ORGANIZING A STATISTICAL PROBLEM A FOUR-STEP PROCESS

STATE: What is the practical question, in the context of the real-world setting?

PLAN: What specific statistical operations does this problem call for?

SOLVE: Make the graphs and carry out the calculations needed for this problem.

CONCLUDE: Give your practical conclusion in the setting of the real-world problem.

CONFIDENCE INTERVALS THE FOUR-STEP PROCESS

STATE: What is the practical question that requires estimating a parameter?

PLAN: Identify the parameter, choose a level of confidence, and select the type of confidence interval that fits your situation.

SOLVE: Carry out the work in two phases:

1. **Check the conditions** for the interval you plan to use.

2. Calculate the **confidence interval.**

CONCLUDE: Return to the practical question to describe your results in this setting.

TESTS OF SIGNIFICANCE THE FOUR-STEP PROCESS

STATE: What is the practical question that requires a statistical test?

PLAN: Identify the parameter, state null and alternative hypotheses, and choose the type of test that fits your situation.

SOLVE: Carry out the test in three phases:

1. **Check the conditions** for the test you plan to use.

2. Calculate the **test statistic.**

3. Find the **P-value.**

CONCLUDE: Return to the practical question to describe your results in this setting.

The Basic Practice of Statistics

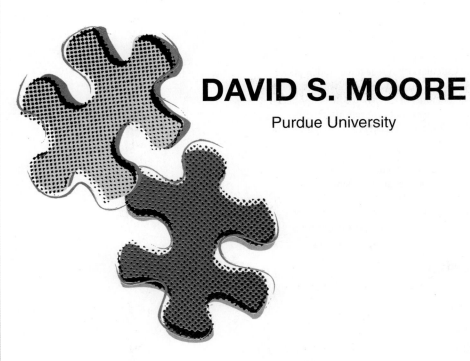

DAVID S. MOORE

Purdue University

W. H. Freeman and Company

New York

Senior Publisher: **Craig Bleyer**
Publisher: **Ruth Baruth**
Senior Media Editor: **Roland Cheyney**
Developmental Editors: **Bruce Kaplan, Shona Burke**
Executive Marketing Manager: **Jennifer Somerville**
Media Editor: **Brian Tedesco**
Associate Editor: **Laura Capuano**
Editorial Assistant: **Katrina Wilhelm**
Photo Editor: **Cecilia Varas**
Photo Researcher: **Elyse Rieder**
Cover and Text Designer: **Vicki Tomaselli**
Cover and Interior Illustrations: **Mark Chickinelli**
Senior Project Editor: **Mary Louise Byrd**
Illustrations: **Aptara**
Production and Illustration Coordinator: **Paul W. Rohloff**
Composition: **Aptara**
Printing and Binding: **Quebecor**

TI-83™ screen shots are used with permission of the publisher © 1996, Texas Instruments Incorporated. TI-83™ Graphic Calculator is a registered trademark of Texas Instruments Incorporated. Minitab is a registered trademark of Minitab, Inc. Microsoft © and Windows © are registered trademarks of the Microsoft Corporation in the United States and other countries. Excel screen shots are reprinted with permission from the Microsoft Corporation. S-PLUS is a registered trademark of the Insightful Corporation.

About the Cover: Completing a jigsaw puzzle makes a meaningful whole out of what seemed like unconnected pieces. Statistics does something similar with data, combining information about the source of the data with graphical displays, numerical summaries, and probability reasoning until meaning emerges from seemed like a jumble. The cover represents how statistics puts everything together.

Library of Congress Control Number: 2008932350

ISBN-13: 978-1-4292-0121-6

ISBN-10: 1-4292-0121-5

Second printing

W. H. Freeman and Company
41 Madison Avenue
New York, NY 10010
Houndmills, Basingstoke RG21 6XS, England
www.whfreeman.com

BRIEF CONTENTS

* Starred material is not required for later parts of the text.

CONTENTS

* Starred material is not required for later parts of the text.

PART V:
Optional Companion Chapters
(AVAILABLE ON THE *BPS* CD AND ONLINE)

About This Book

Welcome to the fifth edition of *The Basic Practice of Statistics* (*BPS*). This book is the cumulation of 40 years of teaching undergraduates and 20 years of writing texts. Previous editions have been very successful, and I think that this new edition is the best yet. In this Preface I describe for instructors the nature and features of the book and the changes in this fifth edition.

BPS is designed to be accessible to college and university students with limited quantitative background—just "algebra" in the sense of being able to read and use simple equations. It is usable with almost any level of technology for calculating and graphing—from a $15 "two-variable statistics" calculator through a graphing calculator or spreadsheet program through full statistical software. Of course, graphs and calculations are less tedious with good technology, so I recommend making available to your students the most effective technology that circumstances permit.

Despite its rather low mathematical level, *BPS* is a "serious" text in the sense that it wants students to do more than master the mechanics of statistical calculations and graphs. Even quite basic statistics is very useful in many fields of study and in everyday life, but only if the student has learned to move from a real world setting to choose and carry out statistical methods and then carry conclusions back to the original setting. These translations require some conceptual understanding of such issues as the distinction between data analysis and inference, the critical role of where the data come from, the reasoning of inference, and the conditions under which we can trust the conclusions of inference. *BPS* tries to teach both the mechanics and the concepts needed for practical statistical work, at a level appropriate for beginners.

Guiding principles

BPS is based on three principles: balanced content, experience with data, and the importance of ideas.

Balanced content. Once upon a time, basic statistics courses taught probability and inference almost exclusively, often preceded by just a week of histograms, means, and medians. Such unbalanced content does not match the actual practice of statistics, where data analysis and design of data production join with probability-based inference to form a coherent science of data. There are also good pedagogical reasons for beginning with data analysis (Chapters 1 to 7), then moving to data production (Chapters 8 and 9), and then to probability (Chapters 10 to 13) and inference (Chapters 14 to 28). In studying data analysis, students learn useful skills immediately and get over some of their fear of statistics. Data analysis is a necessary preliminary to inference in practice, because inference requires clean data. Designed data production is the surest foundation for inference, and the deliberate use of chance in random sampling and randomized comparative experiments motivates the study of probability in a course that emphasizes data-oriented statistics. *BPS* gives a full presentation of basic probability and inference (19 of the 28 chapters) but places it in the context of statistics as a whole.

Experience with data. The study of statistics is supposed to help students work with data in their varied academic disciplines and in their unpredictable later employment. Students learn to work with data by working with data. *BPS* is full of data from many fields of study and from everyday life. Data are more than mere numbers—they are numbers with a context that should play a role in making sense of the numbers and in stating conclusions. Examples and exercises in *BPS*, though intended for beginners, use real data and give enough background to allow students to consider the meaning of their calculations. Exercises often ask for conclusions that are more than a number (or "reject H_0"). Some exercises require judgment in addition to right-or-wrong calculations and conclusions. Statistics, more than mathematics, depends on judgment for effective use. *BPS* begins to develop students' judgment about statistical studies.

The importance of ideas. A first course in statistics introduces many skills, from making a stemplot and calculating a correlation to choosing and carrying out a significance test. In practice (even if not always in the course), calculations and graphs are automated. Moreover, anyone who makes serious use of statistics will need some specific procedures not taught in her college stat course. *BPS* therefore tries to make clear the larger patterns and big ideas of statistics, not in the abstract, but in the context of learning specific skills and working with specific data. Many of the big ideas are summarized in graphical outlines. Three of the most useful appear inside the front cover. Formulas without guiding principles do students little good once the final exam is past, so it is worth the time to slow down a bit and explain the ideas.

These three principles are widely accepted by statisticians concerned about teaching. In fact, statisticians have reached a broad consensus that first courses should reflect how statistics is actually used. As Richard Scheaffer said in discussing a survey paper of mine, "With regard to the content of an introductory statistics course, statisticians are in closer agreement today than at any previous time in my career."[1]* Figure 1 is an outline of the consensus as summarized by the Joint Curriculum Committee of the American Statistical Association and the Mathematical Association of America.[2] I was a member of the ASA/MAA committee, and I agree with their conclusions. More recently, the College Report of the Guidelines for Assessment and Instruction in Statistics Education (GAISE) Project has emphasized exactly the same themes.[3] Fostering active learning is the business of

---

* All notes are collected in the Notes and Data Sources section at the end of the book.

FIGURE 1

1. *EMPHASIZE THE ELEMENTS OF STATISTICAL THINKING:*
    (a) the need for data;
    (b) the importance of data production;
    (c) the omnipresence of variability;
    (d) the measuring and modeling of variability.

2. *INCORPORATE MORE DATA AND CONCEPTS, FEWER RECIPES AND DERIVATIONS. WHER-EVER POSSIBLE, AUTOMATE COMPUTATIONS AND GRAPHICS.* An introductory course should
    (a) rely heavily on *real* (not merely realistic) data;
    (b) emphasize *statistical* concepts, e.g., causation vs. association, experimental vs. observational, and longitudinal vs. cross-sectional studies;
    (c) rely on computers rather than computational recipes;
    (d) treat formal derivations as secondary in importance.

3. *FOSTER ACTIVE LEARNING,* through the following alternatives to lecturing:
    (a) group problem solving and discussion;
    (b) laboratory exercises;
    (c) demonstrations based on class-generated data;
    (d) written and oral presentations;
    (e) projects, either group or individual.

the teacher, though an emphasis on working with data helps. *BPS* is guided by the content emphases of the modern consensus. In the language of the GAISE recommendations, these are: develop statistical thinking, use real data, stress conceptual understanding.

Accessibility

The intent of *BPS* is to be modern *and* accessible. The exposition is straightforward and concentrates on major ideas and skills. One principle of writing for beginners is not to try to tell your students everything you know. Another principle is to offer frequent stopping points. *BPS* presents its content in relatively short chapters, each ending with a summary and two levels of exercises. Within chapters, a few "Apply Your Knowledge" exercises follow each new idea or skill for a quick check of basic mastery—and also to mark off digestible bites of material. Each of the first three parts of the book ends with a review chapter that includes a point-by-point outline of skills learned and many review exercises. (Instructors can choose to cover any or none of the chapters in Parts IV and V, so each of these chapters includes a skills outline.) The review chapters present many additional exercises without the "I just studied that" context, thus asking for another level of learning. I think it is helpful to assign some review exercises. Look at Exercises 21.29 to 21.35

(page 551) for an example of the usefulness of the part reviews. Many instructors will find that the review chapters appear at the right points for pre-examination review.

Technology

Automating calculations increases students' ability to complete problems, reduces their frustration, and helps them concentrate on ideas and problem recognition rather than mechanics. At a minimum, students should have a "two-variable statistics" calculator with functions for correlation and the least-squares regression line as well as for the mean and standard deviation.

Many instructors will take advantage of more elaborate technology, as ASA/MAA and GAISE recommend. And many students who don't use technology in their college statistics course will find themselves using (for example) Excel on the job. *BPS* does not assume or require use of software except in Parts IV and V, where the work is otherwise too tedious. It does accommodate software use and tries to convince students that they are gaining knowledge that will enable them to read and use output from almost any source. There are regular "Using Technology" sections throughout the text. Each of these displays and comments on output from the same three technologies, representing graphing calculators (the Texas Instruments TI-83 or TI-84), spreadsheets (Microsoft Excel), and statistical software (Minitab). The output always concerns one of the main teaching examples, so that students can compare text and output.

A quite different use of technology appears in the interactive applets created to my specifications and available online and on the text CD. These are designed primarily to help in learning statistics rather than in doing statistics. An icon calls attention to comments and exercises based on the applets. I suggest using selected applets for classroom demonstrations even if you do not ask students to work with them. The *Correlation and Regression, Confidence Interval,* and *P-value* applets, for example, convey core ideas more clearly than any amount of chalk and talk.

What's new?

As always, a new edition of *BPS* brings many **new examples and exercises.** There are new data sets provided by researchers from their published work (e.g., Guéguen, Ngai, Suttle, and Vohs in Chapter 24 and earlier). The old favorite Florida manatee regression example returns to Chapters 4, 5, and 23 now that current data are available. Chapter 6 opens with responses of young adults to the survey question "What do you think are the chances you will have much more than a middle-class income at age 30?" Four "Sorry, no chi-square" exercises in Chapter 22 call attention to misuses of the chi-square test, a common source of mistakes even in published reports. These are just a few of a large number of new data settings in this edition.

A new edition is also an opportunity to polish the exposition in ways intended to help students learn. Here are some of the changes:

- Chapters 14 and 15 offer a **revised introduction to inference,** reorganizing material that occupied three chapters in previous editions. Chapter 14 now presents "just the basics" for both confidence intervals and significance tests. The exposition here is shorter and simpler than in past editions, with finer points left for the discussion of conditions, cautions, and planning sample size in Chapter 15. Chapter 14 deemphasizes use of tables to find P-values in order to stress ideas over details.

- There are also substantial improvements in the presentation of **producing data** in Chapters 8 and 9. The distinction between observation and experiment moves to Chapter 9 to introduce experiments. Chapter 8 gains a new discussion of the impact of technology on sample surveys and revised comments on inference from sample to population immediately after the introduction of random sampling.

- Changes in the chapters on **probability** include a new discussion of sources of randomness and an expanded discussion of continuous distributions that compares a histogram of 10,000 random numbers with the idealized uniform distribution (Chapter 10) and more attention to the idea of a population distribution in Chapter 11.

- There is **much rewriting in detail** throughout the book. Among many examples: Chapter Exercises are more carefully graded to place more demanding exercises (often asking use of the four-step process) toward the end; details of the F test for comparing standard deviations have been omitted from Chapter 18, as this test should almost never be used; the Part III Review in Chapter 21 has been expanded to incorporate material that in earlier editions appeared in a short Statistical Thinking Revisited essay, at the end of the text and easy to overlook.

Why did you do that?

There is no single best way to organize our presentation of statistics to beginners. That said, my choices reflect thinking about both content and pedagogy. Here are comments on several "frequently asked questions" about the order and selection of material in *BPS*.

Why does the distinction between population and sample not appear in Part I? This is a sign that there is more to statistics than inference. In fact, statistical inference is appropriate only in rather special circumstances. The chapters in Part I present tools and tactics for describing data—any data. These tools and tactics do not depend on the idea of inference from sample to population. Many data sets in these chapters (for example, the several sets of data about the 50 states) do not lend themselves to inference because they represent an entire population. John Tukey of Bell Labs and Princeton, the philosopher of modern data analysis, insisted that the population-sample distinction be avoided when it is not relevant. He used the word "batch" for data sets in general. I see no need for a special word, but I think Tukey was right.

Why not begin with data production? It is certainly reasonable to do so—the natural flow of a planned study is from design to data analysis to inference. But in their future employment most students will use statistics mainly in settings other than planned research studies. I place the design of data production (Chapters 8 and 9) after data analysis to emphasize that data-analytic techniques apply to any data. One of the primary purposes of statistical designs for producing data is to make inference possible, so the discussion in Chapters 8 and 9 opens Part II and motivates the study of probability.

Why do Normal distributions appear in Part I? Density curves such as the Normal curves are just another tool to describe the distribution of a quantitative variable, along with stemplots, histograms, and boxplots. Professional statistical software offers to make density curves from data just as it offers histograms. I prefer not to suggest that this material is essentially tied to probability, as the traditional order does. And I find it very helpful to break up the indigestible lump of probability that troubles students so much. Meeting Normal distributions early does this and strengthens the "probability distributions are like data distributions" way of approaching probability.

Why not delay correlation and regression until late in the course, as was traditional? *BPS* begins by offering experience working with data and gives a conceptual structure for this nonmathematical but essential part of statistics. Students profit from more experience with data and from seeing the conceptual structure worked out in relations among variables as well as in describing single-variable data. Correlation and least-squares regression are very important descriptive tools, and are often used in settings where there is no population-sample distinction, such as studies of all a firm's employees. Perhaps most important, the *BPS* approach asks students to think about what kind of relationship lies behind the data (confounding, lurking variables, association doesn't imply causation, and so on), without overwhelming them with the demands of formal inference methods. Inference in the correlation and regression setting is a bit complex, demands software, and often comes right at the end of the course. I find that delaying all mention of correlation and regression to that point means that students often don't master the basic uses and properties of these methods. I consider Chapters 4 and 5 (correlation and regression) essential and Chapter 23 (regression inference) optional.

What about probability? Much of the usual formal probability appears in the *optional* Chapters 12 and 13. Chapters 10 and 11 present in a less formal way the ideas of probability and sampling distributions that are needed to understand inference. These two chapters follow a straight line from the idea of probability as long-term regularity, through concrete ways of assigning probabilities, to the central idea of the sampling distribution of a statistic. The law of large numbers and the central limit theorem appear in the context of discussing the sampling distribution of a sample mean. What is left to Chapters 12 and 13 is mostly "general probability rules" (including conditional probability) and the binomial distributions.

I suggest that you omit Chapters 12 and 13 unless you are constrained by external forces. Experienced teachers recognize that students find probability difficult. Research on learning confirms our experience. Even students who can do formally posed probability problems often have a very fragile conceptual grasp of probability ideas. Attempting to present a substantial introduction to probability in a data-oriented statistics course for students who are not mathematically trained is in my opinion unwise. Formal probability does not help these students master the ideas of inference (at least not as much as we teachers often imagine), and it depletes reserves of mental energy that might better be applied to essentially statistical ideas.

Why use the z procedures for a population mean to introduce the reasoning of inference? This is a pedagogical issue, not a question of statistics in practice. Sometime in the golden future we will start with resampling methods. I think that permutation tests make the reasoning of tests clearer than any traditional approach. For now the main choices are z for a mean and z for a proportion.

I find z for means quite a bit more accessible to students. Positively, we can say up front that we are going to explore the reasoning of inference in the overly simple setting described in the box on page 360 titled "Simple conditions for inference about a mean." As this box suggests, exactly Normal population and true simple random sample are as unrealistic as known σ. All the issues of practice—robustness against lack of Normality and application when the data aren't an SRS as well as the need to estimate σ—are put off until, with the reasoning in hand, we discuss the practically useful t procedures. This separation of initial reasoning from messier practice works well.

Negatively, starting with inference for p introduces many side issues: no exact Normal sampling distribution, but a Normal approximation to a discrete distribution; use of $\hat{p}$ in both the numerator and denominator of the test statistic to estimate both the parameter p and $\hat{p}$'s own standard deviation; loss of the direct link between test and confidence interval. Once upon a time we had at least the compensation of developing practically useful procedures. Now the often gross inaccuracy of the traditional z confidence interval for p is better understood. See the following explanation.

Why does the presentation of inference for proportions go beyond the traditional methods? Computational and theoretical work has demonstrated convincingly that the standard confidence intervals for proportions can be trusted only for very large sample sizes. It is hard to abandon old friends, but I think that a look at the graphs in Section 2 of the paper by Brown, Cai, and DasGupta in the May 2001 issue of *Statistical Science* is both distressing and persuasive.[4] The standard intervals often have a true confidence level much less than what was requested, and requiring larger samples encounters a maze of "lucky" and "unlucky" sample sizes until very large samples are reached. Fortunately, there is a simple cure: just add two successes and two failures to your data. I present these "plus four intervals" in Chapters 19 and 20, along with guidelines for use.

Why didn't you cover Topic X? Introductory texts ought not to be encyclopedic. Including each reader's favorite special topic results in a text that is formidable in size and intimidating to students. I chose topics on two grounds: they are the most commonly used in practice, and they are suitable vehicles for learning broader statistical ideas. Students who have completed the core of *BPS*, Chapters 1 to 11 and 14 to 21, will have little difficulty moving on to more elaborate methods. There are of course seven additional chapters in *BPS*, three in this volume and four available on CD and/or online, to begin the next stages of learning.

Acknowledgments

I am grateful to colleagues from two-year and four-year colleges and universities who commented on successive drafts of the manuscript. Special thanks are due to Professor Bradley Hartlaub of Kenyon College. Professor Hartlaub not only read the manuscript with care and offered detailed advice, but is also the author of Chapter 27 on multiple regression, of many two-way ANOVA exercises in Chapter 28, and of some exercises elsewhere.

Special thanks also are due to Professor Sarah Quesen of West Virginia University and Professor Eric Schulz of Walla Walla Community College, who read the manuscript line by line and offered detailed advice. Others who offered comments are:

Brad Bailey,
  North Georgia College and State University
E N Barron,
  Loyola University, Chicago
Jennifer Beineke,
  Western New England College
Diane Benner,
  Harrisburg Area Community College
Zoubir Benzaid,
  University of Wisconsin, Oshkosh
Jennifer Borrello,
  Baylor University
Smiley Cheng,
  University of Manitoba, Winnipeg
Patti Collings,
  Brigham Young University
Tadd Colver,
  Purdue University
James Curl,
  Modesto Junior College
Jonathan Duggins,
  Virginia Tech

Chris Edwards,
  University of Wisconsin, Oshkosh
Margaret Elrich,
  Georgia Perimeter College
Karen Estes,
  St Petersburg College
Eugene Galperin,
  East Stroudsburg University
Mark Gebert,
  Eastern Kentucky University
Kim Gilbert,
  Clayton State University
Aaron Gladish,
  Austin Community College
Ellen Gundlach,
  Purdue University
Arjun Gupta,
  Bowling Green State University
Jeanne Hill,
  Baylor University
Dawn Holmes,
  University of California, Santa Barbara

Patricia Humphrey,
  Georgia Southern University
Thomas Ilvento,
  University of Delaware
Mark Jacobson,
  University of Northern Iowa
Marc Kirschenbaum,
  John Carroll University
Greg Knofczynski,
  Armstrong Atlantic State University
Zhongshan Li,
  Georgia State University
Michael Lichter,
  University of Buffalo
Tom Linton,
  Central College
William Liu,
  Bowling Green—Firelands College
Amy Maddox,
  Grand Rapids Community College
Steve Marsden,
  Glendale Community College
Darcy Mays,
  Virginia Commonwealth University
Andrew McDougall,
  Montclair State University
Bill Meisel,
  *Florida Community College -
  Jacksonville*
Nancy Mendell,
  *State University of New York,
  Stony Brook*
Lynne Nielsen,
  Brigham Young University
Melvin Nyman,
  Alma College
Darlene Olsen,
  Norwich University
Eric Packard,
  Mesa State University
Mary Parker,
  *Austin Community College,
  Rio Grande Campus*
Don Porter,
  Beloit College

Bob Price,
  East Tennessee State University
Asoka Ramanayake,
  University of Wisconsin, Oshkosh
Eric Rdurud,
  St Cloud State University
Christoph Richter,
  Queens University
Scott Richter,
  *University of North Carolina,
  Greensboro*
Corlis Robe,
  East Tennessee State University
Deborah Rumsey,
  Ohio State University
Therese Shelton,
  Southwestern University
Rob Sinn,
  North Georgia College
Eugenia Skirta,
  East Stroudsburg University
Dianna Spence,
  *North George College and State
  University*
Suzhong Tian,
  Husson College
Suzanne Tourville,
  Columbia College
Christopher Tripler,
  Endicott College
Gail Tudor,
  Husson College
Ramin Vakilian,
  *California State University,
  Northridge*
David Vlieger,
  Northwest Missouri State University
Joseph Walker,
  Georgia State University
Steve Waters,
  Pacific Union College
Yuanhui Xiao,
  Georgia State University
Yichuan Zhao,
  Georgia State University

I am also grateful to Craig Bleyer, Ruth Baruth, Bruce Kaplan, Shona Burke, Mary Louise Byrd, Vicki Tomaselli, Pam Bruton, and the other editorial and design professionals who have contributed greatly to the attractiveness of this book.

Finally, I am indebted to the many statistics teachers with whom I have discussed the teaching of our subject over many years; to people from diverse fields with whom I have worked to understand data; and especially to students whose compliments and complaints have changed and improved my teaching. Working with teachers, colleagues in other disciplines, and students constantly reminds me of the importance of hands-on experience with data and of statistical thinking in an era when computer routines quickly handle statistical details.

David S. Moore

For Students

The Basic Practice of Statistics, Fifth Edition, is accompanied by extensive additional materials and alternatives to the printed book. Many of these additional learning resources are available to students at no charge and include interactive statistical applets, interactive exercises, self-quizzers, and data sets. These resources are located on the book's companion Web site and the interactive CD-ROM found at the back of the textbook. Other more extensive materials are available for purchase. These include the electronic alternatives to the printed book: Stats Portal and the eBook, as well as the Online Study Center. Descriptions of all these materials are listed below.

NEW! STATS P▲RTAL

courses.bfwpub.com/bps5e
Text + StatsPortal for BPS, 5e: 1-4292-3093-2

(Access code or online purchase required.)
StatsPortal for The Basic Practice of Statistics, Fifth Edition, is the digital gateway to *BPS*, Fifth Edition, designed to enrich the course and enhance students' study skills through a collection of Web-based tools. *StatsPortal* integrates a rich suite of diagnostic, assessment, tutorial, and enrichment features, enabling students to master statistics at their own pace. It is organized around three main teaching and learning components:

1. Interactive eBook integrates a complete and customizable online version of the text with all of its media resources. Students can quickly search the text, and can personalize the eBook just as they would the print version, with highlighting, bookmarking, and note-taking features. Instructors can add, hide, and reorder content, integrate their own material, and highlight key text.

2. Resources organizes all of *BPS 5e* resources into one location:

Student Resources:

- **StatTutor** Tutorials tied directly to the textbook, containing videos, applets, and animations.

- **Statistical Applets** to help students master key concepts.

- **CrunchIt! Statistical Software**, accessible from any Internet location, offering the basic statistical routines covered in the introductory courses and more.

- **Stats@Work Simulations** put students in the role of consultants, helping them better understand statistics within the context of real-life scenarios.

- **EESEE Case Studies** developed by The Ohio State University Statistics Department teach students to apply their statistical skills by exploring actual case studies, using real data, and answering questions about the study.

- **Podcast Chapter Summary** provides students with a downloadable mp3 version of chapter summaries.

- **Data sets** in ASCII, Excel, JMP, Minitab, TI, SPSS, and S-Plus.

- **Online Tutoring** with SmarThinking is available for homework help from specially-trained, professional educators.

- **Student Study Guide with Selected Solutions** includes explanations of crucial concepts with step-by-step models of important statistical techniques.

- **Statistical Software Manuals** for TI-83/84, Minitab, Excel, JMP, and SPSS provide instruction, examples, and exercises using specific statistical software packages.

- **Interactive Table Reader** allows students to quickly find values in any of the statistical tables used in the course.

- **Tables and Formulas** provide each table and formulas for each chapter.

Resources for Instructors only:

- **Instructor's Guide with Full Solutions** with teaching suggestions, and chapter comments.

- **Test Bank** offering hundreds of multiple choice questions.

- **Lecture PowerPoint slides** offer a detailed lecture presentation for each chapter of BPS.

- **Activities and Projects** offer ideas for projects for Web-based exploration asking students to write critically about statistics.

- **i>clicker Questions** help instructors query students using i>clicker's personal response units in class lectures.

3. Assignment Center (For Instructors only) organizes assignments and guides instructors through an easy-to-create assignment process with access to questions from the Test Bank, Web Quizzes, and Exercises from the text, including many algorithmic problems.

Online Study Center 2.0 for The Basic Practice of Statistics

www.whfreeman.com/osc/bps5e

Text + Online Study Center Access Code: 1-4292-3095-9

(Access code or online purchase required.)

This premium Web-based study alternative helps students pinpoint where their study time should be focused, then provides all the resources needed to improve their comprehension of troublesome areas. Before beginning each chapter, students take a **Self-Test** to assess their knowledge of the material. The OSC then generates a **Detailed Study Plan** linking to the online resources relevant to the questions they answered incorrectly.

The OSC includes all the resources included in StatsPortal, except for the Assignment Center and the eBook. Instructors have access to an easy-to-manage gradebook and all media resources to help them track student progress and prepare lectures or course Web pages.

BPS 5e has an extensive array of free study material:

Book Companion Site at www.whfreeman.com/bps5e, featuring

- Interactive Statistical Applets
- Data sets
- Interactive Exercises and Self-Quizzes to help students prepare for tests.
- Key tables and formulas summary sheet
- All tables from the text in .pdf format for quick, easy reference
- Supplementary Exercises
- Optional Companion chapters 25, 26, 27, and 28, covering nonparametric tests, statistical process control, multiple regression, and two-way analysis of variance, respectively.

Interactive Student CD-ROM

Included with every new copy of *BPS 5e*, the CD contains access to the companion chapters, applets, and datasets also found on the Companion Web site.

Additional materials for students, available for purchase:

Software Manuals

In addition to being a part of StatsPortal, these manuals are available in printed versions through custom publishing. They serve as basic introductions to popular statistical software options and guides to their use with the new edition of *The Basic Practice of Statistics*:

Minitab Manual, 1-4292-2782-6
JMP Manual, 1-4292-2791-5
Excel Manual, 1-4292-27907
SPSS Manual, 1-4292-2785-0
TI 83/84 Manual, 1-4292-2786-9

Study Guide with Selected Solutions, 1-4292-2783-4

Text + Study Guide: 1-4292-3094-0

This Guide offers students explanations of crucial concepts in each section of BPS, plus detailed solutions to key text problems and stepped-through models of important statistical techniques.

Telecourse Study Guide, 1-4292-2460-6

Text + Telecourse Study Guide: 1-4292-3096-7

A study guide for students using the telecourse *Against All Odds: Inside Statistics*.

For Instructors

The Instructor's Web site www.whfreeman.com/bps5e

Password protected, the instructor's Web site features access to all student Web materials on the companion web site, plus:

- Instructor version of EESEE (Electronic Encyclopedia of Statistical Examples and Exercises), with solutions to the exercises in the student version.

- PowerPoint slides containing all textbook figures and tables.

- Lecture PowerPoint slides offering a detailed lecture presentation of statistical concepts covered in each chapter of BPS.

- Full answers to the Supplementary Exercises on the student Web site.

Instructor's Guide with Solutions, 1-4292-2792-3

This printed guide includes full solutions to all exercises and provides video and Internet resources and sample examinations. It also contains brief discussions of the BPS approach for each chapter.

Test Bank, printed, 1-4292-2784-2; CD (Windows and Mac on one disc, 1-4292-2789-3)

The test bank contains hundreds of multiple-choice questions. With the CD version, questions can easily be downloaded, edited, and resequenced.

Enhanced Instructor's Resource CD-ROM, 1-4292-2788-5

Allows instructors to **search** and **export** (by key term or chapter) all the material from the student CD, plus:

- All text images and tables

- Instructor's Guide with full solutions

- PowerPoint files and lecture slides

- Test bank files

Course Management Systems

W. H. Freeman and Company provides course cartridges for Blackboard, WebCT (Campus Edition and Vista), and Angel course management systems. Upon request, we also provide courses for users of Desire2Learn and Moodle.

i>clicker Radio Frequency Classroom Response System
www.iclicker.com

Developed for educators by educators, i>clicker is the easiest-to-use and most flexible classroom response system available.

STATISTICAL THINKING

What genes are active in a tissue? Answering this question can unravel basic questions in biology, distinguish cancer cells from normal cells, and distinguish between closely related types of cancer. To learn the answer, apply the tissue to a "microarray" that contains thousands of snippets of DNA arranged in a grid on a chip about the size of your thumb. As DNA in the tissue binds to the snippets in the array, special recorders pick up spots of light of varying color and intensity across the grid and store what they see as numbers.

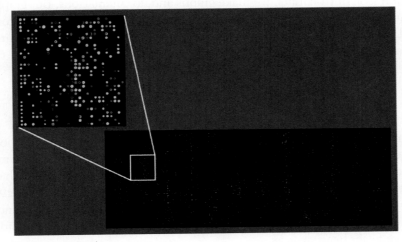

Paphrag at en.wikipedia

What's hot in popular music this week? SoundScan knows. SoundScan collects data electronically from the cash registers in more than 14,000 retail outlets, and also collects data on download sales from Web sites. When you buy a CD or download a digital track, the checkout scanner or Web site is probably telling SoundScan what you bought. SoundScan provides this information to *Billboard Magazine*, MTV, and VH1, as well as to record companies and artists' agents.

Should women take hormones such as estrogen after menopause, when natural production of these hormones ends? In 1992, several major medical organizations said "Yes." In particular, women who took hormones seemed to reduce their risk of a heart attack by 35% to 50%. The risks of taking hormones appeared small compared with the benefits. But in 2002, the National Institutes of Health declared these findings wrong. Use of hormones after menopause immediately plummeted. Both recommendations were based on extensive studies. What happened?

DNA microarrays, SoundScan, and medical studies all produce data (numerical facts), and lots of them. Using data effectively is a large and growing part of most professions. Reacting to data is part of everyday life. That's why statistics is important:

STATISTICS IS THE SCIENCE OF LEARNING FROM DATA

Data are numbers, but they are not "just numbers." **Data are numbers with a context.** The number 10.5, for example, carries no information by itself. But if we hear that a friend's new baby weighed 10.5 pounds at birth, we congratulate her on the healthy size of the child. The context engages our background knowledge and allows us to make judgments. We know that a baby weighing 10.5 pounds is quite large, and that a human baby is unlikely to weigh 10.5 ounces or 10.5 kilograms. The context makes the number informative.

To gain insight from data, we make graphs and do calculations. But graphs and calculations are guided by ways of thinking that amount to educated common sense. Let's begin our study of statistics with an informal look at some principles of statistical thinking.[1]

WHERE THE DATA COME FROM MATTERS

What's behind the flip-flop in the advice offered to women about hormone replacement? The evidence in favor of hormone replacement came from a number of *observational studies* that compared women who were taking hormones with others who were not. But women who choose to take hormones are very different from women who do not: they are richer and better educated and see doctors more often. These women do many things to maintain their health. It isn't surprising that they have fewer heart attacks.

Large and careful observational studies are expensive, but are easier to arrange than careful *experiments*. Experiments don't let women decide what to do. They assign women to either hormone replacement or to dummy pills that look and taste the same as the hormone pills. The assignment is done by a coin toss, so that all kinds of women are equally likely to get either treatment. Part of the difficulty of a good experiment is persuading women to agree to accept the result—invisible to them—of the coin toss. By 2002, several experiments agreed that hormone replacement does *not* reduce the risk of heart attacks, at least for older women. Faced with this better evidence, medical authorities changed their recommendations.[2]

Of course, observational studies are often useful. We can learn from observational studies how chimpanzees behave in the wild, or which popular songs sold best last week, or what percent of workers were unemployed last month. Soundscan's data on popular music and the government's data on employment rate come from *sample surveys*, an important kind of observational study that chooses a part (the sample) to represent a larger whole. Opinion polls interview perhaps 1000 of the 235 million adults in the United States to report the public's views on current issues. Can we trust the results? We'll see that this isn't a simple yes-or-no question. Let's just say that the government's unemployment rate is much more trustworthy than opinion poll results, and not just because the Bureau of Labor Statistics interviews 60,000 people rather than 1000.

We can, however, say right away that some samples *can't* be trusted. The advice columnist Ann Landers once asked her readers, "If you had it to do over again, would you have children?" A few weeks later, her column was headlined "70% OF PARENTS SAY KIDS NOT WORTH IT." Indeed, 70% of the nearly 10,000 parents who wrote in said they would not have children if they could make the choice again. Those 10,000 parents were upset enough with their children to write Ann Landers. Most parents are happy with their kids and don't bother to write. Statistically designed samples, even opinion polls, don't let people choose themselves for the sample. They interview people selected by impersonal chance so that everyone has an equal opportunity to be in the sample. Such a poll showed that 91% of parents *would* have children again. Where data come from matters a lot. If you are careless about how you get your data, you may announce 70% "No" when the truth is close to 90% "Yes."

ALWAYS LOOK AT THE DATA

Yogi Berra said it: "You can observe a lot by just watching." That's a motto for learning from data. *A few carefully chosen graphs are often more instructive than great piles of numbers.* Consider the outcome of the 2000 presidential election in Florida.

Elections don't come much closer: after much recounting, state officials declared that George Bush had carried Florida by 537 votes out of almost 6 million votes cast. Florida's vote decided the election and made George Bush, rather than Al Gore, president. Let's look at some data. Figure 1 (see page xxvi) displays a graph that plots votes for the third-party candidate Pat Buchanan against votes for the Democratic candidate Al Gore in Florida's 67 counties.

What happened in Palm Beach County? The question leaps out from the graph. In this large and heavily Democratic county, a conservative third-party candidate did far better relative to the Democratic candidate than in any other county. The points for the other 66 counties show votes for both candidates increasing together in a roughly straight-line pattern. Both counts go up as county population goes up. Based on this pattern, we would expect Buchanan to receive around 800 votes in Palm Beach County. He actually received more than 3400 votes. That difference determined the election result in Florida and in the nation.

The graph demands an explanation. It turns out that Palm Beach County used a confusing "butterfly" ballot, in which candidate names on both left and right pages led to a voting column in the center (see the illustration on page xxvi). It would be easy for a voter who intended to vote for Gore to in fact cast a vote for Buchanan. The graph is convincing evidence that this in fact happened, more convincing than the complaints of voters who (later) were unsure where their votes ended up.

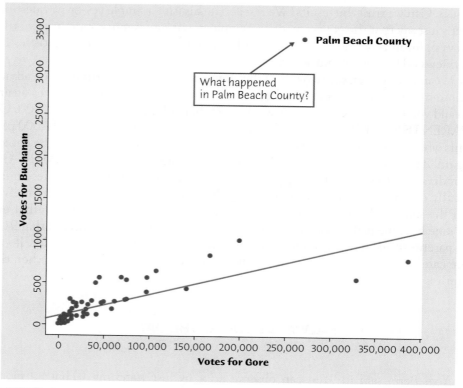

FIGURE 1

Votes in the 2000 presidential election for Al Gore and Patrick Buchanan in Florida's 67 counties. What happened in Palm Beach County?

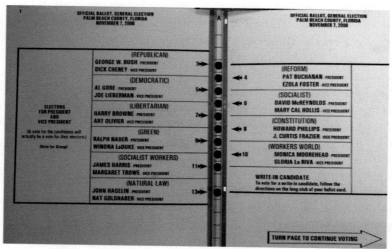

BEWARE THE LURKING VARIABLE

Women who chose hormone replacement after menopause were on the average richer and better educated than those who didn't. No wonder they had fewer heart attacks. Children who play soccer tend to have prosperous and well-educated parents. No wonder they do better in school (on the average) than children who don't play soccer. We can't conclude that hormone replacement reduces heart attacks or that playing soccer increases school grades just because we see these relationships in data. In both examples, education and affluence are *lurking variables*, background factors that help explain the relationships between hormone replacement and good health and between soccer and good grades.

Almost all relationships between two variables are influenced by other variables lurking in the background. To understand the relationship between two variables, you must often look at other variables. Careful statistical studies try to think of and measure possible lurking variables in order to correct for their influence. As the hormone saga illustrates, this doesn't always work well. News reports often just ignore possible lurking variables that might ruin a good headline like "Playing soccer can improve your grades." The habit of asking "What might lie behind this relationship?" is part of thinking statistically.

VARIATION IS EVERYWHERE

The company's sales reps file into their monthly meeting. The sales manager rises. "Congratulations! Our sales were up 2% last month, so we're all drinking champagne this morning. You remember that when sales were down 1% last month I fired half of our reps." This picture is only slightly exaggerated. Many managers overreact to small short-term variations in key figures. Here is Arthur Nielsen, head of the country's largest market research firm, describing his experience:

> *Too many business people assign equal validity to all numbers printed on paper. They accept numbers as representing Truth and find it difficult to work with the concept of probability. They do not see a number as a kind of shorthand for a range that describes our actual knowledge of the underlying condition.*[3]

Business data such as sales and prices vary from month to month for reasons ranging from the weather to a customer's financial difficulties to the inevitable errors in gathering the data. The manager's challenge is to say when there is a real pattern behind the variation. We'll see that statistics provides tools for understanding variation and for seeking patterns behind the screen of variation.

Let's look at some more data. Figure 2 (see page xxviii) plots the average price of a gallon of regular unleaded gasoline each month from January 1990 to July 2008.[4] There certainly is variation! But a close look shows a yearly pattern: gas prices go up during the summer driving season, then down as demand drops in the fall. On

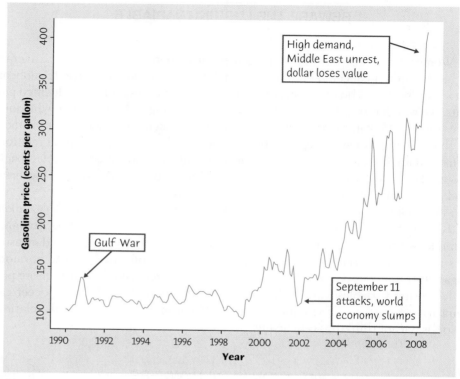

FIGURE 2

Variation is everywhere: the average retail price of regular unleaded gasoline, 1990 to early 2008.

top of this regular pattern we see the effects of international events. For example, prices rose when the 1990 Gulf War threatened oil supplies and dropped when the world economy turned down after the September 11, 2001 terrorist attacks in the United States. The years 2007 and 2008 brought the perfect storm: the ability to produce oil and refine gasoline was overwhelmed by high demand from China and the United States and continued turmoil in the oil-producing areas of the Middle East and Nigeria. Add in a rapid fall in the value of the dollar, and prices at the pump skyrocketed. The data carry an important message: because the United States imports most of its oil, we can't control the price we pay for gasoline.

Variation is everywhere. Individuals vary; repeated measurements on the same individual vary; almost everything varies over time. One reason we need to know some statistics is that statistics helps us deal with variation.

CONCLUSIONS ARE NOT CERTAIN

Cervical cancer is second only to breast cancer as a cause of cancer deaths in women. Almost all cervical cancers are caused by human papillomavirus (HPV).

The first vaccine to protect against the most common varieties of HPV became available in 2006. The Centers for Disease Control and Prevention recommend that all girls be vaccinated at age 11 or 12.

How well does the vaccine work? Doctors rely on experiments (called "clinical trials" in medicine) that give some women the new vaccine and others a dummy vaccine. (This is ethical when it is not yet known whether or not the vaccine is safe and effective.) The conclusion of the most important trial was that an estimated 98% of women up to age 26 who are vaccinated before they are infected with HPV will avoid cervical cancers over a 3-year period.

On the average women who get the vaccine are much less likely to get cervical cancer. But because variation is everywhere, the results are different for different women. Some vaccinated women will get cancer, and many who are not vaccinated will escape. Statistical conclusions are "on the average" statements only. Well then, can we be certain that the vaccine reduces risk on the average? No. We can be very confident, but we can't be certain.

Because variation is everywhere, conclusions are uncertain. Statistics gives us a language for talking about uncertainty that is used and understood by statistically literate people everywhere. In the case of HPV vaccine, the medical journal used that language to tell us that "Vaccine efficiency . . . was 98% (95 percent confidence interval 86% to 100%)."[5] That "98% effective" is, in Arthur Nielsen's words, "shorthand for a range that describes our actual knowledge of the underlying condition." The range is 86% to 100%, and we are 95 percent confident that the truth lies in that range. We will soon learn to understand this language. We can't escape variation and uncertainty. Learning statistics enables us to live more comfortably with these realities.

STATISTICAL THINKING AND YOU

What Lies Ahead in This Book The purpose of *The Basic Practice of Statistics* (BPS) is to give you a working knowledge of the ideas and tools of practical statistics. We will divide practical statistics into three main areas:

1. **Data analysis** concerns methods and strategies for exploring, organizing, and describing data using graphs and numerical summaries. Only organized data can illuminate reality. Only thoughtful exploration of data can defeat the lurking variable. Part I of BPS (Chapters 1 to 7) discusses data analysis.

2. **Data production** provides methods for producing data that can give clear answers to specific questions. Where the data come from really is important. Basic concepts about how to select samples and design experiments are the most influential ideas in statistics. These concepts are the subject of Chapters 8 and 9.

3. **Statistical inference** moves beyond the data in hand to draw conclusions about some wider universe, taking into account that variation is everywhere and that conclusions are uncertain. To describe variation and uncertainty, inference uses the language of probability, introduced in Chapters 10 and 11.

Because we are concerned with practice rather than theory, we need only a limited knowledge of probability. Chapters 12 and 13 offer more probability for those who want it. Chapters 14 and 15 discuss the reasoning of statistical inference. These chapters are the key to the rest of the book. Chapters 17 to 20 present inference as used in practice in the most common settings. Chapters 22 to 24, and the Optional Companion Chapters 25 to 28 on the text CD or online, concern more advanced or specialized kinds of inference.

Because data are numbers with a context, doing statistics means more than manipulating numbers. You must **state** a problem in its real-world context, **plan** your specific statistical work in detail, **solve** the problem by making the necessary graphs and calculations, and **conclude** by explaining what your findings say about the real-world setting. We'll make regular use of this four-step process to encourage good habits that go beyond graphs and calculations to ask, "What do the data tell me?"

Statistics does involve lots of calculating and graphing. The text presents the techniques you need, but you should use technology to automate calculations and graphs as much as possible. Because the big ideas of statistics don't depend on any particular level of access to technology, *BPS* does not require software or a graphing calculator until we reach the more advanced methods in Part IV of the text. Even if you make little use of technology, you should look at the "Using Technology" sections throughout the book. You will see at once that you can read and apply the output from almost any technology used for statistical calculations. The ideas really are more important than the details of how to do the calculations.

Unless you have constant access to software or a graphing calculator, *you will need a basic calculator with some built-in statistical functions.* Specifically, your calculator should find means and standard deviations and calculate correlations and regression lines. Look for a calculator that claims to do "two-variables statistics" or mentions "regression."

Because graphing and calculating are automated in statistical practice, the most important assets you can gain from the study of statistics are an understanding of the big ideas and the beginnings of good judgment in working with data. *BPS* tries to explain the most important ideas of statistics, not just teach methods. Some examples of big ideas that you will meet (one from each of the three areas of statistics) are "always plot your data," "randomized comparative experiments," and "statistical significance."

You learn statistics by doing statistical problems. As you read, you will see several levels of exercises, arranged to help you learn. Short "Apply Your Knowledge" problem sets appear after each major idea. These are straightforward exercises that help you solidify the main points as you read. Be sure you can do these exercises before going on. The end-of-chapter exercises begin with multiple-choice "Check Your Skills" exercises (with all answers in the back of the book). Use them to check your grasp of the basics. The regular "Chapter Exercises" help you combine all the ideas of a chapter. Finally, the three part review chapters (Chapters 7, 16, and 21) look back over major blocks of learning, with many review exercises. At each step you are given less advance knowledge of exactly what statistical ideas and skills the problems will require, so each type of exercise requires more understanding.

The part review chapters (and the individual chapters in Part IV) include point-by-point lists of specific things you should be able to do. Go through that list, and be sure you can say "I can do that" to each item. Then try some of the review exercises.

The key to learning is persistence. The main ideas of statistics, like the main ideas of any important subject, took a long time to discover and take some time to master. The gain will be worth the pain.

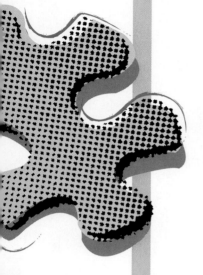

Exploring Data

"What do the data say?" is the first question we ask in any statistical study. *Data analysis* answers this question by open-ended exploration of the data. The tools of data analysis are graphs such as histograms and scatterplots and numerical measures such as means and correlations. At least as important as the tools are principles that organize our thinking as we examine data. The seven chapters in Part I present the principles and tools of statistical data analysis. They equip you with skills that are immediately useful whenever you deal with numbers.

These chapters reflect the strong emphasis on exploring data that characterizes modern statistics. Sometimes we hope to draw conclusions that apply to a setting that goes beyond the data in hand. This is *statistical inference*, the topic of much of the rest of the book. Data analysis is essential if we are to trust the results of inference, but data analysis isn't just preparation for inference. Roughly speaking, you can always do data analysis but inference requires rather special conditions.

One of the organizing principles of data analysis is to first look at one thing at a time and then at relationships. Our presentation follows this principle. In Chapters 1, 2, and 3 you will study *variables and their distributions*. Chapters 4, 5, and 6 concern *relationships among variables*. Chapter 7 reviews this part of the text.

EXPLORING DATA: Variables and Distributions

EXPLORING DATA: Relationships

AP Photo/Mary Altaffer

Picturing Distributions with Graphs

Statistics is the science of data. The volume of data available to us is overwhelming. For example, the Census Bureau's American Community Survey collects data from 3,000,000 housing units each year. Astronomers work with data on tens of millions of galaxies. The checkout scanners at Wal-Mart's 6500 stores in 15 countries record hundreds of millions of transactions every week, all saved to inform both Wal-Mart and its suppliers. The first step in dealing with such a flood of data is to organize our thinking about data. Fortunately, we can do this without looking at millions of data points.

Individuals and variables

Any set of data contains information about some group of *individuals*. The information is organized in *variables*.

INDIVIDUALS AND VARIABLES

Individuals are the objects described by a set of data. Individuals may be people, but they may also be animals or things.

A **variable** is any characteristic of an individual. A variable can take different values for different individuals.

What's that number?

You might think that numbers, unlike words, are universal. Think again. A "billion" in the United States means 1,000,000,000 (nine zeros). In Europe, a "billion" is 1,000,000,000,000 (twelve zeros). OK, those are words that describe numbers. But those commas in big numbers are periods in many other languages. This is so confusing that international standards call for spaces instead, so that an American billion is written 1 000 000 000. And the decimal point of the English-speaking world is the decimal comma in many other languages, so that 3.1416 in the United States becomes 3,1416 in Europe. So what is the number 10,642.389? Depends on where you are.

A college's student data base, for example, includes data about every currently enrolled student. The students are the individuals described by the data set. For each individual, the data contain the values of variables such as date of birth, choice of major, and grade point average. In practice, any set of data is accompanied by background information that helps us understand the data. When you plan a statistical study or explore data from someone else's work, ask yourself the following questions:

1. **Who?** What **individuals** do the data describe? **How many** individuals appear in the data?

2. **What?** How many **variables** do the data contain? What are the **exact definitions** of these variables? In what **unit of measurement** is each variable recorded? Weights, for example, might be recorded in pounds, in thousands of pounds, or in kilograms.

3. **Why? What purpose** do the data have? Do we hope to answer some specific questions? Do we want answers for just these individuals or for some larger group that these individuals are supposed to represent? Are the individuals and variables suitable for the intended purpose?

Some variables, like a person's sex or college major, simply place individuals into categories. Others, like height and grade point average, take numerical values for which we can do arithmetic. It makes sense to give an average income for a company's employees, but it does not make sense to give an "average" sex. We can, however, count the numbers of female and male employees and do arithmetic with these counts.

CATEGORICAL AND QUANTITATIVE VARIABLES

A **categorical variable** places an individual into one of several groups or categories.

A **quantitative variable** takes numerical values for which arithmetic operations such as adding and averaging make sense. The values of a quantitative variable are usually recorded in a **unit of measurement** such as seconds or kilograms.

EXAMPLE 1.1 The American Community Survey

At the Census Bureau Web site, you can view the detailed data collected by the American Community Survey, though of course the identities of people and housing units are protected. If you choose the file of data on people, the *individuals* are the people living in the housing units contacted by the survey. Over 100 variables are recorded for each individual. Figure 1.1 displays a very small part of the data.

Each row records data on one individual. Each column contains the values of one *variable* for all the individuals. Translated from the Census Bureau's abbreviations, the variables are the following:

	A	B	C	D	E	F	G
1	SERIALNO	PWGTP	AGEP	JWMNP	SCHL	SEX	WAGP
2	283	187	66		6	1	24000
3	283	158	66		9	2	0
4	323	176	54	10	12	2	11900
5	346	339	37	10	11	1	6000
6	346	91	27	10	10	2	30000
7	370	234	53	10	13	1	83000
8	370	181	46	15	10	2	74000
9	370	155	18		9	2	0
10	487	233	26		14	2	800
11	487	146	23		12	2	8000
12	511	236	53		9	2	0
13	511	131	53		11	1	0
14	515	213	38		11	2	12500
15	515	194	40		9	1	800
16	515	221	18	20	9	1	2500
17	515	193	11		3	1	

eg01–01.csv

eg01-01

FIGURE 1.1

A spreadsheet displaying data from the American Community Survey, for Example 1.1.

Each row in the spreadsheet contains data on one individual.

SERIALNO   An identifying number for the household.
PWGTP   Weight in pounds.
AGEP   Age in years.
JWMNP   Travel time to work in minutes.
SCHL   Highest level of education. The categories are designated by numbers. For example, 9 = high school graduate, 10 = some college but no degree, and 13 = bachelor's degree.
SEX   Sex, designated by 1 = male and 2 = female.
WAGP   Wage and salary income last year, in dollars.

Look at the highlighted row in Figure 1.1. This individual is a 53-year-old man who weighs 234 pounds, travels 10 minutes to work, has a bachelor's degree, and earned $83,000 last year.

In addition to the household serial number, there are six variables. Education and sex are categorical variables. The values for education and sex are stored as numbers, but these numbers are just labels for the categories and have no units of measurement. The other four variables are quantitative. Their values do have units. These variables are weight in pounds, age in years, travel time in minutes, and income in dollars.

The *purpose* of the American Community Survey is to collect data that represent the entire nation in order to guide government policy and business decisions. To do this, the households contacted are chosen at random from all households in the country. We will see in Chapter 8 why choosing at random is a good idea. ■

Most data tables follow this format—each row is an individual, and each column is a variable. The data set in Figure 1.1 appears in a **spreadsheet** program that has rows and columns ready for your use. Spreadsheets are commonly used to enter and transmit data and to do simple calculations.

spreadsheet

1.1 Fuel economy. Here is a small part of a data set that describes the fuel economy (in miles per gallon) of 2008 model motor vehicles:

Make and model	Vehicle type	Transmission type	Number of cylinders	City mpg	Highway mpg
.					
.					
.					
Aston Martin Vantage	Two-seater	Manual	8	12	19
Honda Civic	Subcompact	Automatic	4	25	36
Toyota Prius	Midsize	Automatic	4	48	45
Chevrolet Impala	Large	Automatic	6	18	29
.					
.					
.					

(a) What are the individuals in this data set?

(b) For each individual, what variables are given? Which of these variables are categorical and which are quantitative?

1.2 Students and TV. You are preparing to study the television-viewing habits of college students. Describe two categorical variables and two quantitative variables that you might measure for each student. Give the units of measurement for the quantitative variables.

Categorical variables: pie charts and bar graphs

exploratory data analysis

Statistical tools and ideas help us examine data in order to describe their main features. This examination is called **exploratory data analysis.** Like an explorer crossing unknown lands, we want first to simply describe what we see. Here are two principles that help us organize our exploration of a set of data.

EXPLORING DATA

1. Begin by examining each variable by itself. Then move on to study the relationships among the variables.

2. Begin with a graph or graphs. Then add numerical summaries of specific aspects of the data.

We will follow these principles in organizing our learning. Chapters 1 to 3 present methods for describing a single variable. We study relationships among several variables in Chapters 4 to 6. In each case, we begin with graphical displays, then add numerical summaries for more complete description.

The proper choice of graph depends on the nature of the variable. To examine a single variable, we usually want to display its *distribution*.

DISTRIBUTION OF A VARIABLE

The **distribution** of a variable tells us what values it takes and how often it takes these values.

The values of a categorical variable are labels for the categories. The **distribution of a categorical variable** lists the categories and gives either the count or the percent of individuals who fall in each category.

EXAMPLE 1.2   **Which major?**

About 1.6 million first-year students enroll in colleges and universities each year. What do they plan to study? Here are data on the percents of first-year students who plan to major in several discipline areas:[1]

Field of study	Percent of students
Arts and humanities	12.8
Biological sciences	7.6
Business	17.4
Education	9.9
Engineering	8.3
Physical sciences	3.1
Professional	14.6
Social science	10.7
Technical	1.2
Other majors	14.1
Total	99.7

It's a good idea to check data for consistency. The percents should add to 100%. In fact, they add to 99.7%. What happened? Each percent is rounded to the nearest tenth. The exact percents would add to 100, but the rounded percents only come close. This is **roundoff error.** Roundoff errors don't point to mistakes in our work, just to the effect of rounding off results. ■

roundoff error

Columns of numbers take time to read. You can use a pie chart or a bar graph to display the distribution of a categorical variable more vividly. Figures 1.2 and 1.3 illustrate these displays for the distribution of intended college majors.

Pie charts show the distribution of a categorical variable as a "pie" whose slices are sized by the counts or percents for the categories. Pie charts are awkward to make by hand, but software will do the job for you. *A pie chart must include all the categories that make up a whole. Use a pie chart only when you want to emphasize each category's relation to the whole.* We need the "Other majors" category in Example 1.2

pie chart

FIGURE 1.2

You can use a pie chart to display the distribution of a categorical variable. Here is a pie chart of the distribution of intended majors of students entering college.

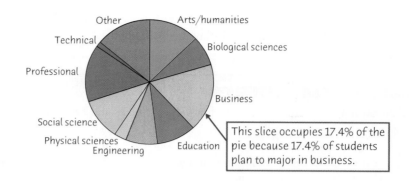

This slice occupies 17.4% of the pie because 17.4% of students plan to major in business.

to complete the whole (all intended majors) and allow us to make the pie chart in Figure 1.2.

bar graph   **Bar graphs** represent each category as a bar. The bar heights show the category counts or percents. Bar graphs are easier to make than pie charts and also easier to read. Figure 1.3 displays two bar graphs of the data on intended majors. The first orders the bars alphabetically by field of study (with "Other" at the end). It is often better to arrange the bars in order of height, as in Figure 1.3(b). This helps us immediately see which majors appear most often.

Bar graphs are more flexible than pie charts. Both graphs can display the distribution of a categorical variable, but a bar graph can also compare any set of quantities that are measured in the same units.

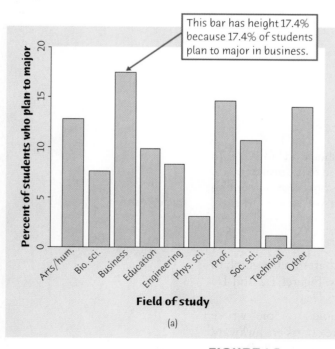

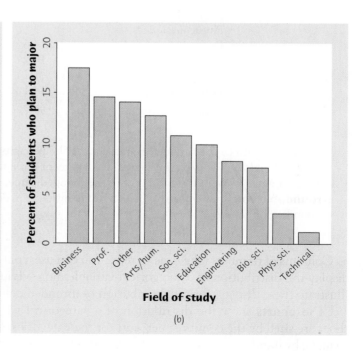

FIGURE 1.3

Bar graphs of the distribution of intended majors of students entering college. In (a), the bars follow the alphabetical order of fields of study. In (b), the same bars appear in order of height.

EXAMPLE 1.3 I love my iPod!

The rating service Arbitron asked adults who used several high-tech devices and services whether they "loved" using them. Here are the percents who said they did:[2]

Device or service	Percent of users who love it
Blackberry or similar device	21
Broadband Internet access	41
Cable TV	20
Digital video recorder	32
High-definition television	34
iPod	45
MP3 player other than iPod	25
Pay TV channels (such as HBO)	16
Satellite radio	33

Michael A. Keller/CORBIS

We can't make a pie chart to display these data. Each percent in the table refers to a different device or service, not to parts of a single whole. Figure 1.4 is a bar graph comparing the nine devices and services. We have again arranged the bars in order of height. ■

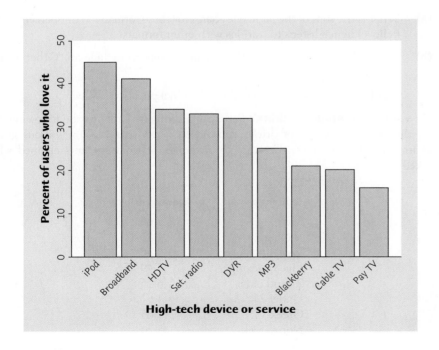

High-tech device or service

FIGURE 1.4

You can use a bar graph to compare quantities that are not part of a whole. This bar graph compares the percents of users who say they "love" using various devices or services, for Example 1.3.

Bar graphs and pie charts are mainly tools for presenting data: they help your audience grasp data quickly. They are of limited use for data analysis because it is easy to understand data on a single categorical variable without a graph. We will move on to quantitative variables, where graphs are essential tools.

1.3 Do you listen to country radio? The rating service Arbitron places U.S. radio stations into more than 50 categories that describe the kind of programs they broadcast. Which formats attract the largest audiences? Here are Arbitron's measurements of the share of the listening audience (aged 12 and over) for the most popular formats:[3]

Format	Audience share
Country	12.6%
News/Talk/Information	10.4%
Adult Contemporary	7.1%
Pop Contemporary Hit	5.5%
Classic Rock	4.7%
Rhythmic Contemporary Hit	4.2%
Urban Contemporary	4.1%
Urban Adult Contemporary	3.4%
Oldies	3.3%
Hot Adult Contemporary	3.2%
Mexican Regional	3.1%

(a)   What is the sum of the audience shares for these formats? What percent of the radio audience listens to stations with other formats?

(b)   Make a bar graph to display these data. Be sure to include an "Other format" category.

(c)   Would it be correct to display these data in a pie chart? Why?

1.4 How much do students drink? Penn State University reports the following data on the average number of drinks consumed "when partying" for various groups of its students.[4] At least, these are the averages of what students claimed when asked.

Student group	Average drinks
Men	6.65
Women	4.31
Live off-campus	6.36
Live on-campus	3.49
21 and older	6.15
Under 21	4.51
Greek	7.65
Non-Greek	5.22

(a)   Explain why it is *not* correct to use a pie chart to display these data.

(b)   Make a bar graph of the data. Notice that because the data contrast groups such as men and women it is better to keep these bars next to each other rather than to arrange the bars in order of height.

1.5 Never on Sunday? Births are not, as you might think, evenly distributed across the days of the week. Here are the average numbers of babies born on each day of the week in 2005:[5]

Day	Births
Sunday	7,374
Monday	11,704
Tuesday	13,169
Wednesday	13,038
Thursday	13,013
Friday	12,664
Saturday	8,459

Present these data in a well-labeled bar graph. Would it also be correct to make a pie chart? Suggest some possible reasons why there are fewer births on weekends.

Quantitative variables: histograms

Quantitative variables often take many values. The distribution tells us what values the variable takes and how often it takes these values. A graph of the distribution is clearer if nearby values are grouped together. The most common graph of the distribution of one quantitative variable is a **histogram**.

histogram

EXAMPLE 1.4 Making a histogram

What percent of your home state's residents were born outside the United States? The country as a whole has 12.5% foreign-born residents, but the states vary from 1.2% in West Virginia to 27.2% in California. Table 1.1 presents the data for all 50 states and the District of Columbia.[6] The *individuals* in this data set are the states. The *variable* is the percent of a state's residents who are foreign-born. It's much easier to see how your state compares with other states from a graph than from the table. To make a histogram of the distribution of this variable, proceed as follows:

Step 1. Choose the classes. Divide the range of the data into classes of equal width. The data in Table 1.1 range from 1.2 to 27.2, so we decide to use these classes:

> percent foreign-born between 0.1 and 5.0
>
> percent foreign-born between 5.1 and 10.0
>
> .
> .
>
> percent foreign-born between 25.1 and 30.0

AP Photo/Mary Altaffer

It is equally correct to use classes 0.0 to 4.9, 5.0 to 9.9, and so on. Just be sure to specify the classes precisely so that each individual falls into exactly one class. Pennsylvania, with 5.1% foreign-born, falls into the second class, but a state with 5.0% would fall into the first.

TABLE 1.1 Percent of state population born outside the United States

STATE	PERCENT	STATE	PERCENT	STATE	PERCENT
Alabama	2.8	Louisiana	2.9	Ohio	3.6
Alaska	7.0	Maine	3.2	Oklahoma	4.9
Arizona	15.1	Maryland	12.2	Oregon	9.7
Arkansas	3.8	Massachusetts	14.1	Pennsylvania	5.1
California	27.2	Michigan	5.9	Rhode Island	12.6
Colorado	10.3	Minnesota	6.6	South Carolina	4.1
Connecticut	12.9	Mississippi	1.8	South Dakota	2.2
Delaware	8.1	Missouri	3.3	Tennessee	3.9
Florida	18.9	Montana	1.9	Texas	15.9
Georgia	9.2	Nebraska	5.6	Utah	8.3
Hawaii	16.3	Nevada	19.1	Vermont	3.9
Idaho	5.6	New Hampshire	5.4	Virginia	10.1
Illinois	13.8	New Jersey	20.1	Washington	12.4
Indiana	4.2	New Mexico	10.1	West Virginia	1.2
Iowa	3.8	New York	21.6	Wisconsin	4.4
Kansas	6.3	North Carolina	6.9	Wyoming	2.7
Kentucky	2.7	North Dakota	2.1	Dist. of Columbia	12.7

Step 2. Count the individuals in each class. Here are the counts:

Class	Count
0.1 to 5.0	20
5.1 to 10.0	13
10.1 to 15.0	10
15.1 to 20.0	5
20.1 to 25.0	2
25.1 to 30.0	1

Check that the counts add to 51, the number of individuals in the data (the 50 states and the District of Columbia).

Step 3. Draw the histogram. Mark the scale for the variable whose distribution you are displaying on the horizontal axis. That's the percent of a state's residents who are foreign-born. The scale runs from 0 to 30 because that is the span of the classes we chose. The vertical axis contains the scale of counts. Each bar represents a class. The base of the bar covers the class, and the bar height is the class count. Draw the bars with no horizontal space between them unless a class is empty, so that its bar has height zero. Figure 1.5 is our histogram. ■

Although histograms resemble bar graphs, their details and uses are different. A histogram displays the distribution of a quantitative variable. The horizontal axis of a histogram is marked in the units of measurement for the variable. A bar

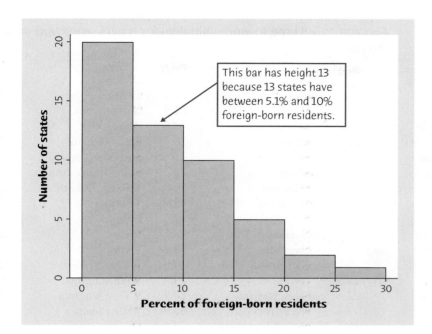

FIGURE 1.5

Histogram of the distribution of the percent of foreign-born residents in the 50 states and the District of Columbia, for Example 1.4.

graph compares the sizes of different quantities. The horizontal axis of a bar graph need not have any measurement scale but simply identifies the quantities being compared. These may be the values of a categorical variable, but they may also be unrelated, like the high-tech devices in Example 1.3. Draw bar graphs with blank space between the bars to separate the quantities being compared. Draw histograms with no space, to indicate that all values of the variable are covered.

Our eyes respond to the *area* of the bars in a histogram.[7] Because the classes are all the same width, area is determined by height and all classes are fairly represented. There is no one right choice of the classes in a histogram. Too few classes will give a "skyscraper" graph, with all values in a few classes with tall bars. Too many will produce a "pancake" graph, with most classes having one or no observations. Neither choice will give a good picture of the shape of the distribution. You must use your judgment in choosing classes to display the shape. Statistics software will choose the classes for you. The software's choice is usually a good one, but you can change it if you want. The histogram function in the *One Variable Statistical Calculator* applet on the text CD and Web site allows you to change the number of classes by dragging with the mouse, so that it is easy to see how the choice of classes affects the histogram.

APPLY YOUR KNOWLEDGE

1.6 Traveling to work. How long must you travel each day to get to work or school? Table 1.2 gives the average travel times to work for workers in each state who are

TABLE 1.2 Average travel time to work (minutes) for adults employed outside the home

STATE	TIME	STATE	TIME	STATE	TIME
Alabama	23.6	Louisiana	25.1	Ohio	22.1
Alaska	17.7	Maine	22.3	Oklahoma	20.0
Arizona	25.0	Maryland	30.6	Oregon	21.8
Arkansas	20.7	Massachusetts	26.6	Pennsylvania	25.0
California	26.8	Michigan	23.4	Rhode Island	22.3
Colorado	23.9	Minnesota	22.0	South Carolina	22.9
Connecticut	24.1	Mississippi	24.0	South Dakota	15.9
Delaware	23.6	Missouri	22.9	Tennessee	23.5
Florida	25.9	Montana	17.6	Texas	24.6
Georgia	27.3	Nebraska	17.7	Utah	20.8
Hawaii	25.5	Nevada	24.2	Vermont	21.2
Idaho	20.1	New Hampshire	24.6	Virginia	26.9
Illinois	27.9	New Jersey	29.1	Washington	25.2
Indiana	22.3	New Mexico	20.9	West Virginia	25.6
Iowa	18.2	New York	30.9	Wisconsin	20.8
Kansas	18.5	North Carolina	23.4	Wyoming	17.9
Kentucky	22.4	North Dakota	15.5	Dist. of Columbia	29.2

at least 16 years old and don't work at home.[8] Make a histogram of the travel times using classes of width 2 minutes starting at 14 minutes. That is, the first bar covers 14.0 to 15.9 minutes, the second covers 16.0 to 17.9 minutes, and so on. (Make this histogram by hand even if you have software, to be sure you understand the process. You may then want to compare your histogram with your software's choice.)

1.7 Choosing classes in a histogram. The data set menu that accompanies the *One Variable Statistical Calculator* applet includes the data on foreign-born residents in the states from Table 1.1. Choose these data, then click on the "Histogram" tab to see a histogram.

(a) How many classes does the applet choose to use? (You can click on the graph outside the bars to get a count of classes.)

(b) Click on the graph and drag to the left. What is the smallest number of classes you can get? What are the lower and upper bounds of each class? (Click on the bar to find out.) Make a rough sketch of this histogram.

(c) Click and drag to the right. What is the greatest number of classes you can get? How many observations does the largest class have?

(d) You see that the choice of classes changes the appearance of a histogram. Drag back and forth until you get the histogram you think best displays the distribution. How many classes did you use?

Interpreting histograms

Making a statistical graph is not an end in itself. *The purpose of graphs is to help us understand the data.* After you make a graph, always ask, "What do I see?" Once you have displayed a distribution, you can see its important features as follows.

EXAMINING A HISTOGRAM

In any graph of data, look for the **overall pattern** and for striking **deviations** from that pattern.

You can describe the overall pattern of a histogram by its **shape, center,** and **spread.**

An important kind of deviation is an **outlier,** an individual value that falls outside the overall pattern.

One way to describe the center of a distribution is by its *midpoint,* the value with roughly half the observations taking smaller values and half taking larger values. For now, we will describe the spread of a distribution by giving the *smallest and largest values.* We will learn better ways to describe center and spread in Chapter 2.

EXAMPLE 1.5 Describing a distribution

Look again at the histogram in Figure 1.5. **Shape:** The distribution has a *single peak* at the left, which represents states in which between 0% and 5% of residents are foreign-born. The distribution is *skewed to the right.* A majority of states have no more than 10% foreign-born residents, but several states have much higher percents, so that the graph extends quite far to the right of its peak. **Center:** Arranging the observations from Table 1.1 in order of size shows that 6.3% (Kansas) is the midpoint of the distribution. There are 25 states with smaller percents foreign-born and 25 with larger. **Spread:** The spread is from 1.2% to 27.2%.

Outliers: Figure 1.5 shows no observations outside the overall single-peaked, right-skewed pattern of the distribution. Figure 1.6 is another histogram of the same distribution, with classes half as wide. Now California, at 27.2%, stands a bit apart to the right of the rest of the distribution. Is California an outlier or just the largest observation in a strongly skewed distribution? Unfortunately, there is no rule. Let's agree to call attention to only strong outliers that suggest something special about an observation—or an error such as typing 10.1 as 101. California is certainly not a strong outlier. ■

Figures 1.5 and 1.6 remind us that interpreting graphs calls for judgment. We also see that *the choice of classes in a histogram can influence the appearance of a distribution.* Because of this, and to avoid worrying about minor details, concentrate on the main features of a distribution. Look for major peaks, not for minor ups and downs, in the bars of the histogram. (For example, don't conclude that Figure 1.6 shows a second peak between 10% and 15%.) Look for clear outliers, not just for the smallest and largest observations. Look for rough *symmetry* or clear *skewness.*

FIGURE 1.6

Another histogram of the distribution of the percent of foreign-born residents, with classes half as wide as in Figure 1.5. Histograms with more classes show more detail but may have a less clear pattern.

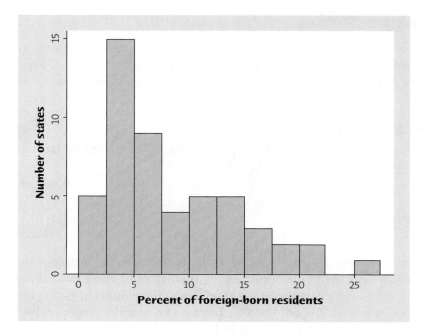

SYMMETRIC AND SKEWED DISTRIBUTIONS

A distribution is **symmetric** if the right and left sides of the histogram are approximately mirror images of each other.

A distribution is **skewed to the right** if the right side of the histogram (containing the half of the observations with larger values) extends much farther out than the left side. It is **skewed to the left** if the left side of the histogram extends much farther out than the right side.

Here are more examples of describing the overall pattern of a histogram.

Courtesy Riverside Publishing

EXAMPLE 1.6   Iowa Test scores

Figure 1.7 displays the scores of all 947 seventh-grade students in the public schools of Gary, Indiana, on the vocabulary part of the Iowa Test of Basic Skills.[9] The distribution is *single-peaked* and *symmetric*. In mathematics, the two sides of symmetric patterns are exact mirror images. Real data are almost never exactly symmetric. We are content to describe Figure 1.7 as symmetric. The center (half above, half below) is close to 7. This is seventh-grade reading level. The scores range from 2.0 (second-grade level) to 12.1 (twelfth-grade level).

Notice that the vertical scale in Figure 1.7 is not the *count* of students but the *percent* of students in each histogram class. A histogram of percents rather than counts is convenient when we want to compare several distributions. To compare Gary with Los Angeles, a much bigger city, we would use percents so that both histograms have the same vertical scale. ■

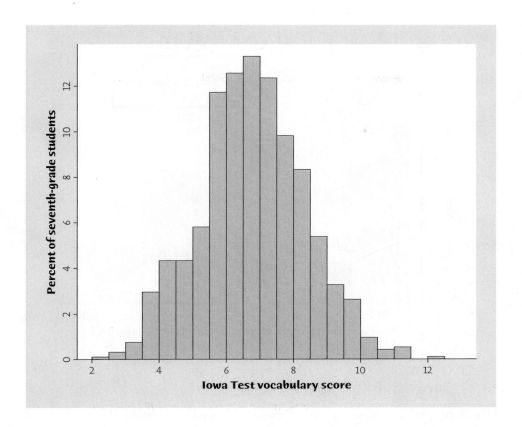

FIGURE 1.7

Histogram of the Iowa Test vocabulary scores of all seventh-grade students in Gary, Indiana, for Example 1.6. This distribution is single-peaked and symmetric.

EXAMPLE 1.7 **Who takes the SAT?**

Depending on where you went to high school, the answer to this question may be "almost everybody" or "almost nobody." Figure 1.8 is a histogram of the percent of high school graduates in each state who took the SAT Reasoning test.[10]

The histogram shows two peaks, a high peak at the left and a lower but broader peak centered in the 60% to 80% class. Several peaks suggest that a distribution mixes several kinds of individuals. That is the case here. There are two major tests of readiness for college, the ACT and the SAT. Most states have a strong preference for one or the other. In some states, many students take the ACT exam and few take the SAT—these states form the peak on the left. In other states, many students take the SAT and few choose the ACT—these states form the broader peak at the right.

Giving the center and spread of this distribution is not very useful. The midpoint falls in the 20% to 40% class, between the two peaks. The story told by the histogram is in the two peaks corresponding to ACT states and SAT states. ■

The overall shape of a distribution is important information about a variable. Some variables have distributions with predictable shapes. Many biological measurements on specimens from the same species and sex—lengths of bird bills, heights of young women—have symmetric distributions. On the other hand,

FIGURE 1.8

Histogram of the percent of high school graduates in each state who took the SAT Reasoning test, for Example 1.7. The graph shows two groups of states: ACT states (where few students take the SAT) at the left and SAT states at the right.

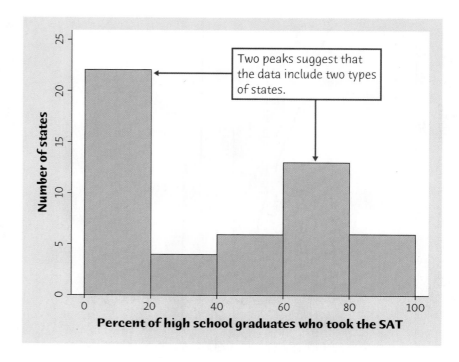

data on people's incomes are usually strongly skewed to the right. There are many moderate incomes, some large incomes, and a few enormous incomes. Many distributions have irregular shapes that are neither symmetric nor skewed. Some data show other patterns, such as the two peaks in Figure 1.8. Use your eyes, describe the pattern you see, and then try to explain the pattern.

APPLY YOUR KNOWLEDGE

1.8 Traveling to work. In Exercise 1.6, you made a histogram of the average travel times to work in Table 1.2. The shape of the distribution is a bit irregular. Is it closer to symmetric or skewed? About where is the center (midpoint) of the data? What is the spread in terms of the smallest and largest values?

1.9 Unmarried women. Figure 1.9 shows the distribution of the state percents of women aged 15 and over who have never been married.

(a) The main body of the distribution is slightly skewed to the right. There is one clear outlier, the District of Columbia. Why is it not surprising that the percent of never-married women is higher in DC than in the 50 states?

(b) The midpoint of the distribution is the 26th state in order of percent of never-married women. In which class does the midpoint fall? About what is the spread (smallest to largest) of the distribution?

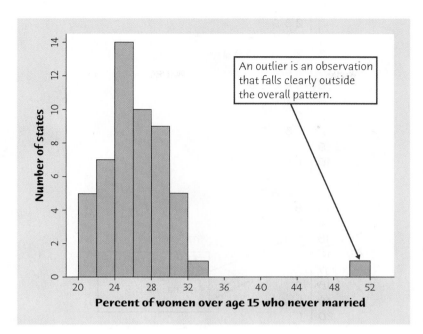

An outlier is an observation that falls clearly outside the overall pattern.

FIGURE 1.9

Histogram of the state percents of women aged 15 and over who have never been married, for Exercise 1.9.

Quantitative variables: stemplots

Histograms are not the only graphical display of distributions. For small data sets, a *stemplot* is quicker to make and presents more detailed information.

> **STEMPLOT**
>
> To make a **stemplot:**
>
> 1. Separate each observation into a **stem,** consisting of all but the final (rightmost) digit, and a **leaf,** the final digit. Stems may have as many digits as needed, but each leaf contains only a single digit.
>
> 2. Write the stems in a vertical column with the smallest at the top, and draw a vertical line at the right of this column. Be sure to include all the stems needed to span the data, even when some will have no leaves.
>
> 3. Write each leaf in the row to the right of its stem, in increasing order out from the stem.

EXAMPLE 1.8 Making a stemplot

Table 1.1 presents the percents of state residents who were born outside the United States. To make a stemplot of these data, take the whole-number part of the percent as the stem and the final digit (tenths) as the leaf. Write stems from 1 for Mississippi, Montana, and West Virginia to 27 for California. Now add leaves. Arizona, 15.1%, has leaf 1 on the 15 stem. Texas, at 15.9%, places leaf 9 on the same stem. These are the

The vital few

Skewed distributions can show us where to concentrate our efforts. Ten percent of the cars on the road account for half of all carbon dioxide emissions. A histogram of CO_2 emissions would show many cars with small or moderate values and a few with very high values. Cleaning up or replacing these cars would reduce pollution at a cost much lower than that of programs aimed at all cars. Statisticians who work at improving quality in industry make a principle of this: distinguish "the vital few" from "the trivial many."

FIGURE 1.10

Stemplot of the percents of foreign-born residents in the states, for Example 1.8. Each stem is a percent and leaves are tenths of a percent.

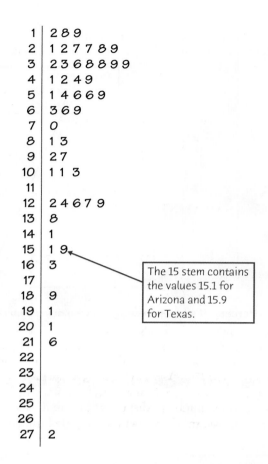

The 15 stem contains the values 15.1 for Arizona and 15.9 for Texas.

only observations on this stem. Arrange the leaves in order, so that 15 | 19 is one row in the stemplot. Figure 1.10 is the complete stemplot for the data in Table 1.1. ■

A stemplot looks like a histogram turned on end. Compare the stemplot in Figure 1.10 with the histograms of the same data in Figures 1.5 and 1.6. The stemplot is like a histogram with many classes. You can choose the classes in a histogram. The classes (the stems) of a stemplot are given to you. All three graphs show a distribution that has one peak and is right-skewed. Figures 1.6 and 1.10 have enough classes to show that California (27.2%) stands slightly apart from the long right tail of the skewed distribution. Histograms are more flexible than stemplots because you can choose the classes. But the stemplot, unlike the histogram, preserves the actual value of each observation. *Stemplots do not work well for large data sets, where each stem must hold a large number of leaves.* Don't try to make a stemplot of a large data set, such as the 947 Iowa Test scores in Figure 1.7.

EXAMPLE 1.9 Pulling wood apart

Student engineers learn that, although handbooks give the strength of a material as a single number, in fact the strength varies from piece to piece. A vital lesson in all fields of study is that "variation is everywhere." Here are data from a typical student laboratory exercise: the load in pounds needed to pull apart pieces of Douglas fir 4 inches long and 1.5 inches square.

33,190	31,860	32,590	26,520	33,280
32,320	33,020	32,030	30,460	32,700
23,040	30,930	32,720	33,650	32,340
24,050	30,170	31,300	28,730	31,920

Courtesy Department of Civil Engineering, University of New Mexico

rounding

A stemplot of these data would have very many stems and no leaves or just one leaf on most stems. So we first **round** the data to the nearest hundred pounds. The rounded data are

332	319	326	265	333	323	330	320	305	327
230	309	327	337	323	241	302	313	287	319

Now we can make a stemplot with the first two digits (thousands of pounds) as stems and the third digit (hundreds of pounds) as leaves. Figure 1.11 is the stemplot. Rotate the stemplot counterclockwise so that it resembles a histogram, with 230 at the left end of the scale. This makes it clear that the distribution is *skewed to the left*. The *midpoint* is around 320 (32,000 pounds) and the *spread* is from 230 to 337. Because of the strong skew, we are reluctant to call the smallest observations outliers. They appear to be part of the long left tail of the distribution. Before using wood like this in construction, we should ask why some pieces are much weaker than the rest. ■

```
23 | 0
24 | 1
25 |
26 | 5
27 |
28 | 7
29 |
30 | 2 5 9
31 | 3 9 9
32 | 0 3 3 6 7 7
33 | 0 2 3 7
```

FIGURE 1.11

Stemplot of the breaking strength of pieces of wood, rounded to the nearest hundred pounds, for Example 1.9. Stems are thousands of pounds and leaves are hundreds of pounds.

Comparing Figures 1.10 (right-skewed) and 1.11 (left-skewed) reminds us that *the direction of skewness is the direction of the long tail, not the direction where most observations are clustered.*

You can also **split stems** in a stemplot to double the number of stems when all the leaves would otherwise fall on just a few stems. Each stem then appears twice. Leaves 0 to 4 go on the upper stem, and leaves 5 to 9 go on the lower stem. If you

splitting stems

split the stems in the stemplot of Figure 1.11, for example, the 32 and 33 stems become

$$
\begin{array}{r|l}
32 & 0\ 3\ 3 \\
32 & 6\ 7\ 7 \\
33 & 0\ 2\ 3 \\
33 & 7
\end{array}
$$

Rounding and splitting stems are matters for judgment, like choosing the classes in a histogram. The wood strength data require rounding but don't require splitting stems. The *One Variable Statistical Calculator* applet on the text CD and Web site allows you to decide whether to split stems, so that it is easy to see the effect.

APPLY YOUR KNOWLEDGE

1.10 Traveling to work. Make a stemplot of the average travel times to work in Table 1.2. Use whole minutes as your stems. Because the stemplot preserves the actual values of the observations, it is easy to find the midpoint (26th of the 51 observations in order) and the spread. What are they?

1.11 Health care spending. Table 1.3 shows the annual spending per person on health care in the world's richer countries.[11] Make a stemplot of the data after rounding to the nearest $100 (so that stems are thousands of dollars and leaves are hundreds of dollars). Split the stems, placing leaves 0 to 4 on the first stem and leaves 5 to 9 on the second stem of the same value. Describe the shape, center, and spread of the distribution. Which country is the high outlier?

TABLE 1.3 Annual spending per person on health care (in U.S. dollars)

COUNTRY	DOLLARS	COUNTRY	DOLLARS	COUNTRY	DOLLARS
Argentina	1067	Hungary	1269	Poland	745
Australia	2874	Iceland	3110	Portugal	1791
Austria	2306	Ireland	2496	Saudi Arabia	578
Belgium	2828	Israel	1911	Singapore	1156
Canada	2989	Italy	2266	Slovakia	777
Croatia	838	Japan	2244	Slovenia	1669
Czech Republic	1302	Korea	1074	South Africa	669
Denmark	2762	Kuwait	567	Spain	1853
Estonia	682	Lithuania	754	Sweden	2704
Finland	2108	Netherlands	2987	Switzerland	3776
France	2902	New Zealand	1893	United Kingdom	2389
Germany	3001	Norway	3809	United States	5711
Greece	1997	Oman	419		

Time plots

Many variables are measured at intervals over time. We might, for example, measure the height of a growing child or the price of a stock at the end of each month. In these examples, our main interest is change over time. To display change over time, make a *time plot*.

TIME PLOT

A **time plot** of a variable plots each observation against the time at which it was measured. Always put time on the horizontal scale of your plot and the variable you are measuring on the vertical scale. Connecting the data points by lines helps emphasize any change over time.

EXAMPLE 1.10   **Water levels in the Everglades**

Water levels in Everglades National Park are critical to the survival of this unique region. The photo shows a water-monitoring station in Shark River Slough, the main path for surface water moving through the "river of grass" that is the Everglades. Figure 1.12 is a time plot of water levels at this station from mid-August 2000 to mid-June 2003.[12] ■

Courtesy U.S. Geological Survey

cycles

When you examine a time plot, look once again for an overall pattern and for strong deviations from the pattern. Figure 1.12 shows strong **cycles,** regular up-and-down movements in water level. The cycles show the effects of Florida's wet season (roughly June to November) and dry season (roughly December to May). Water levels are highest in late fall. In April and May of 2001 and 2002, water levels were less than zero—the water table was below ground level and the surface was dry. If you look closely, you can see year-to-year variation. The dry season in 2003 ended early, with the first-ever April tropical storm. In consequence, the dry-season water level in 2003 never dipped below zero.

Another common overall pattern in a time plot is a **trend,** a long-term upward or downward movement over time. Many economic variables show an upward trend. Incomes, house prices, and (alas) college tuitions tend to move generally upward over time.

trend

Histograms and time plots give different kinds of information about a variable. The time plot in Figure 1.12 presents **time series data** that show the change in water level at one location over time. A histogram displays **cross-sectional data,** such as water levels at many locations in the Everglades at the same time.

time series data
cross-sectional data

APPLY YOUR KNOWLEDGE

1.12 The cost of college. Here are data on the average tuition and fees charged to in-state students by public four-year colleges and universities for the 1976 to 2007

FIGURE 1.12

Time plot of water depth at a monitoring station in Everglades National Park over a period of almost three years, for Example 1.10. The yearly cycles reflect Florida's wet and dry seasons.

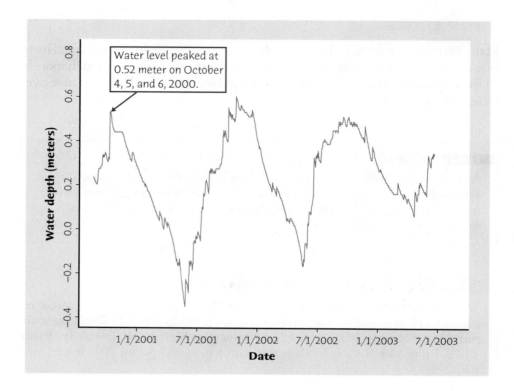

academic years. Because almost any variable measured in dollars increases over time due to inflation (the falling buying power of a dollar), the values are given in "constant dollars," adjusted to have the same buying power that a dollar had in 2007.[13]

Year	Tuition	Year	Tuition	Year	Tuition	Year	Tuition
1976	$2,197	1984	$2,426	1992	$3,444	2000	$4,221
1977	$2,225	1985	$2,532	1993	$3,623	2001	$4,411
1978	$1,986	1986	$2,656	1994	$3,758	2002	$4,715
1979	$1,986	1987	$2,699	1995	$3,802	2003	$5,231
1980	$1,939	1988	$2,721	1996	$3,913	2004	$5,624
1981	$2,018	1989	$2,792	1997	$4,022	2005	$5,814
1982	$2,194	1990	$2,977	1998	$4,131	2006	$5,918
1983	$2,358	1991	$3,187	1999	$4,183	2007	$6,185

(a)   Make a time plot of average tuition and fees.

(b)   What overall pattern does your plot show?

(c)   Some possible deviations from the overall pattern are outliers, periods when charges went down (in 2007 dollars), and periods of particularly rapid increase. Which are present in your plot, and during which years?

CHAPTER 1   SUMMARY

- A data set contains information on a number of **individuals.** Individuals may be people, animals, or things. For each individual, the data give values for one or more **variables.** A variable describes some characteristic of an individual, such as a person's height, sex, or salary.

- Some variables are **categorical** and others are **quantitative.** A categorical variable places each individual into a category, such as male or female. A quantitative variable has numerical values that measure some characteristic of each individual, such as height in centimeters or salary in dollars.

- **Exploratory data analysis** uses graphs and numerical summaries to describe the variables in a data set and the relations among them.

- After you understand the background of your data (individuals, variables, units of measurement), the first thing to do is almost always **plot your data.**

- The **distribution** of a variable describes what values the variable takes and how often it takes these values. **Pie charts** and **bar graphs** display the distribution of a categorical variable. Bar graphs can also compare any set of quantities measured in the same units. **Histograms** and **stemplots** graph the distribution of a quantitative variable.

- When examining any graph, look for an **overall pattern** and for notable **deviations** from the pattern.

- **Shape, center, and spread** describe the overall pattern of the distribution of a quantitative variable. Some distributions have simple shapes, such as **symmetric** or **skewed.** Not all distributions have a simple overall shape, especially when there are few observations.

- **Outliers** are observations that lie outside the overall pattern of a distribution. Always look for outliers and try to explain them.

- When observations on a variable are taken over time, make a **time plot** that graphs time horizontally and the values of the variable vertically. A time plot can reveal **trends, cycles,** or other changes over time.

CHECK YOUR SKILLS

The multiple-choice exercises in "Check Your Skills" ask straightforward questions about basic facts from the chapter. Answers to all of these exercises appear in the back of the book. You should expect all of your answers to be correct.

1.13 Here are the first lines of a professor's data set at the end of a statistics course:

```
Name                Major     Points     Grade

ADVANI, SURA        COMM      397        B
BARTON, DAVID       HIST      323        C
BROWN, ANNETTE      BIOL      446        A
CHIU, SUN           PSYC      405        B
CORTEZ, MARIA       PSYC      461        A
```

The individuals in these data are

(a) the students.   (b) the total points.   (c) the course grades.

1.14 To display the distribution of grades (A, B, C, D, F) in a course, it would be correct to use

(a) a pie chart but not a bar graph.

(b) a bar graph but not a pie chart.

(c) either a pie chart or a bar graph.

1.15 A study of recent college graduates records the sex and total college debt in dollars for 10,000 people a year after they graduate from college.

(a) Sex and college debt are both categorical variables.

(b) Sex and college debt are both quantitative variables.

(c) Sex is a categorical variable and college debt is a quantitative variable.

1.16 A political party's data bank includes the zip codes of past donors, such as

      47906   34236   53075   10010   90210   75204   30304   99709

Zip code is a

(a) quantitative variable.   (b) categorical variable.   (c) unit of measurement.

1.17 Figure 1.9 (page 19) is a histogram of the percent of women in each state aged 15 and over who have never been married. The leftmost bar in the histogram covers percents of never-married women ranging from about

(a) 20% to 24%.   (b) 20% to 22%.   (c) 0% to 20%.

1.18 Here are the amounts of money (cents) in coins carried by 10 students in a statistics class:

      50   35   0   97   76   0   0   87   23   65

To make a stemplot of these data, you would use stems

(a) 0, 1, 2, 3, 4, 5, 6, 7, 8, 9.

(b) 0, 2, 3, 5, 6, 7, 8, 9.

(c) 00, 10, 20, 30, 40, 50, 60, 70, 80, 90.

1.19 The population of the United States is aging, though less rapidly than in other developed countries. Here is a stemplot of the percents of residents aged 65 and older in the 50 states and the District of Columbia. The stems are whole percents and the leaves are tenths of a percent.

```
 6 | 8
 7 |
 8 | 8
 9 | 7 9
10 | 0 8
11 | 1 5 5 6 6
12 | 0 1 2 2 2 3 4 4 4 5 7 8 8 8 9 9 9
13 | 0 1 2 3 3 3 3 3 4 4 4 8 9 9
14 | 0 2 6 6 6
15 | 2 3
16 | 8
```

The outlier is Alaska. What percent of Alaska residents are 65 or older?

(a) 6.8%   (b) 16.8%   (c) 68%

1.20 Ignoring the outlier, the shape of the distribution in Exercise 1.19 is

(a) clearly skewed to the right.

(b) roughly symmetric.

(c) clearly skewed to the left.

1.21 The center of the distribution in Exercise 1.19 is close to

(a) 12.8%.   (b) 12.0%.   (c) 6.8% to 16.8%.

1.22 You look at real estate ads for houses in Naples, Florida. There are many houses ranging from $200,000 to $500,000 in price. The few houses on the water, however, have prices up to $15 million. The distribution of house prices will be

(a) skewed to the left.

(b) roughly symmetric.

(c) skewed to the right.

CHAPTER 1   EXERCISES

1.23 Medical students. Students who have finished medical school are assigned to residencies in hospitals to receive further training in a medical specialty. Here is part of a hypothetical database of students seeking residency positions. USMLE is the student's score on Step 1 of the national medical licensing examination.

Name	Medical school	Sex	Age	USMLE	Specialty sought
Abrams, Laurie	Florida	F	28	238	Family medicine
Brown, Gordon	Meharry	M	25	205	Radiology
Cabrera, Maria	Tufts	F	26	191	Pediatrics
Ismael, Miranda	Indiana	F	32	245	Internal medicine

(a) What individuals does this data set describe?

(b) In addition to the student's name, how many variables does the data set contain? Which of these variables are categorical and which are quantitative?

1.24 Protecting wood. How can we help wood surfaces resist weathering, especially when restoring historic wooden buildings? In a study of this question, researchers prepared wooden panels and then exposed them to the weather. Here are some of the variables recorded. Which of these variables are categorical and which are quantitative?

(a) Type of wood (yellow poplar, pine, cedar)

(b) Type of water repellent (solvent-based, water-based)

(c) Paint thickness (millimeters)

(d) Paint color (white, gray, light blue)

(e) Weathering time (months)

© Photo 24/Age fotostock

1.25 What color is your car? The most popular colors for cars and light trucks change over time. Silver passed green in 2000 to become the most popular color worldwide, then gave way to shades of white in 2007. Here is the distribution of colors for vehicles sold in North America in 2007:[14]

Color	Popularity
White	19%
Silver	18%
Black	16%
Red	13%
Gray	12%
Blue	12%
Beige, brown	5%
Other colors	

Fill in the percent of vehicles that are in other colors. Make a graph to display the distribution of color popularity.

1.26 Buying music online. Young people are more likely than older folk to buy music online. Here are the percents of people in several age groups who bought music online in 2006:[15]

Age group	Bought music online
12 to 17 years	24%
18 to 24 years	21%
25 to 34 years	20%
35 to 44 years	16%
45 to 54 years	10%
55 to 64 years	3%
65 years and over	1%

(a) Explain why it is *not* correct to use a pie chart to display these data.

(b) Make a bar graph of the data.

1.27 Deaths among young people. Among persons aged 15 to 24 years in the United States, the leading causes of death and the number of deaths in 2005 were: accidents, 15,567; homicide, 5359; suicide, 4139; cancer, 1717; heart disease, 1067; congenital defects, 483.[16]

(a) Make a bar graph to display these data.

(b) To make a pie chart, you need one additional piece of information. What is it?

1.28 Hispanic origins. Figure 1.13 is a pie chart prepared by the Census Bureau to show the origin of the more than 43 million Hispanics in the United States in 2006.[17] About what percent of Hispanics are Mexican? Puerto Rican? You see that it is hard to determine numbers from a pie chart. Bar graphs are much easier to use. (The Census Bureau did include the percents in its pie chart.)

**Percent Distribution of Hispanics
by Type: 2006**

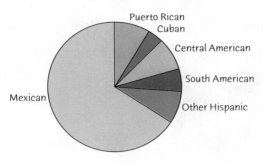

FIGURE 1.13
Pie chart of the national origins of Hispanic residents of the United States, for Exercise 1.28.

1.29 Spam. Email spam is the curse of the Internet. Here is a compilation of the most common types of spam:[18]

Type of spam	Percent
Adult	19
Financial	20
Health	7
Internet	7
Leisure	6
Products	25
Scams	9

Make two bar graphs of these percents, one with bars ordered as in the table (alphabetically) and the other with bars in order from tallest to shortest. Comparisons are easier if you order the bars by height.

1.30 Do adolescent girls eat fruit? We all know that fruit is good for us. Many of us don't eat enough. Figure 1.14 is a histogram of the number of servings of fruit per day claimed by 74 seventeen-year-old girls in a study in Pennsylvania.[19] Describe the shape, center, and spread of this distribution. What percent of these girls ate fewer than two servings per day?

1.31 IQ test scores. Figure 1.15 is a stemplot of the IQ test scores of 78 seventh-grade students in a rural midwestern school.[20]

(a)   Four students had low scores that might be considered outliers. Ignoring these, describe the shape, center, and spread of the distribution. (Notice that it looks roughly bell-shaped.)

(b)   We often read that IQ scores for large populations are centered at 100. What percent of these 78 students have scores above 100?

1.32 Returns on common stocks. The return on a stock is the change in its market price plus any dividend payments made. Total return is usually expressed as a percent

FIGURE 1.14

The distribution of fruit consumption in a sample of 74 seventeen-year-old girls, for Exercise 1.30.

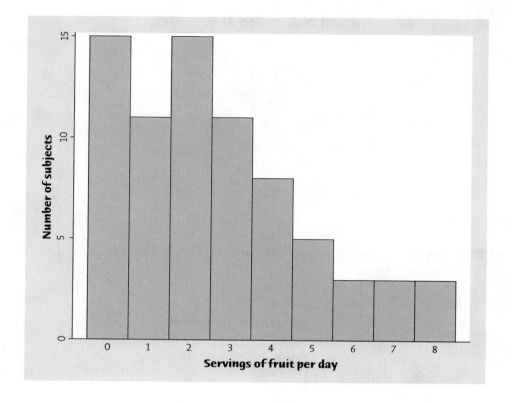

of the beginning price. Figure 1.16 is a histogram of the distribution of the monthly returns for all stocks listed on U.S. markets from January 1985 to September 2007 (273 months).[21] The extreme low outlier is the market crash of October 1987, when stocks lost 23% of their value in one month.

(a) Ignoring the outliers, describe the overall shape of the distribution of monthly returns.

FIGURE 1.15

The distribution of IQ scores for 78 seventh-grade students, for Exercise 1.31.

```
 7 | 2 4
 7 | 7 9
 8 |
 8 | 6 9
 9 | 0 1 3 3
 9 | 6 7 7 8
10 | 0 0 2 2 3 3 3 3 4 4
10 | 5 5 5 6 6 6 7 7 7 8 9
11 | 0 0 0 0 1 1 1 1 2 2 2 2 3 3 3 4 4 4 4
11 | 5 5 6 8 8 9 9 9
12 | 0 0 3 3 4 4
12 | 6 7 7 8 8 8
13 | 0 2
13 | 6
```

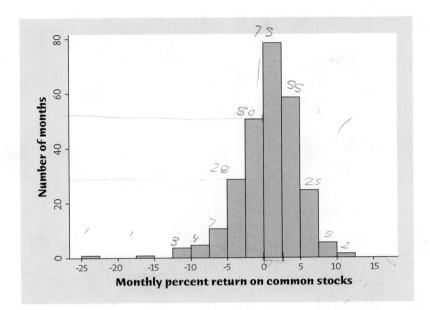

(b) What is the approximate center of this distribution? (For now, take the center to be the value with roughly half the months having lower returns and half having higher returns.)

(c) Approximately what were the smallest and largest monthly returns, leaving out the outliers? (This is one way to describe the spread of the distribution.)

(d) A return less than zero means that stocks lost value in that month. About what percent of all months had returns less than zero?

1.33 Name that variable. A survey of a large college class asked the following questions:

1. Are you female or male? (In the data, male = 0, female = 1.)

2. Are you right-handed or left-handed? (In the data, right = 0, left = 1.)

3. What is your height in inches?

4. How many minutes do you study on a typical weeknight?

Figure 1.17 shows histograms of the student responses, in scrambled order and without scale markings. Which histogram goes with each variable? Explain your reasoning.

1.34 Food oils and health. Fatty acids, despite their unpleasant name, are necessary for human health. Two types of essential fatty acids, called omega-3 and omega-6, are not produced by our bodies and so must be obtained from our food. Food oils, widely used in food processing and cooking, are major sources of these compounds. There is some evidence that a healthy diet should have more omega-3 than omega-6. Table 1.4 gives the ratio of omega-3 to omega-6 in some common food oils.[22] Values greater than 1 show that an oil has more omega-3 than omega-6.

(a) Make a histogram of these data, using classes bounded by the whole numbers from 0 to 6.

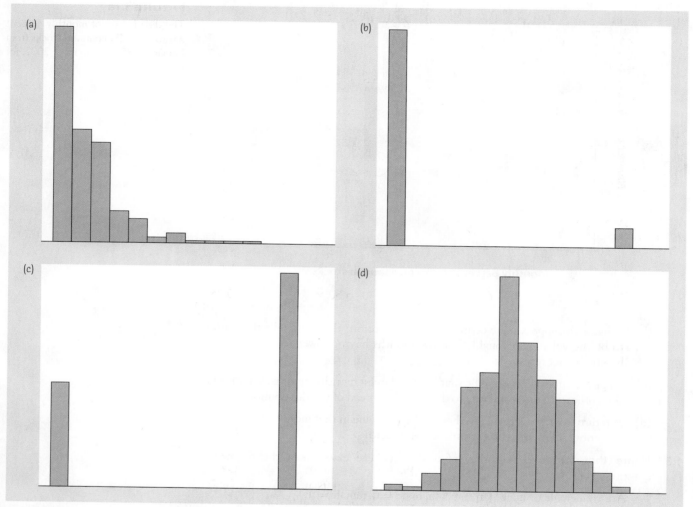

FIGURE 1.17

Histograms of four distributions, for Exercise 1.33.

(b)   What is the shape of the distribution? How many of the 30 food oils have more omega-3 than omega-6? What does this distribution suggest about the possible health effects of modern food oils?

(c)   Table 1.4 contains entries for several fish oils (cod, herring, menhaden, salmon, sardine). How do these values support the idea that eating fish is healthy?

1.35 Where are the doctors? Table 1.5 gives the number of active medical doctors per 100,000 people in each state.[23]

(a)   Why is the number of doctors per 100,000 people a better measure of the availability of health care than a simple count of the number of doctors in a state?

(b)   Make a histogram that displays the distribution of doctors per 100,000 people. Write a brief description of the distribution. Are there any outliers? If so, can you explain them?

TABLE 1.4 Omega-3 fatty acids as a fraction of omega-6 fatty acids
in food oils

OIL	RATIO	OIL	RATIO
Perilla	5.33	Flaxseed	3.56
Walnut	0.20	Canola	0.46
Wheat germ	0.13	Soybean	0.13
Mustard	0.38	Grape seed	0.00
Sardine	2.16	Menhaden	1.96
Salmon	2.50	Herring	2.67
Mayonnaise	0.06	Soybean	0.07
Cod liver	2.00	Rice bran	0.05
Shortening (household)	0.11	Butter	0.64
Shortening (industrial)	0.06	Sunflower	0.03
Margarine	0.05	Corn	0.01
Olive	0.08	Sesame	0.01
Shea nut	0.06	Cottonseed	0.00
Sunflower (oleic)	0.05	Palm	0.02
Sunflower (linoleic)	0.00	Cocoa butter	0.04

TABLE 1.5 Medical doctors per 100,000 people, by state

STATE	DOCTORS	STATE	DOCTORS	STATE	DOCTORS
Alabama	213	Louisiana	264	Ohio	261
Alaska	222	Maine	267	Oklahoma	171
Arizona	208	Maryland	411	Oregon	263
Arkansas	203	Massachusetts	450	Pennsylvania	294
California	259	Michigan	240	Rhode Island	351
Colorado	258	Minnesota	281	South Carolina	230
Connecticut	363	Mississippi	181	South Dakota	219
Delaware	248	Missouri	239	Tennessee	261
Florida	245	Montana	221	Texas	212
Georgia	220	Nebraska	239	Utah	209
Hawaii	310	Nevada	186	Vermont	362
Idaho	169	New Hampshire	260	Virginia	270
Illinois	272	New Jersey	306	Washington	265
Indiana	213	New Mexico	240	West Virginia	229
Iowa	187	New York	389	Wisconsin	254
Kansas	220	North Carolina	253	Wyoming	188
Kentucky	230	North Dakota	242	Dist. of Columbia	798

TABLE 1.6 Carbon dioxide emissions (metric tons per person)

COUNTRY	CO₂	COUNTRY	CO₂	COUNTRY	CO₂
Algeria	2.6	Iran	6.0	Poland	7.8
Argentina	3.6	Iraq	2.9	Romania	4.2
Australia	18.4	Italy	7.8	Russia	10.8
Bangladesh	0.3	Japan	9.5	Saudi Arabia	13.8
Brazil	1.8	Kenya	0.3	South Africa	7.0
Canada	17.0	Korea, North	3.3	Spain	7.9
China	3.9	Korea, South	9.3	Sudan	0.3
Colombia	1.3	Malaysia	5.5	Tanzania	0.1
Congo	0.2	Mexico	3.7	Thailand	3.3
Egypt	2.0	Morocco	1.4	Turkey	3.0
Ethiopia	0.1	Myanmar	0.2	Ukraine	6.3
France	6.2	Nepal	0.1	United Kingdom	8.8
Germany	9.9	Nigeria	0.4	United States	19.6
Ghana	0.3	Pakistan	0.8	Uzbekistan	4.2
India	1.1	Peru	1.0	Venezuela	5.4
Indonesia	1.6	Philippines	0.9	Vietnam	1.0

1.36 Carbon dioxide emissions. Burning fuels in power plants or motor vehicles emits carbon dioxide (CO_2), which contributes to global warming. Table 1.6 displays CO_2 emissions per person from countries with populations of at least 20 million.[24]

(a) Why do you think we choose to measure emissions per person rather than total CO_2 emissions for each country?

(b) Make a stemplot to display the data of Table 1.6. Describe the shape, center, and spread of the distribution. Which countries are outliers?

1.37 Rock sole in the Bering Sea. "Recruitment," the addition of new members to a fish population, is an important measure of the health of ocean ecosystems. The table gives data on the recruitment of rock sole in the Bering Sea from 1973 to 2000.[25] Make a stemplot to display the distribution of yearly rock sole recruitment. (Round to the nearest hundred and split the stems.) Describe the shape, center, and spread of the distribution and any striking deviations that you see.

Sarkis Images/Alamy

Year	Recruitment (millions)	Year	Recruitment (millions)	Year	Recruitment (millions)	Year	Recruitment (millions)
1973	173	1980	1411	1987	4700	1994	505
1974	234	1981	1431	1988	1702	1995	304
1975	616	1982	1250	1989	1119	1996	425
1976	344	1983	2246	1990	2407	1997	214
1977	515	1984	1793	1991	1049	1998	385
1978	576	1985	1793	1992	505	1999	445
1979	727	1986	2809	1993	998	2000	676

1.38 Do women study more than men? We asked the students in a large first-year college class how many minutes they studied on a typical weeknight. Here are the responses of random samples of 30 women and 30 men from the class:

Women					Men				
180	120	180	360	240	90	120	30	90	200
120	180	120	240	170	90	45	30	120	75
150	120	180	180	150	150	120	60	240	300
200	150	180	150	180	240	60	120	60	30
120	60	120	180	180	30	230	120	95	150
90	240	180	115	120	0	200	120	120	180

(a) Examine the data. Why are you not surprised that most responses are multiples of 10 minutes? We eliminated one student who claimed to study 30,000 minutes per night. Are there any other responses you consider suspicious?

(b) Make a **back-to-back stemplot** to compare the two samples. That is, use one set of stems with two sets of leaves, one to the right and one to the left of the stems. (Draw a line on either side of the stems to separate stems and leaves.) Order both sets of leaves from smallest at the stem to largest away from the stem. Report the approximate midpoints of both groups. Does it appear that women study more than men (or at least claim that they do)?

back-to-back stemplot

1.39 Rock sole in the Bering Sea. Make a time plot of the rock sole recruitment data in Exercise 1.37. What does the time plot show that your stemplot in Exercise 1.37 did not show? When you have time series data, a time plot is often needed to understand what is happening.

1.40 Marijuana and traffic accidents. Researchers in New Zealand interviewed 907 drivers at age 21. They had data on traffic accidents and they asked the drivers about marijuana use. Here are data on the numbers of accidents caused by these drivers at age 19, broken down by marijuana use at the same age:[26]

	Marijuana Use per Year			
	Never	**1–10 times**	**11–50 times**	**51 + times**
Drivers	452	229	70	156
Accidents caused	59	36	15	50

(a) Explain carefully why a useful graph must compare *rates* (accidents per driver) rather than *counts* of accidents in the four marijuana use classes.

(b) Make a graph that displays the accident rate for each class. What do you conclude? (You can't conclude that marijuana use *causes* accidents, because risk takers are more likely both to drive aggressively and to use marijuana.)

1.41 Dates on coins. Sketch a histogram for a distribution that is skewed to the left. Suppose that you and your friends emptied your pockets of coins and recorded the year marked on each coin. The distribution of dates would be skewed to the left. Explain why.

1.42 El Niño and the monsoon. The earth is interconnected. For example, it appears that El Niño, the periodic warming of the Pacific Ocean west of South America, affects the monsoon rains that are essential for agriculture in India. Here are the monsoon rains (in millimeters) for the 23 strong El Niño years between 1871 and 2004:[27]

| 628 | 669 | 740 | 651 | 710 | 736 | 717 | 698 | 653 | 604 | 781 | 784 |
| 790 | 811 | 830 | 858 | 858 | 896 | 806 | 790 | 792 | 957 | 872 | |

(a) To make a stemplot of these rainfall amounts, round the data to the nearest 10, so that stems are hundreds of millimeters and leaves are tens of millimeters. Make two stemplots, with and without splitting the stems. Which plot do you prefer?

(b) Describe the shape, center, and spread of the distribution.

(c) The average monsoon rainfall for all years from 1871 to 2004 is about 850 millimeters. What effect does El Niño appear to have on monsoon rains?

1.43 Watch those scales! The impression that a time plot gives depends on the scales you use on the two axes. If you stretch the vertical axis and compress the time axis, change appears to be more rapid. Compressing the vertical axis and stretching the time axis make change appear slower. Make two more time plots of the college tuition data in Exercise 1.12 (page 24), one that makes tuition appear to increase very rapidly and one that shows only a gentle increase. The moral of this exercise is: pay close attention to the scales when you look at a time plot.

FIGURE 1.18

Time plot of the monthly count of new single-family houses started (in thousands) between January 1990 and December 2007, for Exercise 1.44.

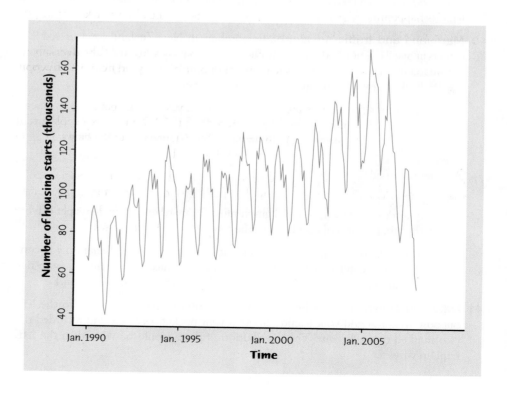

1.44 Housing starts. Figure 1.18 is a time plot of the number of single-family homes started by builders each month from January 1990 to December 2007.[28] The counts are in thousands of homes.

(a) The most notable pattern in this time plot is yearly up-and-down cycles. At what season of the year are housing starts highest? Lowest? The cycles are explained by the weather in the northern part of the country.

(b) Is there a longer-term trend visible in addition to the cycles? If so, describe it.

(c) The big economic news of 2007 was a severe downturn in housing that began in mid-2006. How is this downturn visible in the time plot?

1.45 Alligator bites. Here are data on the number of people bitten by alligators in Florida over a 36-year period:[29]

Year	Number bitten	Year	Number bitten	Year	Number bitten	Year	Number bitten
1972	4	1981	10	1990	17	1999	16
1973	3	1982	7	1991	20	2000	23
1974	4	1983	9	1992	15	2001	25
1975	5	1984	9	1993	19	2002	17
1976	2	1985	7	1994	20	2003	12
1977	14	1986	23	1995	22	2004	13
1978	7	1987	13	1996	13	2005	15
1979	2	1988	18	1997	8	2006	18
1980	5	1989	13	1998	9	2007	18

Millard H. Sharp/Photo Researchers

Make two graphs of these data to illustrate why you should always make a time plot for data collected over time.

(a) Make a histogram of the counts of people bitten by alligators. The distribution has an irregular shape. What is the midpoint of the yearly counts of people bitten?

(b) Make a time plot. There is great variation from year to year, but also an increasing trend. How many of the 22 years from 1986 to 2007 had more people bitten by alligators than your midpoint from (a)? The trend reflects Florida's growing population, which brings more people close to alligators.

1.46 To split or not to split. The data sets in the *One Variable Statistical Calculator* applet on the text CD and Web site include the "pulling wood apart" data from Example 1.9. The applet rounds the data in the same way as Figure 1.11 (page 21). Use the applet to make a stemplot with split stems. Do you prefer this stemplot or that in Figure 1.11? Explain your choice.

Rhona Wise/Icon SMI/Newscom

Describing Distributions with Numbers

We saw in Chapter 1 (page 4) that the American Community Survey asks, among much else, workers' travel times to work. Here are the travel times in minutes for 15 workers in North Carolina, chosen at random by the Census Bureau:[1]

30 20 10 40 25 20 10 60 15 40 5 30 12 10 10

We aren't surprised that most people estimate their travel time in multiples of 5 minutes. Here is a stemplot of these data:

```
0 | 5
1 | 0 0 0 0 2 5
2 | 0 0 5
3 | 0 0
4 | 0 0
5 |
6 | 0
```

The distribution is single-peaked and right-skewed. The longest travel time (60 minutes) may be an outlier. Our goal in this chapter is to describe with numbers the center and spread of this and other distributions.

Measuring center: the mean

The most common measure of center is the ordinary arithmetic average, or *mean*.

THE MEAN $\bar{x}$

To find the **mean** of a set of observations, add their values and divide by the number of observations. If the n observations are $x_1, x_2, \ldots, x_n$, their mean is

$$\bar{x} = \frac{x_1 + x_2 + \cdots + x_n}{n}$$

or, in more compact notation,

$$\bar{x} = \frac{1}{n} \sum x_i$$

The $\sum$ (capital Greek sigma) in the formula for the mean is short for "add them all up." The subscripts on the observations x_i are just a way of keeping the n observations distinct. They do not necessarily indicate order or any other special facts about the data. The bar over the x indicates the mean of all the x-values. Pronounce the mean $\bar{x}$ as "x-bar." This notation is very common. When writers who are discussing data use $\bar{x}$ or $\bar{y}$, they are talking about a mean.

Don't hide the outliers

Data from an airliner's control surfaces, such as the vertical tail rudder, go to cockpit instruments and then to the "black box" flight data recorder. To avoid confusing the pilots, short erratic movements in the data are "smoothed" so that the instruments show overall patterns. When a crash killed 260 people, investigators suspected a catastrophic movement of the tail rudder. But the black box contained only the smoothed data. Sometimes outliers are more important than the overall pattern.

---

resistant measure

EXAMPLE 2.1 **Travel times to work**

The mean travel time of our 15 North Carolina workers is

$$\bar{x} = \frac{x_1 + x_2 + \cdots + x_n}{n}$$
$$= \frac{30 + 20 + \cdots + 10}{15}$$
$$= \frac{337}{15} = 22.5 \text{ minutes}$$

In practice, you can enter the data into your calculator and ask for the mean. You don't have to actually add and divide. But you should know that this is what the calculator is doing.

Notice that only 6 of the 15 travel times are larger than the mean. If we leave out the longest single travel time, 60 minutes, the mean for the remaining 14 people is 19.8 minutes. That one observation raises the mean by 2.7 minutes. ■

Example 2.1 illustrates an important fact about the mean as a measure of center: it is sensitive to the influence of a few extreme observations. These may be outliers, but a skewed distribution that has no outliers will also pull the mean toward its long tail. Because the mean cannot resist the influence of extreme observations, we say that it is not a **resistant measure** of center.

2.1 **Pulling wood apart.** Example 1.9 (page 20) gives the breaking strength in pounds of 20 pieces of Douglas fir. Find the mean breaking strength. How many of the pieces of wood have strengths less than the mean? What feature of the stemplot (Figure 1.11, page 21) explains the fact that the mean is smaller than most of the observations?

2.2 **Health care spending.** Table 1.3 (page 22) gives the annual health care spending per person in 38 richer nations. The United States, at $5711 per person, is a high outlier. Find the mean health care spending in these nations with and without the United States. How much does the one outlier increase the mean?

Measuring center: the median

In Chapter 1, we used the midpoint of a distribution as an informal measure of center. The *median* is the formal version of the midpoint, with a specific rule for calculation.

THE MEDIAN *M*

The **median** M is the midpoint of a distribution, the number such that half the observations are smaller and the other half are larger. To find the median of a distribution:

1. Arrange all observations in order of size, from smallest to largest.
2. If the number of observations n is odd, the median M is the center observation in the ordered list. If the number of observations n is even, the median M is midway between the two center observations in the ordered list.
3. You can always locate the median in the ordered list of observations by counting up $(n + 1)/2$ observations from the start of the list.

Note that the formula $(n + 1)/2$ does *not* give the median, just the location of the median in the ordered list. Medians require little arithmetic, so they are easy to find by hand for small sets of data. Arranging even a moderate number of observations in order is very tedious, however, so that finding the median by hand for larger sets of data is unpleasant. Even simple calculators have an $\overline{x}$ button, but you will need to use software or a graphing calculator to automate finding the median.

EXAMPLE 2.2 **Finding the median: odd *n***

What is the median travel time for our 15 North Carolina workers? Here are the data arranged in order:

5　　10　　10　　10　　10　　12　　15　　**20**　　20　　25　　30　　30　　40　　40　　60

The count of observations $n = 15$ is odd. The bold **20** is the center observation in the ordered list, with 7 observations to its left and 7 to its right. This is the median, M = 20 minutes.

Because $n = 15$, our rule for the location of the median gives

$$\text{location of M} = \frac{n+1}{2} = \frac{16}{2} = 8$$

That is, the median is the 8th observation in the ordered list. It is faster to use this rule than to locate the center by eye. ■

Mitchell Funk/Getty Images

EXAMPLE 2.3   **Finding the median: even *n***

Travel times to work in New York State are (on the average) longer than in North Carolina. Here are the travel times in minutes of 20 randomly chosen New York workers:

10 30 5 25 40 20 10 15 30 20 15 20 85 15 65 15 60 60 40 45

A stemplot not only displays the distribution but makes finding the median easy because it arranges the observations in order:

```
0 | 5
1 | 0 0 5 5 5 5
2 | 0 0 0 5
3 | 0 0
4 | 0 0 5
5 |
6 | 0 0 5
7 |
8 | 5
```

The distribution is single-peaked and right-skewed, with several travel times of an hour or more. There is no center observation, but there is a center pair. These are the bold **20** and **25** in the stemplot, which have 9 observations before them in the ordered list and 9 after them. The median is midway between these two observations:

$$\text{M} = \frac{20 + 25}{2} = 22.5 \text{ minutes}$$

With $n = 20$, the rule for locating the median in the list gives

$$\text{location of M} = \frac{n+1}{2} = \frac{21}{2} = 10.5$$

The location 10.5 means "halfway between the 10th and 11th observations in the ordered list." That agrees with what we found by eye. ■

Comparing the mean and the median

Examples 2.1 and 2.2 illustrate an important difference between the mean and the median. The median travel time (the midpoint of the distribution) is 20 minutes. The mean travel time is higher, 22.5 minutes. The mean is pulled toward the right tail of this right-skewed distribution. The median, unlike the mean, is *resistant*. If the longest travel time were 600 minutes rather than 60 minutes, the mean would increase to more than 58 minutes but the median would not change at all. The outlier just counts as one observation above the center, no matter how far above the center it lies. The mean uses the actual value of each observation and so will chase a single large observation upward. The *Mean and Median* applet is an excellent way to compare the resistance of M and $\bar{x}$.

COMPARING THE MEAN AND THE MEDIAN

The mean and median of a roughly symmetric distribution are close together. If the distribution is exactly symmetric, the mean and median are exactly the same. In a skewed distribution, the mean is usually farther out in the long tail than is the median.[2]

Many economic variables have distributions that are skewed to the right. For example, the median endowment of colleges and universities in 2007 was about $91 million—but the mean endowment was almost $524 million. Most institutions have modest endowments, but a few are very wealthy. Harvard's endowment was almost $35 billion.[3] The few wealthy institutions pull the mean up but do not affect the median. Reports about incomes and other strongly skewed distributions usually give the median ("midpoint") rather than the mean ("arithmetic average"). However, a county that is about to impose a tax of 1% on the incomes of its residents cares about the mean income, not the median. The tax revenue will be 1% of total income, and the total is the mean times the number of residents. The mean and median measure center in different ways, and both are useful. *Don't confuse the "average" value of a variable (the mean) with its "typical" value, which we might describe by the median.*

APPLY YOUR KNOWLEDGE

2.3 New York travel times. Find the mean of the travel times to work for the 20 New York workers in Example 2.3. Compare the mean and median for these data. What general fact does your comparison illustrate?

2.4 House prices. The mean and median selling prices of existing single-family homes sold in 2007 were $218,900 and $265,800.[4] Which of these numbers is the mean and which is the median? Explain how you know.

2.5 Food oils. Table 1.4 (page 33) gives the ratio of two essential fatty acids in 30 food oils. Find the mean and the median for these data. Make a histogram of the data. What feature of the distribution explains why the mean is more than 10 times as large as the median?

© Ryan McVay/Age fotostock

Measuring spread: the quartiles

The mean and median provide two different measures of the center of a distribution. But a measure of center alone can be misleading. The Census Bureau reports that in 2006 the median income of American households was $48,021. Half of all households had incomes below $48,021, and half had higher incomes. The mean was much higher, $66,570, because the distribution of incomes is skewed to the right. But the median and mean don't tell the whole story. The bottom 10% of households had incomes less than $12,000, and households in the top 5% took in more than $174,012.[5] We are interested in the *spread* or *variability* of incomes as well as their center. *The simplest useful numerical description of a distribution requires both a measure of center and a measure of spread.*

One way to measure spread is to give the smallest and largest observations. For example, the travel times of our 15 North Carolina workers range from 5 minutes to 60 minutes. These single observations show the full spread of the data, but they may be outliers. We can improve our description of spread by also looking at the spread of the middle half of the data. The *quartiles* mark out the middle half. Count up the ordered list of observations, starting from the smallest. The *first quartile* lies one-quarter of the way up the list. The *third quartile* lies three-quarters of the way up the list. In other words, the first quartile is larger than 25% of the observations, and the third quartile is larger than 75% of the observations. The second quartile is the median, which is larger than 50% of the observations. That is the idea of quartiles. We need a rule to make the idea exact. The rule for calculating the quartiles uses the rule for the median.

THE QUARTILES Q_1 AND Q_3

To calculate the **quartiles:**

1. Arrange the observations in increasing order and locate the median M in the ordered list of observations.

2. The **first quartile Q_1** is the median of the observations whose position in the ordered list is to the left of the location of the overall median.

3. The **third quartile Q_3** is the median of the observations whose position in the ordered list is to the right of the location of the overall median.

Here are examples that show how the rules for the quartiles work for both odd and even numbers of observations.

EXAMPLE 2.4 **Finding the quartiles: odd *n***

Our North Carolina sample of 15 workers' travel times, arranged in increasing order, is

5   10   10   10   10   12   15   **20**   20   25   30   30   40   40   60

There is an odd number of observations, so the median is the middle one, the bold **20** in the list. The first quartile is the median of the 7 observations to the left of the median. This is the 4th of these 7 observations, so $Q_1 = 10$ minutes. If you want, you can use the rule for the location of the median with $n = 7$:

$$\text{location of } Q_1 = \frac{n+1}{2} = \frac{7+1}{2} = 4$$

The third quartile is the median of the 7 observations to the right of the median, $Q_3 = 30$ minutes. *When there is an odd number of observations, leave out the overall median when you locate the quartiles in the ordered list.*

The quartiles are *resistant* because they are not affected by a few extreme observations. For example, Q_3 would still be 30 if the outlier were 600 rather than 60. ■

EXAMPLE 2.5   **Finding the quartiles: even *n***

Here are the travel times to work of the 20 New Yorkers from Example 2.3, arranged in increasing order:

5   10   10   15   15   15   15   20   20   20 | 25   30   30   40   40   45   60   60   65   85

There is an even number of observations, so the median lies midway between the middle pair, the 10th and 11th in the list. Its value is $M = 22.5$ minutes. We have marked the location of the median by |. The first quartile is the median of the first 10 observations, because these are the observations to the left of the location of the median. Check that $Q_1 = 15$ minutes and $Q_3 = 42.5$ minutes. *When the number of observations is even, include all the observations when you locate the quartiles.* ■

Be careful when, as in these examples, several observations take the same numerical value. Write down all of the observations, arrange them in order, and apply the rules just as if they all had distinct values.

The five-number summary and boxplots

The smallest and largest observations tell us little about the distribution as a whole, but they give information about the tails of the distribution that is missing if we know only the median and the quartiles. To get a quick summary of both center and spread, combine all five numbers.

> **THE FIVE-NUMBER SUMMARY**
>
> The **five-number summary** of a distribution consists of the smallest observation, the first quartile, the median, the third quartile, and the largest observation, written in order from smallest to largest. In symbols, the five-number summary is
>
>   Minimum Q_1 M Q_3 Maximum

These five numbers offer a reasonably complete description of center and spread. The five-number summaries of travel times to work from Examples 2.4 and 2.5 are

North Carolina	5	10	20	30	60
New York	5	15	22.5	42.5	85

The five-number summary of a distribution leads to a new graph, the *boxplot*. Figure 2.1 shows boxplots comparing travel times to work in North Carolina and New York.

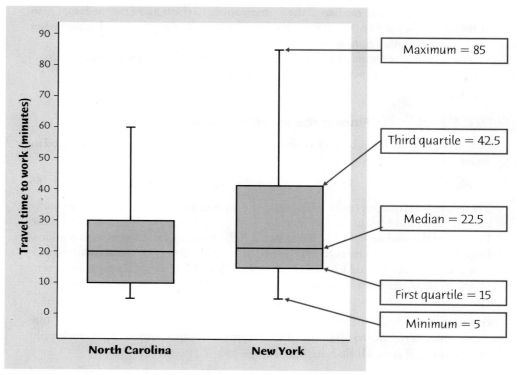

FIGURE 2.1

Boxplots comparing the travel times to work of samples of workers in North Carolina and New York.

> **BOXPLOT**
>
> A **boxplot** is a graph of the five-number summary.
>
> - A central box spans the quartiles Q_1 and Q_3.
> - A line in the box marks the median M.
> - Lines extend from the box out to the smallest and largest observations.

Because boxplots show less detail than histograms or stemplots, they are best used for side-by-side comparison of more than one distribution, as in Figure 2.1. Be sure to include a numerical scale in the graph. When you look at a boxplot, first locate the median, which marks the center of the distribution. Then look at the spread. The span of the central box shows the spread of the middle half of the data, and the extremes (the smallest and largest observations) show the spread of the entire data set. We see from Figure 2.1 that travel times to work are in general a bit longer in New York than in North Carolina. The median, both quartiles, and the maximum are all larger in New York. New York travel times are also more variable, as shown by the span of the box and the spread between the extremes.

Finally, the New York data are more strongly right-skewed. In a symmetric distribution, the first and third quartiles are equally distant from the median. In most distributions that are skewed to the right, on the other hand, the third quartile will be farther above the median than the first quartile is below it. The extremes behave the same way, but remember that they are just single observations and may say little about the distribution as a whole.

APPLY YOUR KNOWLEDGE

2.6 The Dallas Cowboys. The 2007 roster of the Dallas Cowboys professional football team included 9 defensive linemen and 10 offensive linemen. The weights in pounds of the defensive linemen were

$$300 \quad 300 \quad 295 \quad 255 \quad 298 \quad 298 \quad 300 \quad 310 \quad 300$$

and the weights of the offensive linemen were

$$312 \quad 305 \quad 340 \quad 320 \quad 366 \quad 324 \quad 309 \quad 315 \quad 305 \quad 305$$

(a) Make a stemplot of the weights of the defensive linemen and find the five-number summary.

(b) Make a stemplot of the weights of the offensive linemen and find the five-number summary.

(c) Does either group contain one or more clear outliers? Which group of players tends to be heavier?

2.7 Comparing investments. Should you put your money into a fund that buys stocks or a fund that invests in real estate? The answer changes from time to time, and unfortunately we can't look into the future. Looking back into the past, the boxplots in Figure 2.2 compare the daily returns (in percent) on a "total stock market" fund and a real estate fund over a year ending in November 2007.[6]

(a) Read the graph: about what were the highest and lowest daily returns on the stock fund?

(b) Read the graph: the median return was about the same on both investments. About what was the median return?

(c) What is the most important difference between the two distributions?

Kirby Lee/WireImage/Newscom

Spotting suspected outliers*

Look again at the stemplot of travel times to work in New York in Example 2.3. The five-number summary for this distribution is

$$5 \quad 15 \quad 22.5 \quad 42.5 \quad 85$$

How shall we describe the spread of this distribution? The smallest and largest observations are extremes that don't describe the spread of the majority of the data. The distance between the quartiles (the range of the center half of the data) is a more resistant measure of spread. This distance is called the *interquartile range*.

---

* This short section is optional.

Boxplots comparing the distributions of daily returns on two kinds of investment, for Exercise 2.7.

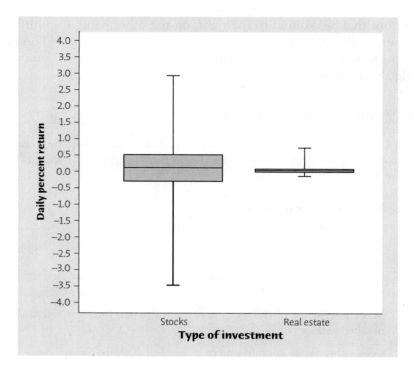

Type of investment

THE INTERQUARTILE RANGE *IQR*

The **interquartile range *IQR*** is the distance between the first and third quartiles,

$$IQR = Q_3 - Q_1$$

For our data on New York travel times, $IQR = 42.5 - 15 = 27.5$ minutes. However, *no single numerical measure of spread, such as IQR, is very useful for describing skewed distributions.* The two sides of a skewed distribution have different spreads, so one number can't summarize them. That's why we give the full five-number summary. The interquartile range is mainly used as the basis for a rule of thumb for identifying suspected outliers.

THE 1.5 × *IQR* RULE FOR OUTLIERS

Call an observation a suspected outlier if it falls more than $1.5 \times IQR$ above the third quartile or below the first quartile.

EXAMPLE 2.6 Using the 1.5 × *IQR* rule

For the New York travel time data, $IQR = 27.5$ and

$$1.5 \times IQR = 1.5 \times 27.5 = 41.25$$

Any values not falling between

$$Q_1 - (1.5 \times IQR) = 15.0 - 41.25 = -26.25 \quad \text{and}$$

$$Q_3 + (1.5 \times IQR) = 42.5 + 41.25 = 83.75$$

are flagged as suspected outliers. Look again at the stemplot in Example 2.3: the only suspected outlier is the longest travel time, 85 minutes. The $1.5 \times IQR$ rule suggests that the three next-longest travel times (60 and 65 minutes) are just part of the long right tail of this skewed distribution. ■

The $1.5 \times IQR$ rule is not a replacement for looking at the data. It is most useful when large volumes of data are scanned automatically.

APPLY YOUR KNOWLEDGE

2.8 Travel time to work. In Example 2.1, we noted the influence of one long travel time of 60 minutes in our sample of 15 North Carolina workers. Does the $1.5 \times IQR$ rule identify this travel time as a suspected outlier?

2.9 Foreign-born residents. Table 1.1 gives the percent of residents in each state who were born outside the United States. In Examples 1.5 and 1.8 we saw that California (27.2%) stands slightly above the rest of the distribution. Is California a suspected outlier by the $1.5 \times IQR$ rule? (Start from the stemplot in Figure 1.10, page 20, which arranges the observations in increasing order.)

Measuring spread: the standard deviation

The five-number summary is not the most common numerical description of a distribution. That distinction belongs to the combination of the mean to measure center and the *standard deviation* to measure spread. The standard deviation and its close relative, the *variance*, measure spread by looking at how far the observations are from their mean.

How much is that house worth?

The town of Manhattan, Kansas, is sometimes called "the little Apple" to distinguish it from that other Manhattan, "the big Apple." A few years ago, a house there appeared in the county appraiser's records valued at $200,059,000. That would be quite a house even on Manhattan Island. As you might guess, the entry was wrong: the true value was $59,500. But before the error was discovered, the county, the city, and the school board had based their budgets on the total appraised value of real estate, which the one outlier jacked up by 6.5%. It can pay to spot outliers before you trust your data.

THE STANDARD DEVIATION s

The **variance** s^2 of a set of observations is an average of the squares of the deviations of the observations from their mean. In symbols, the variance of n observations $x_1, x_2, \ldots, x_n$ is

$$s^2 = \frac{(x_1 - \overline{x})^2 + (x_2 - \overline{x})^2 + \cdots + (x_n - \overline{x})^2}{n - 1}$$

or, more compactly,

$$s^2 = \frac{1}{n-1} \sum (x_i - \overline{x})^2$$

The **standard deviation** s is the square root of the variance s^2:

$$s = \sqrt{\frac{1}{n-1} \sum (x_i - \overline{x})^2}$$

In practice, use software or your calculator to obtain the standard deviation from keyed-in data. Doing an example step-by-step will help you understand how the variance and standard deviation work, however.

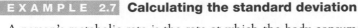

EXAMPLE 2.7 Calculating the standard deviation

A person's metabolic rate is the rate at which the body consumes energy. Metabolic rate is important in studies of weight gain, dieting, and exercise. Here are the metabolic rates of 7 men who took part in a study of dieting. The units are calories per 24 hours. These are the same calories used to describe the energy content of foods.

$$1792 \quad 1666 \quad 1362 \quad 1614 \quad 1460 \quad 1867 \quad 1439$$

The researchers reported $\bar{x}$ and s for these men. First find the mean:

$$\bar{x} = \frac{1792 + 1666 + 1362 + 1614 + 1460 + 1867 + 1439}{7}$$

$$= \frac{11{,}200}{7} = 1600 \text{ calories}$$

Figure 2.3 displays the data as points above the number line, with their mean marked by an asterisk (∗). The arrows mark two of the deviations from the mean. The deviations show how spread out the data are about their mean. They are the starting point for calculating the variance and the standard deviation.

Observations x_i	Deviations $x_i - \bar{x}$	Squared deviations $(x_i - \bar{x})^2$
1792	$1792 - 1600 = \quad 192$	$192^2 = \quad 36{,}864$
1666	$1666 - 1600 = \quad 66$	$66^2 = \quad 4{,}356$
1362	$1362 - 1600 = -238$	$(-238)^2 = \quad 56{,}644$
1614	$1614 - 1600 = \quad 14$	$14^2 = \quad 196$
1460	$1460 - 1600 = -140$	$(-140)^2 = \quad 19{,}600$
1867	$1867 - 1600 = \quad 267$	$267^2 = \quad 71{,}289$
1439	$1439 - 1600 = -161$	$(-161)^2 = \quad 25{,}921$
	sum $= \quad 0$	sum $= 214{,}870$

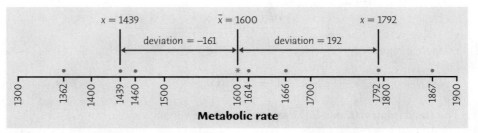

FIGURE 2.3

Metabolic rates for 7 men, with their mean (∗) and the deviations of two observations from the mean, for Example 2.7.

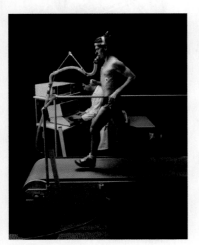

Tom Tracy Photography/Alamy

The variance is the sum of the squared deviations divided by one less than the number of observations:

$$s^2 = \frac{1}{n-1} \sum (x_i - \overline{x})^2 = \frac{214,870}{6} = 35,811.67$$

The standard deviation is the square root of the variance:

$$s = \sqrt{35,811.67} = 189.24 \text{ calories} \blacksquare$$

Notice that the "average" in the variance s^2 divides the sum by one fewer than the number of observations, that is, $n-1$ rather than n. The reason is that the deviations $x_i - \overline{x}$ always sum to exactly 0, so that knowing $n-1$ of them determines the last one. Only $n-1$ of the squared deviations can vary freely, and we average by dividing the total by $n-1$. The number $n-1$ is called the **degrees of freedom** of the variance or standard deviation. Some calculators offer a choice between dividing by n and dividing by $n-1$, so be sure to use $n-1$.

degrees of freedom

More important than the details of hand calculation are the properties that determine the usefulness of the standard deviation:

- s measures *spread about the mean* and should be used only when the mean is chosen as the measure of center.

- s is *always zero or greater than zero*. $s = 0$ only when there is no spread. This happens only when all observations have the same value. Otherwise, $s > 0$. As the observations become more spread out about their mean, s gets larger.

- s has the *same units of measurement as the original observations*. For example, if you measure metabolic rates in calories, both the mean $\overline{x}$ and the standard deviation s are also in calories. This is one reason to prefer s to the variance s^2, which is in squared calories.

- Like the mean $\overline{x}$, s is *not resistant*. A few outliers can make s very large.

The use of squared deviations renders s even more sensitive than $\overline{x}$ to a few extreme observations. For example, the standard deviation of the travel times for the 15 North Carolina workers in Example 2.1 is 15.23 minutes. (Use your calculator or software to verify this.) If we omit the high outlier, the standard deviation drops to 11.56 minutes.

If you feel that the importance of the standard deviation is not yet clear, you are right. We will see in Chapter 3 that the standard deviation is the natural measure of spread for a very important class of symmetric distributions, the Normal distributions. The usefulness of many statistical procedures is tied to distributions of particular shapes. This is certainly true of the standard deviation.

Choosing measures of center and spread

We now have a choice between two descriptions of the center and spread of a distribution: the five-number summary, or $\overline{x}$ and s. Because $\overline{x}$ and s are sensitive to extreme observations, they can be misleading when a distribution is strongly skewed or has outliers. In fact, because the two sides of a skewed distribution have different spreads, no single number such as s describes the spread well. The five-number summary, with its two quartiles and two extremes, does a better job.

CHOOSING A SUMMARY

The five-number summary is usually better than the mean and standard deviation for describing a skewed distribution or a distribution with strong outliers. Use $\overline{x}$ and s only for reasonably symmetric distributions that are free of outliers.

Outliers can greatly affect the values of the mean $\overline{x}$ and the standard deviation s, the most common measures of center and spread. Many more elaborate statistical procedures also can't be trusted when outliers are present. *Whenever you find outliers in your data, try to find an explanation for them.* Sometimes the explanation is as simple as a typing error, such as typing 10.1 as 101. Sometimes a measuring device broke down or a subject gave a frivolous response, like the student in a class survey who claimed to study 30,000 minutes per night. (Yes, that really happened.) In all these cases, you can simply remove the outlier from your data. When outliers are "real data," like the long travel times of some New York workers, you should choose statistical methods that are not greatly disturbed by the outliers. For example, use the five-number summary rather than $\overline{x}$ and s to describe a distribution with extreme outliers. We will meet other examples later in the book.

Remember that a graph gives the best overall picture of a distribution. Numerical measures of center and spread report specific facts about a distribution, but they do not describe its entire shape. Numerical summaries do not disclose the presence of multiple peaks or clusters, for example. Exercise 2.11 shows how misleading numerical summaries can be. **Always plot your data.**

APPLY YOUR KNOWLEDGE

2.10 $\overline{x}$ and s by hand. The air in poultry-processing plants often contains high concentrations of fungus spores, especially during the summer. To measure the presence of spores, air samples are pumped to an agar plate and "colony-forming units (CFUs)" are counted after an incubation period. The CFUs per cubic meter of air in one location for four summer days were 3175, 2526, 1763, and 1090.[7] A graph of only 4 observations gives little information, so we proceed to compute the mean and standard deviation.

(a) Find the mean step-by-step. That is, find the sum of the 4 observations and divide by 4.

(b) Find the standard deviation step-by-step. That is, find the deviations of each observation from the mean, square the deviations, then obtain the variance and the standard deviation. Example 2.7 shows the method.

(c) Now enter the data into your calculator and use the mean and standard deviation buttons to obtain $\bar{x}$ and s. Do the results agree with your hand calculations?

2.11 $\bar{x}$ and *s* are not enough. The mean $\bar{x}$ and standard deviation s measure center and spread but are not a complete description of a distribution. Data sets with different shapes can have the same mean and standard deviation. To demonstrate this fact, use your calculator to find $\bar{x}$ and s for these two small data sets. Then make a stemplot of each and comment on the shape of each distribution.

Data A	9.14	8.14	8.74	8.77	9.26	8.10	6.13	3.10	9.13	7.26	4.74
Data B	6.58	5.76	7.71	8.84	8.47	7.04	5.25	5.56	7.91	6.89	12.50

2.12 Choose a summary. The shape of a distribution is a rough guide to whether the mean and standard deviation are a helpful summary of center and spread. For which of the following distributions would $\bar{x}$ and s be useful? In each case, give a reason for your decision.

(a) Percents of foreign-born residents in the states, Figure 1.5 (page 13).

(b) Iowa Test scores, Figure 1.7 (page 17).

(c) Breaking strength of wood, Figure 1.11 (page 21).

Using technology

Although a calculator with "two-variable statistics" functions will do the basic calculations we need, more elaborate tools are helpful. Graphing calculators and computer software will do calculations and make graphs as you command, freeing you to concentrate on choosing the right methods and interpreting your results. Figure 2.4 displays output describing the travel times to work of 20 people in New York State (Example 2.3). Can you find $\bar{x}$, s, and the five-number summary in each output? The big message of this section is: *once you know what to look for, you can read output from any technological tool.*

The displays in Figure 2.4 come from a Texas Instruments graphing calculator, the Minitab statistical program, and the Microsoft Excel spreadsheet program. Minitab allows you to choose what descriptive measures you want. Excel and the calculator give some things we don't need. Just ignore the extras. Excel's "Descriptive Statistics" menu item doesn't give the quartiles. We used the spreadsheet's separate quartile function to get Q_1 and Q_3.

Texas Instruments Graphing Calculator

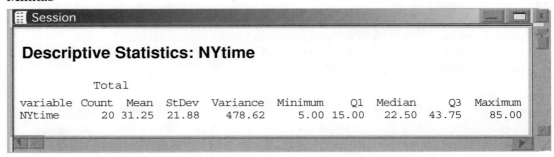

Minitab

Descriptive Statistics: NYtime

		Total							
variable	Count	Mean	StDev	Variance	Minimum	Q1	Median	Q3	Maximum
NYtime	20	31.25	21.88	478.62	5.00	15.00	22.50	43.75	85.00

Microsoft Excel

	A	B	C	D
1	*minutes*			
2				
3	Mean	31.25		
4	Standard Error	4.891924064		
5	Median	22.5	QUARTILE(A2:A21,1)	15
6	Mode	15	QUARTILE(A2:A21,3)	42.5
7	Standard Deviation	21.8773495		
8	Sample Variance	478.6184211		
9	Kurtosis	0.329884126		
10	Skewness	1.040110836		
11	Range	80		
12	Minimum	5		
13	Maximum	85		
14	Sum	625		
15	Count	20		

Sheet4 / Sheet1 / Sheet2 / Sheet

FIGURE 2.4

Output from a graphing calculator, a statistical software package, and a spreadsheet program describing the data on travel times to work in New York State.

EXAMPLE 2.8 **What is the third quartile?**

In Example 2.5, we saw that the quartiles of the New York travel times are $Q_1 = 15$ and $Q_3 = 42.5$. Look at the output displays in Figure 2.4. The calculator and Excel agree with our work. Minitab says that $Q_3 = 43.75$. What happened? *There are several rules for finding the quartiles. Some calculators and software use rules that give results different from ours for some sets of data.* This is true of Minitab and also of Excel, though Excel agrees with our work in this example. Results from the various rules are always close to each other, so that the differences are never important in practice. Our rule is the simplest for hand calculation. ■

Organizing a statistical problem

Most of our examples and exercises have aimed to help you learn basic tools (graphs and calculations) for describing and comparing distributions. You have also learned principles that guide use of these tools, such as "start with a graph" and "look for the overall pattern and striking deviations from the pattern." The data you work with are not just numbers—they describe specific settings such as water depth in the Everglades or travel time to work. Because data come from a specific setting, the final step in examining data is a conclusion for that setting. Water depth in the Everglades has a yearly cycle that reflects Florida's wet and dry seasons. Travel times to work are generally longer in New York than in North Carolina.

As you learn more statistical tools and principles, you will face more complex statistical problems. Although no framework accommodates all the varied issues that arise in applying statistics to real settings, the following four-step thought process gives useful guidance. In particular, the first and last steps emphasize that statistical problems are tied to specific real-world settings and therefore involve more than doing calculations and making graphs.

ORGANIZING A STATISTICAL PROBLEM: A FOUR-STEP PROCESS

STATE: What is the practical question, in the context of the real-world setting?

PLAN: What specific statistical operations does this problem call for?

SOLVE: Make the graphs and carry out the calculations needed for this problem.

CONCLUDE: Give your practical conclusion in the setting of the real-world problem.

To help you master the basics, many exercises will continue to tell you what to do—make a histogram, find the five-number summary, and so on. Real statistical problems don't come with detailed instructions. From now on, especially in the later chapters of the book, you will meet some exercises that are more realistic. Use the four-step process as a guide to solving and reporting these problems. They are marked with the four-step icon, as the following example illustrates.

Art Wolfe/Getty Images

EXAMPLE 2.9 **Comparing tropical flowers**

STATE: Ethan Temeles of Amherst College, with his colleague W. John Kress, studied the relationship between varieties of the tropical flower *Heliconia* on the island of Dominica and the different species of hummingbirds that fertilize the flowers.[8] Over time, the researchers believe, the lengths of the flowers and the forms of the hummingbirds' beaks have evolved to match each other. If that is true, flower varieties fertilized by different hummingbird species should have distinct distributions of length.

Table 2.1 gives length measurements (in millimeters) for samples of three varieties of *Heliconia*, each fertilized by a different species of hummingbird. Do the three varieties display distinct distributions of length? How do the mean lengths compare?

PLAN: Use graphs and numerical descriptions to describe and compare these three distributions of flower length.

SOLVE: We might use boxplots to compare the distributions, but stemplots preserve more detail and work well for data sets of these sizes. Figure 2.5 displays stemplots with the stems lined up for easy comparison. The lengths have been rounded to the nearest tenth of a millimeter. The *bihai* and red varieties have somewhat skewed distributions, so we might choose to compare the five-number summaries. But because the researchers plan to use $\bar{x}$ and s for further analysis, we instead calculate these measures:

Variety	Mean length	Standard deviation
bihai	47.60	1.213
red	39.71	1.799
yellow	36.18	0.975

CONCLUDE: The three varieties differ so much in flower length that there is little overlap among them. In particular, the flowers of *bihai* are longer than either red or yellow. The mean lengths are 47.6 mm for *H. bihai*, 39.7 mm for *H. caribaea* red, and 36.2 mm for *H. caribaea* yellow. ■

TABLE 2.1 Flower lengths (millimeters) for three *Heliconia* varieties

			H. bihai				
47.12	46.75	46.81	47.12	46.67	47.43	46.44	46.64
48.07	48.34	48.15	50.26	50.12	46.34	46.94	48.36
			H. caribaea red				
41.90	42.01	41.93	43.09	41.47	41.69	39.78	40.57
39.63	42.18	40.66	37.87	39.16	37.40	38.20	38.07
38.10	37.97	38.79	38.23	38.87	37.78	38.01	
			H. caribaea yellow				
36.78	37.02	36.52	36.11	36.03	35.45	38.13	37.10
35.17	36.82	36.66	35.68	36.03	34.57	34.63	

	bihai		red		yellow
34		34		34	6 6
35		35		35	2 5 7
36		36		36	0 0 1 5 7 8 8
37		37	4 8 9	37	0 1
38		38	0 0 1 1 2 2 8 9	38	1
39		39	2 6 8	39	
40		40	6 7	40	
41		41	5 7 9 9	41	
42		42	0 2	42	
43		43	1	43	
44		44		44	
45		45		45	
46	3 4 6 7 8 8 9	46		46	
47	1 1 4	47		47	
48	1 2 3 4	48		48	
49		49		49	
50	1 3	50		50	

FIGURE 2.5

Stemplots comparing the distributions of flower lengths from Table 2.1, for Example 2.9. The stems are whole millimeters and the leaves are tenths of a millimeter.

APPLY YOUR KNOWLEDGE

2.13 Logging in the rain forest. "Conservationists have despaired over destruction of tropical rain forest by logging, clearing, and burning." These words begin a report on a statistical study of the effects of logging in Borneo.[9] Charles Cannon of Duke University and his coworkers compared forest plots that had never been logged (Group 1) with similar plots nearby that had been logged 1 year earlier (Group 2) and 8 years earlier (Group 3). All plots were 0.1 hectare in area. Here are the counts of trees for plots in each group:

Group 1	27	22	29	21	19	33	16	20	24	27	28	19
Group 2	12	12	15	9	20	18	17	14	14	2	17	19
Group 3	18	4	22	15	18	19	22	12	12			

To what extent has logging affected the count of trees? Follow the four-step process in reporting your work.

2.14 Diplomatic scofflaws. Until Congress allowed some enforcement in 2002, the thousands of foreign diplomats in New York City could freely violate parking laws. Two economists looked at the number of unpaid parking tickets per diplomat over a five-year period ending when enforcement reduced the problem.[10] They concluded that large numbers of unpaid tickets indicated a "culture of corruption" in a country and lined up well with more elaborate measures of corruption. The data set for 145 countries is too large to print here, but look at the data file *ex02-14.dat* on the text Web site and CD. The first 32 countries in the list (Australia to Trinidad and Tobago) are classified by the World Bank as "developed." The remaining countries (Albania to Zimbabwe) are "developing." The World Bank classification is based only on national income and does not take into account measures of social development.

Give a full description of the distribution of unpaid tickets for both groups of countries and identify any high outliers. Compare the two groups. Does national income alone do a good job of distinguishing countries whose diplomats do and do not obey parking laws?

© James Leynse/CORBIS

CHAPTER 2   SUMMARY

- A numerical summary of a distribution should report at least its **center** and its **spread** or **variability.**

- The **mean** $\bar{x}$ and the **median M** describe the center of a distribution in different ways. The mean is the arithmetic average of the observations, and the median is the midpoint of the values.

- When you use the median to indicate the center of the distribution, describe its spread by giving the **quartiles.** The **first quartile** Q_1 has one-fourth of the observations below it, and the **third quartile** Q_3 has three-fourths of the observations below it.

- The **five-number summary** consisting of the median, the quartiles, and the smallest and largest individual observations provides a quick overall description of a distribution. The median describes the center, and the quartiles and extremes show the spread.

- **Boxplots** based on the five-number summary are useful for comparing several distributions. The box spans the quartiles and shows the spread of the central half of the distribution. The median is marked within the box. Lines extend from the box to the extremes and show the full spread of the data.

- The **variance** s^2 and especially its square root, the **standard deviation s,** are common measures of spread about the mean as center. The standard deviation s is zero when there is no spread and gets larger as the spread increases.

- A **resistant measure** of any aspect of a distribution is relatively unaffected by changes in the numerical value of a small proportion of the total number of observations, no matter how large these changes are. The median and quartiles are resistant, but the mean and the standard deviation are not.

- The mean and standard deviation are good descriptions for symmetric distributions without outliers. They are most useful for the Normal distributions introduced in the next chapter. The five-number summary is a better description for skewed distributions.

- Numerical summaries do not fully describe the shape of a distribution. Always plot your data.

- A statistical problem has a real-world setting. You can organize many problems using the four steps **state, plan, solve,** and **conclude.**

CHECK YOUR SKILLS

2.15 Here are the amounts of money (cents) in coins carried by 10 students in a statistics class:

$$50 \quad 35 \quad 0 \quad 97 \quad 76 \quad 0 \quad 0 \quad 87 \quad 23 \quad 65$$

The mean of these data is

(a) 37.2.   (b) 42.5.   (c) 43.3.

TYSM friend

2.16 The median of the data in Exercise 2.15 is

(a) 35. (b) 42.5. (c) 57.5.

2.17 The five-number summary of the data in Exercise 2.15 is

(a) 0, 0, 42.5, 76, 97.

(b) 0, 29, 57.5, 81.5, 97.

(c) 0, 29, 42.5, 75, 97.

2.18 If a distribution is skewed to the right,

(a) the mean is less than the median.

(b) the mean and median are equal.

(c) the mean is greater than the median.

2.19 What percent of the observations in a distribution lie between the first quartile and the third quartile?

(a) 25% (b) 50% (c) 75%

2.20 To make a boxplot of a distribution, you must know

(a) all of the individual observations.

(b) the mean and the standard deviation.

(c) the five-number summary.

2.21 The standard deviation of the 10 amounts of money in Exercise 2.15 (use your calculator) is

(a) 35.3. (b) 37.2. (c) 43.3.

2.22 What are all the values that a standard deviation s can possibly take?

(a) $0 \leq s$ (b) $0 \leq s \leq 1$ (c) $-1 \leq s \leq 1$

2.23 You have data on the weights in grams of 5 baby pythons. The mean weight is 31.8 and the standard deviation of the weights is 2.39. The correct units for the standard deviation are

(a) no units—it's just a number. (b) grams. (c) grams squared.

2.24 Which of the following is least affected if an extreme high outlier is added to your data?

(a) The median (b) The mean (c) The standard deviation

C H A P T E R 2   E X E R C I S E S

2.25 Incomes of college grads. According to the Census Bureau's 2007 Current Population Survey, the mean and median 2006 income of people at least 25 years old who had a bachelor's degree but no higher degree were $46,453 and $58,886. Which of these numbers is the mean and which is the median? Explain your reasoning.

2.26 Saving for retirement. Retirement seems a long way off and we need money now, so saving for retirement is hard. Among households with an employed person aged 21 to 64, only 63% own a retirement account. The mean value in these accounts is $112,300, but the median value is just $31,600. For people 55 or older, the mean is $222,100 and the median is $64,400.[11] What explains the differences between the two measures of center?

2.27 University endowments. The National Association of College and University Business Officers collects data on college endowments. In 2007, 785 colleges and universities reported the value of their endowments. When the endowment values are arranged in order, what are the positions of the median and the quartiles in this ordered list?

2.28 Pulling wood apart. Example 1.9 (page 20) gives the breaking strengths of 20 pieces of Douglas fir.

(a)   Give the five-number summary of the distribution of breaking strengths. (The stemplot, Figure 1.11, helps because it arranges the data in order, but you should use the unrounded values in numerical work.)

(b)   The stemplot shows that the distribution is skewed to the left. Does the five-number summary show the skew? Remember that only a graph gives a clear picture of the shape of a distribution.

2.29 Comparing tropical flowers. An alternative presentation of the flower length data in Table 2.1 reports the five-number summary and uses boxplots to display the distributions. Do this. Do the boxplots fail to reveal any important information visible in the stemplots in Figure 2.5?

2.30 How much fruit do adolescent girls eat? Figure 1.14 (page 30) is a histogram of the number of servings of fruit per day claimed by 74 seventeen-year-old girls. With a little care, you can find the median and the quartiles from the histogram. What are these numbers? How did you find them?

2.31 Weight of newborns. Here is the distribution of the weight at birth for all babies born in the United States in 2005:[12]

Weight (grams)	Count	Weight (grams)	Count
Less than 500	6,599	3,000 to 3,499	1,596,944
500 to 999	23,864	3,500 to 3,999	1,114,887
1,000 to 1,499	31,325	4,000 to 4,499	289,098
1,500 to 1,999	66,453	4,500 to 4,999	42,119
2,000 to 2,499	210,324	5,000 to 5,499	4,715
2,500 to 2,999	748,042		

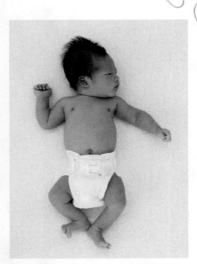

Photodisc Red/Getty Images

(a)   For comparison with other years and with other countries, we prefer a histogram of the *percents* in each weight class rather than the counts. Explain why.

(b)   How many babies were there? Make a histogram of the distribution, using percents on the vertical scale.

(c)   What are the positions of the median and quartiles in the ordered list of all birth weights? In which weight classes do the median and quartiles fall?

2.32 More on study times. In Exercise 1.38 (page 35) you examined the nightly study time claimed by first-year college men and women. The most common methods for formal comparison of two groups use $\bar{x}$ and s to summarize the data.

(a)   What kinds of distributions are best summarized by $\bar{x}$ and s?

(b)   One student in each group claimed to study at least 300 minutes (five hours) per night. How much does removing these observations change $\bar{x}$ and s for each group?

2.33 Making resistance visible. In the *Mean and Median* applet, place three observations on the line by clicking below it: two close together near the center of the line, and one somewhat to the right of these two.

(a)   Pull the single rightmost observation out to the right. (Place the cursor on the point, hold down a mouse button, and drag the point.) How does the mean behave? How does the median behave? Explain briefly why each measure acts as it does.

(b)   Now drag the single rightmost point to the left as far as you can. What happens to the mean? What happens to the median as you drag this point past the other two (watch carefully)?

2.34 Behavior of the median. Place five observations on the line in the *Mean and Median* applet by clicking below it.

(a)   Add one additional observation *without changing the median*. Where is your new point?

(b)   Use the applet to convince yourself that when you add yet another observation (there are now seven in all), the median does not change no matter where you put the seventh point. Explain why this must be true.

2.35 Guinea pig survival times. Here are the survival times in days of 72 guinea pigs after they were injected with infectious bacteria in a medical experiment.[13] Survival times, whether of machines under stress or cancer patients after treatment, usually have distributions that are skewed to the right.

Dorling Kindersley/Getty Images

43	45	53	56	56	57	58	66	67	73	74	79
80	80	81	81	81	82	83	83	84	88	89	91
91	92	92	97	99	99	100	100	101	102	102	102
103	104	107	108	109	113	114	118	121	123	126	128
137	138	139	144	145	147	156	162	174	178	179	184
191	198	211	214	243	249	329	380	403	511	522	598

(a)   Graph the distribution and describe its main features. Does it show the expected right skew?

(b)   Which numerical summary would you choose for these data? Calculate your chosen summary. How does it reflect the skewness of the distribution?

2.36 Never on Sunday: also in Canada? Exercise 1.5 (page 11) gives the number of births in the United States on each day of the week during an entire year. The

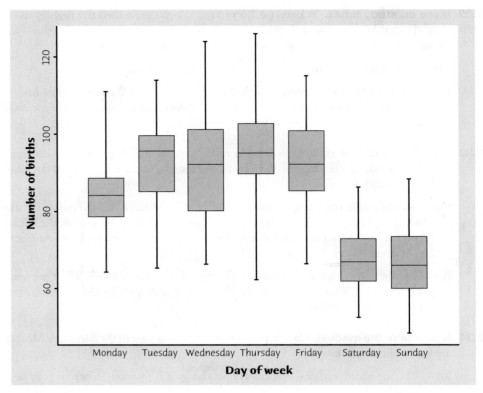

FIGURE 2.6

Boxplots of the distributions of numbers of births in Toronto, Canada, on each day of the week during a year, for Exercise 2.36.

boxplots in Figure 2.6 are based on more detailed data from Toronto, Canada: the number of births on each of the 365 days in a year, grouped by day of the week.[14] Based on these plots, give a more detailed description of how births depend on the day of the week.

2.37 Thinking about means. Table 1.1 (page 12) gives the percent of foreign-born residents in each of the states. For the nation as a whole, 12.5% of residents are foreign-born. Find the mean of the 51 entries in Table 1.1. It is *not* 12.5%. Explain carefully why this happens. (*Hint:* The states with the largest populations are California, Texas, New York, and Florida. Look at their entries in Table 1.1.)

2.38 Thinking about medians. A report says that "the median credit card debt of American households is zero." We know that many households have large amounts of credit card debt. Explain how the median debt can nonetheless be zero.

2.39 A standard deviation contest. This is a standard deviation contest. You must choose four numbers from the whole numbers 0 to 10, with repeats allowed.

(a) Choose four numbers that have the smallest possible standard deviation.

(b)   Choose four numbers that have the largest possible standard deviation.

(c)   Is more than one choice possible in either (a) or (b)? Explain.

2.40 Test your technology. This exercise requires a calculator with a standard deviation button or statistical software on a computer. The observations

$$10,001 \qquad 10,002 \qquad 10,003$$

have mean $\bar{x} = 10{,}002$ and standard deviation $s = 1$. Adding a 0 in the center of each number, the next set becomes

$$100,001 \qquad 100,002 \qquad 100,003$$

The standard deviation remains $s = 1$ as more 0s are added. Use your calculator or software to find the standard deviation of these numbers, adding extra 0s until you get an incorrect answer. How soon did you go wrong? This demonstrates that calculators and software cannot handle an arbitrary number of digits correctly.

2.41 You create the data. Create a set of 5 positive numbers (repeats allowed) that have median 10 and mean 7. What thought process did you use to create your numbers?

2.42 You create the data. Give an example of a small set of data for which the mean is larger than the third quartile.

*Exercises 2.43 to 2.48 ask you to analyze data without having the details outlined for you. The exercise statements give you the **State** step of the four-step process. In your work, follow the **Plan, Solve,** and **Conclude** steps as illustrated in Example 2.9.*

2.43 Athletes' salaries. In 2007, the Boston Red Sox won the World Series for the second time in 4 years. Table 2.2 gives the salaries of the 25 players on the Red Sox World Series roster. Provide the team owner with a full description of the distribution of salaries and a brief summary of its most important features.

TABLE 2.2 Salaries for the 2007 Boston Red Sox World Series team

PLAYER	SALARY	PLAYER	SALARY	PLAYER	SALARY
Josh Beckett	$6,666,667	Jon Lester	$384,000	Jonathan Papelbon	$425,000
Alex Cora	$2,000,000	Javier López	$402,000	Dustin Pedroia	$380,000
Coco Crisp	$3,833,333	Mike Lowell	$9,000,000	Manny Ramirez	$17,016,381
Manny Delcarmen	$380,000	Julio Lugo	$8,250,000	Curt Schilling	$13,000,000
J. D. Drew	$14,400,000	Daisuke Matsuzaka	$6,333,333	Kyle Snyder	$535,000
Jacoby Ellsbury	$380,000	Doug Mirabelli	$750,000	Mike Timlin	$2,800,000
Eric Gagné	$6,000,000	Hideki Okajimi	$1,225,000	Jason Varitek	$11,000,000
Eric Hinske	$5,725,000	David Ortiz	$13,250,000	Kevin Youkilis	$424,000
Bobby Kielty	$2,100,000				

2.44 Returns on stocks. How well have stocks done over the past generation? The Wilshire 5000 index describes the average performance of all U.S. stocks. The average is weighted by the total market value of each company's stock, so think of the

index as measuring the performance of the average investor. Here are the percent returns on the Wilshire 5000 index for the years from 1971 to 2006:

Year	Return	Year	Return	Year	Return
1971	16.19	1983	22.71	1995	36.41
1972	17.34	1984	3.27	1996	21.56
1973	−18.78	1985	31.46	1997	31.48
1974	−27.87	1986	15.61	1998	24.31
1975	37.38	1987	1.75	1999	24.23
1976	26.77	1988	17.59	2000	−10.89
1977	−2.97	1989	28.53	2001	−10.97
1978	8.54	1990	−6.03	2002	−20.86
1979	24.40	1991	33.58	2003	31.64
1980	33.21	1992	9.02	2004	12.48
1981	−3.98	1993	10.67	2005	6.38
1982	20.43	1994	0.06	2006	15.77

What can you say about the distribution of yearly returns on stocks?

2.45 Do good smells bring good business? Businesses know that customers often respond to background music. Do they also respond to odors? Nicolas Guéguen and his colleagues studied this question in a small pizza restaurant in France on Saturday evenings in May. On one evening, a relaxing lavender odor was spread through the restaurant; on another evening, a stimulating lemon odor; a third evening served as a control, with no odor. Table 2.3 shows the amounts (in euros) that customers spent on each of these evenings.[15] Compare the three distributions. What effect did the two odors have on customer spending?

TABLE 2.3 Amount spent (euros) by customers in a restaurant when exposed to odors

NO ODOR									
15.9	18.5	15.9	18.5	18.5	21.9	15.9	15.9	15.9	15.9
15.9	18.5	18.5	18.5	20.5	18.5	18.5	15.9	15.9	15.9
18.5	18.5	15.9	18.5	15.9	18.5	15.9	25.5	12.9	15.9
LEMON ODOR									
18.5	15.9	18.5	18.5	18.5	15.9	18.5	15.9	18.5	18.5
15.9	18.5	21.5	15.9	21.9	15.9	18.5	18.5	18.5	18.5
25.9	15.9	15.9	15.9	18.5	18.5	18.5	18.5		
LAVENDER ODOR									
21.9	18.5	22.3	21.9	18.5	24.9	18.5	22.5	21.5	21.9
21.5	18.5	25.5	18.5	18.5	21.9	18.5	18.5	24.9	21.9
25.9	21.9	18.5	18.5	22.8	18.5	21.9	20.7	21.9	22.5

2.46 Daily activity and obesity. People gain weight when they take in more energy from food than they expend. Table 2.4 compares volunteer subjects who were lean with others who were mildly obese. None of the subjects followed an exercise program.

TABLE 2.4 Time (minutes per day) active and lying down by lean and obese subjects

LEAN SUBJECTS			OBESE SUBJECTS		
SUBJECT	STAND/WALK	LIE	SUBJECT	STAND/WALK	LIE
1	511.100	555.500	11	260.244	521.044
2	607.925	450.650	12	464.756	514.931
3	319.212	537.362	13	367.138	563.300
4	584.644	489.269	14	413.667	532.208
5	578.869	514.081	15	347.375	504.931
6	543.388	506.500	16	416.531	448.856
7	677.188	467.700	17	358.650	460.550
8	555.656	567.006	18	267.344	509.981
9	374.831	531.431	19	410.631	448.706
10	504.700	396.962	20	426.356	412.919

The subjects wore sensors that recorded every move for 10 days. The table shows the average minutes per day spent in activity (standing and walking) and in lying down.[16] Compare the distributions of time spent actively for lean and obese subjects and also the distributions of time spent lying down. How does the behavior of lean and mildly obese people differ?

2.47 Compressing soil. Farmers know that driving heavy equipment on wet soil compresses the soil and hinders the growth of crops. Table 2.5 gives data on the "penetrability" of the same soil at three levels of compression.[17] Penetrability is a measure of the resistance plant roots meet when they grow through the soil. Low penetrability means high resistance. How does increasing compression affect penetrability?

TABLE 2.5 Penetrability of soil at three compression levels

COMPRESSED		INTERMEDIATE		LOOSE	
2.86	3.08	3.14	3.54	3.99	4.11
2.68	2.82	3.38	3.36	4.20	4.30
2.92	2.78	3.10	3.18	3.94	3.96
2.82	2.98	3.40	3.12	4.16	4.03
2.76	3.00	3.38	3.86	4.29	4.89
2.81	2.78	3.14	2.92	4.19	4.12
2.78	2.96	3.18	3.46	4.13	4.00
3.08	2.90	3.26	3.44	4.41	4.34
2.94	3.18	2.96	3.62	3.98	4.27
2.86	3.16	3.02	4.26	4.41	4.91

2.48 Does breast-feeding weaken bones? Breast-feeding mothers secrete calcium into their milk. Some of the calcium may come from their bones, so mothers may lose bone mineral. Researchers compared 47 breast-feeding women with 22 women of similar age who were neither pregnant nor lactating. They measured the percent change in the mineral content of the women's spines over three months. Here are the data:[18]

Breast-feeding women						Other women					
−4.7	−2.5	−4.9	−2.7	−0.8	−5.3	2.4	0.0	0.9	−0.2	1.0	1.7
−8.3	−2.1	−6.8	−4.3	2.2	−7.8	2.9	−0.6	1.1	−0.1	−0.4	0.3
−3.1	−1.0	−6.5	−1.8	−5.2	−5.7	1.2	−1.6	−0.1	−1.5	0.7	−0.4
−7.0	−2.2	−6.5	−1.0	−3.0	−3.6	2.2	−0.4	−2.2	−0.1		
−5.2	−2.0	−2.1	−5.6	−4.4	−3.3						
−4.0	−4.9	−4.7	−3.8	−5.9	−2.5						
−0.3	−6.2	−6.8	1.7	0.3	−2.3						
0.4	−5.3	0.2	−2.2	−5.1							

Do the data show distinctly greater bone mineral loss among the breast-feeding women?

Exercises 2.49 to 2.52 make use of the optional material on the 1.5 × IQR rule for suspected outliers.

2.49 Older Americans. The stemplot in Exercise 1.19 (page 26) displays the distribution of the percents of residents aged 65 and older in the states. Stemplots help you find the five-number summary because they arrange the observations in increasing order.

(a) Give the five-number summary of this distribution.

(b) Which observations does the 1.5 × IQR rule flag as suspected outliers? (The rule flags several observations that are clearly not outliers. The reason is that the center half of the observations are close together, so that IQR is small. This example reminds us to use our eyes, not a rule, to spot outliers.)

2.50 Carbon dioxide emissions. Table 1.6 (page 34) gives carbon dioxide (CO_2) emissions per person for countries with population at least 20 million. A stemplot or histogram shows that the distribution is strongly skewed to the right. The United States and several other countries appear to be high outliers.

(a) Give the five-number summary. Explain why this summary suggests that the distribution is right-skewed.

(b) Which countries are outliers according to the 1.5 × IQR rule? Make a stemplot of the data or look at your stemplot from Exercise 1.36. Do you agree with the rule's suggestions about which countries are and are not outliers?

2.51 Athletes' salaries. Which members of the Boston Red Sox (Table 2.2) have salaries that are suspected outliers by the 1.5 × IQR rule?

2.52 Returns on stocks. The returns on stocks in Exercise 2.44 vary a lot: they range from a loss of more than 27% to a gain of more than 37%. Are any of these years suspected outliers by the 1.5 × IQR rule?

Ios Photos/Newscom

CHAPTER 3

The Normal Distributions

We now have a kit of graphical and numerical tools for describing distributions. What is more, we have a clear strategy for exploring data on a single quantitative variable.

> **EXPLORING A DISTRIBUTION**
>
> 1. Always plot your data: make a graph, usually a histogram or a stemplot.
> 2. Look for the overall pattern (shape, center, spread) and for striking deviations such as outliers.
> 3. Calculate a numerical summary to briefly describe center and spread.

In this chapter, we add one more step to this strategy:

> 4. Sometimes the overall pattern of a large number of observations is so regular that we can describe it by a smooth curve.

Density curves

Figure 3.1 is a histogram of the scores of all 947 seventh-grade students in Gary, Indiana, on the vocabulary part of the Iowa Test of Basic Skills.[1] Scores of many

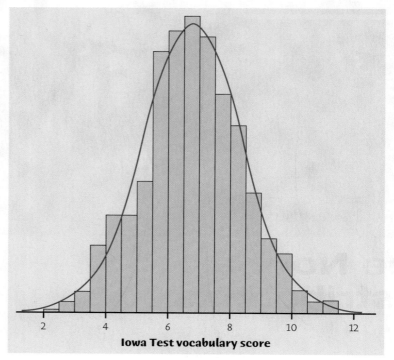

Iowa Test vocabulary score

FIGURE 3.1

Histogram of the Iowa Test vocabulary scores of all seventh-grade students in Gary, Indiana. The smooth curve shows the overall shape of the distribution.

students on this national test have a quite regular distribution. The histogram is symmetric, and both tails fall off smoothly from a single center peak. There are no large gaps or obvious outliers. The smooth curve drawn through the tops of the histogram bars in Figure 3.1 is a good description of the overall pattern of the data.

EXAMPLE 3.1 **From histogram to density curve**

Our eyes respond to the *areas* of the bars in a histogram. The bar areas represent proportions of the observations. Figure 3.2(a) is a copy of Figure 3.1 with the leftmost bars shaded. The area of the shaded bars in Figure 3.2(a) represents the students with vocabulary scores 6.0 or lower. There are 287 such students, who make up the proportion 287/947 = 0.303 of all Gary seventh-graders.

Now look at the curve drawn through the bars. In Figure 3.2(b), the area under the curve to the left of 6.0 is shaded. We can draw histogram bars taller or shorter by adjusting the vertical scale. In moving from histogram bars to a smooth curve, we make a specific choice: adjust the scale of the graph so that *the total area under the curve is exactly 1*. The total area represents the proportion 1, that is, all the observations. We can then interpret areas under the curve as proportions of the observations. The curve is now a *density curve*. The shaded area under the density curve in Figure 3.2(b) represents the proportion of students with score 6.0 or lower. This area is 0.293, only 0.010 away from the actual proportion 0.303. Areas under the density curve give quite good approximations to the actual distribution of the 947 test scores. ■

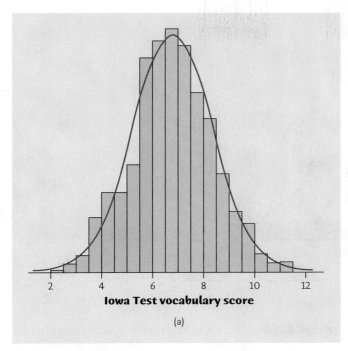

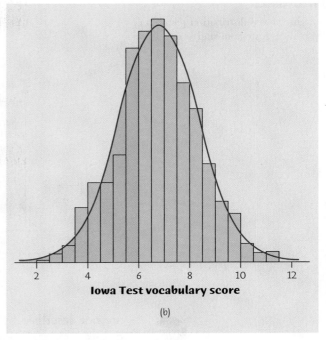

FIGURE 3.2(a)

The proportion of scores less than or equal to 6.0 in the actual data is 0.303.

FIGURE 3.2(b)

The proportion of scores less than or equal to 6.0 from the density curve is 0.293. The density curve is a good approximation to the distribution of the data.

DENSITY CURVE

A **density curve** is a curve that

- is always on or above the horizontal axis, and
- has area exactly 1 underneath it.

A density curve describes the overall pattern of a distribution. The area under the curve and above any range of values is the proportion of all observations that fall in that range.

Density curves, like distributions, come in many shapes. Figure 3.3 shows a strongly skewed distribution, the survival times of guinea pigs from Exercise 2.35 (page 61). The histogram and density curve were both created from the data by software. Both show the overall shape and the "bumps" in the long right tail. The density curve shows a single high peak as a main feature of the distribution. The histogram divides the observations near the peak between two bars, thus reducing the height of the peak. A density curve is often a good description of the overall pattern of a distribution. Outliers, which are deviations from the overall pattern,

FIGURE 3.3

A right-skewed distribution pictured by both a histogram and a density curve.

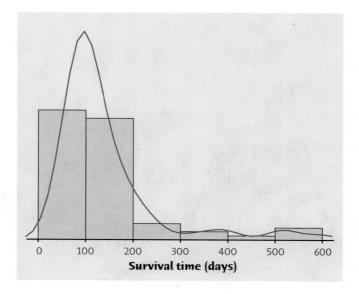

are not described by the curve. *Of course, no set of real data is exactly described by a density curve. The curve is an idealized description that is easy to use and accurate enough for practical use.*

APPLY YOUR KNOWLEDGE

3.1 Sketch density curves. Sketch density curves that describe distributions with the following shapes:

(a) Symmetric, but with two peaks (that is, two strong clusters of observations)

(b) Single peak and skewed to the left

3.2 Accidents on a bike path. Examining the location of accidents on a level, 3-mile bike path shows that they occur uniformly along the length of the path. Figure 3.4 displays the density curve that describes the distribution of accidents.

(a) Explain why this curve satisfies the two requirements for a density curve.

(b) The proportion of accidents that occur in the first mile of the path is the area under the density curve between 0 miles and 1 mile. What is this area?

(c) Sue's property adjoins the bike path between the 0.8 mile mark and the 1.1 mile mark. What proportion of accidents happen in front of Sue's property?

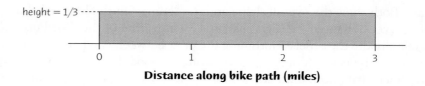

FIGURE 3.4

The density curve for the location of accidents along a 3-mile bike path, for Exercise 3.2.

Describing density curves

Our measures of center and spread apply to density curves as well as to actual sets of observations. The median and quartiles are easy. Areas under a density curve represent proportions of the total number of observations. The median is the point with half the observations on either side. So *the median of a density curve is the equal-areas point*, the point with half the area under the curve to its left and the remaining half of the area to its right. The quartiles divide the area under the curve into quarters. One-fourth of the area under the curve is to the left of the first quartile, and three-fourths of the area is to the left of the third quartile. You can roughly locate the median and quartiles of any density curve by eye by dividing the area under the curve into four equal parts.

Because density curves are idealized patterns, a symmetric density curve is exactly symmetric. The median of a symmetric density curve is therefore at its center. Figure 3.5(a) shows a symmetric density curve with the median marked. It isn't so easy to spot the equal-areas point on a skewed curve. There are mathematical ways of finding the median for any density curve. That's how we marked the median on the skewed curve in Figure 3.5(b).

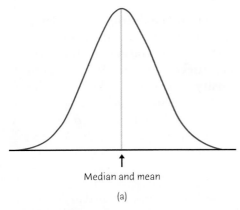

↑
Median and mean

(a)

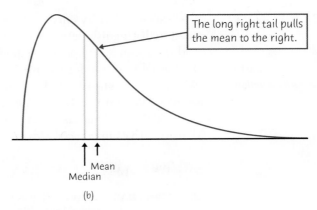

The long right tail pulls the mean to the right.

↑↑ Mean
Median

(b)

FIGURE 3.5(a)

The median and mean of a symmetric density curve both lie at the center of symmetry.

FIGURE 3.5(b)

The median and mean of a right-skewed density curve. The mean is pulled away from the median toward the long tail.

What about the mean? The mean of a set of observations is their arithmetic average. If we think of the observations as weights strung out along a thin rod, the mean is the point at which the rod would balance. This fact is also true of density curves. *The mean is the point at which the curve would balance if made of solid material.* Figure 3.6 illustrates this fact about the mean. A symmetric curve balances at its center because the two sides are identical. *The mean and median of a symmetric density curve are equal,* as in Figure 3.5(a). We know that the mean of a skewed distribution is pulled toward the long tail. Figure 3.5(b) shows how the mean of a skewed density curve is pulled toward the long tail more than is the median. It's hard to locate the balance point by eye on a skewed curve. There are mathematical

FIGURE 3.6
The mean is the balance point of a density curve.

ways of calculating the mean for any density curve, so we are able to mark the mean as well as the median in Figure 3.5(b).

MEDIAN AND MEAN OF A DENSITY CURVE

The **median** of a density curve is the equal-areas point, the point that divides the area under the curve in half.

The **mean** of a density curve is the balance point, at which the curve would balance if made of solid material.

The median and mean are the same for a symmetric density curve. They both lie at the center of the curve. The mean of a skewed curve is pulled away from the median in the direction of the long tail.

Because a density curve is an idealized description of a distribution of data, we need to distinguish between the mean and standard deviation of the density curve and the mean $\bar{x}$ and standard deviation s computed from the actual observations. The usual notation for the **mean of a density curve** is μ (the Greek letter mu). We write the **standard deviation of a density curve** as σ (the Greek letter sigma). We can roughly locate the mean μ of any density curve by eye, as the balance point. There is no easy way to locate the standard deviation σ by eye for density curves in general.

mean μ
standard deviation σ

APPLY YOUR KNOWLEDGE

3.3 Mean and median. What is the mean μ of the density curve pictured in Figure 3.4? (That is, where would the curve balance?) What is the median? (That is, where is the point with area 0.5 on either side?)

3.4 Mean and median. Figure 3.7 displays three density curves, each with three points marked on them. At which of these points on each curve do the mean and the median fall?

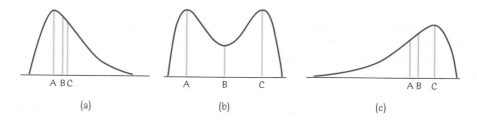

FIGURE 3.7
Three density curves, for Exercise 3.4.

Normal distributions

One particularly important class of density curves has already appeared in Figures 3.1 and 3.2. They are called **Normal curves.** The distributions they describe are called **Normal distributions.** Normal distributions play a large role in statistics, but they are rather special and not at all "normal" in the sense of being usual or

Normal curve
Normal distribution

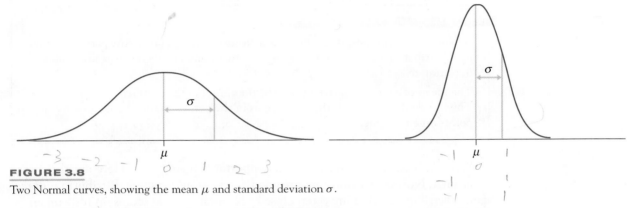

FIGURE 3.8

Two Normal curves, showing the mean μ and standard deviation σ.

average. We capitalize Normal to remind you that these curves are special. Look at the two Normal curves in Figure 3.8. They illustrate several important facts:

- All Normal curves have the same overall shape: symmetric, single-peaked, bell-shaped.

- Any specific Normal curve is completely described by giving its mean μ and its standard deviation σ.

- The mean is located at the center of the symmetric curve and is the same as the median. Changing μ without changing σ moves the Normal curve along the horizontal axis without changing its spread.

- The standard deviation σ controls the spread of a Normal curve. Curves with larger standard deviation are more spread out.

The standard deviation σ is the natural measure of spread for Normal distributions. Not only do μ and σ completely determine the shape of a Normal curve, but we can locate σ by eye on a Normal curve. Here's how. Imagine that you are skiing down a mountain that has the shape of a Normal curve. At first, you descend at an ever-steeper angle as you go out from the peak:

Fortunately, before you find yourself going straight down, the slope begins to grow flatter rather than steeper as you go out and down:

The points at which this change of curvature takes place are located at distance σ on either side of the mean μ. You can feel the change as you run a pencil along a Normal curve, and so find the standard deviation. Remember that *μ and σ alone do not specify the shape of most distributions,* and that the shape of density curves in general does not reveal σ. These are special properties of Normal distributions.

NORMAL DISTRIBUTIONS

A **Normal distribution** is described by a Normal density curve. Any particular Normal distribution is completely specified by two numbers, its mean μ and standard deviation σ.

The mean of a Normal distribution is at the center of the symmetric Normal curve. The standard deviation is the distance from the center to the change-of-curvature points on either side.

Why are the Normal distributions important in statistics? Here are three reasons. First, Normal distributions are good descriptions for some distributions of *real data.* Distributions that are often close to Normal include scores on tests taken by many people (such as Iowa Tests and SAT exams), repeated careful measurements of the same quantity, and characteristics of biological populations (such as lengths of crickets and yields of corn). Second, Normal distributions are good approximations to the results of many kinds of *chance outcomes,* such as the proportion of heads in many tosses of a coin. Third, we will see that many *statistical inference* procedures based on Normal distributions work well for other roughly symmetric distributions. However, many sets of data do not follow a Normal distribution. Most income distributions, for example, are skewed to the right and so are not Normal. Non-Normal data, like nonnormal people, not only are common but are sometimes more interesting than their Normal counterparts.

The 68–95–99.7 rule

Although there are many Normal curves, they all have common properties. In particular, all Normal distributions obey the following rule.

THE 68–95–99.7 RULE

In the Normal distribution with mean μ and standard deviation σ:

■ Approximately **68%** of the observations fall within σ of the mean μ.

■ Approximately **95%** of the observations fall within 2σ of μ.

■ Approximately **99.7%** of the observations fall within 3σ of μ.

Figure 3.9 illustrates the 68–95–99.7 rule. By remembering these three numbers, you can think about Normal distributions without constantly making detailed calculations.

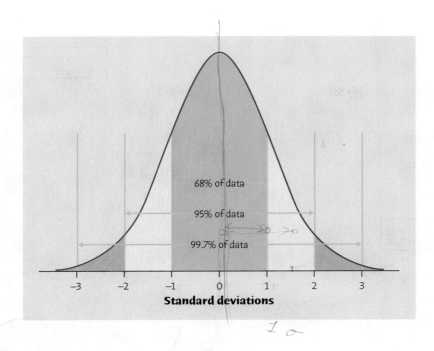

FIGURE 3.9
The 68–95–99.7 rule for Normal distributions.

EXAMPLE 3.2 Iowa Test scores

Figures 3.1 and 3.2 show that the distribution of Iowa Test vocabulary scores for seventh-grade students in Gary, Indiana, is close to Normal. Suppose that the distribution is exactly Normal with mean $\mu = 6.84$ and standard deviation $\sigma = 1.55$. (These are the mean and standard deviation of the 947 actual scores.)

Figure 3.10 applies the 68–95–99.7 rule to the Iowa Test scores. The 95 part of the rule says that 95% of all scores are between

$$\mu - 2\sigma = 6.84 - (2)(1.55) = 6.84 - 3.10 = 3.74$$

and

$$\mu + 2\sigma = 6.84 + (2)(1.55) = 6.84 + 3.10 = 9.94$$

The other 5% of scores are outside this range. Because Normal distributions are symmetric, half of these scores are lower than 3.74 and half are higher than 9.94. That is, 2.5% of the scores are below 3.74 and 2.5% are above 9.94. ■

The 68–95–99.7 rule describes distributions that are exactly Normal. Real data such as the actual Gary scores are never exactly Normal. For one thing, Iowa Test scores are reported only to the nearest tenth. A score can be 9.9 or 10.0, but not 9.94. We use a Normal distribution because it's a good approximation, and because we think the knowledge that the test measures is continuous rather than stopping at tenths.

How well does our work in Example 3.2 describe the actual Iowa Test scores? Well, 900 of the 947 scores are between 3.74 and 9.94. That's 95.04%, very accurate indeed. Of the remaining 47 scores, 20 are below 3.74 and 27 are above 9.94. The tails of the actual data are not quite equal, as they would be in an exactly

The 68–95–99.7 rule applied to the distribution of Iowa Test scores for seventh-grade students in Gary, Indiana, for Example 3.2. The mean and standard deviation are $\mu = 6.84$ and $\sigma = 1.55$.

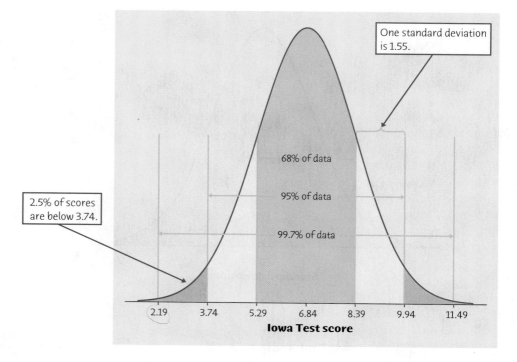

One standard deviation is 1.55.

68% of data

95% of data

99.7% of data

2.5% of scores are below 3.74.

| 2.19 | 3.74 | 5.29 | 6.84 | 8.39 | 9.94 | 11.49 |

Iowa Test score

Normal distribution. Normal distributions often describe real data better in the center of the distribution than in the extreme high and low tails.

EXAMPLE 3.3 **Iowa Test scores**

Look again at Figure 3.10. A score of 5.29 is one standard deviation below the mean. What percent of scores are higher than 5.29? Find the answer by adding areas in the figure. Here is the calculation in pictures:

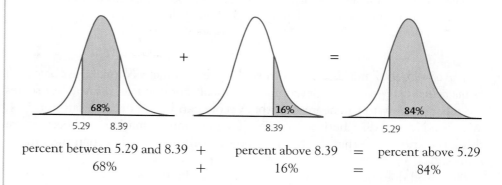

68%		
5.29 8.39	8.39	5.29
16%		
84%		

percent between 5.29 and 8.39 + percent above 8.39 = percent above 5.29
68%   +   16%   =   84%

Be sure you see where the 16% came from. We know that 68% of scores are between 5.29 and 8.39, so 32% of scores are outside that range. These are equally split between the two tails, 16% below 5.29 and 16% above 8.39. ■

Because we will mention Normal distributions often, a short notation is helpful. We abbreviate the Normal distribution with mean μ and standard deviation σ as $N(\mu, \sigma)$. For example, the distribution of Gary Iowa Test scores is approximately $N(6.84, 1.55)$.

APPLY YOUR KNOWLEDGE

3.5 Running a mile. How fast can male college students run a mile? There's lots of variation, of course. During World War II, physical training was required for male students in many colleges, as preparation for military service. That provided an opportunity to collect data on physical performance on a large scale. A study of 12,000 able-bodied male students at the University of Illinois found that their times for the mile run were approximately Normal with mean 7.11 minutes and standard deviation 0.74 minute.[2] Draw a Normal curve on which this mean and standard deviation are correctly located. (*Hint:* Draw an unlabeled Normal curve, locate the points where the curvature changes, then add number labels on the horizontal axis.)

Ios Photos/Newscom

3.6 Running a mile. The times for the mile run of a large group of male college students are approximately Normal with mean 7.11 minutes and standard deviation 0.74 minutes. Use the 68–95–99.7 rule to answer the following questions. (Start by making a sketch like Figure 3.10.)

(a) What range of times covers almost all (99.7%) of this distribution?

(b) What percent of these men run a mile in less than 6.37 minutes?

3.7 Monsoon rains. The summer monsoon brings 80% of India's rainfall and is essential for the country's agriculture. Records going back more than a century show that the amount of monsoon rainfall varies from year to year according to a distribution that is approximately Normal with mean 852 millimeters (mm) and standard deviation 82 mm.[3] Use the 68–95–99.7 rule to answer the following questions.

(a) Between what values do the monsoon rains fall in 95% of all years?

(b) How small are the monsoon rains in the driest 2.5% of all years?

The standard Normal distribution

As the 68–95–99.7 rule suggests, all Normal distributions share many properties. In fact, all Normal distributions are the same if we measure in units of size σ about the mean μ as center. Changing to these units is called *standardizing*. To standardize a value, subtract the mean of the distribution and then divide by the standard deviation.

> **STANDARDIZING AND z-SCORES**
>
> If x is an observation from a distribution that has mean μ and standard deviation σ, the **standardized value** of x is
>
> $$z = \frac{x - \mu}{\sigma}$$
>
> A standardized value is often called a **z-score.**

He said, she said

When *asked* their weight, almost all women say they weigh less than they really do. Heavier men also underreport their weight—but lighter men claim to weigh more than the scale shows. We leave you to ponder the psychology of the two sexes. Just remember that "say so" is no substitute for measuring.

A z-score tells us how many standard deviations the original observation falls away from the mean, and in which direction. Observations larger than the mean are positive when standardized, and observations smaller than the mean are negative.

> **EXAMPLE 3.4** **Standardizing women's heights**
>
> The heights of women aged 20 to 29 are approximately Normal with $\mu = 64$ inches and $\sigma = 2.7$ inches.[4] The standardized height is
>
> $$z = \frac{\text{height} - 64}{2.7}$$
>
> A woman's standardized height is the number of standard deviations by which her height differs from the mean height of all young women. A woman 70 inches tall, for example, has standardized height
>
> $$z = \frac{70 - 64}{2.7} = 2.22$$
>
> or 2.22 standard deviations above the mean. Similarly, a woman 5 feet (60 inches) tall has standardized height
>
> $$z = \frac{60 - 64}{2.7} = -1.48$$
>
> or 1.48 standard deviations less than the mean height. ■

We often standardize observations from symmetric distributions to express them in a common scale. We might, for example, compare the heights of two children of different ages by calculating their z-scores. The standardized heights tell us where each child stands in the distribution for his or her age group.

If the variable we standardize has a Normal distribution, standardizing does more than give a common scale. It makes all Normal distributions into a single distribution, and this distribution is still Normal. Standardizing a variable that has any Normal distribution produces a new variable that has the *standard Normal distribution*.

> **STANDARD NORMAL DISTRIBUTION**
>
> The **standard Normal distribution** is the Normal distribution $N(0, 1)$ with mean 0 and standard deviation 1.
>
> If a variable x has any Normal distribution $N(\mu, \sigma)$ with mean μ and standard deviation σ, then the standardized variable
>
> $$z = \frac{x - \mu}{\sigma}$$
>
> has the standard Normal distribution.

3.8 SAT versus ACT. In 2007, when she was a high school senior, Eleanor scored 680 on the mathematics part of the SAT. The distribution of SAT math scores in 2007 was Normal with mean 515 and standard deviation 114. Gerald took the ACT Assessment mathematics test and scored 27. ACT math scores for 2007 were Normally distributed with mean 21.0 and standard deviation 5.1. Find the standardized scores for both students. Assuming that both tests measure the same kind of ability, who had the higher score?

3.9 Men's and women's heights. The heights of women aged 20 to 29 are approximately Normal with mean 64 inches and standard deviation 2.7 inches. Men the same age have mean height 69.3 inches with standard deviation 2.8 inches. What are the z-scores for a woman 6 feet tall and a man 6 feet tall? Say in simple language what information the z-scores give that the actual heights do not.

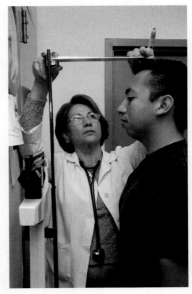

Spencer Grant/PhotoEdit

Finding Normal proportions

Areas under a Normal curve represent proportions of observations from that Normal distribution. There is no formula for areas under a Normal curve. Calculations use either software that calculates areas or a table of areas. Most tables and software calculate one kind of area, *cumulative proportions*. The idea of "cumulative" is "everything that came before." Here is the exact statement.

CUMULATIVE PROPORTIONS

The **cumulative proportion** for a value x in a distribution is the proportion of observations in the distribution that are less than or equal to x.

Cumulative proportion

x

The key to calculating Normal proportions is to match the area you want with areas that represent cumulative proportions. If you make a sketch of the area you want, you will almost never go wrong. Find areas for cumulative proportions either from software or (with an extra step) from a table. The following example shows the method in a picture.

EXAMPLE 3.5 Who qualifies for college sports?

The National Collegiate Athletic Association (NCAA) requires Division II athletes to score at least 820 on the combined mathematics and reading parts of the SAT in order to compete in their first college year. The scores of the 1.5 million high school seniors taking the SAT this year are approximately Normal with mean 1026 and standard deviation 209. What percent of high school seniors qualify for Division II college sports?

Here is the calculation in a picture: the proportion of scores above 820 is the area under the curve to the right of 820. That's the total area under the curve (which is always 1) minus the cumulative proportion up to 820.

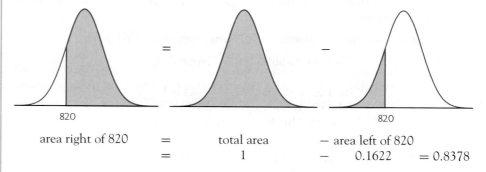

| area right of 820 | = | total area | − area left of 820 |
| | = | 1 | − 0.1622 = 0.8378 |

About 84% of all high school seniors meet the NCAA requirement to compete in Division II college sports. ■

There is *no* area under a smooth curve and exactly over the point 820. Consequently, the area to the right of 820 (the proportion of scores > 820) is the same as the area at or to the right of this point (the proportion of scores ≥ 820). The actual data may contain a student who scored exactly 820 on the SAT. That the proportion of scores exactly equal to 820 is 0 for a Normal distribution is a consequence of the idealized smoothing of Normal distributions for data.

To find the numerical value 0.1622 of the cumulative proportion in Example 3.5 using software, plug in mean 1026 and standard deviation 209 and ask for the cumulative proportion for 820. Software often uses terms such as "cumulative distribution" or "cumulative probability." We will learn in Chapter 10 why the language of probability fits. Here, for example, is Minitab's output:

```
Session                                                    _  □

Cumulative Distribution Function

Normal with mean = 1026 and standard deviation = 209

    x    P ( X <= x )
  820       0.162153
```

The *P* in the output stands for "probability," but we can read it as "proportion of the observations." The *Normal Curve* applet is even handier because it draws pictures as well as finding areas. If you are not using software, you can find cumulative proportions for Normal curves from a table. This requires an extra step.

Using the standard Normal table

The extra step in finding cumulative proportions from a table is that we must first standardize to express the problem in the standard scale of z-scores. This allows us to get by with just one table, a table of *standard Normal cumulative proportions*. Table A in the back of the book gives cumulative proportions for the standard Normal distribution. The pictures at the top of the table remind us that the entries are cumulative proportions, areas under the curve to the left of a value z.

EXAMPLE 3.6 **The standard Normal table**

What proportion of observations on a standard Normal variable z take values less than 1.47?

　　Solution: To find the area to the left of 1.47, locate 1.4 in the left-hand column of Table A, then locate the remaining digit 7 as .07 in the top row. The entry opposite 1.4 and under .07 is 0.9292. This is the cumulative proportion we seek. Figure 3.11 illustrates this area. ■

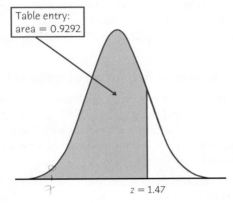

Table entry:
area = 0.9292

$z = 1.47$

FIGURE 3.11

The area under a standard Normal curve to the left of the point $z = 1.47$ is 0.9292. Table A gives areas under the standard Normal curve.

Now that you see how Table A works, let's redo Example 3.5 using the table. We can break Normal calculations using the table into three steps.

EXAMPLE 3.7 **Who qualifies for college sports?**

Scores of high school seniors on the SAT follow the Normal distribution with mean $\mu = 1026$ and standard deviation $\sigma = 209$. What percent of seniors score at least 820?

Step 1. Draw a picture. The picture is exactly as in Example 3.5. It shows that

$$\text{area to the right of } 820 = 1 - \text{area to the left of } 820$$

Step 2. Standardize. Call the SAT score x. Subtract the mean, then divide by the standard deviation, to transform the problem about x into a problem about a standard Normal z:

$$x \geq 820$$

$$\frac{x - 1026}{209} \geq \frac{820 - 1026}{209}$$

$$z \geq -0.99$$

Step 3. Use the table. The picture shows that we need the cumulative proportion for $x = 820$. Step 2 says this is the same as the cumulative proportion for $z = -0.99$. The Table A entry for $z = -0.99$ says that this cumulative proportion is 0.1611. The area to the right of -0.99 is therefore $1 - 0.1611 = 0.8389$. ∎

The area from the table in Example 3.7 (0.8389) is slightly less accurate than the area from software in Example 3.5 (0.8378) because we must round z to two decimal places when we use Table A. The difference is rarely important in practice. Here's the method in outline form.

USING TABLE A TO FIND NORMAL PROPORTIONS

Step 1. State the problem in terms of the observed variable x. **Draw a picture** that shows the proportion you want in terms of cumulative proportions.

Step 2. Standardize x to restate the problem in terms of a standard Normal variable z.

Step 3. Use Table A and the fact that the total area under the curve is 1 to find the required area under the standard Normal curve.

EXAMPLE 3.8 **Who qualifies for an athletic scholarship?**

The NCAA considers a student a "partial qualifier" for Division II athletics if high school grades are satisfactory and the combined SAT score is at least 720. Partial qualifiers can receive athletic scholarships and practice with the team, but they can't compete during their first college year. What proportion of all students who take the SAT would be partial qualifiers?

Step 1. State the problem and draw a picture. Call the SAT score x. The variable x has the $N(1026, 209)$ distribution. What proportion of SAT scores fall between 720 and 820? Here is the picture:

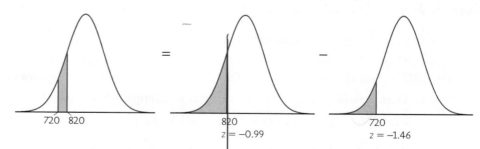

Step 2. Standardize. Subtract the mean, then divide by the standard deviation, to turn x into a standard Normal z:

$$720 \leq x < 820$$

$$\frac{720 - 1026}{209} \leq \frac{x - 1026}{209} < \frac{820 - 1026}{209}$$

$$-1.46 \leq z < -0.99$$

Step 3. Use the table. Follow the picture (we added the z-scores to the picture label to help you):

area between -1.46 and $-0.99 = $ (area left of -0.99) $-$ (area left of -1.46)

$$= 0.1611 - 0.0721 = 0.0890$$

About 9% of high school seniors would be partial qualifiers. ■

Sometimes we encounter a value of z more extreme than those appearing in Table A. For example, the area to the left of $z = -4$ is not given directly in the table. The z-values in Table A leave only area 0.0002 in each tail unaccounted for. For practical purposes, we can act as if there is zero area outside the range of Table A.

APPLY YOUR KNOWLEDGE

3.10 Use the Normal table. Use Table A to find the proportion of observations from a standard Normal distribution that satisfies each of the following statements. In each case, sketch a standard Normal curve and shade the area under the curve that is the answer to the question.

(a) $z < 2.85$ (b) $z > 2.85$ (c) $z > -1.66$ (d) $-1.66 < z < 2.85$

3.11 Monsoon rains. The summer monsoon rains in India follow approximately a Normal distribution with mean 852 millimeters (mm) of rainfall and standard deviation 82 mm.

(a) In the drought year 1987, 697 mm of rain fell. In what percent of all years will India have 697 mm or less of monsoon rain?

(b) "Normal rainfall" means within 20% of the long-term average, or between 683 mm and 1022 mm. In what percent of all years is the rainfall normal?

3.12 Fruit flies. The common fruit fly *Drosophila melanogaster* is the most studied organism in genetic research because it is small, easy to grow, and reproduces rapidly. The length of the thorax (where the wings and legs attach) in a population of male fruit flies is approximately Normal with mean 0.800 millimeters (mm) and standard deviation 0.078 mm.

(a) What proportion of flies have thorax length 0.9 mm or longer?

(b) What proportion have thorax length between 0.9 mm and 1 mm?

Herman Eisenbeiss/Photo Researchers

Finding a value given a proportion

Examples 3.5 to 3.8 illustrate the use of software or Table A to find what proportion of the observations satisfies some condition, such as "SAT score above 820." We may instead want to find the observed value with a given proportion of the observations above or below it. Statistical software will do this directly.

EXAMPLE 3.9 **Find the top 10% using software**

Scores on the SAT reading test in recent years follow approximately the $N(504, 111)$ distribution. How high must a student score in order to place in the top 10% of all students taking the SAT?

We want to find the SAT score x with area 0.1 to its *right* under the Normal curve with mean $\mu = 504$ and standard deviation $\sigma = 111$. That's the same as finding the SAT score x with area 0.9 to its *left*. Figure 3.12 poses the question in graphical form. Most software will tell you x when you plug in mean 504, standard deviation 111, and cumulative proportion 0.9. Here is Minitab's output:

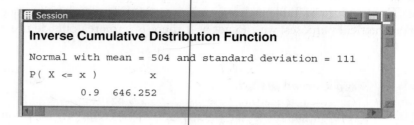

> **Inverse Cumulative Distribution Function**
>
> Normal with mean = 504 and standard deviation = 111
>
> ```
> P( X <= x )        x
>       0.9    646.252
> ```

Minitab gives $x = 646.252$. So scores above 647 are in the top 10%. (Round up because SAT scores can only be whole numbers.) ∎

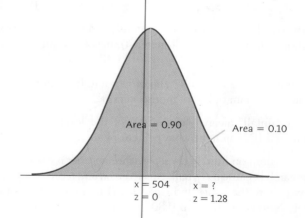

FIGURE 3.12

Locating the point on a Normal curve with area 0.10 to its right, for Examples 3.9 and 3.10.

Area = 0.90   Area = 0.10

$x = 504$ $x = ?$
$z = 0$ $z = 1.28$

Without software, use Table A backward. Find the given proportion in the body of the table and then read the corresponding z from the left column and top row. There are again three steps.

EXAMPLE 3.10 **Find the top 10% using Table A**

Scores on the SAT Verbal test in recent years follow approximately the $N(504, 111)$ distribution. How high must a student score in order to place in the top 10% of all students taking the SAT?

Step 1. State the problem and draw a picture. This step is exactly as in Example 3.9. The picture is Figure 3.12.

Step 2. Use the table. Look in the body of Table A for the entry closest to 0.9. It is 0.8997. This is the entry corresponding to $z = 1.28$. So $z = 1.28$ is the standardized value with area 0.9 to its left.

Step 3. Unstandardize to transform z back to the original x scale. We know that the standardized value of the unknown x is $z = 1.28$. This means that x itself lies 1.28 standard deviations above the mean on this particular Normal curve. That is,

$$x = \text{mean} + (1.28)(\text{standard deviation})$$
$$= 504 + (1.28)(111) = 646.1$$

A student must score at least 647 to place in the highest 10%. ■

EXAMPLE 3.11 Find the first quartile

High levels of cholesterol in the blood increase the risk of heart disease. For 14-year-old boys, the distribution of blood cholesterol is approximately Normal with mean $\mu = 170$ milligrams of cholesterol per deciliter of blood (mg/dl) and standard deviation $\sigma = 30$ mg/dl.[5] What is the first quartile of the distribution of blood cholesterol?

Step 1. State the problem and draw a picture. Call the cholesterol level x. The variable x has the $N(170, 30)$ distribution. The first quartile is the value with 25% of the distribution to its left. Figure 3.13 is the picture.

Step 2. Use the table. Look in the body of Table A for the entry closest to 0.25. It is 0.2514. This is the entry corresponding to $z = -0.67$. So $z = -0.67$ is the standardized value with area 0.25 to its left.

Step 3. Unstandardize. The cholesterol level corresponding to $z = -0.67$ lies 0.67 standard deviations below the mean, so

$$x = \text{mean} - (0.67)(\text{standard deviation})$$
$$= 170 - (0.67)(30) = 149.9$$

The first quartile of blood cholesterol levels in 14-year-old boys is about 150 mg/dl. ■

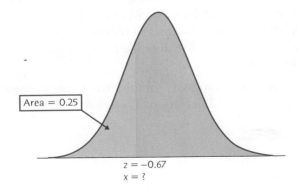

Area = 0.25

z = −0.67
x = ?

FIGURE 3.13

Locating the first quartile of a Normal curve, for Example 3.11.

APPLY YOUR KNOWLEDGE

3.13 Table A. Use Table A to find the value z of a standard Normal variable that satisfies each of the following conditions. (Use the value of z from Table A that comes closest to satisfying the condition.) In each case, sketch a standard Normal curve with your value of z marked on the axis.

(a)   The point z with 20% of the observations falling below it

(b)   The point z with 40% of the observations falling above it

3.14 Fruit flies. The thorax lengths in a population of male fruit flies follow a Normal distribution with mean 0.800 millimeters (mm) and standard deviation 0.078 mm. What are the median and the first and third quartiles of thorax length?

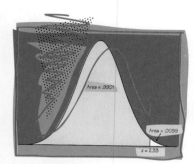

The bell curve?

Does the distribution of human intelligence follow the "bell curve" of a Normal distribution? Scores on IQ tests do roughly follow a Normal distribution. That is because a test score is calculated from a person's answers in a way that is designed to produce a Normal distribution. To conclude that intelligence follows a bell curve, we must agree that the test scores directly measure intelligence. Many psychologists don't think there is one human characteristic that we can call "intelligence" and can measure by a single test score.

C H A P T E R 3   S U M M A R Y

■ We can sometimes describe the overall pattern of a distribution by a **density curve.** A density curve has total area 1 underneath it. An area under a density curve gives the proportion of observations that fall in a range of values.

■ A density curve is an idealized description of the overall pattern of a distribution that smooths out the irregularities in the actual data. We write the **mean of a density curve** as μ and the **standard deviation of a density curve** as σ to distinguish them from the mean $\bar{x}$ and standard deviation s of the actual data.

■ The mean, the median, and the quartiles of a density curve can be located by eye. The **mean** μ is the balance point of the curve. The **median** divides the area under the curve in half. The **quartiles** and the median divide the area under the curve into quarters. The **standard deviation** σ cannot be located by eye on most density curves.

■ The mean and median are equal for symmetric density curves. The mean of a skewed curve is located farther toward the long tail than is the median.

■ The **Normal distributions** are described by a special family of bell-shaped, symmetric density curves, called **Normal curves.** The mean μ and standard deviation σ completely specify a Normal distribution $N(\mu, \sigma)$. The mean is the center of the curve, and σ is the distance from μ to the change-of-curvature points on either side.

■ To **standardize** any observation x, subtract the mean of the distribution and then divide by the standard deviation. The resulting **z-score**

$$z = \frac{x - \mu}{\sigma}$$

says how many standard deviations x lies from the distribution mean.

■ All Normal distributions are the same when measurements are transformed to the standardized scale. In particular, all Normal distributions satisfy the **68−95−99.7 rule,** which describes what percent of observations lie within one, two, and three standard deviations of the mean.

■ If x has the $N(\mu, \sigma)$ distribution, then the **standardized variable** $z = (x - \mu)/\sigma$ has the **standard Normal distribution** $N(0, 1)$ with mean 0 and standard deviation 1. Table A gives the **cumulative proportions** of standard Normal observations that are less than z for many values of z. By standardizing, we can use Table A for any Normal distribution.

CHECK YOUR SKILLS

3.15 Which of these variables is least likely to have a Normal distribution?

    (a) Income per person for 150 different countries

    (b) Lengths of 50 newly hatched pythons

    (c) Heights of 100 white pine trees in a forest

3.16 To completely specify the shape of a Normal distribution, you must give

    (a) the mean and the standard deviation.

    (b) the five-number summary.

    (c) the mean and the median.

3.17 Figure 3.14 shows a Normal curve. The mean of this distribution is

    (a) 0.   (b) 2.   (c) 3.

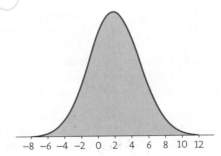

-8 -6 -4 -2 0 2 4 6 8 10 12

FIGURE 3.14
A Normal curve, for Exercises 3.17 and 3.18.

3.18 The standard deviation of the Normal distribution in Figure 3.14 is

    (a) 2.   (b) 3.   (c) 5.

3.19 The length of human pregnancies from conception to birth varies according to a distribution that is approximately Normal with mean 266 days and standard deviation 16 days. 95% of all pregnancies last between

    (a) 250 and 282 days.   (b) 234 and 298 days.   (c) 218 and 314 days.

3.20 The scores of adults on an IQ test are approximately Normal with mean 100 and standard deviation 15. The organization MENSA, which calls itself "the high IQ society," requires an IQ score of 130 or higher for membership. What percent of adults would qualify for membership?

    (a) 95%   (b) 5%   (c) 2.5%

3.21 The scores of adults on an IQ test are approximately Normal with mean 100 and standard deviation 15. Corinne scores 118 on such a test. Her z-score is about

(a) 1.2.   (b) 7.87.   (c) 18.

3.22 The proportion of observations from a standard Normal distribution that take values less than 1.15 is about

(a) 0.1251.   (b) 0.8531.   (c) 0.8749.

3.23 The proportion of observations from a standard Normal distribution that take values larger than -0.75 is about

(a) 0.2266.   (b) 0.7734.   (c) 0.8023.

3.24 The scores of adults on an IQ test are approximately Normal with mean 100 and standard deviation 15. Corinne scores 118 on such a test. She scores higher than what percent of all adults?

(a) About 12%   (b) About 88%   (c) About 98%

C H A P T E R 3   E X E R C I S E S

3.25 Understanding density curves. Remember that it is areas under a density curve, not the height of the curve, that give proportions in a distribution. To illustrate this, sketch a density curve that has a tall, thin peak at 0 on the horizontal axis but has most of its area close to 1 on the horizontal axis without a high peak at 1.

3.26 Daily activity. It appears that people who are mildly obese are less active than leaner people. One study looked at the average number of minutes per day that people spend standing or walking.[6] Among mildly obese people, minutes of activity varied according to the $N(373, 67)$ distribution. Minutes of activity for lean people had the $N(526, 107)$ distribution. Within what limits do the active minutes for 95% of the people in each group fall? Use the 68–95–99.7 rule.

3.27 Low IQ test scores. Scores on the Wechsler Adult Intelligence Scale (WAIS) are approximately Normal with mean 100 and standard deviation 15. People with WAIS scores below 70 are considered mentally retarded when, for example, applying for Social Security disability benefits. According to the 68–95–99.7 rule, about what percent of adults are retarded by this criterion?

3.28 Standard Normal drill. Use Table A to find the proportion of observations from a standard Normal distribution that falls in each of the following regions. In each case, sketch a standard Normal curve and shade the area representing the region.
(a) $z \leq -2.25$ (b) $z \geq -2.25$ (c) $z > 1.77$ (d) $-2.25 < z < 1.77$

3.29 Standard Normal drill.

(a) Find the number z such that the proportion of observations that are less than z in a standard Normal distribution is 0.8.

(b) Find the number z such that 35% of all observations from a standard Normal distribution are greater than z.

3.30 Running a mile. After the physical training required during World War II, the distribution of mile run times for male students at the University of Illinois was approximately Normal with mean 7.11 minutes and standard deviation 0.74 minutes. What proportion of these students could run a mile in 5 minutes or less?

3.31 Acid rain? Emissions of sulfur dioxide by industry set off chemical changes in the atmosphere that result in "acid rain." The acidity of liquids is measured by pH on a scale of 0 to 14. Distilled water has pH 7.0, and lower pH values indicate acidity. Normal rain is somewhat acidic, so acid rain is sometimes defined as rainfall with a pH below 5.0. The pH of rain at one location varies among rainy days according to a Normal distribution with mean 5.4 and standard deviation 0.54. What proportion of rainy days have rainfall with pH below 5.0?

3.32 Runners. In a study of exercise, a large group of male runners walk on a treadmill for 6 minutes. Their heart rates in beats per minute at the end vary from runner to runner according to the N(104, 12.5) distribution. The heart rates for male nonrunners after the same exercise have the N(130, 17) distribution.

(a)   What percent of the runners have heart rates above 130?

(b)   What percent of the nonrunners have heart rates above 130?

3.33 A milling machine. Automated manufacturing operations are quite precise but still vary, often with distributions that are close to Normal. The width in inches of slots cut by a milling machine follows approximately the N(0.8750, 0.0012) distribution. The specifications allow slot widths between 0.8720 and 0.8780 inch. What proportion of slots meet these specifications?

3.34 Making tablets. A pharmaceutical manufacturer forms tablets by compressing a granular material that contains the active ingredient and various fillers. The force in kilograms (kg) applied to the tablets varies a bit, with the N(11.5, 0.2) distribution. The process specifications call for applying a force between 11.2 and 12.2 kg.

(a)   What percent of tablets are subject to a force that meets the specifications?

(b)   The manufacturer adjusts the process so that the mean force is at the center of the specifications, $\mu = 11.7$ kg. The standard deviation remains 0.2 kg. What percent now meet the specifications?

Metalpix/Alamy

Miles per gallon. In its Fuel Economy Guide for 2008 model vehicles, the Environmental Protection Agency gives data on 1152 vehicles. There are a number of outliers, mainly vehicles with very poor gas mileage. If we ignore the outliers, however, the combined city and highway gas mileage of the other 1120 or so vehicles is approximately Normal with mean 18.7 miles per gallon (mpg) and standard deviation 4.3 mpg. Exercises 3.35 to 3.38 concern this distribution.

3.35 In my Chevrolet. The 2008 Chevrolet Malibu with a four-cylinder engine has combined gas mileage 25 mpg. What percent of all vehicles have worse gas mileage than the Malibu?

3.36 The top 10%. How high must a 2008 vehicle's gas mileage be in order to fall in the top 10% of all vehicles? (The distribution omits a few high outliers, mainly hybrid gas-electric vehicles.)

3.37 The middle half. The quartiles of any distribution are the values with cumulative proportions 0.25 and 0.75. They span the middle half of the distribution. What are the quartiles of the distribution of gas mileage?

3.38 Quintiles. The quintiles of any distribution are the values with cumulative proportions 0.20, 0.40, 0.60, and 0.80. What are the quintiles of the distribution of gas mileage?

3.39 What's your percentile? Reports on a student's ACT or SAT usually give the percentile as well as the actual score. The percentile is just the cumulative proportion stated as a percent: the percent of all scores that were lower than this one. In 2007, composite ACT scores were close to Normal with mean 21.2 and standard deviation 5.0.[7] Jacob scored 16. What was his percentile?

3.40 Perfect SAT scores. It is possible to score higher than 1600 on the SAT, but scores 1600 and above are reported as 1600. In 2007 the distribution of SAT scores (combining mathematics and reading) was close to Normal with mean 1021 and standard deviation 211.[8] What proportion of 2007 SAT scores were reported as 1600?

3.41 Heights of men and women. The heights of women aged 20 to 29 follow approximately the $N(64, 2.7)$ distribution. Men the same age have heights distributed as $N(69.3, 2.8)$. What percent of young women are taller than the mean height of young men?

3.42 Heights of men and women. The heights of women aged 20 to 29 follow approximately the $N(64, 2.7)$ distribution. Men the same age have heights distributed as $N(69.3, 2.8)$. What percent of young men are shorter than the mean height of young women?

3.43 A surprising calculation. Changing the mean and standard deviation of a Normal distribution by a moderate amount can greatly change the percent of observations in the tails. Suppose that a college is looking for applicants with SAT math scores 750 and above.

(a) In 2007, the scores of men on the math SAT followed the $N(533, 116)$ distribution. What percent of men scored 750 or better?

(b) Women's SAT math scores that year had the $N(499, 110)$ distribution. What percent of women scored 750 or better? You see that the percent of men above 750 is almost three times the percent of women with such high scores. Why this is true is controversial. (On the other hand, women score higher than men on the new SAT writing test, though by a smaller amount.)

3.44 Grading managers. Some companies "grade on a bell curve" to compare the performance of their managers and professional workers. This forces the use of some low performance ratings so that not all workers are listed as "above average." Ford Motor Company's "performance management process" for a time assigned 10% A grades, 80% B grades, and 10% C grades to the company's managers. Suppose that Ford's performance scores really are Normally distributed. This year, managers with scores less than 25 received C's and those with scores above 475 received A's. What are the mean and standard deviation of the scores?

3.45 Osteoporosis. Osteoporosis is a condition in which the bones become brittle due to loss of minerals. To diagnose osteoporosis, an elaborate apparatus measures bone mineral density (BMD). BMD is usually reported in standardized form. The standardization is based on a population of healthy young adults. The World Health Organization (WHO) criterion for osteoporosis is a BMD 2.5 standard deviations

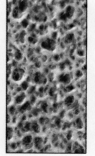
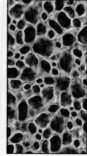

Solid bone matrix   **Weakened bone matrix**

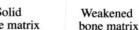

below the mean for young adults. BMD measurements in a population of people similar in age and sex roughly follow a Normal distribution.

(a) What percent of healthy young adults have osteoporosis by the WHO criterion?

(b) Women aged 70 to 79 are of course not young adults. The mean BMD in this age is about -2 on the standard scale for young adults. Suppose that the standard deviation is the same as for young adults. What percent of this older population have osteoporosis?

In later chapters we will meet many statistical procedures that work well when the data are "close enough to Normal." Exercises 3.46 to 3.51 concern data that are mostly close enough to Normal for statistical work. These exercises ask you to do data analysis and Normal calculations to investigate how close to Normal real data are.

3.46 Normal is only approximate: IQ test scores. Here are the IQ test scores of 31 seventh-grade girls in a Midwest school district:[9]

114	100	104	89	102	91	114	114	103	105
108	130	120	132	111	128	118	119	86	72
111	103	74	112	107	103	98	96	112	112	93

(a) We expect IQ scores to be approximately Normal. Make a stemplot to check that there are no major departures from Normality.

(b) Nonetheless, proportions calculated from a Normal distribution are not always very accurate for small numbers of observations. Find the mean $\bar{x}$ and standard deviation s for these IQ scores. What proportions of the scores are within one standard deviation and within two standard deviations of the mean? What would these proportions be in an exactly Normal distribution?

3.47 Normal is only approximate: ACT scores. Scores on the ACT test for the 2007 high school graduating class had mean 21.2 and standard deviation 5.0. In all, 1,300,599 students in this class took the test. Of these, 149,164 had scores higher than 27 and another 50,310 had scores exactly 27. ACT scores are always whole numbers. The exactly Normal $N(21.2, 5.0)$ distribution can include any value, not just whole numbers. What is more, there is *no* area exactly above 27 under the smooth Normal curve. So ACT scores can be only approximately Normal. To illustrate this fact, find

(a) the percent of 2007 ACT scores greater than 27.

(b) the percent of 2007 ACT scores greater than or equal to 27.

(c) the percent of observations from the $N(21.2, 5.0)$ distribution that are greater than 27. (The percent greater than or equal to 27 is the same, because there is no area exactly over 27.)

3.48 Are the data Normal? Acidity of rainfall. Exercise 3.31 concerns the acidity (measured by pH) of rainfall. A sample of 105 rainwater specimens had mean pH 5.43, standard deviation 0.54, and five-number summary 4.33, 5.05, 5.44, 5.79, 6.81.[10]

(a) Compare the mean and median and also the distances of the two quartiles from the median. Does it appear that the distribution is quite symmetric? Why?

(b) If the distribution is really $N(5.43, 0.54)$, what proportion of observations would be less than 5.05? Less than 5.79? Do these proportions suggest that the distribution is close to Normal? Why?

3.49 Are the data Normal? Fruit fly thorax lengths. Here are the lengths in millimeters of the thorax for 49 male fruit flies:[11]

0.64	0.64	0.64	0.68	0.68	0.68	0.72	0.72	0.72	0.72
0.74	0.76	0.76	0.76	0.76	0.76	0.76	0.76	0.76	0.78
0.80	0.80	0.80	0.80	0.80	0.82	0.82	0.84	0.84	0.84
0.84	0.84	0.84	0.84	0.84	0.84	0.84	0.88	0.88	0.88
0.88	0.88	0.88	0.88	0.88	0.92	0.92	0.92	0.94	

(a) Make a histogram of the distribution. Although the result depends a bit on your choice of classes, the distribution appears roughly symmetric with no outliers.

(b) Find the mean, median, standard deviation, and quartiles for these data. Comparing the mean and the median and comparing the distances of the two quartiles from the median suggest that the distribution is quite symmetric. Why?

(c) If the distribution were exactly Normal with the mean and standard deviation you found in (b), what proportion of observations would lie between the two quartiles you found in (b)? What proportion of the actual observations lie between the quartiles (include observations equal to either quartile value). Despite the discrepancy, this distribution is "close enough to Normal" for statistical work in later chapters.

3.50 Are the data Normal? Monsoon rains. Here are the amounts of summer monsoon rainfall (millimeters) for India in the 100 years 1901 to 2000:[12]

722.4	792.2	861.3	750.6	716.8	885.5	777.9	897.5	889.6	935.4
736.8	806.4	784.8	898.5	781.0	951.1	1004.7	651.2	885.0	719.4
866.2	869.4	823.5	863.0	804.0	903.1	853.5	768.2	821.5	804.9
877.6	803.8	976.2	913.8	843.9	908.7	842.4	908.6	789.9	853.6
728.7	958.1	868.6	920.8	911.3	904.0	945.9	874.3	904.2	877.3
739.2	793.3	923.4	885.8	930.5	983.6	789.0	889.6	944.3	839.9
1020.5	810.0	858.1	922.8	709.6	740.2	860.3	754.8	831.3	940.0
887.0	653.1	913.6	748.3	963.0	857.0	883.4	909.5	708.0	882.9
852.4	735.6	955.9	836.9	760.0	743.2	697.4	961.7	866.9	908.8
784.7	785.0	896.6	938.4	826.4	857.3	870.5	873.8	827.0	770.2

(a) Make a histogram of these rainfall amounts. Find the mean and the median.

(b) Although the distribution is reasonably Normal, your work shows some departure from Normality. In what way are the data not Normal?

3.51 Are the data Normal? Soil penetrability. Table 2.5 (page 65) gives data on the penetrability of soil at each of three levels of compression. We might expect the penetrability of specimens of the same soil at the same level of compression to follow a Normal distribution. Make stemplots of the data for loose and for intermediate compression. Does either sample seem roughly Normal? Does either appear distinctly non-Normal? If so, what kind of departure from Normality does your stemplot show?

The Normal Curve applet allows you to do Normal calculations quickly. It is somewhat limited by the number of pixels available for use, so that it can't hit every value exactly. In the exercises below, use the closest available values. In each case, make a sketch of the curve from the applet marked with the values you used to answer the questions.

3.52 How accurate is 68–95–99.7? The 68–95–99.7 rule for Normal distributions is a useful approximation. To see how accurate the rule is, drag one flag across the other so that the applet shows the area under the curve between the two flags.

(a) Place the flags one standard deviation on either side of the mean. What is the area between these two values? What does the 68–95–99.7 rule say this area is?

(b) Repeat for locations two and three standard deviations on either side of the mean. Again compare the 68–95–99.7 rule with the area given by the applet.

3.53 Where are the quartiles? How many standard deviations above and below the mean do the quartiles of any Normal distribution lie? (Use the standard Normal distribution to answer this question.)

3.54 Grading managers. In Exercise 3.44, we saw that Ford Motor Company once graded its managers in such a way that the top 10% received an A grade, the bottom 10% a C, and the middle 80% a B. Let's suppose that performance scores follow a Normal distribution. How many standard deviations above and below the mean do the A/B and B/C cutoffs lie? (Use the standard Normal distribution to answer this question.)

Georgette Douwma/Getty

Scatterplots and Correlation

A medical study finds that short women are more likely to have heart attacks than women of average height, while tall women have the fewest heart attacks. An insurance group reports that heavier cars have fewer deaths per 10,000 vehicles registered than do lighter cars. These and many other statistical studies look at the *relationship between two variables*. Statistical relationships are overall tendencies, not ironclad rules. They allow individual exceptions. Although smokers on the average die younger than nonsmokers, some people live to 90 while smoking three packs a day.

To understand a statistical relationship between two variables, we measure both variables on the same individuals. Often, we must examine other variables as well. To conclude that shorter women have higher risk from heart attacks, for example, the researchers had to eliminate the effect of other variables such as weight and exercise habits. In this and the following chapter we study relationships between variables. One of our main themes is that the relationship between two variables can be strongly influenced by other variables that are lurking in the background.

Explanatory and response variables

We think that car weight helps explain accident deaths and that smoking influences life expectancy. In each of these relationships, the two variables play different roles: one explains or influences the other.

RESPONSE VARIABLE, EXPLANATORY VARIABLE

A **response variable** measures an outcome of a study. An **explanatory variable** may explain or influence changes in a response variable.

You will often find explanatory variables called *independent variables* and response variables called *dependent variables*. The idea behind this language is that the response variable depends on the explanatory variable. Because "independent" and "dependent" have other meanings in statistics that are unrelated to the explanatory-response distinction, we prefer to avoid those words.

It is easiest to identify explanatory and response variables when we actually set values of one variable in order to see how it affects another variable.

EXAMPLE 4.1 **Beer and blood alcohol**

How does drinking beer affect the level of alcohol in our blood? The legal limit for driving in all states is 0.08%. Student volunteers at the Ohio State University drank different numbers of cans of beer. Thirty minutes later, a police officer measured their blood alcohol content. Number of beers consumed is the explanatory variable, and percent of alcohol in the blood is the response variable. ■

When we don't set the values of either variable but just observe both variables, there may or may not be explanatory and response variables. Whether there are depends on how we plan to use the data.

EXAMPLE 4.2 **College debts**

A college student aid officer looks at the findings of the National Student Loan Survey. She notes data on the amount of debt of recent graduates, their current income, and how stressed they feel about college debt. She isn't interested in predictions but is simply trying to understand the situation of recent college graduates. The distinction between explanatory and response variables does not apply.

A sociologist looks at the same data with an eye to using amount of debt and income, along with other variables, to explain the stress caused by college debt. Now amount of debt and income are explanatory variables and stress level is the response variable. ■

In many studies, the goal is to show that changes in one or more explanatory variables actually *cause* changes in a response variable. Other explanatory-response relationships do not involve direct causation. The SAT scores of high school

After you plot your data, think!

The statistician Abraham Wald (1902–1950) worked on war problems during World War II. Wald invented some statistical methods that were military secrets until the war ended. Here is one of his simpler ideas. Asked where extra armor should be added to airplanes, Wald studied the location of enemy bullet holes in planes returning from combat. He plotted the locations on an outline of the plane. As data accumulated, most of the outline filled up. Put the armor in the few spots with no bullet holes, said Wald. That's where bullets hit the planes that didn't make it back.

students help predict the students' future college grades, but high SAT scores certainly don't *cause* high college grades.

Most statistical studies examine data on more than one variable. Fortunately, statistical analysis of several-variable data builds on the tools we used to examine individual variables. The principles that guide our work also remain the same:

- Plot your data. Look for overall patterns and deviations from those patterns.

- Based on what your plot shows, choose numerical summaries for some aspects of the data.

APPLY YOUR KNOWLEDGE

4.1 Explanatory and response variables? You have data on a large group of college students. Here are four pairs of variables measured on these students. For each pair, is it more reasonable to simply explore the relationship between the two variables or to view one of the variables as an explanatory variable and the other as a response variable? In the latter case, which is the explanatory variable and which is the response variable?

(a) Amount of time spent studying for a statistics exam and grade on the exam.

(b) Weight in kilograms and height in centimeters.

(c) Hours per week of extracurricular activities and grade point average.

(d) Score on the SAT Mathematics exam and score on the SAT Critical Reading exam.

4.2 Coral reefs. How sensitive to changes in water temperature are coral reefs? To find out, measure the growth of corals in aquariums where the water temperature is controlled at different levels. Growth is measured by weighing the coral before and after the experiment. What are the explanatory and response variables? Are they categorical or quantitative?

Georgette Douwma/Getty

4.3 Beer and blood alcohol. Example 4.1 describes a study in which college students drank different amounts of beer. The response variable was their blood alcohol content (BAC). BAC for the same amount of beer might depend on other facts about the students. Name two other variables that could influence BAC.

Displaying relationships: scatterplots

The most useful graph for displaying the relationship between two quantitative variables is a *scatterplot*.

EXAMPLE 4.3 State SAT mathematics scores

Figure 1.8 (page 18) reminded us that in some states most high school graduates take the SAT test of readiness for college, and in other states most take the ACT. Who takes a test may influence the average score. Let's follow our four-step process (page 55) to examine this influence.[1]

FIGURE 4.1

Scatterplot of the mean SAT mathematics score in each state against the percent of that state's high school graduates who take the SAT, for Example 4.3. The dotted lines intersect at the point (24, 565), the data for Colorado.

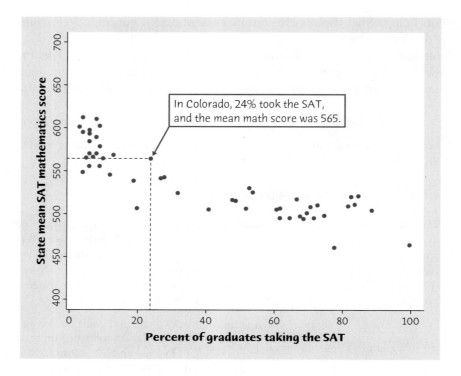

In Colorado, 24% took the SAT, and the mean math score was 565.

STATE: The percent of high school students who take the SAT varies from state to state. Does this fact help explain differences among the states in average SAT mathematics score?

PLAN: Examine the relationship between percent taking the SAT and state mean score on the mathematics part of the SAT. Choose the explanatory and response variables (if any). Make a *scatterplot* to display the relationship between the variables. Interpret the plot to understand the relationship.

SOLVE (MAKE THE PLOT): We suspect that "percent taking" will help explain "mean score." So "percent taking" is the explanatory variable and "mean score" is the response variable. We want to see how mean score changes when percent taking changes, so we put percent taking (the explanatory variable) on the horizontal axis. Figure 4.1 is the scatterplot. Each point represents a single state. In Colorado, for example, 24% took the SAT, and their mean SAT math score was 565. Find 24 on the *x* (horizontal) axis and 565 on the *y* (vertical) axis. Colorado appears as the point (24, 565) above 24 and to the right of 565. ■

SCATTERPLOT

A **scatterplot** shows the relationship between two quantitative variables measured on the same individuals. The values of one variable appear on the horizontal axis, and the values of the other variable appear on the vertical axis. Each individual in the data appears as the point in the plot fixed by the values of both variables for that individual.

Always plot the explanatory variable, if there is one, on the horizontal axis (the
x axis) of a scatterplot. As a reminder, we usually call the explanatory variable x
and the response variable y. If there is no explanatory-response distinction, either
variable can go on the horizontal axis.

APPLY YOUR KNOWLEDGE

4.4 Do heavier people burn more energy? Metabolic rate, the rate at which the body
consumes energy, is important in studies of weight gain, dieting, and exercise. We
have data on the lean body mass and resting metabolic rate for 12 women who are
subjects in a study of dieting. Lean body mass, given in kilograms, is a person's weight
leaving out all fat. Metabolic rate is measured in calories burned per 24 hours, the
same calories used to describe the energy content of foods.

Mass	36.1	54.6	48.5	42.0	50.6	42.0	40.3	33.1	42.4	34.5	51.1	41.2
Rate	995	1425	1396	1418	1502	1256	1189	913	1124	1052	1347	1204

The researchers believe that lean body mass is an important influence on metabolic
rate. Make a scatterplot to examine this belief. (The *Two Variable Statistical Calculator*
Applet provides an easy way to make scatterplots. Click "Data" to enter your data,
then "Scatterplot" to see the plot.)

4.5 Outsourcing by airlines. Airlines have increasingly outsourced the maintenance of
their planes to other companies. Critics say that the maintenance may be less carefully done, so that outsourcing creates a safety hazard. As evidence, they point to
government data on percent of major maintenance outsourced and percent of flight
delays blamed on the airline (often due to maintenance problems):[2]

Airline	Outsource percent	Delay percent	Airline	Outsource percent	Delay percent
AirTran	66	14	Frontier	65	31
Alaska	92	42	Hawaiian	80	70
American	46	26	JetBlue	68	18
America West	76	39	Northwest	76	43
ATA	18	19	Southwest	68	20
Continental	69	20	United	63	27
Delta	48	26	US Airways	77	24

Make a scatterplot that shows how delays depend on outsourcing.

Interpreting scatterplots

To interpret a scatterplot, adapt the strategies of data analysis learned in Chapters
1 and 2 to the new two-variable setting.

EXAMINING A SCATTERPLOT

In any graph of data, look for the **overall pattern** and for striking **deviations** from that pattern.

You can describe the overall pattern of a scatterplot by the **direction, form,** and **strength** of the relationship.

An important kind of deviation is an **outlier,** an individual value that falls outside the overall pattern of the relationship.

EXAMPLE 4.4 **Understanding state SAT scores**

SOLVE (INTERPRET THE PLOT): Figure 4.1 shows a clear *direction:* the overall pattern moves from upper left to lower right. That is, states in which higher percents of high school graduates take the SAT tend to have lower mean SAT mathematics scores. We call this a *negative association* between the two variables.

clusters

The *form* of the relationship is roughly a straight line with a slight curve to the right as it moves down. What is more, most states fall into two distinct **clusters.** As in the histogram in Figure 1.8, the ACT states cluster at the left and the SAT states at the right. In 22 states, fewer than 20% of seniors took the SAT; in another 22 states, more than 50% took the SAT.

The *strength* of a relationship in a scatterplot is determined by how closely the points follow a clear form. The overall relationship in Figure 4.1 is moderately strong: states with similar percents taking the SAT tend to have roughly similar mean SAT math scores.

CONCLUDE: Percent taking explains much of the variation among states in average SAT mathematics score. States in which a higher percent of students take the SAT tend to have lower mean scores because the mean includes a broader group of students. SAT states as a group have lower mean SAT scores than ACT states. So average SAT score says almost nothing about the quality of education in a state. It is foolish to "rank" states by their average SAT scores. ■

POSITIVE ASSOCIATION, NEGATIVE ASSOCIATION

Two variables are **positively associated** when above-average values of one tend to accompany above-average values of the other, and below-average values also tend to occur together.

Two variables are **negatively associated** when above-average values of one tend to accompany below-average values of the other, and vice versa.

Of course, not all relationships have a clear direction that we can describe as positive association or negative association. Exercise 4.8 gives an example that does not have a single direction. Here is an example of a strong positive association with a simple and important form.

TABLE 4.1 Florida boat registrations (thousands) and manatees killed by boats

YEAR	BOATS	MANATEES	YEAR	BOATS	MANATEES	YEAR	BOATS	MANATEES
1977	447	13	1988	675	43	1999	830	82
1978	460	21	1989	711	50	2000	880	78
1979	481	24	1990	719	47	2001	944	81
1980	498	16	1991	681	53	2002	962	95
1981	513	24	1992	679	38	2003	978	73
1982	512	20	1993	678	35	2004	983	69
1983	526	15	1994	696	49	2005	1010	79
1984	559	34	1995	713	42	2006	1024	92
1985	585	33	1996	732	60			
1986	614	33	1997	755	54			
1987	645	39	1998	809	66			

EXAMPLE 4.5   The endangered manatee

STATE: Manatees are large, gentle, slow-moving creatures found along the coast of Florida. Many manatees are injured or killed by boats. Table 4.1 contains data on the number of boats registered in Florida (in thousands) and the number of manatees killed by boats for the years 1977 to 2006.[3] Examine the relationship. Is it plausible that restricting the number of boats would help protect manatees?

linear relationship

PLAN: Make a scatterplot with "boats registered" as the explanatory variable and "manatees killed" as the response variable. Describe the form, direction, and strength of the relationship.

SOLVE: Figure 4.2 is the scatterplot. There is a positive association—more boats goes with more manatees killed. The form of the relationship is **linear.** That is, the overall pattern follows a straight line from lower left to upper right. The relationship is strong because the points don't deviate greatly from a line.

CONCLUDE: As more boats are registered, the number of manatees killed by boats goes up linearly. The Florida Wildlife Commission says that in recent years boats accounted for 24% of manatee deaths and 42% of deaths whose causes could be determined. Although many manatees die from other causes, it appears that fewer boats would mean fewer manatee deaths. ■

Douglas Faulkner/Photo Researchers

As the following chapter will emphasize, *it is wise to always ask what other variables lurking in the background might contribute to the relationship displayed in a scatterplot.* Because both boats registered and manatees killed are recorded year by year, any change in conditions over time might affect the relationship. For example, if boats in Florida have tended to go faster over the years, that might result in more manatees killed by the same number of boats.

FIGURE 4.2

Scatterplot of the number of Florida manatees killed by boats in the years 1977 to 2006 against the number of boats registered in Florida that year, for Example 4.5. There is a strong linear (straight-line) pattern.

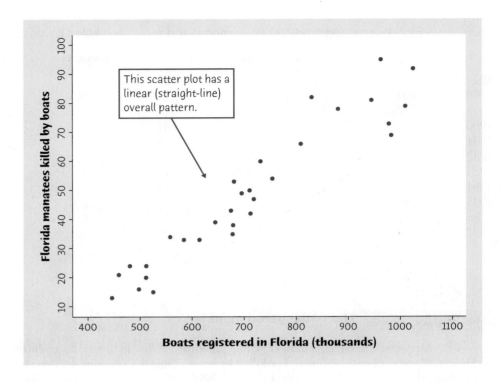

This scatter plot has a linear (straight-line) overall pattern.

APPLY YOUR KNOWLEDGE

4.6 Do heavier people burn more energy? Describe the direction, form, and strength of the relationship between lean body mass and metabolic rate, as displayed in your plot for Exercise 4.4.

4.7 Outsourcing by airlines. Does your plot for Exercise 4.5 show a positive association between maintenance outsourcing and delays caused by the airline? One airline is a high outlier in delay percent. Which airline is this? Aside from the outlier, does the plot show a roughly linear form? Is the relationship very strong?

4.8 Does fast driving waste fuel? How does the fuel consumption of a car change as its speed increases? Here are data for a British Ford Escort. Speed is measured in kilometers per hour, and fuel consumption is measured in liters of gasoline used per 100 kilometers traveled.[4]

Speed	10	20	30	40	50	60	70	80
Fuel	21.00	13.00	10.00	8.00	7.00	5.90	6.30	6.95

Speed	90	100	110	120	130	140	150
Fuel	7.57	8.27	9.03	9.87	10.79	11.77	12.83

(a) Make a scatterplot. (Which is the explanatory variable?)

(b) Describe the form of the relationship. It is not linear. Explain why the form of the relationship makes sense.

(c)   It does not make sense to describe the variables as either positively associated or negatively associated. Why?

(d)   Is the relationship reasonably strong or quite weak? Explain your answer.

Adding categorical variables to scatterplots

The Census Bureau groups the states into four broad regions, named Midwest, Northeast, South, and West. We might ask about regional patterns in SAT exam scores. Figure 4.3 repeats part of Figure 4.1, with an important difference. We have plotted only the Midwest and Northeast groups of states, using the plot symbol "•" for the Midwest states and the symbol "+" for the Northeast states.

The regional comparison is striking. The 9 Northeast states are all SAT states—at least 67% of high school graduates in each of these states take the SAT. The 12 Midwest states are mostly ACT states. In 10 of these states, fewer than 10% of high school graduates take the SAT. One Midwest state is clearly an outlier within the region. Indiana is an SAT state (62% take the SAT) that falls close to the Northeast cluster. Ohio, where 27% take the SAT, also lies outside the Midwest cluster.

Dividing the states into regions introduces a third variable into the scatterplot. "Region" is a categorical variable that has four values, although we plotted data

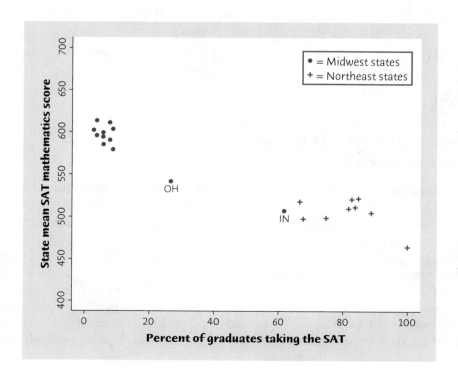

FIGURE 4.3

Mean SAT mathematics score and percent of high school graduates who take the test for only the Midwest (•) and Northeast (+) states.

from only two of the four regions. The two regions are identified by the two different plotting symbols.

CATEGORICAL VARIABLES IN SCATTERPLOTS

To add a categorical variable to a scatterplot, use a different plot color or symbol for each category.

APPLY YOUR KNOWLEDGE

4.9 Do heavier people burn more energy? The study of dieting described in Exercise 4.4 collected data on the lean body mass (in kilograms) and metabolic rate (in calories) for both female and male subjects:

Sex	F	F	F	F	F	F	F	F	F	F
Mass	36.1	54.6	48.5	42.0	50.6	42.0	40.3	33.1	42.4	34.5
Rate	995	1425	1396	1418	1502	1256	1189	913	1124	1052

Sex	F	F	M	M	M	M	M	M	M
Mass	51.1	41.2	51.9	46.9	62.0	62.9	47.4	48.7	51.9
Rate	1347	1204	1867	1439	1792	1666	1362	1614	1460

(a) Make a scatterplot of metabolic rate versus lean body mass for all 19 subjects. Use separate symbols to distinguish women and men.

(b) Does the same overall pattern hold for both women and men? What is the most important difference between women and men?

Measuring linear association: correlation

A scatterplot displays the direction, form, and strength of the relationship between two quantitative variables. Linear (straight-line) relations are particularly important because a straight line is a simple pattern that is quite common. A linear relation is strong if the points lie close to a straight line, and weak if they are widely scattered about a line. Our eyes are not good judges of how strong a linear relationship is. The two scatterplots in Figure 4.4 depict exactly the same data, but the lower plot is drawn smaller in a large field. The lower plot seems to show a stronger linear relationship. Our eyes can be fooled by changing the plotting scales or the amount of space around the cloud of points in a scatterplot.[5] We need to follow our strategy for data analysis by using a numerical measure to supplement the graph. *Correlation* is the measure we use.

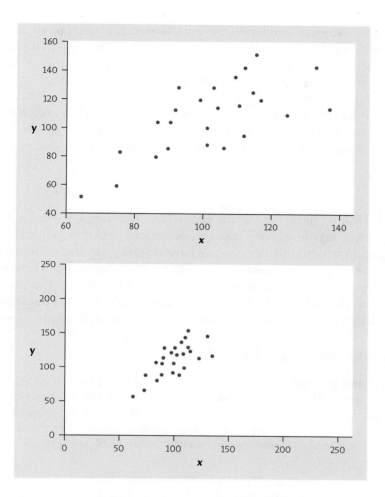

FIGURE 4.4

Two scatterplots of the same data. The straight-line pattern in the lower plot appears stronger because of the surrounding space.

CORRELATION

The **correlation** measures the direction and strength of the linear relationship between two quantitative variables. Correlation is usually written as r.

Suppose that we have data on variables x and y for n individuals. The values for the first individual are x_1 and y_1, the values for the second individual are x_2 and y_2, and so on. The means and standard deviations of the two variables are $\overline{x}$ and s_x for the x-values, and $\overline{y}$ and s_y for the y-values. The correlation r between x and y is

$$r = \frac{1}{n-1}\left[\left(\frac{x_1 - \overline{x}}{s_x}\right)\left(\frac{y_1 - \overline{y}}{s_y}\right) + \left(\frac{x_2 - \overline{x}}{s_x}\right)\left(\frac{y_2 - \overline{y}}{s_y}\right)\right.$$

$$\left. + \cdots + \left(\frac{x_n - \overline{x}}{s_x}\right)\left(\frac{y_n - \overline{y}}{s_y}\right)\right]$$

or, more compactly,

$$r = \frac{1}{n-1}\sum\left(\frac{x_i - \overline{x}}{s_x}\right)\left(\frac{y_i - \overline{y}}{s_y}\right)$$

The formula for the correlation r is a bit complex. It helps us see what correlation is, but in practice you should use software or a calculator that finds r from keyed-in values of two variables x and y. Exercise 4.10 asks you to calculate a correlation step-by-step from the definition to solidify its meaning.

The formula for r begins by standardizing the observations. Suppose, for example, that x is height in centimeters and y is weight in kilograms and that we have height and weight measurements for n people. Then $\overline{x}$ and s_x are the mean and standard deviation of the n heights, both in centimeters. The value

$$\frac{x_i - \overline{x}}{s_x}$$

is the standardized height of the ith person, familiar from Chapter 3. The standardized height says how many standard deviations above or below the mean a person's height lies. Standardized values have no units—in this example, they are no longer measured in centimeters. Standardize the weights also. The correlation r is an average of the products of the standardized height and the standardized weight for all the individuals. Just as in the case of the standard deviation s, the "average" here divides by one fewer than the number of individuals.

APPLY YOUR KNOWLEDGE

Millard H. Sharp/Photo Researchers

4.10 Ebola and gorillas. The deadly Ebola virus is a threat to both people and gorillas in Central Africa. An outbreak in 2002 and 2003 killed 91 of the 95 gorillas in 7 home ranges in the Congo. To study the spread of the virus, measure "distance" by the number of home ranges separating a group of gorillas from the first group infected. Here are data on distance and number of days until deaths began in each later group:[6]

Distance	1	3	4	4	4	5
Days	4	21	33	41	43	46

(a) Make a scatterplot. Which is the explanatory variable? The plot shows a positive linear pattern.

(b) Find the correlation r step-by-step. First find the mean and standard deviation of each variable. Then find the six standardized values for each variable. Finally, use the formula for r. Explain how your value for r matches your graph in (a).

(c) Enter these data into your calculator or software and use the correlation function to find r. Check that you get the same result as in (b), up to roundoff error.

Facts about correlation

The formula for correlation helps us see that r is positive when there is a positive association between the variables. Height and weight, for example, have a positive association. People who are above average in height tend to also be above average in weight. Both the standardized height and the standardized weight are positive. People who are below average in height tend to also have below-average weight. Then both standardized height and standardized weight are negative. In both cases, the products in the formula for r are mostly positive and so r is positive. In the same way, we can see that r is negative when the association between x and y is negative. More detailed study of the formula gives more detailed properties of r. Here is what you need to know in order to interpret correlation.

1. *Correlation makes no distinction between explanatory and response variables.* It makes no difference which variable you call x and which you call y in calculating the correlation.

2. Because r uses the standardized values of the observations, r *does not change when we change the units of measurement of x, y, or both.* Measuring height in inches rather than centimeters and weight in pounds rather than kilograms does not change the correlation between height and weight. The correlation r itself has no unit of measurement; it is just a number.

3. *Positive r indicates positive association between the variables, and negative r indicates negative association.*

4. *The correlation r is always a number between -1 and 1.* Values of r near 0 indicate a very weak linear relationship. The strength of the linear relationship increases as r moves away from 0 toward either -1 or 1. Values of r close to -1 or 1 indicate that the points in a scatterplot lie close to a straight line. The extreme values $r = -1$ and $r = 1$ occur only in the case of a perfect linear relationship, when the points lie exactly along a straight line.

Death from superstition?

Is there a relationship between superstitious beliefs and bad things happening? Apparently there is. Chinese and Japanese people think that the number 4 is unlucky because when pronounced it sounds like the word for "death." Sociologists looked at 15 years' worth of death certificates for Chinese and Japanese Americans and for white Americans. Deaths from heart disease were notably higher on the fourth day of the month among Chinese and Japanese but not among whites. The sociologists think the explanation is increased stress on "unlucky days."

EXAMPLE 4.6 From scatterplot to correlation

The scatterplots in Figure 4.5 illustrate how values of r closer to 1 or -1 correspond to stronger linear relationships. To make the meaning of r clearer, the standard deviations of both variables in these plots are equal, and the horizontal and vertical scales are the same. In general, it is not so easy to guess the value of r from the appearance of a scatterplot. Remember that changing the plotting scales in a scatterplot may mislead our eyes, but it does not change the correlation.

The scatterplots in Figure 4.6 (page 109) show four sets of real data. The patterns are less regular than those in Figure 4.5, but they also illustrate how correlation measures the strength of linear relationships.[7]

(a) This repeats the manatee plot in Figure 4.2. There is a strong positive linear relationship, $r = 0.953$.

(b) Here are the number of named tropical storms each year between 1984 and 2007 plotted against the number predicted before the start of hurricane season by William Gray of Colorado State University. There is a moderate linear relationship, $r = 0.529$.

FIGURE 4.5

How correlation measures the strength of a linear relationship, for Example 4.6. Patterns closer to a straight line have correlations closer to 1 or −1.

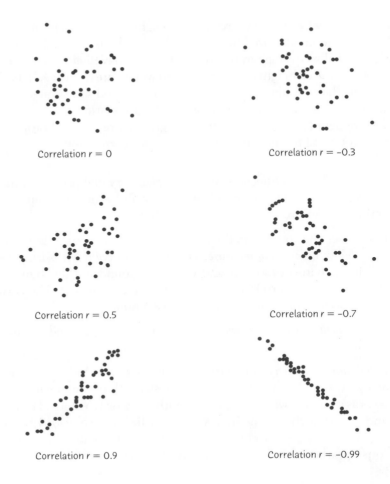

Correlation $r = 0$

Correlation $r = -0.3$

Correlation $r = 0.5$

Correlation $r = -0.7$

Correlation $r = 0.9$

Correlation $r = -0.99$

(c) These data come from an experiment that studied how quickly cuts in the limbs of newts heal. Each point represents the healing rate in micrometers (millionths of a meter) per hour for the two front limbs of the same newt. This relationship is weaker than those in (a) and (b), with $r = 0.358$.

(d) Does last year's stock market performance help predict how stocks will do this year? No. The correlation between last year's percent return and this year's percent return over 56 years is only $r = -0.081$. The scatterplot shows a cloud of points with no visible linear pattern. ■

Describing the relationship between two variables is a more complex task than describing the distribution of one variable. Here are some more facts about correlation, cautions to keep in mind when you use r.

1. *Correlation requires that both variables be quantitative, so that it makes sense to do the arithmetic indicated by the formula for r. We cannot calculate a correlation between the incomes of a group of people and what city they live in, because city is a categorical variable.*

[*Text continues on page 110.*]

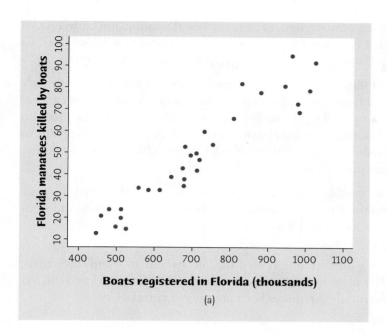

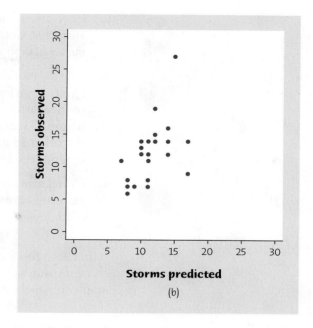

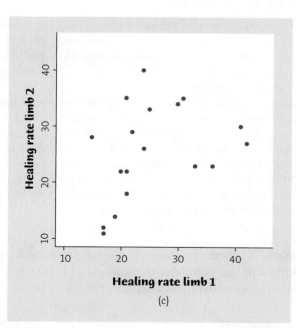

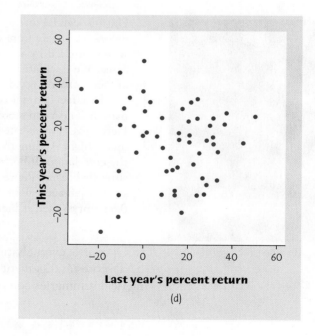

FIGURE 4.6

How correlation measures the strength of a linear relationship, for Example 4.6. Four sets of real data
with (a) $r = 0.953$, (b) $r = 0.529$, (c) $r = 0.358$, and (d) $r = -0.081$.

2. Correlation measures the strength of only the linear relationship between two variables. *Correlation does not describe curved relationships between variables, no matter how strong they are.* Exercise 4.13 illustrates this important fact.

3. *Like the mean and standard deviation, the correlation is not resistant: r is strongly affected by a few outlying observations.* Use r with caution when outliers appear in the scatterplot. Figure 4.6(b) contains an outlier, the disastrous 2005 season, whose 27 named storms included Hurricane Katrina. Adding this one point to the other 23 increases the correlation from 0.475 to 0.529. Because the outlier extends the linear pattern, it increases the correlation.

4. *Correlation is not a complete summary of two-variable data,* even when the relationship between the variables is linear. You should give the means and standard deviations of both x and y along with the correlation.

Because the formula for correlation uses the means and standard deviations, these measures are the proper choice to accompany a correlation. Here is an example in which understanding requires both means and correlation.

Neal Preston/CORBIS

EXAMPLE 4.7 **Scoring figure skaters**

Until a scandal at the 2002 Olympics brought change, figure skating was scored by judges on a scale from 0.0 to 6.0. The scores were often controversial. We have the scores awarded by two judges, Pierre and Elena, to many skaters. How well do they agree? We calculate that the correlation between their scores is r = 0.9. But the mean of Pierre's scores is 0.8 point lower than Elena's mean.

These facts do not contradict each other. They are simply different kinds of information. The mean scores show that Pierre awards lower scores than Elena. But because Pierre gives *every* skater a score about 0.8 point lower than Elena, the correlation remains high. Adding the same number to all values of either x or y does not change the correlation. If both judges score the same skaters, the competition is scored consistently because Pierre and Elena agree on which performances are better. The high r shows their agreement. But if Pierre scores some skaters and Elena others, we must add 0.8 point to each of Pierre's scores to arrive at a fair comparison. ■

Of course, even giving means, standard deviations, and the correlation for state SAT scores and percent taking will not point out the clusters in Figure 4.1. Numerical summaries complement plots of data, but they don't replace them.

APPLY YOUR KNOWLEDGE

4.11 Changing the units. The healing rates plotted in Figure 4.6(c) are measured in micrometers (millionths of a meter) per hour. The correlation between healing rates for the two front limbs of newts is r = 0.358. If the measurements were made in inches per day, would the correlation change? Explain your answer.

4.12 Changing the correlation. Use your calculator, software, or the *Two Variable Statistical Calculator* applet to demonstrate how outliers can affect correlation.

(a) What is the correlation between lean body mass and metabolic rate for the 12 women in Exercise 4.4?

(b) Make a scatterplot of the data with two new points added. Point A: mass 65 kilograms, metabolic rate 1761 calories. Point B: mass 35 kilograms, metabolic rate 1400 calories. Find two new correlations: one for the original data plus Point A, and another for the original data plus Point B.

(c) By looking at your plot, explain why adding Point A makes the correlation stronger (closer to 1) and adding Point B makes the correlation weaker (closer to 0).

4.13 Strong association but no correlation. The gas mileage of an automobile first increases and then decreases as the speed increases. Suppose that this relationship is very regular, as shown by the following data on speed (miles per hour) and mileage (miles per gallon):

Speed	20	30	40	50	60
Mileage	24	28	30	28	24

Make a scatterplot of mileage versus speed. Show that the correlation between speed and mileage is $r = 0$. Explain why the correlation is 0 even though there is a strong relationship between speed and mileage.

CHAPTER 4   SUMMARY

■ To study relationships between variables, we must measure the variables on the same group of individuals.

■ If we think that a variable x may explain or even cause changes in another variable y, we call x an **explanatory variable** and y a **response variable.**

■ A **scatterplot** displays the relationship between two quantitative variables measured on the same individuals. Mark values of one variable on the horizontal axis (x axis) and values of the other variable on the vertical axis (y axis). Plot each individual's data as a point on the graph. Always plot the explanatory variable, if there is one, on the x axis of a scatterplot.

■ Plot points with different colors or symbols to see the effect of a categorical variable in a scatterplot.

■ In examining a scatterplot, look for an overall pattern showing the **direction, form,** and **strength** of the relationship, and then for **outliers** or other deviations from this pattern.

■ **Direction:** If the relationship has a clear direction, we speak of either **positive association** (high values of the two variables tend to occur together) or **negative association** (high values of one variable tend to occur with low values of the other variable).

■ **Form: Linear relationships,** where the points show a straight-line pattern, are an important form of relationship between two variables. Curved relationships and **clusters** are other forms to watch for.

■ **Strength:** The **strength** of a relationship is determined by how close the points in the scatterplot lie to a simple form such as a line.

■ The **correlation** r measures the direction and strength of the linear association between two quantitative variables x and y. Although you can calculate a correlation for any scatterplot, r measures only straight-line relationships.

■ Correlation indicates the direction of a linear relationship by its sign: $r > 0$ for a positive association and $r < 0$ for a negative association. Correlation always satisfies $-1 \leq r \leq 1$ and indicates the strength of a relationship by how close it is to -1 or 1. Perfect correlation, $r = \pm 1$, occurs only when the points on a scatterplot lie exactly on a straight line.

■ Correlation ignores the distinction between explanatory and response variables. The value of r is not affected by changes in the unit of measurement of either variable. Correlation is not resistant, so outliers can greatly change the value of r.

C H E C K   Y O U R   S K I L L S

4.14 You have data for many years on the average price of a barrel of oil and the average retail price of a gallon of unleaded regular gasoline. When you make a scatterplot, the explanatory variable on the x axis

(a) is the price of oil.

(b) is the price of gasoline.

(c) can be either oil price or gasoline price.

4.15 In a scatterplot of the average price of a barrel of oil and the average retail price of a gallon of gasoline, you expect to see

(a) a positive association.

(b) very little association.

(c) a negative association.

4.16 Figure 4.7 is a scatterplot of reading test scores against IQ test scores for 14 fifth-grade children. There is one low outlier in the plot. The IQ and reading scores for this child are

(a) IQ $= 10$, reading $= 124$.

(b) IQ $= 124$, reading $= 72$.

(c) IQ $= 124$, reading $= 10$.

4.17 If we leave out the low outlier, the correlation for the remaining 13 points in Figure 4.7 is closest to

(a) 0.5.   (b) -0.5.   (c) 0.95.

4.18 What are all the values that a correlation r can possibly take?

(a) $r \geq 0$ (b) $0 \leq r \leq 1$ (c) $-1 \leq r \leq 1$

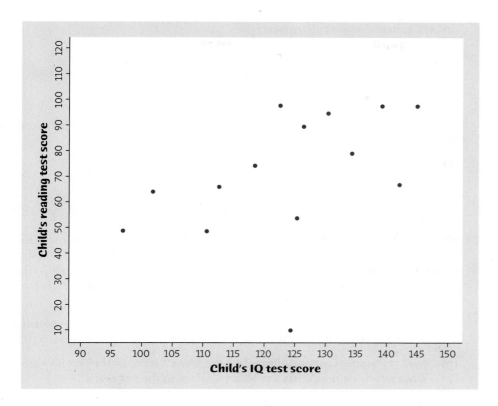

FIGURE 4.7
Scatterplot of reading test score against IQ test score for fifth-grade children, for Exercises 4.16 and 4.17.

4.19 If the correlation between two variables is close to 0, you can conclude that a scatterplot would show

(a)   a strong straight-line pattern.

(b)   a cloud of points with no visible pattern.

(c)   no straight-line pattern, but there might be a strong pattern of another form.

4.20 The points on a scatterplot lie very close to the line whose equation is $y = 4 - 3x$. The correlation between x and y is close to

(a)   −3.   (b)   −1.   (c)   1.

4.21 If women always married men who were 2 years older than themselves, the correlation between the ages of husband and wife would be

(a)   1.   (b)   0.5.   (c)   Can't tell without seeing the data.

4.22 For a biology project, you measure the weight in grams and the tail length in millimeters of a group of mice. The correlation is $r = 0.7$. If you had measured tail length in centimeters instead of millimeters, what would be the correlation? (There are 10 millimeters in a centimeter.)

(a)   0.7/10 = 0.07   (b)   0.7   (c)   (0.7)(10) = 7

4.23 Because elderly people may have difficulty standing to have their heights measured, a study looked at predicting overall height from height to the knee. Here are data (in centimeters) for five elderly men:

Knee height x	57.7	47.4	43.5	44.8	55.2
Height y	192.1	153.3	146.4	162.7	169.1

Use your calculator or software: the correlation between knee height and overall height is about

(a) $r = 0.88$.   (b) $r = 0.09$.   (c) $r = 0.77$.

C H A P T E R 4   E X E R C I S E S

4.24 Scores at the Masters. The Masters is one of the four major golf tournaments. Figure 4.8 is a scatterplot of the scores for the first two rounds of the 2007 Masters for all of the golfers entered. Only the 60 golfers with the lowest two-round total advance to the final two rounds. The plot has a grid pattern because golf scores must be whole numbers.[8]

(a)   Read the graph: What was the lowest score in the first round of play? How many golfers had this low score? What were their scores in the second round?

(b)   Read the graph: Camilo Villegas had the highest score in the second round. What was this score? What was Villegas's score in the first round?

FIGURE 4.8

Scatterplot of the scores in the first two rounds of the 2007 Masters Tournament, for Exercise 4.24.

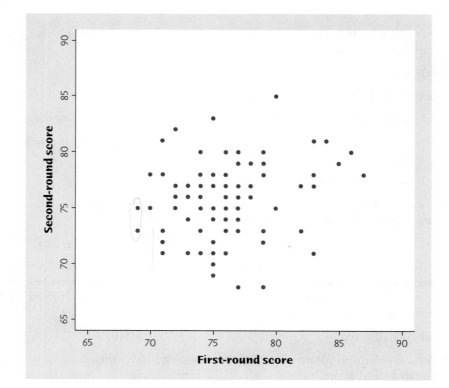

(c) Is the correlation between first-round scores and second-round scores closest to $r = 0.2$, $r = 0.6$, or $r = 0.9$? Explain your choice. Does the graph suggest that knowing a professional golfer's score for one round is much help in predicting his score for another round on the same course?

4.25 Can children estimate their own reading ability? To study this question, investigators asked 60 fifth-grade children to estimate their own reading ability, on a scale from 1 (low) to 5 (high). Figure 4.9 is a scatterplot of the children's estimates (response) against their scores on a reading test (explanatory).[9] Both scores take only whole-number values.

(a) Is there an overall positive association between reading score and self-estimate?

(b) There is one clear outlier. What is this child's self-estimated reading level? Does this appear to over- or underestimate the level as measured by the test?

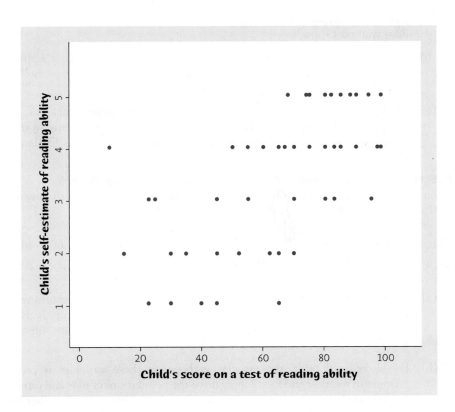

FIGURE 4.9

Scatterplot of children's estimates of their reading ability (on a scale of 1 to 5) against their score on a reading test, for Exercise 4.25.

4.26 Data on dating. A student wonders if tall women tend to date taller men than do short women. She measures herself, her dormitory roommate, and the women in the adjoining rooms; then she measures the next man each woman dates. Here are the data (heights in inches):

Women (x)	66	64	66	65	70	65
Men (y)	72	68	70	68	71	65

66   62   64   62   65   59

(a) Make a scatterplot of these data. Based on the scatterplot, do you expect the correlation to be positive or negative? Near ±1 or not?

(b) Find the correlation r between the heights of the men and women. Do the data show that taller women tend to date taller men?

4.27 Coffee and deforestation. Coffee is a leading export from several developing countries. When coffee prices are high, farmers often clear forest to plant more coffee trees. Here are five years of data on prices paid to coffee growers in Indonesia and the percent of forest area lost in a national park that lies in a coffee-producing region:[10]

Bill Ross/CORBIS

Price (cents per pound)	29	40	54	55	72
Forest lost (percent)	0.49	1.59	1.69	1.82	3.10

(a) Make a scatterplot. Which is the explanatory variable? What kind of pattern does your plot show?

(b) Find the correlation r between coffee price and forest loss. Do your scatterplot and correlation support the idea that higher coffee prices increase the loss of forest?

(c) The price of coffee in international trade is given in dollars and cents. If the prices in the data were translated into the equivalent prices in euros, would the correlation between coffee price and percent of forest loss change? Explain your answer.

4.28 Sparrowhawk colonies. One of nature's patterns connects the percent of adult birds in a colony that return from the previous year and the number of new adults that join the colony. Here are data for 13 colonies of sparrowhawks:[11]

Percent return	74	66	81	52	73	62	52	45	62	46	60	46	38
New adults	5	6	8	11	12	15	16	17	18	18	19	20	20

(a) Plot the count of new adults (response) against the percent of returning birds (explanatory). Describe the direction and form of the relationship. Is the correlation r an appropriate measure of the strength of this relationship? If so, find r.

(b) For short-lived birds, the association between these variables is positive: changes in weather and food supply drive the populations of new and returning birds up or down together. For long-lived territorial birds, on the other hand, the association is negative because returning birds claim their territories in the colony and don't leave room for new recruits. Which type of species is the sparrowhawk?

4.29 Our brains don't like losses. Most people dislike losses more than they like gains. In money terms, people are about as sensitive to a loss of $10 as to a gain of $20. To discover what parts of the brain are active in decisions about gain and loss, psychologists presented subjects with a series of gambles with different odds and different amounts of winnings and losses. From a subject's choices, they constructed a measure of "behavioral loss aversion." Higher scores show greater sensitivity to losses.

Observing brain activity while subjects made their decisions pointed to specific brain regions. Here are data for 16 subjects on behavioral loss aversion and "neural loss aversion," a measure of activity in one region of the brain:[12]

Neural	−50.0	−39.1	−25.9	−26.7	−28.6	−19.8	−17.6	5.5
Behavioral	0.08	0.81	0.01	0.12	0.68	0.11	0.36	0.34
Neural	2.6	20.7	12.1	15.5	28.8	41.7	55.3	155.2
Behavioral	0.53	0.68	0.99	1.04	0.66	0.86	1.29	1.94

(a) Make a scatterplot that shows how behavior responds to brain activity.

(b) Describe the overall pattern of the data. There is one clear outlier.

(c) Find the correlation r between neural and behavioral loss aversion both with and without the outlier. Does the outlier have a strong influence on the value of r? By looking at your plot, explain why adding the outlier to the other data points causes r to increase.

4.30 Sulfur, the ocean, and the sun. Sulfur in the atmosphere affects climate by influencing formation of clouds. The main natural source of sulfur is dimethylsulfide (DMS) produced by small organisms in the upper layers of the oceans. DMS production is in turn influenced by the amount of energy the upper ocean receives from sunlight. Here are monthly data on solar radiation dose (SRD, in watts per square meter) and surface DMS concentration (in nanomolars) for a region in the Mediterranean:[13]

SRD	12.55	12.91	14.34	19.72	21.52	22.41	37.65	48.41
DMS	0.796	0.692	1.744	1.062	0.682	1.517	0.736	0.720
SRD	74.41	94.14	109.38	157.79	262.67	268.96	289.23	
DMS	1.820	1.099	2.692	5.134	8.038	7.280	8.872	

(a) Make a scatterplot that shows how DMS responds to SRD.

(b) Describe the overall pattern of the data. Find the correlation r between DMS and SRD. Because SRD changes with the seasons of the year, the close relationship between SRD and DMS helps explain other seasonal patterns.

4.31 How fast do icicles grow? Japanese researchers measured the growth of icicles in a cold chamber under various conditions of temperature, wind, and water flow.[14] Table 4.2 contains data produced under two sets of conditions. In both cases, there was no wind and the temperature was set at −11°C. Water flowed over the icicle at a higher rate (29.6 milligrams per second) in Run 8905 and at a slower rate (11.9 mg/s) in Run 8903.

(a) Make a scatterplot of the length of the icicle in centimeters versus time in minutes, using separate symbols for the two runs.

(b) What does your plot show about the pattern of growth of icicles? What does it show about the effect of changing the rate of water flow on icicle growth?

TABLE 4.2 Growth of icicles over time

	RUN 8903				RUN 8905		
TIME (min)	LENGTH (cm)	TIME (min)	LENGTH (cm)	TIME (min)	LENGTH (cm)	TIME (min)	LENGTH (cm)
10	0.6	130	18.1	10	0.3	130	10.4
20	1.8	140	19.9	20	0.6	140	11.0
30	2.9	150	21.0	30	1.0	150	11.9
40	4.0	160	23.4	40	1.3	160	12.7
50	5.0	170	24.7	50	3.2	170	13.9
60	6.1	180	27.8	60	4.0	180	14.6
70	7.9			70	5.3	190	15.8
80	10.1			80	6.0	200	16.2
90	10.9			90	6.9	210	17.9
100	12.7			100	7.8	220	18.8
110	14.4			110	8.3	230	19.9
120	16.6			120	9.6	240	21.1

4.32 How many corn plants are too many? How much corn per acre should a farmer plant to obtain the highest yield? Too few plants will give a low yield. On the other hand, if there are too many plants, they will compete with each other for moisture and nutrients, and yields will fall. To find the best planting rate, plant at different rates on several plots of ground and measure the harvest. (Be sure to treat all the plots the same except for the planting rate.) Here are data from such an experiment:[15]

Plants per acre (thousands)	Yield (bushels per acre)			
12	150.1	113.0	118.4	142.6
16	166.9	120.7	135.2	149.8
20	165.3	130.1	139.6	149.9
24	134.7	138.4	156.1	
28	119.0	150.5		

(a) Make a scatterplot of that shows how yield responds to planting rate. Use a scale of yields from 100 to 200 bushels per acre so that the pattern will be clear.

(b) Describe the overall pattern of the relationship. Is it linear? Is there a positive or negative association, or neither? Find the correlation r. Is r a helpful description of this relationship?

(c) Find the mean yield for each of the five planting rates. Plot each mean yield against its planting rate on your scatterplot and connect these five points with lines. This combination of numerical description and graphing makes the relationship clearer. What planting rate would you recommend to a farmer whose conditions were similar to those in the experiment?

4.33 Attracting beetles. To detect the presence of harmful insects in farm fields, we can put up boards covered with a sticky material and examine the insects trapped on the boards. Which colors attract insects best? Experimenters placed six boards of each of four colors at random locations in a field of oats and measured the number of cereal leaf beetles trapped. Here are the data:[16]

Board color	Beetles trapped					
Blue	16	11	20	21	14	7
Green	37	32	20	29	37	32
White	21	12	14	17	13	20
Yellow	45	59	48	46	38	47

Holt Studios International/Alamy

(a) Make a plot of beetles trapped against color (space the four colors equally on the horizontal axis). Which color appears best at attracting beetles?

(b) Does it make sense to speak of a positive or negative association between board color and beetles trapped? Why? Is correlation r a helpful description of the relationship? Why?

4.34 Thinking about correlation. Exercise 4.26 presents data on the heights of women and of the men they date.

(a) If heights were measured in centimeters rather than inches, how would the correlation change? (There are 2.54 centimeters in an inch.)

(b) How would r change if all the men were 6 inches shorter than the heights given in the table? Does the correlation tell us whether women tend to date men taller than themselves?

(c) If every woman dated a man exactly 3 inches taller than herself, what would be the correlation between male and female heights?

4.35 The effect of changing units. Changing the units of measurement can dramatically alter the appearance of a scatterplot. Return to the data on knee height and overall height in Exercise 4.23:

Knee height x	57.7	47.4	43.5	44.8	55.2
Height y	192.1	153.3	146.4	162.7	169.1

Both heights are measured in centimeters. A mad scientist decides to measure knee height in millimeters and height in meters. The same data in these units are

Knee height x	577	474	435	448	552
Height y	1.921	1.533	1.464	1.627	1.691

(a) Make a plot with the x axis extending from 0 to 600 and the y axis from 0 to 250. Plot the original data on these axes. Then plot the new data using a different color or symbol. The two plots look very different.

(b) Nonetheless, the correlation is exactly the same for the two sets of measurements. Why do you know that this is true without doing any calculations? Find the two correlations to verify that they are the same.

4.36 Statistics for investing. Investment reports now often include correlations. Following a table of correlations among mutual funds, a report adds: "Two funds can have perfect correlation, yet different levels of risk. For example, Fund A and Fund B may be perfectly correlated, yet Fund A moves 20% whenever Fund B moves 10%." Write a brief explanation, for someone who knows no statistics, of how this can happen. Include a sketch to illustrate your explanation.

4.37 Statistics for investing. A mutual funds company's newsletter says, "A well-diversified portfolio includes assets with low correlations." The newsletter includes a table of correlations between the returns on various classes of investments. For example, the correlation between municipal bonds and large-cap stocks is 0.50, and the correlation between municipal bonds and small-cap stocks is 0.21.

(a) Rachel invests heavily in municipal bonds. She wants to diversify by adding an investment whose returns do not closely follow the returns on her bonds. Should she choose large-cap stocks or small-cap stocks for this purpose? Explain your answer.

(b) If Rachel wants an investment that tends to increase when the return on her bonds drops, what kind of correlation should she look for?

4.38 Teaching and research. A college newspaper interviews a psychologist about student ratings of the teaching of faculty members. The psychologist says, "The evidence indicates that the correlation between the research productivity and teaching rating of faculty members is close to zero." The paper reports this as "Professor McDaniel said that good researchers tend to be poor teachers, and vice versa." Explain why the paper's report is wrong. Write a statement in plain language (don't use the word "correlation") to explain the psychologist's meaning.

4.39 Sloppy writing about correlation. Each of the following statements contains a blunder. Explain in each case what is wrong.

(a) "There is a high correlation between the gender of American workers and their income."

(b) "We found a high correlation ($r = 1.09$) between students' ratings of faculty teaching and ratings made by other faculty members."

(c) "The correlation between height and weight of the subjects was $r = 0.63$ centimeter."

4.40 Correlation is not resistant. Go to the *Correlation and Regression* applet. Click on the scatterplot to create a group of 10 points in the lower-left corner of the scatterplot with a strong straight-line pattern (correlation about 0.9).

(a) Add one point at the upper right that is in line with the first 10. How does the correlation change?

(b) Drag this last point down until it is opposite the group of 10 points. How small can you make the correlation? Can you make the correlation negative? You see that a single outlier can greatly strengthen or weaken a correlation. Always plot your data to check for outlying points.

4.41 Match the correlation. You are going to use the *Correlation and Regression* applet to make scatterplots with 10 points that have correlation close to 0.7. The lesson is that many patterns can have the same correlation. Always plot your data before you trust a correlation.

(a) Click on the scatterplot to add the first two points. What is the value of the correlation? Why does it have this value?

(b) Make a lower-left to upper-right pattern of 10 points with correlation about $r = 0.7$. (You can drag points up or down to adjust r after you have 10 points.) Make a rough sketch of your scatterplot.

(c) Make another scatterplot with 9 points in a vertical stack at the left of the plot. Add one point far to the right and move it until the correlation is close to 0.7. Make a rough sketch of your scatterplot.

(d) Make yet another scatterplot with 10 points in a curved pattern that starts at the lower left, rises to the right, then falls again at the far right. Adjust the points up or down until you have a quite smooth curve with correlation close to 0.7. Make a rough sketch of this scatterplot also.

*The following exercises ask you to answer questions from data without having the details outlined for you. The exercise statements give you the **State** step of the four-step process. In your work, follow the **Plan, Solve,** and **Conclude** steps of the process, described on page 55.*

4.42 Brighter sunlight? The brightness of sunlight at the earth's surface changes over time depending on whether the earth's atmosphere is more or less clear. Sunlight dimmed between 1960 and 1990. After 1990, air pollution dropped in industrial countries. Did sunlight brighten? Here are data from Boulder, Colorado, averaging over only clear days each year. (Other locations show similar trends.) The response variable is solar radiation in watts per square meter.[17]

Year	1992	1993	1994	1995	1996	1997	1998	1999	2000	2001	2002
Sun	243.2	246.0	248.0	250.3	250.9	250.9	250.0	248.9	251.7	251.4	250.9

4.43 Saving energy with solar panels. We have data from a house in the Midwest that uses natural gas for heating. Will installing solar panels reduce the amount of gas consumed? Gas consumption is higher in cold weather, so the relationship between outside temperature and gas consumption is important. Here are data for 16 consecutive months:[18]

	Nov.	Dec.	Jan.	Feb.	Mar.	Apr.	May	June
Degree-days per day	24	51	43	33	26	13	4	0
Gas used per day	6.3	10.9	8.9	7.5	5.3	4.0	1.7	1.2
	July	Aug.	Sept.	Oct.	Nov.	Dec.	Jan.	Feb.
Degree-days per day	0	1	6	12	30	32	52	30
Gas used per day	1.2	1.2	2.1	3.1	6.4	7.2	11.0	6.9

Paul Glendell/Alamy

Outside temperature is recorded in degree-days, a common measure of demand for heating. A day's degree-days are the number of degrees its average temperature falls below 65°F. Gas used is recorded in hundreds of cubic feet. Here are data for 23 more months after installing solar panels:

Degree-days	19	3	3	0	0	0	8	11	27	46	38	34
Gas used	3.2	2.0	1.6	1.0	0.7	0.7	1.6	3.1	5.1	7.7	7.0	6.1

Degree-days	16	9	2	1	0	2	3	18	32	34	40
Gas used	3.0	2.1	1.3	1.0	1.0	1.0	1.2	3.4	6.1	6.5	7.5

What do the before-and-after data show about the effect of solar panels? (Start by plotting both sets of data on the same plot, using two different plotting symbols.)

4.44 Merlins breeding. Often the percent of an animal species in the wild that survive to breed again is lower following a successful breeding season. This is part of nature's self-regulation to keep population size stable. A study of merlins (small falcons) in northern Sweden observed the number of breeding pairs in an isolated area and the percent of males (banded for identification) who returned the next breeding season. Here are data for nine years:[19]

x Breeding pairs	28	29	29	29	30	32	33	38	38
y Percent return	82	83	70	61	69	58	43	50	47

Investigate the relationship between breeding pairs and percent return.

4.45 Does social rejection hurt? We often describe our emotional reaction to social rejection as "pain." Does social rejection cause activity in areas of the brain that are known to be activated by physical pain? If it does, we really do experience social and physical pain in similar ways. Psychologists first included and then deliberately excluded individuals from a social activity while they measured changes in brain activity. After each activity, the subjects filled out questionnaires that assessed how excluded they felt. Here are data for 13 subjects.[20]

Subject	Social distress	Brain activity	Subject	Social distress	Brain activity
1	1.26	−0.055	8	2.18	0.025
2	1.85	−0.040	9	2.58	0.027
3	1.10	−0.026	10	2.75	0.033
4	2.50	−0.017	11	2.75	0.064
5	2.17	−0.017	12	3.33	0.077
6	2.67	0.017	13	3.65	0.124
7	2.01	0.021			

The explanatory variable is "social distress" measured by each subject's questionnaire score after exclusion relative to the score after inclusion. (So values greater than 1 show the degree of distress caused by exclusion.) The response variable is change in activity in a region of the brain that is activated by physical pain. Discuss what the data show.

TABLE 4.3 Fish supply and wildlife decline in West Africa

YEAR	FISH SUPPLY (kg per person)	BIOMASS CHANGE (percent)	YEAR	FISH SUPPLY (kg per person)	BIOMASS CHANGE (percent)
1971	34.7	2.9	1985	21.3	−5.5
1972	39.3	3.1	1986	24.3	−0.7
1973	32.4	−1.2	1987	27.4	−5.1
1974	31.8	−1.1	1988	24.5	−7.1
1975	32.8	−3.3	1989	25.2	−4.2
1976	38.4	3.7	1990	25.9	0.9
1977	33.2	1.9	1991	23.0	−6.1
1978	29.7	−0.3	1992	27.1	−4.1
1979	25.0	−5.9	1993	23.4	−4.8
1980	21.8	−7.9	1994	18.9	−11.3
1981	20.8	−5.5	1995	19.6	−9.3
1982	19.7	−7.2	1996	25.3	−10.7
1983	20.8	−4.1	1997	22.0	−1.8
1984	21.1	−8.6	1998	21.0	−7.4

4.46 Bushmeat. African peoples often eat "bushmeat," the meat of wild animals. Bushmeat is widely traded in Africa, but its consumption threatens the survival of some animals in the wild. Bushmeat is often not the first choice of consumers—they eat bushmeat when other sources of protein are in short supply. Researchers looked at declines in 41 species of mammals in nature reserves in Ghana and at catches of fish (the primary source of animal protein) in the same region. The data appear in Table 4.3.[21] Fish supply is measured in kilograms per person. The other variable is the percent change in the total "biomass" (weight in tons) for the 41 animal species in six nature reserves. Most of the yearly percent changes in wildlife mass are negative because most years saw fewer wild animals in West Africa. Discuss how the data support the idea that more animals are killed for bushmeat when the fish supply is low.

CHAPTER 5

Regression

Linear (straight-line) relationships between two quantitative variables are easy to understand and quite common. In Chapter 4, we found linear relationships in settings as varied as Florida manatee deaths, icicle growth, and predicting tropical storms. Correlation measures the direction and strength of these relationships. When a scatterplot shows a linear relationship, we would like to summarize the overall pattern by drawing a line on the scatterplot.

Regression lines

A *regression line* summarizes the relationship between two variables, but only in a specific setting: one of the variables helps explain or predict the other. That is, regression describes a relationship between an explanatory variable and a response variable.

REGRESSION LINE

A **regression line** is a straight line that describes how a response variable y changes as an explanatory variable x changes. We often use a regression line to predict the value of y for a given value of x.

EXAMPLE 5.1 Does fidgeting keep you slim?

Why is it that some people find it easy to stay slim? Here, following our four-step process (page 55), is an account of a study that sheds some light on gaining weight.

STATE: Some people don't gain weight even when they overeat. Perhaps fidgeting and other "nonexercise activity" (NEA) explains why. Some people may spontaneously increase nonexercise activity when fed more. Researchers deliberately overfed 16 healthy young adults for 8 weeks. They measured fat gain (in kilograms) and, as an explanatory variable, change in energy use (in calories) from activity other than deliberate exercise—fidgeting, daily living, and the like. Here are the data:[1]

NEA change (cal)	−94	−57	−29	135	143	151	245	355
Fat gain (kg)	4.2	3.0	3.7	2.7	3.2	3.6	2.4	1.3
NEA change (cal)	392	473	486	535	571	580	620	690
Fat gain (kg)	3.8	1.7	1.6	2.2	1.0	0.4	2.3	1.1

Do people with larger increases in NEA tend to gain less fat?

PLAN: Make a scatterplot of the data and examine the pattern. If it is linear, use correlation to measure its strength and draw a regression line on the scatterplot to predict fat gain from change in NEA.

SOLVE: Figure 5.1 is a scatterplot of these data. The plot shows a moderately strong negative linear association with no outliers. The correlation is $r = -0.7786$. The line on the plot is a regression line for predicting fat gain from change in NEA.

FIGURE 5.1

Weight gain after 8 weeks of overeating, plotted against increase in nonexercise activity over the same period, for Example 5.1.

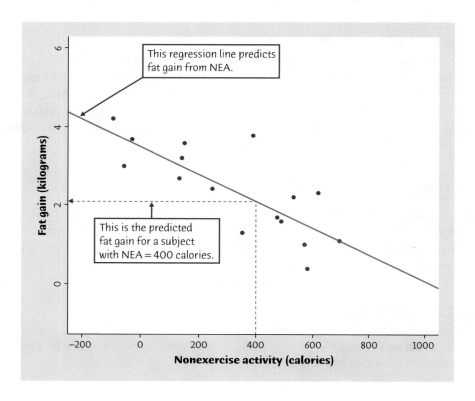

CONCLUDE: People with larger increases in nonexercise activity do indeed gain less fat. To add to this conclusion, we must study regression lines in more detail.

We can, however, already use the regression line to predict fat gain from NEA. Suppose that an individual's NEA increases by 400 calories when she overeats. Go "up and over" on the graph in Figure 5.1. From 400 calories on the x axis, go up to the regression line and then over to the y axis. The graph shows that the predicted gain in fat is a bit more than 2 kilograms. ■

Many calculators and software programs will give you the equation of a regression line from keyed-in data. Understanding and using the line is more important than the details of where the equation comes from.

Regression toward the mean

To "regress" means to go backward. Why are statistical methods for predicting a response from an explanatory variable called "regression"? Sir Francis Galton (1822–1911), who was the first to apply regression to biological and psychological data, looked at examples such as the heights of children versus the heights of their parents. He found that the taller-than-average parents tended to have children who were also taller than average but not as tall as their parents. Galton called this fact "regression toward the mean," and the name came to be applied to the statistical method.

REVIEW OF STRAIGHT LINES

Suppose that y is a response variable (plotted on the vertical axis) and x is an explanatory variable (plotted on the horizontal axis). A straight line relating y to x has an equation of the form

$$y = a + bx$$

In this equation, b is the **slope,** the amount by which y changes when x increases by one unit. The number a is the **intercept,** the value of y when $x = 0$.

EXAMPLE 5.2 **Using a regression line**

Any straight line describing the NEA data has the form

$$\text{fat gain} = a + (b \times \text{NEA change})$$

The line in Figure 5.1 is the regression line with the equation

$$\text{fat gain} = 3.505 - 0.00344 \times \text{NEA change}$$

Be sure you understand the role of the two numbers in this equation:

■ The slope $b = -0.00344$ tells us that fat gained goes down by 0.00344 kilogram for each added calorie of NEA. The slope of a regression line is the *rate of change* in the response as the explanatory variable changes.

■ The intercept, $a = 3.505$ kilograms, is the estimated fat gain if NEA does not change when a person overeats.

The equation of the regression line makes it easy to predict fat gain. If a person's NEA increases by 400 calories when she overeats, substitute $x = 400$ in the equation. The predicted fat gain is

$$\text{fat gain} = 3.505 - (0.00344 \times 400) = 2.13 \text{ kilograms}$$

To **plot the line** on the scatterplot, use the equation to find the predicted y for two values of x, one near each end of the range of x in the data. Plot each y above its x-value and draw the line through the two points. ■

plotting a line

The slope of a regression line is an important numerical description of the relationship between the two variables. Although we need the value of the intercept to draw the line, this value is statistically meaningful only when, as in Example 5.2, the explanatory variable can actually take values close to zero.

The slope $b = -0.00344$ in Example 5.2 is small. This does *not* mean that change in NEA has little effect on fat gain. The size of the slope depends on the units in which we measure the two variables. In this example, the slope is the change in fat gain in kilograms when NEA increases by one calorie. There are 1000 grams in a kilogram. If we measured fat gain in grams, the slope would be 1000 times larger, $b = 3.44$. *You can't say how important a relationship is by looking at the size of the slope of the regression line.*

APPLY YOUR KNOWLEDGE

5.1 City mileage, highway mileage. We expect a car's highway gas mileage to be related to its city gas mileage. Data for all 1198 vehicles in the government's *2008 Fuel Economy Guide* give the regression line

$$\text{highway mpg} = 4.62 + (1.109 \times \text{city mpg})$$

for predicting highway mileage from city mileage.

(a)   What is the slope of this line? Say in words what the numerical value of the slope tells you.

(b)   What is the intercept? Explain why the value of the intercept is not statistically meaningful.

(c)   Find the predicted highway mileage for a car that gets 16 miles per gallon in the city. Do the same for a car with city mileage 28 mpg.

(d)   Draw a graph of the regression line for city mileages between 10 and 50 mpg. (Be sure to show the scales for the x and y axes.)

5.2 What's the line? You use the same bar of soap to shower each morning. The bar weighs 80 grams when it is new. Its weight goes down by 6 grams per day on the average. What is the equation of the regression line for predicting weight from days of use?

The least-squares regression line

In most cases, no line will pass exactly through all the points in a scatterplot. Different people will draw different lines by eye. We need a way to draw a regression line that doesn't depend on our guess as to where the line should go. Because we use the line to predict y from x, the prediction errors we make are errors in y, the vertical direction in the scatterplot. *A good regression line makes the vertical distances of the points from the line as small as possible.*

Figure 5.2 illustrates the idea. This plot shows three of the points from Figure 5.1, along with the line, on an expanded scale. The line passes above one of the points and below two of them. The three prediction errors appear as vertical line

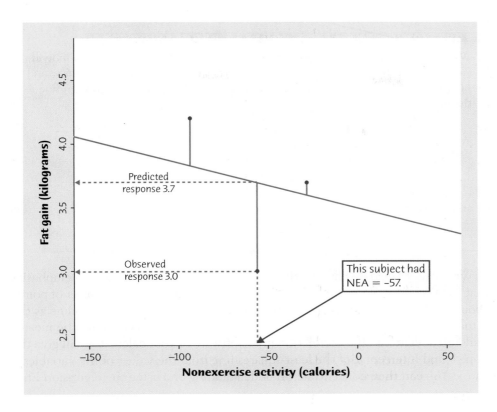

FIGURE 5.2

The least-squares idea. For each observation, find the vertical distance of each point on the scatterplot from a regression line. The least-squares regression line makes the sum of the squares of these distances as small as possible.

segments. For example, one subject had $x = -57$, a decrease of 57 calories in NEA. The line predicts a fat gain of 3.7 kilograms, but the actual fat gain for this subject was 3.0 kilograms. The prediction error is

$$\text{error} = \text{observed response} - \text{predicted response}$$
$$= 3.0 - 3.7 = -0.7 \text{ kilogram}$$

There are many ways to make the collection of vertical distances "as small as possible." The most common is the *least-squares* method.

> **LEAST-SQUARES REGRESSION LINE**
>
> The **least-squares regression line** of y on x is the line that makes the sum of the squares of the vertical distances of the data points from the line as small as possible.

One reason for the popularity of the least-squares regression line is that the problem of finding the line has a simple answer. We can give the equation for the least-squares line in terms of the means and standard deviations of the two variables and the correlation between them.

> ### EQUATION OF THE LEAST-SQUARES REGRESSION LINE
>
> We have data on an explanatory variable x and a response variable y for n individuals. From the data, calculate the means $\overline{x}$ and $\overline{y}$ and the standard deviations s_x and s_y of the two variables, and their correlation r. The least-squares regression line is the line
>
> $$\hat{y} = a + bx$$
>
> with **slope**
>
> $$b = r\frac{s_y}{s_x}$$
>
> and **intercept**
>
> $$a = \overline{y} - b\overline{x}$$

We write $\hat{y}$ (read "y hat") in the equation of the regression line to emphasize that the line gives a *predicted* response $\hat{y}$ for any x. Because of the scatter of points about the line, the predicted response will usually not be exactly the same as the actually *observed* response y. In practice, you don't need to calculate the means, standard deviations, and correlation first. Software or your calculator will give the slope b and intercept a of the least-squares line from the values of the variables x and y. You can then concentrate on understanding and using the regression line.

Using technology

Least-squares regression is one of the most common statistical procedures. Any technology you use for statistical calculations will give you the least-squares line and related information. Figure 5.3 displays the regression output for the data of Examples 5.1 and 5.2 from a graphing calculator, a statistical program, and a spreadsheet program. Each output records the slope and intercept of the least-squares line. The software also provides information that we do not yet need, although we will use much of it later. (In fact, we left out part of the Minitab and Excel outputs.) Be sure that you can locate the slope and intercept on all four outputs. *Once you understand the statistical ideas, you can read and work with almost any software output.*

APPLY YOUR KNOWLEDGE

5.3 **Ebola and gorillas.** An outbreak of the deadly Ebola virus in 2002 and 2003 killed 91 of the 95 gorillas in 7 home ranges in the Congo. To study the spread of the virus, measure "distance" by the number of home ranges separating a group of gorillas from the first group infected. Here are data on distance and number of days until deaths began in each later group:[2]

Distance x	1	3	4	4	4	5
Days y	4	21	33	41	43	46

Texas Instruments Graphing Calculator

FIGURE 5.3

Least-squares regression for the nonexercise activity data: output from a graphing calculator, a statistical program, and a spreadsheet program.

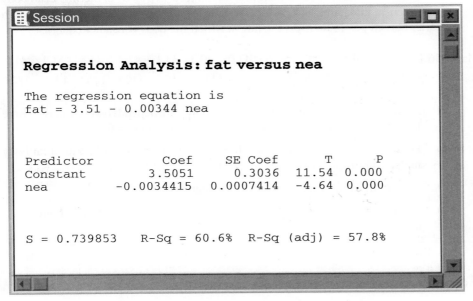

```
LinReg
  y=a+bx
  a=3.505122916
  b=-.003441487
  r²=.6061492049
  r=-.7785558457
```

Minitab

```
┌─────────────────────────────────────────────────────────┐
│ ▦ Session                                      _ □ ✕     │
├─────────────────────────────────────────────────────────┤
│                                                          │
│  Regression Analysis: fat versus nea                     │
│                                                          │
│  The regression equation is                              │
│  fat = 3.51 - 0.00344 nea                                │
│                                                          │
│                                                          │
│  Predictor          Coef     SE Coef       T       P     │
│  Constant         3.5051      0.3036   11.54   0.000     │
│  nea           -0.0034415   0.0007414   -4.64   0.000    │
│                                                          │
│                                                          │
│  S = 0.739853    R-Sq = 60.6%  R-Sq (adj) = 57.8%        │
│                                                          │
└─────────────────────────────────────────────────────────┘
```

Microsoft Excel

	A	B	C	D	E	F
1	SUMMARY OUTPUT					
2						
3	*Regression statistics*					
4	Multiple R	0.778555846				
5	R Square	0.606149205				
6	Adjusted R Square	0.578017005				
7	Standard Error	0.739852874				
8	Observations	16				
9						
10		*Coefficients*	*Standard Error*	*t Stat*	*P-value*	
11	Intercept	3.505122916	0.303616403	11.54458	1.53E-08	
12	nea	-0.003441487	0.00074141	-4.64182	0.000381	
13						

Output / nea data

As you saw in Exercise 4.10 (page 106), there is a linear relationship between distance x and days y.

(a)   Use your calculator to find the mean and standard deviation of both x and y and the correlation r between x and y. Use these basic measures to find the equation of the least-squares line for predicting y from x.

(b)   Enter the data into your software or calculator and use the regression function to find the least-squares line. The result should agree with your work in (a) up to roundoff error.

5.4   Do heavier people burn more energy? We have data on the lean body mass and resting metabolic rate for 12 women who are subjects in a study of dieting. Lean body mass, given in kilograms, is a person's weight leaving out all fat. Metabolic rate, in calories burned per 24 hours, is the rate at which the body consumes energy.

Mass	36.1	54.6	48.5	42.0	50.6	42.0	40.3	33.1	42.4	34.5	51.1	41.2
Rate	995	1425	1396	1418	1502	1256	1189	913	1124	1052	1347	1204

(a)   Make a scatterplot that shows how metabolic rate depends on body mass. There is a quite strong linear relationship, with correlation $r = 0.876$.

(b)   Find the least-squares regression line for predicting metabolic rate from body mass. Add this line to your scatterplot.

(c)   Explain in words what the slope of the regression line tells us.

(d)   Another woman has lean body mass 45 kilograms. What is her predicted metabolic rate?

Facts about least-squares regression

One reason for the popularity of least-squares regression lines is that they have many convenient properties. Here are some facts about least-squares regression lines.

Fact 1. The distinction between explanatory and response variables is essential in regression. Least-squares regression makes the distances of the data points from the line small only in the y direction. If we reverse the roles of the two variables, we get a different least-squares regression line.

EXAMPLE 5.3   **Predicting fat gain, predicting NEA**

Figure 5.4 repeats the scatterplot of the nonexercise activity data in Figure 5.1, but with *two* least-squares regression lines. The solid line is the regression line for predicting fat gain from change in NEA. This is the line that appeared in Figure 5.1.

We might also use the data on these 16 subjects to predict the change in NEA for another subject from that subject's fat gain when overfed for 8 weeks. Now the roles of the variables are reversed: fat gain is the explanatory variable and change in NEA is the response variable. The dashed line in Figure 5.4 is the least-squares line for predicting NEA change from fat gain. The two regression lines are not the same. *In the regression setting, you must know clearly which variable is explanatory.* ■

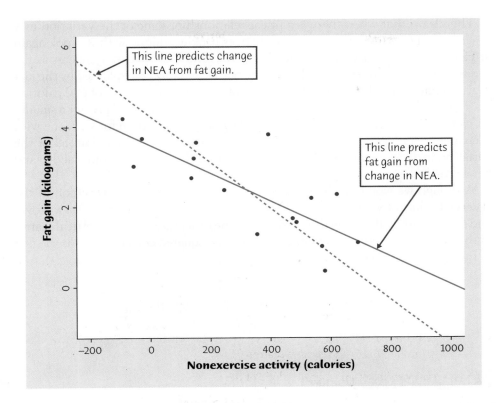

FIGURE 5.4

Two least-squares regression lines for the nonexercise activity data, for Example 5.3. The solid line predicts fat gain from change in nonexercise activity. The dashed line predicts change in nonexercise activity from fat gain.

Fact 2. There is a close connection between correlation and the slope of the least-squares line. The slope is

$$b = r\frac{s_y}{s_x}$$

You see that **the slope and the correlation always have the same sign.** For example, if a scatterplot shows a positive association, then both b and r are positive. The formula for the slope b says more: along the regression line, **a change of one standard deviation in x corresponds to a change of r standard deviations in y.** When the variables are perfectly correlated ($r = 1$ or $r = -1$), the change in the predicted response $\hat{y}$ is the same (in standard deviation units) as the change in x. Otherwise, because $-1 \leq r \leq 1$, the change in $\hat{y}$ (in standard deviation units) is less than the change in x. As the correlation grows less strong, the prediction $\hat{y}$ moves less in response to changes in x.

Fact 3. **The least-squares regression line always passes through the point** $(\overline{x}, \overline{y})$ on the graph of y against x.

Fact 4. The correlation r describes the strength of a straight-line relationship. In the regression setting, this description takes a specific form: **the square of the correlation, r^2, is the fraction of the variation in the values of y that is explained by the least-squares regression of y on x.**

The idea is that when there is a linear relationship, some of the variation in y is accounted for by the fact that as x changes it pulls y along with it. Look again at Figure 5.1, the scatterplot of the NEA data. The variation in y appears as the spread of fat gains from 0.4 to 4.2 kg. Some of this variation is explained by the fact that x (change in NEA) varies from a loss of 94 calories to a gain of 690 calories. As x moves from -94 to 690, it pulls y along the line. You would predict a smaller fat gain for a subject whose NEA increased by 600 calories than for someone with 0 change in NEA. But the straight-line tie of y to x doesn't explain *all* of the variation in y. The remaining variation appears as the scatter of points above and below the line.

Although we won't do the algebra, it is possible to break the variation in the observed values of y into two parts. One part measures the variation in $\hat{y}$ as x moves and pulls $\hat{y}$ with it along the regression line. The other measures the vertical scatter of the data points above and below the line. The squared correlation r^2 is the first of these as a fraction of the whole:

$$r^2 = \frac{\text{variation in } \hat{y} \text{ as } x \text{ pulls it along the line}}{\text{total variation in observed values of } y}$$

EXAMPLE 5.4 **Using r^2**

For the NEA data, $r = -0.7786$ and $r^2 = (-0.7786)^2 = 0.6062$. About 61% of the variation in fat gained is accounted for by the linear relationship with change in NEA. The other 39% is individual variation among subjects that is not explained by the linear relationship.

Figure 4.2 (page 102) shows a stronger linear relationship between boat registrations in Florida and manatees killed by boats. The correlation is $r = 0.953$ and $r^2 = (0.953)^2 = 0.908$. Almost 91% of the year-to-year variation in number of manatees killed by boats is explained by regression on number of boats registered. Only about 9% is variation among years with similar numbers of boats registered. ■

You can find a regression line for any relationship between two quantitative variables, but the usefulness of the line for prediction depends on the strength of the linear relationship. So r^2 is almost as important as the equation of the line in reporting a regression. All of the outputs in Figure 5.3 include r^2, either in decimal form or as a percent. When you see a correlation, square it to get a better feel for the strength of the association. Perfect correlation ($r = -1$ or $r = 1$) means the points lie exactly on a line. Then $r^2 = 1$ and all of the variation in one variable is accounted for by the linear relationship with the other variable. If $r = -0.7$ or $r = 0.7$, $r^2 = 0.49$ and about half the variation is accounted for by the linear relationship. In the r^2 scale, correlation ± 0.7 is about halfway between 0 and ± 1.

Facts 2, 3, and 4 are special properties of least-squares regression. They are not true for other methods of fitting a line to data.

5.5 How useful is regression? Figure 4.8 (page 114) displays the relationship between golfers' scores on the first and second rounds of the 2007 Masters Tournament. The correlation is $r = 0.192$. Exercise 4.30 gives data on solar radiation (SRD) and concentration of dimethylsulfide (DMS) over a region of the Mediterranean. The correlation is $r = 0.969$. Explain in simple language why knowing only these correlations enables you to say that prediction of DMS from SRD by a regression line will be much more accurate than prediction of a golfer's second-round score from his first-round score.

5.6 Growing corn. Exercise 4.32 (page 118) gives data from an agricultural experiment. The purpose of the study was to see how the yield of corn changes as we change the planting rate (plants per acre).

(a) Make a scatterplot of the data. (Use a scale of yields from 100 to 200 bushels per acre.) Find the least-squares regression line for predicting yield from planting rate and add this line to your plot. Why should we *not* use the regression line for prediction in this setting?

(b) What is r^2? What does this value say about the success of the regression line in predicting yield?

Frank Krahmer/Age fotostock

Residuals

One of the first principles of data analysis is to look for an overall pattern and also for striking deviations from the pattern. A regression line describes the overall pattern of a linear relationship between an explanatory variable and a response variable. We see deviations from this pattern by looking at the scatter of the data points about the regression line. The vertical distances from the points to the least-squares regression line are as small as possible, in the sense that they have the smallest possible sum of squares. Because they represent "left-over" variation in the response after fitting the regression line, these distances are called *residuals*.

> **RESIDUALS**
>
> A **residual** is the difference between an observed value of the response variable and the value predicted by the regression line. That is, a residual is the prediction error that remains after we have chosen the regression line:
>
> $$\text{residual} = \text{observed } y - \text{predicted } y$$
> $$= y - \hat{y}$$

EXAMPLE 5.5 I feel your pain

"Empathy" means being able to understand what others feel. To see how the brain expresses empathy, researchers recruited 16 couples in their midtwenties who were married or had been dating for at least two years. They zapped the man's hand with an electrode while the woman watched, and measured the activity in several parts of the

Photodisc Green/Getty Images

woman's brain that would respond to her own pain. Brain activity was recorded as a fraction of the activity observed when the woman herself was zapped with the electrode. The women also completed a psychological test that measures empathy. Will women who score higher in empathy respond more strongly when their partner has a painful experience? Here are data for one brain region:[3]

Subject	1	2	3	4	5	6	7	8
Empathy score	38	53	41	55	56	61	62	48
Brain activity	−0.120	0.392	0.005	0.369	0.016	0.415	0.107	0.506
Subject	9	10	11	12	13	14	15	16
Empathy score	43	47	56	65	19	61	32	105
Brain activity	0.153	0.745	0.255	0.574	0.210	0.722	0.358	0.779

Figure 5.5 is a scatterplot, with empathy score as the explanatory variable x and brain activity as the response variable y. The plot shows a positive association. That is, women who score higher in empathy do indeed react more strongly to their partner's pain. The overall pattern is moderately linear, correlation $r = 0.515$.

The line on the plot is the least-squares regression line of brain activity on empathy score. Its equation is

$$\hat{y} = -0.0578 + 0.00761x$$

For Subject 1, with empathy score 38, we predict

$$\hat{y} = -0.0578 + (0.00761)(38) = 0.231$$

This subject's actual brain activity level was −0.120. The residual is

$$\text{residual} = \text{observed } y - \text{predicted } y$$

$$= -0.120 - 0.231 = -0.351$$

The residual is negative because the data point lies below the regression line. The dashed line segment in Figure 5.5 shows the size of the residual. ∎

There is a residual for each data point. Finding the residuals is a bit unpleasant because you must first find the predicted response for every x. Software or a graphing calculator gives you the residuals all at once. Here are the 16 residuals for the empathy study data, from software:

```
residuals:
-0.3515  -0.2494  -0.3526  -0.3072  -0.1166  -0.1136  0.1231  0.1721
 0.0463   0.0080   0.0084   0.1983   0.4449   0.1369  0.3154  0.0374
```

Because the residuals show how far the data fall from our regression line, examining the residuals helps assess how well the line describes the data. Although residuals can be calculated from any curve fitted to the data, the residuals from the least-squares line have a special property: **the mean of the least-squares residuals is always zero.**

Compare the scatterplot in Figure 5.5 with the *residual plot* for the same data in Figure 5.6. The horizontal line at zero in Figure 5.6 helps orient us. This "residual = 0" line corresponds to the regression line in Figure 5.5.

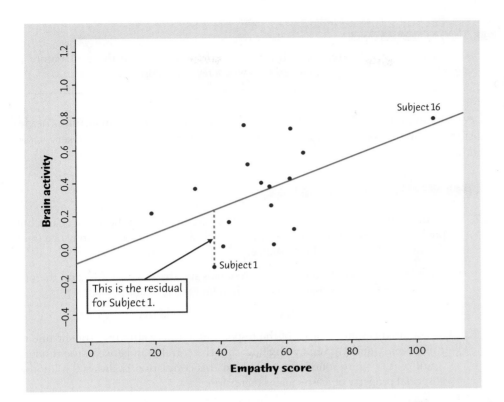

FIGURE 5.5

Scatterplot of activity in a region of the brain that responds to pain versus score on a test of empathy, for Example 5.5. Brain activity is measured as the subject watches her partner experience pain. The line is the least-squares regression line.

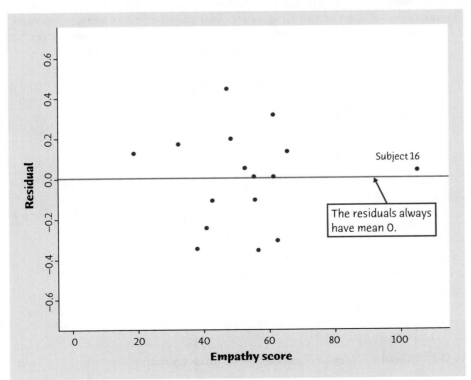

FIGURE 5.6

Residual plot for the data shown in Figure 5.5. The horizontal line at zero residual corresponds to the regression line in Figure 5.5.

> | **RESIDUAL PLOTS** |
> | :--- |
> | A **residual plot** is a scatterplot of the regression residuals against the explanatory variable. Residual plots help us assess how well a regression line fits the data. |

A residual plot in effect turns the regression line horizontal. It magnifies the deviations of the points from the line and makes it easier to see unusual observations and patterns.

APPLY YOUR KNOWLEDGE

5.7 Residuals by hand. In Exercise 5.3 you found the equation of the least-squares line for predicting the number of days y until gorillas in a social group begin to die in an Ebola virus epidemic from the "distance" x from the first group infected.

(a) Use the equation to obtain the 6 residuals step-by-step. That is, find the prediction $\hat{y}$ for each observation and then find the residual $y - \hat{y}$.

(b) Check that (up to roundoff error) the residuals add to 0.

(c) The residuals are the part of the response y left over after the straight-line tie between y and x is removed. Show that the correlation between the residuals and x is 0 (up to roundoff error). That this correlation is always 0 is another special property of least-squares regression.

5.8 Does fast driving waste fuel? Exercise 4.8 (page 102) gives data on the fuel consumption y of a car at various speeds x. Fuel consumption is measured in liters of gasoline per 100 kilometers driven, and speed is measured in kilometers per hour. Software tells us that the equation of the least-squares regression line is

$$\hat{y} = 11.058 - 0.01466x$$

Using this equation we can add the residuals to the original data:

Speed	10	20	30	40	50	60	70	80
Fuel	21.00	13.00	10.00	8.00	7.00	5.90	6.30	6.95
Residual	10.09	2.24	−0.62	−2.47	−3.33	−4.28	−3.73	−2.94
Speed	90	100	110	120	130	140	150	
Fuel	7.57	8.27	9.03	9.87	10.79	11.77	12.83	
Residual	−2.17	−1.32	−0.42	0.57	1.64	2.76	3.97	

(a) Make a scatterplot of the observations and draw the regression line on your plot.

(b) Would you use the regression line to predict y from x? Explain your answer.

(c)   Verify the value of the first residual, for $x = 10$. Verify that the residuals have sum zero (up to roundoff error).

(d)   Make a plot of the residuals against the values of x. Draw a horizontal line at height zero on your plot. How does the pattern of the residuals about this line compare with the pattern of the data points about the regression line in your scatterplot from (a)?

Influential observations

Figures 5.5 and 5.6 show one unusual observation. Subject 16 is an outlier in the x direction, with empathy score 40 points higher than any other subject. Because of its extreme position on the empathy scale, this point has a strong influence on the correlation. Dropping Subject 16 reduces the correlation from $r = 0.515$ to $r = 0.331$. You can see that this point extends the linear pattern in Figure 5.5 and so increases the correlation. We say that Subject 16 is *influential* for calculating the correlation.

INFLUENTIAL OBSERVATIONS

An observation is **influential** for a statistical calculation if removing it would markedly change the result of the calculation.

The result of a statistical calculation may be of little practical use if it depends strongly on a few influential observations.

Points that are outliers in either the x or y direction of a scatterplot are often influential for the correlation. Points that are outliers in the x direction are often influential for the least-squares regression line.

EXAMPLE 5.6   An influential observation?

Subject 16 in Example 5.5 is influential for the correlation between empathy score and brain activity because removing it reduces r from 0.515 to 0.331. Calculating that $r = 0.515$ is not a very useful description of the data, because the value depends so strongly on just one of the 16 subjects.

   Is this observation also influential for the least-squares line? Figure 5.7 shows that it is not. The regression line calculated without Subject 16 (dashed) differs little from the line that uses all of the observations (solid). The reason that the outlier has little influence on the regression line is that it lies close to the dashed regression line calculated from the other observations. ■

To see why points that are outliers in the x direction are often influential for regression, let's try an experiment. Suppose that Subject 16's point in the scatterplot moves straight down. What happens to the regression line? Figure 5.8 gives the

FIGURE 5.7

Subject 16 is an outlier in the *x* direction. The outlier is not influential for least-squares regression, because removing it moves the regression line only a little.

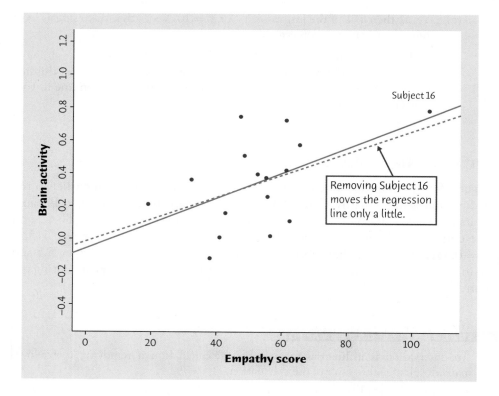

FIGURE 5.8

An outlier in the *x* direction pulls the least-squares line to itself because there are no other observations with similar values of *x* to hold the line in place. When the outlier moves down, the original regression line chases it down. The original regression line is solid, and the final position of the regression line is dashed.

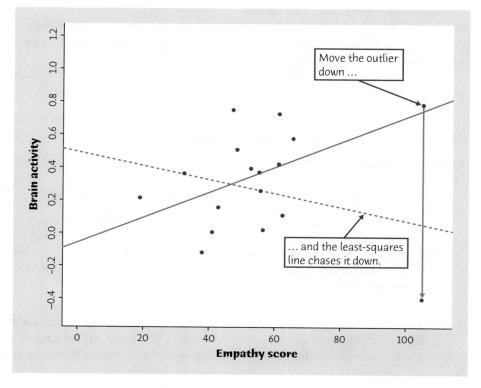

answer. The dashed line is the regression line with the outlier in its new, lower position. Because there are no other points with similar x-values, the line chases the outlier. The *Correlation and Regression* applet allows you to try this experiment yourself—see Exercise 5.9. *An outlier in x pulls the least-squares line toward itself. If the outlier does not lie close to the line calculated from the other observations, it will be influential.*

We did not need the distinction between outliers and influential observations in Chapter 2. A single high salary that pulls up the mean salary $\overline{x}$ for a group of workers is an outlier because it lies far above the other salaries. It is also influential, because the mean changes when it is removed. In the regression setting, however, not all outliers are influential.

APPLY YOUR KNOWLEDGE

5.9 Influence in regression. The *Correlation and Regression* applet allows you to animate Figure 5.8. Click to create a group of 10 points in the lower-left corner of the scatterplot with a strong straight-line pattern (correlation about 0.9). Click the "Show least-squares line" box to display the regression line.

  (a)   Add one point at the upper right that is far from the other 10 points but exactly on the regression line. Why does this outlier have no effect on the line even though it changes the correlation?

  (b)   Now use the mouse to drag this last point straight down. You see that one end of the least-squares line chases this single point, while the other end remains near the middle of the original group of 10. What makes the last point so influential?

5.10 Do heavier people burn more energy? Return to the data of Exercise 5.4 (page 132) on body mass and metabolic rate. We will use these data to illustrate influence.

  (a)   Make a scatterplot of the data that is suitable for predicting metabolic rate from body mass, with two new points added. Point A: mass 42 kilograms, metabolic rate 1500 calories. Point B: mass 70 kilograms, metabolic rate 1400 calories. In which direction is each of these points an outlier?

  (b)   Add three least-squares regression lines to your plot: for the original 12 women, for the original women plus Point A, and for the original women plus Point B. Which new point is more influential for the regression line? Explain in simple language why each new point moves the line in the way your graph shows.

5.11 Outsourcing by airlines. Exercise 4.5 (page 99) gives data for 14 airlines on the percent of major maintenance outsourced and the percent of flight delays blamed on the airline.

  (a)   Make a scatterplot with outsourcing percent as x and delay percent as y. Hawaiian Airlines is a high outlier in the y direction. Because several other airlines have similar values of x, the influence of this outlier is unclear without actual calculation.

(b)   Find the correlation r with and without Hawaiian Airlines. How influential is the outlier for correlation?

(c)   Find the least-squares line for predicting y from x with and without Hawaiian Airlines. Draw both lines on your scatterplot. Use both lines to predict the percent of delays blamed on an airline that has outsourced 76% of its major maintenance. How influential is the outlier for the least-squares line?

Cautions about correlation and regression

Correlation and regression are powerful tools for describing the relationship between two variables. When you use these tools, you must be aware of their limitations. You already know that

■ *Correlation and regression lines describe only linear relationships.* You can do the calculations for any relationship between two quantitative variables, but the results are useful only if the scatterplot shows a linear pattern.

■ *Correlation and least-squares regression lines are not resistant.* Always plot your data and look for observations that may be influential.

Here are two more things to keep in mind when you use correlation and regression.

Beware extrapolation. Suppose that you have data on a child's growth between 3 and 8 years of age. You find a strong linear relationship between age x and height y. If you fit a regression line to these data and use it to predict height at age 25 years, you will predict that the child will be 8 feet tall. Growth slows down and then stops at maturity, so extending the straight line to adult ages is foolish. *Few relationships are linear for all values of x. Don't make predictions far outside the range of x that actually appears in your data.*

> **EXTRAPOLATION**
>
> **Extrapolation** is the use of a regression line for prediction far outside the range of values of the explanatory variable x that you used to obtain the line. Such predictions are often not accurate.

Beware the lurking variable. Another caution is even more important: *the relationship between two variables can often be understood only by taking other variables into account. Lurking variables* can make a correlation or regression misleading.

> **LURKING VARIABLE**
>
> A **lurking variable** is a variable that is not among the explanatory or response variables in a study and yet may influence the interpretation of relationships among those variables.

You should always think about possible lurking variables before you draw conclusions based on correlation or regression.

EXAMPLE 5.7 **Magic Mozart?**

The Kalamazoo (Michigan) Symphony once advertised a "Mozart for Minors" program with this statement: "Question: Which students scored 51 points higher in verbal skills and 39 points higher in math? Answer: Students who had experience in music."[4]

We could as well answer "Students who played soccer." Why? Children with prosperous and well-educated parents are more likely than poorer children to have experience with music and also to play soccer. They are also likely to attend good schools, get good health care, and be encouraged to study hard. These advantages lead to high test scores. Family background is a lurking variable that explains why test scores are related to experience with music. ■

APPLY YOUR KNOWLEDGE

5.12 The endangered manatee. Table 4.1 gives 30 years of data on boats registered in Florida and manatees killed by boats. Figure 4.2 (page 102) shows a strong positive linear relationship. The correlation is $r = 0.953$.

  (a) Find the equation of the least-squares line for predicting manatees killed from thousands of boats registered. Because the linear pattern is so strong, we expect predictions from this line to be quite accurate—but only if conditions in Florida remain similar to those of the past 30 years.

  (b) In 2007, there were 1,027,000 boats registered in Florida. Predict the number of manatees killed by boats in 2007. Explain why we can trust this prediction.

  (c) Predict manatee deaths if there were *no* boats registered in Florida. Explain why the predicted count of deaths is impossible. (We use $x = 0$ to find the intercept of the regression line, but unless the explanatory variable x actually takes values near 0, prediction for $x = 0$ is an example of extrapolation.)

5.13 Is math the key to success in college? A College Board study of 15,941 high school graduates found a strong correlation between how much math minority students took in high school and their later success in college. News articles quoted the head of the College Board as saying that "math is the gatekeeper for success in college."[5] Maybe so, but we should also think about lurking variables. What might lead minority students to take more or fewer high school math courses? Would these same factors influence success in college?

Do left-handers die early?

Yes, said a study of 1000 deaths in California. Left-handed people died at an average age of 66 years; right-handers, at 75 years of age. Should left-handed people fear an early death? No—the lurking variable has struck again. Older people grew up in an era when many natural left-handers were forced to use their right hands. So right-handers are more common among older people, and left-handers are more common among the young. When we look at deaths, the left-handers who die are younger on the average because left-handers in general are younger. Mystery solved.

Association does not imply causation

Thinking about lurking variables leads to the most important caution about correlation and regression. When we study the relationship between two variables, we often hope to show that changes in the explanatory variable *cause* changes in the response variable. *A strong association between two variables is not enough to draw conclusions about cause and effect.* Sometimes an observed association really does reflect cause and effect. A household that heats with natural gas uses more gas in colder months because cold weather requires burning more gas to stay warm. In other cases, an association is explained by lurking variables, and the conclusion that x causes y is either wrong or not proved.

EXAMPLE 5.8   **Does having more cars make you live longer?**

A serious study once found that people with two cars live longer than people who own only one car.[6] Owning three cars is even better, and so on. There is a substantial positive correlation between number of cars x and length of life y.

The basic meaning of causation is that by changing x we can bring about a change in y. Could we lengthen our lives by buying more cars? No. The study used number of cars as a quick indicator of affluence. Well-off people tend to have more cars. They also tend to live longer, probably because they are better educated, take better care of themselves, and get better medical care. The cars have nothing to do with it. There is no cause-and-effect tie between number of cars and length of life. ■

Correlations such as that in Example 5.8 are sometimes called "nonsense correlations." The correlation is real. What is nonsense is the conclusion that changing one of the variables causes changes in the other. A lurking variable—such as personal affluence in Example 5.8—that influences both x and y can create a high correlation even though there is no direct connection between x and y.

ASSOCIATION DOES NOT IMPLY CAUSATION

An association between an explanatory variable x and a response variable y, even if it is very strong, is not by itself good evidence that changes in x actually cause changes in y.

The Super Bowl effect

The Super Bowl is the most-watched TV broadcast in the United States. Data show that on Super Bowl Sunday we consume 3 times as many potato chips as on an average day, and 17 times as much beer. What's more, the number of fatal traffic accidents goes up in the hours after the game ends. Could that be celebration? Or catching up with tasks left undone? Or maybe it's the beer.

EXAMPLE 5.9   **Overweight mothers, overweight daughters**

Overweight parents tend to have overweight children. The results of a study of Mexican American girls aged 9 to 12 years are typical. The investigators measured body mass index (BMI), a measure of weight relative to height, for both the girls and their mothers. People with high BMI are overweight. The correlation between the BMI of daughters and the BMI of their mothers was $r = 0.506$.[7]

Body type is in part determined by heredity. Daughters inherit half their genes from their mothers. There is therefore a direct cause-and-effect link between the BMI of mothers and daughters. But perhaps mothers who are overweight also set an example of little exercise, poor eating habits, and lots of television. Their daughters may pick

Colin Young-Wolff/Alamy

up these habits, so the influence of heredity is mixed up with influences from the girls' environment. Both contribute to the mother-daughter correlation. ■

The lesson of Example 5.9 is more subtle than just "association does not imply causation." *Even when direct causation is present, it may not be the whole explanation for a correlation.* You must still worry about lurking variables. Careful statistical studies try to anticipate lurking variables and measure them. The mother-daughter study did measure TV viewing, exercise, and diet. Elaborate statistical analysis can remove the effects of these variables to come closer to the direct effect of mother's BMI on daughter's BMI. This remains a second-best approach to causation. The best way to get good evidence that x causes y is to do an **experiment** in which we change x and keep lurking variables under control. We will discuss experiments in Chapter 9.

experiment

When experiments cannot be done, explaining an observed association can be difficult and controversial. Many of the sharpest disputes in which statistics plays a role involve questions of causation that cannot be settled by experiment. Do gun control laws reduce violent crime? Does using cell phones cause brain tumors? Has increased free trade widened the gap between the incomes of more educated and less educated American workers? All of these questions have become public issues. All concern associations among variables. And all have this in common: they try to pinpoint cause and effect in a setting involving complex relations among many interacting variables.

EXAMPLE 5.10 Does smoking cause lung cancer?

Despite the difficulties, it is sometimes possible to build a strong case for causation in the absence of experiments. The evidence that smoking causes lung cancer is about as strong as nonexperimental evidence can be.

Doctors had long observed that most lung cancer patients were smokers. Comparison of smokers and "similar" nonsmokers showed a very strong association between smoking and death from lung cancer. Could the association be explained by lurking variables? Might there be, for example, a genetic factor that predisposes people both to nicotine addiction and to lung cancer? Smoking and lung cancer would then be positively associated even if smoking had no direct effect on the lungs. How were these objections overcome? ■

James Leynse/CORBIS

Let's answer this question in general terms: what are the criteria for establishing causation when we cannot do an experiment?

■ *The association is strong.* The association between smoking and lung cancer is very strong.

■ *The association is consistent.* Many studies of different kinds of people in many countries link smoking to lung cancer. That reduces the chance that a lurking variable specific to one group or one study explains the association.

■ *Higher doses are associated with stronger responses.* People who smoke more cigarettes per day or who smoke over a longer period get lung cancer more often. People who stop smoking reduce their risk.

■ *The alleged cause precedes the effect in time.* Lung cancer develops after years of smoking. The number of men dying of lung cancer rose as smoking became more common, with a lag of about 30 years. Lung cancer kills more men than any other form of cancer. Lung cancer was rare among women until women began to smoke. Lung cancer in women rose along with smoking, again with a lag of about 30 years, and has now passed breast cancer as the leading cause of cancer death among women.

■ *The alleged cause is plausible.* Experiments with animals show that tars from cigarette smoke do cause cancer.

Medical authorities do not hesitate to say that smoking causes lung cancer. The U.S. Surgeon General has long stated that cigarette smoking is "the largest avoidable cause of death and disability in the United States."[8] The evidence for causation is overwhelming—but it is not as strong as the evidence provided by well-designed experiments.

APPLY YOUR KNOWLEDGE

5.14 Another reason not to smoke? A stop-smoking booklet says, "Children of mothers who smoked during pregnancy scored nine points lower on intelligence tests at ages three and four than children of nonsmokers." Suggest some lurking variables that may help explain the association between smoking during pregnancy and children's later test scores. The association by itself is not good evidence that mothers' smoking *causes* lower scores.

5.15 Education and income. There is a strong positive association between workers' education and their income. For example, the Census Bureau reports that the median income of young adults (ages 25 to 34) who work full-time increases from $19,956 for those with less than a ninth-grade education, to $29,225 for high school graduates, to $44,125 for holders of a bachelor's degree, and on up for yet more education. In part, this association reflects causation—education helps people qualify for better jobs. Suggest several lurking variables that also contribute. (Ask yourself what kinds of people tend to get more education.)

5.16 To earn more, get married? Data show that men who are married, and also divorced or widowed men, earn quite a bit more than men the same age who have never been married. This does not mean that a man can raise his income by getting married, because men who have never been married are different from married men in many ways other than marital status. Suggest several lurking variables that might help explain the association between marital status and income.

CHAPTER 5 SUMMARY

■ A **regression line** is a straight line that describes how a response variable y changes as an explanatory variable x changes. You can use a regression line to **predict** the value of y for any value of x by substituting this x into the equation of the line.

- The **slope** b of a regression line $\hat{y} = a + bx$ is the rate at which the predicted response $\hat{y}$ changes along the line as the explanatory variable x changes. Specifically, b is the change in $\hat{y}$ when x increases by 1.

- The **intercept** a of a regression line $\hat{y} = a + bx$ is the predicted response $\hat{y}$ when the explanatory variable $x = 0$. This prediction is of no statistical interest unless x can actually take values near 0.

- The most common method of fitting a line to a scatterplot is least squares. The **least-squares regression line** is the straight line $\hat{y} = a + bx$ that minimizes the sum of the squares of the vertical distances of the observed points from the line.

- The least-squares regression line of y on x is the line with slope $b = r s_y / s_x$ and intercept $a = \bar{y} - b\bar{x}$. This line always passes through the point $(\bar{x}, \bar{y})$.

- **Correlation and regression** are closely connected. The correlation r is the slope of the least-squares regression line when we measure both x and y in standardized units. The **square of the correlation** r^2 is the fraction of the variation in one variable that is explained by least-squares regression on the other variable.

- Correlation and regression must be **interpreted with caution. Plot the data** to be sure the relationship is roughly linear and to detect outliers and influential observations. A plot of the **residuals** makes these effects easier to see.

- Look for **influential observations,** individual points that substantially change the correlation or the regression line. Outliers in the x direction are often influential for the regression line.

- Avoid **extrapolation,** the use of a regression line for prediction for values of the explanatory variable far outside the range of the data from which the line was calculated.

- **Lurking variables** may explain the relationship between the explanatory and response variables. Correlation and regression can be misleading if you ignore important lurking variables.

- Most of all, be careful not to conclude that there is a cause-and-effect relationship between two variables just because they are strongly associated. **High correlation does not imply causation.** The best evidence that an association is due to causation comes from an **experiment** in which the explanatory variable is directly changed and other influences on the response are controlled.

C H E C K   Y O U R   S K I L L S

5.17 Figure 5.9 is a scatterplot of reading test scores against IQ test scores for 14 fifth-grade children. The line is the least-squares regression line for predicting reading score from IQ score. If another child in this class has IQ score 110, you predict the reading score to be close to

(a) 50.   (b) 60.   (c) 70.

FIGURE 5.9

IQ test scores and reading test scores for 14 children, for Exercises 5.17 and 5.18.

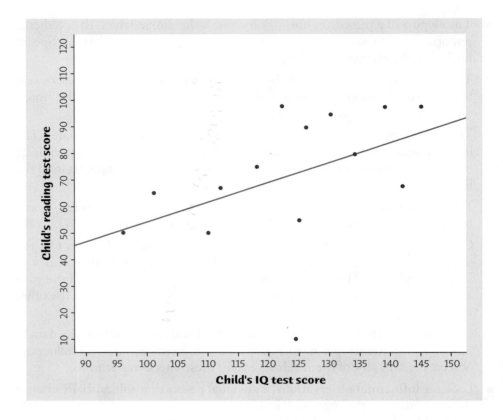

5.18 The slope of the line in Figure 5.9 is closest to

(a) −1.   (b) 0.   (c) 1.

5.19 The points on a scatterplot lie close to the line whose equation is $y = 4 - 3x$. The slope of this line is

(a) 4.   (b) 3.   (c) −3.

5.20 Fred keeps his savings in his mattress. He began with $500 from his mother and adds $100 each year. His total savings y after x years are given by the equation

(a) $y = 500 + 100x$.   (b) $y = 100 + 500x$.   (c) $y = 500 + x$.

5.21 Smokers don't live as long (on the average) as nonsmokers, and heavy smokers don't live as long as light smokers. You regress the age at death of a group of male smokers on the number of packs per day they smoked. The slope of your regression line

(a) will be greater than 0.

(b) will be less than 0.

(c) Can't tell without seeing the data.

5.22 Measurements on young children in Mumbai, India, found this least-squares line for predicting height y from armspan x:[9]

$$\hat{y} = 6.4 + 0.93x$$

All measurements are in centimeters (cm). How much on the average does height increase for each additional centimeter of armspan?

(a) 0.93 cm   (b) 6.4 cm   (c) 7.33 cm

5.23 According to the regression line in Exercise 5.22, the predicted height of a child with armspan 100 cm is about

(a) 106.4 cm.   (b) 99.4 cm.   (c) 93 cm.

5.24 By looking at the equation of the least-squares regression line in Exercise 5.22, you can see that the correlation between height and armspan is

(a) greater than zero.

(b) less than zero.

(c) Can't tell without seeing the data.

5.25 In addition to the regression line in Exercise 5.22, the report on the Mumbai measurements says that $r^2 = 0.95$. This suggests that

(a) although armspan and height are correlated, armspan does not predict height very accurately.

(b) height increases by $\sqrt{0.95} = 0.97$ cm for each additional centimeter of armspan.

(c) prediction of height from armspan will be quite accurate.

5.26 Because elderly people may have difficulty standing to have their heights measured, a study looked at predicting overall height from height to the knee. Here are data (in centimeters) for five elderly men:

Knee height x	57.7	47.4	43.5	44.8	55.2
Height y	192.1	153.3	146.4	162.7	169.1

Use your calculator or software: what is the equation of the least-squares regression line for predicting height from knee height?

(a) $\hat{y} = 2.4 + 44.1x$ (b) $\hat{y} = 44.1 + 2.4x$ (c) $\hat{y} = -2.5 + 0.32x$

C H A P T E R 5 E X E R C I S E S

5.27 Penguins diving. A study of king penguins looked for a relationship between how deep the penguins dive to seek food and how long they stay underwater.[10] For all but the shallowest dives, there is a linear relationship that is different for different penguins. The study report gives a scatterplot for one penguin titled "The relation of dive duration (DD) to depth (D)." Duration DD is measured in minutes and depth D is in meters. The report then says, "The regression equation for this bird is: DD = 2.69 + 0.0138D."

Paul A. Souders/CORBIS

(a) What is the slope of the regression line? Explain in specific language what this slope says about this penguin's dives.

(b) According to the regression line, how long does a typical dive to a depth of 200 meters last?

(c) The dives varied from 40 meters to 300 meters in depth. Plot the regression line from $x = 40$ to $x = 300$.

5.28 Measuring water quality. Biochemical oxygen demand (BOD) measures organic pollutants in water by measuring the amount of oxygen consumed by microorganisms that break down these compounds. BOD is hard to measure accurately. Total organic carbon (TOC) is easy to measure, so it is common to measure TOC and use regression to predict BOD. A typical regression equation for water entering a municipal treatment plant is[11]

$$BOD = -55.43 + 1.507 \, TOC$$

Both BOD and TOC are measured in milligrams per liter of water.

(a) What does the slope of this line say about the relationship between BOD and TOC?

(b) What is the predicted BOD when TOC = 0? Values of BOD less than 0 are impossible. Why do you think the prediction gives an impossible value?

5.29 Does social rejection hurt? Exercise 4.45 (page 122) gives data from a study that shows that social exclusion causes "real pain." That is, activity in an area of the brain that responds to physical pain goes up as distress from social exclusion goes up. A scatterplot shows a moderately strong linear relationship. Figure 5.10 shows Minitab regression output for these data.

(a) What is the equation of the least-squares regression line for predicting brain activity from social distress score? Use the equation to predict brain activity for social distress score 2.0.

FIGURE 5.10

Minitab regression output for a study of the effects of social rejection on brain activity, for Exercise 5.29.

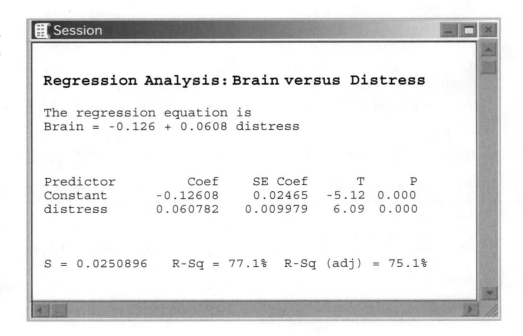

(b) What percent of the variation in brain activity among these subjects is explained by the straight-line relationship with social distress score?

(c) Use the information in Figure 5.10 to find the correlation r between social distress score and brain activity. How do you know whether the sign of r is $+$ or $-$?

5.30 Merlins breeding. Exercise 4.44 (page 122) gives data on the number of breeding pairs of merlins in an isolated area in each of nine years and the percent of males who returned the next year. The data show that the percent returning is lower after successful breeding seasons and that the relationship is roughly linear. Figure 5.11 shows Minitab regression output for these data.

(a) What is the equation of the least-squares regression line for predicting the percent of males that return from the number of breeding pairs? Use the equation to predict the percent of returning males after a season with 30 breeding pairs.

(b) What percent of the year-to-year variation in percent of returning males is explained by the straight-line relationship with number of breeding pairs the previous year?

(c) Use the information in Figure 5.11 to find the correlation r between percent of males that return and number of breeding pairs. How do you know whether the sign of r is $+$ or $-$?

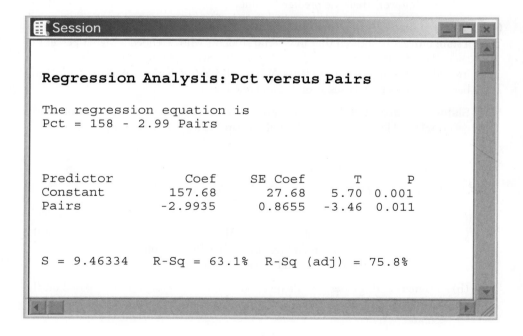

FIGURE 5.11
Minitab regression output for a study of how breeding success affects survival in birds, for Exercise 5.30.

5.31 Husbands and wives. The mean height of American women in their twenties is about 64 inches, and the standard deviation is about 2.7 inches. The mean height of men the same age is about 69.3 inches, with standard deviation about 2.8 inches. Suppose that the correlation between the heights of husbands and wives is about $r = 0.5$.

(a)   What are the slope and intercept of the regression line of the husband's height on the wife's height in young couples?

(b)   Draw a graph of this regression line for heights of wives between 56 and 72 inches. Predict the height of the husband of a woman who is 67 inches tall and plot the wife's height and predicted husband's height on your graph.

(c)   You don't expect this prediction for a single couple to be very accurate. Why not?

5.32 What's my grade? In Professor Friedman's economics course the correlation between the students' total scores prior to the final examination and their final-examination scores is $r = 0.6$. The pre-exam totals for all students in the course have mean 280 and standard deviation 30. The final-exam scores have mean 75 and standard deviation 8. Professor Friedman has lost Julie's final exam but knows that her total before the exam was 300. He decides to predict her final-exam score from her pre-exam total.

(a)   What is the slope of the least-squares regression line of final-exam scores on pre-exam total scores in this course? What is the intercept?

(b)   Use the regression line to predict Julie's final-exam score.

(c)   Julie doesn't think this method accurately predicts how well she did on the final exam. Use r^2 to argue that her actual score could have been much higher (or much lower) than the predicted value.

5.33 Going to class. A study of class attendance and grades among first-year students at a state university showed that in general students who attended a higher percent of their classes earned higher grades. Class attendance explained 16% of the variation in grade index among the students. What is the numerical value of the correlation between percent of classes attended and grade index?

5.34 Sisters and brothers. How strongly do physical characteristics of sisters and brothers correlate? Here are data on the heights (in inches) of 11 adult pairs:[12]

Brother	71	68	66	67	70	71	70	73	72	65	66
Sister	69	64	65	63	65	62	65	64	66	59	62

(a)   Use your calculator or software to find the correlation and the equation of the least-squares line for predicting sister's height from brother's height. Make a scatterplot of the data and add the regression line to your plot.

(b)   Damien is 70 inches tall. Predict the height of his sister Tonya. Based on the scatterplot and the correlation r, do you expect your prediction to be very accurate? Why?

5.35 Keeping water clean. Keeping water supplies clean requires regular measurement of levels of pollutants. The measurements are indirect—a typical analysis involves forming a dye by a chemical reaction with the dissolved pollutant, then passing light through the solution and measuring its "absorbence." To calibrate such measurements, the laboratory measures known standard solutions and uses regression to relate absorbence and pollutant concentration. This is usually done every day. Here is one series of data on the absorbence for different levels of nitrates. Nitrates are measured in milligrams per liter of water.[13]

Nitrates	50	50	100	200	400	800	1200	1600	2000	2000
Absorbence	7.0	7.5	12.8	24.0	47.0	93.0	138.0	183.0	230.0	226.0

(a) Chemical theory says that these data should lie on a straight line. If the correlation is not at least 0.997, something went wrong and the calibration procedure is repeated. Plot the data and find the correlation. Must the calibration be done again?

(b) The calibration process sets nitrate level and measures absorbence. The linear relationship that results is used to estimate the nitrate level in water from a measurement of absorbence. What is the equation of the line used to estimate nitrate level? What is the estimated nitrate level in a water specimen with absorbence 40?

(c) Do you expect estimates of nitrate level from absorbence to be quite accurate? Why?

5.36 Sparrowhawk colonies. One of nature's patterns connects the percent of adult birds in a colony that return from the previous year and the number of new adults that join the colony. Here are data for 13 colonies of sparrowhawks:[14]

Percent return x	74	66	81	52	73	62	52	45	62	46	60	46	38
New adults y	5	6	8	11	12	15	16	17	18	18	19	20	20

Martin B. Withers/Frank Lane Picture Agency/CORBIS

You saw in Exercise 4.28 that there is a moderately strong linear relationship, correlation $r = -0.748$.

(a) Find the least-squares regression line for predicting y from x. Make a scatterplot and draw your line on the plot.

(b) Explain in words what the slope of the regression line tells us.

(c) An ecologist uses the line, based on 13 colonies, to predict how many new birds will join another colony, to which 60% of the adults from the previous year return. What is the prediction?

5.37 Our brains don't like losses. Exercise 4.29 (page 116) describes an experiment that showed a linear relationship between how sensitive people are to monetary losses ("behavioral loss aversion") and activity in one part of their brains ("neural loss aversion").

(a) Make a scatterplot with neural loss aversion as x and behavioral loss aversion as y. One point is a high outlier in both the x and y directions.

(b) Find the least-squares line for predicting y from x, *leaving out the outlier,* and add the line to your plot.

(c) The outlier lies very close to your regression line. Looking at the plot, you now expect that adding the outlier will increase the correlation but will have little effect on the least-squares line. Explain why.

(d)   Find the correlation and the equation of the least-squares line with and without the outlier. Your results verify the expectations from (c).

5.38 Always plot your data! Table 5.1 presents four sets of data prepared by the statistician Frank Anscombe to illustrate the dangers of calculating without first plotting the data.[15]

(a)   Without making scatterplots, find the correlation and the least-squares regression line for all four data sets. What do you notice? Use the regression line to predict y for $x = 10$.

(b)   Make a scatterplot for each of the data sets and add the regression line to each plot.

(c)   In which of the four cases would you be willing to use the regression line to describe the dependence of y on x? Explain your answer in each case.

TABLE 5.1 Four data sets for exploring correlation and regression

Data Set A

x	10	8	13	9	11	14	6	4	12	7	5
y	8.04	6.95	7.58	8.81	8.33	9.96	7.24	4.26	10.84	4.82	5.68

Data Set B

x	10	8	13	9	11	14	6	4	12	7	5
y	9.14	8.14	8.74	8.77	9.26	8.10	6.13	3.10	9.13	7.26	4.74

Data Set C

x	10	8	13	9	11	14	6	4	12	7	5
y	7.46	6.77	12.74	7.11	7.81	8.84	6.08	5.39	8.15	6.42	5.73

Data Set D

x	8	8	8	8	8	8	8	8	8	8	19
y	6.58	5.76	7.71	8.84	8.47	7.04	5.25	5.56	7.91	6.89	12.50

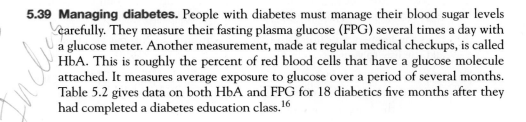

5.39 Managing diabetes. People with diabetes must manage their blood sugar levels carefully. They measure their fasting plasma glucose (FPG) several times a day with a glucose meter. Another measurement, made at regular medical checkups, is called HbA. This is roughly the percent of red blood cells that have a glucose molecule attached. It measures average exposure to glucose over a period of several months. Table 5.2 gives data on both HbA and FPG for 18 diabetics five months after they had completed a diabetes education class.[16]

TABLE 5.2 Two measures of glucose level in diabetics

SUBJECT	HbA (%)	FPG (mg/ml)	SUBJECT	HbA (%)	FPG (mg/ml)	SUBJECT	HbA (%)	FPG (mg/ml)
1	6.1	141	7	7.5	96	13	10.6	103
2	6.3	158	8	7.7	78	14	10.7	172
3	6.4	112	9	7.9	148	15	10.7	359
4	6.8	153	10	8.7	172	16	11.2	145
5	7.0	134	11	9.4	200	17	13.7	147
6	7.1	95	12	10.4	271	18	19.3	255

(a) Make a scatterplot with HbA as the explanatory variable. There is a positive linear relationship, but it is surprisingly weak.

(b) Subject 15 is an outlier in the y direction. Subject 18 is an outlier in the x direction. Find the correlation for all 18 subjects, for all except Subject 15, and for all except Subject 18. Are either or both of these subjects influential for the correlation? Explain in simple language why r changes in opposite directions when we remove each of these points.

5.40 The effect of changing units. The equation of a regression line, unlike the correlation, depends on the units we use to measure the explanatory and response variables. Here are data on knee height and overall height (in centimeters) for five elderly men:

Knee height x	57.7	47.4	43.5	44.8	55.2
Height y	192.1	153.3	146.4	162.7	169.1

(a) Find the equation of the regression line for predicting overall height in centimeters from knee height in centimeters.

(b) A mad scientist decides to measure knee height in millimeters and height in meters. The same data in these units are

Knee height x	577	474	435	448	552
Height y	1.921	1.533	1.464	1.627	1.691

Find the equation of the regression line for predicting overall height in meters from knee height in millimeters.

(c) Use both lines to predict the overall height of a man whose knee height is 50 centimeters, which is the same as 500 millimeters. Use the fact that there are 100 centimeters in a meter to show that the two predictions are the same (up to roundoff error).

5.41 Managing diabetes, continued. Add three regression lines for predicting FPG from HbA to your scatterplot from Exercise 5.39: for all 18 subjects, for all except Subject 15, and for all except Subject 18. Is either Subject 15 or Subject 18 strongly influential for the least-squares line? Explain in simple language what features of the scatterplot explain the degree of influence.

5.42 Do artificial sweeteners cause weight gain? People who use artificial sweeteners in place of sugar tend to be heavier than people who use sugar. Does this mean that artificial sweeteners cause weight gain? Give a more plausible explanation for this association.

5.43 Learning online. Many colleges offer online versions of courses that are also taught in the classroom. It often happens that the students who enroll in the online version do better than the classroom students on the course exams. This does not show that online instruction is more effective than classroom teaching, because the people who sign up for online courses are often quite different from the classroom students. Suggest some differences between online and classroom students that might explain why online students do better.

5.44 Grade inflation and the SAT. The effect of a lurking variable can be surprising when individuals are divided into groups. In recent years, the mean SAT score of all high school seniors has increased. But the mean SAT score has decreased for students at each level of high school grades (A, B, C, and so on). Explain how grade inflation in high school (the lurking variable) can account for this pattern.

5.45 Workers' incomes. Here is another example of the group effect cautioned about in the previous exercise. Explain how, as a nation's population grows older, median income can go down for workers in each age group, yet still go up for all workers.

5.46 Some regression math. Use the equation of the least-squares regression line (box on page 130) to show that the regression line for predicting y from x always passes through the point $(\overline{x}, \overline{y})$. That is, when $x = \overline{x}$, the equation gives $\hat{y} = \overline{y}$.

5.47 Regression to the mean. Figure 4.8 (page 114) displays the relationship between golfers' scores on the first and second rounds of the 2007 Masters Tournament. The least-squares line for predicting second-round scores from first-round scores has equation $\hat{y} = 61.93 + 0.180x$. Find the predicted second-round scores for a player who shot 80 in the first round and for a player who shot 70. The mean second-round score for all players was 75.63. So a player who does well in the first round is predicted to do less well, but still better than average, in the second round. And a player who does poorly in the first is predicted to do better, but still worse than average, in the second.

regression to the mean

(*Comment:* This is **regression to the mean.** If you select individuals with extreme scores on some measure, they tend to have less extreme scores when measured again. That's because their extreme position is partly merit and partly luck. The luck will be different next time. Regression to the mean contributes to lots of "effects." The rookie of the year often doesn't do as well the next year; the best player in an orchestral audition may play less well once hired than the runners-up; a student who feels she needs coaching after the SAT often does better on the next try without coaching.)

5.48 Regression to the mean. We expect that students who do well on the midterm exam in a course will usually also do well on the final exam. Gary Smith of Pomona College looked at the exam scores of all 346 students who took his statistics class over a 10-year period.[17] The least-squares line for predicting final exam score from midterm-exam score was $\hat{y} = 46.6 + 0.41x$. (Both exams have a 100-point scale.)

Octavio scores 10 points above the class mean on the midterm. How many points above the class mean do you predict that he will score on the final? (*Hint:* Use the fact that the least-squares line passes through the point $(\overline{x}, \overline{y})$ and the fact that Octavio's midterm score is $\overline{x} + 10$.) This is another example of regression to the mean: students who do well on the midterm will on the average do less well, but still above average, on the final.

5.49 Is regression useful? In Exercise 4.41 (page 121) you used the *Correlation and Regression* applet to create three scatterplots having correlation about $r = 0.7$ between the horizontal variable x and the vertical variable y. Create three similar scatterplots again, and click the "Show least-squares line" box to display the regression lines. Correlation $r = 0.7$ is considered reasonably strong in many areas of work. Because there is a reasonably strong correlation, we might use a regression line to predict y from x. In which of your three scatterplots does it make sense to use a straight line for prediction?

5.50 Guessing a regression line. In the *Correlation and Regression* applet, click on the scatterplot to create a group of 15 to 20 points from lower left to upper right with a clear positive straight-line pattern (correlation around 0.7). Click the "Draw line" button and use the mouse (right-click and drag) to draw a line through the middle of the cloud of points from lower left to upper right. Note the "thermometer" above the plot. The red portion is the sum of the squared vertical distances from the points in the plot to the least-squares line. The green portion is the "extra" sum of squares for your line—it shows by how much your line misses the smallest possible sum of squares.

(a)   You drew a line by eye through the middle of the pattern. Yet the right-hand part of the bar is probably almost entirely green. What does that tell you?

(b)   Now click the "Show least-squares line" box. Is the slope of the least-squares line smaller (the new line is less steep) or larger (line is steeper) than that of your line? If you repeat this exercise several times, you will consistently get the same result. The least-squares line minimizes the *vertical* distances of the points from the line. It is *not* the line through the "middle" of the cloud of points. This is one reason why it is hard to draw a good regression line by eye.

The following exercises ask you to answer questions from data without having the details outlined for you. The exercise statements give you the **State** *step of the four-step process. In your work, follow the* **Plan, Solve,** *and* **Conclude** *steps of the process, described on page 55.*

5.51 Beavers and beetles. Do beavers benefit beetles? Researchers laid out 23 circular plots, each 4 meters in diameter, in an area where beavers were cutting down cottonwood trees. In each plot, they counted the number of stumps from trees cut by beavers and the number of clusters of beetle larvae. Ecologists think that the new sprouts from stumps are more tender than other cottonwood growth, so that beetles prefer them. If so, more stumps should produce more beetle larvae. Here are the data:[18]

Stumps	2	2	1	3	3	4	3	1	2	5	1	3
Beetle larvae	10	30	12	24	36	40	43	11	27	56	18	40
Stumps	2	1	2	2	1	1	4	1	2	1	4	
Beetle larvae	25	8	21	14	16	6	54	9	13	14	50	

Analyze these data to see if they support the "beavers benefit beetles" idea.

5.52 A computer game. A multimedia statistics learning system includes a test of skill in using the computer's mouse. The software displays a circle at a random location on the computer screen. The subject clicks in the circle with the mouse as quickly as possible. A new circle appears as soon as the subject clicks the old one. Table 5.3 gives data for one subject's trials, 20 with each hand. Distance is the distance from the cursor location to the center of the new circle, in units whose actual size depends on the size of the screen. Time is the time required to click in the new circle, in milliseconds.[19] We suspect that time depends on distance. We also suspect that performance will not be the same with the right and left hands. Analyze the data with a view to predicting performance separately for the two hands.

5.53 Predicting tropical storms. William Gray heads the Tropical Meteorology Project at Colorado State University (well away from the hurricane belt). His forecasts before each year's hurricane season attract lots of attention. Here are data on the number of named Atlantic tropical storms predicted by Dr. Gray and the actual number of storms for the years 1984 to 2007:[20]

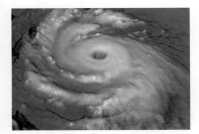

NASA/GSFC

Year	Forecast	Actual	Year	Forecast	Actual	Year	Forecast	Actual
1984	10	12	1992	8	6	2000	12	14
1985	11	11	1993	11	8	2001	12	15
1986	8	6	1994	9	7	2002	11	12
1987	8	7	1995	12	19	2003	14	16
1988	11	12	1996	10	13	2004	14	14
1989	7	11	1997	11	7	2005	15	27
1990	11	14	1998	10	14	2006	17	9
1991	8	8	1999	14	12	2007	17	14

Analyze these data. How accurate are Dr. Gray's forecasts? How many tropical storms would you expect in a year when his preseason forecast calls for 16 storms? What is the effect of the disastrous 2005 season on your answers?

5.54 Climate change. Global warming has many indirect effects on climate. For example, the summer monsoon winds in the Arabian Sea bring rain to India and are critical for agriculture. As the climate warms and winter snow cover in the vast landmass of Europe and Asia decreases, the land heats more rapidly in the summer. This may increase the strength of the monsoon. Here are data on snow cover (in millions of square kilometers) and summer wind stress (in newtons per square meter):[21]

Snow cover	Wind stress	Snow cover	Wind stress	Snow cover	Wind stress
6.6	0.125	16.6	0.111	26.6	0.062
5.9	0.160	18.2	0.106	27.1	0.051
6.8	0.158	15.2	0.143	27.5	0.068
7.7	0.155	16.2	0.153	28.4	0.055
7.9	0.169	17.1	0.155	28.6	0.033
7.8	0.173	17.3	0.133	29.6	0.029
8.1	0.196	18.1	0.130	29.4	0.024

TABLE 5.3 Reaction times in a computer game

TIME	DISTANCE	HAND	TIME	DISTANCE	HAND
115	190.70	right	240	190.70	left
96	138.52	right	190	138.52	left
110	165.08	right	170	165.08	left
100	126.19	right	125	126.19	left
111	163.19	right	315	163.19	left
101	305.66	right	240	305.66	left
111	176.15	right	141	176.15	left
106	162.78	right	210	162.78	left
96	147.87	right	200	147.87	left
96	271.46	right	401	271.46	left
95	40.25	right	320	40.25	left
96	24.76	right	113	24.76	left
96	104.80	right	176	104.80	left
106	136.80	right	211	136.80	left
100	308.60	right	238	308.60	left
113	279.80	right	316	279.80	left
123	125.51	right	176	125.51	left
111	329.80	right	173	329.80	left
95	51.66	right	210	51.66	left
108	201.95	right	170	201.95	left

Analyze these data to uncover the nature and strength of the effect of decreasing snow cover on wind stress.

5.55 Saving energy with solar panels. Exercise 4.43 (page 121) gives monthly data on outside temperature (in degree-days per day) and natural gas consumed for a house in the Midwest both before and after installing solar panels. A cold winter month in this location may average 45 degree-days per day (temperature 20°F). Use before-and-after regression lines to estimate the savings in gas consumption due to solar panels.

5.56 Effects of health care spending. Table 1.3 (page 22) gives United Nations data on annual health care spending per person in 38 richer nations. The United States, at $5711 per person, is a high outlier. The data file *ex05–56.dat* on the text CD and Web site adds more information about these countries: female life expectancy at birth (in years) and infant mortality per 1000 live births.[22] Use these data to investigate the extent to which higher spending on health care increases life expectancy and reduces infant mortality. Would you use regression to predict either outcome from spending? (Two comments to help explain what you find: In addition to the United States, South Africa is an outlier in scatterplots. South Africa is out of place in this group of richer nations, although it does meet the criteria that the United Nations used to make up the list. If the effects of health care spending on overall health seem small, that is in part because the data include only richer nations, most of which spend enough to ensure basic good health for their citizens.)

CHAPTER 6

Two-Way Tables*

We have concentrated on relationships in which at least the response variable is quantitative. Now we will describe relationships between two categorical variables. Some variables—such as sex, race, and occupation—are categorical by nature. Other categorical variables are created by grouping values of a quantitative variable into classes. Published data often appear in grouped form to save space. To analyze categorical data, we use the *counts* or *percents* of individuals that fall into various categories.

EXAMPLE 6.1 I think I'll be rich by age 30

A sample survey of young adults (aged 19 to 25) asked, "What do you think are the chances you will have much more than a middle-class income at age 30?" Table 6.1 shows the responses, omitting a few people who refused to respond or who said they were already rich.[1] This is a **two-way table** because it describes two categorical variables: sex and opinion about becoming rich. Opinion is the **row variable** because each row in the table describes young adults who held one of the five opinions about their chances. Because the opinions have a natural order from "Almost no chance" to

two-way table
row variable

*This material is important in statistics, but it is needed later in this book only for Chapter 22. You may omit it if you do not plan to read Chapter 22 or delay reading it until you reach Chapter 22.

TABLE 6.1 Young adults by sex and chance of getting rich

| | SEX | | |
OPINION	FEMALE	MALE	TOTAL
Almost no chance	96	98	194
Some chance but probably not	426	286	712
A 50-50 chance	696	720	1416
A good chance	663	758	1421
Almost certain	486	597	1083
Total	2367	2459	4826

column variable

"Almost certain," the rows are also in this order. Sex is the **column variable** because each column describes one sex. The entries in the table are the counts of individuals in each opinion-by-sex class. ■

Marginal distributions

How can we best grasp the information contained in Table 6.1? First, *look at the distribution of each variable separately*. The distribution of a categorical variable says how often each outcome occurred. The "Total" column at the right of the table contains the totals for each of the rows. These row totals give the distribution of opinions about becoming rich in the entire group of 4826 young adults: 194 felt that they had almost no chance, 712 thought they had just some chance, and so on.

If the row and column totals are missing, the first thing to do in studying a two-way table is to calculate them. The distributions of opinion alone and sex alone *marginal distribution* are called **marginal distributions** because they appear at the right and bottom margins of the two-way table.

Percents are often more informative than counts. We can display the marginal distribution of opinions in percents by dividing each row total by the table total and converting to a percent.

EXAMPLE 6.2 **Calculating a marginal distribution**

The percent of these young adults who think they are almost certain to be rich by age 30 is

$$\frac{\text{almost certain total}}{\text{table total}} = \frac{1083}{4826} = 0.224 = 22.4\%$$

Do four more such calculations to obtain the marginal distribution of opinion in percents. Here is the complete distribution:

Response	Percent
Almost no chance	$\frac{194}{4826} = 4.0\%$
Some chance	$\frac{712}{4826} = 14.8\%$
A 50-50 chance	$\frac{1416}{4826} = 29.3\%$
A good chance	$\frac{1421}{4826} = 29.4\%$
Almost certain	$\frac{1083}{4826} = 22.4\%$

It seems that many young adults are optimistic about their future income. The total should be 100% because everyone holds one of the five opinions. In fact, the percents add to 99.9% because we rounded each one to the nearest tenth. This is **roundoff error.** ■

roundoff error

Each marginal distribution from a two-way table is a distribution for a single categorical variable. As we saw in Chapter 1, we can use a bar graph or a pie chart to display such a distribution. Figure 6.1 is a bar graph of the distribution of opinion among young adults.

In working with two-way tables, you must calculate lots of percents. Here's a tip to help decide what fraction gives the percent you want. Ask, "What group represents the total of which I want a percent?" The count for that group is the denominator of the fraction that leads to the percent. In Example 6.2, we want a percent "of young adults," so the count of young adults (the table total) is the denominator.

FIGURE 6.1

A bar graph of the distribution of opinions of young adults about becoming rich by age 30. This is one of the marginal distributions for Table 6.1.

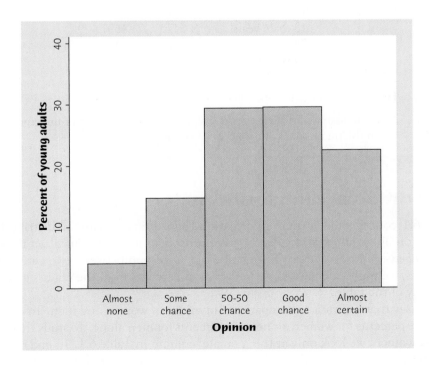

6.1 Attitudes toward recycled products. Recycling is supposed to save resources. Some people think recycled products are lower in quality than other products, a fact that makes recycling less practical. Here are data on attitudes toward coffee filters made of recycled paper among people who had bought these filters and people who had not:[2]

| | Think the quality of the recycled product is | | |
	Higher	The same	Lower
Buyers	20	7	9
Nonbuyers	29	25	43

(a)   How many people does this table describe? How many of these were buyers of coffee filters made of recycled paper?

(b)   Give the marginal distribution of opinion about the quality of recycled filters. What percent of consumers think the quality of the recycled product is the same or higher than the quality of other filters?

6.2 Undergraduates' ages. Here is a two-way table of Census Bureau data describing the age and sex of all American undergraduate college students. The table entries are counts in thousands of students.[3]

Age group	Female	Male
15 to 17 years	116	61
18 to 24 years	5470	4691
25 to 34 years	1319	824
35 years or older	1075	616

Digital Vision/Getty

(a)   How many college undergraduates are there?

(b)   Find the marginal distribution of age group. What percent of undergraduates are in the traditional 18 to 24 college age group?

Conditional distributions

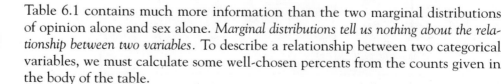

Table 6.1 contains much more information than the two marginal distributions of opinion alone and sex alone. *Marginal distributions tell us nothing about the relationship between two variables.* To describe a relationship between two categorical variables, we must calculate some well-chosen percents from the counts given in the body of the table.

Let's say that we want to compare the opinions of women and men. To do this, compare percents for women alone with percents for men alone. To study the opinions of women, we look only at the "Female" column in Table 6.1. To find the percent *of young women* who think they are almost certain to be rich by age 30, divide the count of such women by the total number of women (the column total):

$$\frac{\text{women who are almost certain}}{\text{column total}} = \frac{486}{2367} = 0.205 = 20.5\%$$

Doing this for all five entries in the "Female" column gives the *conditional distribution* of opinion among women. We use the term "conditional" because this distribution describes only young adults who satisfy the condition that they are female.

MARGINAL AND CONDITIONAL DISTRIBUTIONS

The **marginal distribution** of one of the categorical variables in a two-way table of counts is the distribution of values of that variable among all individuals described by the table.

A **conditional distribution** of a variable is the distribution of values of that variable among only individuals who have a given value of the other variable. There is a separate conditional distribution for each value of the other variable.

EXAMPLE 6.3 **Comparing women and men**

STATE: How do young men and young women differ in their responses to the question "What do you think are the chances you will have much more than a middle-class income at age 30?"

PLAN: Make a two-way table of response by sex. Find the two conditional distributions of response for men alone and for women alone. Compare these two distributions.

SOLVE: Table 6.1 is the two-way table we need. Look first at just the "Female" column to find the conditional distribution for women, then at just the "Male" column to find the conditional distribution for men. Here are the calculations and the two conditional distributions:

Response	Female	Male
Almost no chance	$\frac{96}{2367} = 4.1\%$	$\frac{98}{2459} = 4.0\%$
Some chance	$\frac{426}{2367} = 18.0\%$	$\frac{286}{2459} = 11.6\%$
A 50-50 chance	$\frac{696}{2367} = 29.4\%$	$\frac{720}{2459} = 29.3\%$
A good chance	$\frac{663}{2367} = 28.0\%$	$\frac{758}{2459} = 30.8\%$
Almost certain	$\frac{486}{2367} = 20.5\%$	$\frac{597}{2459} = 24.3\%$

Each set of percents adds to 100% because everyone holds one of the five opinions.

CONCLUDE: Men are somewhat more optimistic about their future income than are women. Men are less likely to say that they have "some chance but probably not" and more likely to say that they have "a good chance" or are "almost certain" to have much more than a middle-class income by age 30. ■

Smiling faces

Women smile more than men. The same data that produce this fact allow us to link smiling to other variables in two-way tables. For example, add as the second variable whether or not the person thinks they are being observed. If yes, that's when women smile more. If no, there's no difference between women and men. Next, take the second variable to be the person's social role (for example, is he or she the boss in an office?). Within each role, there is very little difference in smiling between women and men.

FIGURE 6.2

Minitab output for the two-way table of young adults by sex and chance of getting rich, along with each entry as a percent of its column total. The "Female" and "Male" columns give the conditional distributions of responses for women and men, and the "All" column shows the marginal distribution of responses for all these young adults.

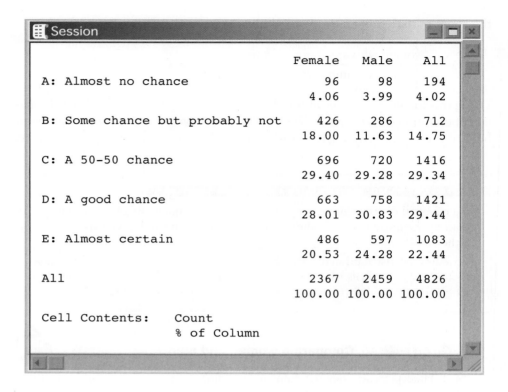

```
 Session                                                      _ □ ×

                                            Female    Male     All
        A: Almost no chance                     96      98     194
                                              4.06    3.99    4.02

        B: Some chance but probably not        426     286     712
                                             18.00   11.63   14.75

        C: A 50-50 chance                      696     720    1416
                                             29.40   29.28   29.34

        D: A good chance                       663     758    1421
                                             28.01   30.83   29.44

        E: Almost certain                      486     597    1083
                                             20.53   24.28   22.44

        All                                   2367    2459    4826
                                            100.00  100.00  100.00

        Cell Contents:     Count
                           % of Column
```

Software will do these calculations for you. Most programs allow you to choose which conditional distributions you want to compare. The output in Figure 6.2 presents the two conditional distributions of opinion, for women and for men, and also the marginal distribution of opinion for all of the young adults. The distributions agree (up to roundoff) with the results in Examples 6.2 and 6.3.

Remember that there are two sets of conditional distributions for any two-way table. Example 6.3 looked at the conditional distributions of opinion for the two sexes. We could also examine the five conditional distributions of sex, one for each of the five opinions, by looking separately at the five rows in Table 6.1. Because the variable "sex" has only two categories, comparing the five conditional distributions amounts to comparing the percents of women among young adults who hold each opinion. Figure 6.3 makes this comparison in a bar graph. The bar heights do *not* add to 100%, because each bar represents a different group of people.

No single graph (such as a scatterplot) portrays the form of the relationship between categorical variables. No single numerical measure (such as the correlation) summarizes the strength of the association. Bar graphs are flexible enough to be helpful, but you must think about what comparisons you want to display. For numerical measures, we rely on well-chosen percents. You must decide which percents you need. Here is a hint: *if there is an explanatory-response relationship, compare the conditional distributions of the response variable for the separate values of the explanatory variable.* If you think that sex influences young adults' opinions about their chances of getting rich by age 30, compare the conditional distributions of opinion for women and for men, as in Example 6.3.

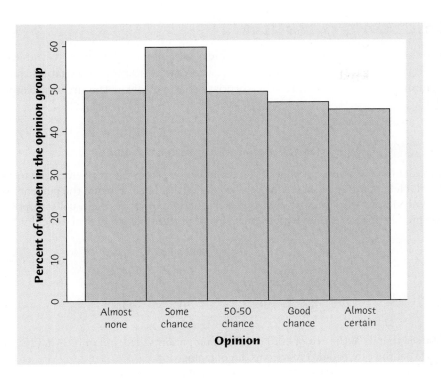

FIGURE 6.3

Bar graph comparing the percents of females among those who hold each opinion about their chances of getting rich by age 30.

APPLY YOUR KNOWLEDGE

6.3 Attitudes toward recycled products. Exercise 6.1 gives data on the opinions of people who have and have not bought coffee filters made from recycled paper. To see the relationship between opinion and experience with the product, find the conditional distributions of opinion (the response variable) for buyers and nonbuyers. What do you conclude?

6.4 Undergraduates' ages. Exercise 6.2 gives Census Bureau data describing the age and sex of all American college undergraduates. We suspect that the percent of women is higher among older students than in the traditional 18 to 24 college age group. Do the data support this suspicion? Follow the four-step process as illustrated in Example 6.3.

6.5 Marginal distributions aren't the whole story. Here are the row and column totals for a two-way table with two rows and two columns:

$$
\begin{array}{cc|c}
a & b & 50 \\
c & d & 50 \\
\hline
60 & 40 & 100
\end{array}
$$

Make up *two different* sets of counts a, b, c, and d for the body of the table that give these same totals. This shows that the relationship between two variables cannot be obtained from the two individual distributions of the variables.

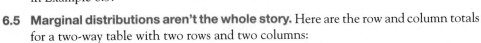

Simpson's paradox

As is the case with quantitative variables, the effects of lurking variables can change or even reverse relationships between two categorical variables. Here is an example that demonstrates the surprises that can await the unsuspecting user of data.

Ashley/Cooper/PICIMPACT/CORBIS

EXAMPLE 6.4 **Do medical helicopters save lives?**

Accident victims are sometimes taken by helicopter from the accident scene to a hospital. Helicopters save time. Do they also save lives? Let's compare the percents of accident victims who die with helicopter evacuation and with the usual transport to a hospital by road. Here are hypothetical data that illustrate a practical difficulty:[4]

	Helicopter	Road
Victim died	64	260
Victim survived	136	840
Total	200	1100

We see that 32% (64 out of 200) of helicopter patients died, but only 24% (260 out of 1100) of the others did. That seems discouraging.

The explanation is that the helicopter is sent mostly to serious accidents, so that the victims transported by helicopter are more often seriously injured. They are more likely to die with or without helicopter evacuation. Here are the same data broken down by the seriousness of the accident:

Serious Accidents			Less Serious Accidents		
	Helicopter	Road		Helicopter	Road
Died	48	60	Died	16	200
Survived	52	40	Survived	84	800
Total	100	100	Total	100	1000

Inspect these tables to convince yourself that they describe the same 1300 accident victims as the original two-way table. For example, 200 (100 + 100) were moved by helicopter, and 64 (48 + 16) of these died.

Among victims of serious accidents, the helicopter saves 52% (52 out of 100) compared with 40% for road transport. If we look only at less serious accidents, 84% of those transported by helicopter survive, versus 80% of those transported by road. Both groups of victims have a higher survival rate when evacuated by helicopter. ■

How can it happen that the helicopter does better for both groups of victims but worse when all victims are lumped together? Examining the data makes the explanation clear. Half the helicopter transport patients are from serious accidents, compared with only 100 of the 1100 road transport patients. So the helicopter carries patients who are more likely to die. The seriousness of the accident was a lurking variable that, until we uncovered it, hid the true relationship between survival and mode of transport to a hospital. Example 6.4 illustrates *Simpson's paradox*.

SIMPSON'S PARADOX

An association or comparison that holds for all of several groups can reverse direction when the data are combined to form a single group. This reversal is called **Simpson's paradox.**

The lurking variable in Simpson's paradox is categorical. That is, it breaks the individuals into groups, as when accident victims are classified as injured in a "serious accident" or a "less serious accident." Simpson's paradox is just an extreme form of the fact that observed associations can be misleading when there are lurking variables.

APPLY YOUR KNOWLEDGE

6.6 **Airline flight delays.** Here are the numbers of flights on time and delayed for two airlines at five airports in one month. Overall on-time percents for each airline are often reported in the news. The airport that flights serve is a lurking variable that can make such reports misleading.[5]

	Alaska Airlines		America West	
	On time	**Delayed**	**On time**	**Delayed**
Los Angeles	497	62	694	117
Phoenix	221	12	4840	415
San Diego	212	20	383	65
San Francisco	503	102	320	129
Seattle	1841	305	201	61

(a) What percent of all Alaska Airlines flights were delayed? What percent of all America West flights were delayed? These are the numbers usually reported.

(b) Now find the percent of delayed flights for Alaska Airlines at each of the five airports. Do the same for America West.

(c) America West did worse at *every one* of the five airports, yet did better overall. That sounds impossible. Explain carefully, referring to the data, how this can happen. (The weather in Phoenix and Seattle lies behind this example of Simpson's paradox.)

6.7 **Which hospital is safer?** To help consumers make informed decisions about health care, the government releases data about patient outcomes in hospitals. You want to compare Hospital A and Hospital B, which serve your community. The table presents data on all patients undergoing surgery in a recent time period. The data include the condition of the patient ("good" or "poor") before the surgery. "Survived" means that the patient lived at least 6 weeks following surgery.

Good Condition		
	Hospital A	**Hospital B**
Died	6	8
Survived	594	592
Total	600	600

Poor Condition		
	Hospital A	**Hospital B**
Died	57	8
Survived	1443	192
Total	1500	200

(a) Compare percents to show that Hospital A has a higher survival rate for both groups of patients.

(b) Combine the data into a single two-way table of outcome ("survived" or "died") by hospital (A or B). The local paper reports just these overall survival rates. Which hospital has the higher rate?

(c) Explain from the data, in language that a reporter can understand, how Hospital B can do better overall even though Hospital A does better for both groups of patients.

CHAPTER 6   SUMMARY

- A **two-way table** of counts organizes data about two categorical variables. Values of the **row variable** label the rows that run across the table, and values of the **column variable** label the columns that run down the table. Two-way tables are often used to summarize large amounts of information by grouping outcomes into categories.

- The **row totals** and **column totals** in a two-way table give the **marginal distributions** of the two individual variables. It is clearer to present these distributions as percents of the table total. Marginal distributions tell us nothing about the relationship between the variables.

- There are two sets of **conditional distributions** for a two-way table: the distributions of the row variable for each fixed value of the column variable, and the distributions of the column variable for each fixed value of the row variable. Comparing one set of conditional distributions is one way to describe the association between the row and the column variables.

- To find the **conditional distribution** of the row variable for one specific value of the column variable, look only at that one column in the table. Find each entry in the column as a percent of the column total.

- **Bar graphs** are a flexible means of presenting categorical data. There is no single best way to describe an association between two categorical variables.

- A comparison between two variables that holds for each individual value of a third variable can be changed or even reversed when the data for all values of the third variable are combined. This is **Simpson's paradox.** Simpson's paradox is an example of the effect of lurking variables on an observed association.

C H E C K   Y O U R   S K I L L S

The National Longitudinal Study of Adolescent Health interviewed several thousand teens (grades 7 to 12). One question asked was "What do you think are the chances you will be married in the next 10 years?" Here is a two-way table of the responses by sex:[6]

Opinion	Female	Male
Almost no chance	119	103
Some chance but probably not	150	171
A 50-50 chance	447	512
A good chance	735	710
Almost certain	1174	756

Comstock Images/Age fotostock

Exercises 6.8 to 6.16 are based on this table.

6.8 How many individuals are described by this table?

(a) 2625   (b) 4877   (c) Need more information

6.9 How many females were among the respondents?

(a) 2625   (b) 4877   (c) Need more information

6.10 The percent of females among the respondents was

(a) about 46%.   (b) about 54%.   (c) about 86%.

6.11 Your percent from the previous exercise is part of

(a) the marginal distribution of sex.

(b) the marginal distribution of opinion about marriage.

(c) the conditional distribution of sex among adolescents with a given opinion.

6.12 What percent of females thought that they were almost certain to be married in the next 10 years?

(a) about 40%   (b) about 45%   (c) about 61%

6.13 Your percent from the previous exercise is part of

(a) the marginal distribution of opinion about marriage.

(b) the conditional distribution of sex among those who thought they were almost certain to be married.

(c) the conditional distribution of opinion about marriage among women.

6.14 What percent of those who thought they were almost certain to be married were female?

(a) about 40%   (b) about 45%   (c) about 61%

6.15 Your percent from the previous exercise is part of

(a) the marginal distribution of opinion about marriage.

(b) the conditional distribution of sex among those who thought they were almost certain to be married.

(c) the conditional distribution of opinion about marriage among women.

6.16 A bar graph showing the conditional distribution of opinion among female respondents would have

(a)   2 bars.   (b)   5 bars.   (c)   10 bars.

6.17 A college looks at the grade point average (GPA) of its full-time and part-time students. Grades in science courses are generally lower than grades in other courses. There are few science majors among part-time students but many science majors among full-time students. The college finds that full-time students who are science majors have higher GPA than part-time students who are science majors. Full-time students who are not science majors also have higher GPA than part-time students who are not science majors. Yet part-time students as a group have higher GPA than full-time students. This finding is

(a)   not possible: if both science and other majors who are full-time have higher GPA than those who are part-time, then all full-time students together must have higher GPA than all part-time students together.

(b)   an example of Simpson's paradox: full-time students do better in both kinds of courses but worse overall because they take more science courses.

(c)   due to comparing two conditional distributions that should not be compared.

C H A P T E R 6   E X E R C I S E S

6.18 **Graduate school for men and women.** The College of Liberal Arts at a large university looks at its graduate students classified by their sex and field of study. Here are the data:[7]

	Female	Male
English	136	89
Foreign languages	61	25
History	35	55
Philosophy	10	54
Political science	29	35

Find the two conditional distributions of field of study, one for women and one for men. Based on your calculations, describe the differences between women and men with a graph and in words.

6.19 **Helping cocaine addicts.** Will giving cocaine addicts an antidepressant drug help them break their addiction? An experiment assigned 24 chronic cocaine users to take the antidepressant drug desipramine, another 24 to take lithium, and another 24 to take a placebo. (Lithium is a standard drug to treat cocaine addiction. A placebo is a dummy pill, used so that the effect of being in the study but not taking any drug can be seen.) After three years, 14 of the 24 subjects in the desipramine group had remained free of cocaine, along with 6 of the 24 in the lithium group and 4 of the 24 in the placebo group.[8]

(a) Make up a two-way table of "Treatment received" by whether or not the subject remained free of cocaine.

(b) Compare the effectiveness of the three treatments in preventing use of cocaine by former addicts. Use percents and draw a bar graph. What do you conclude?

Marital status and job level. *We sometimes hear that getting married is good for your career. Table 6.2 presents data from one of the studies behind this generalization. To avoid gender effects, the investigators looked only at men. The data describe the marital status and the job level of all 8235 male managers and professionals employed by a large manufacturing firm.[9] The firm assigns each position a grade that reflects the value of that particular job to the company. The authors of the study grouped the many job grades into quarters. Grade 1 contains jobs in the lowest quarter of the job grades, and Grade 4 contains those in the highest quarter. Exercises 6.20 to 6.24 are based on these data.*

TABLE 6.2 Marital status and job level

JOB GRADE	MARITAL STATUS				TOTAL
	SINGLE	MARRIED	DIVORCED	WIDOWED	
1	58	874	15	8	955
2	222	3927	70	20	4239
3	50	2396	34	10	2490
4	7	533	7	4	551
Total	337	7730	126	42	8235

6.20 Marginal distributions. Give (in percents) the two marginal distributions, for marital status and for job grade. Do each of your two sets of percents add to exactly 100%? If not, why not?

6.21 Percents. What percent of single men hold Grade 1 jobs? What percent of Grade 1 jobs are held by single men?

6.22 Conditional distribution. Give (in percents) the conditional distribution of job grade among single men. Should your percents add to 100% (up to roundoff error)?

6.23 Marital status and job grade. One way to see the relationship is to look at who holds Grade 1 jobs.

(a) There are 874 married men with Grade 1 jobs, and only 58 single men with such jobs. Explain why these counts by themselves don't describe the relationship between marital status and job grade.

(b) Find the percent of men in each marital status group who have Grade 1 jobs. Then find the percent in each marital group who have Grade 4 jobs. What do these percents say about the relationship?

6.24 Association is not causation. The data in Table 6.2 show that single men are more likely to hold lower-grade jobs than are married men. We should not conclude that single men can help their career by getting married. What lurking variables might help explain the association between marital status and job grade?

6.25 Discrimination? Wabash Tech has two professional schools, business and law. Here are two-way tables of applicants to both schools, categorized by gender and admission decision. (Although these data are made up, similar situations occur in reality.)[10]

Business	Admit	Deny		Law	Admit	Deny
Male	480	120		Male	10	90
Female	180	20		Female	100	200

(a) Make a two-way table of gender by admission decision for the two professional schools together by summing entries in these tables.

(b) From the two-way table, calculate the percent of male applicants who are admitted and the percent of female applicants who are admitted. Wabash admits a higher percent of male applicants.

(c) Now compute separately the percents of male and female applicants admitted by the business school and by the law school. Each school admits a higher percent of female applicants.

(d) This is Simpson's paradox: both schools admit a higher percent of the women who apply, but overall Wabash admits a lower percent of female applicants than of male applicants. Explain carefully, as if speaking to a skeptical reporter, how it can happen that Wabash appears to favor males when each school individually favors females.

6.26 Obesity and health. To estimate the health risks of obesity, we might compare how long obese and nonobese people live. Smoking is a lurking variable that may reduce the gap between the two groups, because smoking tends to both reduce weight and lead to earlier death. So if we ignore smoking, we may underestimate the health risks of obesity. Illustrate Simpson's paradox by a simplified version of this situation: make up two-way tables of obese (yes or no) by early death (yes or no) separately for smokers and nonsmokers such that

■ Obese smokers and obese nonsmokers are both more likely to die earlier than those not obese.

■ But when smokers and nonsmokers are combined into a two-way table of obese by early death, persons who are not obese are more likely to die earlier because more of them are smokers.

*The following exercises ask you to answer questions from data without having the details outlined for you. The exercise statements give you the **State** step of the four-step process. In your work, follow the **Plan, Solve,** and **Conclude** steps of the process as illustrated in Example 6.3.*

6.27 Life at work. The University of Chicago's General Social Survey asked a representative sample of adults this question: "Which of the following statements best describes how your daily work is organized? 1: I am free to decide how my daily work

is organized. 2: I can decide how my daily work is organized, within certain limits. 3: I am not free to decide how my daily work is organized." Here is a two-way table of the responses for three levels of education:[11]

	Highest Degree Completed		
Response	Less than high school	High school	Bachelor's
1	31	161	81
2	49	269	85
3	47	112	14

How does freedom to organize your work depend on level of education?

6.28 Animal testing. "It is right to use animals for medical testing if it might save human lives." The General Social Survey asked 1152 adults to react to this statement. Here is the two-way table of their responses:

Response	Male	Female
Strongly agree	76	59
Agree	270	247
Neither agree nor disagree	87	139
Disagree	61	123
Strongly disagree	22	68

How do the distributions of opinion differ between men and women?

6.29 College degrees. "Colleges and universities across the country are grappling with the case of the mysteriously vanishing male." So said an article in the *Washington Post*. Here are data on the numbers of degrees earned in 2009–2010, as projected by the National Center for Education Statistics. The table entries are counts of degrees in thousands.[12]

	Female	Male
Associate's	447	268
Bachelor's	945	651
Master's	397	251
Professional	49	44
Doctor's	26	25

Briefly contrast the participation of men and women in earning degrees.

6.30 The Mediterranean diet. Cancer of the colon and rectum is less common in the Mediterranean region than in other Western countries. The Mediterranean diet contains little animal fat and lots of olive oil. Italian researchers compared 1953 patients with colon or rectal cancer with a control group of 4154 patients admitted to the same hospitals for unrelated reasons. They estimated consumption of various foods from a detailed interview, then divided the patients into three groups according to their consumption of olive oil. The table presents some of the data.[13]

© The Photo Works

	Olive Oil			
	Low	Medium	High	Total
Colon cancer	398	397	430	1225
Rectal cancer	250	241	237	728
Controls	1368	1377	1409	4154

The researchers conjectured that high olive oil consumption would be more common among patients without cancer than among patients with colon cancer or rectal cancer. What do the data say?

6.31 Do angry people have more heart disease? People who get angry easily tend to have more heart disease. That's the conclusion of a study that followed a random sample of 12,986 people from three locations for about four years. All subjects were free of heart disease at the beginning of the study. The subjects took the Spielberger Trait Anger Scale test, which measures how prone a person is to sudden anger. Here are data for the 8474 people in the sample who had normal blood pressure.[14] CHD stands for "coronary heart disease." This includes people who had heart attacks and those who needed medical treatment for heart disease.

	Low anger	Moderate anger	High anger	Total
CHD	53	110	27	190
No CHD	3057	4621	606	8284
Total	3110	4731	633	8474

Do these data support the study's conclusion about the relationship between anger and heart disease?

6.32 Python eggs. How is the hatching of water python eggs influenced by the temperature of the snake's nest? Researchers placed 104 newly laid eggs in a hot environment, 56 in a neutral environment, and 27 in a cold environment. Hot duplicates the warmth provided by the mother python. Neutral and cold are cooler, as when the mother is absent. The results: 75 of the hot eggs hatched, along with 38 of the neutral eggs and 16 of the cold eggs.[15]

(a) Make a two-way table of "environment temperature" against "hatched or not."

(b) The researchers anticipated that eggs would hatch less well at cooler temperatures. Do the data support that anticipation?

CHAPTER 7

Exploring Data:
Part I Review

Data analysis is the art of describing data using graphs and numerical summaries. The purpose of data analysis is to help us see and understand the most important features of a set of data. Chapter 1 commented on graphs to display distributions: pie charts and bar graphs for categorical variables, histograms and stemplots for quantitative variables. In addition, time plots show how a quantitative variable changes over time. Chapter 2 presented numerical tools for describing the center and spread of the distribution of one variable. Chapter 3 discussed density curves for describing the overall pattern of a distribution, with emphasis on the Normal distributions.

The first STATISTICS IN SUMMARY figure on the next page organizes the big ideas for exploring a quantitative variable. Plot your data, then describe their center and spread using either the mean and standard deviation or the five-number summary. The last step, which makes sense only for some data, is to summarize the data in compact form by using a Normal curve as a description of the overall pattern. The question marks at the last two stages remind us that the usefulness of numerical summaries and Normal distributions depends on what we find when we examine graphs of our data. No short summary does justice to irregular shapes or to data with several distinct clusters.

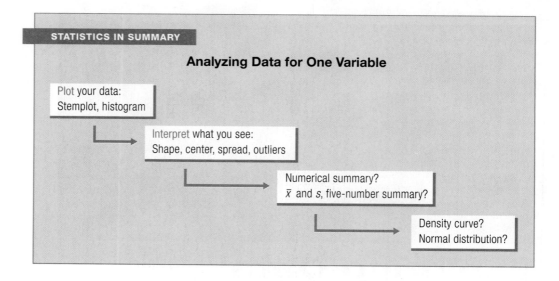

Chapters 4 and 5 applied the same ideas to relationships between two quantitative variables. The second STATISTICS IN SUMMARY figure retraces the big ideas, with details that fit the new setting. Always begin by making graphs of your data. In the case of a scatterplot, we have learned a numerical summary only for data that show a roughly linear pattern on the scatterplot. The summary is then the means and standard deviations of the two variables and their correlation. A regression line drawn on the plot gives a compact description of the overall pattern that we can use for prediction. Once again there are question marks at the last two stages to remind us that correlation and regression describe only straight-line relationships. Chapter 6 shows how to understand relationships between two categorical variables; comparing well-chosen percents is the key.

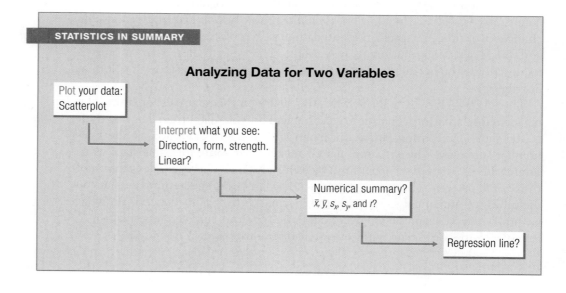

You can organize your work in any open-ended data analysis setting by following the four-step State, Plan, Solve, and Conclude process first introduced in Chapter 2. After we have mastered the extra background needed for statistical inference, this process will also guide practical work on inference later in the book.

PART I   SUMMARY

Here are the most important skills you should have acquired from reading Chapters 1 to 6.

A. Data

1. Identify the individuals and variables in a set of data.
2. Identify each variable as categorical or quantitative. Identify the units in which each quantitative variable is measured.
3. Identify the explanatory and response variables in situations where one variable explains or influences another.

B. Displaying Distributions

1. Recognize when a pie chart can and cannot be used.
2. Make a bar graph of the distribution of a categorical variable, or in general to compare related quantities.
3. Interpret pie charts and bar graphs.
4. Make a histogram of the distribution of a quantitative variable.
5. Make a stemplot of the distribution of a small set of observations. Round leaves or split stems as needed to make an effective stemplot.
6. Make a time plot of a quantitative variable over time. Recognize patterns such as trends and cycles in time plots.

C. Describing Distributions (Quantitative Variable)

1. Look for the overall pattern and for major deviations from the pattern.
2. Assess from a histogram or stemplot whether the shape of a distribution is roughly symmetric, distinctly skewed, or neither. Assess whether the distribution has one or more major peaks.
3. Describe the overall pattern by giving numerical measures of center and spread in addition to a verbal description of shape.
4. Decide which measures of center and spread are more appropriate: the mean and standard deviation (especially for symmetric distributions) or the five-number summary (especially for skewed distributions).
5. Recognize outliers and give plausible explanations for them.

D. Numerical Summaries of Distributions

1. Find the median M and the quartiles Q_1 and Q_3 for a set of observations.
2. Find the five-number summary and draw a boxplot; assess center, spread, symmetry, and skewness from a boxplot.

3. Find the mean $\bar{x}$ and the standard deviation s for a set of observations.

4. Understand that the median is more resistant than the mean. Recognize that skewness in a distribution moves the mean away from the median toward the long tail.

5. Know the basic properties of the standard deviation: $s \geq 0$ always; $s = 0$ only when all observations are identical and increases as the spread increases; s has the same units as the original measurements; s is pulled strongly up by outliers or skewness.

E. Density Curves and Normal Distributions

1. Know that areas under a density curve represent proportions of all observations and that the total area under a density curve is 1.

2. Approximately locate the median (equal-areas point) and the mean (balance point) on a density curve.

3. Know that the mean and median both lie at the center of a symmetric density curve and that the mean moves farther toward the long tail of a skewed curve.

4. Recognize the shape of Normal curves and estimate by eye both the mean and standard deviation from such a curve.

5. Use the 68–95–99.7 rule and symmetry to state what percent of the observations from a Normal distribution fall between two points when both points lie at the mean or one, two, or three standard deviations on either side of the mean.

6. Find the standardized value (z-score) of an observation. Interpret z-scores and understand that any Normal distribution becomes standard Normal $N(0, 1)$ when standardized.

7. Given that a variable has a Normal distribution with a stated mean μ and standard deviation σ, calculate the proportion of values above a stated number, below a stated number, or between two stated numbers.

8. Given that a variable has a Normal distribution with a stated mean μ and standard deviation σ, calculate the point having a stated proportion of all values above it or below it.

F. Scatterplots and Correlation

1. Make a scatterplot to display the relationship between two quantitative variables measured on the same subjects. Place the explanatory variable (if any) on the horizontal scale of the plot.

2. Add a categorical variable to a scatterplot by using a different plotting symbol or color.

3. Describe the direction, form, and strength of the overall pattern of a scatterplot. In particular, recognize positive or negative association and linear (straight-line) patterns. Recognize outliers in a scatterplot.

4. Judge whether it is appropriate to use correlation to describe the relationship between two quantitative variables. Find the correlation r.

5. Know the basic properties of correlation: r measures the direction and strength of only straight-line relationships; r is always a number between -1 and 1; $r = \pm 1$ only for perfect straight-line relationships; r moves away from 0 toward ± 1 as the straight-line relationship gets stronger.

G. Regression Lines

1. Understand that regression requires an explanatory variable and a response variable. Use a calculator or software to find the least-squares regression line of a response variable y on an explanatory variable x from data.

2. Explain what the slope b and the intercept a mean in the equation $\hat{y} = a + bx$ of a regression line.

3. Draw a graph of a regression line when you are given its equation.

4. Use a regression line to predict y for a given x. Recognize extrapolation and be aware of its dangers.

5. Find the slope and intercept of the least-squares regression line from the means and standard deviations of x and y and their correlation.

6. Use r^2, the square of the correlation, to describe how much of the variation in one variable can be accounted for by a straight-line relationship with another variable.

7. Recognize outliers and potentially influential observations from a scatterplot with the regression line drawn on it.

8. Calculate the residuals and plot them against the explanatory variable x. Recognize that a residual plot magnifies the pattern of the scatterplot of y versus x.

H. Cautions about Correlation and Regression

1. Understand that both r and the least-squares regression line can be strongly influenced by a few extreme observations.

2. Recognize possible lurking variables that may explain the observed association between two variables x and y.

3. Understand that even a strong correlation does not mean that there is a cause-and-effect relationship between x and y.

4. Give plausible explanations for an observed association between two variables: direct cause and effect, the influence of lurking variables, or both.

I. Categorical Data (Optional)

1. From a two-way table of counts, find the marginal distributions of both variables by obtaining the row sums and column sums.

2. Express any distribution in percents by dividing the category counts by their total.

3. Describe the relationship between two categorical variables by computing and comparing percents. Often this involves comparing the conditional distributions of one variable for the different categories of the other variable.

4. Recognize Simpson's paradox and be able to explain it.

Driving in Canada

Canada is a civilized and restrained nation, at least in the eyes of Americans. A survey sponsored by the Canada Safety Council suggests that driving in Canada may be more adventurous than expected. Of the Canadian drivers surveyed, 88% admitted to aggressive driving in the past year, and 76% said that sleep-deprived drivers were common on Canadian roads. What really alarms us is the name of the survey: the Nerves of Steel Aggressive Driving Study.

Review exercises help you solidify the basic ideas and skills in Chapters 1 to 6.

7.1 Describing colleges. Popular magazines rank colleges and universities on their academic quality in serving undergraduate students. Below are several variables that might contribute to ranking colleges. Which of these are categorical and which are quantitative?

(a) Percent of freshmen who eventually graduate.

(b) College type: liberal arts college, national university, etc.

(c) Require SAT or ACT for admission (required, recommended, not used)?

(d) Mean faculty salary.

7.2 Data on mice. For a biology project, you measure the tail length in centimeters and weight in grams of 12 mice of the same variety. What units of measurement do each of the following have?

(a) The mean length of the tails.

(b) The first quartile of the tail lengths.

(c) The standard deviation of the tail lengths.

(d) The correlation between tail length and weight.

7.3 What do you think of Microsoft? The Pew Research Center asked a random sample of adults whether they had favorable or unfavorable opinions of a number of major companies. Answers to such questions depend a lot on recent news. Here are the percents with favorable opinions for several of the companies:[1]

Company	Percent favorable
Apple	71
Ben and Jerry's	59
Coors	53
Exxon/Mobil	44
Google	73
Halliburton	25
McDonald's	71
Microsoft	78
Starbucks	64
Wal-Mart	68

Make a graph that displays these data.

7.4 The state of the country. The Pew Research Center reports the following percents of American adults who said "satisfied" when asked, "All in all, are you satisfied or dissatisfied with the way things are going in this country today?"[2]

Year	Percent	Year	Percent	Year	Percent
1988	55	1996	29	2004	38
1990	47	1998	52	2005	36
1992	28	2000	52	2006	30
1994	24	2002	45	2007	30

Make a graph that displays the trend over time for these data.

7.5 How heavy are diamonds? Here are the weights (in milligrams) of 58 diamonds from a nodule carried up to the earth's surface in surrounding rock. This represents a single population of diamonds formed in a single event deep in the earth.[3]

© Gerald Cubitt

13.8	3.7	33.8	11.8	27.0	18.9	19.3	20.8	25.4	23.1	7.8	10.9
9.0	9.0	14.4	6.5	7.3	5.6	18.5	1.1	11.2	7.0	7.6	9.0
9.5	7.7	7.6	3.2	6.5	5.4	7.2	7.8	3.5	5.4	5.1	5.3
3.8	2.1	2.1	4.7	3.7	3.8	4.9	2.4	1.4	0.1	4.7	1.5
2.0	0.1	0.1	1.6	3.5	3.7	2.6	4.0	2.3	4.5		

Make a graph that shows the distribution of weights of diamonds. Describe the shape of the distribution and any outliers. Use numerical measures appropriate for the shape to describe the center and spread.

7.6 Garbage. The formal name for garbage is "municipal solid waste." Here is a breakdown of the materials that make up American municipal solid waste, in millions of tons:[4]

Material	Weight
Food	31.3
Glass	13.2
Metals	19.1
Paper, paperboard	85.3
Plastics	29.5
Rubber, leather, textiles	18.3
Wood	13.9
Yard trimmings	32.4
Other	8.3
Total	251.3

(a) Make a bar graph, ordering the bars from highest to lowest weight.

(b) If you use software, make a pie chart as well. In which graph is it easier to see small differences, such as between glass and wood?

7.7 Recycling. Of the municipal solid waste described in the previous exercise, about 55% is discarded, 32.5% is recovered through recycling or composting, and 12.5% is burned to produce energy. The table presents the percents of several materials in solid waste that are recycled.

Material	Percent recycled
Aluminum cans	45.1
Auto batteries	99.0
Glass containers	25.3
Paper, paperboard	51.6
Plastic bottles	31.0
Steel cans	62.9
Tires	34.9
Yard trimmings	62.0

(a) Make a bar graph, ordering the bars from highest to lowest percent.

(b) Could you also use a pie chart to display these data? Why?

7.8 Nitrogen in diamonds. Scientists made detailed chemical analyses of 24 of the diamonds from Exercise 7.5. The abundances of various elements give clues to how the diamonds were formed. Here are the data on nitrogen content, in parts per million:

487	1430	60	244	196	274	41	54	473	30	98	41
273	94	69	262	120	302	75	242	115	65	311	61

(a) Make a stemplot. What is the overall shape of the distribution?

(b) There is one extreme high outlier. Find the median, mean, and standard deviation with and without this outlier. Which of these measures is least changed when the outlier is removed? (If you recall Chapter 2's discussion of resistant measures, the result is not a surprise.)

7.9 Two species of pine trees. The Aleppo pine and the Torrey pine are widely planted as ornamental trees in Southern California. Here are the lengths (centimeters) of 15 Aleppo pine needles:[5]

10.2 7.2 7.6 9.3 12.1 10.9 9.4 11.3 8.5 8.5 12.8 8.7 9.0 9.0 9.4

Here are the lengths of 18 needles from Torrey pines:

33.7	21.2	26.8	29.7	21.6	21.7	33.7	32.5	23.1
23.7	30.2	29.0	24.2	24.2	25.5	26.6	28.9	29.7

Use five-number summaries and boxplots to compare the two distributions. Given only the length of a needle, do you think you could say which pine species it comes from?

7.10 Genetic engineering for cancer treatment. Here's a new idea for treating advanced melanoma, the most serious kind of skin cancer. Genetically engineer white blood cells to better recognize and destroy cancer cells, then infuse these cells into patients. The subjects in a small initial study were 11 patients whose melanoma had not responded to existing treatments. One question was how rapidly the new cells would multiply after infusion, as measured by the doubling time in days. Here are the doubling times:[6]

1.4 1.0 1.3 1.0 1.3 2.0 0.6 0.8 0.7 0.9 1.9

Graig Tuttle/CORBIS

Another outcome was the increase in the presence of cells that trigger an immune response in the body and so may help fight cancer. Here are the increases, in counts of active cells per 100,000 cells:

$$27 \quad 7 \quad 0 \quad 215 \quad 20 \quad 700 \quad 13 \quad 510 \quad 34 \quad 86 \quad 108$$

Make stemplots of both distributions (use split stems and leave out any extreme outliers). Describe the overall shapes and any outliers. Then give the five-number summary for both. (We can't compare the summaries because the two variables have different scales.)

7.11 Detecting outliers (optional). In Exercise 7.9 you gave five-number summaries for two distributions of lengths of pine needles. Do the data contain any observations that are suspected outliers by the $1.5 \times IQR$ rule?

7.12 Detecting outliers (optional). In Exercise 7.10 you gave five-number summaries for the distributions of two responses to a new cancer treatment. Do the data contain any observations that are suspected outliers by the $1.5 \times IQR$ rule?

7.13 Distribution shapes. Biologists commonly act as if measurements on many individuals from the same species follow a Normal distribution. They therefore use the mean $\bar{x}$ and standard deviation s as numerical summaries.

(a) Make stemplots of the lengths of needles of Aleppo pines and of Torrey pines from Exercise 7.9. Which distribution appears "more Normal" and why? (The distribution of a small number of observations often has an irregular shape even when, as here, more data would show that the overall distribution is close to Normal.)

(b) Find the mean and median length for both species. What fact about the distributions explains why the mean and median are close together for both species?

7.14 Weights aren't Normal. The heights of people of the same sex and similar ages follow a Normal distribution reasonably closely. Weights, on the other hand, are not Normally distributed. The weights of women aged 20 to 29 have mean 141.7 pounds and median 133.2 pounds. The first and third quartiles are 118.3 pounds and 157.3 pounds. What can you say about the shape of the weight distribution? Why?

7.15 Pine needles. The lengths of needles from Aleppo pines follow approximately the Normal distribution with mean 9.6 centimeters (cm) and standard deviation 1.6 cm. According to the 68–95–99.7 rule, what range of lengths covers the center 95% of Aleppo pine needles? What percent of needles are less than 6.4 cm long?

7.16 Body mass index. Your body mass index (BMI) is your weight in kilograms divided by the square of your height in meters. Many online BMI calculators allow you to enter weight in pounds and height in inches. High BMI is a common but controversial indicator of overweight or obesity. A study by the National Center for Health Statistics found that the BMI of American young women (ages 20 to 29) is approximately Normal with mean 26.8 and standard deviation 7.4.[7]

(a) People with BMI less than 18.5 are often classed as "underweight." What percent of young women are underweight by this criterion?

(b) People with BMI over 30 are often classed as "obese." What percent of young women are obese by this criterion?

7.17 The Medical College Admission Test. Almost all medical schools in the United States require applicants to take the Medical College Admission Test (MCAT). The scores of applicants on the biological sciences part of the MCAT in 2007 were approximately Normal with mean 9.6 and standard deviation 2.2. For applicants who actually entered medical school, the mean score was 10.6 and the standard deviation was 1.7.[8]

(a) What percent of all applicants had scores higher than 13?

(b) What percent of those who entered medical school had scores between 8 and 12?

7.18 Breaking bolts. Mechanical measurements on supposedly identical objects usually vary. The variation often follows a Normal distribution. The stress required to break a type of bolt varies Normally with mean 75 kilopounds per square inch (ksi) and standard deviation 8.3 ksi.

(a) What percent of these bolts will withstand a stress of 90 ksi without breaking?

(b) What range covers the middle 50% of breaking strengths for these bolts?

Soap in the shower. *From Rex Boggs in Australia comes an unusual data set: before showering in the morning, he weighed the bar of soap in his shower stall. The weight goes down as the soap is used. The data appear below (weights in grams). Notice that Mr. Boggs forgot to weigh the soap on some days. Exercises 7.19 to 7.21 are based on the soap data set.*

Day	Weight	Day	Weight	Day	Weight
1	124	8	84	16	27
2	121	9	78	18	16
5	103	10	71	19	12
6	96	12	58	20	8
7	90	13	50	21	6

7.19 Scatterplot and correlation. Plot the weight of the bar of soap against day. Is the overall pattern roughly linear? Based on your scatterplot, is the correlation between day and weight close to 1, positive but not close to 1, close to 0, negative but not close to −1, or close to −1? Explain your answer. Then find the correlation r to verify what you concluded from the graph.

7.20 Regression. Find the equation of the least-squares regression line for predicting soap weight from day.

(a) What is the equation? Explain what it tells us about the rate at which the soap lost weight.

(b) Mr. Boggs did not measure the weight of the soap on Day 4. Use the regression equation to predict that weight.

(c) Add the regression line to your scatterplot from the previous exercise.

7.21 Prediction? Use the regression equation in the previous exercise to predict the weight of the soap after 30 days. Why is it clear that your answer makes no sense? What's wrong with using the regression line to predict weight after 30 days?

Beer in South Dakota

Take a break from doing exercises to apply your math to beer cans in South Dakota. A newspaper there reported that every year an average of 650 beer cans per mile are tossed onto the state's highways. South Dakota has about 83,000 miles of roads. How many beer cans is that in all? The Census Bureau says that there are about 780,000 people in South Dakota. How many beer cans does each man, woman, and child in the state toss on the road each year? That's pretty impressive. Maybe the paper got its numbers wrong.

7.22 Growing icicles. Table 4.2 (page 118) gives data on the growth of icicles over time. Let's look again at Run 8903, for which a slower flow of water produces faster growth.

(a) How can you tell from a calculation, without drawing a scatterplot, that the pattern of growth is very close to a straight line?

(b) What is the equation of the least-squares regression line for predicting an icicle's length from time in minutes under these conditions?

(c) Predict the length of an icicle after one full day. This prediction can't be trusted. Why not?

7.23 Thin monkeys, fat monkeys. Animals and people that take in more energy than they expend will get fatter. Here are data on 12 rhesus monkeys: 6 lean monkeys (4% to 9% body fat) and 6 obese monkeys (13% to 44% body fat). The data report the energy expended in 24 hours (kilojoules per minute) and the lean body mass (kilograms, leaving out fat) for each monkey.[9]

Lean		Obese	
Mass	Energy	Mass	Energy
6.6	1.17	7.9	0.93
7.8	1.02	9.4	1.39
8.9	1.46	10.7	1.19
9.8	1.68	12.2	1.49
9.7	1.06	12.1	1.29
9.3	1.16	10.8	1.31

(a) What is the mean lean body mass of the lean monkeys? Of the obese monkeys? Because animals with higher lean mass usually expend more energy, we can't directly compare energy expended.

(b) Instead, look at how energy expended is related to body mass. Make a scatterplot of energy versus mass, using different plot symbols for lean and obese monkeys. Then add to the plot two regression lines, one for lean monkeys and one for obese monkeys. What do these lines suggest about the monkeys?

7.24 The end of smoking? The number of adult Americans who smoke continues to drop. Here are estimates of the percent of adults (aged 18 and over) who were smokers in the years between 1965 and 2006:[10]

Year	1965	1974	1979	1983	1987	1990	1993	1997	2000	2002	2006
Smokers	41.9	37.0	33.3	31.9	28.6	25.3	24.8	24.6	23.1	22.5	20.8

(a) Make a scatterplot of these data. There is a strong negative linear association. Find the least-squares regression line for predicting percent of smokers from year and add the line to your plot.

(b) According to your regression line, how much did smoking decline per year during this period, on the average? What percent of the observed variation in percent of adults who smoke can be explained by linear change over time?

(c)   One of the government's national health objectives was to reduce smoking to no more than 12% of adults by 2010. Use your regression line to predict the percent of adults who smoke in 2010. Did it appear that this health objective would be met?

7.25 Squirrels and their food supply. That animal species produce more offspring when their supply of food goes up isn't surprising. That some animals appear able to anticipate unusual food abundance is more surprising. Red squirrels eat seeds from pine cones, a food source that occasionally has very large crops (called seed masting). Here are data on an index of the abundance of pine cones and average number of offspring per female over 16 years:[11]

© Don Johnston

Cone index x	0.00	2.02	0.25	3.22	4.68	0.31	3.37	3.09
Offspring ŷ	1.49	1.10	1.29	2.71	4.07	1.29	3.36	2.41
Cone index x	2.44	4.81	1.88	0.31	1.61	1.88	0.91	1.04
Offspring ŷ	1.97	3.41	1.49	2.02	3.34	2.41	2.15	2.12

Describe the relationship with both a graph and numerical measures, then summarize in words. What is striking is that the offspring are conceived in the spring, *before* the cones mature in the fall to feed the new young squirrels through the winter.

7.26 The end of smoking, continued. Use your regression line from Exercise 7.24 to predict the percent of adults who will smoke in 2050. Why is your result impossible? Why was it foolish to use the regression line for this prediction?

7.27 Monkey calls. The usual way to study the brain's response to sounds is to have subjects listen to "pure tones." The response to recognizable sounds may differ. To compare responses, researchers anesthetized macaque monkeys. They fed

TABLE 7.1 Neuron response to tones and monkey calls

NEURON	TONE	CALL	NEURON	TONE	CALL	NEURON	TONE	CALL
1	474	500	14	145	42	26	71	134
2	256	138	15	141	241	27	68	65
3	241	485	16	129	194	28	59	182
4	226	338	17	113	123	29	59	97
5	185	194	18	112	182	30	57	318
6	174	159	19	102	141	31	56	201
7	176	341	20	100	118	32	47	279
8	168	85	21	74	62	33	46	62
9	161	303	22	72	112	34	41	84
10	150	208	23	20	193	35	26	203
11	19	66	24	21	129	36	28	192
12	20	54	25	26	135	37	31	70
13	35	103						

pure tones and also monkey calls directly to their brains by inserting electrodes. Response to the stimulus was measured by the firing rate (electrical spikes per second) of neurons in various areas of the brain. Table 7.1 contains the responses for 37 neurons.[12]

(a) One important finding is that responses to monkey calls are generally stronger than responses to pure tones. For how many of the 37 neurons is this true?

(b) We might expect some neurons to have strong responses to any stimulus and others to have consistently weak responses. There would then be a strong relationship between tone response and call response. Make a scatterplot of monkey call response against pure tone response (explanatory variable). Find the correlation r between tone and call responses. How strong is the linear relationship?

7.28 Remember what you ate. How well do people remember their past diet? Data are available for 91 people who were asked about their diet when they were 18 years old. Researchers asked them at about age 55 to describe their eating habits at age 18. For each subject, the researchers calculated the correlation between actual intakes of many foods at age 18 and the intakes the subjects now remember. The median of the 91 correlations was $r = 0.217$. The authors say, "We conclude that memory of food intake in the distant past is fair to poor."[13] Explain why $r = 0.217$ points to this conclusion.

7.29 Statistics for investing. Joe's retirement plan invests in stocks through an "index fund" that follows the behavior of the stock market as a whole, as measured by the S&P 500 stock index. Joe wants to buy a mutual fund that does not track the index closely. He reads that monthly returns from Fidelity Technology Fund have correlation $r = 0.77$ with the S&P 500 index and that Fidelity Real Estate Fund has correlation $r = 0.37$ with the index.

(a) Which of these funds has the closer relationship to returns from the stock market as a whole? How do you know?

(b) Does the information given tell Joe anything about which fund has had higher returns?

7.30 Moving in step? One reason to invest abroad is that markets in different countries don't move in step. When American stocks go down, foreign stocks may go up. So an investor who holds both bears less risk. That's the theory. Now we read: "The correlation between changes in American and European share prices has risen from 0.4 in the mid-1990s to 0.8 in 2000."[14] Explain to an investor who knows no statistics why this fact reduces the protection provided by buying European stocks.

7.31 Interpreting correlation. The same article that claims that the correlation between changes in stock prices in Europe and the United States is 0.8 goes on to say: "Crudely, that means that movements on Wall Street can explain 80% of price movements in Europe." Is this true? What is the correct percent explained if $r = 0.8$?

7.32 Weeds among the corn. Lamb's-quarter is a common weed that interferes with the growth of corn. An agriculture researcher planted corn at the same rate in 16 small plots of ground, then weeded the plots by hand to allow a fixed number of

Scott Camazine/Photo Researchers

lamb's-quarter plants to grow in each meter of corn row. No other weeds were allowed to grow. Here are the yields of corn (bushels per acre) in each of the plots:[15]

Weeds per meter	Corn yield	Weeds per meter	Corn yield	Weeds per meter	Corn yield	Weeds per meter	Corn yield
0	166.7	1	166.2	3	158.6	9	162.8
0	172.2	1	157.3	3	176.4	9	142.4
0	165.0	1	166.7	3	153.1	9	162.8
0	176.9	1	161.1	3	156.0	9	162.4

(a)   What are the explanatory and response variables in this experiment?

(b)   Make side-by-side stemplots of the yields, after rounding to the nearest bushel. Give the mean yield for each group (using the unrounded data). What do you conclude about the effect of this weed on corn yield?

(c)   With only 4 observations in each group, it isn't surprising that the "3 weeds"and "9 weeds" groups each have a possible outlier. Find the median yield for each group. How does switching from means to medians affect your conclusion?

7.33 Weeds among the corn, continued. We can also use regression to analyze the data on weeds and corn yield. The advantage of regression over the side-by-side comparison in the previous exercise is that we can use the fitted line to draw conclusions for counts of weeds other than the ones the researcher actually used.

(a)   Make a scatterplot of corn yield against weeds per meter. Find the least-squares regression line and add it to your plot. What does the slope of the fitted line tell us about the effect of lamb's-quarter on corn yield?

(b)   Predict the yield for corn grown under these conditions with 6 lamb's-quarter plants per meter of row. (The small number of observations and possible outliers make this prediction unreliable.)

7.34 Catalog shopping (optional). What is the most important reason that students buy from catalogs? The answer may differ for different groups of students. Here are counts for samples of American and East Asian students at a large midwestern university:[16]

Reason	American	Asian
Save time	29	10
Easy	28	11
Low price	17	34
Live far from stores	11	4
No pressure to buy	10	3
Other	20	7
Total	115	69

(a)   Give the marginal distribution of reasons for all students, in percents.

(b)   Give the two conditional distributions of reasons, for American and for East Asian students. Make a graph of each conditional distribution.

(c)   What are the most important differences between the two groups of students?

S U P P L E M E N T A R Y   E X E R C I S E S

*Supplementary exercises apply the skills you have learned in ways that require more thought or more elaborate use of technology. Some of these exercises ask you to follow the **Plan, Solve,** and **Conclude** steps of the four-step process introduced on page 55.*

7.35 The Mississippi River. Table 7.2 gives the volume of water discharged by the Mississippi River into the Gulf of Mexico for each year from 1954 to 2001.[17] The units are cubic kilometers of water—the Mississippi is a big river.

TABLE 7.2 Yearly discharge (cubic kilometers of water) of the Mississippi River

YEAR	DISCHARGE	YEAR	DISCHARGE	YEAR	DISCHARGE	YEAR	DISCHARGE
1954	290	1966	410	1978	560	1990	680
1955	420	1967	460	1979	800	1991	700
1956	390	1968	510	1980	500	1992	510
1957	610	1969	560	1981	420	1993	900
1958	550	1970	540	1982	640	1994	640
1959	440	1971	480	1983	770	1995	590
1960	470	1972	600	1984	710	1996	670
1961	600	1973	880	1985	680	1997	680
1962	550	1974	710	1986	600	1998	690
1963	360	1975	670	1987	450	1999	580
1964	390	1976	420	1988	420	2000	390
1965	500	1977	430	1989	630	2001	580

(a) Make a graph of the distribution of water volume. Describe the overall shape of the distribution and any outliers.

(b) Based on the shape of the distribution, do you expect the mean to be close to the median, clearly less than the median, or clearly greater than the median? Why? Find the mean and the median to check your answer.

(c) Based on the shape of the distribution, does it seem reasonable to use $\overline{x}$ and s to describe the center and spread of this distribution? Why? Find $\overline{x}$ and s if you think they are a good choice. Otherwise, find the five-number summary.

7.36 More on the Mississippi River. The data in Table 7.2 are a time series. Make a time plot that shows how the volume of water in the Mississippi changed between 1954 and 2001. What does the time plot reveal that the histogram from the previous exercise does not? It is a good idea to always make a time plot of time series data because a histogram cannot show changes over time.

Falling through the ice. *The Nenana Ice Classic is an annual contest to guess the exact time in the spring thaw when a tripod erected on the frozen Tanana River near Nenana, Alaska, will fall through the ice. The 2007 jackpot prize was $303,000. The contest has*

2006 Bill Watkins/Alaska Stock.com

TABLE 7.3 Days from April 20 for the Tanana River tripod to fall

YEAR	DAY	YEAR	DAY	YEAR	DAY	YEAR	DAY	YEAR	DAY	YEAR	DAY
1917	11	1933	19	1949	25	1965	18	1981	11	1997	11
1918	22	1934	11	1950	17	1966	19	1982	21	1998	1
1919	14	1935	26	1951	11	1967	15	1983	10	1999	10
1920	22	1936	11	1952	23	1968	19	1984	20	2000	12
1921	22	1937	23	1953	10	1969	9	1985	23	2001	19
1922	23	1938	17	1954	17	1970	15	1986	19	2002	18
1923	20	1939	10	1955	20	1971	19	1987	16	2003	10
1924	22	1940	1	1956	12	1972	21	1988	8	2004	5
1925	16	1941	14	1957	16	1973	15	1989	12	2005	9
1926	7	1942	11	1958	10	1974	17	1990	5	2006	13
1927	23	1943	9	1959	19	1975	21	1991	12	2007	8
1928	17	1944	15	1960	13	1976	13	1992	25		
1929	16	1945	27	1961	16	1977	17	1993	4		
1930	19	1946	16	1962	23	1978	11	1994	10		
1931	21	1947	14	1963	16	1979	11	1995	7		
1932	12	1948	24	1964	31	1980	10	1996	16		

been run since 1917. Table 7.3 gives simplified data that record only the date on which the tripod fell each year. The earliest date so far is April 20. To make the data easier to use, the table gives the date each year in days starting with April 20. That is, April 20 is 1, April 21 is 2, and so on. Exercises 7.37 to 7.39 concern these data.[18]

7.37 When does the ice break up? We have 91 years of data on the date of ice breakup on the Tanana River. Describe the distribution of the breakup date with both a graph or graphs and appropriate numerical summaries. What is the median date (month and day) for ice breakup?

7.38 Global warming? Because of the high stakes, the falling of the tripod has been carefully observed for many years. If the date the tripod falls has been getting earlier, that may be evidence for the effects of global warming.

(a) Make a time plot of the date the tripod falls against year.

(b) There is a great deal of year-to-year variation. Fitting a regression line to the data may help us see the trend. Fit the least-squares line and add it to your time plot. What do you conclude?

(c) There is much variation about the line. Give a numerical description of how much of the year-to-year variation in ice breakup time is accounted for by the time trend represented by the regression line. (This simple example is typical of more complex evidence for the effects of global warming: large year-to-year variation requires many years of data to see a trend.)

7.39 More on global warming. Side-by-side boxplots offer a different look at the data. Group the data into periods of roughly equal length: 1917 to 1939, 1940 to 1962,

1963 to 1985, and 1986 to 2007. Make boxplots to compare ice breakup dates in these four time periods. Write a brief description of what the plots show.

7.40 Big government? The data file *ex07-40.dat* on the text CD and Web site contains the percent of gross domestic product (GDP, the total value of all goods and services a country produces) taken by the government in 82 countries. For example, the government share of GDP is 12.28% in Canada and 10.54% in the United States.[19]

(a) Make a stemplot or a histogram to display the distribution of government share of GDP.

(b) There are several high outliers. What countries are these? (In the most extreme case, the government took more than the total annual GDP!) What is the overall shape of the distribution if you ignore the outliers?

(c) Based on your work in (b), give a numerical summary of the center and spread of the distribution, omitting the outliers.

(d) Some Americans complain about big government and heavy taxes. Where does the United States (10.54%) stand in this international comparison?

7.41 Cicadas as fertilizer? Every 17 years, swarms of cicadas emerge from the ground in the eastern United States, live for about six weeks, then die. (There are several "broods," so we experience cicada eruptions more often than every 17 years.) There are so many cicadas that their dead bodies can serve as fertilizer and increase plant growth. In an experiment, a researcher added 10 cicadas under some plants in a natural plot of American bellflowers in a forest, leaving other plants undisturbed. One of the response variables was the size of seeds produced by the plants. Here are data (seed mass in milligrams) for 39 cicada plants and 33 undisturbed (control) plants:[20]

Cicada plants				Control plants			
0.237	0.277	0.241	0.142	0.212	0.188	0.263	0.253
0.109	0.209	0.238	0.277	0.261	0.265	0.135	0.170
0.261	0.227	0.171	0.235	0.203	0.241	0.257	0.155
0.276	0.234	0.255	0.296	0.215	0.285	0.198	0.266
0.239	0.266	0.296	0.217	0.178	0.244	0.190	0.212
0.238	0.210	0.295	0.193	0.290	0.253	0.249	0.253
0.218	0.263	0.305	0.257	0.268	0.190	0.196	0.220
0.351	0.245	0.226	0.276	0.246	0.145	0.247	0.140
0.317	0.310	0.223	0.229	0.241			
0.192	0.201	0.211					

Alastair Shay; Papilio/CORBIS

Describe and compare the two distributions. Do the data support the idea that dead cicadas can serve as fertilizer?

7.42 A big toe problem. Hallux abducto valgus (call it HAV) is a deformation of the big toe that is not common in youth and often requires surgery. Doctors used X-rays to measure the angle (in degrees) of deformity in 38 consecutive patients under the age of 21 who came to a medical center for surgery to correct HAV.[21] The angle is a measure of the seriousness of the deformity. The data appear in Table 7.4 as "HAV angle." Describe the distribution of the angle of deformity among young patients needing surgery for this condition.

TABLE 7.4 Angle of deformity (degrees) for two types of foot deformity

HAV ANGLE	MA ANGLE	HAV ANGLE	MA ANGLE	HAV ANGLE	MA ANGLE
28	18	21	15	16	10
32	16	17	16	30	12
25	22	16	10	30	10
34	17	21	7	20	10
38	33	23	11	50	12
26	10	14	15	25	25
25	18	32	12	26	30
18	13	25	16	28	22
30	19	21	16	31	24
26	10	22	18	38	20
28	17	20	10	32	37
13	14	18	15	21	23
20	20	26	16		

7.43 Prey attract predators. Here is one way in which nature regulates the size of animal populations: high population density attracts predators, who remove a higher proportion of the population than when the density of the prey is low. One study looked at kelp perch and their common predator, the kelp bass. The researcher set up four large circular pens on sandy ocean bottom in Southern California. He chose young perch at random from a large group and placed 10, 20, 40, and 60 perch in the four pens. Then he dropped the nets protecting the pens, allowing bass to swarm in, and counted the perch left after 2 hours. Here are data on the proportions of perch eaten in four repetitions of this setup:[22]

Perch	Proportion killed			
10	0.0	0.1	0.3	0.3
20	0.2	0.3	0.3	0.6
40	0.075	0.3	0.6	0.725
60	0.517	0.55	0.7	0.817

Do the data support the principle that "more prey attract more predators, who drive down the number of prey"?

7.44 Predicting foot problems. Metatarsus adductus (call it MA) is a turning in of the front part of the foot that is common in adolescents and usually corrects itself. Table 7.4 gives the severity of MA ("MA angle") as well. Doctors speculate that the severity of MA can help predict the severity of HAV. Describe the relationship between MA and HAV. Do you think the data confirm the doctors' speculation? Why or why not?

7.45 Change in the Serengeti. Long-term records from the Serengeti National Park in Tanzania show interesting ecological relationships. When wildebeest are more abundant, they graze the grass more heavily, so there are fewer fires and more trees grow. Lions feed more successfully when there are more trees, so the lion population increases. Here are data on one part of this cycle, wildebeest abundance (in thousands of animals) and the percent of the grass area that burned in the same year:[23]

Gallo Image—Anthony Bannister/Getty Images

Wildebeest (1000s)	Percent burned	Wildebeest (1000s)	Percent burned	Wildebeest (1000s)	Percent burned
396	56	360	88	1147	32
476	50	444	88	1173	31
698	25	524	75	1178	24
1049	16	622	60	1253	24
1178	7	600	56	1249	53
1200	5	902	45		
1302	7	1440	21		

To what extent do these data support the claim that more wildebeest reduce the percent of grasslands that burn? How rapidly does burned area decrease as the number of wildebeest increases? Include a graph and suitable calculations.

7.46 Casting aluminum. In casting metal parts, molten metal flows through a "gate" into a die that shapes the part. The gate velocity (the speed at which metal is forced through the gate) plays a critical role in die casting. A firm that casts cylindrical aluminum pistons examined 12 types formed from the same alloy. How does the piston wall thickness (inches) influence the gate velocity (feet per second) chosen by the skilled workers who do the casting? If there is a clear pattern, it can be used to direct new workers or to automate the process. Analyze these data and report your findings.[24]

Thickness	Velocity	Thickness	Velocity	Thickness	Velocity
0.248	123.8	0.524	228.6	0.697	145.2
0.359	223.9	0.552	223.8	0.752	263.1
0.366	180.9	0.628	326.2	0.806	302.4
0.400	104.8	0.697	302.4	0.821	302.4

7.47 How are schools doing? (optional) The nonprofit group Public Agenda conducted telephone interviews with parents of high school children. Interviewers chose equal numbers of black, Hispanic, and non-Hispanic white parents at random. One question asked was "Are the high schools in your state doing an excellent, good, fair or poor job, or don't you know enough to say?" The survey results[25] are presented in the table.

Opinion	Black parents	Hispanic parents	White parents
Excellent	12	34	22
Good	69	55	81
Fair	75	61	60
Poor	24	24	24
Don't know	22	28	14
Total	202	202	201

Write a brief analysis of these results that focuses on the relationship between parent group and opinions about schools.

7.48 Influence: hot mutual funds? Investment advertisements always warn that "past performance does not guarantee future results." Here is an example that shows why you should pay attention to this warning. Stocks fell sharply in 2002, then rose sharply in 2003. The table below gives the percent returns from 23 Fidelity Investments "sector funds" in these two years. Sector funds invest in narrow segments of the stock market. They often rise and fall faster than the market as a whole.

2002 return	2003 return	2002 return	2003 return	2002 return	2003 return
−17.1	23.9	−0.7	36.9	−37.8	59.4
−6.7	14.1	−5.6	27.5	−11.5	22.9
−21.1	41.8	−26.9	26.1	−0.7	36.9
−12.8	43.9	−42.0	62.7	64.3	32.1
−18.9	31.1	−47.8	68.1	−9.6	28.7
−7.7	32.3	−50.5	71.9	−11.7	29.5
−17.2	36.5	−49.5	57.0	−2.3	19.1
−11.4	30.6	−23.4	35.0		

(a) Make a scatterplot of 2003 return (response) against 2002 return (explanatory). The funds with the best performance in 2002 tend to have the worst performance in 2003. Fidelity Gold Fund, the only fund with a positive return in both years, is an extreme outlier.

(b) To demonstrate that correlation is not resistant, find r for all 23 funds and then find r for the 22 funds other than Gold. Explain from Gold's position in your plot why omitting this point makes r more negative.

(c) Find the equations of two least-squares lines for predicting 2003 return from 2002 return, one for all 23 funds and one omitting Fidelity Gold Fund. Add both lines to your scatterplot. Starting with the least-squares idea, explain why adding Fidelity Gold Fund to the other 22 funds moves the line in the direction that your graph shows.

7.49 Influence: monkey calls. Table 7.1 (page 188) contains data on the response of 37 monkey neurons to pure tones and to monkey calls. You made a scatterplot of these data in Exercise 7.27.

(a) Find the least-squares line for predicting a neuron's call response from its pure tone response. Add the line to your scatterplot. Mark on your plot the point (call it A) with the largest residual (either positive or negative) and also the point (call it B) that is an outlier in the x direction.

(b) How influential are each of these points for the correlation r?

(c) How influential are each of these points for the regression line?

7.50 Influence: bushmeat. Table 4.3 (page 123) gives data on fish catches in a region of West Africa and the percent change in the biomass (total weight) of 41 animals in nature reserves. It appears that years with smaller fish catches see greater declines in animals, probably because local people turn to "bushmeat" when other sources of protein are not available. The next year (1999) had a fish catch of 23.0 kilograms per person and animal biomass change of −22.9%.

(a) Make a scatterplot that shows how change in animal biomass depends on fish catch. Be sure to include the additional data point. Describe the overall pattern. The added point is a low outlier in the y direction.

(b) Find the correlation between fish catch and change in animal biomass both with and without the outlier. The outlier is influential for correlation. Explain from your plot why adding the outlier makes the correlation smaller.

(c) Find the least-squares line for predicting change in animal biomass from fish catch both with and without the additional data point for 1999. Add both lines to your scatterplot from (a). The outlier is not influential for the least-squares line. Explain from your plot why this is true.

From Exploration to Inference

The purpose of statistics is to gain understanding from data. We can seek understanding in different ways, depending on the circumstances. We have studied one approach to data, *exploratory data analysis*, in some detail. Now we move from data analysis toward *statistical inference*. Both types of reasoning are essential to effective work with data. Here is a brief sketch of the differences between them.

EXPLORATORY DATA ANALYSIS	STATISTICAL INFERENCE
Purpose is unrestricted exploration of the data, searching for interesting patterns.	Purpose is to answer specific questions, posed before the data were produced.
Conclusions apply only to the individuals and circumstances for which we have data in hand.	Conclusions apply to a larger group of individuals or a broader class of circumstances.
Conclusions are informal, based on what we see in the data.	Conclusions are formal, backed by a statement of our confidence in them.

Our journey toward inference begins in Chapters 8 and 9, which describe statistical designs for *producing data* by samples and experiments. The conclusions of inference use the language of *probability*, the mathematics of chance. Chapters 10 and 11 present the ideas we need, and the optional Chapters 12 and 13 add more detail. Armed with designs for producing trustworthy data, data analysis to examine the data, and the language of probability, we are prepared to understand the big ideas of inference in Chapters 14 and 15. These chapters are the foundation for the discussion of inference in practice that occupies the rest of the book.

Manoj Shah/Getty

PART II

John Elk III/Alamy

Producing Data: Sampling

Statistics, the science of data, provides ideas and tools that we can use in many settings. Sometimes we have data that describe a group of individuals and want to learn what the data say. That's the job of exploratory data analysis. Sometimes we have specific questions but no data to answer them. To get sound answers, we must *produce data* in a way that is designed to answer our questions.

Suppose our question is "What percent of college students think that people should not obey laws that violate their personal values?" To answer the question, we interview undergraduate college students. We can't afford to ask all students, so we put the question to a *sample* chosen to represent the entire student *population*. How shall we choose a sample that truly represents the opinions of the entire population? Statistical designs for choosing samples are the topic of this chapter. We will see that

- a sound statistical design is necessary if we are to trust data from a sample;

- in sampling from large human populations, however, "practical problems" can overwhelm even sound designs; and

- the impact of technology (particularly cell phones and the Web) is making it harder to produce trustworthy national data by sampling.

Population versus sample

A political scientist wants to know what percent of college-age adults consider themselves conservatives. An automaker hires a market research firm to learn what percent of adults aged 18 to 35 recall seeing television advertisements for a new gas-electric hybrid car. Government economists inquire about average household income. In all these cases, we want to gather information about a large group of individuals. Time, cost, and inconvenience forbid contacting every individual. So we gather information about only part of the group in order to draw conclusions about the whole.

POPULATION, SAMPLE, SAMPLING DESIGN

The **population** in a statistical study is the entire group of individuals about which we want information.

A **sample** is a part of the population from which we actually collect information. We use a sample to draw conclusions about the entire population.

A **sampling design** describes exactly how to choose a sample from the population.

Pay careful attention to the details of the definitions of "population" and "sample." Look at Exercise 8.1 right now to check your understanding.

We often draw conclusions about a whole on the basis of a sample. Everyone has tasted a sample of ice cream and ordered a cone on the basis of that taste. But ice cream is uniform, so that the single taste represents the whole. Choosing a representative sample from a large and varied population is not so easy. The first *sample survey* step in planning a **sample survey** is to say exactly *what population* we want to describe. The second step is to say exactly *what we want to measure*, that is, to give exact definitions of our variables. These preliminary steps can be complicated, as the following example illustrates.

EXAMPLE 8.1 The Current Population Survey

The most important government sample survey in the United States is the monthly Current Population Survey (CPS). The CPS contacts about 60,000 households each month. It produces the monthly unemployment rate and much other economic and social information. (See Figure 8.1.) To measure unemployment, we must first specify

FIGURE 8.1

The home page of the Current Population Survey at the Bureau of Labor Statistics.

the population we want to describe. Which age groups will we include? Will we include illegal immigrants or people in prisons? The CPS defines its population as all U.S. residents (legal or not) 16 years of age and over who are civilians and are not in an institution such as a prison. The unemployment rate announced in the news refers to this specific population.

The second question is harder: what does it mean to be "unemployed"? Someone who is not looking for work—for example, a full-time student—should not be called unemployed just because she is not working for pay. If you are chosen for the CPS sample, the interviewer first asks whether you are available to work and whether you actually looked for work in the past four weeks. If not, you are neither employed nor unemployed—you are not in the labor force.

If you are in the labor force, the interviewer goes on to ask about employment. If you did any work for pay or in your own business during the week of the survey, you are employed. If you worked at least 15 hours in a family business without pay, you are employed. You are also employed if you have a job but didn't work because of vacation, being on strike, or other good reason. An unemployment rate of 4.7% means that 4.7% of the sample was unemployed, using the exact CPS definitions of both "labor force" and "unemployed." ∎

The final step in planning a sample survey is the sampling design. We will now introduce basic statistical designs for sampling.

APPLY YOUR KNOWLEDGE

8.1 Sampling students. A political scientist wants to know how college students feel about the Social Security system. She obtains a list of the 3456 undergraduates at her college and mails a questionnaire to 250 students selected at random. Only 104 questionnaires are returned.

(a) What is the population in this study? Be careful: what group does she *want information about?*

(b) What is the sample? Be careful: from what group does she *actually obtain information?*

8.2 Student archaeologists. An archaeological dig turns up large numbers of pottery shards, broken stone implements, and other artifacts. Students working on the project classify each artifact and assign it a number. The counts in different categories are important for understanding the site, so the project director chooses 2% of the artifacts at random and checks the students' work. What are the population and the sample here?

John ELK III/Alamy

8.3 Customer satisfaction. A department store mails a customer satisfaction survey to people who make credit card purchases at the store. This month, 45,000 people made credit card purchases. Surveys are mailed to 1000 of these people, chosen at random, and 137 people return the survey form. What is the population for this survey? What is the sample from which information was actually obtained?

How to sample badly

How can we choose a sample that we can trust to represent the population? A sampling design is a specific method for choosing a sample from the population. The easiest—but not the best—design just chooses individuals close at hand. If we are interested in finding out how many people have jobs, for example, we might go to a shopping mall and ask people passing by if they are employed. A sample selected by taking the members of the population that are easiest to reach is called *convenience sample* a **convenience sample.** Convenience samples often produce unrepresentative data.

> **EXAMPLE 8.2** **Sampling at the mall**
>
> A sample of mall shoppers is fast and cheap. But people at shopping malls tend to be more prosperous than typical Americans. They are also more likely to be teenagers or retired. Moreover, unless interviewers are carefully trained, they tend to question well-dressed, respectable people and avoid poorly dressed or tough-looking individuals. In short, mall interviews will not contact a sample that is representative of the entire population. ■

Interviews at shopping malls will almost surely overrepresent middle-class and retired people and underrepresent the poor. This will happen almost every time we take such a sample. That is, it is a systematic error caused by a bad sampling design, not just bad luck on one sample. This is *bias:* the outcomes of mall surveys will repeatedly miss the truth about the population in the same ways.

> **BIAS**
>
> The design of a statistical study is **biased** if it systematically favors certain outcomes.

> **EXAMPLE 8.3** **Online polls**
>
> The CNN evening commentator Lou Dobbs doesn't like illegal immigration. One of his broadcasts in 2007 was largely devoted to attacking a proposal by the governor of New York State to offer driver's licenses to illegal immigrants as a public safety measure. During the show, Mr. Dobbs invited his viewers to go to `loudobbs.com` to vote on the question "Would you be more or less likely to vote for a presidential candidate who supports giving drivers' licences to illegal aliens?" We aren't surprised that 97% of the 7350 people who voted by the end of the broadcast said "Less likely." ■

The `loudobbs.com` poll was biased because people chose whether or not to participate. Most who voted were viewers of Lou Dobbs's program who had just heard him denounce the governor's idea. *People who take the trouble to respond to an open invitation are usually not representative of any clearly defined population.* That's true of the people who bother to respond to write-in, call-in, or online polls in general. Polls like these are examples of *voluntary response sampling.*

> ### VOLUNTARY RESPONSE SAMPLE
>
> A **voluntary response sample** consists of people who choose themselves by responding to a broad appeal. Voluntary response samples are biased because people with strong opinions are most likely to respond.

APPLY YOUR KNOWLEDGE

8.4 Sampling on campus. You see a woman student standing in front of the student center, now and then stopping other students to ask them questions. She says that she is collecting student opinions for a class assignment. Explain why this sampling method is almost certainly biased.

8.5 More sampling on campus. Your college wants to gather student opinion about parking for students on campus. It isn't practical to contact all students.

(a) Give an example of a way to choose a sample of students that is poor practice because it depends on voluntary response.

(b) Give another example of a bad way to choose a sample that doesn't use voluntary response.

Simple random samples

In a voluntary response sample, people choose whether to respond. In a convenience sample, the interviewer makes the choice. In both cases, personal choice produces bias. The statistician's remedy is to allow impersonal chance to choose the sample. A sample chosen by chance rules out both favoritism by the sampler and self-selection by respondents. Choosing a sample by chance attacks bias by giving all individuals an equal chance to be chosen. Rich and poor, young and old, black and white, all have the same chance to be in the sample.

The simplest way to use chance to select a sample is to place names in a hat (the population) and draw out a handful (the sample). This is the idea of *simple random sampling*.

> ### SIMPLE RANDOM SAMPLE
>
> A **simple random sample (SRS)** of size n consists of n individuals from the population chosen in such a way that every set of n individuals has an equal chance to be the sample actually selected.

An SRS not only gives each individual an equal chance to be chosen but also gives every possible sample an equal chance to be chosen. There are other random sampling designs that give each individual, but not each sample, an equal chance. Exercise 8.41 describes one such design.

When you think of an SRS, picture drawing names from a hat to remind yourself that an SRS doesn't favor any part of the population. That's why an SRS is a better method of choosing samples than convenience or voluntary response sampling. But writing names on slips of paper and drawing them from a hat is slow and inconvenient. That's especially true if, like the Current Population Survey, we must draw a sample of size 60,000. In practice, samplers use software. The *Simple Random Sample* applet makes the choosing of an SRS very fast. If you don't use the applet or other software, you can randomize by using a *table of random digits*. In fact, software for choosing samples starts by generating random digits, so using a table just does by hand what the software does more quickly.

RANDOM DIGITS

A **table of random digits** is a long string of the digits 0, 1, 2, 3, 4, 5, 6, 7, 8, 9 with these two properties:

1. Each entry in the table is equally likely to be any of the 10 digits 0 through 9.
2. The entries are independent of each other. That is, knowledge of one part of the table gives no information about any other part.

Table B at the back of the book is a table of random digits. Table B begins with the digits 19223950340575628713. To make the table easier to read, the digits appear in groups of five and in numbered rows. The groups and rows have no meaning—the table is just a long list of randomly chosen digits. There are two steps in using the table to choose a simple random sample.

USING TABLE B TO CHOOSE AN SRS

Label: Give each member of the population a numerical label of the *same length*.

Table: To choose an SRS, read from Table B successive groups of digits of the length you used as labels. Your sample contains the individuals whose labels you find in the table.

Are these random digits really random?

Not a chance. The random digits in Table B were produced by a computer program. Computer programs do exactly what you tell them to do. Give the program the same input and it will produce exactly the same "random" digits. Of course, clever people have devised computer programs that produce output that *looks* like random digits. These are called "pseudo-random numbers," and that's what Table B contains. Pseudo-random numbers work fine for statistical randomizing, but they have hidden nonrandom patterns that can mess up more refined uses.

You can label up to 100 items with two digits: 01, 02, ..., 99, 00. Up to 1000 items can be labeled with three digits, and so on. Always use the shortest labels that will cover your population. As standard practice, we recommend that you begin with label 1 (or 01 or 001, as needed). Reading groups of digits from the table gives all individuals the same chance to be chosen because all labels of the same length have the same chance to be found in the table. For example, any pair of digits in the table is equally likely to be any of the 100 possible labels 01, 02, ..., 99, 00. Ignore any group of digits that was not used as a label or that duplicates a label already in the sample. You can read digits from Table B in any order—across a row, down a column, and so on—because the table has no order. As standard practice, we recommend reading across rows.

EXAMPLE 8.4 Sampling spring break resorts

A campus newspaper plans a major article on spring break destinations. The authors intend to call 4 randomly chosen resorts at each destination to ask about their attitudes toward groups of students as guests. Here are the resorts listed in one city:

01	Aloha Kai	08	Captiva	15	Palm Tree	22	Sea Shell
02	Anchor Down	09	Casa del Mar	16	Radisson	23	Silver Beach
03	Banana Bay	10	Coconuts	17	Ramada	24	Sunset Beach
04	Banyan Tree	11	Diplomat	18	Sandpiper	25	Tradewinds
05	Beach Castle	12	Holiday Inn	19	Sea Castle	26	Tropical Breeze
06	Best Western	13	Lime Tree	20	Sea Club	27	Tropical Shores
07	Cabana	14	Outrigger	21	Sea Grape	28	Veranda

Robert Daly/Getty Images

Label: Because two digits are needed to label the 28 resorts, all labels will have two digits. We have added labels 01 to 28 in the list of resorts. Always say how you labeled the members of the population. To sample from the 1240 resorts in a major vacation area, you would label the resorts 0001, 0002, ..., 1239, 1240.

Table: To use the *Simple Random Sample* applet, just enter 28 in the "Population =" box and 4 in the "Select a sample" box, click "Reset," and click "Sample." Figure 8.2 shows the result of one sample.

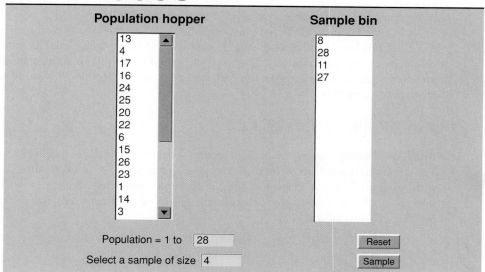

FIGURE 8.2

The *Simple Random Sample* applet used to choose an SRS of size $n = 4$ from a population of size 28.

To use Table B, read two-digit groups until you have chosen four resorts. Starting at line 130 (any line will do), we find

69051 64817 87174 09517 84534 06489 87201 97245

Because the labels are two digits long, read successive two-digit groups from the table. Ignore groups not used as labels, like the initial 69. Also ignore any repeated labels, like the second and third 17s in this row, because you can't choose the same resort twice. Your sample contains the resorts labeled 05, 16, 17, and 20. These are Beach Castle, Radisson, Ramada, and Sea Club. ■

We can trust results from an SRS, as well as from other types of random samples that we will meet later, because the use of impersonal chance avoids bias. Online polls and mall interviews also produce samples. We can't trust results from these samples, because they are chosen in ways that invite bias. *The first question to ask about any sample is whether it was chosen at random.*

EXAMPLE 8.5 The future of the environment

"Do you think the condition of the environment for the next generation will be better, worse, or about the same as it is now?" When the *New York Times* and CBS News asked this question of 1052 adults, 57% said "worse" and just 11% said "better." Can we trust the opinions of this sample to fairly represent the opinions of all adults? Here's part of the statement by the *Times* on "How the Poll Was Conducted":[1]

The latest New York Times/CBS News poll is based on telephone interviews conducted April 20 through April 24 with 1,052 adults throughout the United States.

The sample of telephone exchanges called was randomly selected by a computer from a complete list of more than 42,000 active residential exchanges across the country. The exchanges were chosen so as to ensure that each region of the country was represented in proportion to its population.

Within each exchange, random digits were added to form a complete telephone number, thus permitting access to listed and unlisted numbers alike. Within each household, one adult was designated by a random procedure to be the respondent for the survey.

random digit dialing

This is a good description of the most common method for choosing national samples, called **random digit dialing.** We'll come back to random digit dialing and its problems later, but this statement is a good start toward gaining our confidence. We know the size of the sample, when the poll was taken, and the comforting word "random" appears three times. ■

APPLY YOUR KNOWLEDGE

8.6 Apartment living. You are planning a report on apartment living in a college town. You decide to select three apartment complexes at random for in-depth interviews with residents. Use the *Simple Random Sample* applet, other software, or Table B to select a simple random sample of three of the following apartment complexes. If you use Table B, start at line 117.

Ashley Oaks	Country View	Mayfair Village
Bay Pointe	Country Villa	Nobb Hill
Beau Jardin	Crestview	Pemberly Courts
Bluffs	Del-Lynn	Peppermill
Brandon Place	Fairington	Pheasant Run
Briarwood	Fairway Knolls	River Walk
Brownstone	Fowler	Sagamore Ridge
Burberry Place	Franklin Park	Salem Courthouse
Cambridge	Georgetown	Village Square
Chauncey Village	Greenacres	Waterford Court
Country Squire	Lahr House	Williamsburg

8.7 **Minority managers.** A firm wants to understand the attitudes of its minority managers toward its system for assessing management performance. Below is a list of all the firm's managers who are members of minority groups. Use the *Simple Random Sample* applet, other software, or Table B at line 139 to choose six to be interviewed in detail about the performance appraisal system.

Abdulhamid	Duncan	Huang	Puri
Agarwal	Fernandez	Kim	Richards
Baxter	Fleming	Lumumba	Rodriguez
Bonds	Gates	Mourning	Santiago
Brown	Gomez	Nguyen	Shen
Castro	Gupta	Peters	Vargas
Chavez	Hernandez	Peña	Wang

8.8 **Sampling gravestones.** The local genealogical society in Coles County, Illinois, has compiled records on all 55,914 gravestones in cemeteries in the county for the years 1825 to 1985. Historians plan to use these records to learn about African Americans in Coles County's history. They first choose an SRS of 395 records to check their accuracy by visiting the actual gravestones.[2]

(a) How would you label the 55,914 records?

(b) Use Table B, beginning at line 120, to choose the first five records for the SRS.

© The Photo Works

Inference about the population

The purpose of a sample is to give us information about a larger population. The process of drawing conclusions about a population on the basis of sample data is called **inference** because we *infer* information about the population from what we *know* about the sample.

inference

Inference from convenience samples or voluntary response samples would be misleading because these methods of choosing a sample are biased. We are almost certain that the sample does *not* fairly represent the population. *The first reason to rely on random sampling is to eliminate bias in selecting samples from the list of available individuals.*

Nonetheless, it is unlikely that results from a random sample are exactly the same as for the entire population. Sample results, like the unemployment rate obtained from the monthly Current Population Survey, are only estimates of the truth about the population. If we select two samples at random from the same population, we will almost certainly draw different individuals. So the sample results will differ somewhat, just by chance. Properly designed samples avoid systematic bias, but their results are rarely exactly correct and they vary from sample to sample.

Why can we trust random samples? The big idea is that the results of random sampling don't change haphazardly from sample to sample. Because we deliberately use chance, the results obey the laws of probability that govern chance behavior. These laws allow us to say how likely it is that sample results are close to the truth about the population. *The second reason to use random sampling is that the laws of probability allow trustworthy inference about the population.* Results from random samples come with a margin of error that sets bounds on the size of the likely error. How to do this is part of the technique of statistical inference. We will describe the reasoning in Chapter 14 and present details throughout the rest of the book.

One point is worth making now: *larger random samples give more accurate results than smaller samples*. By taking a very large sample, you can be confident that the sample result is very close to the truth about the population. The Current Population Survey contacts about 60,000 households, so it estimates the national unemployment rate very accurately. Opinion polls that contact 1000 or 1500 people give less accurate results. Of course, only samples chosen by chance carry this guarantee. Lou Dobbs's online sample tells us little about overall American public opinion even though 7350 people clicked a response.

APPLY YOUR KNOWLEDGE

8.9 Ask more people. Just before a presidential election, a national opinion-polling firm increases the size of its weekly sample from the usual 1500 people to 4000 people. Why do you think the firm does this?

8.10 Sampling Pentecostals. Pentecostals are among the fastest-growing Christian groups in many countries. The Pew Forum on Religion and Public Life surveyed Pentecostal Christians in 10 countries and compared their opinions with those of the general population. In South Korea, random samples by Gallup Korea had margins of error (more detail in later chapters) of ±4% for the general public and ±9% for Pentecostals.[3] What do you think explains the fact that estimates for Pentecostals were less accurate?

Golfing at random

Random drawings give everyone the same chance to be chosen, so they offer a fair way to decide who gets a scarce good—like a round of golf. Lots of golfers want to play the famous Old Course at St. Andrews, Scotland. Some can reserve in advance, at considerable expense. Most must hope that chance favors them in the daily random drawing for tee times. At the height of the summer season, only 1 in 6 wins the right to pay $250 for a round.

Other sampling designs

Random sampling, the use of chance to select the sample, is the essential principle of statistical sampling. Designs for random sampling from large populations spread out over a wide area are usually more complex than an SRS. For example, it is common to sample important groups within the population separately, then combine these samples. This is the idea of a *stratified random sample*.

> ### STRATIFIED RANDOM SAMPLE
>
> To select a **stratified random sample,** first classify the population into groups of similar individuals, called **strata.** Then choose a separate SRS in each stratum and combine these SRSs to form the full sample.

Choose the strata based on facts known before the sample is taken. For example, a population of election districts might be divided into urban, suburban, and rural strata. A stratified design can produce more precise information than an SRS of the same size by taking advantage of the fact that individuals in the same stratum are similar to one another.

EXAMPLE 8.6 **Seat belt use in Hawaii**

Each state conducts an annual survey of seat belt use by drivers, following guidelines set by the federal government. The guidelines require random sampling. Seat belt use is observed at randomly chosen road locations at random times during daylight hours. The locations are not an SRS of all locations in the state but rather a stratified sample using the state's counties as strata.

In Hawaii, the counties are the islands that make up the state's territory. The seat belt survey sample consists of 135 road locations in the four most populated islands: 66 in Oahu, 24 in Maui, 23 in Hawaii, and 22 in Kauai. The sample sizes on the islands are proportional to the amount of road traffic.[4] ■

Ryan McVay/Photo Disc/Getty Images

Most large-scale sample surveys use **multistage samples.** For example, the opinion poll described in Example 8.5 has three stages: choose a random sample of telephone exchanges (stratified by region of the country), then an SRS of household telephone numbers within each exchange, then a random adult in each household.

multistage sample

Analysis of data from sampling designs more complex than an SRS takes us beyond basic statistics. But the SRS is the building block of more elaborate designs, and analysis of other designs differs more in complexity of detail than in fundamental concepts.

APPLY YOUR KNOWLEDGE

8.11 Sampling metro Chicago. Cook County, Illinois, has the second-largest population of any county in the United States (after Los Angeles County, California). Cook County has 30 suburban townships and an additional 8 townships that make up the city of Chicago. The suburban townships are

Barrington	Elk Grove	Maine	Orland	Riverside
Berwyn	Evanston	New Trier	Palatine	Schaumburg
Bloom	Hanover	Niles	Palos	Stickney
Bremen	Lemont	Northfield	Proviso	Thornton
Calumet	Leyden	Norwood Park	Rich	Wheeling
Cicero	Lyons	Oak Park	River Forest	Worth

The Chicago townships are

Hyde Park	Lake	North Chicago	South Chicago
Jefferson	Lake View	Rogers Park	West Chicago

Because city and suburban areas may differ, the first stage of a multistage sample chooses a stratified sample of 6 suburban townships and 4 of the more heavily populated Chicago townships. Use Table B or software to choose this sample. (If you use Table B, assign labels in alphabetical order and start at line 101 for the suburbs and at line 110 for Chicago.)

8.12 Academic dishonesty. A study of academic dishonesty among college students used a two-stage sampling design. The first stage chose a sample of 30 colleges and universities. Then the study authors mailed questionnaires to a stratified sample of 200 seniors, 100 juniors, and 100 sophomores at each school.[5] One of the schools chosen has 1127 freshmen, 989 sophomores, 943 juniors, and 895 seniors. You have alphabetical lists of the students in each class. Explain how you would assign labels for stratified sampling. Then use software or Table B, starting at line 122, to select the first 5 students in the sample from each stratum.

Cautions about sample surveys

Random selection eliminates bias in the choice of a sample from a list of the population. When the population consists of human beings, however, accurate information from a sample requires more than a good sampling design.

To begin, we need an accurate and complete list of the population. Because such a list is rarely available, most samples suffer from some degree of *undercoverage*. A sample survey of households, for example, will miss not only homeless people but prison inmates and students in dormitories. An opinion poll conducted by calling landline telephone numbers will miss households that have only cell phones as well as households without a phone. The results of national sample surveys therefore have some bias if the people not covered differ from the rest of the population.

A more serious source of bias in most sample surveys is *nonresponse*, which occurs when a selected individual cannot be contacted or refuses to cooperate. Nonresponse to sample surveys often exceeds 50%, even with careful planning and several callbacks. Because nonresponse is higher in urban areas, most sample surveys substitute other people in the same area to avoid favoring rural areas in the final sample. If the people contacted differ from those who are rarely at home or who refuse to answer questions, some bias remains.

UNDERCOVERAGE AND NONRESPONSE

Undercoverage occurs when some groups in the population are left out of the process of choosing the sample.

Nonresponse occurs when an individual chosen for the sample can't be contacted or refuses to participate.

EXAMPLE 8.7 How bad is nonresponse?

The Census Bureau's American Community Survey (ACS) has the lowest nonresponse rate of any poll we know: only about 1% of the households in the sample refuse to respond; the overall nonresponse rate, including "never at home" and other causes, is just 2.5%.[6] This monthly survey of about 250,000 households replaces the "long form" that in the past was sent to some households in the every-ten-years national census. Participation in the ACS is mandatory, and the Census Bureau follows up by telephone and then in person if a household fails to return the mail questionnaire.

The University of Chicago's General Social Survey (GSS) is the nation's most important social science survey. (See Figure 8.3.) The GSS contacts its sample in person, and it is run by a university. Despite these advantages, its most recent survey had a 30% rate of nonresponse.

FIGURE 8.3
The home page of the General Social Survey at the University of Chicago's National Opinion Research Center. The GSS has tracked opinions about a wide variety of issues since 1972.

What about opinion polls by news media and opinion-polling firms? We don't know their rates of nonresponse because they won't say. That itself is a bad sign. The Pew Research Center for the People and the Press imitated a careful random digit dialing survey and published the results: over 5 days, the survey reached 76% of the households in its chosen sample, but "because of busy schedules, skepticism and outright refusals, interviews were completed in just 38% of households that were reached." Combining households that could not be contacted with those who did not complete the interview gave a nonresponse rate of 73%.[7] ∎

In addition, the behavior of the respondent or of the interviewer can cause **response bias** in sample results. People know that they should take the trouble to vote, for example, so many who didn't vote in the last election will tell an interviewer that they did. The race or sex of the interviewer can influence responses to questions about race relations or attitudes toward feminism. Answers to questions that ask respondents to recall past events are often inaccurate because of faulty memory. For example, many people "telescope" events in the past, bringing them forward in memory to more recent time periods. "Have you visited a dentist in the last 6 months?" will often draw a "Yes" from someone who last visited a dentist 8 months ago.[8] Careful training of interviewers and careful supervision to avoid variation among the interviewers can reduce response bias. Good interviewing technique is another aspect of a well-done sample survey.

response bias

The **wording of questions** is the most important influence on the answers given to a sample survey. Confusing or leading questions can introduce strong bias, and changes in wording can greatly change a survey's outcome. Even the order in which questions are asked matters. Here are some examples.[9]

wording effects

EXAMPLE 8.8 What was that question?

How do Americans feel about illegal immigrants? "Should illegal immigrants be prosecuted and deported for being in the U.S. illegally, or shouldn't they?" Asked this question in an opinion poll, 69% favored deportation. But when the very same sample was asked whether illegal immigrants who have worked in the United States for two years "should be given a chance to keep their jobs and eventually apply for legal status," 62% said that they should. Different questions give quite different impressions of attitudes toward illegal immigrants.

What about government help for the poor? Only 13% think we are spending too much on "assistance to the poor," but 44% think we are spending too much on "welfare." ■

EXAMPLE 8.9 Are you happy?

Ask a sample of college students these two questions:

"How happy are you with your life in general?" (Answers on a scale of 1 to 5)

"How many dates did you have last month?"

The correlation between answers is $r = -0.012$ when asked in this order. It appears that dating has little to do with happiness. Reverse the order of the questions, however, and $r = 0.66$. Asking a question that brings dating to mind makes dating success a big factor in happiness. ■

Don't trust the results of a sample survey until you have read the exact questions asked. The amount of nonresponse and the date of the survey are also important. Good statistical design is a part, but only a part, of a trustworthy survey.

APPLY YOUR KNOWLEDGE

8.13 Ring-no-answer. A common form of nonresponse in telephone surveys is "ring-no-answer." That is, a call is made to an active number but no one answers. The Italian National Statistical Institute looked at nonresponse to a government survey of households in Italy during the periods January 1 to Easter and July 1 to August 31. All calls were made between 7 and 10 P.M., but 21.4% gave "ring-no-answer" in one period versus 41.5% "ring-no-answer" in the other period.[10] Which period do you think had the higher rate of no answers? Why? Explain why a high rate of nonresponse makes sample results less reliable.

8.14 Question wording. In 2000, when the federal budget showed a large surplus, the Pew Research Center asked two questions of random samples of adults. Both questions stated that Social Security would be "fixed." Here are two questions about using the remaining surplus:

> Question A: *Should the money be used for a tax cut, or should it be used to fund new government programs?*
> Question B: *Should the money be used for a tax cut, or should it be spent on programs for education, the environment, health care, crime-fighting and military defense?*

One of these questions drew 60% favoring a tax cut. The other drew only 22%. Which wording pulls respondents toward a tax cut? Why?

The impact of technology

A few national sample surveys, including the General Social Survey and the government's American Community Survey and Current Population Survey, interview some or all of their subjects in person. This is expensive and time-consuming, so most national surveys contact subjects by telephone using the random digit dialing (RDD) method described in Example 8.5 (page 208). Technology, especially the spread of cell phones, is making traditional RDD methods outdated.

First, *call screening* is now common. A large majority of American households have answering machines or caller ID, and many use these methods to screen their calls. Calls from polling organizations are rarely returned.

More seriously, the number of *cell-phone-only households* is increasing rapidly. Already by mid-2007, 14% of American households had a cell phone but no landline phone. Even if the United States and Canada don't approach the 52% of households in Finland that have no landline phone, it's clear that RDD reaching only landline numbers is in trouble. Can surveys just add cell phone numbers? Not easily. Federal regulations require hand dialing of cell phone numbers, ruling out computerized RDD sampling and adding expense. A cell phone can be anywhere, so stratifying by location becomes difficult. And a cell phone user may be driving or otherwise unable to talk safely.

People who screen calls and people who have only a cell phone tend to be younger than the general population. In fact, one projection claims that by the end of 2009 more than 40% of American adults under age 30 will have no landline phone. So RDD surveys may be biased. Careful surveys weight their responses to reduce bias. For example, if a sample contains too few young adults, the responses of the young adults who do respond are given extra weight. Somewhat surprisingly, detailed studies showed that as of 2006 the bias due to call screening and omitting cell phone numbers was quite small. But response rates are steadily dropping and cell phone only is steadily growing. The future of RDD landline telephone surveys is not promising.[11]

One alternative is to use *Web surveys* rather than telephone surveys. It's important to distinguish professional Web surveys from the overwhelming number of voluntary response online surveys that are intended to be entertaining rather than to give trustworthy information about a clearly defined population. Undercoverage is a serious problem for even careful Web surveys because (as of 2007) almost a quarter of Americans lack Internet access and only about half have broadband access. People without Internet access are more likely to be poor, elderly, minority, or rural than the overall population, so the potential for bias in a Web survey is clear. There is no easy way to choose a random sample even from people with Web access, because there is no technology that generates personal email addresses at random in the way that RDD generates residential telephone numbers. Even if such technology existed, etiquette and regulations aimed at spammers would prevent mass emailing. For the present, Web surveys work well only for restricted populations, for example, surveying students at your university using the school's list of student email addresses.[12] Here is an example of a successful Web survey.

Do not call!

People who do sample surveys hate telemarketing. We all get so many unwanted sales pitches by phone that many people hang up before learning that the caller is conducting a survey rather than selling vinyl siding. You can eliminate calls from commercial telemarketers by placing your phone number on the National Do Not Call Registry. Just go to www.donotcall.gov to sign up.

AB/Getty Images

EXAMPLE 8.10 **Doctors and placebos**

A placebo is a dummy treatment like a salt pill that has no direct effect on a patient but may bring about a response because patients expect it to. Do academic physicians who maintain private practices sometimes give their patients placebos? A Web survey of doctors in internal medicine departments at Chicago-area medical schools was possible because almost all the doctors had listed email addresses.

Send an email to each doctor explaining the purpose of the study, promising anonymity, and giving an individual Web link for response. In all, 231 of 443 doctors responded. The response rate was helped by the fact that the email came from a team at a medical school. Result: 45% said they sometimes used placebos in their clinical practice.[13] ■

APPLY YOUR KNOWLEDGE

8.15 Let's go polling. Use Google or your favorite search engine to search the Web for "web polling software." Choose one of the sites that offer software that allows you to conduct your own online opinion polls. (At the time of writing, **www.pollmonkey.com** was a favorite, but things change quickly on the Web.) Briefly describe some attractive features that the software offers. (For example, you would like to list the answer choices in random order, so that the same choice is not always in the first position). Despite these features, all such polls share a fatal weakness. What is this?

C H A P T E R   8   S U M M A R Y

■ A **sample survey** selects a **sample** from the **population** of all individuals about which we desire information. We base conclusions about the population on data from the sample. It is important to specify exactly what population you are interested in and what variables you will measure.

■ The **design** of a sample describes the method used to select the sample from the population. **Random sampling** designs use chance to select a sample.

■ The basic random sampling design is a **simple random sample (SRS).** An SRS gives every possible sample of a given size the same chance to be chosen.

■ Choose an SRS by labeling the members of the population and using **random digits** to select the sample. Software can automate this process.

■ To choose a **stratified random sample,** classify the population into **strata,** groups of individuals that are similar in some way that is important to the response. Then choose a separate SRS from each stratum.

■ Failure to use random sampling often results in **bias,** or systematic errors in the way the sample represents the population. **Voluntary response samples,** in which the respondents choose themselves, are particularly prone to large bias.

■ In human populations, even random samples can suffer from bias due to **undercoverage** or **nonresponse,** from **response bias,** or from misleading results due to **poorly worded questions.** Sample surveys must deal expertly with these potential problems in addition to using a random sampling design.

■ Most national sample surveys are carried out by telephone, using **random digit dialing** to choose residential telephone numbers at random. Call screening is increasing nonresponse to such surveys, and the rise of cell-phone-only households is increasing undercoverage.

8.16 An opinion poll contacts 1161 adults and asks them, "Which political party do you think has better ideas for leading the country in the twenty-first century?" In all, 696 of the 1161 say, "The Democrats." The sample in this setting is

(a)   all 235 million adults in the United States.

(b)   the 1161 people interviewed.

(c)   the 696 people who chose the Democrats.

8.17 A committee on community relations in a college town plans to survey local businesses about the importance of students as customers. From telephone book listings, the committee chooses 150 businesses at random. Of these, 73 return the questionnaire mailed by the committee. The population for this study is

(a)   all businesses in the college town.

(b)   the 150 businesses chosen.

(c)   the 73 businesses that returned the questionnaire.

8.18 The Web portal AOL places opinion poll questions next to many of its news stories. Simply click your response to join the sample. One of the questions in January 2008 was "Do you plan to diet this year?" More than 30,000 people responded, with 68% saying "Yes." You can conclude that

(a)   about 68% of Americans planned to diet in 2008.

(b)   the poll uses voluntary response, so the results tell us little about the population of all adults.

(c)   the sample is too small to draw any conclusion.

8.19 You must choose an SRS of 10 of the 440 retail outlets in New York that sell your company's products. How would you label this population in order to use Table B?

(a)   001, 002, 003, ..., 439, 440

(b)   000, 001, 002, ..., 439, 440

(c)   1, 2, ..., 439, 440

8.20 You are using the table of random digits to choose a simple random sample of 6 students from a class of 30 students. You label the students 01 to 30 in alphabetical order. Go to line 133 of Table B. Your sample contains the students labeled

(a)   45, 74, 04, 18, 07, 65.

(b)   04, 18, 07, 13, 02, 07.

(c)   04, 18, 07, 13, 02, 05.

8.21 You want to choose an SRS of 5 of the 7200 salaried employees of a corporation. You label the employees 0001 to 7200 in alphabetical order. Using line 111 of Table B, your sample contains the employees labeled

(a)   6694, 5130, 0041, 2712, 3827.

(b)   6694, 0513, 0929, 7004, 1271.

(c)   8148, 6694, 8760, 5130, 9297.

8.22 Archaeologists plan to examine a sample of 2-meter-square plots near an ancient Greek city for artifacts visible in the ground. They choose separate samples of plots from floodplain, coast, foothills, and high hills. What kind of sample is this?

(a)   A simple random sample.

(b)   A stratified random sample.

(c)   A voluntary response sample.

8.23 A sample of households in a community is selected at random from the telephone directory. In this community, 4% of households have no telephone, 10% have only cell phones, and another 25% have unlisted telephone numbers. The sample will certainly suffer from

(a)   nonresponse.

(b)   undercoverage.

(c)   false responses.

8.24 The Gallup Poll asked a random sample of adults, "Do you have enough time to do what you want to do?" In the entire sample, 53% said "No." But 62% of parents of children younger than age 18 said "No." Which of these two sample percents will be more accurate as an estimate of the truth about the population?

(a)   The result for the entire sample is more accurate because it comes from a larger sample.

(b)   The result for parents is more accurate because it's easier to estimate a result for a smaller population.

(c)   Both are equally accurate because both come from the same sample.

CHAPTER 8 EXERCISES

In all exercises asking for an SRS, you may use Table B, the Simple Random Sample applet, or other software.

8.25 Are you feeling stressed? A Gallup Poll asked, "In general, how often do you experience stress in your daily life—never, rarely, sometimes, or frequently?" Gallup's report said, "Results are based on telephone interviews with 1,027 national adults, aged 18 and older, conducted Dec. 6–9, 2007."[14] What is the population for this sample survey? What is the sample?

8.26 Sampling stuffed envelopes. A large retailer prepares its customers' monthly credit card bills using an automatic machine that folds the bills, stuffs them into envelopes, and seals the envelopes for mailing. Are the envelopes completely sealed?

Inspectors choose 40 envelopes from the 1000 stuffed each hour for visual inspection. What is the population for this sample survey? What is the sample?

8.27 Do you trust the Internet? You want to ask a sample of college students the question "How much do you trust information about health that you find on the Internet—a great deal, somewhat, not much, or not at all?" You try out this and other questions on a pilot group of 10 students chosen from your class. The class members are

Anderson	Deng	Glaus	Nguyen	Samuels
Arroyo	De Ramos	Helling	Palmiero	Shen
Batista	Drasin	Husain	Percival	Tse
Bell	Eckstein	Johnson	Prince	Velasco
Burke	Fernandez	Kim	Puri	Wallace
Cabrera	Fullmer	Molina	Richards	Washburn
Calloway	Gandhi	Morgan	Rider	Zabidi
Delluci	Garcia	Murphy	Rodriguez	Zhao

Choose an SRS of 10 students. If you use Table B, start at line 117.

8.28 Sampling telephone area codes. There are approximately 341 active telephone area codes covering Canada, the United States, and some Caribbean areas. (More are created regularly.) You want to choose an SRS of 25 of these area codes for a study of available telephone numbers. Label the codes 001 to 341 and use the *Simple Random Sample* applet or other software to choose your sample. (If you use Table B, start at line 129 and choose only the first 5 codes in the sample.)

8.29 Sampling the forest. To gather data on a 1200-acre pine forest in Louisiana, the U.S. Forest Service laid a grid of 1410 equally spaced circular plots over a map of the forest. A ground survey visited a sample of 10% of these plots.[15]

(a) How would you label the plots?

(b) Choose the first 5 plots in an SRS of 141 plots. (If you use Table B, start at line 105.)

8.30 Sampling pharmacists. All pharmacists in the Canadian province of Ontario are required to be members of the Ontario College of Pharmacists. The membership list contains 7500 names.

(a) How would you label the names in order to select an SRS?

(b) Use software or Table B, starting at line 142, to select an SRS of 10 Ontario pharmacists.

8.31 Random digits. In using Table B repeatedly to choose random samples, you should not always begin at the same place, such as line 101. Why not?

8.32 Random digits. Which of the following statements are true of a table of random digits, and which are false? Briefly explain your answers.

(a) There are exactly four 0s in each row of 40 digits.

(b) Each pair of digits has chance 1/100 of being 00.

(c) The digits 0000 can never appear as a group, because this pattern is not random.

8.33 Movie viewing. An opinion poll calls 2000 randomly chosen residential telephone numbers, then asks to speak with an adult member of the household. The interviewer asks, "How many movies have you watched in a movie theater in the past 12 months?"

(a)   What population do you think the poll has in mind?

(b)   In all, 831 people respond. What is the rate (percent) of nonresponse?

(c)   What source of response error is likely for the question asked?

8.34 Online polls. Example 8.3 reports an online poll in which 97% of the respondents opposed issuing driver's licenses to illegal immigrants. National random samples taken at the same time showed about 70% of the respondents opposed to such licenses. Explain briefly to someone who knows no statistics why the random samples report public opinion more reliably than the online poll.

8.35 Nonresponse. Academic sample surveys, unlike commercial polls, often discuss nonresponse. A survey of drivers began by randomly sampling all listed residential telephone numbers in the United States. Of 45,956 calls to these numbers, 5029 were completed.[16] What was the rate of nonresponse for this sample? (Only one call was made to each number. Nonresponse would be lower if more calls were made.)

8.36 Running red lights. The sample described in the previous exercise produced a list of 5024 licensed drivers. The investigators then chose an SRS of 880 of these drivers to answer questions about their driving habits.

(a)   How would you assign labels to the 5024 drivers? Use Table B, starting at line 104, to choose the first 5 drivers in the sample.

(b)   One question asked was "Recalling the last ten traffic lights you drove through, how many of them were red when you entered the intersections?" Of the 880 respondents, 171 admitted that at least one light had been red. A practical problem with this survey is that people may not give truthful answers. What is the likely direction of the bias: do you think more or fewer than 171 of the 880 respondents really ran a red light? Why?

8.37 Seat belt use. A study in El Paso, Texas, looked at seat belt use by drivers. Drivers were observed at randomly chosen convenience stores. After they left their cars, they were invited to answer questions that included questions about seat belt use. In all, 75% said they always used seat belts, yet only 61.5% were wearing seat belts when they pulled into the store parking lots.[17] Explain the reason for the bias observed in responses to the survey. Do you expect bias in the same direction in most surveys about seat belt use?

8.38 Sampling at a party. At a party there are 30 students over age 21 and 20 students under age 21. You choose at random 3 of those over 21 and separately choose at random 2 of those under 21 to interview about attitudes toward alcohol. You have given every student at the party the same chance to be interviewed: what is that chance? Why is your sample not an SRS?

8.39 Sampling at a party. At a large block party there are 290 men and 110 women. You want to ask opinions about how to improve the next party. To be sure that women's opinions are adequately represented, you decide to choose a stratified random sample of 20 men and 20 women. Explain how you will assign labels to the names of the people at the party. Give the labels of the first 3 men and the first 3 women in your sample. If you use Table B, start at line 130.

8.40 Sampling Amazon forests. Stratified samples are widely used to study large areas of forest. Based on satellite images, a forest area in the Amazon basin is divided into 14 types. Foresters studied the four most commercially valuable types: alluvial climax forests of quality levels 1, 2, and 3, and mature secondary forest. They divided the area of each type into large parcels, chose parcels of each type at random, and counted tree species in a 20- by 25-meter rectangle randomly placed within each parcel selected. Here is some detail:

© Age fotostock/SuperStock

Forest type	Total parcels	Sample size
Climax 1	36	4
Climax 2	72	7
Climax 3	31	3
Secondary	42	4

Choose the stratified sample of 18 parcels. Be sure to explain how you assigned labels to parcels. If you use Table B, start at line 102.

8.41 Systematic random samples. *Systematic random samples* go through a list of the population at fixed intervals from a randomly chosen starting point. For example, a study of dating among college students chose a systematic sample of 200 single male students at a university as follows.[18] Start with a list of all 9000 single male students. Because $9000/200 = 45$, choose one of the first 45 names on the list at random and then every 45th name after that. For example, if the first name chosen is at position 23, the systematic sample consists of the names at positions, 23, 68, 113, 158, and so on up to 8978.

(a) Use Table B to choose a systematic random sample of 5 addresses from a list of 200. Enter the table at line 120.

(b) Like an SRS, a systematic sample gives all individuals the same chance to be chosen. Explain why this is true, then explain carefully why a systematic sample is nonetheless *not* an SRS.

8.42 Why random digit dialing is common. The list of individuals from which a sample is actually selected is called the *sampling frame*. Ideally, the frame should list every individual in the population, but in practice this is often difficult. A frame that leaves out part of the population is a common source of undercoverage.

(a) Suppose that a sample of households in a community is selected at random from the telephone directory. What households are omitted from this frame? What types of people do you think are likely to live in these households? These people will probably be underrepresented in the sample.

(b) It is usual in telephone surveys to use random digit dialing equipment that selects the last four digits of a telephone number at random after being given the exchange (the first three digits), as described in Example 8.5 (page 208). Which of the households you mentioned in your answer to (a) will be included in the sampling frame by random digit dialing?

8.43 Regulating guns. The National Gun Policy Survey asked respondents' opinions about government regulation of firearms. A report from the survey says, "Participating households were identified through random digit dialing; the respondent in each household was selected by the most-recent-birthday method."[19]

(a) What is random digit dialing? Why is it a practical method for obtaining (almost) an SRS of households with landline phones?

(b) The survey wants the opinion of an individual adult. Several adults may live in a household. In that case, the survey interviewed the adult with the most recent birthday. Why is this preferable to simply interviewing the person who answers the phone?

8.44 Wording survey questions. Comment on each of the following as a potential sample survey question. Is the question clear? Is it slanted toward a desired response?

(a) "Some cell phone users have developed brain cancer. Should all cell phones come with a warning label explaining the danger of using cell phones?"

(b) "Do you agree that a national system of health insurance should be favored because it would provide health insurance for everyone and would reduce administrative costs?"

(c) "In view of the negative externalities in parent labor force participation and pediatric evidence associating increased group size with morbidity of children in day care, do you support government subsidies for day care programs?"

8.45 Your own bad questions. Write your own examples of bad sample survey questions.

(a) Write a biased question designed to get one answer rather than another.

(b) Write a question to which many people may not give truthful answers.

8.46 Canada's national health care. The Ministry of Health in the Canadian province of Ontario wants to know whether the national health care system is achieving its goals in the province. Much information about health care comes from patient records, but that source doesn't allow us to compare people who use health services with those who don't. So the Ministry of Health conducted the Ontario Health Survey, which interviewed a random sample of 61,239 people who live in Ontario.[20]

(a) What is the population for this sample survey? What is the sample?

(b) The survey found that 76% of males and 86% of females in the sample had visited a general practitioner at least once in the past year. Do you think these estimates are close to the truth about the entire population? Why?

8.47 Polling Hispanics. A New York Times News Service article on a poll concerned with the opinions of Hispanics includes this paragraph:

> The poll was conducted by telephone from July 13 to 27, with 3,092 adults nationwide, 1,074 of whom described themselves as Hispanic. It has a margin of sampling error of plus or minus three percentage points for the entire poll and plus or minus four percentage points for Hispanics. Sample sizes for most Hispanic nationalities, like Cubans or Dominicans, were too small to break out the results separately.[21]

(a) Why is the "margin of sampling error" larger for Hispanics than for all 3092 respondents?

(b) Why would a very small sample size prevent a responsible news organization from breaking out results for Cubans separately?

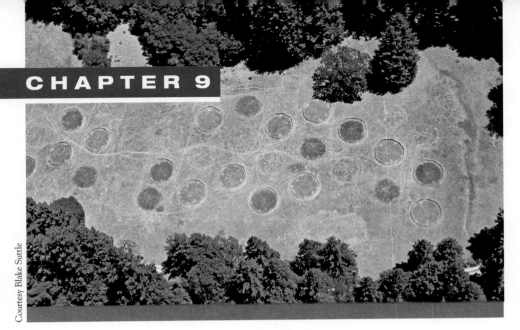

Producing Data: Experiments

A sample survey aims to gather information about a population without disturbing the population in the process. Sample surveys are one kind of *observational study*. Other observational studies observe the behavior of animals in the wild or the interactions between teacher and students in the classroom. This chapter is about statistical designs for *experiments*, a quite different way to produce data.

Observation versus experiment

In contrast to observational studies, experiments don't just observe individuals or ask them questions. They actively impose some treatment in order to observe the response. Experiments can answer questions such as "Does aspirin reduce the chance of a heart attack?" and "Do a majority of college students prefer Pepsi to Coke when they taste both without knowing which they are drinking?"

OBSERVATION VERSUS EXPERIMENT

An **observational study** observes individuals and measures variables of interest but does not attempt to influence the responses. The purpose of an observational study is to describe some group or situation.

> An **experiment,** on the other hand, deliberately imposes some treatment on individuals in order to observe their responses. The purpose of an experiment is to study whether the treatment causes a change in the response.

An observational study, even one based on a statistical sample, is a poor way to gauge the effect of a treatment. To see the response to a change, we must actually impose the change. *When our goal is to understand cause and effect, experiments are the only source of fully convincing data.* For this reason, the distinction between observation and experiment is one of the most important in statistics.

You just don't understand

A sample survey of journalists and scientists found quite a communications gap. Journalists think that scientists are arrogant, while scientists think that journalists are ignorant. We won't take sides, but here is one interesting result from the survey: 82% of the scientists agree that the "media do not understand statistics well enough to explain new findings" in medicine and other fields.

EXAMPLE 9.1 **Drink a little, but not a lot**

Many observational studies show that people who drink a moderate amount of alcohol have less heart disease than people who drink no alcohol or who drink heavily.[1] ("Moderate" means one or two drinks a day for men and one drink a day for women.) Is this *association* good reason to think that moderate drinking actually *causes* less heart disease? People who choose to drink in moderation are, as a group, different from both heavy drinkers and abstainers. They are more likely to maintain a healthy weight, get enough sleep, and exercise regularly. Moderate drinkers may be healthier because of these healthy habits rather than because of the effect of alcohol on health.

It is easy to imagine an experiment that would settle the issue of whether moderate drinking really *causes* reduced heart disease. Choose half of a large group of adults at random to be the "treatment" group. The remaining half becomes the "control" group. Require the treatment group to have one alcoholic drink every day. Require the control group to abstain from alcohol. Follow both groups for a decade. This experiment isolates the effect of alcohol. Of course, it isn't practical to carry out such an experiment. ■

The point of Example 9.1 is the contrast between observing people who choose for themselves what to drink and an experiment that requires some people to drink and others to abstain. When we simply observe people's drinking choices, the effect of moderate drinking is *confounded* with (mixed up with) the characteristics of people who choose to drink in moderation. These characteristics are lurking variables (see page 143) that make it hard to see the true relationship between the explanatory and response variables. Figure 9.1 shows the confounding in picture form.

CONFOUNDING

Two variables (explanatory variables or lurking variables) are **confounded** when their effects on a response variable cannot be distinguished from each other.

Observational studies of the effect of one variable on another often fail because the explanatory variable is confounded with lurking variables. Well-designed experiments take steps to prevent confounding.

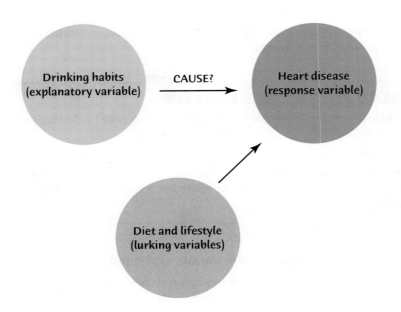

FIGURE 9.1
Confounding: We can't distinguish the effects of drinking habits from the effects of overall diet and lifestyle.

APPLY YOUR KNOWLEDGE

9.1 Cell phones and brain cancer. A study of cell phones and the risk of brain cancer looked at a group of 469 people who have brain cancer. The investigators matched each cancer patient with a person of the same sex, age, and race who did not have brain cancer, then asked about use of cell phones.[2] Result: "Our data suggest that use of handheld cellular telephones is not associated with risk of brain cancer." Is this an observational study or an experiment? Why? What are the explanatory and response variables?

9.2 Teaching economics. An educational software company wants to compare the effectiveness of its computer animation for teaching about supply and demand curves with that of a textbook presentation. The company tests the economic knowledge of a number of first-year college students, then divides them into two groups. One group uses the animation, and the other studies the text. The company retests all the students and compares the increase in economic understanding in the two groups. Is this an experiment? Why or why not? What are the explanatory and response variables?

9.3 Effects of binge drinking. A common definition of "binge drinking" is 5 or more drinks at one setting for men, and 4 or more for women. An observational study finds that students who binge have lower average GPA than those who don't. Suggest some lurking variables that may be confounded with binge drinking. The possibility of confounding means that we can't conclude that binge drinking *causes* lower GPA.

Subjects, factors, treatments

A study is an experiment when we actually do something to people, animals, or objects in order to observe the response. Because the purpose of an experiment is to reveal the response of one variable to changes in other variables, the distinction between explanatory and response variables is essential. Here is the basic vocabulary of experiments.

SUBJECTS, FACTORS, TREATMENTS

The **individuals** studied in an experiment are often called **subjects,** particularly when they are people.

The explanatory variables in an experiment are often called **factors.**

A **treatment** is any specific experimental condition applied to the subjects. If an experiment has more than one factor, a treatment is a combination of specific values of each factor.

EXAMPLE 9.2 Foster care versus orphanages

Do abandoned children placed in foster homes do better than similar children placed in an institution? The Bucharest Early Intervention Project found that the answer is a clear "Yes." The *subjects* were 136 young children abandoned at birth and living in orphanages in Bucharest, Romania. Half of the children, chosen at random, were placed in foster homes. The other half remained in the orphanages. The experiment compared these two *treatments*. There is a single *factor*, foster versus institutional care. The *response variables* included measures of mental and physical development.[3] (Foster care was not easily available in Romania at the time and so was paid for by the study. See Exercise 15 on page 259 in the Data Ethics essay for ethical questions concerning this experiment.) ■

EXAMPLE 9.3 Effects of TV advertising

What are the effects of repeated exposure to an advertising message? The answer may depend both on the length of the ad and on how often it is repeated. An experiment investigated this question using undergraduate students as *subjects*. All subjects viewed a 40-minute television program that included ads for a digital camera. Some subjects saw a 30-second commercial; others, a 90-second version. The same commercial was shown either 1, 3, or 5 times during the program.

This experiment has two *factors*: length of the commercial, with 2 values, and repetitions, with 3 values. The 6 combinations of one value of each factor form 6 *treatments*. Figure 9.2 shows the layout of the treatments. After viewing, all of the subjects answered questions about their recall of the ad, their attitude toward the camera, and their intention to purchase it. These are the *response variables*. ■

Examples 9.2 and 9.3 illustrate the advantages of experiments over observational studies. In an experiment, we can study the effects of the specific treatments we are interested in. By assigning subjects to treatments, we can avoid confounding. For example, observational studies of the effects of foster homes versus institutions on the development of children have often been biased because healthier or more alert children tend to be placed in homes. The random assignment in Example 9.2 eliminates bias in placing the children. Moreover, we can control the

```
                         Factor B
                        Repetitions
                                                      ┌─────────────────────────────┐
                 1 time    3 times   5 times          │ Subjects assigned to Treatment│
            ┌────────────┬──────────┬──────────┐      │ 3 see a 30-second ad five     │
     30     │            │          │          │      │ times during the program.     │
   seconds  │     1      │    2     │    3 ─────┼──────└─────────────────────────────┘
            │            │          │          │
 Factor A   ├────────────┼──────────┼──────────┤
  Length    │            │          │          │
     90     │     4      │    5     │    6     │
   seconds  │            │          │          │
            └────────────┴──────────┴──────────┘
```

FIGURE 9.2

The treatments in the experimental design of Example 9.3. Combinations of values of the two factors form six treatments.

environment of the subjects to hold constant factors that are of no interest to us, such as the specific product advertised in Example 9.3.

Another advantage of experiments is that we can study the combined effects of several factors simultaneously. The interaction of several factors can produce effects that could not be predicted from looking at the effect of each factor alone. Perhaps longer commercials increase interest in a product, and more commercials also increase interest, but if we both make a commercial longer and show it more often, viewers get annoyed and their interest in the product drops. The two-factor experiment in Example 9.3 will help us find out.

APPLY YOUR KNOWLEDGE

For each of the following experiments, identify the subjects, the factors, the treatments, and the response variables.

9.4 Ginkgo extract and the post-lunch dip. The post-lunch dip is the drop in mental alertness after a midday meal. Does an extract of the leaves of the ginkgo tree reduce the post-lunch dip? Assign healthy people aged 18 to 40 to take either ginkgo extract or a placebo pill. After lunch, ask them to read seven pages of random letters and place an X over every *e*. Count the number of misses.

9.5 Growing in the shade. Ability to grow in shade may help pines found in the dry forests of Arizona resist drought. How well do these pines grow in shade? Plant pine seedlings in a greenhouse in either full light, light reduced to 25% of normal by shade cloth, or light reduced to 5% of normal. At the end of the study, dry the young trees and weigh them.

9.6 Exercise and heart rate. A student project measured the increase in the heart rates of fellow students when they stepped up and down for three minutes to the beat of a metronome. The step was either 5.75 inches or 11.5 inches high and the metronome beat was either 14, 21, or 28 steps per minute. Five students stepped at each combination of height and speed. (Use a diagram like Figure 9.2 to display the factors and treatments.)

Howard Bjornson/Getty Images

How to experiment badly

Experiments are the preferred method for examining the effect of one variable on another. By imposing the specific treatment of interest and controlling other influences, we can pin down cause and effect. Statistical designs are often essential for effective experiments. To see why, let's look at an example in which an experiment suffers from confounding just as observational studies do.

> **EXAMPLE 9.4** **An uncontrolled experiment**
>
> A college regularly offers a review course to prepare candidates for the Graduate Management Admission Test (GMAT), which is required by most graduate business schools. This year, it offers only an online version of the course. The average GMAT score of students in the online course is 10% higher than the longtime average for those who took the classroom review course. Is the online course more effective?
>
> This experiment has a very simple design. A group of subjects (the students) were exposed to a treatment (the online course), and the outcome (GMAT scores) was observed. Here is the design:
>
> $$\text{Subjects} \longrightarrow \text{Online course} \longrightarrow \text{GMAT scores}$$
>
> A closer look at the GMAT review course showed that the students in the online review course were quite different from the students who in past years took the classroom course. In particular, they were older and more likely to be employed. An online course appeals to these mature people, but we can't compare their performance with that of the undergraduates who previously dominated the course. The online course might even be less effective than the classroom version. The effect of online versus in-class instruction is confounded with the effect of lurking variables. As a result of confounding, the experiment is biased in favor of the online course. ■

Most laboratory experiments use a design like that in Example 9.4:

$$\text{Subjects} \longrightarrow \text{Treatment} \longrightarrow \text{Measure response}$$

In the controlled environment of the laboratory, simple designs often work well. Field experiments and experiments with living subjects are exposed to more variable conditions and deal with more variable subjects. *Outside the laboratory, uncontrolled experiments often yield worthless results because of confounding with lurking variables.*

APPLY YOUR KNOWLEDGE

9.7 Reducing unemployment. Will cash bonuses speed the return to work of unemployed people? A state department of labor notes that last year 68% of people who filed claims for unemployment insurance found a new job within 15 weeks. As an experiment, the state offers $500 to people filing unemployment claims if they find a job within 15 weeks. The percent who do so increases to 77%. Suggest some conditions that might make it easier or harder to find a job this year as opposed to last year. Confounding with these lurking variables makes it impossible to say whether the bonus really caused the increase.

Randomized comparative experiments

The remedy for the confounding in Example 9.4 is to do a *comparative experiment* in which some students are taught in the classroom and other, similar students take the course online. The first group is called a **control group.** Most well-designed experiments compare two or more treatments. Part of the design of an experiment is a description of the factors (explanatory variables) and the layout of the treatments, with comparison as the leading principle.

control group

Comparison alone isn't enough to produce results we can trust. If the treatments are given to groups that differ markedly when the experiment begins, bias will result. For example, if we allow students to elect online or classroom instruction, students who are older and employed are likely to sign up for the online course. Personal choice will bias our results in the same way that volunteers bias the results of online opinion polls. The solution to the problem of bias in sampling is random selection, and the same is true in experiments. The subjects assigned to any treatment should be chosen at random from the available subjects.

RANDOMIZED COMPARATIVE EXPERIMENT

An experiment that uses both comparison of two or more treatments and random assignment of subjects to treatments is a **randomized comparative experiment.**

EXAMPLE 9.5 **On-campus versus online**

The college decides to compare the progress of 25 on-campus students taught in the classroom with that of 25 students taught the same material online. Select the students who will be taught online by taking a simple random sample of size 25 from the 50 available subjects. The remaining 25 students form the control group. They will receive classroom instruction. The result is a randomized comparative experiment with two groups. Figure 9.3 outlines the design in graphical form.

The selection procedure is exactly the same as it is for sampling. **Label:** Label the 50 students 01 to 50. **Table:** Go to the table of random digits and read successive

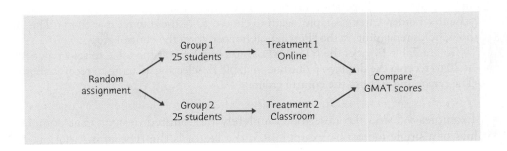

FIGURE 9.3

Outline of a randomized comparative experiment to compare online and classroom instruction, for Example 9.5.

two-digit groups. The first 25 labels encountered select the online group. As usual, ignore repeated labels and groups of digits not used as labels. For example, if you begin at line 125 in Table B, the first five students chosen are those labeled 21, 49, 37, 18, and 44. Software such as the *Simple Random Sample* applet makes it particularly easy to choose treatment groups at random. ■

The design in Example 9.5 is *comparative* because it compares two treatments (the two instructional settings). It is *randomized* because the subjects are assigned to the treatments by chance. This "flowchart" outline in Figure 9.3 presents all the essentials: randomization, the sizes of the groups and which treatment they receive, and the response variable. There are, as we will see later, statistical reasons for generally using treatment groups about equal in size. We call designs like that in Figure 9.3 *completely randomized*.

COMPLETELY RANDOMIZED DESIGN

In a **completely randomized** experimental design, all the subjects are allocated at random among all the treatments.

Completely randomized designs can compare any number of treatments. Here is an example that compares three treatments.

EXAMPLE 9.6 Conserving energy

Many utility companies have introduced programs to encourage energy conservation among their customers. An electric company considers placing small digital displays in households to show current electricity use and what the cost would be if this use continued for a month. Will the displays reduce electricity use? Would cheaper methods work almost as well? The company decides to conduct an experiment.

One cheaper approach is to give customers a chart and information about monitoring their electricity use from their outside meter. The experiment compares these two approaches (display, chart) and also a control. The control group of customers receives information about energy conservation but no help in monitoring electricity use. The response variable is total electricity used in a year. The company finds 60 single-family residences in the same city willing to participate, so it assigns 20 residences at random to each of the three treatments. Figure 9.4 outlines the design.

To use the *Simple Random Sample* applet, set the population labels as 1 to 60 and the sample size to 20 and click "Reset" and "Sample." The 20 households chosen receive the displays. The "Population hopper" now contains the 40 remaining households, in scrambled order. Click "Sample" again to choose 20 of these to receive charts. The 20 households remaining in the "Population hopper" form the control group.

To use Table B, label the 60 households 01 to 60. Enter the table to select an SRS of 20 to receive the displays. Continue in Table B, selecting 20 more to receive charts. The remaining 20 form the control group. ■

Examples 9.5 and 9.6 describe completely randomized designs that compare values of a single factor. In Example 9.5, the factor is the type of instruction. In

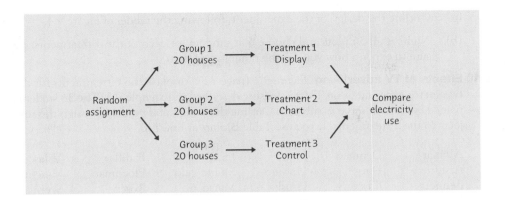

FIGURE 9.4
Outline of a completely randomized design comparing three energy-saving programs, for Example 9.6.

Example 9.6, it is the method used to encourage energy conservation. Completely randomized designs can have more than one factor. The advertising experiment of Example 9.3 has two factors: the length and the number of repetitions of a television commercial. Their combinations form the six treatments outlined in Figure 9.2. A completely randomized design assigns subjects at random to these six treatments. Once the layout of treatments is set, the randomization needed for a completely randomized design is tedious but straightforward.

APPLY YOUR KNOWLEDGE

9.8 Evaluating your own performance. Undergraduate music students often don't evaluate their own performances accurately. Can small-group discussions help? The subjects were 29 students preparing for the end-of-semester performance that is an important part of their grade. Assign 15 students to the treatment: videotape a practice performance, ask the student to evaluate it, then have the student discuss the tape with a small group of other students. The remaining 14 students form a control group who watch and evaluate their tapes alone. At the end of the semester, the discussion-group students evaluated their final performance more accurately.[4]

(a)   Outline the design of this experiment, following the model of Figure 9.3.

(b)   Carry out the random assignment of 15 students to the treatment group, using the *Simple Random Sample* applet, other software, or Table B, starting at line 132.

9.9 More rain for California? The changing climate will probably bring more rain to California, but we don't know whether the additional rain will come during the winter wet season or extend into the long dry season in spring and summer. Kenwyn Suttle of the University of California at Berkeley and his coworkers carried out a randomized controlled experiment to study the effects of more rain in either season. They randomly assigned plots of open grassland to 3 treatments: added water equal to 20% of annual rainfall either during January to March (winter) or during April to June (spring), and no added water (control). Thirty-six circular plots of area 70 square meters were available (see the photo), of which 18 were used for this study. One response variable was total plant biomass, in grams per square meter, produced in a plot over a year.[5]

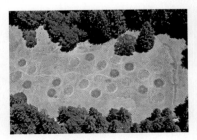

Courtesy Blake Suttle

(a)   Outline the design of the experiment, following the model of Figure 9.4.

(b)   Number all 36 plots and choose 6 at random for each of the 3 treatments. Be sure to explain how you did the random selection.

9.10 Effects of TV advertising. Figure 9.2 (page 227) displays the 6 treatments for the two-factor experiment on TV advertising described in Example 9.3. The 36 students named below will serve as subjects. Outline the design and randomly assign the subjects to the 6 treatments. If you use Table B, start at line 130.

Alomar	Denman	Han	Liang	Padilla	Valasco
Asihiro	Durr	Howard	Maldonado	Plochman	Vaughn
Bennett	Edwards	Hruska	Marsden	Rosen	Wei
Bikalis	Farouk	Imrani	Montoya	Solomon	Wilder
Chao	Fleming	James	O'Brian	Trujillo	Willis
Clemente	George	Kaplan	Ogle	Tullock	Zhang

The logic of randomized comparative experiments

Randomized comparative experiments are designed to give good evidence that differences in the treatments actually *cause* the differences we see in the response. The logic is as follows:

- Random assignment of subjects forms groups that should be similar in all respects before the treatments are applied. Exercise 9.50 uses the *Simple Random Sample* applet to demonstrate this.

- Comparative design ensures that influences other than the experimental treatments operate equally on all groups.

- Therefore, differences in average response must be due either to the treatments or to the play of chance in the random assignment of subjects to the treatments.

That "either-or" deserves more thought. In Example 9.5, we cannot say that *any* difference between the average GMAT scores of students enrolled online and in the classroom must be caused by a difference in the effectiveness of the two types of instruction. There would be some difference even if both groups received the same instruction, because of variation among students in background and study habits. Chance assigns students to one group or the other, and this creates a chance difference between the groups. We would not trust an experiment with just one student in each group, for example. The results would depend too much on which group got lucky and received the stronger student. If we assign many subjects to each group, however, the effects of chance will average out and there will be little difference in the average responses in the two groups unless the treatments themselves cause a difference. "Use enough subjects to reduce chance variation" is the third big idea of statistical design of experiments.

PRINCIPLES OF EXPERIMENTAL DESIGN

The basic principles of statistical design of experiments are

1. **Control** the effects of lurking variables on the response, most simply by comparing two or more treatments.

2. **Randomize**—use chance to assign subjects to treatments.

3. **Use enough subjects** in each group to reduce chance variation in the results.

We hope to see a difference in the responses so large that it is unlikely to happen just because of chance variation. We can use the laws of probability, which describe chance behavior, to learn if the treatment effects are larger than we would expect to see if only chance were operating. If they are, we call them *statistically significant*.

STATISTICAL SIGNIFICANCE

An observed effect so large that it would rarely occur by chance is called **statistically significant.**

If we observe statistically significant differences among the groups in a randomized comparative experiment, we have good evidence that the treatments actually caused these differences. You will often see the phrase "statistically significant" in reports of investigations in many fields of study. The great advantage of randomized comparative experiments is that they can produce data that give good evidence for a cause-and-effect relationship between the explanatory and response variables. We know that in general a strong association does not imply causation. A statistically significant association in data from a well-designed experiment *does* imply causation.

APPLY YOUR KNOWLEDGE

9.11 Prayer and meditation. You read in a magazine that "nonphysical treatments such as meditation and prayer have been shown to be effective in controlled scientific studies for such ailments as high blood pressure, insomnia, ulcers, and asthma." Explain in simple language what the article means by "controlled scientific studies." Why can such studies in principle provide good evidence that, for example, meditation is an effective treatment for high blood pressure?

9.12 Conserving energy. Example 9.6 describes an experiment to learn whether providing households with digital displays or charts will reduce their electricity consumption. An executive of the electric company objects to including a control group. He says: "It would be simpler to just compare electricity use last year (before the display or chart was provided) with consumption in the same period this year. If households use less electricity this year, the display or chart must be working." Explain clearly why this design is inferior to that in Example 9.6.

9.13 Arsenic and lung cancer. Arsenic is frequently found both in the natural environment and in food. A study of the relationship between arsenic in drinking water

What's news?

Randomized comparative experiments provide the best evidence for medical advances. Do newspapers care? Maybe not. University researchers looked at 1192 articles in medical journals, of which 7% were turned into stories by the two newspapers examined. Of the journal articles, 37% concerned observational studies and 25% described randomized experiments. Among the articles publicized by the newspapers, 58% were observational studies and only 6% were randomized experiments. Conclusion: the newspapers want exciting stories, especially bad-news stories, whether or not the evidence is good.

Digital Vision/Getty Images

and deaths from lung cancer measured arsenic levels in drinking water in 138 villages in Taiwan and examined death certificates to identify lung cancer deaths. The study summary says that "arsenic levels above 0.64 mg/l were associated with a significant increase in the mortality of lung cancer in both genders, but no significant effect was observed at lower levels."[6]

(a)   Explain why this is an observational study rather than an experiment.

(b)   The word "significant" in the conclusion has its statistical meaning, not its everyday meaning. Restate the study conclusion without using the word "significant" in a way that is clear to readers who know no statistics.

Cautions about experimentation

The logic of a randomized comparative experiment depends on our ability to treat all the subjects identically in every way except for the actual treatments being compared. Good experiments therefore require careful attention to details to ensure that all subjects really are treated identically.

If some subjects in a medical experiment take a pill each day and a control group takes no pill, the subjects are not treated identically. Many medical experiments are therefore "placebo-controlled." A study of the effects of taking vitamin E on heart disease is typical. All of the subjects receive the same medical attention during the several years of the experiment. All of them take a pill every day, vitamin E in the treatment group and a placebo in the control group. A **placebo** is a dummy treatment. Many patients respond favorably to any treatment, even a placebo, perhaps because they trust the doctor. The response to a dummy treatment is called the *placebo effect*. If the control group did not take any pills, the effect of vitamin E in the treatment group would be confounded with the placebo effect, the effect of simply taking pills.

In addition, such studies are usually *double-blind*. The subjects don't know whether they are taking vitamin E or a placebo. Neither do the medical personnel who work with them. The double-blind method avoids unconscious bias by, for example, a doctor who is convinced that a vitamin must be better than a placebo. In many medical studies, only the statistician who does the randomization knows which treatment each patient is receiving.

Scratch my furry ears

Rats and rabbits, specially bred to be uniform in their inherited characteristics, are the subjects in many experiments. Animals, like people, are quite sensitive to how they are treated. This can create opportunities for hidden bias. For example, human affection can change the cholesterol level of rabbits. Choose some rabbits at random and regularly remove them from their cages to have their heads scratched by friendly people. Leave other rabbits unloved. All the rabbits eat the same diet, but the rabbits that receive affection have lower cholesterol.

lack of realism

> **DOUBLE-BLIND EXPERIMENTS**
>
> In a **double-blind** experiment, neither the subjects nor the people who interact with them know which treatment each subject is receiving.

Placebo controls and the double-blind method are more ways to eliminate possible confounding. But even well-designed experiments often face another problem: **lack of realism.** Practical constraints may mean that the subjects or treatments or setting of an experiment don't realistically duplicate the conditions we really want to study. Here are two examples.

EXAMPLE 9.7   Response to advertising

The study of television advertising in Example 9.3 showed a 40-minute video to students who knew an experiment was going on. We can't be sure that the results apply to everyday television viewers. Many behavioral science experiments use as subjects students or other volunteers who know they are subjects in an experiment. That's not a realistic setting. ■

EXAMPLE 9.8   Center brake lights

Do those high center brake lights, required on all cars sold in the United States since 1986, really reduce rear-end collisions? Randomized comparative experiments with fleets of rental and business cars, done before the lights were required, showed that the third brake light reduced rear-end collisions by as much as 50%. Alas, requiring the third light in all cars led to only a 5% drop.

    What happened? Most cars did not have the extra brake light when the experiments were carried out, so it caught the eye of following drivers. Now that almost all cars have the third light, they no longer capture attention. ■

©Image 100/CORBIS

Lack of realism can limit our ability to apply the conclusions of an experiment to the settings of greatest interest. Most experimenters want to generalize their conclusions to some setting wider than that of the actual experiment. *Statistical analysis of an experiment cannot tell us how far the results will generalize.* Nonetheless, the randomized comparative experiment, because of its ability to give convincing evidence for causation, is one of the most important ideas in statistics.

APPLY YOUR KNOWLEDGE

9.14 Testosterone for older men. As men age, their testosterone levels gradually decrease. This may cause a reduction in lean body mass, an increase in fat, and other undesirable changes. Do testosterone supplements reverse some of these effects? A study in the Netherlands assigned 237 men aged 60 to 80 with low or low-normal testosterone levels to either a testosterone supplement or a placebo. The report in the *Journal of the American Medical Association* described the study as a "double-blind, randomized, placebo-controlled trial."[7] Explain each of these terms to someone who knows no statistics.

9.15 Does meditation reduce anxiety? An experiment that claimed to show that meditation reduces anxiety proceeded as follows. The experimenter interviewed the subjects and rated their level of anxiety. Then the subjects were randomly assigned to two groups. The experimenter taught one group how to meditate and they meditated daily for a month. The other group was simply told to relax more. At the end of the month, the experimenter interviewed all the subjects again and rated their anxiety level. The meditation group now had less anxiety. Psychologists said that the results were suspect because the ratings were not blind. Explain what this means and how lack of blindness could bias the reported results.

Matched pairs and other block designs

Completely randomized designs are the simplest statistical designs for experiments. They illustrate clearly the principles of control, randomization, and adequate number of subjects. However, completely randomized designs are often inferior to more elaborate statistical designs. In particular, matching the subjects in various ways can produce more precise results than simple randomization.

matched pairs design

One common design that combines matching with randomization is the **matched pairs design.** A matched pairs design compares just two treatments. Choose pairs of subjects that are as closely matched as possible. Use chance to decide which subject in a pair gets the first treatment. The other subject in that pair gets the other treatment. That is, the random assignment of subjects to treatments is done within each matched pair, not for all subjects at once. Sometimes each "pair" in a matched pairs design consists of just one subject, who gets both treatments one after the other. Each subject serves as his or her own control. The *order* of the treatments can influence the subject's response, so we randomize the order for each subject.

Royalty Free/CORBIS

EXAMPLE 9.9 **Cell phones and driving**

Does talking on a hands-free cell phone distract drivers? Undergraduate students "drove" in a high-fidelity driving simulator equipped with a hands-free cell phone. The car ahead brakes: how quickly does the subject react? Let's compare two designs for this experiment. There are 40 student subjects available.

In a *completely randomized design*, all 40 subjects are assigned at random, 20 to simply drive and the other 20 to talk on the cell phone while driving. In the *matched pairs design* that was actually used, all subjects drive both with and without using the cell phone. The two drives are on separate days to reduce carryover effects. The *order* of the two treatments is assigned at random: 20 subjects are chosen to drive first with the phone, and the remaining 20 drive first without the phone.[8]

Some subjects naturally react faster than others. The completely randomized design relies on chance to distribute the faster subjects roughly evenly between the two groups. The matched pairs design compares each subject's reaction time with and without the cell phone. This makes it easier to see the effects of using the phone. ■

Matched pairs designs use the principles of comparison of treatments and randomization. However, the randomization is not complete—we do not randomly assign all the subjects at once to the two treatments. Instead, we randomize only within each matched pair. This allows matching to reduce the effect of variation among the subjects. Matched pairs are one kind of *block design*, with each pair forming a *block*.

BLOCK DESIGN

A **block** is a group of individuals that are known before the experiment to be similar in some way that is expected to affect the response to the treatments.

In a **block design,** the random assignment of individuals to treatments is carried out separately within each block.

A block design combines the idea of creating equivalent treatment groups by matching with the principle of forming treatment groups at random. Blocks are another form of *control*. They control the effects of some outside variables by bringing those variables into the experiment to form the blocks. Here are some typical examples of block designs.

EXAMPLE 9.10 **Men, women, and advertising**

Women and men respond differently to advertising. An experiment to compare the effectiveness of three advertisements for the same product will want to look separately at the reactions of men and women, as well as assess the overall response to the ads.

A *completely randomized design* considers all subjects, both men and women, as a single pool. The randomization assigns subjects to three treatment groups without regard to their sex. This ignores the differences between men and women. A *block design* considers women and men separately. Randomly assign the women to three groups, one to view each advertisement. Then separately assign the men at random to three groups. Figure 9.5 outlines this improved design. ■

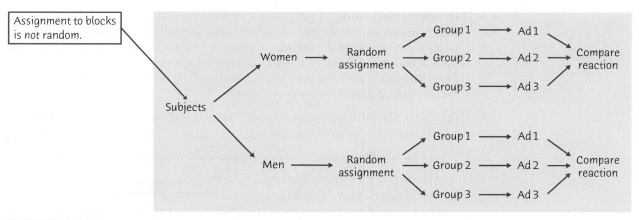

FIGURE 9.5

Outline of a block design, for Example 9.10. The blocks consist of male and female subjects. The treatments are three advertisements for the same product.

EXAMPLE 9.11 **Comparing welfare policies**

A social policy experiment will assess the effect on family income of several proposed new welfare systems and compare them with the present welfare system. Because the future income of a family is strongly related to its present income, the families who agree to participate are divided into blocks of similar income levels. The families in each block are then allocated at random among the welfare systems. ■

A block design allows us to draw separate conclusions about each block, for example, about men and women in Example 9.10. Blocking also allows more precise overall conclusions, because the systematic differences between men and women can be removed when we study the overall effects of the three advertisements.

The idea of blocking is an important additional principle of statistical design of experiments. A wise experimenter will form blocks based on the most important unavoidable sources of variability among the subjects. Randomization will then average out the effects of the remaining variation and allow an unbiased comparison of the treatments.

Like the design of samples, the design of complex experiments is a job for experts. Now that we have seen a bit of what is involved, we will concentrate for the most part on completely randomized experiments.

APPLY YOUR KNOWLEDGE

9.16 Comparing hand strength. Is the right hand generally stronger than the left in right-handed people? You can crudely measure hand strength by placing a bathroom scale on a shelf with the end protruding, then squeezing the scale between the thumb below and the four fingers above it. The reading of the scale shows the force exerted. Describe the design of a matched pairs experiment to compare the strength of the right and left hands, using 10 right-handed people as subjects. (You need not actually do the randomization.)

9.17 How long did I work? A psychologist wants to know if the difficulty of a task influences our estimate of how long we spend working at it. She designs two sets of mazes that subjects can work through on a computer. One set has easy mazes and the other has hard mazes. Subjects work until told to stop (after 6 minutes, but subjects do not know this). They are then asked to estimate how long they worked. The psychologist has 30 students available to serve as subjects.

(a) Describe the design of a completely randomized experiment to learn the effect of difficulty on estimated time.

(b) Describe the design of a matched pairs experiment using the same 30 subjects.

9.18 Technology for teaching statistics. The Brigham Young University statistics department is performing randomized comparative experiments to compare teaching methods. Response variables include students' final-exam scores and a measure of their attitude toward statistics. One study compares two levels of technology for large lectures: standard (overhead projectors and chalk) and multimedia. The individuals in the study are the 8 lectures in a basic statistics course. There are four instructors, each of whom teaches two lectures. Because the lecturers differ, their lectures form four blocks.[9] Suppose the lectures and lecturers are as follows:

Lecture	Lecturer	Lecture	Lecturer
1	Hilton	5	Tolley
2	Christensen	6	Hilton
3	Hadfield	7	Tolley
4	Hadfield	8	Christensen

Outline a block design and do the randomization that your design requires.

C H A P T E R   9   S U M M A R Y

■ We can produce data intended to answer specific questions by **observational studies** or **experiments.** Sample surveys that select a part of a population of interest to represent the whole are one type of observational study. **Experiments,** unlike observational studies, actively impose some treatment on the subjects of the experiment.

■ Variables are **confounded** when their effects on a response can't be distinguished from each other. Observational studies and uncontrolled experiments often fail to show that changes in an explanatory variable actually cause changes in a response variable because the explanatory variable is confounded with lurking variables.

■ In an experiment, we impose one or more **treatments** on individuals, often called **subjects.** Each treatment is a combination of values of the explanatory variables, which we call **factors.**

■ The **design** of an experiment describes the choice of treatments and the manner in which the subjects are assigned to the treatments. The basic principles of statistical design of experiments are **control** and **randomization** to combat bias and **using enough subjects** to reduce chance variation.

■ The simplest form of control is **comparison.** Experiments should compare two or more treatments in order to avoid confounding of the effect of a treatment with other influences, such as lurking variables.

■ **Randomization** uses chance to assign subjects to the treatments. Randomization creates treatment groups that are similar (except for chance variation) before the treatments are applied. Randomization and comparison together prevent **bias,** or systematic favoritism, in experiments.

■ You can carry out randomization by using software or by giving numerical labels to the subjects and using a **table of random digits** to choose treatment groups.

■ Applying each treatment to many subjects reduces the role of chance variation and makes the experiment more sensitive to differences among the treatments.

■ Good experiments require attention to detail as well as good statistical design. Many behavioral and medical experiments are **double-blind.** Some give a **placebo** to a control group. **Lack of realism** in an experiment can prevent us from generalizing its results.

■ In addition to comparison, a second form of control is to restrict randomization by forming **blocks** of individuals that are similar in some way that is important to the response. Randomization is then carried out separately within each block.

■ **Matched pairs** are a common form of blocking for comparing just two treatments. In some matched pairs designs, each subject receives both treatments in a random order. In others, the subjects are matched in pairs as closely as possible, and each subject in a pair receives one of the treatments.

9.19 The Nurses' Health Study has interviewed a sample of more than 100,000 female registered nurses every two years since 1976. The study finds that "light-to-moderate drinkers had a significantly lower risk of death" than either nondrinkers or heavy drinkers. The Nurses' Health Study is

(a) an observational study.

(b) an experiment.

(c) Can't tell without more information.

9.20 What electrical changes occur in muscles as they get tired? Student subjects hold their arms above their shoulders until they have to drop them. Meanwhile, the electrical activity in their arm muscles is measured. This is

(a) an observational study.

(b) an uncontrolled experiment.

(c) a randomized comparative experiment.

9.21 Can changing diet reduce high blood pressure? Vegetarian diets and low-salt diets are both promising. Men with high blood pressure are assigned at random to four diets: (1) normal diet with unrestricted salt; (2) vegetarian with unrestricted salt; (3) normal with restricted salt; and (4) vegetarian with restricted salt. This experiment has

(a) one factor, the choice of diet.

(b) two factors, normal/vegetarian diet and unrestricted/restricted salt.

(c) four factors, the four diets being compared.

9.22 In the experiment of the previous exercise, the 240 subjects are labeled 001 to 240. Software assigns an SRS of 60 subjects to Diet 1, an SRS of 60 of the remaining 180 to Diet 2, and an SRS of 60 of the remaining 120 to Diet 3. The 60 who are left get Diet 4. This is a

(a) completely randomized design.

(b) block design, with four blocks.

(c) matched pairs design.

9.23 An important response variable in the experiment described in Exercise 9.21 must be

(a) the amount of salt in the subject's diet.

(b) which of the four diets a subject is assigned to.

(c) change in blood pressure after 8 weeks on the assigned diet.

9.24 A medical experiment compares an antidepression medicine with a placebo for relief of chronic headaches. There are 36 headache patients available to serve as subjects. To choose 18 patients to receive the medicine, you would

(a) assign labels 01 to 36 and use Table B to choose 18.

(b)   assign labels 01 to 18, because only 18 need to be chosen.

(c)   assign the first 18 who signed up to get the medicine.

9.25 The Community Intervention Trial for Smoking Cessation asked whether a community-wide advertising campaign would reduce smoking. The researchers located 11 pairs of communities, each pair similar in location, size, economic status, and so on. One community in each pair participated in the advertising campaign and the other did not. This is

(a)   an observational study.

(b)   a matched pairs experiment.

(c)   a completely randomized experiment.

9.26 To decide which community in each pair in the previous exercise should get the advertising campaign, it is best to

(a)   toss a coin.

(b)   choose the community that will help pay for the campaign.

(c)   choose the community with a mayor who will participate.

9.27 A marketing class designs two videos advertising an expensive Mercedes sports car. They test the videos by asking fellow students to view both (in random order) and say which makes them more likely to buy the car. Mercedes should be reluctant to agree that the video favored in this study will sell more cars because

(a)   the study used a matched pairs design instead of a completely randomized design.

(b)   results from students may not generalize to the older and richer customers who might buy a Mercedes.

(c)   this is an observational study, not an experiment.

C H A P T E R   9   E X E R C I S E S

In all exercises that require randomization, you may use Table B, the Simple Random Sample applet, or other software. See Example 9.6 for directions on using the applet for more than two treatment groups.

9.28 Alcohol and heart attacks. Many studies have found that people who drink alcohol in moderation have lower risk of heart attacks than either nondrinkers or heavy drinkers. Does alcohol consumption also improve survival after a heart attack? One study followed 1913 people who were hospitalized after severe heart attacks. In the year before their heart attacks, 47% of these people did not drink, 36% drank moderately, and 17% drank heavily. After four years, fewer of the moderate drinkers had died.[10]

(a)   Is this an observational study or an experiment? Why? What are the explanatory and response variables?

(b) Suggest some lurking variables that may be confounded with the drinking habits of the subjects. The possible confounding makes it difficult to conclude that drinking habits explain death rates.

9.29 Reducing nonresponse. How can we reduce the rate of refusals in telephone surveys? Most people who answer at all listen to the interviewer's introductory remarks and then decide whether to continue. One study made telephone calls to randomly selected households to ask opinions about the next election. In some calls, the interviewer gave her name, in others she identified the university she was representing, and in still others she identified both herself and the university. The study recorded what percent of each group of interviews was completed. Is this an observational study or an experiment? Why? What are the explanatory and response variables?

9.30 Samples versus experiments. Give an example of a question about college students, their behavior, or their opinions that would best be answered by

(a) a sample survey.

(b) an experiment.

9.31 Observation versus experiment. Observational studies had suggested that vitamin E reduces the risk of heart disease. Careful experiments, however, showed that vitamin E has no effect. According to a commentary in the *Journal of the American Medical Association*:

> Thus, vitamin E enters the category of therapies that were promising in epidemiologic and observational studies but failed to deliver in adequately powered randomized controlled trials. As in other studies, the "healthy user" bias must be considered, ie, the healthy lifestyle behaviors that characterize individuals who care enough about their health to take various supplements are actually responsible for the better health, but this is minimized with the rigorous trial design.[11]

A friend who knows no statistics asks you to explain this.

(a) What is the difference between observational studies and experiments?

(b) What is a "randomized controlled trial"? (We'll discuss "adequately powered" in Chapter 15.)

(c) How does "healthy user bias" explain how people who take vitamin E supplements have better health in observational studies but not in controlled experiments?

9.32 Attitudes toward homeless people. Negative attitudes toward poor people are common. Are attitudes more negative when a person is homeless? To find out, read to subjects a description of a poor person. There are two versions. One begins

> Jim is a 30-year-old single man. He is currently living in a small single-room apartment.

The other description begins

> Jim is a 30-year-old single man. He is currently homeless and lives in a shelter for homeless people.

After reading the description, ask subjects what they believe about Jim and what they think should be done to help him. The subjects are 544 adults interviewed by telephone.[12] Outline the design of this experiment.

9.33 Getting teachers to come to school. Elementary schools in rural India are usually small, with a single teacher. The teachers often fail to show up for work. Here is an idea for improving attendance: give the teacher a digital camera with a tamper-proof time and date stamp and ask a student to take a photo of the teacher and class at the beginning and end of the day. Offer the teacher better pay for good attendance, verified by the photos. Will this work? A randomized comparative experiment started with 120 rural schools in Rajasthan and assigned 60 to this treatment and 60 to a control group. Random checks for teacher attendance showed that 21% of teachers in the treatment group were absent, as opposed to 42% in the control group.[13]

(a) Outline the design of this experiment.

(b) Label the schools and choose the first 10 schools for the treatment group. If you use Table B, start at line 108.

9.34 Marijuana and work. How does smoking marijuana affect willingness to work? Canadian researchers persuaded young adult men who used marijuana to live for 98 days in a "planned environment." The men earned money by weaving belts. They used their earnings to pay for meals and other consumption and could keep any money left over. One group smoked two potent marijuana cigarettes every evening. The other group smoked two weak marijuana cigarettes. All subjects could buy more cigarettes but were given strong or weak cigarettes depending on their group. Did the weak and strong groups differ in work output and earnings?[14]

(a) Outline the design of this experiment.

(b) Here are the names of the 20 subjects. Use software or Table B at line 131 to carry out the randomization your design requires.

Abate	Dubois	Gutierrez	Lucero	Rosen
Afifi	Engel	Huang	McNeill	Thompson
Brown	Fluharty	Iselin	Morse	Travers
Cheng	Gerson	Kaplan	Quinones	Ullmann

9.35 The benefits of red wine. Some people think that red wine protects moderate drinkers from heart disease better than other alcoholic beverages. This calls for a randomized comparative experiment. The subjects were healthy men aged 35 to 65. They were randomly assigned to drink red wine (9 subjects), drink white wine (9 subjects), drink white wine and also take polyphenols from red wine (6 subjects), take polyphenols alone (9 subjects), or drink vodka and lemonade (6 subjects).[15] Outline the design of the experiment and randomly assign the 39 subjects to the 5 groups. If you use Table B, start at line 107.

9.36 Fabric finishing. A maker of fabric for clothing is setting up a new line to "finish" the raw fabric. The line will use either metal rollers or natural-bristle rollers to raise the surface of the fabric; a dyeing cycle time of either 30 minutes or 40 minutes; and a temperature of either 150°C or 175°C. An experiment will compare all combinations of these choices. Three specimens of fabric will be subjected to each treatment and scored for quality.

(a) What are the factors and the treatments? How many individuals (fabric specimens) does the experiment require?

Tim Elliott/FeaturePics

(b) Outline a completely randomized design for this experiment. (You need not actually do the randomization.)

9.37 Relieving headaches. Doctors identify "chronic tension-type headaches" as headaches that occur almost daily for at least six months. Can antidepressant medications or stress management training reduce the number and severity of these headaches? Are both together more effective than either alone?

(a) Use a diagram like Figure 9.2 to display the treatments in a design with two factors: "medication, yes or no" and "stress management, yes or no." Then outline the design of a completely randomized experiment to compare these treatments.

(b) The headache sufferers named below have agreed to participate in the study. Randomly assign the subjects to the treatments. If you use the *Simple Random Sample* applet or other software, assign all the subjects. If you use Table B, start at line 130 and assign subjects to only the first treatment group.

Abbott	Decker	Herrera	Lucero	Richter
Abdalla	Devlin	Hersch	Masters	Riley
Alawi	Engel	Hurwitz	Morgan	Samuels
Broden	Fuentes	Irwin	Nelson	Smith
Chai	Garrett	Jiang	Nho	Suarez
Chuang	Gill	Kelley	Ortiz	Upasani
Cordoba	Glover	Kim	Ramdas	Wilson
Custer	Hammond	Landers	Reed	Xiang

Treating sinus infections. *Sinus infections are common, and doctors commonly treat them with antibiotics. Another treatment is to spray a steroid solution into the nose. A well-designed clinical trial found that these treatments, alone or in combination, do not reduce the severity or the length of sinus infections.*[16] *Exercises 9.38 to 9.40 concern this trial.*

9.38 Experimental design. The clinical trial was a completely randomized experiment that assigned 240 patients at random among 4 treatments as follows:

	Antibiotic pill	Placebo pill
Steroid spray	53	64
Placebo spray	60	63

(a) Outline the design of the experiment.

(b) How will you label the 240 subjects?

(c) Explain briefly how you would do the random assignment of patients to treatments. Assign the first 5 patients who will receive the first treatment.

9.39 Describing the design. The report of this study in the *Journal of the American Medical Association* describes it as a "double-blind, randomized, placebo-controlled factorial trial." "Factorial" means that the treatments are formed from more than one factor. What are the factors? What do "double-blind" and "placebo-controlled" mean?

9.40 Checking the randomization. If the random assignment of patients to treatments did a good job of eliminating bias, possible lurking variables such as smoking history, asthma, and hay fever should be similar in all 4 groups. After recording and comparing many such variables, the investigators said that "all showed no significant difference between groups." Explain to someone who knows no statistics what "no significant difference" means. Does it mean that the presence of all these variables was exactly the same in all four treatment groups?

9.41 Frappuccino light? Here's the opening of a Starbucks press release: "Starbucks Corp. on Monday said it would roll out a line of blended coffee drinks intended to tap into the growing popularity of reduced-calorie and reduced-fat menu choices for Americans." You wonder if Starbucks customers like the new "Mocha Frappuccino Light" as well as the regular mocha Frappuccino coffee.

(a) Describe a matched pairs design to answer this question. Be sure to include proper blinding of your subjects.

(b) You have 20 regular Starbucks customers on hand. Use the *Simple Random Sample* applet or Table B at line 141 to do the randomization that your design requires.

9.42 Growing trees faster. The concentration of carbon dioxide (CO_2) in the atmosphere is increasing rapidly due to our use of fossil fuels. Because green plants use CO_2 to fuel photosynthesis, more CO_2 may cause trees to grow faster. An elaborate apparatus allows researchers to pipe extra CO_2 to a 30-meter circle of forest. We want to compare the growth in base area of trees in treated and untreated areas to see if extra CO_2 does in fact increase growth. We can afford to treat three circular areas.[17]

(a) Describe the design of a completely randomized experiment using six well-separated 30-meter circular areas in a pine forest. Sketch the circles and carry out the randomization your design calls for.

(b) Areas within the forest may differ in soil fertility. Describe a matched pairs design using three pairs of circles that will reduce the extra variation due to different fertility. Sketch the circles and carry out the randomization your design calls for.

9.43 Athletes taking oxygen. We often see players on the sidelines of a football game inhaling oxygen. Their coaches think this will speed their recovery. We might measure recovery from intense exertion as follows: Have a football player run 100 yards three times in quick succession. Then allow three minutes to rest before running 100 yards again. Time the final run. Because players vary greatly in speed, you plan a matched pairs experiment using 25 football players as subjects. Discuss the design of such an experiment to investigate the effect of inhaling oxygen during the rest period.

9.44 Protecting ultramarathon runners. An ultramarathon, as you might guess, is a footrace longer than the 26.2 miles of a marathon. Runners commonly develop respiratory infections after an ultramarathon. Will taking 600 milligrams of vitamin C daily reduce these infections? Researchers randomly assigned ultramarathon runners to receive either vitamin C or a placebo. Separately, they also randomly assigned these treatments to a group of nonrunners the same age as the runners. All subjects were watched for 14 days after the big race to see if infections developed.[18]

Wade Payne/AP Photos

(a) What is the name for this experimental design?

(b) Use a diagram to outline the design.

9.45 Wine, beer, or spirits? There is good evidence that moderate alcohol use improves health. Some people think that red wine is better for your health than other alcoholic drinks. You have recruited 300 adults aged 45 to 65 who are willing to follow your orders about alcohol consumption over the next five years. You want to compare the effects on heart disease of moderate drinking of just red wine, just beer, or just spirits. Outline the design of a completely randomized experiment to do this. (No such experiment has been done because subjects aren't willing to have their drinking regulated for years.)

9.46 Wine, beer, or spirits? Women as a group develop heart disease much later than men. We can improve the completely randomized design of Exercise 9.45 by using women and men as blocks. Your 300 subjects include 120 women and 180 men. Outline a block design for comparing wine, beer, and spirits. Be sure to say how many subjects you will put in each group in your design.

9.47 Quick randomizing. Here's a quick and easy way to randomize. You have 100 subjects, 50 women and 50 men. Toss a coin. If it's heads, assign all the men to the treatment group and all the women to the control group. If the coin comes up tails, assign all the women to treatment and all the men to control. This gives every individual subject a 50-50 chance of being assigned to treatment or control. Why isn't this a good way to randomly assign subjects to treatment groups?

9.48 Do antioxidants prevent cancer? People who eat lots of fruits and vegetables have lower rates of colon cancer than those who eat little of these foods. Fruits and vegetables are rich in "antioxidants" such as vitamins A, C, and E. Will taking antioxidants help prevent colon cancer? A medical experiment studied this question with 864 people who were at risk of colon cancer. The subjects were divided into four groups: daily beta-carotene, daily vitamins C and E, all three vitamins every day, or daily placebo. After four years, the researchers were surprised to find no significant difference in colon cancer among the groups.[19]

(a) What are the explanatory and response variables in this experiment?

(b) Outline the design of the experiment. Use your judgment in choosing the group sizes.

(c) The study was double-blind. What does this mean?

(d) What does "no significant difference" mean in describing the outcome of the study?

(e) Suggest some lurking variables that could explain why people who eat lots of fruits and vegetables have lower rates of colon cancer. The experiment suggests that these variables, rather than the antioxidants, may be responsible for the observed benefits of fruits and vegetables.

9.49 An herb for depression? Does the herb Saint-John's-wort relieve major depression? Here are some excerpts from the report of a study of this issue.[20] The study concluded that the herb is no more effective than a placebo.

(a) "Design: Randomized, double-blind, placebo-controlled clinical trial...." A clinical trial is a medical experiment using actual patients as subjects. Explain the meaning of each of the other terms in this description.

(b) "Participants...were randomly assigned to receive either Saint-John's-wort extract ($n = 98$) or placebo ($n = 102$).... The primary outcome measure was the rate of change in the Hamilton Rating Scale for Depression over the treatment period." Based on this information, use a diagram to outline the design of this clinical trial.

9.50 Randomization avoids bias. Suppose that the 25 even-numbered students among the 50 students available for the comparison of on-campus and online instruction (Example 9.5) are older, employed students. We hope that randomization will distribute these students roughly equally between the on-campus and online groups. Use the *Simple Random Sample* applet to take 20 samples of size 25 from the 50 students. (Be sure to click "Reset" after each sample.) Record the counts of even-numbered students in each of your 20 samples. You see that there is considerable chance variation but no systematic bias in favor of one or the other group in assigning the older students. Larger samples from a larger population will on the average do an even better job of creating two similar groups.

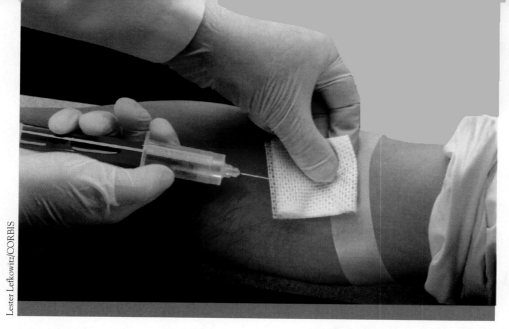

Commentary: Data Ethics*

The production and use of data, like all human endeavors, raise ethical questions. We won't discuss the telemarketer who begins a telephone sales pitch with "I'm conducting a survey." Such deception is clearly unethical. It enrages legitimate survey organizations, which find the public less willing to talk with them. Neither will we discuss those few researchers who, in the pursuit of professional advancement, publish fake data. There is no ethical question here—faking data to advance your career is just wrong. It will end your career when uncovered. But just how honest must researchers be about real, unfaked data? Here is an example that suggests the answer is "More honest than they often are."

EXAMPLE 1 The whole truth?

Papers reporting scientific research are supposed to be short, with no extra baggage. Brevity, however, can allow researchers to avoid complete honesty about their data. Did they choose their subjects in a biased way? Did they report data on only some of their subjects? Did they try several statistical analyses and report only the ones that looked best? The statistician John Bailar screened more than 4000 medical papers in more than a decade as consultant to the *New England Journal of Medicine*. He says,

*This short essay concerns a very important topic, but the material is not needed to read the rest of the book.

"When it came to the statistical review, it was often clear that critical information was lacking, and the gaps nearly always had the practical effect of making the authors' conclusions look stronger than they should have."[1] The situation is no doubt worse in fields that screen published work less carefully. ■

The most complex issues of data ethics arise when we collect data from people. The ethical difficulties are more severe for experiments that impose some treatment on people than for sample surveys that simply gather information. Trials of new medical treatments, for example, can do harm as well as good to their subjects. Here are some basic standards of data ethics that must be obeyed by all studies that gather data from human subjects, both observational studies and experiments.

BASIC DATA ETHICS

All planned studies must be reviewed in advance by an **institutional review board** charged with protecting the safety and well-being of the subjects.

All individuals who are subjects in a study must give their **informed consent** before data are collected.

All individual data must be kept **confidential.** Only statistical summaries for groups of subjects may be made public.

The law requires that studies carried out or funded by the federal government obey these principles.[2] But neither the law nor the consensus of experts is completely clear about the details of their application.

Institutional review boards

The purpose of an institutional review board is not to decide whether a proposed study will produce valuable information or whether it is statistically sound. The board's purpose is, in the words of one university's board, "to protect the rights and welfare of human subjects (including patients) recruited to participate in research activities." The board reviews the plan of the study and can require changes. It reviews the consent form to ensure that subjects are informed about the nature of the study and about any potential risks. Once research begins, the board monitors the study's progress at least once a year.

The most pressing issue concerning institutional review boards is whether their workload has become so large that their effectiveness in protecting subjects drops. When the government temporarily stopped human subject research at Duke University Medical Center in 1999 due to inadequate protection of subjects, more than 2000 studies were going on. That's a lot of review work. There are shorter review procedures for projects that involve only minimal risks to subjects, such as most sample surveys. When a board is overloaded, there is a temptation to put more proposals in the minimal risk category to speed the work.

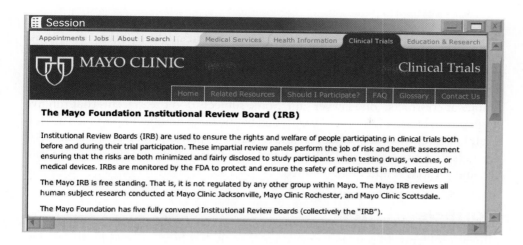

The Web page of the Mayo Clinic's institutional review board. It begins by describing the job of such boards.

Informed consent

Both words in the phrase "informed consent" are important, and both can be controversial. Subjects must be *informed* in advance about the nature of a study and any risk of harm it may bring. In the case of a sample survey, physical harm is not possible. The subjects should be told what kinds of questions the survey will ask and about how much of their time it will take. Experimenters must tell subjects the nature and purpose of the study and outline possible risks. Subjects must then *consent* in writing.

Bernardo Bucci/CORBIS

EXAMPLE 2 Who can consent?

Are there some subjects who can't give informed consent? It was once common, for example, to test new vaccines on prison inmates who gave their consent in return for good-behavior credit. Now we worry that prisoners are not really free to refuse, and the law forbids almost all medical research in prisons.

Children can't give fully informed consent, so the usual procedure is to ask their parents. A study of new ways to teach reading is about to start at a local elementary school, so the study team sends consent forms home to parents. Many parents don't return the forms. Can their children take part in the study because the parents did not say "No," or should we allow only children whose parents returned the form and said "Yes"?

What about research into new medical treatments for people with mental disorders? What about studies of new ways to help emergency room patients who may be unconscious? In most cases, there is not time to get the consent of the family. Does the principle of informed consent bar realistic trials of new treatments for unconscious patients?

These are questions without clear answers. Reasonable people differ strongly on all of them. There is nothing simple about informed consent.[3] ■

The difficulties of informed consent do not vanish even for capable subjects. Some researchers, especially in medical trials, regard consent as a barrier to getting patients to participate in research. They may not explain all possible risks; they may not point out that there are other therapies that might be better than those being studied; they may be too optimistic in talking with patients even when the consent form has all the right details. On the other hand, mentioning every possible risk leads to very long consent forms that really are barriers. "They are like rental car contracts," one lawyer said. Some subjects don't read forms that run five or six printed pages. Others are frightened by the large number of possible (but unlikely) disasters that might happen and so refuse to participate. Of course, unlikely disasters sometimes happen. When they do, lawsuits follow and the consent forms become yet longer and more detailed.

Confidentiality

Ethical problems do not disappear once a study has been cleared by the review board, has obtained consent from its subjects, and has actually collected data about the subjects. It is important to protect the subjects' privacy by keeping all data about individuals confidential. The report of an opinion poll may say what percent of the 1200 respondents felt that legal immigration should be reduced. It may not report what *you* said about this or any other issue.

anonymity
Confidentiality is not the same as **anonymity.** Anonymity means that subjects are anonymous—their names are not known even to the director of the study. Anonymity is rare in statistical studies. Even where it is possible (mainly in surveys conducted by mail), anonymity prevents any follow-up to improve nonresponse or inform subjects of results.

Any breach of confidentiality is a serious violation of data ethics. The best practice is to separate the identity of the subjects from the rest of the data at once. Sample surveys, for example, use the identification only to check on who did or did not respond. In an era of advanced technology, however, it is no longer enough to be sure that each individual set of data protects people's privacy. The government, for example, maintains a vast amount of information about citizens in many separate data bases—census responses, tax returns, Social Security information, data from surveys such as the Current Population Survey, and so on. Many of these data bases can be searched by computers for statistical studies. A clever computer search of several data bases might be able, by combining information, to identify you and learn a great deal about you even if your name and other identification have been removed from the data available for search. A colleague from Germany once remarked that "female full professor of statistics with PhD from the United States" was enough to identify her among all the 83 million residents of Germany. Privacy and confidentiality of data are hot issues among statisticians in the computer age.

> ### EXAMPLE 3 Uncle Sam knows
>
> Citizens are required to give information to the government. Think of tax returns and Social Security contributions. The government needs these data for administrative purposes—to see if you paid the right amount of tax and how large a Social Security benefit you are owed when you retire. Some people feel that individuals should be able

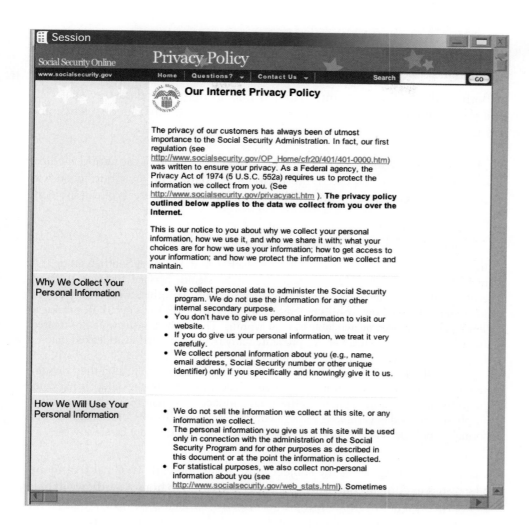

The privacy policy of the government's Social Security Administration Web site.

to forbid any other use of their data, even with all identification removed. This would prevent using government records to study, say, the ages, incomes, and household sizes of Social Security recipients. Such a study could well be vital to debates on reforming Social Security. ▪

Clinical trials

Clinical trials are experiments that study the effectiveness of medical treatments on actual patients. Medical treatments can harm as well as heal, so clinical trials spotlight the ethical problems of experiments with human subjects. Here are the starting points for a discussion:

- Randomized comparative experiments are the only way to see the true effects of new treatments. Without them, risky treatments that are no more effective than placebos will become common.

- Clinical trials produce great benefits, but most of these benefits go to future patients. The trials also pose risks, and these risks are borne by the subjects of the trial. So we must balance future benefits against present risks.

- Both medical ethics and international human rights standards say that "the interests of the subject must always prevail over the interests of science and society."

The quoted words are from the 1964 Helsinki Declaration of the World Medical Association, the most respected international standard. The most outrageous examples of unethical experiments are those that ignore the interests of the subjects.

EXAMPLE 4 **The Tuskegee study**

In the 1930s, syphilis was common among black men in the rural South, a group that had almost no access to medical care. The Public Health Service Tuskegee study recruited 399 poor black sharecroppers with syphilis and 201 others without the disease in order to observe how syphilis progressed when no treatment was given. Beginning in 1943, penicillin became available to treat syphilis. The study subjects were not treated. In fact, the Public Health Service prevented any treatment until word leaked out and forced an end to the study in the 1970s.

The Tuskegee study is an extreme example of investigators following their own interests and ignoring the well-being of their subjects. A 1996 review said, "It has come to symbolize racism in medicine, ethical misconduct in human research, paternalism by physicians, and government abuse of vulnerable people." In 1997, President Clinton formally apologized to the surviving participants in a White House ceremony.[4] ■

Because "the interests of the subject must always prevail," medical treatments can be tested in clinical trials only when there is reason to hope that they will help the patients who are subjects in the trials. Future benefits aren't enough to justify experiments with human subjects. Of course, if there is already strong evidence that a treatment works and is safe, it is unethical *not* to give it. Here are the words of Dr. Charles Hennekens of the Harvard Medical School, who directed the large clinical trial that showed that aspirin reduces the risk of heart attacks:

There's a delicate balance between when to do or not do a randomized trial. On the one hand, there must be sufficient belief in the agent's potential to justify exposing half the subjects to it. On the other hand, there must be sufficient doubt about its efficacy to justify withholding it from the other half of subjects who might be assigned to placebos.[5]

Why is it ethical to give a control group of patients a placebo? Well, we know that placebos often work. Moreover, placebos have no harmful side effects. So in the state of balanced doubt described by Dr. Hennekens, the placebo group may be getting a better treatment than the drug group. If we *knew* which treatment was better, we would give it to everyone. When we don't know, it is ethical to try both and compare them.

Behavioral and social science experiments

When we move from medicine to the behavioral and social sciences, the direct risks to experimental subjects are less acute, but so are the possible benefits to the subjects. Consider, for example, the experiments conducted by psychologists in their study of human behavior.

EXAMPLE 5   **Psychologists in the men's room**

Psychologists observe that people have a "personal space" and are uneasy if others come too close to them. We don't like strangers to sit at our table in a coffee shop if other tables are available, and we see people move apart in elevators if there is room to do so. Americans tend to require more personal space than people in most other cultures. Can violations of personal space have physical, as well as emotional, effects?

Investigators set up shop in a men's public restroom. They blocked off urinals to force men walking in to use either a urinal next to an experimenter (treatment group) or a urinal separated from the experimenter (control group). Another experimenter, using a periscope from a toilet stall, measured how long the subject took to start urinating and how long he continued.[6] ■

David Pollack/CORBIS

This personal space experiment illustrates the difficulties facing those who plan and review behavioral studies.

■ There is no risk of harm to the subjects, although they would certainly object to being watched through a periscope. What should we protect subjects from when physical harm is unlikely? Possible emotional harm? Undignified situations? Invasion of privacy?

■ What about informed consent? The subjects did not even know they were participating in an experiment. Many behavioral experiments rely on hiding the true purpose of the study. The subjects would change their behavior if told in advance what the investigators were looking for. Subjects are asked to consent on the basis of vague information. They receive full information only after the experiment.

The "Ethical Principles" of the American Psychological Association require consent unless a study merely observes behavior in a public place. They allow deception only when it is necessary to the study, does not hide information that might influence a subject's willingness to participate, and is explained to subjects as soon as possible. The personal space study (from the 1970s) does not meet current ethical standards.

We see that the basic requirement for informed consent is understood differently in medicine and psychology. Here is an example of another setting with yet another interpretation of what is ethical. The subjects get no information and give no consent. They don't even know that an experiment may be sending them to jail for the night.

EXAMPLE 6 Reducing domestic violence

How should police respond to domestic violence calls? In the past, the usual practice was to remove the offender and order him to stay out of the household overnight. Police were reluctant to make arrests because the victims rarely pressed charges. Women's groups argued that arresting offenders would help prevent future violence even if no charges were filed. Is there evidence that arrest will reduce future offenses? That's a question that experiments have tried to answer.

A typical domestic violence experiment compares two treatments: arrest the suspect and hold him overnight, or warn the suspect and release him. When police officers reach the scene of a domestic violence call, they calm the participants and investigate. Weapons or death threats require an arrest. If the facts permit an arrest but do not require it, an officer radios headquarters for instructions. The person on duty opens the next envelope in a file prepared in advance by a statistician. The envelopes contain the treatments in random order. The police either arrest the suspect or warn and release him, depending on the contents of the envelope. The researchers then watch police records and visit the victim to see if the domestic violence reoccurs.

Such experiments show that arresting domestic violence suspects does reduce their future violent behavior.[7] As a result of this evidence, arrest has become the common police response to domestic violence. ■

The domestic violence experiments shed light on an important issue of public policy. Because there is no informed consent, the ethical rules that govern clinical trials and most social science studies would forbid these experiments. They were cleared by review boards because, in the words of one domestic violence researcher, "These people became subjects by committing acts that allow the police to arrest them. You don't need consent to arrest someone."

DISCUSSION EXERCISES

Most of these exercises pose issues for discussion. There are no right or wrong answers, but there are more and less thoughtful answers.

1. **Minimal risk?** You are a member of your college's institutional review board. You must decide whether several research proposals qualify for lighter review because they involve only minimal risk to subjects. Federal regulations say that "minimal risk" means the risks are no greater than "those ordinarily encountered in daily life or during the performance of routine physical or psychological examinations or tests." That's vague. Which of these do you think qualifies as "minimal risk"?

   (a) Draw a drop of blood by pricking a finger in order to measure blood sugar.

   (b) Draw blood from the arm for a full set of blood tests.

   (c) Insert a tube that remains in the arm, so that blood can be drawn regularly.

2. **Who reviews?** Government regulations require that institutional review boards consist of at least five people, including at least one scientist, one nonscientist, and one person from outside the institution. Most boards are larger, but many contain just one outsider.

    (a)   Why should review boards contain people who are not scientists?

    (b)   Do you think that one outside member is enough? How would you choose that member? (For example, would you prefer a medical doctor? A member of the clergy? An activist for patients' rights?)

3. Getting consent. A researcher suspects that traditional religious beliefs tend to be associated with an authoritarian personality. She prepares a questionnaire that measures authoritarian tendencies and also asks many religious questions. Write a description of the purpose of this research to be read by subjects in order to obtain their informed consent. You must balance the conflicting goals of not deceiving the subjects as to what the questionnaire will tell about them and of not biasing the sample by scaring off religious people.

4. No consent needed? In which of the circumstances below would you allow collecting personal information without the subjects' consent?

    (a)   A government agency takes a random sample of income tax returns to obtain information on the average income of people in different occupations. Only the incomes and occupations are recorded from the returns, not the names.

    (b)   A social psychologist attends public meetings of a religious group to study the behavior patterns of members.

    (c)   The social psychologist pretends to be converted to membership in a religious group and attends private meetings to study the behavior patterns of members.

5. Studying your blood. Long ago, doctors drew a blood specimen from you as part of treating minor anemia. Unknown to you, the sample was stored. Now researchers plan to use stored samples from you and many other people to look for genetic factors that may influence anemia. It is no longer possible to ask your consent. Modern technology can read your entire genetic makeup from the blood sample.

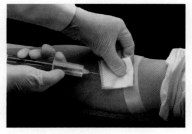

Lester Lefkowitz/CORBIS

    (a)   Do you think it violates the principle of informed consent to use your blood sample if your name is on it but you were not told that it might be saved and studied later?

    (b)   Suppose that your identity is not attached. The blood sample is known only to come from (say) "a 20-year-old white female being treated for anemia." Is it now OK to use the sample for research?

    (c)   Perhaps we should use biological materials such as blood samples only from patients who have agreed to allow the material to be stored for later use in research. It isn't possible to say in advance what kind of research, so this falls short of the usual standard for informed consent. Is it nonetheless acceptable, given complete confidentiality and the fact that using the sample can't physically harm the patient?

6. Anonymous? Confidential? One of the most important nongovernment surveys in the United States is the National Opinion Research Center's General Social Survey. The GSS regularly monitors public opinion on a wide variety of political and social issues. Interviews are conducted in person in the subject's home. Are a subject's responses to GSS questions anonymous, confidential, or both? Explain your answer.

7. **Anonymous? Confidential?** Texas A&M, like many universities, offers screening for HIV, the virus that causes AIDS. Students may choose either anonymous or confidential screening. An announcement says, "Persons who sign up for screening will be assigned a number so that they do not have to give their name." They can learn the results of the test by telephone, still without giving their name. Does this describe the *anonymous* or the *confidential* screening? Why?

8. **Political polls.** The presidential election campaign is in full swing, and the candidates have hired polling organizations to take sample surveys to find out what the voters think about the issues. What information should the pollsters be required to give out?

   (a) What does the standard of informed consent require the pollsters to tell potential respondents?

   (b) The standards accepted by polling organizations also require giving respondents the name and address of the organization that carries out the poll. Why do you think this is required?

   (c) The polling organization usually has a professional name such as "Samples Incorporated," so respondents don't know that the poll is being paid for by a political party or candidate. Would revealing the sponsor to respondents bias the poll? Should the sponsor always be announced whenever poll results are made public?

9. **Making poll results public.** Some people think that the law should require that all political poll results be made public. Otherwise, the possessors of poll results can use the information to their own advantage. They can act on the information, release only selected parts of it, or time the release for best effect. A candidate's organization replies that they are paying for the poll in order to gain information for their own use, not to amuse the public. Do you favor requiring complete disclosure of political poll results? What about other private surveys, such as market research surveys of consumer tastes?

10. **Student subjects.** Students taking Psychology 001 are required to serve as experimental subjects. Students in Psychology 002 are not required to serve, but they are given extra credit if they do so. Students in Psychology 003 are required either to sign up as subjects or to write a term paper. Serving as an experimental subject may be educational, but current ethical standards frown on using "dependent subjects" such as prisoners or charity medical patients. Students are certainly somewhat dependent on their teachers. Do you object to any of these course policies? If so, which ones, and why?

11. **The Willowbrook hepatitis studies.** In the 1960s, children entering the Willowbrook State School, an institution for the mentally retarded, were deliberately infected with hepatitis. The researchers argued that almost all children in the institution quickly became infected anyway. The studies showed for the first time that two strains of hepatitis existed. This finding contributed to the development of effective vaccines. Despite these valuable results, the Willowbrook studies are now considered an example of unethical research. Explain why, according to current ethical standards, useful results are not enough to allow a study.

12. **Unequal benefits.** Researchers on aging proposed to investigate the effect of supplemental health services on the quality of life of older people. Eligible patients on the rolls of a large medical clinic were to be randomly assigned to treatment and control

groups. The treatment group would be offered hearing aids, dentures, transportation, and other services not available without charge to the control group. The review board felt that providing these services to some but not other persons in the same institution raised ethical questions. Do you agree?

13. **How many have HIV?** Researchers from Yale, working with medical teams in Tanzania, wanted to know how common infection with HIV, the virus that causes AIDS, is among pregnant women in that African country. To do this, they planned to test blood samples drawn from pregnant women.

    Yale's institutional review board insisted that the researchers get the informed consent of each woman and tell her the results of the test. This is the usual procedure in developed nations. The Tanzanian government did not want to tell the women why blood was drawn or tell them the test results. The government feared panic if many people turned out to have an incurable disease for which the country's medical system could not provide care. The study was canceled. Do you think that Yale was right to apply its usual standards for protecting subjects?

14. **AIDS trials in Africa.** The drug programs that treat AIDS in rich countries are very expensive, so some African nations cannot afford to give them to large numbers of people. Yet AIDS is more common in parts of Africa than anywhere else. "Short-course" drug programs that are much less expensive might help, for example, in preventing infected pregnant women from passing the infection to their unborn children. Is it ethical to compare a short-course program with a placebo in a clinical trial? Some say no: this is a double standard, because in rich countries the full drug program would be the control treatment. Others say yes: the intent is to find treatments that are practical in Africa, and the trial does not withhold any treatment that subjects would otherwise receive. What do you think?

15. **Abandoned children in Romania.** The study described in Example 9.2 randomly assigned abandoned children in Romanian orphanages to move to foster homes or to remain in an orphanage. All of the children would otherwise have remained in an orphanage. The foster care was paid for by the study. There was no informed consent because the children had been abandoned and had no adult to speak for them. The experiment was considered ethical because "people who cannot consent can be protected by enrolling them only in minimal-risk research, whose risks do not exceed those of everyday life," and because the study "aimed to produce results that would primarily benefit abandoned, institutionalized children."[8] Do you agree?

16. **Asking teens about sex.** The Centers for Disease Control and Prevention, in a survey of teenagers, asked the subjects if they were sexually active. Those who said "Yes" were then asked, "How old were you when you had sexual intercourse for the first time?" Should consent of parents be required to ask minors about sex, drugs, and other such issues, or is consent of the minors themselves enough? Give reasons for your opinion.

17. **Deceiving subjects.** Students sign up to be subjects in a psychology experiment. When they arrive, they are told that interviews are running late and are taken to a waiting room. The experimenters then stage a theft of a valuable object left in the waiting room. Some subjects are alone with the thief, and others are in pairs—these are the treatments being compared. Will the subject report the theft?

The students had agreed to take part in an unspecified study, and the true nature of the experiment is explained to them afterward. Do you think this study is ethically OK?

18. **Deceiving subjects.** A psychologist conducts the following experiment: she measures the attitude of subjects toward cheating, then has them play a game rigged so that winning without cheating is impossible. The computer that organizes the game also records—unknown to the subjects—whether or not they cheat. Then attitude toward cheating is retested.

Subjects who cheat tend to change their attitudes to find cheating more acceptable. Those who resist the temptation to cheat tend to condemn cheating more strongly on the second test of attitude. These results confirm the psychologist's theory.

This experiment tempts subjects to cheat. The subjects are led to believe that they can cheat secretly when in fact they are observed. Is this experiment ethically objectionable? Explain your position.

Darrell Walker/HWMS/
Icon SMI/Newscom

CHAPTER 10

Introducing Probability

Why is probability, the mathematics of chance behavior, needed to understand statistics, the science of data? Let's look at a typical sample survey.

EXAMPLE 10.1 Do you lotto?

What proportion of all adults bought a lottery ticket in the past 12 months? We don't know, but we do have results from the Gallup Poll. Gallup took a random sample of 1523 adults. The poll found that 868 of the people in the sample bought tickets. The proportion who bought tickets was

$$\text{sample proportion} = \frac{868}{1523} = 0.57 \quad \text{(that is, 57\%)}$$

Because all adults had the same chance to be among the chosen 1523, it seems reasonable to use this 57% as an estimate of the unknown proportion in the population. It's a *fact* that 57% of the sample bought lottery tickets—we know because Gallup asked them. We don't know what percent of all adults bought tickets, but we *estimate* that about 57% did. This is a basic move in statistics: use a result from a sample to estimate something about a population. ■

What if Gallup took a second random sample of 1523 adults? The new sample would have different people in it. It is almost certain that there would not be exactly 868 positive responses. That is, Gallup's estimate of the proportion of adults who bought a lottery ticket will vary from sample to sample. Could it happen that one random sample finds that 57% of adults recently bought a lottery ticket and a second random sample finds that only 37% had done so? *Random samples eliminate bias from the act of choosing a sample, but they can still be wrong because of the variability that results when we choose at random.* If the variation when we take repeat samples from the same population is too great, we can't trust the results of any one sample.

This is where we need facts about probability to make progress in statistics. Because Gallup uses chance to choose its samples, the laws of probability govern the behavior of the samples. Gallup says that the probability is 0.95 that an estimate from one of their samples comes within ±3 percentage points of the truth about the population of all adults. The first step toward understanding this statement is to understand what "probability 0.95" means. Our purpose in this chapter is to understand the language of probability, but without going into the mathematics of probability theory.

The idea of probability

To understand why we can trust random samples and randomized comparative experiments, we must look closely at chance behavior. The big fact that emerges is this: **chance behavior is unpredictable in the short run but has a regular and predictable pattern in the long run.**

Toss a coin, or choose a random sample. The result can't be predicted in advance, because the result will vary when you toss the coin or choose the sample repeatedly. But there is still a regular pattern in the results, a pattern that emerges clearly only after many repetitions. This remarkable fact is the basis for the idea of probability.

SuperStock

EXAMPLE 10.2 **Coin tossing**

When you toss a coin, there are only two possible outcomes, heads or tails. Figure 10.1 shows the results of tossing a coin 5000 times twice. For each number of tosses from 1 to 5000, we have plotted the proportion of those tosses that gave a head. Trial A (solid red line) begins tail, head, tail, tail. You can see that the proportion of heads for Trial A starts at 0 on the first toss, rises to 0.5 when the second toss gives a head, then falls to 0.33 and 0.25 as we get two more tails. Trial B, on the other hand, starts with five straight heads, so the proportion of heads is 1 until the sixth toss.

The proportion of tosses that produce heads is quite variable at first. Trial A starts low and Trial B starts high. As we make more and more tosses, however, the proportion of heads for both trials gets close to 0.5 and stays there. If we made yet a third trial at tossing the coin a great many times, the proportion of heads would again settle down to 0.5 in the long run. This is the intuitive idea of probability. Probability 0.5 means "occurs half the time in a very large number of trials." The probability 0.5 appears as a horizontal line on the graph. ■

We might suspect that a coin has probability 0.5 of coming up heads just because the coin has two sides. But we can't be sure. In fact, spinning a penny on a flat

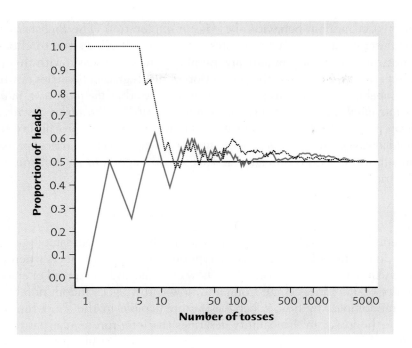

FIGURE 10.1

The proportion of tosses of a coin that give a head changes as we make more tosses. Eventually, however, the proportion approaches 0.5, the probability of a head. This figure shows the results of two trials of 5000 tosses each.

surface, rather than tossing the coin, gives heads probability about 0.45 rather than 0.5.[1] The idea of probability is empirical. That is, it is based on observation rather than theorizing. Probability describes what happens in very many trials, and we must actually observe many trials to pin down a probability. In the case of tossing a coin, some diligent people have in fact made thousands of tosses.

EXAMPLE 10.3 Some coin tossers

The French naturalist Count Buffon (1707–1788) tossed a coin 4040 times. Result: 2048 heads, or proportion 2048/4040 = 0.5069 for heads.

Around 1900, the English statistician Karl Pearson heroically tossed a coin 24,000 times. Result: 12,012 heads, a proportion of 0.5005.

While imprisoned by the Germans during World War II, the South African mathematician John Kerrich tossed a coin 10,000 times. Result: 5067 heads, a proportion of 0.5067. ■

RANDOMNESS AND PROBABILITY

We call a phenomenon **random** if individual outcomes are uncertain but there is nonetheless a regular distribution of outcomes in a large number of repetitions.

The **probability** of any outcome of a random phenomenon is the proportion of times the outcome would occur in a very long series of repetitions.

The best way to understand randomness is to observe random behavior, as in Figure 10.1. You can do this with physical devices like coins, but computer simulations (imitations) of random behavior allow faster exploration. The *Probability* applet is a computer simulation that animates Figure 10.1. It allows you to choose the probability of a head and simulate any number of tosses of a coin with that probability. Experience shows that the proportion of heads gradually settles down close to the probability. Equally important, it also shows that *the proportion in a small or moderate number of tosses can be far from the probability. Probability describes* only *what happens in the long run.* Of course, we can never observe a probability exactly. We could always continue tossing the coin, for example. Mathematical probability is an idealization based on imagining what would happen in an indefinitely long series of trials.

The search for randomness*

Random numbers are valuable. They are used to choose random samples, to shuffle the cards in online poker games, to encrypt our credit card numbers when we buy online, and as part of simulations of the flow of traffic and the spread of epidemics. Where does randomness come from, and how can we get random numbers? We defined randomness by how it behaves: unpredictable in the short run, regular pattern in the long run. Probability describes the long-run regular pattern. That many things are random in this sense is an observed fact about the world. Not all these things are "really" random. Here's a quick tour of how to find random behavior and get random numbers.

The easiest way to get random numbers is from a *computer program*. Of course, a computer program just does what it is told to do. Run the program again and you get exactly the same result. The random numbers in Table B, the outcomes of the *Probability* applet, and the random numbers that shuffle cards for online poker come from computer programs, so they aren't "really" random. Clever computer programs produce outcomes that look random even though they really aren't. These *pseudorandom numbers* are more than good enough for choosing samples and shuffling cards. But they may have hidden patterns that can distort scientific simulations.

You might think that *physical devices such as coins and dice* produce really random outcomes. But a tossed coin obeys the laws of physics. If we knew all the inputs of the toss (forces, angles, and so on), then we could say in advance whether the outcome will be heads or tails. The outcome of a toss is predictable rather than random. Why do the results of tossing a coin *look* random? The outcomes are extremely sensitive to the inputs, so that very small changes in the forces you apply when you toss a coin change the outcome from heads to tails and back again. In practice, the outcomes are not predictable. Probability is a lot more useful than physics for describing coin tosses.

We call a phenomenon with "small changes in, big changes out" behavior *chaotic*. If we can feed chaotic behavior into a computer, we can do better than pseudorandom numbers. Coins and dice are awkward, but you can go the the Web site **random.org** to get random numbers from radio noise in the atmosphere, a chaotic phenomenon that is easy to feed to a computer.

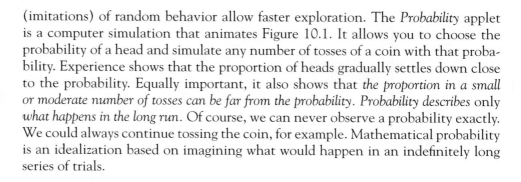

Does God play dice?

Few things in the world are truly random in the sense that no amount of information will allow us to predict the outcome. But according to the branch of physics called quantum mechanics, randomness does rule events inside individual atoms. Although Albert Einstein helped quantum theory get started, he always insisted that nature must have some fixed reality, not just probabilities. "I shall never believe that God plays dice with the world," said the great scientist. A century after Einstein's first work on quantum theory, it appears that he was wrong.

*This short discussion is optional.

Is anything really random? As far as current science can say, behavior inside atoms really is random—that is, there isn't any way to predict behavior in advance no matter how much information we have. It was this "really, truly random" idea that Einstein disliked as he watched the new science of quantum mechanics emerge. You can go to the HotBits Web site **www.fourmilab.ch/hotbits** to get really, truly random numbers generated from the radioactive decay of atoms.

APPLY YOUR KNOWLEDGE

10.1 **Texas hold 'em.** In the popular Texas hold 'em variety of poker, players make their best five-card poker hand by combining the two cards they are dealt with three of five cards available to all players. You read in a book on poker that if you hold a pair (two cards of the same rank) in your hand, the probability of getting four of a kind is 88/1000. Explain carefully what this means. In particular, explain why it does *not* mean that if you play 1000 such hands, exactly 88 will be four of a kind.

10.2 **Probability says ...** Probability is a measure of how likely an event is to occur. Match one of the probabilities that follow with each statement of likelihood given. (The probability is usually a more exact measure of likelihood than is the verbal statement.)

0   0.01   0.3   0.6   0.99   1

(a) This event is impossible. It can never occur.

(b) This event is certain. It will occur on every trial.

(c) This event is very unlikely, but it will occur once in a while in a long sequence of trials.

(d) This event will occur more often than not.

10.3 **Random digits.** The table of random digits (Table B) was produced by a random mechanism that gives each digit probability 0.1 of being a 0.

(a) What proportion of the first 50 digits in the table are 0s? This proportion is an estimate, based on 50 repetitions, of the true probability, which we know is 0.1.

(b) The *Probability* applet can imitate random digits. Set the probability of heads in the applet to 0.1. Check "Show true probability" to show this value on the graph. A head stands for a 0 in the random digit table and a tail stands for any other digit. Simulate 200 digits (keep clicking "Toss" to get 40 at a time—don't click "Reset"). If you kept going forever, presumably you would get 10% heads. What was the result of your 200 tosses?

10.4 **The long run but not the short run.** Our intuition about chance behavior is not very accurate. In particular, we tend to expect that the long-run pattern described by probability will show up in the short run as well. For example, we tend to think that tossing a coin 20 times will give close to 10 heads.

(a) Set the probability of heads in the *Probability* applet to 0.5 and the number of tosses to 20. Click "Toss" to simulate 20 tosses of a balanced coin. What was the proportion of heads?

Cut and Deal Ltd./Alamy

(b) Click "Reset" and toss again. The simulation is fast, so do it 25 times and keep a record of the proportion of heads in each set of 20 tosses. Make a stemplot of your results. You see that the result of tossing a coin 20 times is quite variable and need not be very close to the probability 0.5 of heads.

Probability models

Gamblers have known for centuries that the fall of coins, cards, and dice displays clear patterns in the long run. The idea of probability rests on the observed fact that the average result of many thousands of chance outcomes can be known with near certainty. How can we give a mathematical description of long-run regularity?

To see how to proceed, think first about a very simple random phenomenon, tossing a coin once. When we toss a coin, we cannot know the outcome in advance. What *do* we know? We are willing to say that the outcome will be either heads or tails. We believe that each of these outcomes has probability 1/2. This description of coin tossing has two parts:

- a list of possible outcomes

- a probability for each outcome

Such a description is the basis for all *probability models*. Here is the basic vocabulary we use.

PROBABILITY MODELS

The **sample space** S of a random phenomenon is the set of all possible outcomes.

An **event** is an outcome or a set of outcomes of a random phenomenon. That is, an event is a subset of the sample space.

A **probability model** is a mathematical description of a random phenomenon consisting of two parts: a sample space S and a way of assigning probabilities to events.

A sample space S can be very simple or very complex. When we toss a coin once, there are only two outcomes, heads and tails. The sample space is $S = \{H, T\}$. When Gallup draws a random sample of 1523 adults, the sample space contains all possible choices of 1523 of the 235 million adults in the country. This S is extremely large. Each member of S is a possible sample, which explains the term *sample space*.

EXAMPLE 10.4 **Rolling dice**

Rolling two dice is a common way to lose money in casinos. There are 36 possible outcomes when we roll two dice and record the up-faces in order (first die, second die).

FIGURE 10.2

Figure 10.2 displays these outcomes. They make up the sample space S. "Roll a 5" is an event, call it A, that contains four of these 36 outcomes:

$$A = \{ \qquad \qquad \qquad \qquad \}$$

How can we assign probabilities to this sample space? We can find the actual probabilities for two specific dice only by actually tossing the dice many times, and even then only approximately. So we will give a probability model that assumes ideal, perfectly balanced dice. This model will be quite accurate for carefully made casino dice and less accurate for the cheap dice that come with a board game.

If the dice are perfectly balanced, all 36 outcomes in Figure 10.2 will be *equally likely*. That is, each of the 36 outcomes will come up on one thirty-sixth of all rolls in the long run. So each outcome has probability 1/36. There are 4 outcomes in the event A ("roll a 5"), so this event has probability 4/36. In this way we can assign a probability to any event. So we have a complete probability model. ■

EXAMPLE 10.5 **Rolling dice and counting the spots**

Gamblers care only about the total number of spots on the up-faces of the dice. The sample space for rolling two dice and counting the spots is

$$S = \{2, 3, 4, 5, 6, 7, 8, 9, 10, 11, 12\}$$

Comparing this S with Figure 10.2 reminds us that *we can change S by changing the detailed description of the random phenomenon we are describing.*

What are the probabilities for this new sample space? The 11 possible outcomes are *not* equally likely, because there are six ways to roll a 7 and only one way to roll a 2 or a 12. That's the key: each outcome in Figure 10.2 has probability 1/36. So "roll a 7" has probability 6/36 because this event contains 6 of the 36 outcomes. Similarly, "roll a 2" has probability 1/36, and "roll a 5" (4 outcomes from Figure 10.2) has probability 4/36. Here is the complete probability model:

Spots	2	3	4	5	6	7	8	9	10	11	12
Probability	1/36	2/36	3/36	4/36	5/36	6/36	5/36	4/36	3/36	2/36	1/36

■

10.5   Sample space. Choose a student at random from a large statistics class. Describe a sample space S for each of the following. (In some cases you may have some freedom in specifying S.)

(a)   Is the student male or female?

(b)   What is the student's height in inches?

(c)   Ask how much money in coins (not bills) the student is carrying.

(d)   Record the student's letter grade at the end of the course.

10.6   Role-playing games. Computer games in which the players take the roles of characters are very popular. They go back to earlier tabletop games such as Dungeons & Dragons. These games use many different types of dice. A four-sided die has faces with 1, 2, 3, and 4 spots.

(a)   What is the sample space for rolling the die twice (spots on first and second rolls)? Follow the example of Figure 10.2.

(b)   What is the assignment of probabilities to outcomes in this sample space? Assume that the die is perfectly balanced, and follow the method of Example 10.4.

10.7   Role-playing games. The intelligence of a character in a game is determined by rolling the four-sided die twice and adding 1 to the sum of the spots. Start with your work in the previous exercise to give a probability model (sample space and probabilities of outcomes) for the character's intelligence. Follow the method of Example 10.5.

Probability rules

In Examples 10.4 and 10.5 we found probabilities for tossing dice. As random phenomena go, dice are pretty simple. Even so, we had to assume idealized perfectly balanced dice. In most situations, it isn't easy to give a "correct" probability model. We can make progress by listing some facts that must be true for *any* assignment of probabilities. These facts follow from the idea of probability as "the long-run proportion of repetitions on which an event occurs."

1.   **Any probability is a number between 0 and 1.** Any proportion is a number between 0 and 1, so any probability is also a number between 0 and 1. An event with probability 0 never occurs, and an event with probability 1 occurs on every trial. An event with probability 0.5 occurs in half the trials in the long run.

2.   **All possible outcomes together must have probability 1.** Because some outcome must occur on every trial, the sum of the probabilities for all possible outcomes must be exactly 1.

3.   **If two events have no outcomes in common, the probability that one or the other occurs is the sum of their individual probabilities.** If one event occurs

in 40% of all trials, a different event occurs in 25% of all trials, and the two can never occur together, then one or the other occurs on 65% of all trials because 40% + 25% = 65%.

4. **The probability that an event does not occur is 1 minus the probability that the event does occur.** If an event occurs in (say) 70% of all trials, it fails to occur in the other 30%. The probability that an event occurs and the probability that it does not occur always add to 100%, or 1.

We can use mathematical notation to state Facts 1 to 4 more concisely. Capital letters near the beginning of the alphabet denote events. If A is any event, we write its probability as $P(A)$. Here are our probability facts in formal language. As you apply these rules, remember that they are just another form of intuitively true facts about long-run proportions.

Equally likely?

A game of bridge begins by dealing all 52 cards in the deck to the four players, 13 to each. If the deck is well shuffled, all of the immense number of possible hands will be equally likely. But don't expect the hands that appear in newspaper bridge columns to reflect the equally likely probability model. Writers on bridge choose "interesting" hands, especially those that lead to high bids that are rare in actual play.

PROBABILITY RULES

Rule 1. The probability $P(A)$ of any event A satisfies $0 \le P(A) \le 1$.

Rule 2. If S is the sample space in a probability model, then $P(S) = 1$.

Rule 3. Two events A and B are **disjoint** if they have no outcomes in common and so can never occur together. If A and B are disjoint,

$$P(A \text{ or } B) = P(A) + P(B)$$

This is the **addition rule for disjoint events.**

Rule 4. For any event A,

$$P(A \text{ does not occur}) = 1 - P(A)$$

The addition rule extends to more than two events that are disjoint in the sense that no two have any outcomes in common. If events A, B, and C are disjoint, the probability that one of these events occurs is $P(A) + P(B) + P(C)$.

EXAMPLE 10.6 **Using the probability rules**

We already used the addition rule, without calling it by that name, to find the probabilities in Example 10.5. The event "roll a 5" contains the four disjoint outcomes displayed in Example 10.4, so the addition rule (Rule 3) says that its probability is

$$P(\text{roll a } 5) = P(\boxdot\ \boxdot) + P(\boxdot\ \boxdot) + P(\boxdot\ \boxdot) + P(\boxdot\ \boxdot)$$

$$= \frac{1}{36} + \frac{1}{36} + \frac{1}{36} + \frac{1}{36}$$

$$= \frac{4}{36} = 0.111$$

Image Source/Alamy

Check that the probabilities in Example 10.5, found using the addition rule, are all between 0 and 1 and add to exactly 1. That is, this probability model obeys Rules 1 and 2.

What is the probability of rolling anything other than a 5? By Rule 4,

$$P(\text{roll does not give a 5}) = 1 - P(\text{roll a 5})$$

$$= 1 - 0.111 = 0.889$$

Our model assigns probabilities to individual outcomes. To find the probability of an event, just add the probabilities of the outcomes that make up the event. For example:

$$P(\text{outcome is odd}) = P(3) + P(5) + P(7) + P(9) + P(11)$$

$$= \frac{2}{36} + \frac{4}{36} + \frac{6}{36} + \frac{4}{36} + \frac{2}{36}$$

$$= \frac{18}{36} = \frac{1}{2} \quad ■$$

APPLY YOUR KNOWLEDGE

10.8 Preparing for the GMAT. In many settings, the "rules of probability" are just basic facts about percents. A company that offers courses to prepare students for the Graduate Management Admission Test (GMAT) has the following information about its customers: 20% are currently undergraduate students in business; 15% are undergraduate students in other fields of study; 60% are college graduates who are currently employed; and 5% are college graduates who are not employed.

(a) What percent of customers are currently undergraduates? Which rule of probability did you use to find the answer?

(b) What percent of customers are not undergraduate business students? Which rule of probability did you use to find the answer?

10.9 Overweight? Although the rules of probability are just basic facts about percents or proportions, we need to be able to use the language of events and their probabilities. Choose an American adult at random. Define two events:

  $A =$ the person chosen is obese
  $B =$ the person chosen is overweight, but not obese

According to the National Center for Health Statistics, $P(A) = 0.32$ and $P(B) = 0.34$.

(a) Explain why events A and B are disjoint.

(b) Say in plain language what the event "A or B" is. What is $P(A \text{ or } B)$?

(c) If C is the event that the person chosen has normal weight or less, what is $P(C)$?

10.10 Languages in Canada. Canada has two official languages, English and French. Choose a Canadian at random and ask, "What is your mother tongue?" Here is the distribution of responses, combining many separate languages from the broad Asia/Pacific region:[2]

Language	English	French	Asian/Pacific	Other
Probability	0.63	0.22	0.06	?

(a)   What probability should replace "?" in the distribution?

(b)   What is the probability that a Canadian's mother tongue is not English?

(c)   What is the probability that a Canadian's mother tongue is a language other than English or French?

Discrete probability models

Examples 10.4, 10.5, and 10.6 illustrate one way to assign probabilities to events: assign a probability to every individual outcome, then add these probabilities to find the probability of any event. This idea works well when there are only a finite (fixed and limited) number of outcomes.

DISCRETE PROBABILITY MODEL

A probability model with a finite sample space is called **discrete.**

To assign probabilities in a discrete model, list the probabilities of all the individual outcomes. These probabilities must be numbers between 0 and 1 that add to exactly 1. The probability of any event is the sum of the probabilities of the outcomes making up the event.

EXAMPLE 10.7   Benford's law

Faked numbers in tax returns, invoices, or expense account claims often display patterns that aren't present in legitimate records. Some patterns, like too many round numbers, are obvious and easily avoided by a clever crook. Others are more subtle. It is a striking fact that the first digits of numbers in legitimate records often follow a model known as Benford's law.[3] Call the first digit of a randomly chosen record X for short. Benford's law gives this probability model for X (note that a first digit can't be 0):

First digit X	1	2	3	4	5	6	7	8	9
Probability	0.301	0.176	0.125	0.097	0.079	0.067	0.058	0.051	0.046

Check that the probabilities of the outcomes sum to exactly 1. This is therefore a legitimate discrete probability model. Investigators can detect fraud by comparing the first digits in records such as invoices paid by a business with these probabilities.

The probability that a first digit is equal to or greater than 6 is

$$P(X \geq 6) = P(X = 6) + P(X = 7) + P(X = 8) + P(X = 9)$$
$$= 0.067 + 0.058 + 0.051 + 0.046 = 0.222$$

This is less than the probability that a record has first digit 1,

$$P(X = 1) = 0.301$$

Fraudulent records tend to have too few 1s and too many higher first digits.

Note that the probability that a first digit is greater than or equal to 6 is not the same as the probability that a first digit is strictly greater than 6. The latter probability is

$$P(X > 6) = 0.058 + 0.051 + 0.046 = 0.155$$

The outcome $X = 6$ is included in "greater than or equal to" and is not included in "strictly greater than." ∎

APPLY YOUR KNOWLEDGE

10.11 Rolling a die. Figure 10.3 displays several discrete probability models for rolling a die. We can learn which model is actually *accurate* for a particular die only by rolling the die many times. However, some of the models are not *legitimate*. That is, they do not obey the rules. Which are legitimate and which are not? In the case of the illegitimate models, explain what is wrong.

FIGURE 10.3

Four assignments of probabilities to the six faces of a die, for Exercise 10.11.

Outcome	Probability			
	Model 1	Model 2	Model 3	Model 4
⚀	1/7	1/3	1/3	1
⚁	1/7	1/6	1/6	1
⚂	1/7	1/6	1/6	2
⚃	1/7	0	1/6	1
⚄	1/7	1/6	1/6	1
⚅	1/7	1/6	1/6	2

10.12 Benford's law. The first digit of a randomly chosen expense account claim follows Benford's law (Example 10.7). Consider the events

$$A = \{\text{first digit is 7 or greater}\}$$
$$B = \{\text{first digit is odd}\}$$

(a) What outcomes make up the event A? What is $P(A)$?

(b) What outcomes make up the event B? What is $P(B)$?

(c)   What outcomes make up the event "*A or B*"? What is $P(A \text{ or } B)$? Why is this probability not equal to $P(A) + P(B)$?

10.13 Working out. Choose a person aged 19 to 25 years at random and ask, "In the past seven days, how many times did you go to an exercise or fitness center or work out?" Call the response X for short. Based on a large sample survey, here is a probability model for the answer you will get:[4]

Days	0	1	2	3	4	5	6	7
Probability	0.68	0.05	0.07	0.08	0.05	0.04	0.01	0.02

(a)   Verify that this is a legitimate discrete probability model.

(b)   Describe the event $X < 7$ in words. What is $P(X < 7)$?

(c)   Express the event "worked out at least once" in terms of X. What is the probability of this event?

Continuous probability models

When we use the table of random digits to select a digit between 0 and 9, the discrete probability model assigns probability 1/10 to each of the 10 possible outcomes. Suppose that we want to choose a number at random between 0 and 1, allowing *any* number between 0 and 1 as the outcome. Software random number generators will do this. For example, here is the result of asking software to produce 5 random numbers between 0 and 1:

0.2893511   0.3213787   0.5816462   0.9787920   0.4475373

The sample space is now an entire interval of numbers:

$$S = \{\text{all numbers between 0 and 1}\}$$

Call the outcome of the random number generator Y for short. How can we assign probabilities to such events as $\{0.3 \le Y \le 0.7\}$? As in the case of selecting a random digit, we would like all possible outcomes to be equally likely. But we cannot assign probabilities to each individual value of Y and then add them, because there are infinitely many possible values.

We use a new way of assigning probabilities directly to events—as *areas under a density curve*. Any density curve has area exactly 1 underneath it, corresponding to total probability 1. We met density curves as models for data in Chapter 3 (page 69).

Really random digits

For purists, the RAND Corporation long ago published a book titled *One Million Random Digits*. The book lists 1,000,000 digits that were produced by a very elaborate physical randomization and really are random. An employee of RAND once told me that this is not the most boring book that RAND has ever published.

CONTINUOUS PROBABILITY MODEL

A **continuous probability model** assigns probabilities as areas under a density curve. The area under the curve and above any range of values is the probability of an outcome in that range.

EXAMPLE 10.8 **Random numbers**

The random number generator will spread its output uniformly across the entire interval from 0 to 1 as we allow it to generate a long sequence of numbers. Figure 10.4 is a histogram of 10,000 random numbers. They are quite uniform, but not exactly so. The bar heights would all be exactly equal (1000 numbers for each bar) if the 10,000 numbers were exactly uniform. In fact, the counts vary from a low of 960 to a high of 1022.

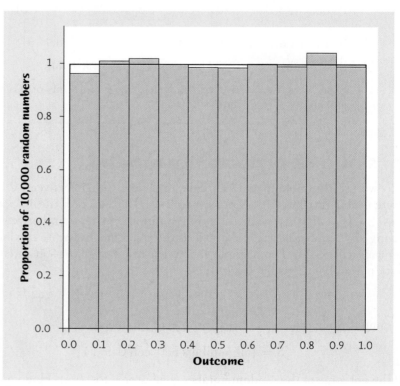

FIGURE 10.4

The probability model for the outcomes of a software random number generator, for Example 10.8. Compare the histogram of 10,000 actual outcomes with the uniform density curve that spreads probability evenly between 0 and 1.

uniform distribution

As in Chapter 3, we have adjusted the histogram scale so that the total area of the bars is exactly 1. Now we can add the density curve that describes the distribution of perfectly random numbers. This density curve also appears in Figure 10.4. It has height 1 over the interval from 0 to 1. This is the density curve of a **uniform distribution.** It is the continuous probability model for the results of generating very many random numbers. Like the probability models for perfectly balanced coins and dice, the density curve is an idealized description of the outcomes of a perfectly uniform random number generator. It is a good approximation for software outcomes, but even 10,000 tries isn't enough for actual outcomes to look exactly like the idealized model. ■

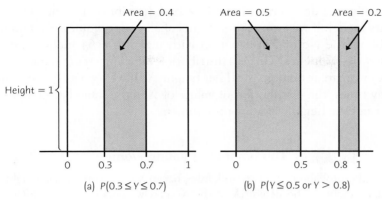

FIGURE 10.5

Probability as area under a density curve. The uniform density curve spreads probability evenly between 0 and 1.

The uniform density curve has height 1 over the interval from 0 to 1. The area under the curve is 1, and the probability of any event is the area under the curve and above the event in question. Figure 10.5 illustrates finding probabilities as areas under the density curve. The probability that the random number generator produces a number between 0.3 and 0.7 is

$$P(0.3 \leq Y \leq 0.7) = 0.4$$

because the area under the density curve and above the interval from 0.3 to 0.7 is 0.4. The height of the curve is 1 and the area of a rectangle is the product of height and length, so the probability of any interval of outcomes is just the length of the interval. Similarly,

$$P(Y \leq 0.5) = 0.5$$
$$P(Y > 0.8) = 0.2$$
$$P(Y \leq 0.5 \text{ or } Y > 0.8) = 0.7$$

The last event consists of two nonoverlapping intervals, so the total area above the event is found by adding two areas, as illustrated by Figure 10.5(b). This assignment of probabilities obeys all of our rules for probability.

The probability model for a continuous random variable assigns probabilities to intervals of outcomes rather than to individual outcomes. In fact, *all continuous probability models assign probability 0 to every individual outcome*. Only intervals of values have positive probability. To see that this is true, consider a specific outcome such as $P(Y = 0.8)$. The probability of any interval is the same as its length. The point 0.8 has no length, so its probability is 0. Put another way, $P(Y > 0.8)$ and $P(Y \geq 0.8)$ are both 0.2 because that is the area in Figure 10.5(b) between 0.8 and 1.

We can use any density curve to assign probabilities. The density curves that are most familiar to us are the Normal curves. **Normal distributions are continuous probability models** as well as descriptions of data. There is a close connection between a Normal distribution as an idealized description for data and a Normal probability model. If we look at the heights of all young women, we find that they closely follow the Normal distribution with mean $\mu = 64$ inches and standard deviation $\sigma = 2.7$ inches. This is a distribution for a large set of data. Now choose one young woman at random. Call her height X. If we repeat the random choice very many times, the distribution of values of X is the same Normal distribution that describes the heights of all young women.

EXAMPLE 10.9 **The heights of young women**

What is the probability that a randomly chosen young woman has height between 68 and 70 inches? The height X of the woman we choose has the $N(64, 2.7)$ distribution. We want $P(68 \leq X \leq 70)$. This is the area under the Normal curve in Figure 10.6. Software or the *Normal Curve* applet will give us the answer at once: $P(68 \leq X \leq 70) = 0.0561$.

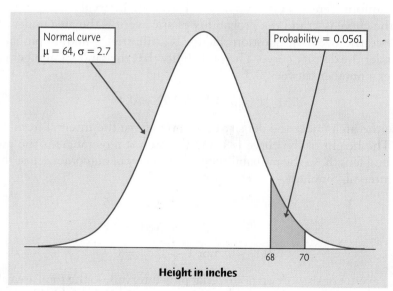

Normal curve
$\mu = 64, \sigma = 2.7$

Probability = 0.0561

68   70

Height in inches

FIGURE 10.6
The probability in Example 10.9 as an area under a Normal curve.

We can also find the probability by standardizing and using Table A, the table of standard Normal probabilities. We will reserve capital Z for a standard Normal variable.

$$P(68 \leq X \leq 70) = P\left(\frac{68 - 64}{2.7} \leq \frac{X - 64}{2.7} \leq \frac{70 - 64}{2.7}\right)$$

$$= P(1.48 \leq Z \leq 2.22)$$

$$= 0.9868 - 0.9306 = 0.0562$$

(As usual, there is a small roundoff error.) The calculation is the same as those we did in Chapter 3. Only the language of probability is new. ■

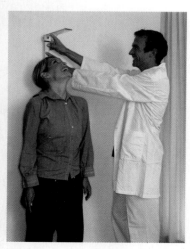

APPLY YOUR KNOWLEDGE

10.14 Random numbers. Let Y be a random number between 0 and 1 produced by the idealized random number generator described in Example 10.8 and Figure 10.4. Find the following probabilities:

(a) $P(Y \leq 0.4)$

(b) $P(Y < 0.4)$

(c) $P(0.3 \leq Y \leq 0.5)$

10.15 Adding random numbers. Generate two random numbers between 0 and 1 and take X to be their sum. The sum X can take any value between 0 and 2. The density curve of X is the triangle shown in Figure 10.7.

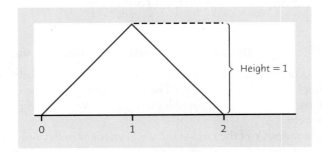

FIGURE 10.7

The density curve for the sum of two random numbers, for Exercise 10.15. This density curve spreads probability between 0 and 2.

(a) Verify by geometry that the area under this curve is 1.

(b) What is the probability that X is less than 1? (Sketch the density curve, shade the area that represents the probability, then find that area. Do this for (c) also.)

(c) What is the probability that X is less than 0.5?

10.16 Iowa Test scores. The Normal distribution with mean $\mu = 6.8$ and standard deviation $\sigma = 1.6$ is a good description of the Iowa Test vocabulary scores of seventh-grade students in Gary, Indiana. This is a continuous probability model for the score of a randomly chosen student. Figure 3.1 (page 68) pictures the density curve. Call the score of a randomly chosen student X for short.

(a) Write the event "the student chosen has a score of 10 or higher" in terms of X.

(b) Find the probability of this event.

Random variables

Examples 10.7 to 10.9 use a shorthand notation that is often convenient. In Example 10.9, we let X stand for the result of choosing a woman at random and measuring her height. We know that X would take a different value if we made another random choice. Because its value changes from one random choice to another, we call the height X a *random variable*.

> **RANDOM VARIABLE**
>
> A **random variable** is a variable whose value is a numerical outcome of a random phenomenon.
>
> The **probability distribution** of a random variable X tells us what values X can take and how to assign probabilities to those values.

We usually denote random variables by capital letters near the end of the alphabet, such as X or Y. Of course, the random variables of greatest interest to us are outcomes such as the mean $\bar{x}$ of a random sample, for which we will keep the familiar notation. There are two main types of random variables, corresponding to two types of probability models: *discrete* and *continuous*.

EXAMPLE 10.10 **Discrete and continuous random variables**

discrete random variable

The first digit X in Example 10.7 is a random variable whose possible values are the whole numbers {1, 2, 3, 4, 5, 6, 7, 8, 9}. The distribution of X assigns a probability to each of these outcomes. Random variables that have a finite list of possible outcomes are called **discrete**.

Compare the output Y of the random number generator in Example 10.8. The values of Y fill the entire interval of numbers between 0 and 1. The probability distribution of Y is given by its density curve, shown in Figure 10.4. Random variables that can take on any value in an interval, with probabilities given as areas under a density curve, are called **continuous**. ■

continuous random variable

APPLY YOUR KNOWLEDGE

10.17 Grades in a statistics course. North Carolina State University posts the grade distributions for its courses online.[5] Students in Statistics 101 in the Fall 2007 semester received 26% A's, 42% B's, 20% C's, 10% D's, and 2% F's. Choose a Statistics 101 student at random. To "choose at random" means to give every student the same chance to be chosen. The student's grade on a four-point scale (with A = 4) is a discrete random variable X with this probability distribution:

Value of X	0	1	2	3	4
Probability	0.02	0.10	0.20	0.42	0.26

(a) Say in words what the meaning of $P(X \geq 3)$ is. What is this probability?

(b) Write the event "the student got a grade poorer than C" in terms of values of the random variable X. What is the probability of this event?

10.18 Running a mile. A study of 12,000 able-bodied male students at the University of Illinois found that their times for the mile run were approximately Normal with mean 7.11 minutes and standard deviation 0.74 minute.[6] Choose a student at random from this group and call his time for the mile Y.

(a)   Say in words what the meaning of $P(Y \geq 8)$ is. What is this probability?

(b)   Write the event "the student could run a mile in less than 6 minutes" in terms of values of the random variable Y. What is the probability of this event?

Personal probability*

We began our discussion of probability with one idea: the probability of an outcome of a random phenomenon is the proportion of times that outcome would occur in a very long series of repetitions. This idea ties probability to actual outcomes. It allows us, for example, to estimate probabilities by simulating random phenomena. Yet we often meet another, quite different, idea of probability.

> **EXAMPLE 10.11** **Joe and the Chicago Cubs**
>
> Joe sits staring into his beer as his favorite baseball team, the Chicago Cubs, loses another game. The Cubbies have some good young players, so let's ask Joe, "What's the chance that the Cubs will go to the World Series next year?" Joe brightens up. "Oh, about 10%," he says.
>
> Does Joe assign probability 0.10 to the Cubs' appearing in the World Series? The outcome of next year's pennant race is certainly unpredictable, but we can't reasonably ask what would happen in many repetitions. Next year's baseball season will happen only once and will differ from all other seasons in players, weather, and many other ways. If probability measures "what would happen if we did this many times," Joe's 0.10 is not a probability. Probability is based on data about many repetitions of the same random phenomenon. Joe is giving us something else, his personal judgment. ■

Although Joe's 0.10 isn't a probability in our usual sense, it gives useful information about Joe's opinion. More seriously, a company asking, "How likely is it that building this plant will pay off within five years?" can't employ an idea of probability based on many repetitions of the same thing. The opinions of company officers and advisers are nonetheless useful information, and these opinions can be expressed in the language of probability. These are *personal probabilities*.

What are the odds?

Gamblers often express chance in terms of *odds* rather than probability. Odds of A to B against an outcome means that the probability of that outcome is $B/(A+B)$. So "odds of 5 to 1" is another way of saying "probability 1/6." A probability is always between 0 and 1, but odds range from 0 to infinity. Although odds are mainly used in gambling, they give us a way to make very small probabilities clearer. "Odds of 999 to 1" may be easier to understand than "probability 0.001."

> **PERSONAL PROBABILITY**
>
> A **personal probability** of an outcome is a number between 0 and 1 that expresses an individual's judgment of how likely the outcome is.

Rachel's opinion about the Cubs may differ from Joe's, and the opinions of several company officers about the new plant may differ. Personal probabilities are indeed personal: they vary from person to person. Moreover, a personal probability can't be called right or wrong. If we say, "In the long run, this coin will come up

*This short section is optional.

heads 60% of the time," we can find out if we are right by actually tossing the coin several thousand times. If Joe says, "I think the Cubs have a 10% chance of going to the World Series next year," that's just Joe's opinion. Why think of personal probabilities as probabilities? Because *any set of personal probabilities that makes sense obeys the same basic Rules 1 to 4 that describe any legitimate assignment of probabilities to events.* If Joe thinks there's a 10% chance that the Cubs will go to the World Series, he must also think that there's a 90% chance that they won't go. There is just one set of rules of probability, even though we now have two interpretations of what probability means.

APPLY YOUR KNOWLEDGE

10.19 Will you have an accident? The probability that a randomly chosen driver will be involved in an accident in the next year is about 0.2. This is based on the proportion of millions of drivers who have accidents. "Accident" includes things like crumpling a fender in your own driveway, not just highway accidents.

    (a)   What do you think is your own probability of being in an accident in the next year? This is a personal probability.

    (b)   Give some reasons why your personal probability might be a more accurate prediction of your "true chance" of having an accident than the probability for a random driver.

    (c)   Almost everyone says their personal probability is lower than the random driver probability. Why do you think this is true?

10.20 Winning the ACC tournament. The annual Atlantic Coast Conference men's basketball tournament has temporarily taken Joe's mind off the Chicago Cubs. He says to himself, "I think that Duke has probability 0.2 of winning. Clemson's probability is half of Duke's and North Carolina's probability is twice Duke's."

    (a)   What are Joe's personal probabilities for Clemson and North Carolina?

    (b)   What is Joe's personal probability that one of the 9 teams other than Clemson, Duke, and North Carolina will win the tournament?

CHAPTER 10 SUMMARY

■ A **random phenomenon** has outcomes that we cannot predict but that nonetheless have a regular distribution in very many repetitions.

■ The **probability** of an event is the proportion of times the event occurs in many repeated trials of a random phenomenon.

■ A **probability model** for a random phenomenon consists of a sample space S and an assignment of probabilities P.

■ The **sample space** S is the set of all possible outcomes of the random phenomenon. Sets of outcomes are called **events.** P assigns a number $P(A)$ to an event A as its probability.

- Any assignment of probability must obey the rules that state the basic properties of probability:
  1. $0 \leq P(A) \leq 1$ for any event A.
  2. $P(S) = 1$.
  3. **Addition rule:** Events A and B are **disjoint** if they have no outcomes in common. If A and B are disjoint, then $P(A \text{ or } B) = P(A) + P(B)$.
  4. For any event A, $P(A \text{ does not occur}) = 1 - P(A)$.

- When a sample space S contains finitely many possible values, a **discrete probability model** assigns each of these values a probability between 0 and 1 such that the sum of all the probabilities is exactly 1. The probability of any event is the sum of the probabilities of all the values that make up the event.

- A sample space can contain all values in some interval of numbers. A **continuous probability model** assigns probabilities as areas under a **density curve.** The probability of any event is the area under the curve above the values that make up the event.

- A **random variable** is a variable taking numerical values determined by the outcome of a random phenomenon. The **probability distribution** of a random variable X tells us what the possible values of X are and how probabilities are assigned to those values.

- A random variable X and its distribution can be **discrete** or **continuous.** A **discrete random variable** has finitely many possible values. Its distribution gives the probability of each value. A **continuous random variable** takes all values in some interval of numbers. A density curve describes the probability distribution of a continuous random variable.

CHECK YOUR SKILLS

10.21 You read in a book on poker that the probability of being dealt three of a kind in a five-card poker hand is 1/50. This means that

(a) if you deal thousands of poker hands, the fraction of them that contain three of a kind will be very close to 1/50.

(b) if you deal 50 poker hands, exactly 1 of them will contain three of a kind.

(c) if you deal 10,000 poker hands, exactly 200 of them will contain three of a kind.

10.22 A basketball player shoots 8 free throws during a game. The sample space for counting the number she makes is

(a) $S =$ any number between 0 and 1.

(b) $S =$ whole numbers 0 to 8.

(c) $S =$ all sequences of 8 hits or misses, like HMMHHHMH.

Here is the probability model for the blood type of a randomly chosen person in the United States. Exercises 10.23 to 10.26 use this information.

Blood type	O	A	B	AB
Probability	0.45	0.40	0.11	?

10.23 This probability model is

(a) continuous.   (b) discrete.   (c) equally likely.

10.24 The probability that a randomly chosen American has type AB blood must be

(a) any number between 0 and 1   (b) 0.04.   (c) 0.4.

10.25 Maria has type B blood. She can safely receive blood transfusions from people with blood types O and B. What is the probability that a randomly chosen American can donate blood to Maria?

(a) 0.11   (b) 0.44   (c) 0.56

10.26 What is the probability that a randomly chosen American does not have type O blood?

(a) 0.55   (b) 0.45   (c) 0.04

10.27 In a table of random digits such as Table B, each digit is equally likely to be any of 0, 1, 2, 3, 4, 5, 6, 7, 8, or 9. What is the probability that a digit in the table is a 0?

(a) 1/9   (b) 1/10   (c) 9/10

10.28 In a table of random digits such as Table B, each digit is equally likely to be any of 0, 1, 2, 3, 4, 5, 6, 7, 8, or 9. What is the probability that a digit in the table is 7 or greater?

(a) 7/10   (b) 4/10   (c) 3/10

10.29 Choose an American household at random and let the random variable X be the number of cars (including SUVs and light trucks) they own. Here is the probability model if we ignore the few households that own more than 5 cars:

Number of cars X	0	1	2	3	4	5
Probability	0.09	0.36	0.35	0.13	0.05	0.02

A housing company builds houses with two-car garages. What percent of households have more cars than the garage can hold?

(a) 20%   (b) 45%   (c) 55%

10.30 Choose a common fruit fly *Drosophila melanogaster* at random. Call the length of the thorax (where the wings and legs attach) Y. The random variable Y has the Normal distribution with mean $\mu = 0.800$ millimeter (mm) and standard deviation $\sigma = 0.078$ mm. The probability $P(Y > 1)$ that the fly you choose has a thorax more than 1 mm long is about

(a) 0.995.   (b) 0.5.   (c) 0.005.

C H A P T E R 1 0   E X E R C I S E S

10.31 Sample space. In each of the following situations, describe a sample space S for the random phenomenon.

(a) A basketball player shoots four free throws. You record the sequence of hits and misses.

(b) A basketball player shoots four free throws. You record the number of baskets she makes.

10.32 Probability models? In each of the following situations, state whether or not the given assignment of probabilities to individual outcomes is legitimate, that is, satisfies the rules of probability. If not, give specific reasons for your answer.

Darrell Walker/HWMS/Icon SMI/Newscom

(a) Roll a die and record the count of spots on the up-face:

$P(1) = 0$ $P(2) = 1/6$ $P(3) = 1/3$ $P(4) = 1/3$ $P(5) = 1/6$ $P(6) = 0$

(b) Deal a card from a shuffled deck:

$P(\text{clubs}) = 12/52$ $P(\text{diamonds}) = 12/52$ $P(\text{hearts}) = 12/52$
$P(\text{spades}) = 16/52$

(c) Choose a college student at random and record sex and enrollment status:

$P(\text{female full-time}) = 0.56$ $P(\text{male full-time}) = 0.44$
$P(\text{female part-time}) = 0.24$ $P(\text{male part-time}) = 0.17$

10.33 Education among young adults. Choose a young adult (aged 25 to 29) at random. The probability is 0.13 that the person chosen did not complete high school, 0.29 that the person has a high school diploma but no further education, and 0.30 that the person has at least a bachelor's degree.

(a) What must be the probability that a randomly chosen young adult has some education beyond high school but does not have a bachelor's degree?

(b) What is the probability that a randomly chosen young adult has at least a high school education?

10.34 Land in Canada. Canada's national statistics agency, Statistics Canada, says that the land area of Canada is 9,094,000 square kilometers. Of this land, 4,176,000 square kilometers are forested. Choose a square kilometer of land in Canada at random.

(a) What is the probability that the area you choose is forested?

(b) What is the probability that it is not forested?

10.35 Foreign-language study. Choose a student in grades 9 to 12 at random and ask if he or she is studying a language other than English. Here is the distribution of results:

Language	Spanish	French	German	All others	None
Probability	0.26	0.09	0.03	0.03	0.59

(a) Explain why this is a legitimate probability model.

(b) What is the probability that a randomly chosen student is studying a language other than English?

(c) What is the probability that a randomly chosen student is studying French, German, or Spanish?

10.36 Car colors. Choose a new car or light truck at random and note its color. Here are the probabilities of the most popular colors for vehicles made in North America in 2007:[7]

Color	White	Silver	Black	Red	Gray	Blue
Probability	0.19	0.18	0.16	0.13	0.12	0.12

(a) What is the probability that the vehicle you choose has any color other than the six listed?

(b) What is the probability that a randomly chosen vehicle is neither silver nor white?

10.37 Drawing cards. You are about to draw a card at random (that is, all choices have the same probability) from a set of 7 cards. Although you can't see the cards, here they are:

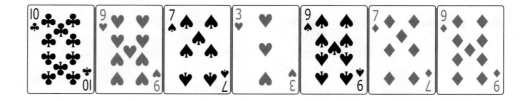

(a) What is the probability that you draw a 9?

(b) What is the probability that you draw a red 9?

(c) What is the probability that you do not draw a 7?

10.38 Loaded dice. There are many ways to produce crooked dice. To *load* a die so that 6 comes up too often and 1 (which is opposite 6) comes up too seldom, add a bit of lead to the filling of the spot on the 1 face. If a die is loaded so that 6 comes up with probability 0.2 and the probabilities of the 2, 3, 4, and 5 faces are not affected, what is the assignment of probabilities to the six faces?

10.39 A door prize. A party host gives a door prize to one guest chosen at random. There are 48 men and 42 women at the party. What is the probability that the prize goes to a woman? Explain how you arrived at your answer.

10.40 Race and ethnicity. The Census Bureau allows each person to choose from a long list of races. That is, in the eyes of the Census Bureau, you belong to whatever race you say you belong to. "Hispanic/Latino" is a separate category; Hispanics may be of any race. If we choose a resident of the United States at random, the Census Bureau gives these probabilities:[8]

	Hispanic	Not Hispanic
Asian	0.001	0.044
Black	0.006	0.124
White	0.139	0.674
Other	0.003	0.009

(a) Verify that this is a legitimate assignment of probabilities.

(b) What is the probability that a randomly chosen American is Hispanic?

(c) Non-Hispanic whites are the historical majority in the United States. What is the probability that a randomly chosen American is not a member of this group?

Choose at random a young adult aged 19 to 22 years. Ask their age and where they live (with their parents, in their own place, or in some other place such as a college dormitory). Here is the probability model for the 12 possible answers:[9]

	Age in Years			
	19	20	21	22
With parents	0.11	0.13	0.11	0.11
Own place	0.04	0.09	0.13	0.16
Other	0.03	0.04	0.03	0.02

Exercises 10.41 to 10.43 use this probability model.

10.41 Where do young people live?

(a) Why is this a legitimate discrete probability model?

(b) What is the probability that the person chosen is a 19-year-old who lives in his or her own place?

(c) What is the probability that the person is 19 years old?

(d) What is the probability that the person chosen lives in his or her own place?

10.42 Where do young people live, continued.

(a) List the outcomes that make up the event

$$A = \{\text{The person chosen is } either \text{ 19 years old } or \text{ lives}$$
$$\text{in his or her own place, or both}\}$$

(b) What is $P(A)$? Explain carefully why $P(A)$ is not the sum of the probabilities you found in parts (c) and (d) of the previous exercise.

10.43 Where do young people live, continued.

(a) What is the probability that the person chosen is 21 years old or older?

(b) What is the probability that the person chosen does not live with his or her parents?

10.44 Spelling errors. Spell-checking software catches "nonword errors" that result in a string of letters that is not a word, as when "the" is typed as "teh." When undergraduates are asked to type a 250-word essay (without spell-checking), the number X of nonword errors has the following distribution:

Value of X	0	1	2	3	4
Probability	0.1	0.2	0.3	0.3	0.1

(a) Is the random variable X discrete or continuous? Why?

(b) Write the event "at least one nonword error" in terms of X. What is the probability of this event?

(c) Describe the event $X \leq 2$ in words. What is its probability? What is the probability that $X < 2$?

10.45 First digits again. A crook who never heard of Benford's law might choose the first digits of his faked invoices so that all of 1, 2, 3, 4, 5, 6, 7, 8, and 9 are equally likely. Call the first digit of a randomly chosen fake invoice W for short.

(a) Write the probability distribution for the random variable W.

(b) Find $P(W \geq 6)$ and compare your result with the Benford's law probability from Example 10.7.

10.46 Who goes to Paris? Abby, Deborah, Mei-Ling, Sam, and Roberto work in a firm's public relations office. Their employer must choose two of them to attend a conference in Paris. To avoid unfairness, the choice will be made by drawing two names from a hat. (This is an SRS of size 2.)

(a) Write down all possible choices of two of the five names. This is the sample space.

(b) The random drawing makes all choices equally likely. What is the probability of each choice?

(c) What is the probability that Mei-Ling is chosen?

(d) What is the probability that neither of the two men (Sam and Roberto) is chosen?

M. Konarzewska/Stock.xchng

10.47 Birth order. A couple plans to have three children. There are 8 possible arrangements of girls and boys. For example, GGB means the first two children are girls and the third child is a boy. All 8 arrangements are (approximately) equally likely.

(a) Write down all 8 arrangements of the sexes of three children. What is the probability of any one of these arrangements?

(b) Let X be the number of girls the couple has. What is the probability that $X = 2$?

(c) Starting from your work in (a), find the distribution of X. That is, what values can X take, and what are the probabilities for each value?

10.48 Unusual dice. Nonstandard dice can produce interesting distributions of outcomes. You have two balanced, six-sided dice. One is a standard die, with faces having 1, 2, 3, 4, 5, and 6 spots. The other die has three faces with 0 spots and three faces with 6 spots. Find the probability distribution for the total number of spots Y on the up-faces when you roll these two dice. (*Hint:* Start with a picture like Figure 10.2 for the possible up-faces. Label the three 0 faces on the second die 0a, 0b, 0c in your picture, and similarly distinguish the three 6 faces.)

10.49 Random numbers. Many random number generators allow users to specify the range of the random numbers to be produced. Suppose that you specify that the random number Y can take any value between 0 and 2. Then the density curve of the outcomes has constant height between 0 and 2, and height 0 elsewhere.

(a) Is the random variable Y discrete or continuous? Why?

(b) What is the height of the density curve between 0 and 2? Draw a graph of the density curve.

(c) Use your graph from (b) and the fact that probability is area under the curve to find $P(Y \leq 1)$.

10.50 More random numbers. Find these probabilities as areas under the density curve you sketched in Exercise 10.49.

(a) $P(0.5 < Y < 1.3)$

(b) $P(Y \geq 0.8)$

10.51 Did you vote? A sample survey contacted an SRS of 663 registered voters in Oregon shortly after an election and asked respondents whether they had voted. Voter records show that 56% of registered voters had actually voted. We will see later that in this situation the proportion of the sample who voted (call this proportion V) has approximately the Normal distribution with mean $\mu = 0.56$ and standard deviation $\sigma = 0.019$.

(a) If the respondents answer truthfully, what is $P(0.52 \leq V \leq 0.60)$? This is the probability that the sample proportion V estimates the population proportion 0.56 within plus or minus 0.04.

(b) In fact, 72% of the respondents said they had voted ($V = 0.72$). If respondents answer truthfully, what is $P(V \geq 0.72)$? This probability is so small that it is good evidence that some people who did not vote claimed that they did vote.

10.52 Friends. How many close friends do you have? Suppose that the number of close friends adults claim to have varies from person to person with mean $\mu = 9$ and standard deviation $\sigma = 2.5$. An opinion poll asks this question of an SRS of 1100 adults. We will see later that in this situation the sample mean response $\bar{x}$ has approximately the Normal distribution with mean 9 and standard deviation 0.075. What is $P(8.9 \leq \bar{x} \leq 9.1)$, the probability that the sample result $\bar{x}$ estimates the population truth $\mu = 9$ to within ± 0.1?

10.53 Playing Pick 4. The Pick 4 games in many state lotteries announce a four-digit winning number each day. Each of the 10,000 possible numbers 0000 to 9999 has the same chance of winning. You win if your choice matches the winning digits. Suppose your chosen number is 5974.

(a) What is the probability that the winning number matches your number exactly?

(b)   What is the probability that the winning number has the same digits as your number *in any order?*

10.54 Nickels falling over. You may feel that it is obvious that the probability of a head in tossing a coin is about 1/2 because the coin has two faces. Such opinions are not always correct. Stand a nickel on edge on a hard, flat surface. Pound the surface with your hand so that the nickel falls over. What is the probability that it falls with heads upward? Make at least 50 trials to estimate the probability of a head.

10.55 What probability doesn't say. The idea of probability is that the *proportion* of heads in many tosses of a balanced coin eventually gets close to 0.5. But does the actual *count* of heads get close to one-half the number of tosses? Let's find out. Set the "Probability of heads" in the *Probability* applet to 0.5 and the number of tosses to 40. You can extend the number of tosses by clicking "Toss" again to get 40 more. Don't click "Reset" during this exercise.

(a)   After 40 tosses, what is the proportion of heads? What is the count of heads? What is the difference between the count of heads and 20 (one-half the number of tosses)?

(b)   Keep going to 120 tosses. Again record the proportion and count of heads and the difference between the count and 60 (half the number of tosses).

(c)   Keep going. Stop at 240 tosses and again at 480 tosses to record the same facts. Although it may take a long time, the laws of probability say that the proportion of heads will always get close to 0.5 and also that the difference between the count of heads and half the number of tosses will always grow without limit.

10.56 Shaq's free throws. The basketball player Shaquille O'Neal makes about half of his free throws over an entire season. Use the *Probability* applet or statistical software to simulate 100 free throws shot by a player who has probability 0.5 of making each shot. (In most software, the key phrase to look for is "Bernoulli trials." This is the technical term for independent trials with Yes/No outcomes. Our outcomes here are "Hit" and "Miss.")

(a)   What percent of the 100 shots did he hit?

(b)   Examine the sequence of hits and misses. How long was the longest run of shots made? Of shots missed? (Sequences of random outcomes often show runs longer than our intuition thinks likely.)

10.57 Simulating an opinion poll. An opinion poll showed that about 65% of the American public have a favorable opinion of the software company Microsoft. Suppose that this is exactly true. Choosing a person at random then has probability 0.65 of getting one who has a favorable opinion of Microsoft. Use the *Probability* applet or statistical software to simulate choosing many people at random. (In most software, the key phrase to look for is "Bernoulli trials." This is the technical term for independent trials with Yes/No outcomes. Our outcomes here are "Favorable" or not.)

(a)   Simulate drawing 20 people, then 80 people, then 320 people. What proportion have a favorable opinion of Microsoft in each case? We expect (but because of chance variation we can't be sure) that the proportion will be closer to 0.65 in longer runs of trials.

(b) Simulate drawing 20 people 10 times and record the percents in each sample who have a favorable opinion of Microsoft. Then simulate drawing 320 people 10 times and again record the 10 percents. Which set of 10 results is less variable? We expect the results of samples of size 320 to be more predictable (less variable) than the results of samples of size 20. That is "long-run regularity" showing itself.

Bruce Coleman/Alamy

Sampling Distributions

What is the average income of American households? Each March, the government's Current Population Survey asks detailed questions about income. The 98,105 households contacted in March 2007 had a mean "total money income" of $66,570 in 2006.[1] (The median income was of course lower, $48,201.) That $66,570 describes the sample, but we use it to estimate the mean income of all households. This is an example of statistical inference: we use information from a sample to infer something about a wider population.

Because the results of random samples and randomized comparative experiments include an element of chance, we can't guarantee that our inferences are correct. What we can guarantee is that our methods usually give correct answers. The reasoning of statistical inference rests on asking, "How often would this method give a correct answer if I used it very many times?" If our data come from random sampling or randomized comparative experiments, the laws of probability answer the question "What would happen if we did this many times?" This chapter presents some facts about probability that help answer this question.

Parameters and statistics

As we begin to use sample data to draw conclusions about a wider population, we must take care to keep straight whether a number describes a sample or a population. Here is the vocabulary we use.

PARAMETER, STATISTIC

A **parameter** is a number that describes the population. In statistical practice, the value of a parameter is not known because we cannot examine the entire population.

A **statistic** is a number that can be computed from the sample data without making use of any unknown parameters. In practice, we often use a statistic to estimate an unknown parameter.

EXAMPLE 11.1 Household earnings

The mean income of the sample of 98,105 households contacted by the Current Population Survey was $\overline{x} = \$66{,}570$. The number \$66,570 is a *statistic* because it describes this one Current Population Survey sample. The population that the poll wants to draw conclusions about is all 116 million U.S. households. The *parameter* of interest is the mean income of all of these households. We don't know the value of this parameter. ■

Remember **s** and **p**: statistics come from **s**amples, and **p**arameters come from **p**opulations. As long as we were just doing data analysis, the distinction between population and sample was not important. Now, however, it is essential. The notation we use must reflect this distinction. We write μ (the Greek letter mu) for the **mean of a population.** This is a fixed parameter that is unknown when we use a sample for inference. The **mean of the sample** is the familiar $\overline{x}$, the average of the observations in the sample. This is a statistic that would almost certainly take a different value if we chose another sample from the same population. The sample mean $\overline{x}$ from a sample or an experiment is an estimate of the mean μ of the underlying population.

population mean μ
sample mean $\overline{x}$

APPLY YOUR KNOWLEDGE

11.1 **Effects of caffeine.** How does caffeine affect our bodies? In a matched pairs experiment, subjects pushed a button as quickly as they could after taking a caffeine pill and also after taking a placebo pill. The mean pushes per minute were **283** for the placebo and **311** for caffeine. Is each of the boldface numbers a parameter or a statistic?

11.2 **Florida voters.** Florida has played a key role in recent presidential elections. Voter registration records show that **41%** of Florida voters are registered as Democrats and **37%** as Republicans. (Most of the others did not choose a party.) To test a random digit dialing device, you use it to call 250 randomly chosen residential telephones in Florida. Of the registered voters contacted, **33%** are registered Democrats. Is each of the boldface numbers a parameter or a statistic?

11.3   Ancient projectile points. Most of what we know about North America before Columbus comes from artifacts such as fragments of clay pottery and stone projectile points. Locations and cultures can be distinguished by the types of artifacts found. At one site in North Carolina, **82%** of the projectile points unearthed came from the Middle Archaic period (6000 to 3000 B.C.) and the remaining **18%** from the Late Archaic period (3000 to 1000 B.C.). Is each of the boldface numbers a parameter or a statistic?

Courtesy of Padre Island Seashore/National Park Service

Statistical estimation and the law of large numbers

Statistical inference uses sample data to draw conclusions about the entire population. Because good samples are chosen randomly, statistics such as $\bar{x}$ are random variables. We can describe the behavior of a sample statistic by a probability model that answers the question "What would happen if we did this many times?" Here is an example that will lead us toward the probability ideas most important for statistical inference.

EXAMPLE 11.2 Does this wine smell bad?

Sulfur compounds such as dimethyl sulfide (DMS) are sometimes present in wine. DMS causes "off-odors" in wine, so winemakers want to know the odor threshold, the lowest concentration of DMS that the human nose can detect. Different people have different thresholds, so we start by asking about the mean threshold μ in the population of all adults. The number μ is a parameter that describes this population.

To estimate μ, we present tasters with both natural wine and the same wine spiked with DMS at different concentrations to find the lowest concentration at which they identify the spiked wine. Here are the odor thresholds (measured in micrograms of DMS per liter of wine) for 10 randomly chosen subjects:

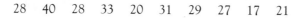

28   40   28   33   20   31   29   27   17   21

The mean threshold for these subjects is $\bar{x} = 27.4$. It seems reasonable to use the sample result $\bar{x} = 27.4$ to estimate the unknown μ. An SRS should fairly represent the population, so the mean $\bar{x}$ of the sample should be somewhere near the mean μ of the population. Of course, we don't expect $\bar{x}$ to be exactly equal to μ. We realize that if we choose another SRS, the luck of the draw will probably produce a different $\bar{x}$. ■

Enigma/Alamy

If $\bar{x}$ is rarely exactly right and varies from sample to sample, why is it nonetheless a reasonable estimate of the population mean μ? Here is one answer: *if we keep on taking larger and larger samples, the statistic $\bar{x}$ is guaranteed to get closer and closer to the parameter μ.* We have the comfort of knowing that if we can afford to keep on measuring more subjects, eventually we will estimate the mean odor threshold of all adults very accurately. This remarkable fact is called the *law of large numbers*. It is remarkable because it holds for *any* population, not just for some special class such as Normal distributions.

High-tech gambling

There are twice as many slot machines as bank ATMs in the United States. Once upon a time, you put in a coin and pulled the lever to spin three wheels, each with 20 symbols. No longer. Now the machines are video games with flashy graphics and outcomes produced by random number generators. Machines can accept many coins at once, can pay off on a bewildering variety of outcomes, and can be networked to allow common jackpots. Gamblers still search for systems, but in the long run the law of large numbers guarantees the house its 5% profit.

> **LAW OF LARGE NUMBERS**
>
> Draw observations at random from any population with finite mean μ. As the number of observations drawn increases, the mean $\bar{x}$ of the observed values gets closer and closer to the mean μ of the population.

The law of large numbers can be proved mathematically starting from the basic laws of probability. The behavior of $\bar{x}$ is similar to the idea of probability. In the long run, the *proportion* of outcomes taking any value gets close to the probability of that value, and the *average* outcome gets close to the population mean. Figure 10.1 (page 263) shows how proportions approach probability in one example. Here is an example of how sample means approach the population mean.

EXAMPLE 11.3 **The law of large numbers in action**

In fact, the distribution of odor thresholds among all adults has mean 25. The mean $\mu = 25$ is the true value of the parameter we seek to estimate. Figure 11.1 shows how the sample mean $\bar{x}$ of an SRS drawn from this population changes as we add more subjects to our sample.

The first subject in Example 11.2 had threshold 28, so the line in Figure 11.1 starts there. The mean for the first two subjects is

$$\bar{x} = \frac{28 + 40}{2} = 34$$

FIGURE 11.1

The law of large numbers in action: as we take more observations, the sample mean $\bar{x}$ always approaches the mean μ of the population.

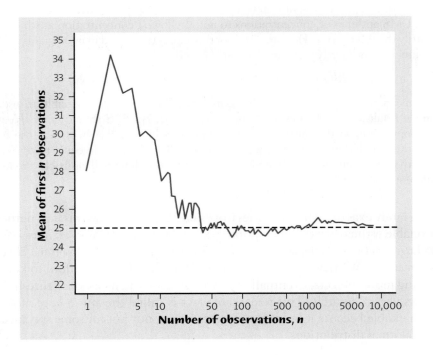

This is the second point on the graph. At first, the graph shows that the mean of the sample changes as we take more observations. Eventually, however, the mean of the observations gets close to the population mean $\mu = 25$ and settles down at that value.

If we started over, again choosing people at random from the population, we would get a different path from left to right in Figure 11.1. The law of large numbers says that whatever path we get will always settle down at 25 as we draw more and more people. ■

The *Law of Large Numbers* applet animates Figure 11.1 in a different setting. You can use the applet to watch $\overline{x}$ change as you average more observations until it eventually settles down at the mean μ.

The law of large numbers is the foundation of such business enterprises as gambling casinos and insurance companies. The winnings (or losses) of a gambler on a few plays are uncertain—that's why some people find gambling exciting. In Figure 11.1, the mean of even 100 observations is not yet very close to μ. It is only *in the long run* that the mean outcome is predictable. The house plays tens of thousands of times. So the house, unlike individual gamblers, can count on the long-run regularity described by the law of large numbers. The average winnings of the house on tens of thousands of plays will be very close to the mean of the distribution of winnings. Needless to say, this mean guarantees the house a profit. That's why gambling can be a business.

APPLY YOUR KNOWLEDGE

11.4   The law of large numbers made visible. Roll two balanced dice and count the spots on the up faces. The probability model appears in Example 10.5 (page 267). You can see that this distribution is symmetric with 7 as its center, so it's no surprise that the mean is $\mu = 7$. This is the population mean for the idealized population that contains the results of rolling two dice forever. The law of large numbers says that the average $\overline{x}$ from a finite number of rolls gets closer and closer to 7 as we do more and more rolls.

(a)   Click "More dice" once in the *Law of Large Numbers* applet to get two dice. Click "Show mean" to see the mean 7 on the graph. Leaving the number of rolls at 1, click "Roll dice" three times. How many spots did each roll produce? What is the average for the three rolls? You see that the graph displays at each point the average number of spots for all rolls up to the last one. This is exactly like Figure 11.1.

(b)   Set the number of rolls to 100 and click "Roll dice." The applet rolls the two dice 100 times. The graph shows how the average count of spots changes as we make more rolls. That is, the graph shows $\overline{x}$ as we continue to roll the dice. Sketch (or print out) the final graph.

(c)   Repeat your work from (b). Click "Reset" to start over, then roll two dice 100 times. Make a sketch of the final graph of the mean $\overline{x}$ against the number of rolls. Your two graphs will often look very different. What they have in common is that the average eventually gets close to the population mean

$\mu = 7$. The law of large numbers says that this will *always* happen if you keep on rolling the dice.

11.5 **Insurance.** The idea of insurance is that we all face risks that are unlikely but carry high cost. Think of a fire destroying your home. Insurance spreads the risk: we all pay a small amount, and the insurance policy pays a large amount to those few of us whose homes burn down. An insurance company looks at the records for millions of homeowners and sees that the mean loss from fire in a year is $\mu = \$250$ per person. (Most of us have no loss, but a few lose their homes. The $250 is the average loss.) The company plans to sell fire insurance for $250 plus enough to cover its costs and profit. Explain clearly why it would be unwise to sell only 12 policies. Then explain why selling thousands of such policies is a safe business.

Sampling distributions

The law of large numbers assures us that if we measure enough subjects, the statistic $\overline{x}$ will eventually get very close to the unknown parameter μ. But the odor threshold study in Example 11.2 had just 10 subjects. What can we say about estimating μ by $\overline{x}$ from a sample of 10 subjects? Put this one sample in the context of all such samples by asking, "What would happen if we took many samples of 10 subjects from this population?" Here's how to answer this question:

- Take a large number of samples of size 10 from the population.

- Calculate the sample mean $\overline{x}$ for each sample.

- Make a histogram of the values of $\overline{x}$.

- Examine the shape, center, and spread of the distribution displayed in the histogram.

simulation   In practice it is too expensive to take many samples from a large population such as all adult U.S. residents. But we can imitate many samples by using software. Using software to imitate chance behavior is called **simulation.**

EXAMPLE 11.4 **What would happen in many samples?**

Extensive studies have found that the DMS odor threshold of adults follows roughly a Normal distribution with mean $\mu = 25$ micrograms per liter and standard deviation $\sigma = 7$ micrograms per liter. We call this the *population distribution* of odor threshold.

Figure 11.2 illustrates the process of choosing many samples and finding the sample mean threshold $\overline{x}$ for each one. Follow the flow of the figure from the population at the left, to choosing an SRS and finding the $\overline{x}$ for this sample, to collecting together the $\overline{x}$'s from many samples. The first sample has $\overline{x} = 26.42$. The second sample contains a different 10 people, with $\overline{x} = 24.28$, and so on. The histogram at the right of the figure shows the distribution of the values of $\overline{x}$ from 1000 separate SRSs of size 10. This histogram displays the *sampling distribution* of the statistic $\overline{x}$. ■

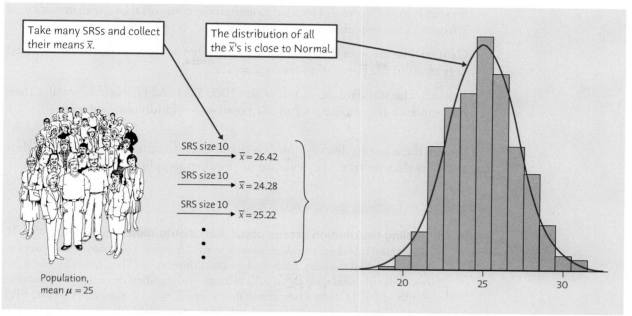

FIGURE 11.2

The idea of a sampling distribution: take many samples from the same population, collect the $\bar{x}$'s from all the samples, and display the distribution of the $\bar{x}$'s. The histogram shows the results of 1000 samples.

POPULATION DISTRIBUTION, SAMPLING DISTRIBUTION

The **population distribution** of a variable is the distribution of values of the variable among all the individuals in the population.

The **sampling distribution** of a statistic is the distribution of values taken by the statistic in all possible samples of the same size from the same population.

Be careful: The population distribution describes the *individuals* that make up the population. A sampling distribution describes how a *statistic* varies in many samples from the population.

Strictly speaking, the sampling distribution is the ideal pattern that would emerge if we looked at all possible samples of size 10 from our population. A distribution obtained from a fixed number of trials, like the 1000 trials in Figure 11.2, is only an approximation to the sampling distribution. One of the uses of probability theory in statistics is to obtain sampling distributions without simulation. The interpretation of a sampling distribution is the same, however, whether we obtain it by simulation or by the mathematics of probability.

We can use the tools of data analysis to describe any distribution. Let's apply those tools to Figure 11.2. What can we say about the shape, center, and spread of this distribution?

- **Shape:** It looks Normal! Detailed examination confirms that the distribution of $\bar{x}$ from many samples is very close to Normal.

- **Center:** The mean of the 1000 $\bar{x}$'s is 24.95. That is, the distribution is centered very close to the population mean $\mu = 25$.

- **Spread:** The standard deviation of the 1000 $\bar{x}$'s is 2.217, notably smaller than the standard deviation $\sigma = 7$ of the population of individual subjects.

Although these results describe just one simulation of a sampling distribution, they reflect facts that are true whenever we use random sampling.

APPLY YOUR KNOWLEDGE

11.6   **Sampling distribution versus population distribution.** During World War II, 12,000 able-bodied male undergraduates at the University of Illnois participated in required physical training. Each student ran a timed mile. Their times followed the Normal distribution with mean 7.11 minutes and standard deviation 0.74 minute. An SRS of 100 of these students has mean time $\bar{x} = 7.15$ minutes. A second SRS of size 100 has mean $\bar{x} = 6.97$ minutes. After many SRSs, the many values of the sample mean $\bar{x}$ follow the Normal distribution with mean 7.11 minutes and standard deviation 0.074 minute.

   (a)   What is the population? What values does the population distribution describe? What is this distribution?

   (b)   What values does the sampling distribution of $\bar{x}$ describe? What is the sampling distribution?

11.7   **Generating a sampling distribution.** Let's illustrate the idea of a sampling distribution in the case of a very small sample from a very small population. The population is the scores of 10 students on an exam:

Student	0	1	2	3	4	5	6	7	8	9
Score	82	62	80	58	72	73	65	66	74	62

The parameter of interest is the mean score μ in this population. The sample is an SRS of size $n = 4$ drawn from the population. Because the students are labeled 0 to 9, a single random digit from Table B chooses one student for the sample.

   (a)   Find the mean of the 10 scores in the population. This is the population mean μ.

   (b)   Use the first digits in row 116 of Table B to draw an SRS of size 4 from this population. What are the four scores in your sample? What is their mean $\bar{x}$? This statistic is an estimate of μ.

   (c)   Repeat this process 9 more times, using the first digits in rows 117 to 125 of Table B. Make a histogram of the 10 values of $\bar{x}$. You are constructing the sampling distribution of $\bar{x}$. Is the center of your histogram close to μ?

The sampling distribution of $\bar{x}$

Figure 11.2 suggests that when we choose many SRSs from a population, the sampling distribution of the sample means is centered at the mean of the original population and is less spread out than the distribution of individual observations. Here are the facts.

MEAN AND STANDARD DEVIATION OF A SAMPLE MEAN[2]

Suppose that $\bar{x}$ is the mean of an SRS of size n drawn from a large population with mean μ and standard deviation σ. Then the sampling distribution of $\bar{x}$ has **mean** μ and **standard deviation** $\sigma/\sqrt{n}$.

These facts about the mean and the standard deviation of the sampling distribution of $\bar{x}$ are true for *any* population, not just for some special class such as Normal distributions. They have important implications for statistical inference:

- The mean of the statistic $\bar{x}$ is always equal to the mean μ of the population. That is, the sampling distribution of $\bar{x}$ is centered at μ. In repeated sampling, $\bar{x}$ will sometimes fall above the true value of the parameter μ and sometimes below, but there is no systematic tendency to overestimate or underestimate the parameter. This makes the idea of lack of bias in the sense of "no favoritism" more precise. Because the mean of $\bar{x}$ is equal to μ, we say that the statistic $\bar{x}$ is an **unbiased estimator** of the parameter μ.

  unbiased estimator

- An unbiased estimator is "correct on the average" in many samples. How close the estimator falls to the parameter in most samples is determined by the spread of the sampling distribution. If individual observations have standard deviation σ, then sample means $\bar{x}$ from samples of size n have standard deviation $\sigma/\sqrt{n}$. That is, **averages are less variable than individual observations.**

- Not only is the standard deviation of the distribution of $\bar{x}$ smaller than the standard deviation of individual observations, but it gets smaller as we take larger samples. **The results of large samples are less variable than the results of small samples.**

The upshot of all this is that we can trust the sample mean from a large random sample to estimate the population mean accurately. If the sample size n is large, the standard deviation of $\bar{x}$ is small, and almost all samples will give values of $\bar{x}$ that lie very close to the true parameter μ. *However, the standard deviation of the sampling distribution gets smaller only at the rate $\sqrt{n}$. To cut the standard deviation of $\bar{x}$ in half, we must take four times as many observations, not just twice as many.* So very accurate estimates may be expensive.

We have described the center and spread of the sampling distribution of a sample mean $\bar{x}$, but not its shape. The shape of the sampling distribution depends on the shape of the population distribution. In one important case there is a simple relationship between the two distributions: if the population distribution is Normal, then so is the sampling distribution of the sample mean.

Sample size matters

The new thing in baseball is using statistics to evaluate players, with new measures of performance to help decide which players are worth the high salaries they demand. This challenges traditional subjective evaluation of young players and the usefulness of traditional measures such as batting average. But success has led many major league teams to hire statisticians. The statisticians say that sample size matters in baseball also: the 162-game regular season is long enough for the better teams to come out on top, but 5-game and 7-game playoff series are so short that luck has a lot to say about who wins.

> **SAMPLING DISTRIBUTION OF A SAMPLE MEAN**
>
> If individual observations have the $N(\mu, \sigma)$ distribution, then the sample mean $\bar{x}$ of an SRS of size n has the $N(\mu, \sigma/\sqrt{n})$ distribution.

EXAMPLE 11.5 Population distribution, sampling distribution

If we measure the DMS odor thresholds of individual adults, the values follow the Normal distribution with mean $\mu = 25$ micrograms per liter and standard deviation $\sigma = 7$ micrograms per liter. This is the population distribution of odor threshold.

Take many SRSs of size 10 from this population and find the sample mean $\bar{x}$ for each sample, as in Figure 11.2. The sampling distribution describes how the values of $\bar{x}$ vary among samples. That sampling distribution is also Normal, with mean $\mu = 25$ and standard deviation

$$\frac{\sigma}{\sqrt{n}} = \frac{7}{\sqrt{10}} = 2.2136$$

Figure 11.3 contrasts these two Normal distributions. Both are centered at the population mean, but sample means are much less variable than individual observations.

The smaller variation of sample means shows up in probability calculations. You can show (using software or standardizing and using Table A) that about 52% of all adults have odor thresholds between 20 and 30. But almost 98% of means of samples of size 10 lie in this range. ■

FIGURE 11.3

The distribution of single observations (the population distribution) compared with the sampling distribution of the means $\bar{x}$ of 10 observations, for Example 11.5. Both have the same mean, but averages are less variable than individual observations.

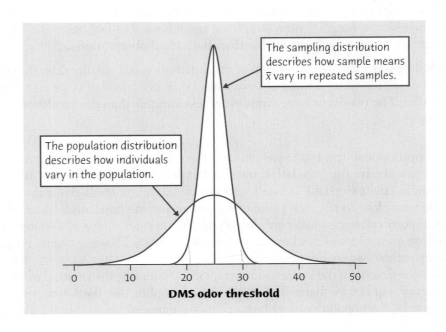

The sampling distribution describes how sample means $\bar{x}$ vary in repeated samples.

The population distribution describes how individuals vary in the population.

DMS odor threshold

11.8 A sample of young men. A government sample survey plans to measure the blood cholesterol level of an SRS of men aged 20 to 34. The researchers will report the mean $\overline{x}$ from their sample as an estimate of the mean cholesterol level μ in this population.

(a)   Explain to someone who knows no statistics what it means to say that $\overline{x}$ is an "unbiased" estimator of μ.

(b)   The sample result $\overline{x}$ is an unbiased estimator of the population truth μ no matter what size SRS the study uses. Explain to someone who knows no statistics why a large sample gives more trustworthy results than a small sample.

11.9 Larger sample, more accurate estimate. Suppose that in fact the blood cholesterol level of all men aged 20 to 34 follows the Normal distribution with mean $\mu = 188$ milligrams per deciliter (mg/dl) and standard deviation $\sigma = 41$ mg/dl.

(a)   Choose an SRS of 100 men from this population. What is the sampling distribution of $\overline{x}$? What is the probability that $\overline{x}$ takes a value between 185 and 191 mg/dl? This is the probability that $\overline{x}$ estimates μ within ± 3 mg/dl.

(b)   Choose an SRS of 1000 men from this population. Now what is the probability that $\overline{x}$ falls within ± 3 mg/dl of μ? The larger sample is much more likely to give an accurate estimate of μ.

11.10 Measurements in the lab. Juan makes a measurement in a chemistry laboratory and records the result in his lab report. The standard deviation of students' lab measurements is $\sigma = 10$ milligrams. Juan repeats the measurement 3 times and records the mean $\overline{x}$ of his 3 measurements.

(a)   What is the standard deviation of Juan's mean result? (That is, if Juan kept on making 3 measurements and averaging them, what would be the standard deviation of all his $\overline{x}$'s?)

(b)   How many times must Juan repeat the measurement to reduce the standard deviation of $\overline{x}$ to 5? Explain to someone who knows no statistics the advantage of reporting the average of several measurements rather than the result of a single measurement.

The central limit theorem

The facts about the mean and standard deviation of $\overline{x}$ are true no matter what the shape of the population distribution may be. But what is the shape of the sampling distribution when the population distribution is not Normal? *It is a remarkable fact that as the sample size increases, the distribution of $\overline{x}$ changes shape: it looks less like that of the population and more like a Normal distribution.* When the sample is large enough, the distribution of $\overline{x}$ is very close to Normal. This is true no matter what shape the population distribution has, as long as the population has a finite standard deviation σ. This famous fact of probability theory is called the *central limit theorem*. It is much more useful than the fact that the distribution of $\overline{x}$ is exactly Normal if the population is exactly Normal.

What was that probability again?

Wall Street uses fancy mathematics to predict the probabilities that fancy investments will go wrong. The probabilities are always too low—sometimes because something was assumed to be Normal but was not. Probability predictions in other areas also go wrong. In mid-September 2007, the New York Mets had probability 0.998 of making the National League playoffs, or so an elaborate calculation said. Then the Mets lost 12 of their final 17 games, the Phillies won 13 of their final 17, and the Mets were out. Maybe next year?

> ### CENTRAL LIMIT THEOREM
>
> Draw an SRS of size n from any population with mean μ and finite standard deviation σ. The **central limit theorem** says that when n is large, the sampling distribution of the sample mean $\bar{x}$ is approximately Normal:
>
> $$\bar{x} \text{ is approximately } N\left(\mu, \frac{\sigma}{\sqrt{n}}\right)$$
>
> The central limit theorem allows us to use Normal probability calculations to answer questions about sample means from many observations even when the population distribution is not Normal.

More general versions of the central limit theorem say that the distribution of any sum or average of many small random quantities is close to Normal. This is true even if the quantities are correlated with each other (as long as they are not too highly correlated) and even if they have different distributions (as long as no one random quantity is so large that it dominates the others). The central limit theorem suggests why the Normal distributions are common models for observed data. Any variable that is a sum of many small influences will have approximately a Normal distribution.

How large a sample size n is needed for $\bar{x}$ to be close to Normal depends on the population distribution. More observations are required if the shape of the population distribution is far from Normal. Here are two examples in which the population is far from Normal.

EXAMPLE 11.6 The central limit theorem in action

In March 2007, the Current Population Survey contacted 98,105 households. Figure 11.4(a) is a histogram of the earnings of the 61,742 households that had earned income greater than zero in 2006.[3] As we expect, the distribution of earned incomes is strongly skewed to the right and very spread out. The right tail of the distribution is even longer than the histogram shows because there are too few high incomes for their bars to be visible on this scale. In fact, we cut off the earnings scale at $400,000 to save space—a few households earned even more than $400,000. The mean earnings for these 61,742 households was $69,750.

Regard these 61,742 households as a population with mean $\mu = \$69,750$. Take an SRS of 100 households. The mean earnings in this sample is $\bar{x} = \$66,807$. That's less than the mean of the population. Take another SRS of size 100. The mean for this sample is $\bar{x} = \$70,820$. That's higher than the mean of the population. *What would happen if we did this many times?* Figure 11.4(b) is a histogram of the mean earnings for 500 samples, each of size 100. The scales in Figures 11.4(a) and 11.4(b) are the same, for easy comparison. Although the distribution of individual earnings is skewed and very spread out, the distribution of sample means is roughly symmetric and much less spread out.

Figure 11.4(c) zooms in on the center part of the histogram in Figure 11.4(b) to more clearly show its shape. Although $n = 100$ is not a very large sample size and the population distribution is extremely skewed, we can see that the distribution of sample means is close to Normal. ■

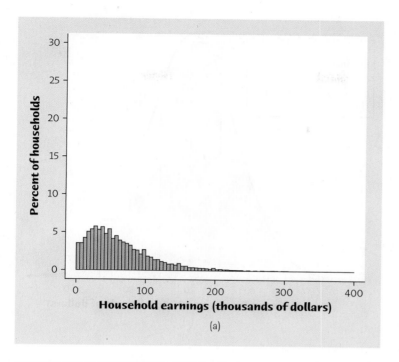

(a)

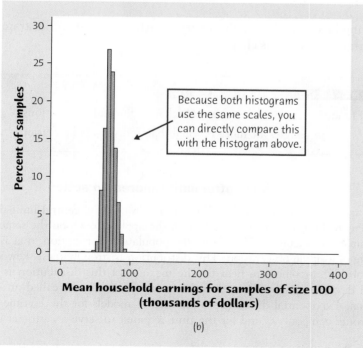

Because both histograms use the same scales, you can directly compare this with the histogram above.

(b)

FIGURE 11.4

The central limit theorem in action, for Example 11.6. **(a)** The distribution of earned income in a population of 61,742 households. **(b)** The distribution of the mean earnings for 500 SRSs of 100 households each from this population. (*Continued*)

FIGURE 11.4

(*Continued*) **(c)** The distribution of the sample means in more detail: the shape is close to Normal.

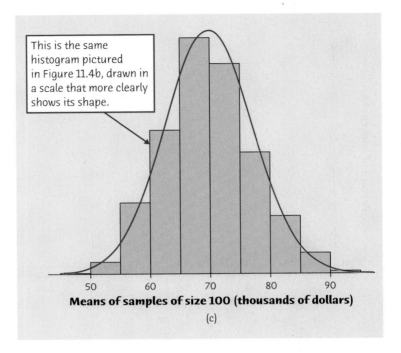

This is the same histogram pictured in Figure 11.4b, drawn in a scale that more clearly shows its shape.

Means of samples of size 100 (thousands of dollars)

(c)

Comparing Figure 11.4(a) with Figures 11.4(b) and 11.4(c) illustrates the two most important ideas of this chapter.

THINKING ABOUT SAMPLE MEANS

Means of random samples are **less variable** than individual observations.

Means of random samples are **more Normal** than individual observations.

EXAMPLE 11.7 **The central limit theorem in action**

The *Central Limit Theorem* applet allows you to watch the central limit theorem in action. Figure 11.5 presents snapshots from the applet, drawn on the same scales for easy comparison. Figure 11.5(a) shows the population distribution, that is, the density curve of a single observation. This distribution is strongly right-skewed, and the most probable outcomes are near 0. The mean μ of this distribution is 1, and its standard deviation σ is also 1. This particular distribution is called an *exponential distribution*. Exponential distributions are used as models for the lifetime in service of electronic components and for the time required to serve a customer or repair a machine.

Figures 11.5(b), (c), and (d) are the density curves of the sample means of 2, 10, and 25 observations from this population. As n increases, the shape becomes more Normal. The mean remains at $\mu = 1$, and the standard deviation decreases, taking the value $1/\sqrt{n}$. The density curve for 10 observations is still somewhat skewed to the right but already resembles a Normal curve having $\mu = 1$ and $\sigma = 1/\sqrt{10} = 0.32$. The density

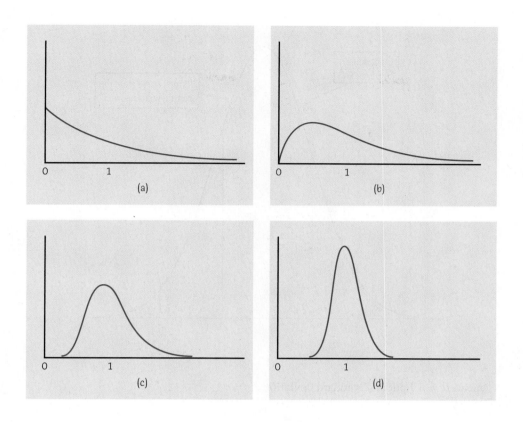

FIGURE 11.5

The central limit theorem in action, for Example 11.7. The distribution of sample means $\bar{x}$ from a strongly non-Normal population becomes more Normal as the sample size increases. **(a)** The distribution of 1 observation. **(b)** The distribution of $\bar{x}$ for 2 observations. **(c)** The distribution of $\bar{x}$ for 10 observations. **(d)** The distribution of $\bar{x}$ for 25 observations.

curve for $n = 25$ is yet more Normal. The contrast between the shapes of the population distribution and of the distribution of the mean of 10 or 25 observations is striking. ■

Let's use Normal calculations based on the central limit theorem to answer a question about the very non-Normal distribution in Figure 11.5(a).

EXAMPLE 11.8 **Maintaining air conditioners**

STATE: The time (in hours) that a technician requires to perform preventive maintenance on an air-conditioning unit is governed by the exponential distribution whose density curve appears in Figure 11.5(a). The mean time is $\mu = 1$ hour and the standard deviation is $\sigma = 1$ hour. Your company has a contract to maintain 70 of these units in an apartment building. You must schedule technicians' time for a visit to this building. Is it safe to budget an average of 1.1 hours for each unit? Or should you budget an average of 1.25 hours?

PLAN: We can treat these 70 air conditioners as an SRS from all units of this type. What is the probability that the average maintenance time for 70 units exceeds 1.1 hours? That the average time exceeds 1.25 hours?

SOLVE: The central limit theorem says that the sample mean time $\bar{x}$ spent working on 70 units has approximately the Normal distribution with mean equal to the population

FIGURE 11.6

The exact distribution (dotted) and the Normal approximation from the central limit theorem (solid) for the average time needed to maintain an air conditioner, for Example 11.8. The probability we want is the area to the right of 1.1.

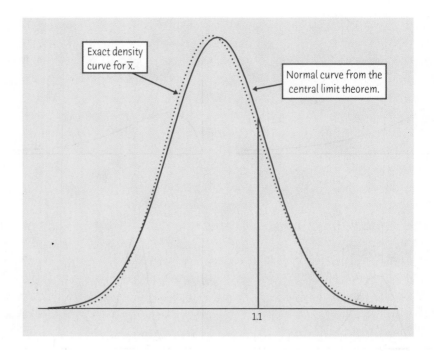

Exact density curve for $\bar{x}$.

Normal curve from the central limit theorem.

1.1

mean $\mu = 1$ hour and standard deviation

$$\frac{\sigma}{\sqrt{70}} = \frac{1}{\sqrt{70}} = 0.12 \text{ hour}$$

The distribution of $\bar{x}$ is therefore approximately $N(1, 0.12)$. This Normal curve is the solid curve in Figure 11.6.

Using this Normal distribution, the probabilities we want are

$$P(\bar{x} > 1.10 \text{ hours}) = 0.2014$$

$$P(\bar{x} > 1.25 \text{ hours}) = 0.0182$$

Software gives these probabilities immediately, or you can standardize and use Table A. For example,

$$P(\bar{x} > 1.10) = P\left(\frac{\bar{x} - 1}{0.12}\right) > P\left(\frac{1.10 - 1}{0.12}\right)$$

$$= P(Z > 0.83) = 1 - 0.7967 = 0.2033$$

with the usual roundoff error. Don't forget to use standard deviation 0.12 in your software or when you standardize $\bar{x}$.

CONCLUDE: If you budget 1.1 hours per unit, there is a 20% chance that the technicians will not complete the work in the building within the budgeted time. This chance drops to 2% if you budget 1.25 hours. You therefore budget 1.25 hours per unit. ■

Using more mathematics, we can start with the exponential distribution and find the actual density curve of $\bar{x}$ for 70 observations. This is the dotted curve in

Figure 11.6. You can see that the solid Normal curve is a good approximation. The exactly correct probability for 1.1 hours is an area to the right of 1.1 under the dotted density curve. It is 0.1977. The central limit theorem Normal approximation 0.2014 is off by only about 0.004.

APPLY YOUR KNOWLEDGE

11.11 What does the central limit theorem say? Asked what the central limit theorem says, a student replies, "As you take larger and larger samples from a population, the histogram of the sample values looks more and more Normal." Is the student right? Explain your answer.

11.12 Detecting gypsy moths. The gypsy moth is a serious threat to oak and aspen trees. A state agriculture department places traps throughout the state to detect the moths. When traps are checked periodically, the mean number of moths trapped is only 0.5, but some traps have several moths. The distribution of moth counts is discrete and strongly skewed, with standard deviation 0.7.

Bruce Coleman/Alamy

(a) What are the mean and standard deviation of the average number of moths $\bar{x}$ in 50 traps?

(b) Use the central limit theorem to find the probability that the average number of moths in 50 traps is greater than 0.6.

11.13 More on insurance. An insurance company knows that in the entire population of millions of homeowners, the mean annual loss from fire is $\mu = \$250$ and the standard deviation of the loss is $\sigma = \$1000$. The distribution of losses is strongly right-skewed: most policies have $0 loss, but a few have large losses. If the company sells 10,000 policies, can it safely base its rates on the assumption that its average loss will be no greater than $275? Follow the four-step process as illustrated in Example 11.8.

CHAPTER 11 SUMMARY

- A **parameter** in a statistical problem is a number that describes a population, such as the **population mean μ.** To estimate an unknown parameter, use a **statistic** calculated from a sample, such as the **sample mean $\bar{x}$.**

- The **law of large numbers** states that the actually observed mean outcome $\bar{x}$ must approach the mean μ of the population as the number of observations increases.

- The **population distribution** of a variable describes the values of the variable for all individuals in a population.

- The **sampling distribution** of a statistic describes the values of the statistic in all possible samples of the same size from the same population.

- When the sample is an SRS from the population, the **mean** of the sampling distribution of the sample mean $\bar{x}$ is the same as the population mean μ. That is, $\bar{x}$ is an **unbiased estimator** of μ.

- The **standard deviation** of the sampling distribution of $\overline{x}$ is $\sigma/\sqrt{n}$ for an SRS of size n if the population has standard deviation σ. That is, averages are less variable than individual observations.

- When the sample is an SRS from a population that has a Normal distribution, the sample mean $\overline{x}$ also has a Normal distribution.

- Choose an SRS of size n from any population with mean μ and finite standard deviation σ. The **central limit theorem** states that when n is large the sampling distribution of $\overline{x}$ is approximately Normal. That is, averages are more Normal than individual observations. We can use the $N(\mu, \sigma/\sqrt{n})$ distribution to calculate approximate probabilities for events involving $\overline{x}$.

CHECK YOUR SKILLS

11.14 The Bureau of Labor Statistics announces that last month it interviewed all members of the labor force in a sample of 60,000 households; **4.9%** of the people interviewed were unemployed. The boldface number is a

(a) sampling distribution.　　(b) parameter.　　(c) statistic.

11.15 A study of voting chose 663 registered voters at random shortly after an election. Of these, 72% said they had voted in the election. Election records show that only **56%** of registered voters voted in the election. The boldface number is a

(a) sampling distribution.　　(b) parameter.　　(c) statistic.

11.16 Annual returns on the more than 5000 common stocks available to investors vary a lot. In a recent year, the mean return was 8.3% and the standard deviation of returns was 28.5%. The law of large numbers says that

(a) you can get an average return higher than the mean 8.3% by investing in a large number of stocks.

(b) as you invest in more and more stocks chosen at random, your average return on these stocks gets ever closer to 8.3%.

(c) if you invest in a large number of stocks chosen at random, your average return will have approximately a Normal distribution.

11.17 Scores on the mathematics part of the SAT exam in a recent year were roughly Normal with mean 515 and standard deviation 114. You choose an SRS of 100 students and average their SAT math scores. If you do this many times, the mean of the average scores you get will be close to

(a) 515.　　(b) $515/100 = 5.15$.　　(c) $515/\sqrt{100} = 51.5$.

11.18 Scores on the mathematics part of the SAT exam in a recent year were roughly Normal with mean 515 and standard deviation 114. You choose an SRS of 100 students and average their SAT math scores. If you do this many times, the standard deviation of the average scores you get will be close to

(a) 114.　　(b) $114/100 = 1.14$.　　(c) $114/\sqrt{100} = 11.4$.

11.19 A newborn baby has extremely low birth weight (ELBW) if it weighs less than 1000 grams. A study of the health of such children in later years examined a random sample of 219 children. Their mean weight at birth was $\bar{x} = 810$ grams. This sample mean is an *unbiased estimator* of the mean weight μ in the population of all ELBW babies. This means that

(a) in many samples from this population, the mean of the many values of $\bar{x}$ will be equal to μ.

(b) as we take larger and larger samples from this population, $\bar{x}$ will get closer and closer to μ.

(c) in many samples from this population, the many values of $\bar{x}$ will have a distribution that is close to Normal.

11.20 The number of hours a light bulb burns before failing varies from bulb to bulb. The distribution of burnout times is strongly skewed to the right. The central limit theorem says that

(a) as we look at more and more bulbs, their average burnout time gets ever closer to the mean μ for all bulbs of this type.

(b) the average burnout time of a large number of bulbs has a distribution of the same shape (strongly skewed) as the distribution for individual bulbs.

(c) the average burnout time of a large number of bulbs has a distribution that is close to Normal.

11.21 The length of human pregnancies from conception to birth varies according to a distribution that is approximately Normal with mean 266 days and standard deviation 16 days. The probability that the average pregnancy length for 6 randomly chosen women exceeds 270 days is about

(a) 0.40.   (b) 0.27.   (c) 0.07.

C H A P T E R 1 1   E X E R C I S E S

11.22 Testing glass. How well materials conduct heat matters when designing houses. As a test of a new measurement process, 10 measurements are made on pieces of glass known to have conductivity **1**. The average of the 10 measurements is **1.09**. Is each of the boldface numbers a parameter or a statistic? Explain your answer.

11.23 Small classes in school. The Tennessee STAR experiment randomly assigned children to regular or small classes during their first four years of school. When these children reached high school, **40.2%** of blacks from small classes took the ACT or SAT college entrance exams. Only **31.7%** of blacks from regular classes took one of these exams. Is each of the boldface numbers a parameter or a statistic? Explain your answer.

11.24 Roulette. A roulette wheel has 38 slots, of which 18 are black, 18 are red, and 2 are green. When the wheel is spun, the ball is equally likely to come to rest in any of the slots. One of the simplest wagers chooses red or black. A bet of $1 on red returns $2 if the ball lands in a red slot. Otherwise, the player loses his dollar. When gamblers bet on red or black, the two green slots belong to the house. Because the probability of winning $2 is 18/38, the mean payoff from a $1 bet is twice 18/38, or 94.7 cents. Explain what the law of large numbers tells us about what will happen if a gambler makes very many bets on red.

Gandee Vasan/Getty Images

11.25 Lightning strikes. The number of lightning strikes on a square kilometer of open ground in a year has mean 6 and standard deviation 2.4. (These values are typical of much of the United States.) The National Lightning Detection Network uses automatic sensors to watch for lightning in a sample of 10 square kilometers. What are the mean and standard deviation of $\bar{x}$, the mean number of strikes per square kilometer?

11.26 Heights of male students. To estimate the mean height μ of male students on your campus, you will measure an SRS of students. Heights of people of the same sex and similar ages are close to Normal. You know from government data that the standard deviation of the heights of young men is about 2.8 inches. Suppose that (unknown to you) the mean height of all male students is 70 inches.

(a) If you choose one student at random, what is the probability that he is between 69 and 71 inches tall?

(b) You measure 25 students. What is the sampling distribution of their average height $\bar{x}$?

(c) What is the probability that the mean height of your sample is between 69 and 71 inches?

11.27 Glucose testing. Shelia's doctor is concerned that she may suffer from gestational diabetes (high blood glucose levels during pregnancy). There is variation both in the actual glucose level and in the blood test that measures the level. A patient is classified as having gestational diabetes if the glucose level is above 140 milligrams per deciliter (mg/dl) one hour after having a sugary drink. Shelia's measured glucose level one hour after the sugary drink varies according to the Normal distribution with $\mu = 125$ mg/dl and $\sigma = 10$ mg/dl.

(a) If a single glucose measurement is made, what is the probability that Shelia is diagnosed as having gestational diabetes?

(b) If measurements are made on 4 separate days and the mean result is compared with the criterion 140 mg/dl, what is the probability that Shelia is diagnosed as having gestational diabetes?

11.28 Durable press fabrics. "Durable press" cotton fabrics are treated to improve their recovery from wrinkles after washing. Unfortunately, the treatment also reduces the strength of the fabric. The breaking strength of untreated fabric is Normally distributed with mean 58 pounds and standard deviation 2.3 pounds. The same type of fabric after treatment has Normally distributed breaking strength with mean 30 pounds and standard deviation 1.6 pounds.[4] A clothing manufacturer tests an SRS of 5 specimens of each fabric.

(a) What is the probability that the mean breaking strength of the 5 untreated specimens exceeds 50 pounds?

(b) What is the probability that the mean breaking strength of the 5 treated specimens exceeds 50 pounds?

11.29 Glucose testing, continued. Shelia's measured glucose level one hour after a sugary drink varies according to the Normal distribution with $\mu = 125$ mg/dl and $\sigma = 10$ mg/dl. What is the level L such that there is probability only 0.05 that the mean glucose level of 4 test results falls above L? (*Hint:* This requires a backward Normal calculation. See page 83 in Chapter 3 if you need to review.)

11.30 Pollutants in auto exhausts. The level of nitrogen oxides (NOX) in the exhaust of cars of a particular model varies Normally with mean 0.2 grams per mile (g/mi) and standard deviation 0.05 g/mi. Government regulations call for NOX emissions no higher than 0.3 g/mi.

(a) What is the probability that a single car of this model fails to meet the NOX requirement?

(b) A company has 25 cars of this model in its fleet. What is the probability that the average NOX level $\overline{x}$ of these cars is above the 0.3 g/mi limit?

11.31 Auto accidents. The number of accidents per week at a hazardous intersection varies with mean 2.2 and standard deviation 1.4. This distribution takes only whole-number values, so it is certainly not Normal.

(a) Let $\overline{x}$ be the mean number of accidents per week at the intersection during a year (52 weeks). What is the approximate distribution of $\overline{x}$ according to the central limit theorem?

(b) What is the approximate probability that $\overline{x}$ is less than 2?

(c) What is the approximate probability that there are fewer than 100 accidents at the intersection in a year? (*Hint:* Restate this event in terms of $\overline{x}$.)

11.32 Pollutants in auto exhausts, continued. The level of nitrogen oxides (NOX) in the exhaust of cars of a particular model varies Normally with mean 0.2 g/mi and standard deviation 0.05 g/mi. A company has 25 cars of this model in its fleet. What is the level L such that the probability that the average NOX level $\overline{x}$ for the fleet is greater than L is only 0.01? (*Hint:* This requires a backward Normal calculation. See page 83 in Chapter 3 if you need to review.)

11.33 Returns on stocks. Andrew plans to retire in 40 years. He plans to invest part of his retirement funds in stocks, so he seeks out information on past returns. He learns that over the entire 20th century, the real (that is, adjusted for inflation) annual returns on U.S. common stocks had mean 8.7% and standard deviation 20.2%.[5] The distribution of annual returns on common stocks is roughly symmetric, so the mean return over even a moderate number of years is close to Normal. What is the probability (assuming that the past pattern of variation continues) that the mean annual return on common stocks over the next 40 years will exceed 10%? What is the probability that the mean return will be less than 5%? Follow the four-step process as illustrated in Example 11.8.

11.34 Airline passengers get heavier. In response to the increasing weight of airline passengers, the Federal Aviation Administration in 2003 told airlines to assume that passengers average 190 pounds in the summer, including clothing and carry-on baggage. But passengers vary, and the FAA did not specify a standard deviation. A reasonable standard deviation is 35 pounds. Weights are not Normally distributed, especially when the population includes both men and women, but they are not very non-Normal. A commuter plane carries 19 passengers. What is the approximate probability that the total weight of the passengers exceeds 4000 pounds? Use the four-step process to guide your work. (*Hint:* To apply the central limit theorem, restate the problem in terms of the mean weight.)

11.35 Sampling male students. To estimate the mean height μ of male students on your campus, you will measure an SRS of students. You know from government

Alan Hicks/Getty Images

Jeff Greenberg/The Image Works

data that heights of young men are approximately Normal with standard deviation about 2.8 inches. How large an SRS must you take to reduce the standard deviation of the sample mean to one-half inch?

11.36 Sampling male students, continued. To estimate the mean height μ of male students on your campus, you will measure an SRS of students. You know from government data that heights of young men are approximately Normal with standard deviation about 2.8 inches. You want your sample mean $\bar{x}$ to estimate μ with an error of no more than one-half inch in either direction.

(a)   What standard deviation must $\bar{x}$ have so that 99.7% of all samples give an $\bar{x}$ within one-half inch of μ? (Use the 68–95–99.7 rule.)

(b)   How large an SRS do you need to reduce the standard deviation of $\bar{x}$ to the value you found in part (a)?

11.37 Playing the numbers. The numbers racket is a well-entrenched illegal gambling operation in most large cities. One version works as follows: you choose one of the 1000 three-digit numbers 000 to 999 and pay your local numbers runner a dollar to enter your bet. Each day, one three-digit number is chosen at random and pays off $600. The mean payoff for the population of thousands of bets is $\mu = 60$ cents. Joe makes one bet every day for many years. Explain what the law of large numbers says about Joe's results as he keeps on betting.

11.38 Playing the numbers: a gambler gets chance outcomes. The law of large numbers tells us what happens in the long run. Like many games of chance, the numbers racket has outcomes so variable—one three-digit number wins $600 and all others win nothing—that gamblers never reach "the long run." Even after many bets, their average winnings may not be close to the mean. For the numbers racket, the mean payout for single bets is $0.60 (60 cents) and the standard deviation of payouts is about $18.96. If Joe plays 350 days a year for 40 years, he makes 14,000 bets.

(a)   What are the mean and standard deviation of the average payout $\bar{x}$ that Joe receives from his 14,000 bets?

(b)   The central limit theorem says that his average payout is approximately Normal with the mean and standard deviation you found in part (a). What is the approximate probability that Joe's average payout per bet is between $0.50 and $0.70? You see that Joe's average may not be very close to the mean $0.60 even after 14,000 bets.

11.39 Playing the numbers: the house has a business. Unlike Joe (see the previous exercise) the operators of the numbers racket can rely on the law of large numbers. It is said that the New York City mobster Casper Holstein took as many as 25,000 bets per day in the Prohibition era. That's 150,000 bets in a week if he takes Sunday off. Casper's mean winnings per bet are $0.40 (he pays out 60 cents of each dollar bet to people like Joe and keeps the other 40 cents.) His standard deviation for single bets is about $18.96, the same as Joe's.

(a)   What are the mean and standard deviation of Casper's average winnings $\bar{x}$ on his 150,000 bets?

(b)   According to the central limit theorem, what is the approximate probability that Casper's average winnings per bet are between $0.30 and $0.50? After only a week, Casper can be pretty confident that his winnings will be quite close to $0.40 per bet.

11.40 Can we trust the central limit theorem? The central limit theorem says that "when n is large" we can act as if the distribution of a sample mean $\overline{x}$ is close to Normal. How large a sample we need depends on how far the population distribution is from being Normal. Example 11.8 shows that we can trust this Normal approximation for quite moderate sample sizes even when the population has a strongly skewed continuous distribution.

The central limit theorem requires much larger samples for Joe's bets with his local numbers racket. The population of individual bets has a discrete distribution with only 2 possible outcomes: $600 (probability 0.001) and $0 (probability 0.999). This distribution has mean $\mu = 0.6$ and standard deviation about $\sigma = 18.96$. With more math and good software we can find exact probabilities for Joe's average winnings.

(a) If Joe makes 14,000 bets, the exact probability $P(0.5 \leq \overline{x} \leq 0.7) = 0.4961$. How accurate was your Normal approximation from part (b) of Exercise 11.38?

(b) If Joe makes only 3500 bets, $P(0.5 \leq \overline{x} \leq 0.7) = 0.4048$. How accurate is the Normal approximation for this probability?

(c) If Joe and his buddies make 150,000 bets, $P(0.5 \leq \overline{x} \leq 0.7) = 0.9629$. How accurate is the Normal approximation?

11.41 What's the mean? Suppose that you roll three balanced dice. We wonder what the mean number of spots on the up-faces of the three dice is. The law of large numbers says that we can find out by experience: roll three dice many times, and the average number of spots will eventually approach the true mean. Set up the *Law of Large Numbers* applet to roll three dice. Don't click "Show mean" yet. Roll the dice until you are confident you know the mean quite closely, then click "Show mean" to verify your discovery. What is the mean? Make a rough sketch of the path the averages $\overline{x}$ followed as you kept adding more rolls.

Barry Austin Photography/Getty

General Rules of Probability*

Probability models can describe the flow of traffic through a highway system, a telephone interchange, or a computer processor; the genetic makeup of populations; the energy states of subatomic particles; the spread of epidemics or rumors; and the rate of return on risky investments. Although we are interested in probability mainly because it is the foundation for statistical inference, the mathematics of chance is important in many fields of study. Our introduction to probability in Chapter 10 concentrated on basic ideas and facts. Now we look at some further details. With more probability at our command, we can model more complex random phenomena.

Although we won't emphasize the math, everything in this chapter (and much more) follows from the first three of the four rules we met in Chapter 10. Here they are again.

PROBABILITY RULES

Rule 1. For any event A, $0 \leq P(A) \leq 1$.

Rule 2. If S is the sample space, $P(S) = 1$.

---

*This more advanced chapter introduces some of the mathematics of probability. The material is not needed to read the rest of the book.

> **PROBABILITY RULES (CONTINUED)**
>
> **Rule 3. Addition rule:** If A and B are **disjoint** events,
>
> $$P(A \text{ or } B) = P(A) + P(B)$$
>
> **Rule 4.** For any event A,
>
> $$P(A \text{ does not occur}) = 1 - P(A)$$

Independence and the multiplication rule

Rule 3, the addition rule for disjoint events, describes the probability that *one or the other* of two events A and B occurs in the special situation when A and B cannot occur together. Now we will describe the probability that *both* events A and B occur, again only in a special situation.

FIGURE 12.1

Venn diagram showing disjoint events A and B.

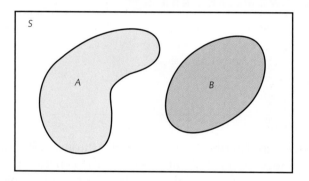

You may find it helpful to draw a picture to display relations among several events. A picture like Figure 12.1 that shows the sample space S as a rectangular area and events as areas within S is called a **Venn diagram.** The events A and B in Figure 12.1 are disjoint because they do not overlap. The Venn diagram in Figure 12.2 illustrates two events that are not disjoint. The event $\{A \text{ and } B\}$ appears as the overlapping area that is common to both A and B. Can we find the probability

Venn diagram

FIGURE 12.2

Venn diagram showing events A and B that are not disjoint. The event $\{A \text{ and } B\}$ consists of outcomes common to A and B.

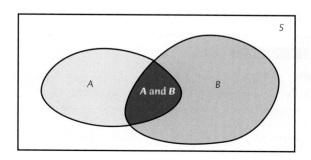

$P(A \text{ and } B)$ that both events occur if we know the individual probabilities $P(A)$ and $P(B)$?

EXAMPLE 12.1 **Can you taste PTC?**

That molecule in the diagram is PTC, a substance with an unusual property: 70% of people find that it has a bitter taste and the other 30% can't taste it at all. The difference is genetic, depending on a single gene. Ask two people chosen at random to taste PTC. We are interested in the events

$$A = \{\text{first person can taste PTC}\}$$
$$B = \{\text{second person can taste PTC}\}$$

We know that $P(A) = 0.7$ and $P(B) = 0.7$. What is the probability $P(A \text{ and } B)$ that both can taste PTC?

We can think our way to the answer. The first person chosen can taste PTC in 70% of all samples and then the second person can taste it in 70% of those samples. We will get two tasters in 70% of 70% of all samples. That's $P(A \text{ and } B) = 0.7 \times 0.7 = 0.49$. ∎

The argument in Example 12.1 works because knowing that the first person can taste PTC tells us nothing about the second person. The probability is still 0.7 that the second person can taste PTC whether or not the first person can. We say that the events "first person can taste PTC" and "second person can taste PTC" are **independent.** Now we have another rule of probability.

independent events

> **MULTIPLICATION RULE FOR INDEPENDENT EVENTS**
>
> Two events A and B are **independent** if knowing that one occurs does not change the probability that the other occurs. If A and B are independent,
>
> $$P(A \text{ and } B) = P(A)P(B)$$

EXAMPLE 12.2 **Independent or not?**

To use this multiplication rule, we must decide whether events are independent. In Example 12.1, we think that the ability of one randomly chosen person to taste PTC tells us nothing about whether or not a second person, also randomly chosen, can taste PTC. That's independence. But if the two people are members of the same family, the fact that ability to taste PTC is inherited warns us that they are not independent.

Independence is clearest in artificial settings such as games of chance. Because a coin has no memory and most coin tossers cannot influence the fall of the coin, it is safe to assume that successive coin tosses are independent. On the other hand, the colors of successive cards dealt from the same deck are not independent. A standard 52-card deck contains 26 red and 26 black cards. For the first card dealt from a shuffled deck, the probability of a red card is $26/52 = 0.50$. Once we see that the first card is red, we know that there are only 25 reds among the remaining 51 cards. The probability that the second card is red is therefore only $25/51 = 0.49$. Knowing the outcome of the first deal changes the probabilities for the second. ∎

Condemned by independence

Assuming independence when it isn't true can lead to disaster. Several mothers in England were convicted of murder simply because two of their children had died in their cribs with no visible cause. An "expert witness" for the prosecution said that the probability of an unexplained crib death in a nonsmoking middle-class family is 1/8500. He then multiplied 1/8500 by 1/8500 to claim that there is only a 1 in 73 million chance that two children in the same family could have died naturally. This is nonsense: it assumes that crib deaths are independent, and data suggest that they are not. Some common genetic or environmental cause, not murder, probably explains the deaths.

The multiplication rule extends to collections of more than two events, provided that all are independent. Independence of events A, B, and C means that no information about any one or any two can change the probability of the remaining events. Independence is often assumed in setting up a probability model when the events we are describing seem to have no connection.

EXAMPLE 12.3 Surviving?

During World War II, the British found that the probability that a bomber is lost through enemy action on a mission over occupied Europe was 0.05. The probability that the bomber returns safely from a mission was therefore 0.95. It is reasonable to assume that missions are independent. Take A_i to be the event that a bomber survives its ith mission. The probability of surviving 2 missions is

$$P(A_1 \text{ and } A_2) = P(A_1)P(A_2)$$
$$= (0.95)(0.95) = 0.9025$$

The multiplication rule also applies to more than two independent events, so the probability of surviving 3 missions is

$$P(A_1 \text{ and } A_2 \text{ and } A_3) = P(A_1)P(A_2)P(A_3)$$
$$= (0.95)(0.95)(0.95) = 0.8574$$

The probability of surviving 20 missions is only

$$P(A_1 \text{ and } A_2 \text{ and } \ldots \text{ and } A_{20}) = P(A_1)P(A_2)\cdots P(A_{20})$$
$$= (0.95)(0.95)\cdots(0.95)$$
$$= (0.95)^{20} = 0.3585$$

The tour of duty for an airman was 30 missions. ■

If two events A and B are independent, the event that A does not occur is also independent of B, and so on. For example, choose two people at random and ask if they can taste PTC. Because 70% can taste PTC and 30% cannot, the probability that the first person is a taster and the second is not is $(0.7)(0.3) = 0.21$.

EXAMPLE 12.4 Rapid HIV testing

STATE: Many people who come to clinics to be tested for HIV, the virus that causes AIDS, don't come back to learn the test results. Clinics now use "rapid HIV tests" that give a result while the client waits. In a clinic in Malawi, for example, use of rapid tests increased the percent of clients who learned their test results from 69% to 99.7%.

The trade-off for fast results is that rapid tests are less accurate than slower laboratory tests. Applied to people who have no HIV antibodies, one rapid test has probability about 0.004 of producing a false positive (that is, of falsely indicating that antibodies are present).[1] If a clinic tests 200 people who are free of HIV antibodies, what is the chance that at least one false positive will occur?

PLAN: It is reasonable to assume that the test results for different individuals are independent. We have 200 independent events, each with probability 0.004. What is the probability that at least one of these events occurs?

SOLVE: "At least one" combines many outcomes. It is much easier to use the fact that

$$P(\text{at least one positive}) = 1 - P(\text{no positives})$$

and find $P(\text{no positives})$ first.

The probability of a negative result for any one person is $1 - 0.004 = 0.996$. To find the probability that all 200 people tested have negative results, use the multiplication rule:

$$P(\text{no positives}) = P(\text{all 200 negative})$$

$$= (0.996)(0.996) \cdots (0.996)$$

$$= 0.996^{200} = 0.4486$$

The probability we want is therefore

$$P(\text{at least one positive}) = 1 - 0.4486 = 0.5514$$

CONCLUDE: The probability is greater than 1/2 that at least one of the 200 people will test positive for HIV, even though no one has the virus. ■

The multiplication rule $P(A \text{ and } B) = P(A)P(B)$ holds if A and B are independent but not otherwise. The addition rule $P(A \text{ or } B) = P(A) + P(B)$ holds if A and B are disjoint but not otherwise. Resist the temptation to use these simple rules when the circumstances that justify them are not present. *You must also be careful not to confuse disjointness and independence.* If A and B are disjoint, then the fact that A occurs tells us that B cannot occur—look again at Figure 12.1. So disjoint events are not independent. Unlike disjointness, we cannot picture independence in a Venn diagram, because it involves the probabilities of the events rather than just the outcomes that make up the events.

APPLY YOUR KNOWLEDGE

12.1 Older college students. Government data show that 8% of adults are full-time college students and that 30% of adults are age 55 or older. Nonetheless, we can't conclude that, because $(0.08)(0.30) = 0.024$, about 2.4% of adults are college students 55 or older. Why not?

12.2 Common names. The Census Bureau says that the 10 most common names in the United States are (in order) Smith, Johnson, Williams, Brown, Jones, Miller, Davis, Garcia, Rodriguez, and Wilson. These names account for 9.6% of all U.S. residents. Out of curiosity, you look at the authors of the textbooks for your current courses. There are 9 authors in all. Would you be surprised if none of the names of these authors were among the 10 most common? (Assume that authors' names are independent and follow the same probability distribution as the names of all residents.)

12.3 Lost Internet sites. Internet sites often vanish or move, so that references to them can't be followed. In fact, 13% of Internet sites referenced in major scientific journals are lost within two years after publication.[2] If a paper contains seven Internet references, what is the probability that all seven are still good two years later? What specific assumptions did you make in order to calculate this probability?

The general addition rule

We know that if A and B are disjoint events, then $P(A \text{ or } B) = P(A) + P(B)$. If events A and B are *not* disjoint, they can occur together. The probability that one or the other occurs is then *less* than the sum of their probabilities. As Figure 12.3 illustrates, outcomes common to both are counted twice when we add probabilities, so we must subtract this probability once. Here is the addition rule for any two events, disjoint or not.

FIGURE 12.3

The general addition rule: for any events A and B, $P(A \text{ or } B) = P(A) + P(B) - P(A \text{ and } B)$.

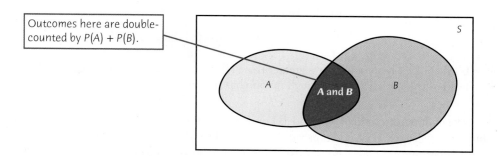

Outcomes here are double-counted by $P(A) + P(B)$.

ADDITION RULE FOR ANY TWO EVENTS

For any two events A and B,

$$P(A \text{ or } B) = P(A) + P(B) - P(A \text{ and } B)$$

If A and B are disjoint, the event $\{A \text{ and } B\}$ that both occur contains no outcomes and therefore has probability 0. So the general addition rule includes Rule 3, the addition rule for disjoint events.

EXAMPLE 12.5 **Motor vehicle sales**

Motor vehicles sold in the United States (ignoring heavy trucks) are classified as either cars or light trucks and as either domestic or imported. "Light trucks" include SUVs and minivans. "Domestic" means made in Canada, Mexico, or the United States, so that a Toyota made in Canada counts as domestic.

In a recent year, 77% of the new vehicles sold to individuals were domestic, 52% were light trucks, and 44% were domestic light trucks.[3] Choose a vehicle sale at random. Then

$$P(\text{domestic or light truck}) = P(\text{domestic}) + P(\text{light truck}) - P(\text{domestic light truck})$$
$$= 0.77 + 0.52 - 0.44 = 0.85$$

Michael Newman/PhotoEdit

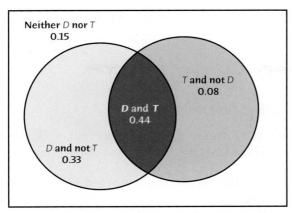

D = vehicle is domestic   T = vehicle is a light truck

FIGURE 12.4

Venn diagram and probabilities for motor vehicle sales, for Example 12.5.

That is, 85% of vehicles sold were either domestic or light trucks. A vehicle is an imported car if it is *neither* domestic nor a light truck. So

$$P(\text{imported car}) = 1 - 0.85 = 0.15 \blacksquare$$

Venn diagrams clarify events and their probabilities because you can just think of adding and subtracting areas. Figure 12.4 shows all the events formed from "domestic" and "truck" in Example 12.5. The four probabilities that appear in the figure add to 1 because they refer to four disjoint events that make up the entire sample space. All of these probabilities come from the information in Example 12.5. For example, the probability that a randomly chosen vehicle sale is a domestic car ("*D* and not *T*" in the figure) is

$$P(\text{domestic car}) = P(\text{domestic}) - P(\text{domestic light truck})$$
$$= 0.77 - 0.44 = 0.33$$

APPLY YOUR KNOWLEDGE

12.4 College degrees. Of all college degrees awarded in the United States, 50% are bachelor's degrees, 59% are earned by women, and 29% are bachelor's degrees earned by women. Make a Venn diagram and use it to answer these questions.

(a) What percent of all degrees are earned by men?

(b) What percent of all degrees are bachelor's degrees earned by men?

12.5 Distance learning. A study of the students taking distance learning courses at a university finds that they are mostly older students not living in the university town. Choose a distance learning student at random. Let *A* be the event that the student is 25 years old or older and *B* the event that the student is local. The study finds that $P(A) = 0.7$, $P(B) = 0.25$, and $P(A \text{ and } B) = 0.05$.

(a) Make a Venn diagram similar to Figure 12.4 showing the events {*A* and *B*}, {*A* and not *B*}, {*B* and not *A*}, and {neither *A* nor *B*}.

Barry Austin Photography/Getty

(b)   Describe each of these events in words.

(c)   Find the probabilities of all four events and add the probabilities to your Venn diagram.

Conditional probability

The probability we assign to an event can change if we know that some other event has occurred. This idea is the key to many applications of probability.

> **EXAMPLE 12.6** **Trucks among imported motor vehicles**
>
> Figure 12.4, based on the information in Example 12.5, gives the following probabilities for a randomly chosen light motor vehicle sold at retail in the United States:
>
> | | Domestic | Imported | Total |
> |---|---|---|---|
> | Light truck | 0.44 | 0.08 | 0.52 |
> | Car | 0.33 | 0.15 | 0.48 |
> | Total | 0.77 | 0.23 | 1 |
>
> The four probabilities in the body of the table add to 1 because they describe all vehicles sold. We obtain the "Total" row and column from these probabilities by the addition rule. For example, the probability that a randomly chosen vehicle is a light truck is
>
> $$P(\text{truck}) = P(\text{truck and domestic}) + P(\text{truck and imported})$$
>
> $$= 0.44 + 0.08 = 0.52$$
>
> Now we are told that the vehicle chosen is imported. That is, it is one of the 23% in the "Imported" column of the table. The probability that a vehicle is a light truck, *given the information that it is imported*, is the proportion of trucks in the "Imported" column,
>
> $$P(\text{truck} \mid \text{imported}) = \frac{0.08}{0.23} = 0.35$$
>
> *conditional probability*
>
> This is a **conditional probability.** You can read the bar | as "given the information that." ■

Although 52% of all vehicles sold are trucks, only 35% of imported vehicles are trucks. It's common sense that knowing that one event (the vehicle is imported) occurs often changes the probability of another event (the vehicle is a truck). The example also shows how we should define conditional probability. The idea of a conditional probability $P(B \mid A)$ of one event B given that another event A occurs is the proportion *of all occurrences of* A for which B also occurs.

> **CONDITIONAL PROBABILITY**
>
> When $P(A) > 0$, the **conditional probability** of B given A is
>
> $$P(B \mid A) = \frac{P(A \text{ and } B)}{P(A)}$$

The conditional probability $P(B \mid A)$ makes no sense if the event A can never occur, so we require that $P(A) > 0$ whenever we talk about $P(B \mid A)$. *Be sure to keep in mind the distinct roles of the events A and B in $P(B \mid A)$.* Event A represents the information we are given, and B is the event whose probability we are calculating. Here is an example that emphasizes this distinction.

EXAMPLE 12.7 Imports among trucks

What is the conditional probability that a randomly chosen vehicle is imported, *given the information that it is a truck?* Using the definition of conditional probability,

$$P(\text{imported} \mid \text{truck}) = \frac{P(\text{imported and truck})}{P(\text{truck})}$$

$$= \frac{0.08}{0.52} = 0.15$$

Only 15% of trucks sold are imports. ■

Be careful not to confuse the two different conditional probabilities

$$P(\text{truck} \mid \text{imported}) = 0.35$$

$$P(\text{imported} \mid \text{truck}) = 0.15$$

The first answers the question "What proportion of imports are trucks?" The second answers "What proportion of trucks are imports?"

APPLY YOUR KNOWLEDGE

12.6 College degrees. In the setting of Exercise 12.4, what is the conditional probability that a degree is earned by a woman, given that it is a bachelor's degree?

12.7 Distance learning. In the setting of Exercise 12.5, what is the conditional probability that a student is local, given that he or she is less than 25 years old?

12.8 Computer games. Here is the distribution of computer games sold by type of game:[4]

Game type	Probability
Strategy	0.354
Role playing	0.139
Family entertainment	0.127
Shooters	0.109
Children's	0.057
Other	0.214

What is the conditional probability that a computer game is a role-playing game, given that it is not a strategy game?

The general multiplication rule

The definition of conditional probability reminds us that in principle all probabilities, including conditional probabilities, can be found from the assignment of probabilities to events that describes a random phenomenon. More often, however, conditional probabilities are part of the information given to us in a probability model. The definition of conditional probability then turns into a rule for finding the probability that both of two events occur.

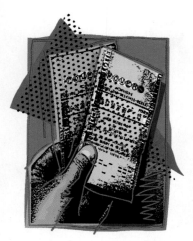

Winning the lottery twice

In 1986, Evelyn Marie Adams won the New Jersey lottery for the second time, adding $1.5 million to her previous $3.9 million jackpot. The *New York Times* claimed that the odds of one person winning the big prize twice were 1 in 17 trillion. Nonsense, said two statisticians in a letter to the *Times*. The chance that Evelyn Marie Adams would win twice is indeed tiny, but it is almost certain that *someone* among the millions of lottery players would win two jackpots. Sure enough, Robert Humphries won his second Pennsylvania lottery jackpot ($6.8 million total) in 1988.

MULTIPLICATION RULE FOR ANY TWO EVENTS

The probability that both of two events A and B happen together can be found by

$$P(A \text{ and } B) = P(A)P(B \mid A)$$

Here $P(B \mid A)$ is the conditional probability that B occurs, given the information that A occurs.

In words, this rule says that for both of two events to occur, first one must occur and then, given that the first event has occurred, the second must occur. This is just common sense expressed in the language of probability, as the following example illustrates.

EXAMPLE 12.8　Teens with online profiles

The Pew Internet and American Life Project finds that 93% of teenagers (ages 12 to 17) use the Internet, and that 55% of online teens have posted a profile on a social networking site.[5] What percent of teens are online *and* have posted a profile?

Use the multiplication rule:

$$P(\text{online}) = 0.93$$

$$P(\text{profile} \mid \text{online}) = 0.55$$

$$P(\text{online and have profile}) = P(\text{online}) \times P(\text{profile} \mid \text{online})$$

$$= (0.93)(0.55) = 0.5115$$

That is, about 51% of all teens use the Internet and have a profile on a social networking site.

You should think your way through this: if 93% of teens are online and 55% *of these* have posted a profile, then 55% of 93% are both online and have a profile. ■

We can extend the multiplication rule to find the probability that all of several events occur. The key is to condition each event on the occurrence of *all* of the preceding events. So for any three events A, B, and C,

$$P(A \text{ and } B \text{ and } C) = P(A)P(B \mid A)P(C \mid \text{both } A \text{ and } B)$$

Here is an example of the extended multiplication rule.

EXAMPLE 12.9 **Fundraising by telephone**

STATE: A charity raises funds by calling a list of prospective donors to ask for pledges. It is able to talk with 40% of the names on its list. Of those the charity reaches, 30% make a pledge. But only half of those who pledge actually make a contribution. What percent of the donor list contributes?

PLAN: Express the information we are given in terms of events and their probabilities:

If A = {the charity reaches a prospect}　　then　$P(A) = 0.4$
If B = {the prospect makes a pledge}　　then　$P(B \mid A) = 0.3$
If C = {the prospect makes a contribution}　then　$P(C \mid \text{both } A \text{ and } B) = 0.5$

We want to find $P(A \text{ and } B \text{ and } C)$.

SOLVE: Use the multiplication rule:

$$P(A \text{ and } B \text{ and } C) = P(A)P(B \mid A)P(C \mid \text{both } A \text{ and } B)$$

$$= 0.4 \times 0.3 \times 0.5 = 0.06$$

CONCLUDE: Only 6% of the prospective donors make a contribution. ■

As Example 12.9 illustrates, formulating a problem in the language of probability is often the key to success in applying probability ideas.

APPLY YOUR KNOWLEDGE

12.9 At the gym. Suppose that 10% of adults belong to health clubs, and 40% of these health club members go to the club at least twice a week. What percent of all adults go to a health club at least twice a week? Write the information given in terms of probabilities and use the general multiplication rule.

12.10 Teens online. We saw in Example 12.8 that 93% of teenagers are online and that 55% of online teens have posted a profile on a social networking site. Of online teens with a profile, 76% have placed comments on a friend's blog. What percent of all teens are online, have a profile, and comment on a friend's blog? Define events and probabilities and follow the pattern of Example 12.9.

12.11 The probability of a flush. A poker player holds a flush when all 5 cards in the hand belong to the same suit (clubs, diamonds, hearts, or spades). We will find the probability of a flush when 5 cards are dealt. Remember that a deck contains 52 cards, 13 of each suit, and that when the deck is well shuffled, each card dealt is equally likely to be any of those that remain in the deck.

(a) Concentrate on spades. What is the probability that the first card dealt is a spade? What is the conditional probability that the second card is a spade, given that the first is a spade? (*Hint:* How many cards remain? How many of these are spades?)

(b) Continue to count the remaining cards to find the conditional probabilities of a spade on the third, the fourth, and the fifth card, given in each case that all previous cards are spades.

Cut and Deal Ltd./Alamy

(c)   The probability of being dealt 5 spades is the product of the 5 probabilities you have found. Why? What is this probability?

(d)   The probability of being dealt 5 hearts or 5 diamonds or 5 clubs is the same as the probability of being dealt 5 spades. What is the probability of being dealt a flush?

Independence again

The conditional probability $P(B \mid A)$ is generally not equal to the unconditional probability $P(B)$. That's because the occurrence of event A generally gives us some additional information about whether or not event B occurs. If knowing that A occurs gives no additional information about B, then A and B are independent events. The precise definition of independence is expressed in terms of conditional probability.

INDEPENDENT EVENTS

Two events A and B that both have positive probability are **independent** if

$$P(B \mid A) = P(B)$$

We now see that the multiplication rule for independent events, $P(A \text{ and } B) = P(A)P(B)$, is a special case of the general multiplication rule, $P(A \text{ and } B) = P(A)P(B \mid A)$, just as the addition rule for disjoint events is a special case of the general addition rule. We rarely use the definition of independence because most often independence is part of the information given to us in a probability model.

APPLY YOUR KNOWLEDGE

12.12 Independent? The Clemson University Fact Book for 2007 shows that 123 of the university's 338 assistant professors were women, along with 76 of the 263 associate professors and 73 of the 375 full professors.

(a)   What is the probability that a randomly chosen Clemson professor is a woman?

(b)   What is the conditional probability that a randomly chosen professor is a woman, given that the person chosen is a full professor?

(c)   Are the rank and gender of Clemson professors independent? How do you know?

Tree diagrams

Probability models often have several stages, with probabilities at each stage conditional on the outcomes of earlier states. These models require us to combine several of the basic rules into a more elaborate calculation. Here is an example.

EXAMPLE 12.10 **Who visits YouTube?**

STATE: Video sharing sites, led by YouTube, are popular destinations on the Internet. Let's look only at adult Internet users, age 18 and over. About 27% of adult Internet users are 18 to 29 years old, another 45% are 30 to 49 years old, and the remaining 28% are 50 and over. The Pew Internet and American Life Project finds that 70% of Internet users aged 18 to 29 have visited a video sharing site, along with 51% of those aged 30 to 49 and 26% of those 50 or older. What percent of all adult Internet users visit video sharing sites?

PLAN: To use the tools of probability, restate all of these percents as probabilities. If we choose an online adult at random,

$$P(\text{age 18 to 29}) = 0.27$$

$$P(\text{age 30 to 49}) = 0.45$$

$$P(\text{age 50 and older}) = 0.28$$

These three probabilities add to 1 because all adult Internet users are in one of the three age groups. The percents of each group who visit video sharing sites are *conditional* probabilities:

$$P(\text{video yes} \mid \text{age 18 to 29}) = 0.70$$

$$P(\text{video yes} \mid \text{age 30 to 49}) = 0.51$$

$$P(\text{video yes} \mid \text{age 50 and older}) = 0.26$$

We want to find the unconditional probability $P(\text{video yes})$.

SOLVE: The **tree diagram** in Figure 12.5 organizes this information. Each segment in the tree is one stage of the problem. Each complete branch shows a path through the two stages. The probability written on each segment is the conditional probability of an

Jim Craigmyle/CORBIS

tree diagram

FIGURE 12.5

Tree diagram for use of the Internet and video sharing sites such as YouTube, for Example 12.10. The three disjoint paths to the outcome that an adult Internet user visits video sharing sites are colored red.

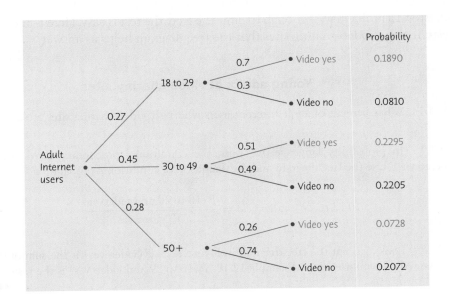

Internet user following that segment, given that he or she has reached the node from which it branches.

Starting at the left, an Internet user falls into one of the three age groups. The probabilities of these groups mark the leftmost segments in the tree. Look at age 18 to 29, the top branch. The two segments going out from the "18 to 29" branch point carry the conditional probabilities

$$P(\text{video yes} \mid \text{age 18 to 29}) = 0.70$$

$$P(\text{video no} \mid \text{age 18 to 29}) = 0.30$$

The full tree shows the probabilities for all three age groups.

Now use the multiplication rule. The probability that a randomly chosen Internet user is an 18- to 29-year-old who visits video sharing sites is

$$P(\text{18 to 29 and video yes}) = P(\text{18 to 29})P(\text{video yes} \mid \text{18 to 29})$$

$$= (0.27)(0.70) = 0.1890$$

This probability appears at the end of the topmost branch. The multiplication rule says that the probability of any complete branch in the tree is the product of the probabilities of the segments in that branch.

There are three disjoint paths to "video yes," one for each of the three age groups. These paths are colored red in Figure 12.5. Because the three paths are disjoint, the probability that an adult Internet user visits video-sharing sites is the sum of their probabilities:

$$P(\text{video yes}) = (0.27)(0.70) + (0.45)(0.51) + (0.28)(0.26)$$

$$= 0.1890 + 0.2295 + 0.0728 = 0.4913$$

CONCLUDE: About 49% of all adult Internet users have visited a video sharing site. ∎

It takes longer to explain a tree diagram than it does to use it. Once you have understood a problem well enough to draw the tree, the rest is easy. Here is another question about video-sharing sites that the tree diagram helps us answer.

EXAMPLE 12.11 **Young adults at video sharing sites**

STATE: What percent of adult Internet users who visit video sharing sites are age 18 to 29?

PLAN: In probability language, we want the conditional probability $P(\text{18 to 29} \mid \text{video yes})$. Use the tree diagram and the definition of conditional probability:

$$P(\text{18 to 29} \mid \text{video yes}) = \frac{P(\text{18 to 29 and video yes})}{P(\text{video yes})}$$

SOLVE: Look again at the tree diagram in Figure 12.5. $P(\text{video yes})$ is the sum of the three red probabilities, as in Example 12.10. $P(\text{18 to 29 and video yes})$ is the result of following just the top branch in the tree diagram. So

$$P(\text{18 to 29} \mid \text{video yes}) = \frac{P(\text{18 to 29 and video yes})}{P(\text{video yes})}$$

$$= \frac{0.1890}{0.4913} = 0.3847$$

CONCLUDE: About 38% of adults who visit video sharing sites are between 18 and 29 years old. Compare this conditional probability with the original information (unconditional) that 27% of adult Internet users are between 18 and 29 years old. Knowing that a person visits video sharing sites increases the probability that he or she is young. ■

Examples 12.10 and 12.11 illustrate a common setting for tree diagrams. Some outcome (such as visiting video sharing sites) has several sources (such as the three age groups). Starting from

- the probability of each source, and

- the conditional probability of the outcome given each source

the tree diagram leads to the overall probability of the outcome. Example 12.10 does this. You can then use the probability of the outcome and the definition of conditional probability to find the conditional probability of one of the sources, given that the outcome occurred. Example 12.11 shows how.

APPLY YOUR KNOWLEDGE

12.13 Peanut and tree nut allergies. About 1% of the American population is allergic to peanuts or tree nuts.[6] Choose 5 individuals at random and let the random variable X be the number in this sample who are allergic to peanuts or tree nuts. The possible values X can take are 0, 1, 2, 3, 4, and 5. Make a five-stage tree diagram of the outcomes (allergic or not allergic) for the 5 individuals and use it to find the probability distribution of X.

12.14 Testing for HIV. Enzyme immunoassay tests are used to screen blood specimens for the presence of antibodies to HIV, the virus that causes AIDS. Antibodies indicate the presence of the virus. The test is quite accurate but is not always correct. Here are approximate probabilities of positive and negative test results when the blood tested does and does not actually contain antibodies to HIV:[7]

	Test Result	
	Positive	**Negative**
Antibodies present	0.9985	0.0015
Antibodies absent	0.0060	0.9940

Suppose that 1% of a large population carries antibodies to HIV in their blood.

Politically correct

In 1950, the Soviet mathematician B. V. Gnedenko (1912–1995) wrote *The Theory of Probability*, a text that was popular around the world. The introduction contains a mystifying paragraph that begins, "We note that the entire development of probability theory shows evidence of how its concepts and ideas were crystallized in a severe struggle between materialistic and idealistic conceptions." It turns out that "materialistic" is jargon for "Marxist-Leninist." It was good for the health of Soviet scientists in the Stalin era to add such statements to their books.

(a)   Draw a tree diagram for selecting a person from this population (outcomes: antibodies present or absent) and testing his or her blood (outcomes: test positive or negative).

(b)   What is the probability that the test is positive for a randomly chosen person from this population?

12.15   Peanut and tree nut allergies. Continue your work from Exercise 12.13. What is the conditional probability that exactly 1 of the people will be allergic to peanuts or tree nuts, given that at least 1 of the 5 people suffers from one of these allergies?

12.16   False HIV positives. Continue your work from Exercise 12.14. What is the probability that a person has the antibody, given that the test is positive? (Your result illustrates a fact that is important when considering proposals for widespread testing for HIV, illegal drugs, or agents of biological warfare: if the condition being tested is uncommon in the population, most positives will be false positives.)

C H A P T E R 1 2   S U M M A R Y

■ Events A and B are **disjoint** if they have no outcomes in common. In that case, $P(A \text{ or } B) = P(A) + P(B)$.

■ The **conditional probability** $P(B \mid A)$ of an event B given an event A is defined by

$$P(B \mid A) = \frac{P(A \text{ and } B)}{P(A)}$$

when $P(A) > 0$. In practice, we most often find conditional probabilities from directly available information rather than from the definition.

■ Events A and B are **independent** if knowing that one event occurs does not change the probability we would assign to the other event; that is, $P(B \mid A) = P(B)$. In that case, $P(A \text{ and } B) = P(A)P(B)$.

■ Any assignment of probability obeys these rules:

Addition rule for disjoint events: If events A, B, C, ... are all disjoint in pairs, then

$P(\text{at least one of these events occurs}) = P(A) + P(B) + P(C) + \cdots$

Multiplication rule for independent events: If events A, B, C, ... are independent, then

$P(\text{all of these events occur}) = P(A)P(B)P(C) \cdots$

General addition rule: For any two events A and B,

$P(A \text{ or } B) = P(A) + P(B) - P(A \text{ and } B)$

General multiplication rule: For any two events A and B,

$P(A \text{ and } B) = P(A)P(B \mid A)$

■ **Tree diagrams** organize probability models that have several stages.

CHECK YOUR SKILLS

12.17 An instant lottery game gives you probability 0.02 of winning on any one play. Plays are independent of each other. If you play 3 times, the probability that you win on none of your plays is about

(a) 0.98.   (b) 0.94.   (c) 0.000008.

12.18 The probability that you win on one or more of your 3 plays of the game in the previous exercise is about

(a) 0.02.   (b) 0.06.   (c) 0.999992.

12.19 An athlete suspected of having used steroids is given two tests that operate independently of each other. Test A has probability 0.9 of being positive if steroids have been used. Test B has probability 0.8 of being positive if steroids have been used. What is the probability that *neither* test is positive if steroids have been used?

(a) 0.72   (b) 0.38   (c) 0.02

Accidents, suicide, and murder are the leading causes of death for young adults. Here are the counts of violent deaths in a recent year among people 20 to 24 years of age:

	Female	Male
Accidents	1818	6457
Homicide	457	2870
Suicide	345	2152

Exercises 12.20 to 12.23 are based on this table.

12.20 Choose a violent death in this age group at random. The probability that the victim was male is about

(a) 0.81.   (b) 0.78.   (c) 0.19.

12.21 The conditional probability that the victim was male, given that the death was accidental, is about

(a) 0.81.   (b) 0.78.   (c) 0.56.

12.22 The conditional probability that the death was accidental, given that the victim was male, is about

(a) 0.81.   (b) 0.78.   (c) 0.56.

12.23 Let A be the event that a victim of violent death was a woman and B the event that the death was a suicide. The proportion of suicides among violent deaths of women is expressed in probability notation as

(a) $P(A \text{ and } B)$.   (b) $P(A \mid B)$.   (c) $P(B \mid A)$.

12.24 Choose an American adult at random. The probability that you choose a woman is 0.52. The probability that the person you choose has never married is 0.25. The probability that you choose a woman who has never married is 0.11. The probability that the person you choose is either a woman or never married (or both) is therefore about

(a) 0.77.   (b) 0.66.   (c) 0.13.

12.25 Of people who died in the United States in recent years, 86% were white, 12% were black, and 2% were Asian. (This ignores a small number of deaths among other races.) Diabetes caused 2.8% of deaths among whites, 4.4% among blacks, and 3.5% among Asians. The probability that a randomly chosen death is a white who died of diabetes is about

(a) 0.107.   (b) 0.030.   (c) 0.024.

12.26 Using the information in the previous exercise, the probability that a randomly chosen death was due to diabetes is about

(a) 0.107.   (b) 0.030.   (c) 0.024.

CHAPTER 12 EXERCISES

12.27 Playing the lottery. New York State's "Quick Draw" lottery moves right along. Players choose between one and ten numbers from the range 1 to 80; 20 winning numbers are displayed on a screen every four minutes. If you choose just one number, your probability of winning is 20/80, or 0.25. Lester plays one number 8 times as he sits in a bar. What is the probability that all 8 bets lose?

12.28 Universal blood donors. People with type O-negative blood are universal donors. That is, any patient can receive a transfusion of O-negative blood. Only 7.2% of the American population have O-negative blood. If 10 people appear at random to give blood, what is the probability that at least 1 of them is a universal donor?

12.29 Playing the slots. Slot machines are now video games, with outcomes determined by random number generators. In the old days, slot machines were like this: you pull the lever to spin three wheels; each wheel has 20 symbols, all equally likely to show when the wheel stops spinning; the three wheels are independent of each other. Suppose that the middle wheel has 9 cherries among its 20 symbols, and the left and right wheels have 1 cherry each.

Peter Dazeley/Getty

(a) You win the jackpot if all three wheels show cherries. What is the probability of winning the jackpot?

(b) There are three ways that the three wheels can show two cherries and one symbol other than a cherry. Find the probability of each of these ways.

(c) What is the probability that the wheels stop with exactly two cherries showing among them?

12.30 A random walk on Wall Street? The "random walk" theory of stock prices holds that price movements in disjoint time periods are independent of each other. Suppose that we record only whether the price is up or down each year, and that the probability that our portfolio rises in price in any one year is 0.65. (This probability is approximately correct for a portfolio containing equal dollar amounts of all common stocks listed on the New York Stock Exchange.)

(a) What is the probability that our portfolio goes up for three consecutive years?

(b) What is the probability that the portfolio's value moves in the same direction (either up or down) for three consecutive years?

12.31 Getting into college. Ramon has applied to both Princeton and Stanford. He thinks the probability that Princeton will admit him is 0.4, the probability that Stanford will admit him is 0.5, and the probability that both will admit him is 0.2. Make a Venn diagram. Then answer these questions.

(a)   What is the probability that neither university admits Ramon?

(b)   What is the probability that he gets into Stanford but not Princeton?

(c)   Are admission to Princeton and admission to Stanford independent events?

12.32 Tendon surgery. You have torn a tendon and are facing surgery to repair it. The surgeon explains the risks to you: infection occurs in 3% of such operations, the repair fails in 14%, and both infection and failure occur together in 1%. What percent of these operations succeed and are free from infection? Follow the four-step process in your answer.

12.33 Screening job applicants. A company retains a psychologist to assess whether job applicants are suited for assembly-line work. The psychologist classifies applicants as one of A (well suited), B (marginal), or C (not suited). The company is concerned about the event D that an employee leaves the company within a year of being hired. Data on all people hired in the past five years give these probabilities:

$$P(A) = 0.4 \qquad P(B) = 0.3 \qquad P(C) = 0.3$$
$$P(A \text{ and } D) = 0.1 \qquad P(B \text{ and } D) = 0.1 \qquad P(C \text{ and } D) = 0.2$$

Sketch a Venn diagram of the events A, B, C, and D and mark on your diagram the probabilities of all combinations of psychological assessment and leaving (or not) within a year. What is $P(D)$, the probability that an employee leaves within a year?

12.34 Foreign-language study. Choose a student in grades 9 to 12 at random and ask if he or she is studying a language other than English. Here is the distribution of results:

Language	Spanish	French	German	All others	None
Probability	0.26	0.09	0.03	0.03	0.59

What is the conditional probability that a student is studying Spanish, given that he or she is studying some language other than English?

12.35 Income tax returns. Here is the distribution of the adjusted gross income (in thousands of dollars) reported on individual federal income tax returns in 2005:

Income	<25	25–49	50–99	100–499	≥500
Probability	0.431	0.248	0.215	0.100	0.006

(a)   What is the probability that a randomly chosen return shows an adjusted gross income of $50,000 or more?

(b)   Given that a return shows an income of at least $50,000, what is the conditional probability that the income is at least $100,000?

Mehmetali Uslu/Stock.xchng

12.36 Mike's pizza. You work at Mike's pizza shop. You have the following information about the 7 pizzas in the oven: 3 of the 7 have thick crust, and of these one has only sausage and 2 have only mushrooms; the remaining 4 pizzas have regular crust, and of these 2 have only sausage and 2 have only mushrooms. Choose a pizza at random from the oven.

(a) Are the events {getting a thick crust pizza} and {getting a pizza with mushrooms} independent? Explain.

(b) You add an eighth pizza to the oven. This pizza has thick crust with only cheese. Now are the events {getting a thick crust pizza} and {getting a pizza with mushrooms} independent? Explain.

12.37 Geometric probability. Choose a point at random in the square with sides $0 \le x \le 1$ and $0 \le y \le 1$. This means that the probability that the point falls in any region within the square is equal to the area of that region. Let X be the x coordinate and Y the y coordinate of the point chosen. Find the conditional probability $P(Y < 1/2 \mid Y > X)$. (*Hint:* Draw a diagram of the square and the events $Y < 1/2$ and $Y > X$.)

12.38 A probability teaser. Suppose (as is roughly correct) that each child born is equally likely to be a boy or a girl and that the sexes of successive children are independent. If we let BG mean that the older child is a boy and the younger child is a girl, then each of the combinations BB, BG, GB, GG has probability 0.25. Ashley and Brianna each have two children.

(a) You know that at least one of Ashley's children is a boy. What is the conditional probability that she has two boys?

(b) You know that Brianna's older child is a boy. What is the conditional probability that she has two boys?

12.39 College degrees. A striking trend in higher education is that more women than men reach each level of attainment. Here are the counts (in thousands) of earned degrees in the United States in the 2010–2011 academic year, classified by level and by the sex of the degree recipient:[8]

	Bachelor's	Master's	Professional	Doctorate	Total
Female	986	411	52	32	1481
Male	693	260	45	27	1025
Total	1679	671	97	59	2506

(a) If you choose a degree recipient at random, what is the probability that the person you choose is a woman?

(b) What is the conditional probability that you choose a woman, given that the person chosen received a doctorate?

(c) Are the events "choose a woman" and "choose a doctoral degree recipient" independent? How do you know?

12.40 College degrees. Exercise 12.39 gives the counts (in thousands) of earned degrees in the United States in the 2010–2011 academic year. Use these data to answer the following questions.

(a) What is the probability that a randomly chosen degree recipient is a man?

(b) What is the conditional probability that the person chosen received a bachelor's degree, given that he is a man?

(c) Use the multiplication rule to find the probability of choosing a male bachelor's degree recipient. Check your result by finding this probability directly from the table of counts.

12.41 Deer and pine seedlings. As suburban gardeners know, deer will eat almost anything green. In a study of pine seedlings at an environmental center in Ohio, researchers noted how deer damage varied with how much of the seedling was covered by thorny undergrowth:[9]

| | Deer Damage | |
Thorny Cover	Yes	No
None	60	151
<1/3	76	158
1/3 to 2/3	44	177
>2/3	29	176

Peter Skinner/Photo Researchers

(a) What is the probability that a randomly selected seedling was damaged by deer?

(b) What are the conditional probabilities that a randomly selected seedling was damaged, given each level of cover?

(c) Does knowing about the amount of thorny cover on a seedling change the probability of deer damage? If so, cover and damage are not independent.

12.42 Deer and pine seedlings. In the setting of Exercise 12.41, what percent of the trees that were damaged by deer were less than 1/3 covered by thorny plants?

12.43 Deer and pine seedlings. In the setting of Exercise 12.41, what percent of the trees that were not damaged by deer were more than 2/3 covered by thorny plants?

Julie is graduating from college. She has studied biology, chemistry, and computing and hopes to use her science background in crime investigation. Late one night she thinks about some jobs for which she has applied. Let A, B, and C be the events that Julie is offered a job by

   A = the Connecticut Office of the Chief Medical Examiner
   B = the New Jersey Division of Criminal Justice
   C = the federal Disaster Mortuary Operations Response Team

Julie writes down her personal probabilities for being offered these jobs:

$P(A) = 0.6$ $P(B) = 0.4$ $P(C) = 0.2$
$P(A \text{ and } B) = 0.1$ $P(A \text{ and } C) = 0.05$ $P(B \text{ and } C) = 0.05$
$P(A \text{ and } B \text{ and } C) = 0$

Make a Venn diagram of the events A, B, and C. As in Figure 12.4, mark the probabilities of every intersection involving these events. Use this diagram for Exercises 12.44 to 12.46.

12.44 Will Julie get a job offer? What is the probability that Julie is offered at least one of the three jobs?

12.45 Will Julie get just these offers? What is the probability that Julie is offered both the Connecticut and New Jersey jobs, but not the federal job?

12.46 Julie's conditional probabilities. If Julie is offered the federal job, what is the conditional probability that she is also offered the New Jersey job? If Julie is offered the New Jersey job, what is the conditional probability that she is also offered the federal job?

12.47 The geometric distributions. You are tossing a pair of balanced dice in a board game. Tosses are independent. You land in a danger zone that requires you to roll doubles (both faces show the same number of spots) before you are allowed to play again. How long will you wait to play again?

(a) What is the probability of rolling doubles on a single toss of the dice? (If you need review, the possible outcomes appear in Figure 10.2 (page 267). All 36 outcomes are equally likely.)

(b) What is the probability that you do not roll doubles on the first toss, but you do on the second toss?

(c) What is the probability that the first two tosses are not doubles and the third toss is doubles? This is the probability that the first doubles occurs on the third toss.

(d) Now you see the pattern. What is the probability that the first doubles occurs on the fourth toss? On the fifth toss? Give the general result: what is the probability that the first doubles occurs on the kth toss?

(*Comment:* The distribution of the number of trials to the first success is called a *geometric distribution.* In this problem you have found geometric distribution probabilities when the probability of a success on each trial is 1/6. The same idea works for any probability of success.)

12.48 Winning at tennis. A player serving in tennis has two chances to get a serve into play. If the first serve is out, the player serves again. If the second serve is also out, the player loses the point. Here are probabilities based on four years of the Wimbledon Championship:[10]

$$P(\text{1st serve in}) = 0.59$$

$$P(\text{win point} \mid \text{1st serve in}) = 0.73$$

$$P(\text{2nd serve in} \mid \text{1st serve out}) = 0.86$$

$$P(\text{win point} \mid \text{1st serve out and 2nd serve in}) = 0.59$$

Make a tree diagram for the results of the two serves and the outcome (win or lose) of the point. (The branches in your tree have different numbers of stages depending on the outcome of the first serve.) What is the probability that the serving player wins the point?

12.49 Urban voters. The voters in a large city are 40% white, 40% black, and 20% Hispanic. (Hispanics may be of any race in official statistics, but here we are speaking of political blocks.) A black mayoral candidate anticipates attracting 30% of the white vote, 90% of the black vote, and 50% of the Hispanic vote. Draw a tree diagram with probabilities for the race (white, black, or Hispanic) and vote (for or against the candidate) of a randomly chosen voter. What percent of the overall vote does the candidate expect to get? Use the four-step process to guide your work.

12.50 Winning at tennis, continued. Based on your work in Exercise 12.48, in what percent of points won by the server was the first serve in? (Write this as a conditional probability and use the definition of conditional probability.)

12.51 Where do the votes come from? In the election described in Exercise 12.49, what percent of the candidate's votes come from black voters? (Write this as a conditional probability and use the definition of conditional probability.)

12.52 Lactose intolerance. Lactose intolerance causes difficulty digesting dairy products that contain lactose (milk sugar). It is particularly common among people of African and Asian ancestry. In the United States (ignoring other groups and people who consider themselves to belong to more than one race), 82% of the population is white, 14% is black, and 4% is Asian. Moreover, 15% of whites, 70% of blacks, and 90% of Asians are lactose intolerant.[11]

(a) What percent of the entire population is lactose intolerant?

(b) What percent of people who are lactose intolerant are Asian?

12.53 Fundraising by telephone. Tree diagrams can organize problems having more than two stages. Figure 12.6 shows probabilities for a charity calling potential donors by telephone.[12] Each person called is either a recent donor, a past donor, or a new prospect. At the next stage, the person called either does or does not pledge to contribute, with conditional probabilities that depend on the donor class the person belongs to. Finally, those who make a pledge either do or don't actually make a contribution.

(a) What percent of calls result in a contribution?

(b) What percent of those who contribute are recent donors?

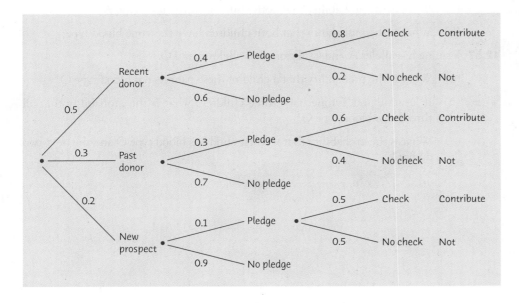

FIGURE 12.6

Tree diagram for fundraising by telephone, for Exercise 12.53. The three stages are the type of prospect called, whether or not the person makes a pledge, and whether or not a person who pledges actually makes a contribution.

Mendelian inheritance. *Some traits of plants and animals depend on inheritance of a single gene. This is called Mendelian inheritance, after Gregor Mendel (1822–1884). Exercises 12.54 to 12.57 are based on the following information about Mendelian inheritance of blood type.*

Each of us has an ABO blood type, which describes whether two characteristics called A and B are present. Every human being has two blood type alleles (gene forms), one inherited from our mother and one from our father. Each of these alleles can be A, B, or O. Which two we inherit determines our blood type. Here is a table that shows what our blood type is for each combination of two alleles:

Alleles inherited	Blood type
A and A	A
A and B	AB
A and O	A
B and B	B
B and O	B
O and O	O

We inherit each of a parent's two alleles with probability 0.5. We inherit independently from our mother and father.

12.54 Rachel and Jonathan both have alleles A and B.

(a) What blood types can their children have?

(b) What is the probability that their next child has each of these blood types?

12.55 Sarah and David both have alleles B and O.

(a) What blood types can their children have?

(b) What is the probability that their next child has each of these blood types?

12.56 Isabel has alleles A and O. Carlos has alleles A and B. They have two children.

(a) What is the probability that both children have blood type A?

(b) What is the probability that both children have the same blood type?

12.57 Jasmine has alleles A and O. Tyrone has alleles B and O.

(a) What is the probability that a child of these parents has blood type O?

(b) If Jasmine and Tyrone have three children, what is the probability that all three have blood type O?

(c) What is the probability that the first child has blood type O and the next two do not?

Binomial Distributions*

A basketball player shoots 5 free throws. How many does she make? A sample survey dials 1200 residential phone numbers at random. How many live people answer the phone? You plant 10 dogwood trees. How many live through the winter? In all these situations, we want a probability model for a *count* of successful outcomes.

The binomial setting and binomial distributions

The distribution of a count depends on how the data are produced. Here is a common situation.

THE BINOMIAL SETTING

1. There are a fixed number n of observations.
2. The n observations are all **independent.** That is, knowing the result of one observation does not change the probabilities we assign to other observations.
3. Each observation falls into one of just two categories, which for convenience we call "success" and "failure."
4. The probability of a success, call it p, is the same for each observation.

---

*This more advanced chapter concerns a special topic in probability. The material is not needed to read the rest of the book.

Think of tossing a coin n times as an example of the binomial setting. Each toss gives either heads or tails. Knowing the outcome of one toss doesn't change the probability of a head on any other toss, so the tosses are independent. If we call heads a success, then p is the probability of a head and remains the same as long as we toss the same coin. For tossing a coin, p is close to 0.5. If we spin the coin on a flat surface rather than toss it, p is not equal to 0.5. The number of heads we count is a discrete random variable X. The distribution of X is called a *binomial distribution*.

BINOMIAL DISTRIBUTION

The count X of successes in the binomial setting has the **binomial distribution** with parameters n and p. The parameter n is the number of observations, and p is the probability of a success on any one observation. The possible values of X are the whole numbers from 0 to n.

The binomial distributions are an important class of discrete probability models. *Pay attention to the binomial setting, because not all counts have binomial distributions.*

EXAMPLE 13.1 Blood types

Genetics says that children receive genes from their parents independently. Each child of a particular pair of parents has probability 0.25 of having type O blood. If these parents have 5 children, the number who have type O blood is the count X of successes in 5 independent observations with probability 0.25 of a success on each observation. So X has the binomial distribution with $n = 5$ and $p = 0.25$. ■

EXAMPLE 13.2 Counting boys

Here is set of genetic examples that require more thought.

Choose two births at random from the last year's births at a large hospital and count the number of boys (0, 1, or 2). The genders of children born to different mothers are surely independent. The probability that a randomly chosen birth in Canada and the United States is a boy is about 0.52. (Why it is not 0.5 is something of a mystery.) So the count of boys has a binomial distribution with $n = 2$ and $p = 0.52$.

Next, observe successive births at a large hospital and let X be the number of births until the first boy is born. Births are independent and each has probability 0.52 of being a boy. Yet X is *not* binomial, because there is no fixed number of observations. "Count observations until the first success" is a different setting than "count the number of successes in a fixed number of observations."

Finally, choose at random a family with exactly two children and count the number of boys. Careful study of such families shows that the count of boys is *not* binomial: the probability of exactly 1 boy is too high.[1] Families are less likely to have a third child if the first two are a boy and a girl, so when we look at families that stopped at two children, "one of each" is more common than if we look at randomly chosen births. The sexes of successive children in two-child families are *not independent,* because the parents' choices interfere with the genetics. ■

Binomial distributions in statistical sampling

The binomial distributions are important in statistics when we wish to make inferences about the proportion p of "successes" in a population. Here is a typical example.

EXAMPLE 13.3 **Choosing an SRS of CDs**

A music distributor inspects an SRS of 10 CDs from a shipment of 10,000 music CDs. Suppose that (unknown to the distributor) 10% of the CDs in the shipment have defective copy-protection schemes that will harm personal computers. Count the number X of bad CDs in the sample.

This is not quite a binomial setting. Removing one CD changes the proportion of bad CDs remaining in the shipment. So the probability that the second CD chosen is bad changes when we know whether the first is good or bad. But removing one CD from a shipment of 10,000 changes the makeup of the remaining 9999 CDs very little. In practice, the distribution of X is very close to the binomial distribution with $n = 10$ and $p = 0.1$. ∎

Was he good or was he lucky?

When a baseball player hits .300, everyone applauds. A .300 hitter gets a hit in 30% of times at bat. Could a .300 year just be luck? Typical major leaguers bat about 500 times a season and hit about .260. A hitter's successive tries seem to be independent, so we have a binomial distribution. From this model, we can calculate or simulate the probability of hitting .300. It is about 0.025. Out of 100 run-of-the-mill major league hitters, two or three each year will bat .300 because they were lucky.

Example 13.3 shows how we can use the binomial distributions in the statistical setting of selecting an SRS. When the population is much larger than the sample, a count of successes in an SRS of size n has approximately the binomial distribution with n equal to the sample size and p equal to the proportion of successes in the population.

SAMPLING DISTRIBUTION OF A COUNT

Choose an SRS of size n from a population with proportion p of successes. When the population is much larger than the sample, the count X of successes in the sample has approximately the binomial distribution with parameters n and p.

APPLY YOUR KNOWLEDGE

In each of Exercises 13.1 to 13.3, X is a count. Does X have a binomial distribution? Give your reasons in each case.

13.1 **Random digit dialing.** When an opinion poll calls residential telephone numbers at random, only 20% of the calls reach a live person. You watch the random dialing machine make 15 calls. X is the number that reach a live person.

13.2 **Random digit dialing.** When an opinion poll calls residential telephone numbers at random, only 20% of the calls reach a live person. You watch the random dialing machine make calls. X is the number of calls until the first live person answers.

13.3   Computer instruction. A student studies binomial distributions using computer-assisted instruction. After the lesson, the computer presents 10 problems. The student solves each problem and enters her answer. The computer gives additional instruction between problems if the answer is wrong. The count X is the number of problems that the student gets right.

13.4   Teens feel the heat. Opinion polls find that 63% of American teens say that their parents put at least some pressure on them to get into a good college.[2] If you take an SRS of 1000 teens, what is the approximate distribution of the number in your sample who say they feel at least some pressure from their parents to get into a good college?

Binomial probabilities

We can find a formula for the probability that a binomial random variable takes any value by adding probabilities for the different ways of getting exactly that many successes in n observations. Here is an example that illustrates the idea.

> **EXAMPLE 13.4 Inheriting blood type**
>
> The blood types of successive children born to the same parents are independent and have fixed probabilities that depend on the genetic makeup of the parents. Each child born to a particular set of parents has probability 0.25 of having blood type O. If these parents have 5 children, what is the probability that exactly 2 of them have type O blood?
>
> The count of children with type O blood is a binomial random variable X with $n = 5$ tries and probability $p = 0.25$ of a success on each try. We want $P(X = 2)$. ■

Because the method doesn't depend on the specific example, let's use "S" for success and "F" for failure for short. Do the work in two steps.

Step 1. Find the probability that a specific 2 of the 5 tries, say the first and the third, give successes. This is the outcome SFSFF. Because tries are independent, the multiplication rule for independent events applies. The probability we want is

$$P(\text{SFSFF}) = P(S)P(F)P(S)P(F)P(F)$$

$$= (0.25)(0.75)(0.25)(0.75)(0.75)$$

$$= (0.25)^2(0.75)^3$$

Step 2. Observe that *any one arrangement* of 2 S's and 3 F's has this same probability. This is true because we multiply together 0.25 twice and 0.75 three times whenever we have 2 S's and 3 F's. The probability that $X = 2$ is the probability of getting 2 S's and 3 F's in any arrangement whatsoever. Here are all the possible arrangements:

$$\text{SSFFF} \quad \text{SFSFF} \quad \text{SFFSF} \quad \text{SFFFS} \quad \text{FSSFF}$$
$$\text{FSFSF} \quad \text{FSFFS} \quad \text{FFSSF} \quad \text{FFSFS} \quad \text{FFFSS}$$

What looks random?

Toss a coin six times and record heads (H) or tails (T) on each toss. Which of these outcomes is more probable: HTHTTH or TTTHHH? Almost everyone says that HTHTTH is more probable, because TTTHHH does not "look random." In fact, both are equally probable. That heads has probability 0.5 says that about half of a very long sequence of tosses will be heads. It doesn't say that heads and tails must come close to alternating in the short run. The coin doesn't know what past outcomes were, and it can't try to create a balanced sequence.

There are 10 of them, all with the same probability. The overall probability of 2 successes is therefore

$$P(X = 2) = 10(0.25)^2(0.75)^3 = 0.2637$$

The pattern of this calculation works for any binomial probability. To use it, we must count the number of arrangements of k successes in n observations. We use the following fact to do the counting without actually listing all the arrangements.

BINOMIAL COEFFICIENT

The number of ways of arranging k successes among n observations is given by the **binomial coefficient**

$$\binom{n}{k} = \frac{n!}{k!\,(n-k)!}$$

for $k = 0, 1, 2, \ldots, n$.

The formula for binomial coefficients uses the **factorial** notation. For any positive whole number n, its factorial $n!$ is

factorial

$$n! = n \times (n-1) \times (n-2) \times \cdots \times 3 \times 2 \times 1$$

In addition, we define $0! = 1$.

The larger of the two factorials in the denominator of a binomial coefficient will cancel much of the $n!$ in the numerator. For example, the binomial coefficient we need for Example 13.4 is

$$\binom{5}{2} = \frac{5!}{2!\,3!}$$

$$= \frac{(5)(4)(3)(2)(1)}{(2)(1) \times (3)(2)(1)}$$

$$= \frac{(5)(4)}{(2)(1)} = \frac{20}{2} = 10$$

The binomial coefficient $\binom{5}{2}$ *is not related to the fraction* $\dfrac{5}{2}$. A helpful way to remember its meaning is to read it as "5 choose 2." Binomial coefficients have many uses, but we are interested in them only as an aid to finding binomial probabilities. The binomial coefficient $\binom{n}{k}$ counts the number of different ways in which k successes can be arranged among n observations. The binomial probability $P(X = k)$

is this count multiplied by the probability of any one specific arrangement of the k successes. Here is the result we seek.

BINOMIAL PROBABILITY

If X has the binomial distribution with n observations and probability p of success on each observation, the possible values of X are 0, 1, 2, ..., n. If k is any one of these values,

$$P(X = k) = \binom{n}{k} p^k (1 - p)^{n-k}$$

EXAMPLE 13.5 Inspecting CDs

The number X of CDs with defective copy protection in Example 13.3 has approximately the binomial distribution with $n = 10$ and $p = 0.1$.

The probability that the sample contains no more than 1 defective CD is

$$P(X \leq 1) = P(X = 1) + P(X = 0)$$

$$= \binom{10}{1}(0.1)^1(0.9)^9 + \binom{10}{0}(0.1)^0(0.9)^{10}$$

$$= \frac{10!}{1!\,9!}(0.1)(0.3874) + \frac{10!}{0!\,10!}(1)(0.3487)$$

$$= (10)(0.1)(0.3874) + (1)(1)(0.3487)$$

$$= 0.3874 + 0.3487 = 0.7361$$

This calculation uses the facts that $0! = 1$ and that $a^0 = 1$ for any number a other than 0. We see that about 74% of all samples will contain no more than 1 bad CD. In fact, 35% of the samples will contain no bad CDs. A sample of size 10 cannot be trusted to alert the distributor to the presence of unacceptable CDs in the shipment. ■

Using technology

The binomial probability formula is awkward to use unless the number of observations n is quite small. You can find tables of binomial probabilities $P(X = k)$ and cumulative probabilities $P(X \leq k)$ for selected values of n and p, but the most efficient way to do binomial calculations is to use technology.

Figure 13.1 shows output for the calculation in Example 13.5 from a graphing calculator, a statistical program, and a spreadsheet program. We asked all three to give cumulative probabilities. The calculator and Minitab have menu entries for binomial cumulative probabilities. Excel has no menu entry, but the worksheet function BINOMDIST is available. All of the outputs agree with the result 0.7361 of Example 13.5.

Texas Instruments Graphing Calculator

FIGURE 13.1

The binomial probability $P(X \le 1)$ for Example 13.5: output from a graphing calculator, a statistical program, and a spreadsheet program.

```
binomcdf(10,0.1,
1)
                .7361
```

Minitab

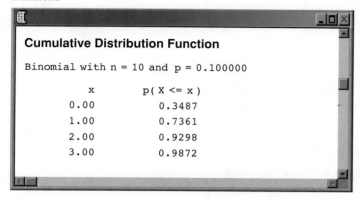

Cumulative Distribution Function

Binomial with n = 10 and p = 0.100000

x	p(X <= x)
0.00	0.3487
1.00	0.7361
2.00	0.9298
3.00	0.9872

Microsoft Excel

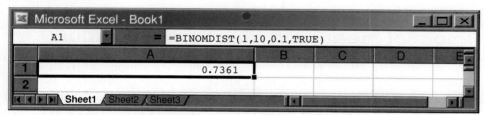

Microsoft Excel - Book1

A1   =   =BINOMDIST(1,10,0.1,TRUE)

A	B	C	D	E
0.7361				

Sheet1 / Sheet2 / Sheet3 /

APPLY YOUR KNOWLEDGE

13.5 **Proofreading.** Typing errors in a text are either nonword errors (as when "the" is typed as "teh") or word errors that result in a real but incorrect word. Spell-checking software will catch nonword errors but not word errors. Human proofreaders catch 70% of word errors. You ask a fellow student to proofread an essay in which you have deliberately made 10 word errors.

(a) If the student matches the usual 70% rate, what is the distribution of the number of errors caught? What is the distribution of the number of errors missed?

(b) Missing 3 or more out of 10 errors seems a poor performance. What is the probability that a proofreader who catches 70% of word errors misses exactly 3 out of 10? If you use software, also find the probability of missing 3 or more out of 10.

13.6 **Random digit dialing.** When an opinion poll calls residential telephone numbers at random, only 20% of the calls reach a live person. You watch the random digit dialing machine make 15 calls.

(a)   What is the probability that exactly 3 calls reach a person?

(b)   What is the probability that at most 3 calls reach a person?

(c)   What is the probability that at least 3 calls reach a person?

(d)   What is the probability that less than 3 calls reach a person?

(e)   What is the probability that more than 3 calls reach a person?

13.7   **Google does binomial.** Point your Web browser to www.google.com. Instead of searching the Web or looking for images, you can request a calculation in the Search box.

(a)   Enter 5 choose 2 and click Search. What does Google return?

(b)   You see that Google calculates the binomial coefficient "5 choose 2." What are the values of the binomial coefficients for "500 choose 2" and "500 choose 100"? We expect that there are more ways to choose 100 than to choose 2, but how many more may be a surprise. That 10^{107} in Google's answer means a 1 followed by 107 zeros.

(c)   Google also does binomial probabilities. Enter (10 choose 1)*0.1* 0.9^9 to find the first binomial probability in Example 13.5. What is Google's answer with all its decimal places?

Binomial mean and standard deviation

If a count X has the binomial distribution based on n observations with probability p of success, what is its mean μ? That is, in very many repetitions of the binomial setting, what will be the average count of successes? We can guess the answer. If a basketball player makes 80% of her free throws, the mean number made in 10 tries should be 80% of 10, or 8. In general, the mean of a binomial distribution should be $\mu = np$. Here are the facts.

BINOMIAL MEAN AND STANDARD DEVIATION

If a count X has the binomial distribution with number of observations n and probability of success p, the **mean** and **standard deviation** of X are

$$\mu = np$$

$$\sigma = \sqrt{np(1 - p)}$$

Remember that these short formulas are good only for binomial distributions. They can't be used for other distributions.

EXAMPLE 13.6 **Inspecting CDs**

Continuing Example 13.5, the count X of bad CDs is binomial with $n = 10$ and $p = 0.1$. The histogram in Figure 13.2 displays this probability distribution. (Because

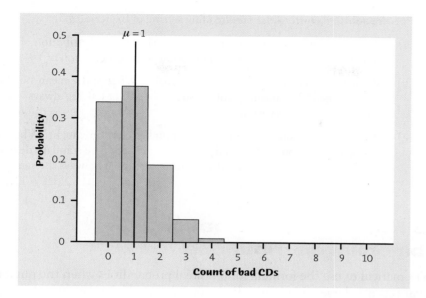

FIGURE 13.2

Probability histogram for the binomial distribution with $n = 10$ and $p = 0.1$, for Example 13.6.

probabilities are long-run proportions, using probabilities as the heights of the bars shows what the distribution of X would be in very many repetitions.) The distribution is strongly skewed. Although X can take any whole-number value from 0 to 10, the probabilities of values larger than 5 are so small that they do not appear in the histogram.

The mean and standard deviation of the binomial distribution in Figure 13.2 are

$$\mu = np$$

$$= (10)(0.1) = 1$$

$$\sigma = \sqrt{np(1-p)}$$

$$= \sqrt{(10)(0.1)(0.9)} = \sqrt{0.9} = 0.9487$$

The mean is marked on the probability histogram in Figure 13.2. ■

APPLY YOUR KNOWLEDGE

13.8   Random digit dialing. When an opinion poll calls residential telephone numbers at random, only 20% of the calls reach a live person. You watch the random digit dialing machine make 15 calls.

(a)   What is the mean number of calls that reach a person?

(b)   What is the standard deviation σ of the count of calls that reach a person?

(c)   If calls are made to New York City rather than nationally, the probability that a call reaches a person is only $p = 0.08$. How does this new p affect the standard deviation? What would be the standard deviation if $p = 0.01$? What does your work show about the behavior of the standard deviation of a binomial distribution as the probability of a success gets closer to 0?

Randomness turns silver to bronze

After many charges of favoritism by judges, the rules for scoring international figure skating competitions changed in 2004. The big change is that 12 judges score all performances, then scores from 3 judges chosen at random are dropped for each part of the program. So there are $\binom{12}{9} = 220$ possible panels of 9 judges for (say) the "Free Skate" and these panels will have slightly different scores. Result: at the 2006 World Figure Skating Championships, the Russian pair Maria Petrova and Alexei Tikhonov received the bronze medal when the consensus of all 12 judges would have given them the silver medal. Perhaps the system needs another change.

13.9   Proofreading. Return to the proofreading setting of Exercise 13.5.

(a)   If X is the number of word errors missed, what is the distribution of X? If Y is the number of word errors caught, what is the distribution of Y?

(b)   What is the mean number of errors caught? What is the mean number of errors missed? The mean counts of successes and of failures always add to n, the number of observations.

(c)   What is the standard deviation of the number of errors caught? What is the standard deviation of the number of errors missed? The standard deviations of the count of successes and the count of failures are always the same.

The Normal approximation to binomial distributions

It isn't practical to use the formula for binomial probabilities when the number of observations n is large. (Look at part (b) of Exercise 13.7 to see why.) Software or a graphing calculator will handle many problems that are beyond the reach of hand calculation. If technology does not rescue you, there is another alternative: *as the number of observations n gets larger, the binomial distribution gets close to a Normal distribution.* When n is large, we can use Normal probability calculations to approximate binomial probabilities. Here are the facts.

> **NORMAL APPROXIMATION FOR BINOMIAL DISTRIBUTIONS**
>
> Suppose that a count X has the binomial distribution with n observations and success probability p. When n is large, the distribution of X is approximately Normal, $N(np, \sqrt{np(1-p)})$.
>
> As a rule of thumb, we will use the Normal approximation when n is so large that $np \geq 10$ and $n(1-p) \geq 10$.

The Normal approximation is easy to remember because it says to act as if X is Normal with exactly the same mean and standard deviation as the binomial. The accuracy of the Normal approximation improves as the sample size n increases. It is most accurate for any fixed n when p is close to 1/2 and least accurate when p is near 0 or 1. This is why the rule of thumb in the box depends on p as well as n.

EXAMPLE 13.7 **Attitudes toward shopping**

Sample surveys show that fewer people enjoy shopping than in the past. A survey asked a nationwide random sample of 2500 adults if they agreed or disagreed that "I like buying new clothes, but shopping is often frustrating and time-consuming."[3] The population that the poll wants to draw conclusions about is all U.S. residents aged 18 and over. Suppose that in fact 60% of all adult U.S. residents would say "Agree" if asked the same question. What is the probability that 1520 or more of the sample agree? ■

Erica Shires/zefa/CORBIS

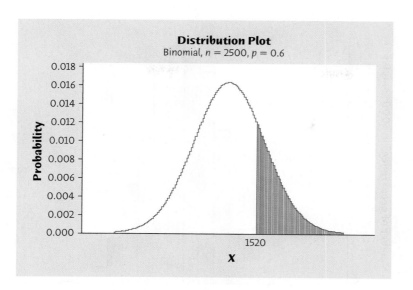

FIGURE 13.3

Probability histogram for the binomial distribution with $n = 2500$, $p = 0.6$. The bars at and above 1520 are shaded to highlight the probability of getting at least 1520 successes. The shape of this binomial probability distribution closely resembles a Normal curve.

Because there are about 235 million adults in the United States, the responses of 2500 randomly chosen adults are very close to independent. So the number in our sample who agree that shopping is frustrating is a random variable X having the binomial distribution with $n = 2500$ and $p = 0.6$. To find the probability $P(X \geq 1520)$ that at least 1520 of the people in the sample find shopping frustrating, we must add the binomial probabilities of all outcomes from $X = 1520$ to $X = 2500$. Figure 13.3 is a probability histogram of this binomial distribution, from Minitab. As the Normal approximation suggests, the shape of the distribution looks Normal. The probability we want is the sum of the heights of the shaded bars. Here are three ways to find this probability.

1. Use technology. Statistical software can find the exact binomial probability. In most cases, software finds cumulative probabilities $P(X \leq x)$. So start by writing

$$P(X \geq 1520) = 1 - P(X \leq 1519)$$

Here is Minitab's answer for $P(X \leq 1519)$:

```
Binomial with n = 2500 and p = 0.6

    x       P(X <= x)
  1519       0.786861
```

The probability we want is $1 - 0.786861 = 0.213139$, correct to 6 decimal places.

2. Simulate a large number of samples. Figure 13.4 displays a histogram of the counts X from 5000 samples of size 2500 when the truth about the population is $p = 0.6$. The simulated distribution, like the exact distribution in Figure 13.3, looks Normal. Because 1085 of these 5000 samples have X at least 1520,

FIGURE 13.4

Histogram of 5000 simulated binomial counts ($n = 2500$, $p = 0.6$).

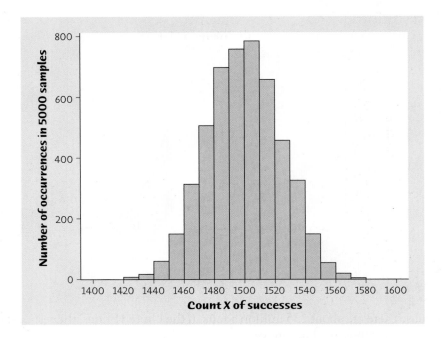

the probability estimated from the simulation is

$$P(X \geq 1520) = \frac{1085}{5000} = 0.2170$$

This estimate misses the true probability by about 0.004. The law of large numbers says that the results of such simulations always get closer to the true probability as we simulate more and more samples.

3. Both of the previous methods require software. We can avoid the need for software by using the Normal approximation.

EXAMPLE 13.8 **Normal calculation of a binomial probability**

Act as though the count X has the Normal distribution with the same mean and standard deviation as the binomial distribution:

$$\mu = np = (2500)(0.6) = 1500$$

$$\sigma = \sqrt{np(1 - p)} = \sqrt{(2500)(0.6)(0.4)} = 24.49$$

Standardizing X gives a standard Normal variable Z. The probability we want is

$$P(X \geq 1520) = P\left(\frac{X - 1500}{24.49} \geq \frac{1520 - 1500}{24.49}\right)$$

$$= P(Z \geq 0.82)$$

$$= 1 - 0.7939 = 0.2061$$

The Normal approximation 0.2061 misses the true probability by about 0.007. ■

The *Normal Approximation to Binomial* applet shows in visual form how well the Normal approximation fits the binomial distribution for any n and p. You can slide n and watch the approximation get better. Whether or not the Normal approximation is satisfactory depends on how accurate your calculations need to be. For most statistical purposes, great accuracy is not required. Our rule of thumb for use of the Normal approximation reflects this judgment.

APPLY YOUR KNOWLEDGE

13.10 Using Benford's law. According to Benford's law (Example 10.7, page 271) the probability that the first digit of the amount of a randomly chosen invoice is a 1 or a 2 is 0.477. You examine 90 invoices from a vendor and find that 29 have first digits 1 or 2. If Benford's law holds, the count of 1s and 2s will have the binomial distribution with $n = 90$ and $p = 0.477$. Too few 1s and 2s suggests fraud. What is the approximate probability of 29 or fewer if the invoices follow Benford's law? Do you suspect that the invoice amounts are not genuine?

13.11 College admissions. A small liberal arts college would like to have an entering class of 415 students next year. Past experience shows that about 27% of the students admitted will decide to attend. The college therefore plans to admit 1535 students. Suppose that students make their decisions independently and that the probability is 0.27 that a randomly chosen student will accept the offer of admission.

    (a)   What are the mean and standard deviation of the number of students who accept the admissions offer from this college?

    (b)   Use the Normal approximation: what is the approximate probability that the college gets more students than they want?

    (c)   Use software to compute the exact probability that the college gets more students than they want. How good is the approximation in part (b)?

13.12 Checking for survey errors. One way of checking the effect of undercoverage, nonresponse, and other sources of error in a sample survey is to compare the sample with known facts about the population. About 12% of American adults are black. The number X of blacks in random samples of 1500 adults should therefore vary with the binomial ($n = 1500$, $p = 0.12$) distribution.

    (a)   What are the mean and standard deviation of X?

    (b)   Use the Normal approximation to find the probability that the sample will contain between 165 and 195 blacks. Be sure to check that you can safely use the approximation.

C H A P T E R 1 3   S U M M A R Y

■ A count X of successes has a **binomial distribution** in the **binomial setting:** there are n observations; the observations are independent of each other; each observation results in a success or a failure; each observation has the same probability p of a success.

■ The binomial distribution with n observations and probability p of success gives a good approximation to the sampling distribution of the count of successes in an SRS of size n from a large population containing proportion p of successes.

■ If X has the binomial distribution with parameters n and p, the possible values of X are the whole numbers $0, 1, 2, \ldots, n$. The **binomial probability** that X takes any of these values is

$$P(X = k) = \binom{n}{k} p^k (1 - p)^{n-k}$$

Binomial probabilities in practice are best found using software.

■ The **binomial coefficient**

$$\binom{n}{k} = \frac{n!}{k!\,(n - k)!}$$

counts the number of ways k successes can be arranged among n observations. Here the **factorial $n!$** is

$$n! = n \times (n - 1) \times (n - 2) \times \cdots \times 3 \times 2 \times 1$$

for positive whole numbers n, and $0! = 1$.

■ The **mean** and **standard deviation** of a binomial count X are

$$\mu = np$$

$$\sigma = \sqrt{np(1 - p)}$$

■ The **Normal approximation** to the binomial distribution says that if X is a count having the binomial distribution with parameters n and p, then when n is large, X is approximately $N(np, \sqrt{np(1-p)}\,)$. Use this approximation only when $np \geq 10$ and $n(1 - p) \geq 10$.

CHECK YOUR SKILLS

13.13 Joe reads that 1 out of 4 eggs contains salmonella bacteria. So he never uses more than 3 eggs in cooking. If eggs do or don't contain salmonella independently of each other, the number of contaminated eggs when Joe uses 3 chosen at random has the distribution

(a) binomial with $n = 4$ and $p = 1/4$.

(b) binomial with $n = 3$ and $p = 1/4$.

(c) binomial with $n = 3$ and $p = 1/3$.

13.14 In the previous exercise, the probability that at least 1 of Joe's 3 eggs contains salmonella is about

(a) 0.68.   (b) 0.58.   (c) 0.30.

13.15 In a group of 10 college students, 4 are business majors. You choose 3 of the 10 students at random and ask their major. The distribution of the number of business majors you choose is

(a)   binomial with $n = 10$ and $p = 0.4$.

(b)   binomial with $n = 3$ and $p = 0.4$.

(c)   not binomial.

13.16 If a basketball player makes 5 free throws and misses 2 free throws during a game, in how many ways can you arrange the sequence of hits and misses?

(a) $\binom{7}{5} = 42$ (b) $\binom{7}{5} = 21$ (c) $\binom{5}{2} = 10$

13.17 A basketball player makes 70% of her free throws. She takes 7 free throws in a game. If the shots are independent of each other, the probability that she makes the first 5 and misses the last 2 is about

(a)   0.635.   (b)   0.318.   (c)   0.015.

13.18 A basketball player makes 70% of her free throws. She takes 7 free throws in a game. If the shots are independent of each other, the probability that she makes 5 out of the 7 shots is about

(a)   0.635.   (b)   0.318.   (c)   0.015.

Each entry in a table of random digits like Table B has probability 0.1 of being a 0, and digits are independent of each other. Exercises 13.19 to 13.21 use this setting.

13.19 The probability of finding exactly 4 0s in a line 40 digits long is about

(a)   0.0000225.   (b)   0.0225.   (c)   0.2059.

13.20 The mean number of 0s in a line 40 digits long is

(a)   4.   (b)   3.098.   (c)   0.4.

13.21 Ten lines in the table contain 400 digits. The count of 0s in these lines is approximately Normal with

(a)   mean 40 and standard deviation 36.

(b)   mean 40 and standard deviation 6.

(c)   mean 36 and standard deviation 6.

C H A P T E R 1 3   E X E R C I S E S

13.22 Binomial setting? In each situation below, is it reasonable to use a binomial distribution for the random variable X? Give reasons for your answer in each case.

(a)   An auto manufacturer chooses one car from each hour's production for a detailed quality inspection. One variable recorded is the count X of finish defects (dimples, ripples, etc.) in the car's paint.

(b)   The pool of potential jurors for a murder case contains 100 persons chosen at random from the adult residents of a large city. Each person in the pool is asked whether he or she opposes the death penalty; X is the number who say "Yes."

(c)   Joe buys a ticket in his state's Pick 3 lottery game every week; X is the number of times in a year that he wins a prize.

13.23 Binomial setting? A binomial distribution will be approximately correct as a model for one of these two sports settings and not for the other. Explain why by briefly discussing both settings.

(a)   A National Football League kicker has made 80% of his field goal attempts in the past. This season he attempts 20 field goals. The attempts differ widely in distance, angle, wind, and so on.

(b)   A National Basketball Association player has made 80% of his free-throw attempts in the past. This season he takes 150 free throws. Basketball free throws are always attempted from 15 feet away with no interference from other players.

David Bergman/CORBIS

13.24 Testing ESP. In a test for ESP (extrasensory perception), a subject is told that cards the experimenter can see but he cannot contain either a star, a circle, a wave, or a square. As the experimenter looks at each of 20 cards in turn, the subject names the shape on the card. A subject who is just guessing has probability 0.25 of guessing correctly on each card.

(a)   The count of correct guesses in 20 cards has a binomial distribution. What are n and p?

(b)   What is the mean number of correct guesses in many repetitions of the experiment?

(c)   What is the probability of exactly 5 correct guesses?

13.25 Random stock prices. A believer in the random walk theory of stock markets thinks that an index of stock prices has probability 0.65 of increasing in any year. Moreover, the change in the index in any given year is not influenced by whether it rose or fell in earlier years. Let X be the number of years among the next 5 years in which the index rises.

(a) X has a binomial distribution. What are n and p?

(b)   What are the possible values that X can take?

(c)   Find the probability of each value of X. Draw a probability histogram for the distribution of X. (See Figure 13.2 for an example of a probability histogram.)

(d)   What are the mean and standard deviation of this distribution? Mark the location of the mean on your histogram.

13.26 On the Web. What kinds of Web sites do males aged 18 to 34 visit? About 50% of male Internet users in this age group visit an auction site such as eBay at least once a month.[4] Interview a random sample of 12 male Internet users aged 18 to 34.

(a)   What is the distribution of the number who have visited an online auction site in the past month?

(b)   What is the probability that exactly 8 of the 12 have visited an auction site in the past month? If you have software, also find the probability that at least 8 of the 12 have visited an auction site in the past month.

13.27 The pill. Many women take oral contraceptives to prevent pregnancy. Under ideal conditions, 1% of women taking the pill become pregnant within one year. In typical use, however, 5% become pregnant.[5] Choose at random 20 women taking the pill. How many become pregnant in the next year?

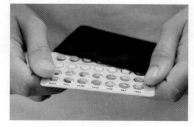

(a) Explain why this is a binomial setting.

(b) What is the probability that at least one of the women becomes pregnant under ideal conditions? What is the probability in typical use?

© CORBIS Premium RF/Alamy

13.28 On the Web, continued. A study of Internet usage interviews a random sample of 500 men aged 18 to 34. Based on the information in Exercise 13.26, what is the probability that at least 235 of the men in the sample visit an online auction site at least once a month? (Check that the Normal approximation is permissible and use it to find this probability. If your software allows, find the exact binomial probability and compare the two results.)

13.29 The pill, continued. A study of the effectiveness of oral contraceptives interviews a random sample of 500 women who are taking the pill.

(a) Based on the information about typical use in Exercise 13.27, what is the probability that at least 25 of these women become pregnant in the next year? (Check that the Normal approximation is permissible and use it to find this probability. If your software allows, find the exact binomial probability and compare the two results.)

(b) We can't use the Normal approximation to the binomial distributions to find this probability under ideal conditions as described in Exercise 13.27. Why not?

13.30 Hitting the fairway. One statistic used to assess professional golfers is driving accuracy, the percent of drives that land in the fairway. Driving accuracy for PGA Tour professionals ranges from about 40% to about 75%. Tiger Woods hits the fairway about 60% of the time.[6]

(a) Tiger hits 14 drives in a round. What assumptions must you make in order to use a binomial distribution for the count X of fairways he hits? Which of these assumptions is least realistic?

(b) Assuming that a binomial distribution can be used, what is the most likely number of fairways that Tiger hits in a round in which he hits 14 drives?

13.31 Genetics. According to genetic theory, the blossom color in the second generation of a certain cross of sweet peas should be red or white in a 3:1 ratio. That is, each plant has probability 3/4 of having red blossoms, and the blossom colors of separate plants are independent.

(a) What is the probability that exactly 6 out of 8 of these plants have red blossoms?

(b) What is the mean number of red-blossomed plants when 80 plants of this type are grown from seeds?

(c) What is the probability of obtaining at least 60 red-blossomed plants when 80 plants are grown from seeds? Use the Normal approximation. If your software allows, find the exact binomial probability and compare the two results.

© blickwinkel/Alamy

13.32 False positives in testing for HIV. A rapid test for the presence in the blood of antibodies to HIV, the virus that causes AIDS, gives a positive result with probability about 0.004 when a person who is free of HIV antibodies is tested. A clinic tests 1000 people who are all free of HIV antibodies.

(a) What is the distribution of the number of positive tests?

(b) What is the mean number of positive tests?

(c) You cannot safely use the Normal approximation for this distribution. Explain why.

13.33 High school equivalency. The Census Bureau says that 21% of Americans aged 18 to 24 do not have a high school diploma. A vocational school wants to attract young people who may enroll in order to achieve high school equivalency. The school mails an advertising flyer to 25,000 persons between the ages of 18 and 24.

(a) If the mailing list can be considered a random sample of the population, what is the mean number of high school dropouts who will receive the flyer?

(b) What is the approximate probability that at least 5000 dropouts will receive the flyer?

13.34 Survey demographics. According to the Census Bureau, 13% of American adults (age 18 and over) are Hispanic. An opinion poll plans to contact an SRS of 1200 adults.

(a) What is the mean number of Hispanics in such samples? What is the standard deviation?

(b) According to the 68–95–99.7 rule, what range will include the counts of Hispanics in 95% of all such samples?

(c) How large a sample is required to make the mean number of Hispanics at least 200?

13.35 Multiple-choice tests. Here is a simple probability model for multiple-choice tests. Suppose that each student has probability p of correctly answering a question chosen at random from a universe of possible questions. (A strong student has a higher p than a weak student.) Answers to different questions are independent.

(a) Jodi is a good student for whom $p = 0.75$. Use the Normal approximation to find the probability that Jodi scores between 70% and 80% on a 100-question test.

(b) If the test contains 250 questions, what is the probability that Jodi will score between 70% and 80%? You see that Jodi's score on the longer test is more likely to be close to her "true score."

13.36 Is this coin balanced? While he was a prisoner of war during World War II, John Kerrich tossed a coin 10,000 times. He got 5067 heads. If the coin is perfectly balanced, the probability of a head is 0.5. Is there reason to think that Kerrich's coin was not balanced? To answer this question, find the probability that tossing a balanced coin 10,000 times would give a count of heads at least this far from 5000 (that is, at least 5067 heads or no more than 4933 heads.)

13.37 Binomial variation. Never forget that probability describes only what happens in the long run. Example 13.5 concerns the count of bad CDs in inspection samples of size 10. The count has the binomial distribution with $n = 10$ and $p = 0.1$. The *Probability* applet simulates inspecting a lot of CDs if you set the probability of heads to 0.1, toss 10 times, and let each head stand for a bad CD.

(a) The mean number of bad CDs in a sample is 1. Click "Toss" and "Reset" repeatedly to simulate 20 samples. How many bad CDs did you find in each sample? How close to the mean 1 is the average number of bad CDs in these samples?

(b) Example 13.5 shows that the probability of exactly 1 bad CD is 0.3874. How close to the probability is the proportion of the 20 lots that have exactly 1 bad CD?

Whooping cough. *Whooping cough (pertussis) is a highly contagious bacterial infection that was a major cause of childhood deaths before the development of vaccines. About 80% of unvaccinated children who are exposed to whooping cough will develop the infection, as opposed to only about 5% of vaccinated children. Exercises 13.38 to 13.41 are based on this information.*

13.38 Vaccination at work. A group of 20 children at a nursery school are exposed to whooping cough by playing with an infected child.

(a) If all 20 have been vaccinated, what is the mean number of new infections? What is the probability that no more than 2 of the 20 children develop infections?

(b) If none of the 20 have been vaccinated, what is the mean number of new infections? What is the probability that 18 or more of the 20 children develop infections?

13.39 A whooping cough outbreak. In 2007, Bob Jones University ended its fall semester a week early because of a whooping cough outbreak; 158 students were isolated and another 1200 given antibiotics as a precaution.[7] Authorities react strongly to whooping cough outbreaks because the disease is so contagious. Because the effect of childhood vaccination often wears off by late adolescence, treat the Bob Jones students as if they were unvaccinated. It appears that about 1400 students were exposed. What is the probability that at least 75% of these students develop infections if not treated? (Fortunately, whooping cough is much less serious after infancy.)

13.40 A mixed group: means. A group of 20 children at a nursery school are exposed to whooping cough by playing with an infected child. Of these children 17 have been vaccinated and 3 have not.

(a) What is the distribution of the number of new infections among the 17 vaccinated children? What is the mean number of new infections?

(b) What is the distribution of the number of new infections among the 3 unvaccinated children? What is the mean number of new infections?

(c) Add your means from parts (a) and (b). This is the mean number of new infections among all 20 exposed children.

13.41 A mixed group: probabilities. We would like to find the probability that exactly 2 of the 20 exposed children in the previous exercise develop whooping cough.

(a) One way to get 2 infections is to get 1 among the 17 vaccinated children and 1 among the 3 unvaccinated children. Find the probability of exactly 1 infection among the 17 vaccinated children. Find the probability of exactly 1 infection among the 3 unvaccinated children. These events are independent: what is the probability of exactly 1 infection in each group?

(b) Write down all the ways in which 2 infections can be divided between the two groups of children. Follow the pattern of part (a) to find the probability of each of these possibilities. Add all of your results (including the result of part (a)) to obtain the probability of exactly 2 infections among the 20 children.

13.42 Estimating π from random numbers. Kenyon College student Eric Newman used basic geometry to evaluate software random number generators as part of a summer research project. He generated 2000 independent random points (X, Y) in the unit square. (That is, X and Y are independent random numbers between 0 and 1, each having the density function illustrated in Figure 10.4 (page 274). The probability that (X, Y) falls in any region within the unit square is the area of the region.)[8]

(a) Sketch the unit square, the region of possible values for the point (X, Y).

(b) The set of points (X, Y) where $X^2 + Y^2 < 1$ describes a circle of radius 1. Add this circle to your sketch in part (a) and label the intersection of the two regions A.

(c) Let T be the total number of the 2000 points that fall into the region A. T follows a binomial distribution. Identify n and p. (*Hint:* Recall that the area of a circle is πr^2.)

(d) What are the mean and standard deviation of T?

(e) Explain how Eric used a random number generator and the facts above to estimate π.

13.43 The continuity correction. One reason why the Normal approximation may fail to give accurate estimates of binomial probabilities is that the binomial distributions are discrete and the Normal distributions are continuous. That is, counts take only whole number values but Normal variables can take any value. We can improve the Normal approximation by treating each whole number count as if it occupied the interval from 0.5 below the number to 0.5 above the number. For example, approximate a binomial probability $P(X \geq 10)$ by finding the Normal probability $P(X \geq 9.5)$. Be careful: binomial $P(X > 10)$ is approximated by Normal $P(X \geq 10.5)$.

We saw in Exercise 13.30 that Tiger Woods hits the fairway in 60% of his drives. We will assume that his drives are independent and that each has probability 0.6 of hitting the fairway. Tiger drives 25 times. The exact binomial probability that he hits 15 or more fairways is 0.5858.

(a) Show that this setting satisfies the rule of thumb for use of the Normal approximation (just barely).

(b) What is the Normal approximation to $P(X \geq 15)$?

(c) What is the Normal approximation using the continuity correction? That's a lot closer to the true binomial probability.

Introduction to Inference

After we have selected a sample, we know the responses of the individuals in the sample. The usual reason for taking a sample is not to learn about the individuals in the sample but to *infer* from the sample data some conclusion about the wider population that the sample represents.

STATISTICAL INFERENCE

Statistical inference provides methods for drawing conclusions about a population from sample data.

Because a different sample might lead to different conclusions, we can't be certain that our conclusions are correct. Statistical inference uses the language of probability to say how trustworthy our conclusions are. This chapter introduces the two most common types of inference, *confidence intervals* for estimating the value of a population parameter and *tests of significance* for assessing the evidence for a claim about a population. Both types of inference are based on the sampling distributions of statistics. That is, both use probability to say what would happen if we applied the inference method many times.

**IN THIS CHAPTER
WE COVER...**

- The reasoning of statistical estimation
- Margin of error and confidence level
- Confidence intervals for a population mean
- The reasoning of tests of significance
- Stating hypotheses
- *P*-value and statistical significance
- Tests for a population mean
- Significance from a table

This chapter presents the basic reasoning of statistical inference. To make the reasoning as clear as possible, we start with a setting that is too simple to be realistic. Here is the setting for our work in this chapter.

SIMPLE CONDITIONS FOR INFERENCE ABOUT A MEAN

1. We have an SRS from the population of interest. There is no nonresponse or other practical difficulty.

2. The variable we measure has an exactly Normal distribution $N(\mu, \sigma)$ in the population.

3. We don't know the population mean μ. But we do know the population standard deviation σ.

The conditions that we have a perfect SRS, that the population is exactly Normal, and that we know the population σ are all unrealistic. Chapter 15 begins to move from the "simple conditions" toward the reality of statistical practice. Later chapters deal with inference in fully realistic settings.

The reasoning of statistical estimation

Body mass index (BMI) is used to screen for possible weight problems. It is calculated as weight divided by the square of height, measuring weight in kilograms and height in meters. Many online BMI calculators allow you to enter weight in pounds and height in inches. Adults with BMI less than 18.5 are considered underweight and those with BMI greater than 25 may be overweight. For data about BMI, we turn to the National Health and Nutrition Examination Survey (NHANES), a continuing government sample survey that monitors the health of the American population.

EXAMPLE 14.1 Body mass index of young women

The most recent NHANES report gives data for 654 women aged 20 to 29 years.[1] The mean BMI of these 654 women was $\overline{x} = 26.8$. On the basis of this sample, we want to estimate the mean BMI μ in the population of all 18 million women in this age group.

To match the "simple conditions," we will treat the NHANES sample as an SRS from a Normal population with standard deviation $\sigma = 7.5$. ■

Here is the reasoning of statistical estimation in a nutshell:

1. To estimate the unknown population mean BMI μ, use the mean $\overline{x} = 26.8$ of the random sample. We don't expect $\overline{x}$ to be exactly equal to μ, so we want to say how accurate this estimate is.

2. We know the sampling distribution of $\overline{x}$. In repeated samples, $\overline{x}$ has the Normal distribution with mean μ and standard deviation $\sigma/\sqrt{n}$. So the

Thomas Northcut/Getty

average BMI $\bar{x}$ of an SRS of 654 young women has standard deviation

$$\frac{\sigma}{\sqrt{n}} = \frac{7.5}{\sqrt{654}} = 0.3 \ \text{(rounded off)}$$

3. The 95 part of the 68–95–99.7 rule for Normal distributions says that $\bar{x}$ is within 0.6 (that's two standard deviations) of the mean μ in 95% of all samples. That is, for 95% of all samples of size 654, the distance between the sample mean $\bar{x}$ and the population mean μ is less than 0.6. So if we estimate that μ lies somewhere in the interval from $\bar{x} - 0.6$ to $\bar{x} + 0.6$, we'll be right for 95% of all possible samples. For this particular sample, this interval is

$$\bar{x} - 0.6 = 26.8 - 0.6 = 26.2$$

to

$$\bar{x} + 0.6 = 26.8 + 0.6 = 27.4$$

4. Because we got the interval 26.2 to 27.4 from a method that captures the population mean for 95% of all possible samples, we say that we are 95% *confident* that the mean BMI μ of all young women is some value in that interval, no lower than 26.2 and no higher than 27.4.

   The big idea is that the sampling distribution of $\bar{x}$ tells us how close to μ the sample mean $\bar{x}$ is likely to be. Statistical estimation just turns that information around to say how close to $\bar{x}$ the unknown population mean μ is likely to be. We call the interval of numbers between the values $\bar{x} \pm 0.6$ a 95% *confidence interval* for μ.

Ranges are for statistics?

Many people like to think that statistical estimates are exact. The Nobel Prize–winning economist Daniel McFadden tells a story of his time on the Council of Economic Advisers. Presented with a range of forecasts for economic growth, President Lyndon Johnson replied: "Ranges are for cattle; give me one number."

APPLY YOUR KNOWLEDGE

14.1   Number skills of young men. The National Assessment of Educational Progress (NAEP) gave a test of basic arithmetic and the ability to apply it in everyday life to a sample of 840 men 21 to 25 years of age.[2] Scores range from 0 to 500; for example, someone with a score of 325 can determine the price of a meal from a menu. The mean score for these 840 young men was $\bar{x} = 272$. We want to estimate the mean score μ in the population of all young men. Consider the NAEP sample as an SRS from a Normal population with standard deviation $\sigma = 60$.

   (a)   If we take many samples, the sample mean $\bar{x}$ varies from sample to sample according to a Normal distribution with mean equal to the unknown mean score μ in the population. What is the standard deviation of this sampling distribution?

   (b)   According to the 95 part of the 68–95–99.7 rule, 95% of all values of $\bar{x}$ fall within _____ on either side of the unknown mean μ. What is the missing number?

   (c)   What is the 95% confidence interval for the population mean score μ based on this one sample?

Margin of error and confidence level

The 95% confidence interval for the mean BMI of young women, based on the NHANES sample, is $\overline{x} \pm 0.6$. Once we have the sample results in hand, we know that for this sample $\overline{x} = 26.8$, so that our confidence interval is 26.8 ± 0.6. Most confidence intervals have a form similar to this,

$$\text{estimate} \pm \text{margin of error}$$

margin of error

The estimate ($\overline{x} = 26.8$ in our example) is our guess for the value of the unknown parameter. The **margin of error** ± 0.6 shows how accurate we believe our guess is, based on the variability of the estimate. We have a 95% confidence interval because the interval $\overline{x} \pm 0.6$ catches the unknown parameter in 95% of all possible samples.

CONFIDENCE INTERVAL

A **level C confidence interval** for a parameter has two parts:

■ An interval calculated from the data, usually of the form

$$\text{estimate} \pm \text{margin of error}$$

■ A **confidence level** C, which gives the probability that the interval will capture the true parameter value in repeated samples. That is, the confidence level is the success rate for the method.

Users can choose the confidence level, usually 90% or higher because we usually want to be quite sure of our conclusions. The most common confidence level is 95%.

INTERPRETING A CONFIDENCE INTERVAL

The confidence level is the success rate of the method that produces the interval. We don't know whether the 95% confidence interval from a particular sample is one of the 95% that capture μ or one of the unlucky 5% that miss.

To say that we are **95% confident** that the unknown μ lies between 26.2 and 27.4 is shorthand for **"We got these numbers using a method that gives correct results 95% of the time."**

EXAMPLE 14.2 **Statistical estimation in pictures**

Figures 14.1 and 14.2 illustrate the behavior of confidence intervals. Study these figures carefully. If you understand what they say, you have mastered one of the big ideas of statistics.

Figure 14.1 illustrates the behavior of the interval $\overline{x} \pm 0.6$ for the mean BMI of young women. Starting with the population, imagine taking many SRSs of 654 young women. The first sample has $\overline{x} = 26.8$, the second has $\overline{x} = 27.0$, the third has

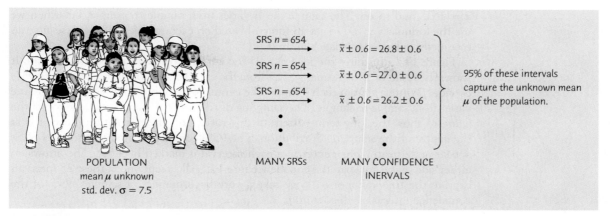

FIGURE 14.1

To say that $\bar{x} \pm 0.6$ is a 95% confidence interval for the population mean μ is to say that, in repeated samples, 95% of these intervals capture μ.

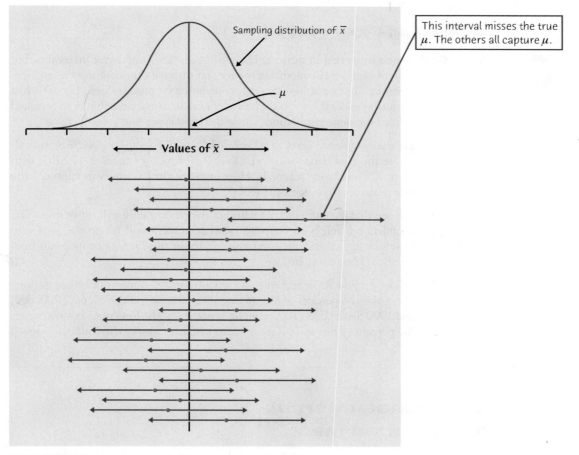

FIGURE 14.2

Twenty-five samples from the same population gave these 95% confidence intervals. In the long run, 95% of all samples give an interval that contains the population mean μ.

$\bar{x} = 26.2$, and so on. The sample mean varies from sample to sample, but when we use the formula $\bar{x} \pm 0.6$ to get an interval based on each sample, 95% *of these intervals capture the unknown population mean μ.*

Figure 14.2 illustrates the idea of a 95% confidence interval in a different form. It shows the result of drawing many SRSs from the same population and calculating a 95% confidence interval from each sample. The center of each interval is at $\bar{x}$ and therefore varies from sample to sample. The sampling distribution of $\bar{x}$ appears at the top of the figure to show the long-term pattern of this variation. The population mean μ is at the center of the sampling distribution. The 95% confidence intervals from 25 SRSs appear underneath. The center $\bar{x}$ of each interval is marked by a dot. The arrows on either side of the dot span the confidence interval. All except one of these 25 intervals capture the true value of μ. If we take a very large number of samples, 95% of the confidence intervals will contain μ. ■

The *Confidence Interval* applet animates Figure 14.2. You can use the applet to watch confidence intervals from one sample after another capture or fail to capture the true parameter.

APPLY YOUR KNOWLEDGE

14.2   Confidence intervals in action. The idea of an 80% confidence interval is that in 80% of all samples the method produces an interval that captures the true parameter value. That's not high enough confidence for practical use, but 80% hits and 20% misses make it easy to see how a confidence interval behaves in repeated samples from the same population. Go to the *Confidence Interval* applet.

(a)   Set the confidence level to 80%. Click "Sample" to choose an SRS and calculate the confidence interval. Do this 10 times to simulate 10 SRSs with their 10 confidence intervals. How many of the 10 intervals captured the true mean μ? How many missed?

(b)   You see that we can't predict whether the next sample will hit or miss. The confidence level, however, tells us what percent will hit in the long run. Reset the applet and click "Sample 50" to get the confidence intervals from 50 SRSs. How many hit?

(c)   Click "Sample 50" repeatedly and write down the number of hits each time. What was the percent of hits among 100, 200, 300, 400, 500, 600, 700, 800, and 1000 SRSs? Even 1000 samples is not truly "the long run," but we expect the percent of hits in 1000 samples to be fairly close to the confidence level, 80%.

Confidence intervals for a population mean

In the setting of Example 14.1 we outlined the reasoning that leads to a 95% confidence interval for the unknown mean μ of a population. Now we will reduce the reasoning to a formula.

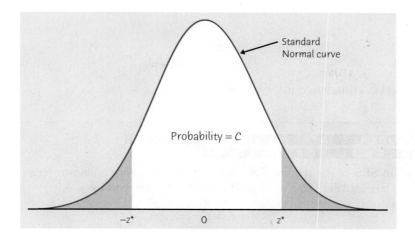

FIGURE 14.3

The critical value z^* is the number that catches central probability C under a standard Normal curve between $-z^*$ and z^*.

To find a 95% confidence interval for the mean BMI of young women, we first caught the central 95% of the Normal sampling distribution by going out two standard deviations in both directions from the mean. To find a level C confidence interval, we first catch the central area C under the Normal sampling distribution. Because all Normal distributions are the same in the standard scale, we can obtain everything we need from the standard Normal curve.

Figure 14.3 shows how the central area C under a standard Normal curve is marked off by two points z^* and $-z^*$. Numbers like z^* that mark off specified areas are called **critical values** of the standard Normal distribution. Values of z^* for many choices of C appear at the bottom of Table C in the back of the book, in the row labeled z^*. Here are the entries for the most common confidence levels:

critical value

Confidence level C	90%	95%	99%
Critical value z^*	1.645	1.960	2.576

You see that for $C = 95\%$ the table gives $z^* = 1.960$. This is a bit more precise than the approximate value $z^* = 2$ based on the 68–95–99.7 rule. You can of course use software to find critical values z^*, as well as the entire confidence interval.

Figure 14.3 shows that there is area C under the standard Normal curve between $-z^*$ and z^*. So *any* Normal curve has area C within z^* standard deviations on either side of its mean. The Normal sampling distribution of $\overline{x}$ has area C within $z^*\sigma/\sqrt{n}$ on either side of the population mean μ because it has mean μ and standard deviation $\sigma/\sqrt{n}$. If we start at $\overline{x}$ and go out $z^*\sigma/\sqrt{n}$ in both directions, we get an interval that contains the population mean μ in a proportion C of all samples. This interval is

$$\text{from} \quad \overline{x} - z^*\frac{\sigma}{\sqrt{n}} \quad \text{to} \quad \overline{x} + z^*\frac{\sigma}{\sqrt{n}}$$

or

$$\overline{x} \pm z^* \frac{\sigma}{\sqrt{n}}$$

It is a level C confidence interval for μ.

CONFIDENCE INTERVAL FOR THE MEAN OF A NORMAL POPULATION

Draw an SRS of size n from a Normal population having unknown mean μ and known standard deviation σ. A level C **confidence interval for μ** is

$$\overline{x} \pm z^* \frac{\sigma}{\sqrt{n}}$$

The critical value z^* is illustrated in Figure 14.3 and found in the z^* row of Table C.

The steps in finding a confidence interval mirror the overall four-step process for organizing statistical problems.

CONFIDENCE INTERVALS: THE FOUR-STEP PROCESS

STATE: What is the practical question that requires estimating a parameter?

PLAN: Identify the parameter, choose a level of confidence, and select the type of confidence interval that fits your situation.

SOLVE: Carry out the work in two phases:

1. **Check the conditions** for the interval you plan to use.
2. Calculate the **confidence interval.**

CONCLUDE: Return to the practical question to describe your results in this setting.

EXAMPLE 14.3 Healing of skin wounds

STATE: Biologists studying the healing of skin wounds measured the rate at which new cells closed a razor cut made in the skin of an anesthetized newt. Here are data from 18 newts, measured in micrometers (millionths of a meter) per hour:[3]

29	27	34	40	22	28	14	35	26
35	12	30	23	18	11	22	23	33

This is one of several sets of measurements made under different conditions. We want to estimate the mean healing rate for comparison with rates under other conditions.

PLAN: We will estimate the mean rate μ for all newts of this species by giving a 95% confidence interval. The confidence interval just introduced fits this situation.

Royalty Free/CORBIS

SOLVE: We should start by checking the conditions for inference. For this example, we will first find the interval and then discuss how statistical practice deals with conditions that are never perfectly satisfied.

The mean of the sample is $\bar{x} = 25.67$. As part of the "simple conditions," suppose that from past experience with this species of newts we know that the standard deviation of healing rates is $\sigma = 8$ micrometers per hour. For 95% confidence, the critical value is $z^* = 1.960$. A 95% confidence interval for μ is therefore

$$\bar{x} \pm z^* \frac{\sigma}{\sqrt{n}} = 25.67 \pm 1.960 \frac{8}{\sqrt{18}}$$

$$= 25.67 \pm 3.70$$

$$= 21.97 \text{ to } 29.37$$

CONCLUDE: We are 95% confident that the mean healing rate for all newts of this species is between 21.97 and 29.37 micrometers per hour. ∎

In practice, the first part of the *Solve* step is to check the conditions for inference. The "simple conditions" are as follows:

1. **SRS:** We don't have an actual SRS from the population of all newts of this species. Scientists usually act as if animal subjects are SRSs from their species or genetic type if there is nothing special about how the subjects were obtained. This study was a randomized comparative experiment in which these 18 newts were assigned at random from a larger group of newts to get one of the treatments being compared.

2. **Normal distribution:** The biologists expect from past experience that measurements like this on several animals of the same species under the same conditions will follow approximately a Normal distribution. We can't look at the population, but we can examine the sample. Figure 14.4 is a stemplot, with split stems. The shape is irregular, but there are no outliers or strong skewness. Shapes like this often occur in small samples from Normal populations, so we have no reason to doubt that the population distribution is Normal.

3. **Known σ:** It really is unrealistic to suppose that we know that $\sigma = 8$. We will see in Chapter 17 that it is easy to do away with the need to know σ.

As this discussion suggests, inference methods are often used when conditions like SRS and Normal population are not exactly satisfied. In this introductory chapter, we act as though the "simple conditions" are satisfied. In reality, wise use of inference requires judgment. Chapter 15 and the later chapters on each inference method will give you a better basis for judgment.

```
1 | 1 2 4
1 | 8
2 | 2 2 3 3
2 | 6 7 8 9
3 | 0 3 4
3 | 5 5
4 | 0
```

FIGURE 14.4

Stemplot of the healing rates in Example 14.3.

APPLY YOUR KNOWLEDGE

14.3 Find a critical value. The critical value z^* for confidence level 97.5% is not in Table C. Use software or Table A of standard Normal probabilities to find z^*. Include in your answer a sketch like Figure 14.3 with $C = 0.975$ and your critical value z^* marked on the axis.

14.4 **Measuring conductivity.** The National Institute of Standards and Technology (NIST) supplies "standard materials" whose physical properties are supposed to be known. For example, you can buy from NIST a liquid whose electrical conductivity is supposed to be 5. (The units for conductivity are microsiemens per centimeter. Distilled water has conductivity 0.5.) Of course, no measurement is exactly correct. NIST knows the variability of its measurements very well, so it is quite realistic to assume that the population of all measurements of the same liquid has the Normal distribution with mean μ equal to the true conductivity and standard deviation $\sigma = 0.2$. Here are 6 measurements on the same standard liquid, which is supposed to have conductivity 5:

$$5.32 \quad 4.88 \quad 5.10 \quad 4.73 \quad 5.15 \quad 4.75$$

NIST wants to give the buyer of this liquid a 90% confidence interval for its true conductivity. What is this interval? Follow the four-step process as illustrated in Example 14.3.

14.5 **IQ test scores.** Here are the IQ test scores of 31 seventh-grade girls in a Midwest school district:[4]

114	100	104	89	102	91	114	114	103	105
108	130	120	132	111	128	118	119	86	72
111	103	74	112	107	103	98	96	112	112	93

(a) These 31 girls are an SRS of all seventh-grade girls in the school district. Suppose that the standard deviation of IQ scores in this population is known to be $\sigma = 15$. We expect the distribution of IQ scores to be close to Normal. Make a stemplot of the distribution of these 31 scores (split the stems) to verify that there are no major departures from Normality. You have now checked the "simple conditions" to the extent possible.

(b) Estimate the mean IQ score for all seventh-grade girls in the school district, using a 99% confidence interval. Follow the four-step process as illustrated in Example 14.3.

The reasoning of tests of significance

Confidence intervals are one of the two most common types of statistical inference. Use a confidence interval when your goal is to estimate a population parameter. The second common type of inference, called *tests of significance*, has a different goal: to assess the evidence provided by data about some claim concerning a population. Here is the reasoning of statistical tests in a nutshell.

> **EXAMPLE 14.4** **I'm a great free-throw shooter**
>
> I claim that I make 80% of my basketball free throws. To test my claim, you ask me to shoot 20 free throws. I make only 8 of the 20. "Aha!" you say. "Someone who makes 80% of his free throws would almost never make only 8 out of 20. So I don't believe your claim."

Your reasoning is based on asking what would happen if my claim were true and we repeated the sample of 20 free throws many times—I would almost never make as few as 8. This outcome is so unlikely that it gives strong evidence that my claim is not true.

You can say how strong the evidence against my claim is by giving the probability that I would make as few as 8 out of 20 free throws if I really make 80% in the long run. This probability is 0.0001. I would make as few as 8 of 20 only once in 10,000 tries in the long run if my claim to make 80% were true. The small probability convinces you that my claim is false. ■

The *Reasoning of a Statistical Test* applet animates Example 14.4. You can ask a player to shoot free throws until the data do (or don't) convince you that he makes fewer than 80%. Significance tests use an elaborate vocabulary, but the basic idea is simple: *an outcome that would rarely happen if a claim were true is good evidence that the claim is not true*.

The reasoning of statistical tests, like that of confidence intervals, is based on asking what would happen if we repeated the sample or experiment many times. We will act as if the "simple conditions" listed on page 360 are true: we have a perfect SRS from an exactly Normal population with standard deviation σ known to us. Here is an example we will explore.

EXAMPLE 14.5 **Sweetening colas**

Diet colas use artificial sweeteners to avoid sugar. These sweeteners gradually lose their sweetness over time. Manufacturers therefore test new colas for loss of sweetness before marketing them. Trained tasters sip the cola along with drinks of standard sweetness and score the cola on a "sweetness score" of 1 to 10. The cola is then stored for a month at high temperature to imitate the effect of four months' storage at room temperature. Each taster scores the cola again after storage. This is a matched pairs experiment. Our data are the differences (score before storage minus score after storage) in the tasters' scores. The bigger these differences, the bigger the loss of sweetness.

Ramin/Talaie/CORBIS

Suppose we know that for any cola, the sweetness loss scores vary from taster to taster according to a Normal distribution with standard deviation $\sigma = 1$. The mean μ for all tasters measures loss of sweetness and is different for different colas.

Here are the sweetness losses for a new cola, as measured by 10 trained tasters:

$$2.0 \quad 0.4 \quad 0.7 \quad 2.0 \quad -0.4 \quad 2.2 \quad -1.3 \quad 1.2 \quad 1.1 \quad 2.3$$

Most are positive. That is, most tasters found a loss of sweetness. But the losses are small, and two tasters (the negative scores) thought the cola gained sweetness. The average sweetness loss is given by the sample mean $\bar{x} = 1.02$. Are these data good evidence that the cola lost sweetness in storage? ■

The reasoning is the same as in Example 14.4. We make a claim and ask if the data give evidence *against* it. We seek evidence that there *is* a sweetness loss, so the claim we test is that there is *not* a loss. In that case, the mean loss for the population of all trained testers would be $\mu = 0$.

FIGURE 14.5

If the cola does not lose sweetness in storage, the mean score $\bar{x}$ for 10 tasters will have this sampling distribution. The actual result for one cola was $\bar{x} = 0.3$. That could easily happen just by chance. Another cola had $\bar{x} = 1.02$. That's so far out on the Normal curve that it is good evidence that this cola did lose sweetness.

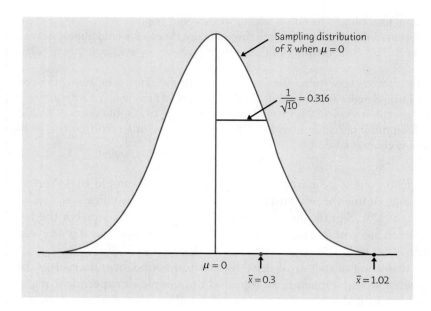

- If the claim that $\mu = 0$ is true, the sampling distribution of $\bar{x}$ from 10 tasters is Normal with mean $\mu = 0$ and standard deviation

$$\frac{\sigma}{\sqrt{n}} = \frac{1}{\sqrt{10}} = 0.316$$

Figure 14.5 shows this sampling distribution. We can judge whether any observed $\bar{x}$ is surprising by locating it on this distribution.

- For a cola already on the market, 10 tasters had mean loss $\bar{x} = 0.3$. It is clear from Figure 14.5 that an $\bar{x}$ this large could easily occur just by chance when the population mean is $\mu = 0$. That 10 tasters found $\bar{x} = 0.3$ is not evidence that this cola loses sweetness.

- The taste test for the new cola produced $\bar{x} = 1.02$. That's way out on the Normal curve in Figure 14.5—so far out that *an observed value this large would rarely occur just by chance if the true μ were 0*. This observed value is good evidence that the true μ is in fact greater than 0, that is, that the cola lost sweetness. The manufacturer must reformulate the cola and try again.

<hr>

APPLY YOUR KNOWLEDGE

14.6 **Student attitudes.** The Survey of Study Habits and Attitudes (SSHA) is a psychological test that measures students' study habits and attitude toward school. Scores range from 0 to 200. The mean score for college students is about 115, and the standard deviation is about 30. A teacher suspects that the mean μ for older students is higher than 115. She gives the SSHA to an SRS of 25 students who are at least 30 years old. Suppose we know that scores in the population of older students are Normally distributed with standard deviation $\sigma = 30$.

(a) We seek evidence *against* the claim that $\mu = 115$. What is the sampling distribution of the mean score $\bar{x}$ of a sample of 25 students if the claim is true? Draw the density curve of this distribution. (Sketch a Normal curve, then mark on the axis the values of the mean and 1, 2, and 3 standard deviations on either side of the mean.)

(b) Suppose that the sample data give $\bar{x} = 118.6$. Mark this point on the axis of your sketch. In fact, the result was $\bar{x} = 125.8$. Mark this point on your sketch. Using your sketch, explain in simple language why one result is good evidence that the mean score of all older students is greater than 115 and why the other outcome is not.

14.7 **Measuring conductivity.** The National Institute of Standards and Technology (NIST) supplies a "standard liquid" whose electrical conductivity is supposed to be exactly 5. Is there reason to think that the true conductivity of a shipment of this liquid is not 5? To find out, NIST measures the conductivity 6 times. Repeated measurements of the same thing vary, which is why NIST makes 6 measurements. These measurements are an SRS from the population of all possible measurements. This population has a Normal distribution with mean μ equal to the true conductivity and standard deviation $\sigma = 0.2$.

(a) We seek evidence *against* the claim that $\mu = 5$. What is the sampling distribution of the mean $\bar{x}$ in many samples of 6 measurements if the claim is true? Make a sketch of the Normal curve for this distribution. (Draw a Normal curve, then mark on the axis the values of the mean and 1, 2, and 3 standard deviations on either side of the mean.)

(b) Suppose that the sample mean is $\bar{x} = 4.98$. Mark this value on the axis of your sketch. Another shipment of liquid has $\bar{x} = 4.7$ for 6 measurements. Mark this value on the axis as well. Explain in simple language why one result is good evidence that the true conductivity differs from 5 and why the other result gives no reason to doubt that 5 is correct.

Stating hypotheses

A statistical test starts with a careful statement of the claims we want to compare. In Example 14.5, we saw that the taste test data are not plausible if the new cola loses no sweetness. Because the reasoning of tests looks for evidence *against* a claim, we start with the claim we seek evidence against, such as "no loss of sweetness."

NULL AND ALTERNATIVE HYPOTHESES

The claim tested by a statistical test is called the **null hypothesis.** The test is designed to assess the strength of the evidence *against* the null hypothesis. Usually the null hypothesis is a statement of "no effect" or "no difference."

The claim about the population that we are trying to find evidence *for* is the **alternative hypothesis.** The alternative hypothesis is **one-sided** if it states that a parameter is *larger than* or *smaller than* the null hypothesis value. It is **two-sided** if it states that the parameter is *different from* the null value (it could be either smaller or larger).

We abbreviate the null hypothesis as H_0 and the alternative hypothesis as H_a. *Hypotheses always refer to a population, not to a particular outcome. Be sure to state H_0 and H_a in terms of population parameters.* Because H_a expresses the effect that we hope to find evidence *for*, it is sometimes easier to begin by stating H_a and then set up H_0 as the statement that the hoped-for effect is not present.

In Example 14.5, we are seeking evidence *for* loss in sweetness. The null hypothesis says "no loss" on the average in a large population of tasters. The alternative hypothesis says "there is a loss." So the hypotheses are

$$H_0: \mu = 0$$
$$H_a: \mu > 0$$

The alternative hypothesis is *one-sided* because we are interested only in whether the cola *lost* sweetness.

EXAMPLE 14.6   Studying job satisfaction

Does the job satisfaction of assembly workers differ when their work is machine-paced rather than self-paced? Assign workers either to an assembly line moving at a fixed pace or to a self-paced setting. All subjects work in both settings, in random order. This is a matched pairs design. After two weeks in each work setting, the workers take a test of job satisfaction. The response variable is the difference in satisfaction scores, self-paced minus machine-paced.

The parameter of interest is the mean μ of the differences in scores in the population of all assembly workers. The null hypothesis says that there is no difference between self-paced and machine-paced work, that is,

$$H_0: \mu = 0$$

The authors of the study wanted to know if the two work conditions have different levels of job satisfaction. They did not specify the direction of the difference. The alternative hypothesis is therefore *two-sided*:

$$H_a: \mu \neq 0 \; \blacksquare$$

The hypotheses should express the hopes or suspicions we have **before** *we see the data. It is cheating to first look at the data and then frame hypotheses to fit what the data show.* For example, the data for the study in Example 14.6 showed that the workers were more satisfied with self-paced work, but this should not influence the choice of H_a. If you do not have a specific direction firmly in mind in advance, use a two-sided alternative.

APPLY YOUR KNOWLEDGE

14.8 Student attitudes. State the null and alternative hypotheses for the study of older students' attitudes described in Exercise 14.6. (Is the alternative hypothesis one-sided or two-sided?)

14.9 Measuring conductivity. State the null and alternative hypotheses for the study of electrical conductivity described in Exercise 14.7. (Is the alternative hypothesis one-sided or two-sided?)

14.10 Grading a teaching assistant. The examinations in a large accounting class are scaled after grading so that the mean score is 50. The professor thinks that one teaching assistant is a poor teacher and suspects that his students have a lower mean score than the class as a whole. The TA's students this semester can be considered a sample from the population of all students in the course, so the professor compares their mean score with 50. State the hypotheses H_0 and H_a.

14.11 Women's heights. The average height of 18-year-old American women is 64.2 inches. You wonder whether the mean height of this year's female graduates from your local high school is different from the national average. You measure an SRS of 78 female graduates and find that $\bar{x} = 63.1$ inches. What are your null and alternative hypotheses?

14.12 Stating hypotheses. In planning a study of the birth weights of babies whose mothers did not see a doctor before delivery, a researcher states the hypotheses as

$$H_0: \bar{x} = 1000 \text{ grams}$$
$$H_a: \bar{x} < 1000 \text{ grams}$$

What's wrong with this?

Honest hypotheses?

Chinese and Japanese, for whom the number 4 is unlucky, die more often on the fourth day of the month than on other days. The authors of a study did a statistical test of the claim that the fourth day has more deaths than other days and found good evidence in favor of this claim. Can we trust this? Not if the authors looked at all days, picked the one with the most deaths, then made "this day is different" the claim to be tested. A critic raised that issue, and the authors replied: No, we had day 4 in mind in advance, so our test was legitimate.

P-value and statistical significance

The idea of stating a null hypothesis that we want to find evidence *against* seems odd at first. It may help to think of a criminal trial. The defendant is "innocent until proven guilty." That is, the null hypothesis is innocence and the prosecution must try to provide convincing evidence against this hypothesis. That's exactly how statistical tests work, though in statistics we deal with evidence provided by data and use a probability to say how strong the evidence is.

The probability that measures the strength of the evidence against a null hypothesis is called a *P-value*. Statistical tests generally work like this:

TEST STATISTIC AND *P*-VALUE

A **test statistic** calculated from the sample data measures how far the data diverge from what we would expect if the null hypothesis H_0 were true. Large values of the statistic show that the data are not consistent with H_0.

The probability, computed assuming that H_0 is true, that the test statistic would take a value as extreme or more extreme than that actually observed is called the **P-value** of the test. The smaller the P-value, the stronger the evidence against H_0 provided by the data.

Small P-values are evidence against H_0 because they say that the observed result would be unlikely to occur if H_0 were true. Large P-values fail to give evidence against H_0. Statistical software will give you the P-value of a test when you enter your null and alternative hypotheses and your data. So your most important task is to understand what a P-value says.

EXAMPLE 14.7 **Sweetening colas: one-sided _P_-value**

The study of sweetness loss in Example 14.5 tests the hypotheses

$$H_0: \mu = 0$$
$$H_a: \mu > 0$$

Because the alternative hypothesis says that $\mu > 0$, values of $\overline{x}$ greater than 0 favor H_a over H_0. The test statistic compares the observed $\overline{x}$ with the hypothesized value $\mu = 0$. For now, let's concentrate on the P-value.

The discussion on page 370 compares two colas, though Example 14.5 gives actual data only for one. For the first cola, the 10 tasters found mean sweetness loss $\overline{x} = 0.3$. For the second, the data gave $\overline{x} = 1.02$. *The P-value for each test is the probability of getting an $\overline{x}$ this large when the mean sweetness loss is really $\mu = 0$.*

The shaded area in Figure 14.6 shows the P-value when $\overline{x} = 0.3$. The Normal curve is the sampling distribution of $\overline{x}$ when the null hypothesis $H_0: \mu = 0$ is true. A Normal probability calculation (Exercise 14.13) shows that the P-value is $P(\overline{x} \geq 0.3) = 0.1711$.

A value as large as $\overline{x} = 0.3$ would appear just by chance in 17% of all samples when $H_0: \mu = 0$ is true. So observing $\overline{x} = 0.3$ is not strong evidence against H_0. On the other hand, you can calculate that the probability that $\overline{x}$ is 1.02 or larger when in fact $\mu = 0$ is only 0.0006. We would very rarely observe a mean sweetness loss of 1.02 or larger if H_0 were true. This small P-value provides strong evidence against H_0 and in favor of the alternative $H_a: \mu > 0$. ■

Figure 14.6 is actually the output of the *P-Value of a Test of Significance* applet, along with the information we entered into the applet. This applet automates the work of finding P-values for samples of size 50 or smaller under the "simple conditions" for inference about a mean.

FIGURE 14.6

The one-sided P-value for the cola with mean sweetness loss $\overline{x} = 0.3$ in Example 14.7. The figure shows both the input and the output for the P-value applet.

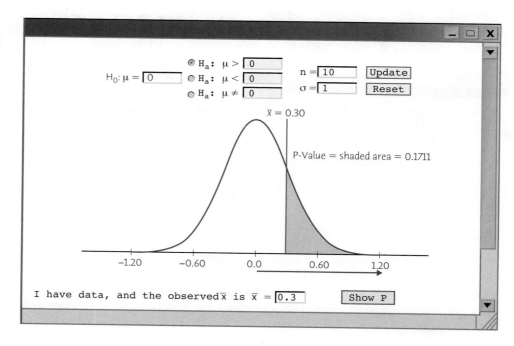

The alternative hypothesis sets the direction that counts as evidence against H_0. In Example 14.7, only large positive values count because the alternative is one-sided on the high side. If the alternative is two-sided, both directions count.

EXAMPLE 14.8 Job satisfaction: two-sided *P*-value

The study of job satisfaction in Example 14.6 requires that we test

$$H_0: \mu = 0$$

$$H_a: \mu \neq 0$$

Suppose we know that differences in job satisfaction scores (self-paced minus machine-paced) in the population of all workers follow a Normal distribution with standard deviation $\sigma = 60$.

Data from 18 workers give $\bar{x} = 17$. That is, these workers prefer the self-paced environment on the average. Because the alternative is two-sided, the *P*-value is the probability of getting an $\bar{x}$ at least as far from $\mu = 0$ *in either direction* as the observed $\bar{x} = 17$.

Enter the information for this example into the *P-Value of a Test of Significance* applet and click "Show P." Figure 14.7 shows the applet output as well as the information we entered. The *P*-value is the sum of the two shaded areas under the Normal curve. It is $P = 0.2302$. Values as far from 0 as $\bar{x} = 17$ (in either direction) would happen 23% of the time when the true population mean is $\mu = 0$. An outcome that would occur so often when H_0 is true is not good evidence against H_0. ∎

The conclusion of Example 14.8 is *not* that H_0 is true. The study looked for evidence against $H_0: \mu = 0$ and failed to find strong evidence. That is all we can

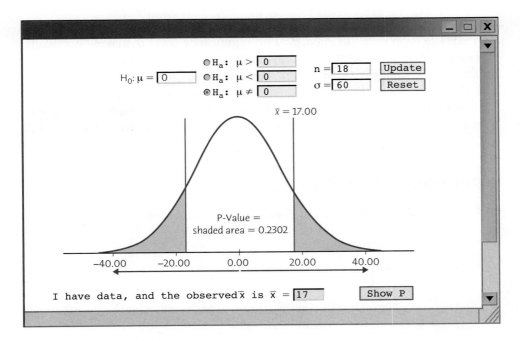

FIGURE 14.7

The two-sided *P*-value for Example 14.8. The figure shows both the input and the output for the *P-value* applet. Note that the *P*-value is the shaded area under the curve, not the unshaded area.

say. No doubt the mean μ for the population of all assembly workers is not exactly equal to 0. A large enough sample would give evidence of the difference, even if it is very small. Tests of significance assess the evidence against H_0. If the evidence is strong, we can confidently reject H_0 in favor of the alternative. *Failing to find evidence against H_0 means only that the data are consistent with H_0, not that we have clear evidence that H_0 is true.*

In Examples 14.7 and 14.8, we decided that P-value $P = 0.0006$ was strong evidence against the null hypothesis and that P-values $P = 0.1711$ and $P = 0.2302$ did not give convincing evidence. There is no rule for how small a P-value we should require to reject H_0—it's a matter of judgment and depends on the specific circumstances.

Nonetheless, we can compare a P-value with some fixed values that are in common use as standards for evidence against H_0. The most common fixed values are 0.05 and 0.01. If $P \leq 0.05$, there is no more than 1 chance in 20 that a sample would give evidence this strong just by chance when H_0 is actually true. If $P \leq 0.01$, we have a result that in the long run would happen no more than once per 100 samples if H_0 were true. These fixed standards for P-values are called ***significance level*** **significance levels.** We use α, the Greek letter alpha, to stand for a significance level.

STATISTICAL SIGNIFICANCE

If the P-value is as small or smaller than α, we say that the data are **statistically significant at level α.**

"Significant" in the statistical sense does not mean "important." It means simply "not likely to happen just by chance." The significance level α makes "not likely" more exact. Significance at level 0.01 is often expressed by the statement "The results were significant ($P < 0.01$)." Here P stands for the P-value. The actual P-value is more informative than a statement of significance because it allows us to assess significance at any level we choose. For example, a result with $P = 0.03$ is significant at the $\alpha = 0.05$ level but is not significant at the $\alpha = 0.01$ level.

APPLY YOUR KNOWLEDGE

14.13 Sweetening colas: find the *P*-value. The P-value for the first cola in Example 14.7 is the probability (taking the null hypothesis $\mu = 0$ to be true) that $\overline{x}$ takes a value at least as large as 0.3.

(a)   What is the sampling distribution of $\overline{x}$ when $\mu = 0$? This distribution appears in Figure 14.6.

(b)   Do a Normal probability calculation to find the P-value. Your result should agree with Example 14.7 up to roundoff error.

14.14 Job satisfaction: find the *P*-value. The *P*-value in Example 14.8 is the probability (taking the null hypothesis $\mu = 0$ to be true) that $\bar{x}$ takes a value at least as far from 0 as 17.

(a) What is the sampling distribution of $\bar{x}$ when $\mu = 0$? This distribution appears in Figure 14.7.

(b) Do a Normal probability calculation to find the *P*-value. Your result should agree with Example 14.8 up to roundoff error.

14.15 Protecting long-distance runners. A randomized comparative experiment compared vitamin C with a placebo as protection against respiratory infections after running a very long distance. The report of the study said:[5]

> Sixty-eight percent of the runners in the placebo group reported the development of symptoms of upper respiratory tract infection after the race; this was significantly more (P < 0.01) than that reported by the vitamin C–supplemented group (33%).

John Lund/Sam Diephuis/Photolibrary

(a) Explain to someone who knows no statistics why "significantly more" means there is good reason to think that vitamin C works.

(b) Now explain more exactly: what does $P < 0.01$ mean?

14.16 Student attitudes. Exercise 14.6 describes a study of the attitudes of older college students. You stated the null and alternative hypotheses in Exercise 14.8.

(a) One sample of 25 students had mean SSHA score $\bar{x} = 118.6$. Enter this $\bar{x}$, along with the other required information, into the *P-Value of a Test of Significance* applet. What is the *P*-value? Is this outcome statistically significant at the $\alpha = 0.05$ level? At the $\alpha = 0.01$ level?

(b) Another sample of 25 students had $\bar{x} = 125.8$. Use the applet to find the *P*-value for this outcome. Is it statistically significant at the $\alpha = 0.05$ level? At the $\alpha = 0.01$ level?

(c) Explain briefly why these *P*-values tell us that one outcome is strong evidence against the null hypothesis and that the other outcome is not.

14.17 Measuring conductivity. Exercise 14.7 describes 6 measurements of the electrical conductivity of a liquid. You stated the null and alternative hypotheses in Exercise 14.9.

(a) One set of measurements has mean conductivity $\bar{x} = 4.98$. Enter this $\bar{x}$, along with the other required information, into the *P-Value of a Test of Significance* applet. What is the *P*-value? Is this outcome statistically significant at the $\alpha = 0.05$ level? At the $\alpha = 0.01$ level?

(b) Another set of measurements has $\bar{x} = 4.7$. Use the applet to find the *P*-value for this outcome. Is it statistically significant at the $\alpha = 0.05$ level? At the $\alpha = 0.01$ level?

(c) Explain briefly why these *P*-values tell us that one outcome is strong evidence against the null hypothesis and that the other outcome is not.

Tests for a population mean

We have used tests for hypotheses about the mean μ of a population, under the "simple conditions," to introduce tests of significance. The big idea is the reasoning of a test: *data that would rarely occur if the null hypothesis H₀ were true provide evidence that H₀ is not true.* The *P*-value gives us a probability to measure "would rarely occur." In practice, the steps in carrying out a significance test mirror the overall four-step process for organizing realistic statistical problems.

Significance strikes down a new drug

The pharmaceutical company Pfizer spent $1 billion developing a new cholesterol-fighting drug. The final test for its effectiveness was a clinical trial with 15,000 subjects. To enforce double-blindness, only an independent group of experts saw the data during the trial. Three years into the trial, the monitors declared that there was a statistically significant excess of deaths and of heart problems in the group assigned to the new drug. Pfizer ended the trial. There went $1 billion.

> **TESTS OF SIGNIFICANCE: THE FOUR-STEP PROCESS**
>
> **STATE:** What is the practical question that requires a statistical test?
>
> **PLAN:** Identify the parameter, state null and alternative hypotheses, and choose the type of test that fits your situation.
>
> **SOLVE:** Carry out the test in three phases:
>
> 1. **Check the conditions** for the test you plan to use.
> 2. Calculate the **test statistic.**
> 3. Find the **P-value.**
>
> **CONCLUDE:** Return to the practical question to describe your results in this setting.

Once you have stated your question, formulated hypotheses, and checked the conditions for your test, you or your software can find the test statistic and *P*-value by following a rule. Here is the rule for the test we have used in our examples.

> **z TEST FOR A POPULATION MEAN**
>
> Draw an SRS of size n from a Normal population that has unknown mean μ and known standard deviation σ. To **test the null hypothesis that μ has a specified value,**
>
> $$H_0: \mu = \mu_0$$
>
> calculate the **one-sample z test statistic**
>
> $$z = \frac{\bar{x} - \mu_0}{\sigma/\sqrt{n}}$$
>
> In terms of a variable Z having the standard Normal distribution, the *P*-value for a test of H_0 against
>
> $$H_a: \mu > \mu_0 \quad \text{is} \quad P(Z \geq z)$$
>
>

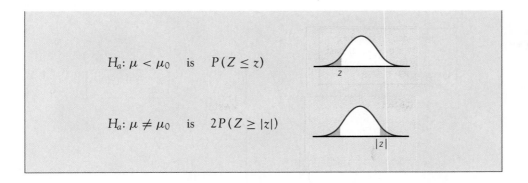

$H_a: \mu < \mu_0$ is $P(Z \leq z)$

$H_a: \mu \neq \mu_0$ is $2P(Z \geq |z|)$

As promised, the test statistic z measures how far the observed sample mean $\overline{x}$ deviates from the hypothesized population value μ_0. The measurement is in the familiar standard scale obtained by dividing by the standard deviation of $\overline{x}$. So we have a common scale for all z tests, and the 68–95–99.7 rule helps us see at once if $\overline{x}$ is far from μ_0. The pictures that illustrate the P-value look just like Figures 4.6 and 4.7 except that they are in the standard scale.

EXAMPLE 14.9 Executives' blood pressures

STATE: The National Center for Health Statistics reports that the systolic blood pressure for males 35 to 44 years of age has mean 128 and standard deviation 15. The medical director of a large company looks at the medical records of 72 executives in this age group and finds that the mean systolic blood pressure in this sample is $\overline{x} = 126.07$. Is this evidence that the company's executives have a different mean blood pressure from the general population?

PLAN: The null hypothesis is "no difference" from the national mean $\mu_0 = 128$. The alternative is two-sided, because the medical director did not have a particular direction in mind before examining the data. So the hypotheses about the unknown mean μ of the executive population are

$$H_0: \mu = 128$$
$$H_a: \mu \neq 128$$

We know that the one-sample z test is appropriate for these hypotheses under the "simple conditions."

SOLVE: As part of the "simple conditions," suppose we know that executives' blood pressures follow a Normal distribution with standard deviation $\sigma = 15$. Software can now calculate z and P for you. Going ahead by hand, the **test statistic** is

$$z = \frac{\overline{x} - \mu_0}{\sigma/\sqrt{n}} = \frac{126.07 - 128}{15/\sqrt{72}}$$
$$= -1.09$$

To help find the **P-value,** sketch the standard Normal curve and mark on it the observed value of z. Figure 14.8 shows that the P-value is the probability that a standard Normal

ImageState/Alamy

FIGURE 14.8

The P-value for the two-sided test in Example 14.9. The observed value of the test statistic is $z = -1.09$.

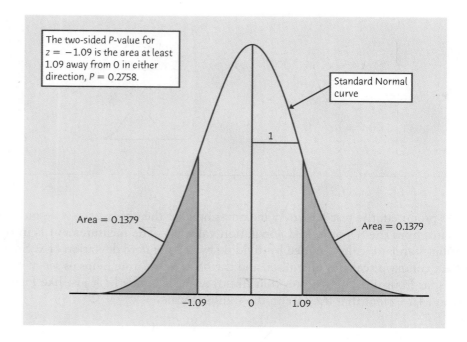

The two-sided P-value for $z = -1.09$ is the area at least 1.09 away from 0 in either direction, $P = 0.2758$.

Standard Normal curve

Area = 0.1379

Area = 0.1379

−1.09   0   1.09

variable Z takes a value at least 1.09 away from zero. From Table A or software, this probability is

$$P = 2P(Z < -1.09) = (2)(0.1379) = 0.2758$$

CONCLUDE: More than 27% of the time, an SRS of size 72 from the general male population would have a mean blood pressure at least as far from 128 as that of the executive sample. The observed $\bar{x} = 126.07$ is therefore not good evidence that executives differ from other men. ■

In this chapter we are acting as if the "simple conditions" stated on page 360 are true. In practice, you must verify these conditions.

1. **SRS:** The most important condition is that the 72 executives in the sample are an SRS from the population of all middle-aged male executives in the company. We should check this requirement by asking how the data were produced. If medical records are available only for executives with recent medical problems, for example, the data are of little value for our purpose because of the obvious health bias. It turns out that all executives are given a free annual medical exam, and that the medical director selected 72 exam results at random.

2. **Normal distribution:** We should also examine the distribution of the 72 observations to look for signs that the population distribution is not Normal.

3. **Known σ:** It really is unrealistic to suppose that we know that $\sigma = 15$. We will see in Chapter 17 that it is easy to do away with the need to know σ.

APPLY YOUR KNOWLEDGE

14.18 The z statistic. Published reports of research work are terse. They often report just a test statistic and P-value. For example, the conclusion of Example 14.9 might be stated as "($z = -1.09$, $P = 0.2758$)." Find the values of the one-sample z statistic needed to complete these conclusions:

(a)   For the first cola in Example 14.7, ($z = ?$, $P = 0.1711$).

(b)   For the second cola in Example 14.7, ($z = ?$, $P = 0.0006$).

(c)   For Example 14.8, ($z = ?$, $P = 0.2302$).

14.19 Measuring conductivity. Here are 6 measurements of the electrical conductivity of a liquid:

$$5.32 \quad 4.88 \quad 5.10 \quad 4.73 \quad 5.15 \quad 4.75$$

The liquid is supposed to have conductivity 5. Do the measurements give good evidence that the true conductivity is not 5?

   The 6 measurements are an SRS from the population of all results we would get if we kept measuring conductivity forever. This population has a Normal distribution with mean equal to the true conductivity of the liquid and standard deviation 0.2. Use this information to carry out a test, following the four-step process as illustrated in Example 14.9.

14.20 Reading a computer screen. Does the use of fancy type fonts slow down the reading of text on a computer screen? Adults can read four paragraphs of text in an average time of 22 seconds in the common Times New Roman font. Ask 25 adults to read this text in the ornate font named Gigi. Here are their times:[6]

23.2	21.2	28.9	27.7	29.1	27.3	16.1	22.6	25.6
34.2	23.9	26.8	20.5	34.3	21.4	32.6	26.2	34.1
31.5	24.6	23.0	28.6	24.4	28.1	41.3		

Suppose that reading times are Normal with $\sigma = 6$ seconds. Is there good evidence that the mean reading time for Gigi is greater than 22 seconds? Follow the four-step process as illustrated in Example 14.9.

Significance from a table

Statistics in practice uses technology (graphing calculator or software) to get P-values quickly and accurately. In the absence of suitable technology, you can get approximate P-values quickly by comparing the value of your test statistic with critical values from a table. For the z statistic, the table is Table C, the same table we used for confidence intervals.

   Look at the bottom row of critical values in Table C, labeled z^*. At the top of the table, you see the confidence level C for each z^*. At the bottom of the table, you see both the one-sided and two-sided P-values for each z^*. Values of a test statistic z that are farther out than a z^* (in the direction given by the alternative hypothesis) are significant at the level that matches z^*.

Robert Daly/Getty Images

SIGNIFICANCE FROM A TABLE OF CRITICAL VALUES

To find the approximate P-value for any z statistic, compare z (ignoring its sign) with the critical values z^* at the bottom of Table C. If z falls between two values of z^*, the P-value falls between the two corresponding values of P in the "One-sided P" or the "Two-sided P" row of Table C.

EXAMPLE 14.10 Is it significant?

z^*	2.054	2.326
One-sided P	.02	.01

The z statistic for a one-sided test is $z = 2.13$. How significant is this result? Compare $z = 2.13$ with the z^* row in Table C. It lies between $z^* = 2.054$ and $z^* = 2.326$. So the P-value lies between the corresponding entries in the "One-sided P" row, which are $P = 0.02$ and $P = 0.01$. This z *is* significant at the $\alpha = 0.02$ level and *is not* significant at the $\alpha = 0.01$ level.

Figure 14.9 illustrates the situation. The shaded area under the Normal curve is the P-value for $z = 2.13$. You can see that P falls between the areas to the right of the two critical values, for $P = 0.02$ and $P = 0.01$.

z^*	1.036	1.282
Two-sided P	.30	.20

The z statistic in Example 14.9 is $z = -1.09$. The alternative hypothesis is two-sided. Compare $z = -1.09$ (ignoring the minus sign) with the z^* row in Table C. It lies between $z^* = 1.036$ and $z^* = 1.282$. So the P-value lies between the matching entries in the "Two-sided P" row, $P = 0.30$ and $P = 0.20$. This is enough to conclude that the data do not provide good evidence against the null hypothesis. ■

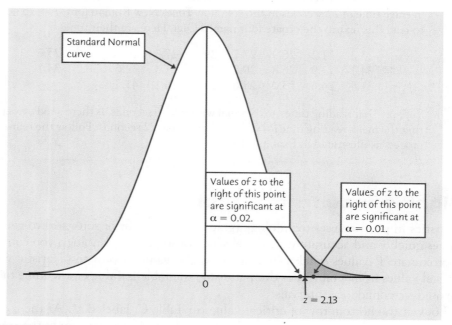

FIGURE 14.9

Is it significant? The test statistic value $z = 2.13$ falls between the critical values required for significance at the $\alpha = 0.02$ and $\alpha = 0.01$ levels. So the test *is* significant at $\alpha = 0.02$ and *is not* significant at $\alpha = 0.01$.

14.21 Significance from a table. A test of H_0: $\mu = 1$ against H_a: $\mu > 1$ has test statistic $z = 1.776$. Is this test significant at the 5% level ($\alpha = 0.05$)? Is it significant at the 1% level ($\alpha = 0.01$)?

14.22 Significance from a table. A test of H_0: $\mu = 1$ against H_a: $\mu \neq 1$ has test statistic $z = 1.776$. Is this test significant at the 5% level ($\alpha = 0.05$)? Is it significant at the 1% level?

14.23 Testing a random number generator. A random number generator is supposed to produce random numbers that are uniformly distributed on the interval from 0 to 1. If this is true, the numbers generated come from a population with $\mu = 0.5$ and $\sigma = 0.2887$. A command to generate 100 random numbers gives outcomes with mean $\overline{x} = 0.4365$. Assume that the population σ remains fixed. We want to test

$$H_0: \mu = 0.5$$
$$H_a: \mu \neq 0.5$$

(a)   Calculate the value of the z test statistic.

(b)   Use Table C: is z significant at the 5% level ($\alpha = 0.05$)?

(c)   Use Table C: is z significant at the 1% level ($\alpha = 0.01$)?

(d)   Between which two Normal critical values z^* in the bottom row of Table C does z lie? Between what two numbers does the P-value lie? Does the test give good evidence against the null hypothesis?

CHAPTER 14 SUMMARY

■ A **confidence interval** uses sample data to estimate an unknown population parameter with an indication of how accurate the estimate is and of how confident we are that the result is correct.

■ Any confidence interval has two parts: an interval calculated from the data and a confidence level C. The **interval** often has the form

estimate ± margin of error

■ The **confidence level** is the success rate of the method that produces the interval. That is, C is the probability that the method will give a correct answer. If you use 95% confidence intervals often, in the long run 95% of your intervals will contain the true parameter value. You do not know whether or not a 95% confidence interval calculated from a particular set of data contains the true parameter value.

■ A level C **confidence interval for the mean** μ of a Normal population with known standard deviation σ, based on an SRS of size n, is given by

$$\overline{x} \pm z^* \frac{\sigma}{\sqrt{n}}$$

■ The **critical value** z^* is chosen so that the standard Normal curve has area C between $-z^*$ and z^*.

■ A **test of significance** assesses the evidence provided by data against a **null hypothesis** H_0 in favor of an **alternative hypothesis** H_a.

■ Hypotheses are always stated in terms of population parameters. Usually H_0 is a statement that no effect is present, and H_a says that a parameter differs from its null value in a specific direction (**one-sided alternative**) or in either direction (**two-sided alternative**).

■ The essential reasoning of a significance test is as follows. Suppose for the sake of argument that the null hypothesis is true. If we repeated our data production many times, would we often get data as inconsistent with H_0 as the data we actually have? Data that would rarely occur if H_0 were true provide evidence against H_0.

■ A test is based on a **test statistic** that measures how far the sample outcome is from the value stated by H_0.

■ The **P-value** of a test is the probability, computed supposing H_0 to be true, that the test statistic will take a value at least as extreme as that actually observed. Small P-values indicate strong evidence against H_0. To calculate a P-value we must know the sampling distribution of the test statistic when H_0 is true.

■ If the P-value is as small or smaller than a specified value α, the data are **statistically significant** at significance level α.

■ **Significance tests for the null hypothesis H_0: $\mu = \mu_0$** concerning the unknown mean μ of a population are based on the **one-sample z test statistic**

$$z = \frac{\bar{x} - \mu_0}{\sigma/\sqrt{n}}$$

■ The z test assumes an SRS of size n from a Normal population with known population standard deviation σ. P-values can be obtained either with computations from the standard Normal distribution or by using technology (applet or software).

CHECK YOUR SKILLS

14.24 To give a 96% confidence interval for a population mean μ, you would use the critical value

    (a) $z^* = 1.960$.   (b) $z^* = 2.054$.   (c) $z^* = 2.326$.

14.25 You use software to carry out a test of significance. The program tells you that the P-value is $P = 0.031$. This result is

    (a) not significant at either $\alpha = 0.05$ or $\alpha = 0.01$.

    (b) significant at $\alpha = 0.05$ but not at $\alpha = 0.01$.

    (c) significant at both $\alpha = 0.05$ and $\alpha = 0.01$.

14.26 The z statistic for a one-sided test is $z = 2.433$. This test is

    (a)   not significant at either $\alpha = 0.05$ or $\alpha = 0.01$.

    (b)   significant at $\alpha = 0.05$ but not at $\alpha = 0.01$.

    (c)   significant at both $\alpha = 0.05$ and $\alpha = 0.01$.

Use the following information for Exercises 14.27 through 14.30. A laboratory scale is known to have a standard deviation of $\sigma = 0.001$ gram in repeated weighings. Scale readings in repeated weighings are Normally distributed, with mean equal to the true weight of the specimen. Three weighings of a specimen on this scale give 3.412, 3.416, and 3.414 grams.

Spencer Grant/Photo Edit

14.27 A 95% confidence interval for the true weight of this specimen is

    (a) 3.414 ± 0.00113.   (b) 3.414 ± 0.00065.   (c) 3.414 ± 0.00196.

14.28 You want a 99% confidence interval for the true weight of this specimen. The margin of error for this interval will be

    (a)   smaller than the margin of error for 95% confidence.

    (b)   greater than the margin of error for 95% confidence.

    (c)   about the same as the margin of error for 95% confidence.

14.29 The z statistic for testing H_0: $\mu = 3.41$ based on these 3 measurements is

    (a) $z = 0.004$.   (b) $z = 4$.   (c) $z = 6.928$.

14.30 Another specimen is weighed 8 times on this scale. The average weight is 4.1602 grams. A 99% confidence interval for the true weight of this specimen is

    (a) 4.1602 ± 0.00032.   (b) 4.1602 ± 0.00069.   (c) 4.1602 ± 0.00091.

14.31 Experiments on learning in animals sometimes measure how long it takes mice to find their way through a maze. The mean time is 18 seconds for one particular maze. A researcher thinks that a loud noise will cause the mice to complete the maze faster. She measures how long each of 10 mice takes with a noise as stimulus. The sample mean is $\bar{x} = 16.5$ seconds. The null hypothesis for the significance test is

    (a) H_0: $\mu = 18$.   (b) H_0: $\mu = 16.5$.   (c) H_0: $\mu < 18$.

14.32 The alternative hypothesis for the test in Exercise 14.31 is

    (a) H_a: $\mu \neq 18$.   (b) H_a: $\mu < 18$.   (c) H_a: $\mu = 16.5$.

14.33 You use software to carry out a test of significance. The program tells you that the P-value is $P = 0.031$. This means that

    (a)   the probability that the null hypothesis is true is 0.031.

    (b)   the value of the test statistic is 0.031.

    (c)   a test statistic as extreme as these data give would happen with probability 0.031 if the null hypothesis were true.

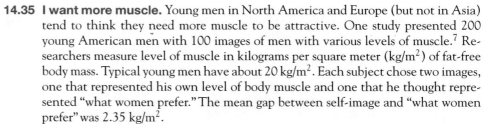

In all exercises that call for P-values, give the actual value if you use software or the P-value applet. Otherwise, use Table C to give values between which P must fall.

14.34 Student study times. A class survey in a large class for first-year college students asked, "About how many minutes do you study on a typical weeknight?" The mean response of the 269 students was $\overline{x} = 137$ minutes. Suppose that we know that the study time follows a Normal distribution with standard deviation $\sigma = 65$ minutes in the population of all first-year students at this university.

(a) Use the survey result to give a 99% confidence interval for the mean study time of all first-year students.

(b) What condition not yet mentioned must be met for your confidence interval to be valid?

14.35 I want more muscle. Young men in North America and Europe (but not in Asia) tend to think they need more muscle to be attractive. One study presented 200 young American men with 100 images of men with various levels of muscle.[7] Researchers measure level of muscle in kilograms per square meter (kg/m^2) of fat-free body mass. Typical young men have about $20\ kg/m^2$. Each subject chose two images, one that represented his own level of body muscle and one that he thought represented "what women prefer." The mean gap between self-image and "what women prefer" was $2.35\ kg/m^2$.

Suppose that the "muscle gap" in the population of all young men has a Normal distribution with standard deviation $2.5\ kg/m^2$. Give a 90% confidence interval for the mean amount of muscle young men think they should add to be attractive to women. (They are wrong: women actually prefer a level close to that of typical men.)

14.36 An outlier strikes. There were actually 270 responses to the class survey in Exercise 14.34. One student claimed to study 30,000 minutes per night. We know he's joking, so we left out this value. If we did a calculation without looking at the data, we would get $\overline{x} = 248$ minutes for all 270 students. Now what is the 99% confidence interval for the population mean? (Continue to use $\sigma = 65$.) Compare the new interval with that in Exercise 14.34. The message is clear: always look at your data, because outliers can greatly change your result.

14.37 Explaining confidence. A student reads that a 95% confidence interval for the mean body mass index (BMI) of young American women is 26.8 ± 0.6. Asked to explain the meaning of this interval, the student says, "95% of all young women have BMI between 26.2 and 27.4." Is the student right? Explain your answer.

14.38 Explaining confidence. You ask another student to explain the confidence interval for mean BMI described in the previous exercise. The student answers, "We can be 95% confident that future samples of young women will have mean BMI between 26.2 and 27.4." Is this explanation correct? Explain your answer.

14.39 Explaining confidence. Here is an explanation from the Associated Press concerning one of its opinion polls. Explain briefly but clearly in what way this explanation is incorrect.

For a poll of 1,600 adults, the variation due to sampling error is no more than three percentage points either way. The error margin is said to be valid at the 95 percent confidence level. This means that, if the same questions were repeated in 20 polls, the results of at least 19 surveys would be within three percentage points of the results of this survey.

14.40 Student study times. Exercise 14.34 describes a class survey in which students claimed to study an average of $\bar{x} = 137$ minutes on a typical weeknight. Regard these students as an SRS from the population of all first-year students at this university. Does the study give good evidence that students claim to study more than 2 hours per night on the average?

(a) State null and alternative hypotheses in terms of the mean study time in minutes for the population.

(b) What is the value of the test statistic z?

(c) What is the P-value of the test? Can you conclude that students do claim to study more than two hours per weeknight on the average?

14.41 I want more muscle. If young men thought that their own level of muscle was about what women prefer, the mean "muscle gap" in the study described in Exercise 14.35 would be 0. We suspect (before seeing the data) that young men think women prefer more muscle than they themselves have.

(a) State null and alternative hypotheses for testing this suspicion.

(b) What is the value of the test statistic z?

(c) You can tell just from the value of z that the evidence in favor of the alternative is very strong (that is, the P-value is very small). Explain why this is true.

14.42 Hotel managers' personalities. Successful hotel managers must have personality characteristics often thought of as feminine (such as "compassionate") as well as those often thought of as masculine (such as "forceful"). The Bem Sex-Role Inventory (BSRI) is a personality test that gives separate ratings for female and male stereotypes, both on a scale of 1 to 7. A sample of 148 male general managers of three-star and four-star hotels had mean BSRI femininity score $\bar{y} = 5.29$.[8] The mean score for the general male population is $\mu = 5.19$. Do hotel managers on the average differ significantly in femininity score from men in general? Assume that the standard deviation of scores in the population of all male hotel managers is the same as the $\sigma = 0.78$ for the adult male population.

(a) State null and alternative hypotheses in terms of the mean femininity score μ for male hotel managers.

(b) Find the z test statistic.

(c) What is the P-value for your z? What do you conclude about male hotel managers?

14.43 Is this what *P* means? When asked to explain the meaning of "the P-value was $P = 0.03$," a student says, "This means there is only probability 0.03 that the null hypothesis is true." Explain what $P = 0.03$ really means in a way that makes it clear that the student's explanation is wrong.

14.44 How to show that you are rich. Every society has its own marks of wealth and prestige. In ancient China, it appears that owning pigs was such a mark. Evidence comes from examining burial sites. The skulls of sacrificed pigs tend to appear along with expensive ornaments, which suggests that the pigs, like the ornaments, signal the wealth and prestige of the person buried. A study of burials from around 3500 B.C. concluded that "there are striking differences in grave goods between burials with pig skulls and burials without them.... A test indicates that the two samples of total artifacts are significantly different at the 0.01 level."[9] Explain clearly why "significantly different at the 0.01 level" gives good reason to think that there really is a systematic difference between burials that contain pig skulls and those that lack them.

Alastair Shay; Papilio/CORBIS

14.45 Cicadas as fertilizer? Every 17 years, swarms of cicadas emerge from the ground in the eastern United States, live for about six weeks, then die. There are so many cicadas that their dead bodies can serve as fertilizer. In an experiment, a researcher added cicadas under some plants in a natural plot of bellflowers on the forest floor, leaving other plants undisturbed. "In this experiment, cicada-supplemented bellflowers from a natural field population produced foliage with 12% greater nitrogen content relative to controls ($P = 0.031$)."[10] A colleague who knows no statistics says that an increase of 12% isn't a lot—maybe it's just an accident due to natural variation among the plants. Explain in simple language how "$P = 0.031$" answers this objection.

14.46 Forests and windstorms. Does the destruction of large trees in a windstorm change forests in any important way? Here is the conclusion of a study that found that the answer is "No":

> We found surprisingly little divergence between treefall areas and adjacent control areas in the richness of woody plants ($P = 0.62$), in total stem densities ($P = 0.98$), or in population size or structure for any individual shrub or tree species.[11]

The two P-values refer to null hypotheses that say "no change" in measurements between treefall and control areas. Explain clearly why these values provide no evidence of change.

14.47 5% versus 1%. Sketch the standard Normal curve for the z test statistic and mark off areas under the curve to show why a value of z that is significant at the 1% level in a one-sided test is always significant at the 5% level. If z is significant at the 5% level, what can you say about its significance at the 1% level?

14.48 The wrong alternative. One of your friends is comparing movie ratings by female and male students for a class project. She starts with no expectations as to which sex will rate a movie more highly. After seeing that women rate a particular movie more highly than men, she tests a one-sided alternative about the mean ratings,

$$H_0: \mu_F = \mu_M$$
$$H_a: \mu_F > \mu_M$$

She finds $z = 2.1$ with one-sided P-value $P = 0.0179$.

(a)   Explain why your friend should have used the two-sided alternative hypothesis.

(b)   What is the correct P-value for $z = 2.1$?

14.49 The wrong P. The report of a study of seat belt use by drivers says, "Hispanic drivers were not significantly more likely than White/non-Hispanic drivers to overreport safety belt use (27.4 vs. 21.1%, respectively; $z = 1.33$, $P > 1.0$.)"[12] How do you know that the P-value given is incorrect? What is the correct one-sided P-value for test statistic $z = 1.33$?

*Exercises 14.50 to 14.56 ask you to answer questions from data. Assume that the "simple conditions" hold in each case. The exercise statements give you the **State** step of the four-step process. In your work, follow the **Plan, Solve,** and **Conclude** steps, illustrated in Example 14.3 for a confidence interval and in Example 14.9 for a test of significance.*

14.50 Pulling wood apart. How heavy a load (pounds) is needed to pull apart pieces of Douglas fir 4 inches long and 1.5 inches square? Here are data from students doing a laboratory exercise:

33,190	31,860	32,590	26,520	33,280
32,320	33,020	32,030	30,460	32,700
23,040	30,930	32,720	33,650	32,340
24,050	30,170	31,300	28,730	31,920

(a)   We are willing to regard the wood pieces prepared for the lab session as an SRS of all similar pieces of Douglas fir. Engineers also commonly assume that characteristics of materials vary Normally. Make a graph to show the shape of the distribution for these data. Does the Normality condition appear safe? Suppose that the strength of pieces of wood like these follows a Normal distribution with standard deviation 3000 pounds.

(b)   Give a 90% confidence interval for the mean load required to pull the wood apart.

14.51 Bone loss by nursing mothers. Breast-feeding mothers secrete calcium into their milk. Some of the calcium may come from their bones, so mothers may lose bone mineral. Researchers measured the percent change in mineral content of the spines of 47 mothers during three months of breast-feeding.[13] Here are the data:

−4.7	−2.5	−4.9	−2.7	−0.8	−5.3	−8.3	−2.1	−6.8	−4.3
2.2	−7.8	−3.1	−1.0	−6.5	−1.8	−5.2	−5.7	−7.0	−2.2
−6.5	−1.0	−3.0	−3.6	−5.2	−2.0	−2.1	−5.6	−4.4	−3.3
−4.0	−4.9	−4.7	−3.8	−5.9	−2.5	−0.3	−6.2	−6.8	1.7
0.3	−2.3	0.4	−5.3	0.2	−2.2	−5.1			

(a)   The researchers are willing to consider these 47 women as an SRS from the population of all nursing mothers. Suppose that the percent change in this population has standard deviation $\sigma = 2.5\%$. Make a stemplot of the data to see that they appear to follow a Normal distribution quite closely. (Don't forget that you need both a 0 and a −0 stem because there are both positive and negative values.)

(b)   Use a 99% confidence interval to estimate the mean percent change in the population.

14.52 **Pulling wood apart.** Exercise 14.50 gives data on the pounds of load needed to pull apart pieces of Douglas fir. The data are a random sample from a Normal distribution with standard deviation 3000 pounds.

(a) Is there significant evidence at the $\alpha = 0.10$ level against the hypothesis that the mean is 32,000 pounds for the two-sided alternative?

(b) Is there significant evidence at the $\alpha = 0.10$ level against the hypothesis that the mean is 31,500 pounds for the two-sided alternative?

14.53 **Bone loss by nursing mothers.** Exercise 14.51 gives the percent change in the mineral content of the spine for 47 mothers during three months of nursing a baby. As in that exercise, suppose that the percent change in the population of all nursing mothers has a Normal distribution with standard deviation $\sigma = 2.5\%$. Do these data give good evidence that on the average nursing mothers lose bone mineral?

14.54 **This wine stinks.** Sulfur compounds cause "off-odors" in wine, so winemakers want to know the odor threshold, the lowest concentration of a compound that the human nose can detect. The odor threshold for dimethyl sulfide (DMS) in trained wine tasters is about 25 micrograms per liter of wine ($\mu g/l$). The untrained noses of consumers may be less sensitive, however. Here are the DMS odor thresholds for 10 untrained students:

31   31   43   36   23   34   32   30   20   24

(a) Assume that the standard deviation of the odor threshold for untrained noses is known to be $\sigma = 7 \ \mu g/l$. Briefly discuss the other two "simple conditions," using a stemplot to verify that the distribution is roughly symmetric with no outliers.

(b) Give a 95% confidence interval for the mean DMS odor threshold among all students.

14.55 **Eye grease.** Athletes performing in bright sunlight often smear black eye grease under their eyes to reduce glare. Does eye grease work? In one study, 16 student subjects took a test of sensitivity to contrast after 3 hours facing into bright sun, both with and without eye grease. This is a matched pairs design. Here are the differences in sensitivity, with eye grease minus without eye grease:[14]

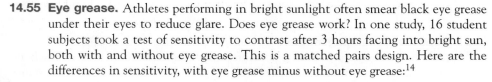

0.07	0.64	−0.12	−0.05	−0.18	0.14	−0.16	0.03
0.05	0.02	0.43	0.24	−0.11	0.28	0.05	0.29

We want to know whether eye grease increases sensitivity on the average.

(a) What are the null and alternative hypotheses? Say in words what mean μ your hypotheses concern.

(b) Suppose that the subjects are an SRS of all young people with normal vision, that contrast differences follow a Normal distribution in this population, and that the standard deviation of differences is $\sigma = 0.22$. Carry out a test of significance.

CORBIS/Veer

14.56 **This wine stinks.** Are untrained students less sensitive on the average than trained tasters in detecting "off-odors" in wine? Exercise 14.54 gives the lowest levels of dimethyl sulfide (DMS) that 10 students could detect. The units are micrograms of DMS per liter of wine ($\mu g/l$). Assume that the odor threshold for untrained noses is Normally distributed with $\sigma = 7\ \mu g/l$. Is there evidence that the mean threshold for untrained tasters is greater than $25\ \mu g/l$?

14.57 Tests from confidence intervals. A confidence interval for the population mean μ tells us which values of μ are plausible (those inside the interval) and which values are not plausible (those outside the interval) at the chosen level of confidence. You can use this idea to carry out a test of any null hypothesis H_0: $\mu = \mu_0$ starting with a confidence interval: *reject H_0 if μ_0 is outside the interval and fail to reject if μ_0 is inside the interval.*

The alternative hypothesis is always two-sided, H_a: $\mu \neq \mu_0$, because the confidence interval extends in both directions from $\overline{x}$. A 95% confidence interval leads to a test at the 5% significance level because the interval is wrong 5% of the time. In general, confidence level C leads to a test at significance level $\alpha = 1 - C$.

(a) In Example 14.9, a medical director found mean blood pressure $\overline{x} = 126.07$ for an SRS of 72 executives. The standard deviation of the blood pressures of all executives is $\sigma = 15$. Give a 90% confidence interval for the mean blood pressure μ of all executives.

(b) The hypothesized value $\mu_0 = 128$ falls *inside* this confidence interval. Carry out the z test for H_0: $\mu = 128$ against the two-sided alternative. Show that the test is *not significant* at the 10% level.

(c) The hypothesized value $\mu_0 = 129$ falls *outside* this confidence interval. Carry out the z test for H_0: $\mu = 129$ against the two-sided alternative. Show that the test *is significant* at the 10% level.

14.58 Tests from confidence intervals. A 95% confidence interval for a population mean is 31.5 ± 3.4. Use the method described in the previous exercise to answer these questions.

(a) With a two-sided alternative, can you reject the null hypothesis that $\mu = 34$ at the 5% ($\alpha = 0.05$) significance level? Why?

(b) With a two-sided alternative, can you reject the null hypothesis that $\mu = 35$ at the 5% significance level? Why?

Thinking about Inference

To this point, we have met just two procedures for statistical inference. Both concern inference about the mean μ of a population when the "simple conditions" (page 360) are true: the data are an SRS, the population has a Normal distribution, and we know the standard deviation σ of the population. Under these conditions, a confidence interval for the mean μ is

$$\overline{x} \pm z^* \frac{\sigma}{\sqrt{n}}$$

To test a hypothesis $H_0: \mu = \mu_0$ we use the one-sample z statistic:

$$z = \frac{\overline{x} - \mu_0}{\sigma/\sqrt{n}}$$

We call these **z procedures** because they both start with the one-sample z statistic and use the standard Normal distribution.

In later chapters we will modify these procedures for inference about a population mean to make them useful in practice. We will also introduce procedures for confidence intervals and tests in most of the settings we met in learning to explore data. There are libraries—both of books and of software—full of more elaborate statistical techniques. The reasoning of confidence intervals and tests is the same, no matter how elaborate the details of the procedure are.

z procedures

There is a saying among statisticians that "mathematical theorems are true; statistical methods are effective when used with judgment." That the one-sample z statistic has the standard Normal distribution when the null hypothesis is true is a mathematical theorem. Effective use of statistical methods requires more than knowing such facts. It requires even more than understanding the underlying reasoning. This chapter begins the process of helping you develop the judgment needed to use statistics in practice. That process will continue in examples and exercises through the rest of this book.

Conditions for inference in practice

Any confidence interval or significance test can be trusted only under specific conditions. It's up to you to understand these conditions and judge whether they fit your problem. With that in mind, let's look back at the "simple conditions" for the z procedures.

The final "simple condition," that we know the standard deviation σ of the population, is rarely satisfied in practice. The z procedures are therefore of little practical use. Fortunately, it's easy to remove the "known σ" condition. Chapter 17 shows how. The first two "simple conditions" (SRS, Normal population) are harder to escape. In fact, they represent the kinds of conditions needed if we are to trust almost any statistical inference. As you plan inference, you should always ask "Where did the data come from?" and you must often also ask "What is the shape of the population distribution?" This is the point where knowing mathematical facts gives way to the need for judgment.

Where did the data come from? *The most important requirement for any inference procedure is that the data come from a process to which the laws of probability apply.* Inference is most reliable when the data come from a random sample or a randomized comparative experiment. Random samples use chance to choose respondents. Randomized comparative experiments use chance to assign subjects to treatments. The deliberate use of chance ensures that the laws of probability apply to the outcomes, and this in turn ensures that statistical inference makes sense.

> **WHERE THE DATA COME FROM MATTERS**
>
> When you use statistical inference, you are acting as if your data are a random sample or come from a randomized comparative experiment.

If your data don't come from a random sample or a randomized comparative experiment, your conclusions may be challenged. To answer the challenge, you must usually rely on subject-matter knowledge, not on statistics. It is common to apply statistical inference to data that are not produced by random selection. When you see such a study, ask whether the data can be trusted as a basis for the conclusions of the study.

EXAMPLE 15.1 **The psychologist and the sociologist**

A psychologist is interested in how our visual perception can be fooled by optical illusions. Her subjects are students in Psychology 101 at her university. Most psychologists would agree that it's safe to treat the students as an SRS of all people with normal vision. There is nothing special about being a student that changes visual perception.

A sociologist at the same university uses students in Sociology 101 to examine attitudes toward poor people and antipoverty programs. Students as a group are younger than the adult population as a whole. Even among young people, students as a group come from more prosperous and better-educated homes. Even among students, this university isn't typical of all campuses. Even on this campus, students in a sociology course may have opinions that are quite different from those of engineering students. The sociologist can't reasonably act as if these students are a random sample from any interesting population. ■

Our first examples of inference, using the z procedures, act as if the data are an SRS from the population of interest. Let's look back at the examples in Chapter 14.

EXAMPLE 15.2 **Is it really an SRS?**

The NHANES survey that produced the BMI data for Example 14.1 used a complex multistage sample design, so it's a bit oversimplified to treat the BMI data as coming from an SRS from the population of young women.[1] The overall effect of the NHANES sample is close to an SRS. Nonetheless, professional statisticians would use more complex inference procedures to match the more complex design of the sample.

The 18 newts for the skin-healing study in Example 14.3 were chosen from a laboratory population of newts to receive one of several treatments being compared in a randomized comparative experiment. Recall that each treatment group in a completely randomized experiment is an SRS of the available subjects. Scientists usually act as if the available animal subjects are an SRS from their species or genetic type if there is nothing special about where the subjects came from. We can treat these newts as an SRS from this species.

The cola taste test in Example 14.5 uses scores from 10 tasters. All were examined to be sure that they have no medical condition that interferes with normal taste and then carefully trained to score sweetness using a set of standard drinks. We are willing to take their scores as an SRS from the population of trained tasters.

The medical director who examined executives' blood pressures in Example 14.9 actually chose an SRS from the medical records of all executives in this company. ■

These examples are typical. One is an actual SRS, two are situations in which common practice is to act as if the sample were an SRS, and in the remaining example procedures that assume an SRS can be used for a quick analysis of data from a more complex random sample. *There is no simple rule for deciding when you can act as if a sample is an SRS. Pay attention to these cautions:*

Don't touch the plants

We know that confounding can distort inference. We don't always recognize how easy it is to confound data. Consider the innocent scientist who visits plants in the field once a week to measure their size. To measure the plants, he has to touch them. A study of six plant species found that one touch a week significantly increased leaf damage by insects in two species and significantly decreased damage in another species.

Really wrong numbers

By now you know that "statistics" that don't come from properly designed studies are often dubious and sometimes just made up. It's rare to find wrong numbers that anyone can see are wrong, but it does happen. A German physicist claimed that 2006 was the first year since 1441 with more than one Friday the 13th. Sorry: Friday the 13th occurred in February and August of 2004, which is a bit more recent than 1441.

■ *Practical problems such as nonresponse in samples or dropouts from an experiment can hinder inference even from a well-designed study.* The NHANES survey has about an 80% response rate. This is much higher than opinion polls and most other national surveys, so by realistic standards NHANES data are quite trustworthy. (NHANES uses advanced methods to try to correct for nonresponse, but these methods work a lot better when response is high to start with.)

■ *Different methods are needed for different designs.* The z procedures aren't correct for random sampling designs more complex than an SRS. Later chapters give methods for some other designs, but we won't discuss inference for really complex designs like that used by NHANES. Always be sure that you (or your statistical consultant) know how to carry out the inference your design calls for.

■ *There is no cure for fundamental flaws like voluntary response surveys or uncontrolled experiments.* Look back at the bad examples in Chapters 8 and 9 and steel yourself to just ignore data from such studies.

What is the shape of the population distribution? Most statistical inference procedures require some conditions on the shape of the population distribution. Many of the most basic methods of inference are designed for Normal populations. That's the case for the z procedures and also for the more practical procedures for inference about means that we will meet in Chapters 17 and 18. Fortunately, this condition is less essential than where the data come from.

This is true because the z procedures and many other procedures designed for Normal distributions are based on Normality of the sample mean $\overline{x}$, not Normality of individual observations. The central limit theorem tells us that $\overline{x}$ is more Normal than the individual observations and that $\overline{x}$ becomes more Normal as the size of the sample increases. In practice, the z procedures are reasonably accurate for any roughly symmetric distribution for samples of even moderate size. If the sample is large, $\overline{x}$ will be close to Normal even if individual measurements are strongly skewed, as Figures 11.4 (page 303) and 11.5 (page 305) illustrate. Later chapters give practical guidelines for specific inference procedures.

There is one important exception to the principle that the shape of the population is less critical than how the data were produced. Outliers can distort the results of inference. *Any inference procedure based on sample statistics like the sample mean $\overline{x}$ that are not resistant to outliers can be strongly influenced by a few extreme observations.*

We rarely know the shape of the population distribution. In practice we rely on previous studies and on data analysis. Sometimes long experience suggests that our data are likely to come from a roughly Normal distribution, or not. For example, heights of people of the same sex and similar ages are close to Normal, but weights are not. Always explore your data before doing inference. When the data are chosen at random from a population, the shape of the data distribution mirrors the shape of the population distribution. Make a stemplot or histogram of your data and look to see whether the shape is roughly Normal. Remember that small samples have a lot of chance variation, so that Normality is hard to judge from just a few observations. Always look for outliers and try to correct them or justify their removal before performing the z procedures or other inference based on statistics like $\bar{x}$ that are not resistant.

When outliers are present or the data suggest that the population is strongly non-Normal, consider alternative methods that don't require Normality and are not sensitive to outliers. Some of these methods appear in Chapter 25 (available online and on the text CD).

APPLY YOUR KNOWLEDGE

15.1   **Rate that movie.** A professor interested in the opinions of college-age adults about a new hit movie asks the 25 students in her course on documentary filmmaking to rate the entertainment value of the movie on a scale of 0 to 5. Which of the following is the most important reason why a confidence interval for the mean rating by all college-age adults based on these data is of little use? Comment briefly on each reason to explain your answer.

(a) The course is small, so the margin of error will be large.

(b) Many of the students in the course will probably refuse to respond.

(c) The students in the course can't be considered a random sample from the population of all college-age adults.

15.2   **Running red lights.** A survey of licensed drivers inquired about running red lights. One question asked, "Of every ten motorists who run a red light, about how many do you think will be caught?" The mean result for 880 respondents was $\bar{x} = 1.92$ and the standard deviation was $s = 1.83.$[2] For this large sample, s will be close to the population standard deviation σ, so suppose we know that $\sigma = 1.83$.

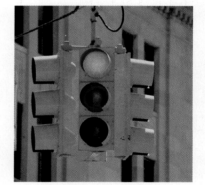

(a) Give a 95% confidence interval for the mean opinion in the population of all licensed drivers.

(b) The distribution of responses is skewed to the right rather than Normal. This will not strongly affect the z confidence interval for this sample. Why not?

(c) The 880 respondents are an SRS from completed calls among 45,956 calls to randomly chosen residential telephone numbers listed in telephone directories. Only 5029 of the calls were completed. This information gives two reasons to suspect that the sample may not represent all licensed drivers. What are these reasons?

Ilene MacDonald/Alamy

15.3   **Sampling shoppers.** A marketing consultant observes 50 consecutive shoppers at a supermarket, recording how much each shopper spends in the store. Suggest some reasons why it may be risky to act as if 50 consecutive shoppers at a particular time are an SRS of all shoppers at this store.

How confidence intervals behave

The z confidence interval $\bar{x} \pm z^*\sigma/\sqrt{n}$ for the mean of a Normal population illustrates several important properties that are shared by all confidence intervals in common use. The user chooses the confidence level, and the margin of error follows from this choice. We would like high confidence and also a small margin of error. High confidence says that our method almost always gives correct answers. A small margin of error says that we have pinned down the parameter quite precisely. The factors that influence the margin of error of the z confidence interval are typical of most confidence intervals.

How do we get a small margin of error? The margin of error for the z confidence interval is

$$\text{margin of error} = z^* \frac{\sigma}{\sqrt{n}}$$

This expression has z^* and σ in the numerator and $\sqrt{n}$ in the denominator. Therefore, the margin of error gets smaller when

- z^* gets smaller. Smaller z^* is the same as lower confidence level C (look again at Figure 14.3 on page 365). *There is a trade-off between the confidence level and the margin of error. To obtain a smaller margin of error from the same data, you must be willing to accept lower confidence.*

- σ is smaller. The standard deviation σ measures the variation in the population. You can think of the variation among individuals in the population as noise that obscures the average value μ. It is easier to pin down μ when σ is small.

- n gets larger. Increasing the sample size n reduces the margin of error for any confidence level. Larger samples thus allow more precise estimates. However, *because n appears under a square root sign, we must take four times as many observations in order to cut the margin of error in half.*

EXAMPLE 15.3 **Changing the margin of error**

In Example 14.3 (page 366), biologists measured the rate of healing of the skin of 18 newts. The data gave $\bar{x} = 25.67$ micrometers per hour and we know that $\sigma = 8$ micrometers per hour. The 95% confidence interval for the mean healing rate for all newts is

$$\bar{x} \pm z^* \frac{\sigma}{\sqrt{n}} = 25.67 \pm 1.960 \frac{8}{\sqrt{18}}$$

$$= 25.67 \pm 3.70$$

The 90% confidence interval based on the same data replaces the 95% critical value $z^* = 1.960$ by the 90% critical value $z^* = 1.645$. This interval is

$$\bar{x} \pm z^* \frac{\sigma}{\sqrt{n}} = 25.67 \pm 1.645 \frac{8}{\sqrt{18}}$$

$$= 25.67 \pm 3.10$$

Lower confidence results in a smaller margin of error, ± 3.10 in place of ± 3.70. You can calculate that the margin of error for 99% confidence is larger, ± 4.86. Figure 15.1 compares these three confidence intervals.

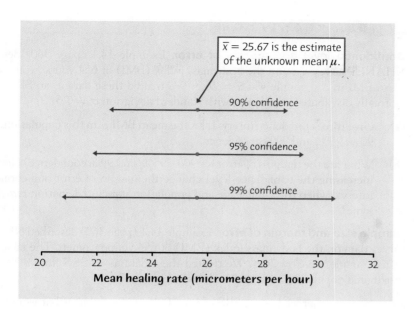

FIGURE 15.1

The lengths of three confidence intervals for Example 15.3. All three are centered at the estimate $\bar{x} = 25.67$. When the data and the sample size remain the same, higher confidence requires a larger margin of error.

If we had a sample of only 9 newts, you can check that the margin of error for 95% confidence increases from ± 3.70 to ± 5.23. Cutting the sample size in half does *not* double the margin of error, because the sample size n appears under a square root sign. ▪

What does the margin of error include? The most important caution about confidence intervals in general is a consequence of the use of a sampling distribution. A sampling distribution shows how a statistic such as $\bar{x}$ varies in repeated random sampling. This variation causes *random sampling error* because the statistic misses the true parameter by a random amount. No other source of variation or bias in the sample data influences the sampling distribution. So *the margin of error in a confidence interval ignores everything except the sample-to-sample variation due to choosing the sample randomly.*

THE MARGIN OF ERROR DOESN'T COVER ALL ERRORS

The margin of error in a confidence interval covers only random sampling errors.

Practical difficulties such as undercoverage and nonresponse are often more serious than random sampling error. The margin of error does not take such difficulties into account.

Recall from Chapter 8 that national opinion polls often have response rates less than 50% and that even small changes in the wording of questions can strongly influence results. In such cases, the announced margin of error is probably unrealistically small. And of course there is no way to assign a meaningful margin of error to results from voluntary response or convenience samples, because there is no random selection. Look carefully at the details of a study before you trust a confidence interval.

15.4 **Confidence level and margin of error.** Example 14.1 (page 360) described NHANES survey data on the body mass index (BMI) of 654 young women. The mean BMI in the sample was $\bar{x} = 26.8$. We treated these data as an SRS from a Normally distributed population with standard deviation $\sigma = 7.5$.

(a) Give three confidence intervals for the mean BMI μ in this population, using 90%, 95%, and 99% confidence.

(b) What are the margins of error for 90%, 95%, and 99% confidence? How does increasing the confidence level change the margin of error of a confidence interval when the sample size and population standard deviation remain the same?

15.5 **Sample size and margin of error.** Example 14.1 (page 360) described NHANES survey data on the body mass index (BMI) of 654 young women. The mean BMI in the sample was $\bar{x} = 26.8$. We treated these data as an SRS from a Normally distributed population with standard deviation $\sigma = 7.5$.

(a) Suppose that we had an SRS of just 100 young women. What would be the margin of error for 95% confidence?

(b) Find the margins of error for 95% confidence based on SRSs of 400 young women and 1600 young women.

(c) Compare the three margins of error. How does increasing the sample size change the margin of error of a confidence interval when the confidence level and population standard deviation remain the same?

15.6 **Is your food safe?** "Do you feel confident or not confident that the food available at most grocery stores is safe to eat?" When a Gallup Poll asked this question, 87% of the sample said they were confident.[3] Gallup announced the poll's margin of error for 95% confidence as ±3 percentage points. Which of the following sources of error are included in this margin of error?

(a) Gallup dialed landline telephone numbers at random and so missed all people without landline phones, including people whose only phone is a cell phone.

(b) Some people whose numbers were chosen never answered the phone in several calls or answered but refused to participate in the poll.

(c) There is chance variation in the random selection of telephone numbers.

© Jupiterimages/Age fotostock

How significance tests behave

Significance tests are widely used in most areas of statistical work. New pharmaceutical products require significant evidence of effectiveness and safety. Courts inquire about statistical significance in hearing class action discrimination cases. Marketers want to know whether a new package design will significantly increase sales. Medical researchers want to know whether a new therapy performs significantly better. In all these uses, statistical significance is valued because it points to an effect that is unlikely to occur simply by chance. Here are some points to keep in mind when you use or interpret significance tests.

How small a *P* is convincing? The purpose of a test of significance is to describe the degree of evidence provided by the sample against the null hypothesis. The P-value does this. But how small a P-value is convincing evidence against the null hypothesis? This depends mainly on two circumstances:

■ *How plausible is H_0?* If H_0 represents an assumption that the people you must convince have believed for years, strong evidence (small P) will be needed to persuade them.

■ *What are the consequences of rejecting H_0?* If rejecting H_0 in favor of H_a means making an expensive changeover from one type of product packaging to another, you need strong evidence that the new packaging will boost sales.

These criteria are a bit subjective. Different people will often insist on different levels of significance. Giving the P-value allows each of us to decide individually if the evidence is sufficiently strong.

Users of statistics have often emphasized standard levels of significance such as 10%, 5%, and 1%. For example, courts have tended to accept 5% as a standard in discrimination cases.[4] This emphasis reflects the time when tables of critical values rather than software dominated statistical practice. The 5% level ($\alpha = 0.05$) is particularly common. *There is no sharp border between "significant" and "not significant," only increasingly strong evidence as the P-value decreases. There is no practical distinction between the P-values 0.049 and 0.051. It makes no sense to treat $P \leq 0.05$ as a universal rule for what is significant.*

Significance depends on the alternative hypothesis You may have noticed that the P-value for a one-sided test is one-half the P-value for the two-sided test of the same null hypothesis based on the same data. The two-sided P-value combines two equal areas, one in each tail of a Normal curve. The one-sided P-value is just one of these areas, in the direction specified by the alternative hypothesis. It makes sense that the evidence against H_0 is stronger when the alternative is one-sided, because it is based on the data *plus* information about the direction of possible deviations from H_0. If you lack this added information, always use a two-sided alternative hypothesis.

Significance depends on sample size A sample survey shows that significantly fewer students are heavy drinkers at colleges that ban alcohol on campus. "Significantly fewer" is not enough information to decide whether there is an important difference in drinking behavior at schools that ban alcohol. *How important an effect is depends on the size of the effect as well as on its statistical significance.* If the number of heavy drinkers is only 1% less at colleges that ban alcohol than at other colleges, this is not an important effect even if it is statistically significant. In fact, the sample survey found that 38% of students at colleges that ban alcohol

Should tests be banned?

Significance tests don't tell us how large or how important an effect is. Research in psychology has emphasized tests, so much so that some think their weaknesses should ban them from use. The American Psychological Association asked a group of experts. They said: Use anything that sheds light on your study. Use more data analysis and confidence intervals. But: "The task force does not support any action that could be interpreted as banning the use of null hypothesis significance testing or P-values in psychological research and publication."

are "heavy episodic drinkers" compared with 48% at other colleges.[5] That difference is large enough to be important. (Of course, this observational study doesn't prove that an alcohol ban directly reduces drinking; it may be that colleges that ban alcohol attract more students who don't want to drink heavily.)

Such examples remind us to always look at the size of an effect (like 38% versus 48%) as well as its significance. They also raise a question: can a tiny effect really be highly significant? Yes. The behavior of the z test statistic is typical. The statistic is

$$z = \frac{\bar{x} - \mu_0}{\sigma/\sqrt{n}}$$

The numerator measures how far the sample mean deviates from the hypothesized mean μ_0. Larger values of the numerator give stronger evidence against H_0: $\mu = \mu_0$. The denominator is the standard deviation of $\bar{x}$. It measures how much random variation we expect. There is less variation when the number of observations n is large. So z gets larger (more significant) when the estimated effect $\bar{x} - \mu_0$ gets larger *or* when the number of observations n gets larger. Significance depends both on the size of the effect we observe *and* on the size of the sample. Understanding this fact is essential to understanding significance tests.

SAMPLE SIZE AFFECTS STATISTICAL SIGNIFICANCE

Because large random samples have small chance variation, very small population effects can be highly significant if the sample is large.

Because small random samples have a lot of chance variation, even large population effects can fail to be significant if the sample is small.

Statistical significance does not tell us whether an effect is large enough to be important. That is, **statistical significance is not the same thing as practical significance.**

Keep in mind that statistical significance means "the sample showed an effect larger than would often occur just by chance." The extent of chance variation changes with the size of the sample, so the size of the sample does matter. Exercise 15.8 demonstrates in detail how increasing the sample size drives down the P-value. Here is another example.

EXAMPLE 15.4 **It's significant. Or not. So what?**

We are testing the hypothesis of no correlation between two variables. With 1000 observations, an observed correlation of only $r = 0.08$ is significant evidence at the 1% level that the correlation in the population is not zero but positive. *The small P-value does not mean there is a strong association, only that there is strong evidence of some association.* The true population correlation is probably quite close to the observed sample value, $r = 0.08$. We might well conclude that for practical purposes we can ignore the association between these variables, even though we are confident (at the 1% level) that the correlation is positive.

On the other hand, if we have only 10 observations, a correlation of $r = 0.5$ is not significantly greater than zero even at the 5% level. Small samples vary so much that a large r is needed if we are to be confident that we aren't just seeing chance variation at work. So a small sample will often fall short of significance even if the true population correlation is quite large. ■

Beware of multiple analyses Statistical significance ought to mean that you have found an effect that you were looking for. The reasoning behind statistical significance works well if you decide what effect you are seeking, design a study to search for it, and use a test of significance to weigh the evidence you get. In other settings, significance may have little meaning.

Edward Bock/CORBIS

EXAMPLE 15.5 Cell phones and brain cancer

Might the radiation from cell phones be harmful to users? Many studies have found little or no connection between using cell phones and various illnesses. Here is part of a news account of one study:

> A hospital study that compared brain cancer patients and a similar group without brain cancer found no statistically significant association between cell phone use and a group of brain cancers known as gliomas. But when 20 types of glioma were considered separately an association was found between phone use and one rare form. Puzzlingly, however, this risk appeared to decrease rather than increase with greater mobile phone use.[6]

Think for a moment. Suppose that the 20 null hypotheses (no association) for these 20 significance tests are all true. Then each test has a 5% chance of being significant at the 5% level. That's what $\alpha = 0.05$ means: results this extreme occur 5% of the time just by chance when the null hypothesis is true. Because 5% is 1/20, we expect about 1 of 20 tests to give a significant result just by chance. That's what the study observed. ■

Running one test and reaching the 5% level of significance is reasonably good evidence that you have found something. Running 20 tests and reaching that level only once is not. The caution about multiple analyses applies to confidence intervals as well. A single 95% confidence interval has probability 0.95 of capturing the true parameter each time you use it. The probability that all of 20 confidence intervals will capture their parameters is much less than 95%. If you think that multiple tests or intervals may have discovered an important effect, you need to gather new data to do inference about that specific effect.

APPLY YOUR KNOWLEDGE

15.7 Is it significant? In the absence of special preparation SAT mathematics (SATM) scores in recent years have varied Normally with mean $\mu = 518$ and $\sigma = 114$. Fifty students go through a rigorous training program designed to raise their SATM scores by improving their mathematics skills. Either by hand or by using the

P-Value of a Test of Significance applet, carry out a test of

$$H_0: \mu = 518$$

$$H_a: \mu > 518$$

(with $\sigma = 114$) in each of the following situations:

(a)   The students' average score is $\bar{x} = 544$. Is this result significant at the 5% level?

(b)   The average score is $\bar{x} = 545$. Is this result significant at the 5% level?

The difference between the two outcomes in (a) and (b) is of no importance. Beware attempts to treat $\alpha = 0.05$ as sacred.

15.8 Detecting acid rain. Emissions of sulfur dioxide by industry set off chemical changes in the atmosphere that result in "acid rain." The acidity of liquids is measured by pH on a scale of 0 to 14. Distilled water has pH 7.0, and lower pH values indicate acidity. Normal rain is somewhat acidic, so acid rain is sometimes defined as rainfall with a pH below 5.0. Suppose that pH measurements of rainfall on different days in a Canadian forest follow a Normal distribution with standard deviation $\sigma = 0.5$. A sample of n days finds that the mean pH is $\bar{x} = 4.8$. Is this good evidence that the mean pH μ for all rainy days is less than 5.0? The answer depends on the size of the sample.

Either by hand or using the *P-Value of a Test of Significance* applet, carry out three tests of

$$H_0: \mu = 5.0$$

$$H_a: \mu < 5.0$$

Use $\sigma = 0.5$ and $\bar{x} = 4.8$ in all three tests. But use three different sample sizes, $n = 5$, $n = 15$, and $n = 40$.

(a)   What are the P-values for the three tests? *The P-value of the same result $\bar{x} = 4.8$ gets smaller (more significant) as the sample size increases.*

(b)   For each test, sketch the Normal curve for the sampling distribution of $\bar{x}$ when H_0 is true. This curve has mean 5.0 and standard deviation $0.5/\sqrt{n}$. Mark the observed $\bar{x} = 4.8$ on each curve. (If you use the applet, you can just copy the curves displayed by the applet.) *The same result $\bar{x} = 4.8$ gets more extreme on the sampling distribution as the sample size increases.*

15.9 Confidence intervals help. Give a 95% confidence interval for the mean pH μ for each sample size in the previous exercise. The intervals, unlike the P-values, give a clear picture of what mean pH values are plausible for each sample.

15.10 Searching for ESP. A researcher looking for evidence of extrasensory perception (ESP) tests 500 subjects. Four of these subjects do significantly better ($P < 0.01$) than random guessing.

(a)   You can't conclude that these four people have ESP. Why not?

(b)   What should the researcher now do to test whether any of these four subjects have ESP?

Planning studies: sample size for confidence intervals

A wise user of statistics never plans a sample or an experiment without at the same time planning the inference. The number of observations is a critical part of planning a study. Larger samples give smaller margins of error in confidence intervals and make significance tests better able to detect effects in the population. But taking observations costs both time and money. How many observations are enough? We will look at this question first for confidence intervals and then for tests. Planning a confidence interval is much simpler than planning a test. It is also more useful, because estimation is generally more informative than testing. The section on planning tests is therefore optional.

You can arrange to have both high confidence and a small margin of error by taking enough observations. The margin of error of the z confidence interval for the mean of a Normally distributed population is $m = z^*\sigma/\sqrt{n}$. To obtain a desired margin of error m, put in the value of z^* for your desired confidence level, and solve for the sample size n. Here is the result.

SAMPLE SIZE FOR DESIRED MARGIN OF ERROR

The z confidence interval for the mean of a Normal population will have a specified margin of error m when the sample size is

$$n = \left(\frac{z^*\sigma}{m}\right)^2$$

Notice that it is the size of the sample that determines the margin of error. The size of the population does not influence the sample size we need. (This is true as long as the population is much larger than the sample.)

EXAMPLE 15.6 How many observations?

Example 14.3 (page 366) reports a study of the healing rate of cuts in the skin of newts. We know that the population standard deviation is $\sigma = 8$ micrometers per hour. We want to estimate the mean healing rate μ for this species of newts within ± 3 micrometers per hour with 90% confidence. How many newts must we measure?

The desired margin of error is $m = 3$. For 90% confidence, Table C gives $z^* = 1.645$. Therefore,

$$n = \left(\frac{z^*\sigma}{m}\right)^2 = \left(\frac{1.645 \times 8}{3}\right)^2 = 19.2$$

Because 19 newts will give a slightly larger margin of error than desired, and 20 newts a slightly smaller margin of error, we must measure 20 newts. *Always round up to the next higher whole number when finding n.* ■

15.11 Body mass index of young women. Example 14.1 (page 360) assumed that the body mass index (BMI) of all American young women follows a Normal distribution with standard deviation $\sigma = 7.5$. How large a sample would be needed to estimate the mean BMI μ in this population to within ± 1 with 95% confidence?

15.12 Number skills of young men. Suppose that scores of men aged 21 to 25 years on the quantitative part of the National Assessment of Educational Progress (NAEP) test follow a Normal distribution with standard deviation $\sigma = 60$. You want to estimate the mean score within ± 10 with 90% confidence. How large an SRS of scores must you choose?

Planning studies: the power of a statistical test*

How large a sample should we take when we plan to carry out a test of significance? We know that if our sample is too small, even large effects in the population will often fail to give statistically significant results. Here are the questions we must answer to decide how many observations we need:

Significance level. How much protection do we want against getting a significant result from our sample when there really is no effect in the population?

Effect size. How large an effect in the population is important in practice?

Power. How confident do we want to be that our study will detect an effect of the size we think is important?

The three boldface terms are statistical shorthand for three pieces of information. *Power* is a new idea.

EXAMPLE 15.7 Sweetening colas: planning a study

Let's illustrate typical answers to these questions in the example of testing a new cola for loss of sweetness in storage (Example 14.5, page 369). Ten trained tasters rated the sweetness on a 10-point scale before and after storage, so that we have each taster's judgment of loss of sweetness. From experience, we know that sweetness loss scores vary from taster to taster according to a Normal distribution with standard deviation about $\sigma = 1$. To see if the taste test gives reason to think that the cola does lose sweetness, we will test

$$H_0: \mu = 0$$

$$H_a: \mu > 0$$

Are 10 tasters enough, or should we use more?

---

* Power calculations are important in planning studies, but this more advanced material is not needed to read the rest of the book.

Significance level. Requiring significance at the 5% level is enough protection against declaring there is a loss in sweetness when in fact there is no change if we could look at the entire population. This means that when there is no change in sweetness in the population, 1 out of 20 samples of tasters will wrongly find a significant loss.

Effect size. A mean sweetness loss of 0.8 point on the 10-point scale will be noticed by consumers and so is important in practice.

Power. We want to be 90% confident that our test will detect a mean loss of 0.8 point in the population of all tasters. We agreed to use significance at the 5% level as our standard for detecting an effect. So we want probability at least 0.9 that a test at the $\alpha = 0.05$ level will reject the null hypothesis H_0: $\mu = 0$ when the true population mean is $\mu = 0.8$. ■

The probability that the test successfully detects a sweetness loss of the specified size is the *power* of the test. You can think of tests with high power as being highly sensitive to deviations from the null hypothesis. In Example 15.7, we decided that we want power 90% when the truth about the population is that $\mu = 0.8$.

POWER

The **power** of a test against a specific alternative is the probability that the test will reject H_0 at a chosen significance level α when the specified alternative value of the parameter is true.

For most statistical tests, calculating power is a job for comprehensive statistical software. The z test is easier, but we will nonetheless skip the details. The two following examples illustrate two approaches: an applet that shows the meaning of power, and statistical software.

EXAMPLE 15.8 **Finding power: use an applet**

Finding the power of the z test is less challenging than most other power calculations because it requires only a Normal distribution probability calculation. The *Power of a Test* applet does this and illustrates the calculation with Normal curves. Enter the information from Example 15.7 into the applet: hypotheses, significance level $\alpha = 0.05$, alternative value $\mu = 0.8$, standard deviation $\sigma = 1$, and sample size $n = 10$. Click "Update." The applet output appears in Figure 15.2.

The power of the test against the specific alternative $\mu = 0.8$ is 0.808. That is, the test will reject H_0 about 81% of the time when this alternative is true. So 10 observations are too few to give power 90%. ■

The two Normal curves in Figure 15.2 show the sampling distribution of $\overline{x}$ under the null hypothesis $\mu = 0$ (top) and also under the specific alternative $\mu = 0.8$ (bottom). The curves have the same shape because σ does not change. The top curve is centered at $\mu = 0$ and the bottom curve at $\mu = 0.8$. The shaded region at the right of the top curve has area 0.05. It marks off values of $\overline{x}$ that are statistically

FIGURE 15.2

Output from the *Power of a Test* applet for Example 15.8, along with the information entered into the applet. The top curve shows the behavior of $\bar{x}$ when the null hypothesis is true ($\mu = 0$). The bottom curve shows the distribution of $\bar{x}$ when $\mu = 0.8$.

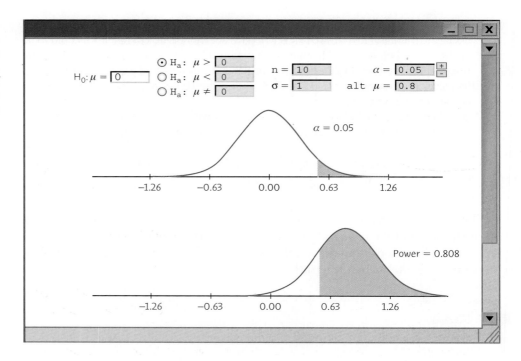

significant at the $\alpha = 0.05$ level. The lower curve shows the probability of these same values when $\mu = 0.8$. This area is the power, 0.808.

The applet will find the power for any given sample size. It's more helpful in practice to turn the process around and learn what sample size we need to achieve a given power. Statistical software will do this, but usually doesn't show the helpful Normal curves that are part of the applet's output.

EXAMPLE 15.9 Finding power: use software

We asked Minitab to find the number of observations needed for the one-sided z test to have power 0.9 against several specific alternatives at the 5% significance level when the population standard deviation is $\sigma = 1$. Here is the table that results:

Difference	Sample Size	Target Power	Actual Power
0.1	857	0.9	0.900184
0.2	215	0.9	0.901079
0.3	96	0.9	0.902259
0.4	54	0.9	0.902259
0.5	35	0.9	0.905440
0.6	24	0.9	0.902259
0.7	18	0.9	0.907414
0.8	14	0.9	0.911247
0.9	11	0.9	0.909895
1.0	9	0.9	0.912315

In this output, "Difference" is the difference between the null hypothesis value $\mu = 0$ and the alternative we want to detect. This is the effect size. The "Sample Size" column shows the smallest number of observations needed for power 0.9 against each effect size.

We see again that our earlier sample of 10 tasters is not large enough to be 90% confident of detecting (at the 5% significance level) an effect of size 0.8. If we want power 90% against effect size 0.8, we need at least 14 tasters. The actual power with 14 tasters is 0.911247.

Statistical software, unlike the applet, will do power calculations for most of the tests in this book. ∎

Fish, fishermen, and power

Are the stocks of cod in the ocean off eastern Canada declining? Studies over many years failed to find significant evidence of a decline. These studies had low power—that is, they might fail to find a decline even if one was present. When it became clear that the cod were vanishing, quotas on fishing ravaged the economy in parts of Canada. If the earlier studies had high power, they would likely have seen the decline. Quick action might have reduced the economic and environmental costs.

The table in Example 15.9 makes it clear that smaller effects require larger samples to reach 90% power. Here is an overview of influences on "How large a sample do I need?"

- If you insist on a smaller significance level (such as 1% rather than 5%), you will need a larger sample. A smaller significance level requires stronger evidence to reject the null hypothesis.

- If you insist on higher power (such as 99% rather than 90%), you will need a larger sample. Higher power gives a better chance of detecting an effect when it is really there.

- At any significance level and desired power, a two-sided alternative requires a larger sample than a one-sided alternative.

- At any significance level and desired power, detecting a small effect requires a larger sample than detecting a large effect.

Planning a serious statistical study always requires an answer to the question "How large a sample do I need?" If you intend to test the hypothesis $H_0: \mu = \mu_0$ about the mean μ of a population, you need at least a rough idea of the size of the population standard deviation σ and of how big a deviation $\mu - \mu_0$ of the population mean from its hypothesized value you want to be able to detect. More elaborate settings, such as comparing the mean effects of several treatments, require more elaborate advance information. You can leave the details to experts, but you should understand the idea of power and the factors that influence how large a sample you need.

To calculate the power of a test, we act as if we are interested in a fixed level of significance such as $\alpha = 0.05$. That's essential to do a power calculation, but remember that in practice we think in terms of P-values rather than a fixed level α. To effectively plan a statistical test we must find the power for several significance levels and for a range of sample sizes and effect sizes to get a full picture of how the test will behave.

Type I and Type II errors in significance tests We can assess the performance of a test by giving two probabilities: the significance level α and the power for an alternative that we want to be able to detect. The significance level of a test is the probability of reaching the *wrong* conclusion when the null hypothesis is true. The power for a specific alternative is the probability of reaching the *right* conclusion when that alternative is true. We can just as well describe the test by giving the probabilities of being *wrong* under both conditions.

TYPE I AND TYPE II ERRORS

If we reject H_0 when in fact H_0 is true, this is a **Type I error.**

If we fail to reject H_0 when in fact H_a is true, this is a **Type II error.**

The **significance level** α of any fixed level test is the probability of a Type I error.

The **power** of a test against any alternative is 1 minus the probability of a Type II error for that alternative.

The possibilities are summed up in Figure 15.3. If H_0 is true, our conclusion is correct if we fail to reject H_0 and is a Type I error if we reject H_0. If H_a is true, our conclusion is either correct or a Type II error. Only one error is possible at one time.

		Truth about the population	
		H_0 true	H_a true
Conclusion based on sample	Reject H_0	Type I error	Correct conclusion
	Fail to reject H_0	Correct conclusion	Type II error

FIGURE 15.3

The two types of error in testing hypotheses.

EXAMPLE 15.10 **Calculating error probabilities**

Because the probabilities of the two types of error are just a rewording of significance level and power, we can see from Figure 15.2 what the error probabilities are for the test in Example 15.7.

$$P\,(\text{Type I error}) = P\,(\text{reject } H_0 \text{ when in fact } \mu = 0)$$

$$= \text{significance level } \alpha = 0.05$$

$$P\,(\text{Type II error}) = P\,(\text{fail to reject } H_0 \text{ when in fact } \mu = 0.8)$$

$$= 1 - \text{power} = 1 - 0.808 = 0.192$$

The two Normal curves in Figure 15.2 are used to find the probabilities of a Type I error (top curve, $\mu = 0$) and of a Type II error (bottom curve, $\mu = 0.8$). ■

APPLY YOUR KNOWLEDGE

15.13 What is power? You manufacture and sell a liquid product whose electrical conductivity is supposed to be 5. You plan to make 6 measurements of the conductivity of each lot of product. You know that the standard deviation of your measurements is $\sigma = 0.2$. If the product meets specifications, the mean of many measurements will be 5. You will therefore test

$$H_0: \mu = 5$$
$$H_a: \mu \neq 5$$

If the true conductivity is 5.1, the liquid is not suitable for its intended use. You learn that the power of your test at the 5% significance level against the alternative $\mu = 5.1$ is 0.23.

(a) Explain in simple language what "power = 0.23" means.

(b) Explain why the test you plan will not adequately protect you against selling a liquid with conductivity 5.1.

15.14 Thinking about power. Answer these questions in the setting of the previous exercise about measuring the conductivity of a liquid.

(a) You could get higher power against the same alternative with the same α by changing the number of measurements you make. Should you make more measurements or fewer to increase power?

(b) If you decide to use $\alpha = 0.10$ in place of $\alpha = 0.05$, with no other changes in the test, will the power increase or decrease?

(c) If you shift your interest to the alternative $\mu = 5.2$ with no other changes, will the power increase or decrease?

15.15 How power behaves. In the setting of Exercise 15.13, use the *Power of a Test* applet to find the power in each of the following circumstances. Be sure to set the applet to the two-sided alternative.

(a) Standard deviation $\sigma = 0.2$, significance level $\alpha = 0.05$, alternative $\mu = 5.1$, and sample sizes $n = 6$, $n = 12$, and $n = 24$. How does increasing the sample size with no other changes affect the power?

(b) Standard deviation $\sigma = 0.2$, significance level $\alpha = 0.05$, sample size $n = 6$, and alternatives $\mu = 5.1$, $\mu = 5.2$, and $\mu = 5.3$. How do alternatives more distant from the hypothesis (larger effect sizes) affect the power?

(c) Standard deviation $\sigma = 0.2$, sample size $n = 6$, alternative $\mu = 5.1$, and significance levels $\alpha = 0.05$, $\alpha = 0.10$, and $\alpha = 0.25$. (Click the + and − buttons to change α.) How does increasing the desired significance level affect the power?

15.16 How power behaves. Another approach to improving the unsatisfactory power of the test in Exercise 15.13 is to improve the measurement process. That is, use a measurement process that is less variable. Use the *Power of a Test* applet to find the power of the test in Exercise 15.13 in each of these circumstances: significance level $\alpha = 0.05$, alternative $\mu = 5.1$, sample size $n = 6$, and $\sigma = 0.2$, $\sigma = 0.1$, and

$\sigma = 0.05$. How does decreasing the variability of the population of measurements affect the power?

15.17 **Two types of error.** Your company markets a computerized medical diagnostic program used to evaluate thousands of people. The program scans the results of routine medical tests (pulse rate, blood tests, etc.) and refers the case to a doctor if there is evidence of a medical problem. The program makes a decision about each person.

(a) What are the two hypotheses and the two types of error that the program can make? Describe the two types of error in terms of "false positive" and "false negative" test results.

(b) The program can be adjusted to decrease one error probability, at the cost of an increase in the other error probability. Which error probability would you choose to make smaller, and why? (This is a matter of judgment. There is no single correct answer.)

CHAPTER 15 SUMMARY

■ A specific confidence interval or test is correct only under specific conditions. The most important conditions concern the method used to produce the data. Other factors such as the shape of the population distribution may also be important.

■ Whenever you use statistical inference, you are acting as if your data are a random sample or come from a randomized comparative experiment.

■ Always do data analysis before inference to detect outliers or other problems that would make inference untrustworthy.

■ Other things being equal, the margin of error of a confidence interval gets smaller as
  – the confidence level C decreases,
  – the population standard deviation σ decreases,
  – the sample size n increases.

■ The margin of error in a confidence interval accounts for only the chance variation due to random sampling. In practice, errors due to nonresponse or undercoverage are often more serious.

■ There is no universal rule for how small a P-value in a test of significance is convincing evidence against the null hypothesis. Beware of placing too much weight on traditional significance levels such as $\alpha = 0.05$.

■ Very small effects can be highly significant (small P) when a test is based on a large sample. A statistically significant effect need not be practically important. Plot the data to display the effect you are seeking, and use confidence intervals to estimate the actual values of parameters.

■ On the other hand, lack of significance does not imply that H_0 is true. Even a large effect can fail to be significant when a test is based on a small sample.

- Many tests run at once will probably produce some significant results by chance alone, even if all the null hypotheses are true.

- When you plan a statistical study, plan the inference as well. In particular, ask what sample size you need for successful inference.

- The z confidence interval for a Normal mean has specified margin of error m when the sample size is

$$n = \left(\frac{z^*\sigma}{m}\right)^2$$

Here z^* is the critical value for the desired level of confidence. Always round n up when you use this formula.

- The **power** of a significance test measures its ability to detect an alternative hypothesis. The power against a specific alternative is the probability that the test will reject H_0 at a particular level α when that alternative is true.

- Increasing the size of the sample increases the power of a significance test. You can use statistical software to find the sample size needed to achieve a desired power.

CHECK YOUR SKILLS

15.18 The most important condition for sound conclusions from statistical inference is usually

(a) that the data can be thought of as a random sample from the population of interest.

(b) that the population distribution is exactly Normal.

(c) that the data contain no outliers.

15.19 The coach of a college men's basketball team records the resting heart rates of the 15 team members. You should not trust a confidence interval for the mean resting heart rate of all male students at this college based on these data because

(a) with only 15 observations, the margin of error will be large.

(b) heart rates may not have a Normal distribution.

(c) the members of the basketball team can't be considered a random sample of all students.

15.20 You turn your Web browser to the online Harris Interactive poll. Based on 6748 responses, the poll reports that 16% of U.S. adults sometimes use the Internet to make telephone calls.[7] You should refuse to calculate a 95% confidence interval based on this sample because

(a) the poll was taken a week ago.

(b) inference from a voluntary response sample can't be trusted.

(c) the sample is too large.

15.21 Many sample surveys use well-designed random samples but half or more of the original sample can't be contacted or refuse to take part. Any errors due to this nonresponse

(a)   have no effect on the accuracy of confidence intervals.

(b)   are included in the announced margin of error.

(c)   are in addition to the random variation accounted for by the announced margin of error.

15.22 A writer in a medical journal says: "An uncontrolled experiment in 17 women found a significantly improved mean clinical symptom score after treatment. Methodologic flaws make it difficult to interpret the results of this study." The writer is skeptical about the significant improvement because

(a)   there is no control group, so the improvement might be due to the placebo effect or to the fact that many medical conditions improve over time.

(b)   the P-value given was $P = 0.03$, which is too large to be convincing.

(c)   the response variable might not have an exactly Normal distribution in the population.

15.23 Vigorous exercise helps people live several years longer (on the average). Whether mild activities like slow walking extend life is not clear. Suppose that the added life expectancy from regular slow walking is just 2 months. A statistical test is more likely to find a significant increase in mean life if

(a)   it is based on a very large random sample.

(b)   it is based on a very small random sample.

(c)   The size of the sample doesn't have any effect on the significance of the test.

CORBIS/Age fotostock

15.24 A laboratory scale is known to have a standard deviation of $\sigma = 0.001$ gram in repeated weighings. Scale readings in repeated weighings are Normally distributed, with mean equal to the true weight of the specimen. How many times must you weigh a specimen on this scale in order to get a margin of error no larger than ± 0.0005 with 95% confidence?

(a)   4 times   (b)   15 times   (c)   16 times

15.25 A medical experiment compared the herb echinacea with a placebo for preventing colds. One response variable was "volume of nasal secretions" (if you have a cold, you blow your nose a lot). Take the average volume of nasal secretions in people without colds to be $\mu = 1$. An increase to $\mu = 3$ indicates a cold. The significance level of a test of $H_0: \mu = 1$ versus $H_a: \mu > 1$ is defined as

(a)   the probability that the test rejects H_0 when $\mu = 1$ is true.

(b)   the probability that the test rejects H_0 when $\mu = 3$ is true.

(c)   the probability that the test fails to reject H_0 when $\mu = 3$ is true.

15.26 (**Optional**) The power of the test in the previous exercise against the specific alternative $\mu = 3$ is defined as

(a)   the probability that the test rejects H_0 when $\mu = 1$ is true.

(b) the probability that the test rejects H_0 when $\mu = 3$ is true.

(c) the probability that the test fails to reject H_0 when $\mu = 3$ is true.

15.27 (Optional) The power of a test is important in practice because power

(a) describes how well the test performs when the null hypothesis is actually true.

(b) describes how sensitive the test is to violations of conditions such as Normal population distribution.

(c) describes how well the test performs when the null hypothesis is actually not true.

C H A P T E R 1 5   E X E R C I S E S

15.28 Hotel managers. In Exercise 14.42 (page 387) you carried out a test of significance based on data from 148 general managers of three-star and four-star hotels. Before you trust your results, you would like more information about the data. What facts would you most like to know?

15.29 Color blindness in Africa. An anthropologist claims that color blindness is less common in societies that live by hunting and gathering than in settled agricultural societies. He tests a number of adults in two populations in Africa, one of each type. The proportion of color-blind people is significantly lower ($P < 0.05$) in the hunter-gatherer population. What additional information would you want to help you decide whether you believe the anthropologist's claim?

15.30 Hotel managers. In Exercise 14.42 (page 387) you carried out a test of significance based on the BSRI femininity scores of 148 male general managers of three-star and four-star hotels. You now realize that a confidence interval for the mean score of male hotel managers would be more informative than a test. You would be satisfied to estimate the mean to within ± 0.2 with 99% confidence. The standard deviation of the scores is probably close to the value $\sigma = 0.78$ for the adult male population. How large an SRS of hotel managers do you need?

15.31 Sampling at the mall. A market researcher chooses at random from women entering a large suburban shopping mall. One outcome of the study is a 95% confidence interval for the mean of "the highest price you would pay for a pair of casual shoes."

(a) Explain why this confidence interval does not give useful information about the population of all women.

(b) Explain why it may give useful information about the population of women who shop at large suburban malls.

15.32 Pulling wood apart. You want to estimate the mean load needed to pull apart the pieces of wood in Exercise 14.50 (page 389) to within ± 1000 pounds with 95% confidence. How large a sample is needed?

15.33 Sensitive questions. The National AIDS Behavioral Surveys found that 170 individuals in its random sample of 2673 adult heterosexuals said they had multiple sexual partners in the past year. That's 6.36% of the sample. Why is this estimate likely to be biased? Does the margin of error of a 95% confidence interval for the proportion of all adults with multiple partners allow for this bias?

15.34 College degrees. At the Census Bureau Web site www.census.gov you can find the percent of adults in each state who have at least a bachelor's degree. It makes no sense to find $\bar{x}$ for these data and use it to get a confidence interval for the mean percent μ in all 50 states. Why not?

15.35 An outlier strikes. You have data on an SRS of recent graduates from your college that shows how long each student took to complete a bachelor's degree. The data contain one high outlier. Will this outlier have a greater effect on a confidence interval for mean completion time if your sample is small or large? Why?

15.36 Can we trust this interval? Here are data on the percent change in the total mass (in tons) of wildlife in several West African game preserves in the years 1971 to 1999:[8]

1971	1972	1973	1974	1975	1976	1977	1978	1979	1980
2.9	3.1	−1.2	−1.1	−3.3	3.7	1.9	−0.3	−5.9	−7.9
1981	1982	1983	1984	1985	1986	1987	1988	1989	1990
−5.5	−7.2	−4.1	−8.6	−5.5	−0.7	−5.1	−7.1	−4.2	0.9
1991	1992	1993	1994	1995	1996	1997	1998	1999	
−6.1	−4.1	−4.8	−11.3	−9.3	−10.7	−1.8	−7.4	−22.9	

Software gives the 95% confidence interval for the mean annual percent change as −6.66% to −2.55%. There are several reasons why we might not trust this interval.

(a) Examine the distribution of the data. What feature of the distribution throws doubt on the validity of statistical inference?

(b) Plot the percents against year. What trend do you see in this time series? Explain why a trend over time casts doubt on the condition that years 1971 to 1999 can be treated as an SRS from a larger population of years.

15.37 When to use pacemakers. A medical panel prepared guidelines for when cardiac pacemakers should be implanted in patients with heart problems. The panel reviewed a large number of medical studies to judge the strength of the evidence supporting each recommendation. For each recommendation, they ranked the evidence as level A (strongest), B, or C (weakest). Here, in scrambled order, are the panel's descriptions of the three levels of evidence.[9] Which is A, which B, and which C? Explain your ranking.

Layne Kennedy/CORBIS

Evidence was ranked as level _____ when data were derived from a limited number of trials involving comparatively small numbers of patients or from well-designed data analysis of nonrandomized studies or observational data registries.

Evidence was ranked as level _____ if the data were derived from multiple randomized clinical trials involving a large number of individuals.

Evidence was ranked as level _____ when consensus of expert opinion was the primary source of recommendation.

15.38 What is significance good for? Which of the following questions does a test of significance answer? Briefly explain your replies.

(a) Is the sample or experiment properly designed?

(b) Is the observed effect due to chance?

(c) Is the observed effect important?

15.39 Why are larger samples better? Statisticians prefer large samples. Describe briefly the effect of increasing the size of a sample (or the number of subjects in an experiment) on each of the following:

(a) The margin of error of a 95% confidence interval.

(b) The P-value of a test, when H_0 is false and all facts about the population remain unchanged as n increases.

(c) (Optional) The power of a fixed level α test, when α, the alternative hypothesis, and all facts about the population remain unchanged.

15.40 Predicting success of trainees. What distinguishes managerial trainees who eventually become executives from those who don't succeed and leave the company? We have abundant data on past trainees—data on their personalities and goals, their college preparation and performance, even their family backgrounds and their hobbies. Statistical software makes it easy to perform dozens of significance tests on these dozens of variables to see which ones best predict later success. We find that future executives are significantly more likely than washouts to have an urban or suburban upbringing and an undergraduate degree in a technical field. Explain clearly why using these "significant" variables to select future trainees is not wise.

15.41 A test goes wrong. Software can generate samples from (almost) exactly Normal distributions. Here is a random sample of size 5 from the Normal distribution with mean 10 and standard deviation 2:

$$6.47 \quad 7.51 \quad 10.10 \quad 13.63 \quad 9.91$$

These data match the conditions for a z test better than real data will: the population is very close to Normal and has known standard deviation $\sigma = 2$, and the population mean is $\mu = 10$. Test the hypotheses

$$H_0: \mu = 8$$
$$H_a: \mu \neq 8$$

(a) What are the z statistic and its P-value? Is the test significant at the 5% level?

(b) We know that the null hypothesis does not hold, but the test failed to give strong evidence against H_0. Explain why this is not surprising.

15.42 Helping welfare mothers. A study compares two groups of mothers with young children who were on welfare two years ago. One group attended a voluntary training program that was offered free of charge at a local vocational school and was advertised in the local news media. The other group did not choose to attend the training program. The study finds a significant difference ($P < 0.01$) between the proportions of the mothers in the two groups who are still on welfare. The difference is not only significant but quite large. The report says that with 95% confidence the percent of the nonattending group still on welfare is 21% ± 4% higher than that of the group who attended the program. You are on the staff of a member of Congress who is interested in the plight of welfare mothers and who asks you about the report.

(a)   Explain in simple language what "a significant difference ($P < 0.01$)" means.

(b)   Explain clearly and briefly what "95% confidence" means.

(c)   Is this study good evidence that requiring job training of all welfare mothers would greatly reduce the percent who remain on welfare for several years?

15.43 How far do rich parents take us? How much education children get is strongly associated with the wealth and social status of their parents. In social science jargon, this is "socioeconomic status," or SES. But the SES of parents has little influence on whether children who have graduated from college go on to yet more education. One study looked at whether college graduates took the graduate admissions tests for business, law, and other graduate programs. The effects of the parents' SES on taking the LSAT test for law school were "both statistically insignificant and small."

(a)   What does "statistically insignificant" mean?

(b)   Why is it important that the effects were small in size as well as insignificant?

The following exercises concern the optional material on the power of a test.

15.44 The first child has higher IQ. Does the birth order of a family's children influence their IQ scores? A careful study of 241,310 Norwegian 18- and 19-year-olds found that firstborn children scored 2.3 points higher on the average than second children in the same family. This difference was highly significant ($P < 0.001$). A commentator said, "One puzzle highlighted by these latest findings is why certain other within-family studies have failed to show equally consistent results. Some of these previous null findings, which have all been obtained in much smaller samples, may be explained by inadequate statistical power."[10] Explain in simple language why tests having low power often fail to give evidence against a null hypothesis even when the hypothesis is really false.

15.45 How valium works. Valium is a common antidepressant and sedative. A study investigated how valium works by comparing its effect on sleep in 7 genetically modified mice and 8 normal control mice. There was no significant difference between the two groups. The authors say that this lack of significance "is related to the large inter-individual variability that is also reflected in the low power (20%) of the test."[11]

(a)   Explain exactly what power 20% against a specific alternative means.

(b)   Explain in simple language why tests having low power often fail to give evidence against a null hypothesis even when the null hypothesis is really false.

(c)   What fact about this experiment most likely explains the low power?

15.46 Treating knee pain. An article in the *New England Journal of Medicine* describes a double-blind randomized clinical trial that compared a type of surgery for knee pain due to arthritis with a placebo surgery. The experiment found no significant difference between the treatment and the placebo. According to the article, "The trial was designed to have 90 percent power, with a two-sided type I error of 0.04, to detect a moderate effect size (0.55) between the groups."[12]

(a) What fixed significance level was used in calculating the power?

(b) Explain to someone who knows no statistics why power 90% means that the experiment would probably have been significant if the surgery performed better than the placebo.

15.47 Power. In Exercise 15.41, a sample from a Normal population with mean $\mu = 10$ and standard deviation $\sigma = 2$ failed to reject the null hypothesis $H_0: \mu = 8$ at the $\alpha = 0.05$ significance level. Enter the information from this example into the *Power of a Test* applet. (Don't forget that the alternative hypothesis is two-sided.) What is the power of the test against the alternative $\mu = 10$? Because the power is not high, it isn't surprising that the sample in Exercise 15.41 failed to reject H_0.

15.48 Finding power by hand. Even though software is used in practice to calculate power, doing the work by hand builds your understanding. Return to the test in Example 15.7. There are $n = 10$ observations from a population with standard deviation $\sigma = 1$ and unknown mean μ. We will test

$$H_0: \mu = 0$$
$$H_a: \mu > 0$$

with fixed significance level $\alpha = 0.05$. Find the power against the alternative $\mu = 0.8$ by following these steps.

(a) The z test statistic is

$$z = \frac{\bar{x} - \mu_0}{\sigma/\sqrt{n}} = \frac{\bar{x} - 0}{1/\sqrt{10}} = 3.162\bar{x}$$

(Remember that you won't know the numerical value of $\bar{x}$ until you have data.) What values of z lead to rejecting H_0 at the 5% significance level?

(b) Starting from your result in (a), what values of $\bar{x}$ lead to rejecting H_0? The area above these values is shaded under the top curve in Figure 15.2.

(c) The power is the probability that you observe any of these values of $\bar{x}$ when $\mu = 0.8$. This is the shaded area under the bottom curve in Figure 15.2. What is this probability?

15.49 Finding power by hand: two-sided test. The previous exercise shows how to calculate the power of a one-sided z test. Power calculations for two-sided tests follow the same outline. We will find the power of a test based on 6 measurements of the conductivity of a liquid, reported in Exercise 15.13. The hypotheses are

$$H_0: \mu = 5$$
$$H_a: \mu \neq 5$$

The population of all measurements is Normal with standard deviation $\sigma = 0.2$, and the alternative we hope to be able to detect is $\mu = 5.1$. (If you used the *Power of a Test* applet for Exercise 15.15, the two Normal curves for $n = 6$ illustrate parts (a) and (b) below.)

(a) Write the z test statistic in terms of the sample mean $\bar{x}$. For what values of z does this two-sided test reject H_0 at the 5% significance level?

(b) Restate your result from part (a): what values of $\bar{x}$ lead to rejection of H_0?

(c)   Now suppose that $\mu = 5.1$. What is the probability of observing an $\overline{x}$ that leads to rejection of H_0? This is the power of the test.

15.50 Error probabilities. You read that a statistical test at significance level $\alpha = 0.05$ has power 0.78. What are the probabilities of Type I and Type II errors for this test?

15.51 Power. You read that a statistical test at the $\alpha = 0.01$ level has probability 0.14 of making a Type II error when a specific alternative is true. What is the power of the test against this alternative?

15.52 Find the error probabilities. You have an SRS of size $n = 9$ from a Normal distribution with $\sigma = 1$. You wish to test

$$H_0: \mu = 0$$

$$H_a: \mu > 0$$

You decide to reject H_0 if $\overline{x} > 0$ and to accept H_0 otherwise.

(a)   Find the probability of a Type I error. That is, find the probability that the test rejects H_0 when in fact $\mu = 0$.

(b)   Find the probability of a Type II error when $\mu = 0.3$. This is the probability that the test accepts H_0 when in fact $\mu = 0.3$.

(c)   Find the probability of a Type II error when $\mu = 1$.

15.53 Two types of error. Go to the *Statistical Significance* applet. This applet carries out tests at a fixed significance level. When you arrive, the applet is set for the cola-tasting test of Example 15.7. That is, the hypotheses are

$$H_0: \mu = 0$$

$$H_a: \mu > 0$$

We have an SRS of size 10 from a Normal population with standard deviation $\sigma = 1$, and we will do a test at level $\alpha = 0.05$. At the bottom of the screen, a button allows you to choose a value of the mean μ and then to generate samples from a population with that mean.

(a)   Set $\mu = 0$, so that the null hypothesis is true. Each time you click the button, a new sample appears. If the sample $\overline{x}$ lands in the colored region, that sample rejects H_0 at the 5% level. Click 100 times rapidly, keeping track of how many samples reject H_0. Use your results to estimate the probability of a Type I error. If you kept clicking forever, what probability would you get?

(b)   Now set $\mu = 0.8$. Example 15.8 shows that the test has power 0.808 against this alternative. Click 100 times rapidly, keeping track of how many samples fail to reject H_0. Use your results to estimate the probability of a Type II error. If you kept clicking forever, what probability would you get?

© David Paynter/Age fotostock

CHAPTER 16

From Exploration to Inference: Part II Review

In Part I of this book, you mastered **data analysis,** the use of graphs and numerical summaries to organize and explore any set of data. Part II has introduced designs for data production, probability, and the reasoning of statistical inference. Parts III and IV will deal in detail with practical inference.

Designs for producing data are essential if the data are intended to represent some wider population or process. Figures 16.1 and 16.2 display the big ideas visually. Random sampling and randomized comparative experiments are perhaps the most important statistical inventions of the 20th century. Both were slow to

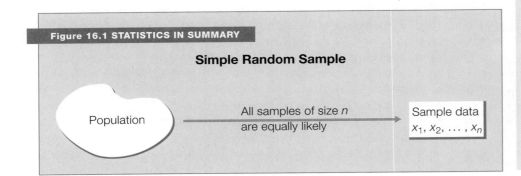

Figure 16.1 STATISTICS IN SUMMARY

Simple Random Sample

Population → All samples of size n are equally likely → Sample data $x_1, x_2, \ldots, x_n$

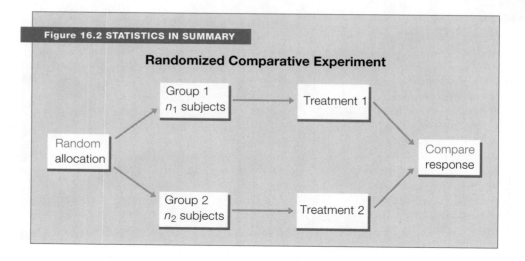

gain acceptance, and you will still see many voluntary response samples and uncontrolled experiments. You should now understand good designs for producing data and also why bad designs often produce data that are worthless for inference. The deliberate use of chance in producing data is a central idea in statistics. It not only reduces bias but allows us to use **probability,** the mathematics of chance, as the basis for inference. Fortunately, we need only some basic facts about probability in order to understand statistical inference.

Statistical inference draws conclusions about a population on the basis of sample data and uses probability to indicate how reliable the conclusions are. A confidence interval estimates an unknown parameter. A significance test shows how strong the evidence is for some claim about a parameter.

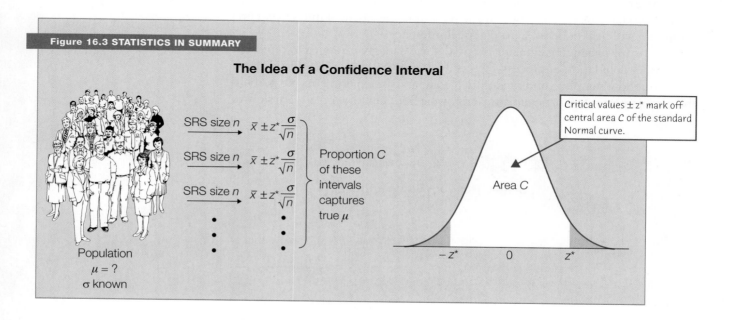

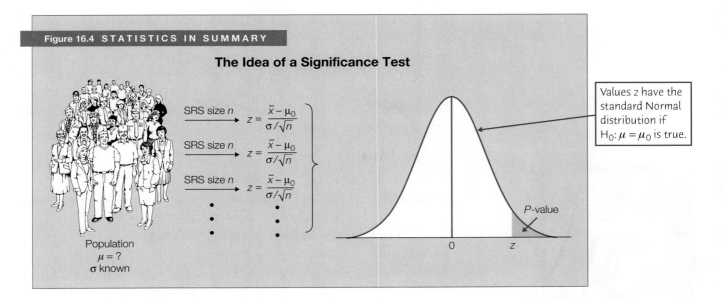

Figure 16.4 STATISTICS IN SUMMARY

The Idea of a Significance Test

The probabilities in both confidence intervals and tests tell us what would happen if we used the method for the interval or test very many times.

■ A confidence level is the success rate of the method for a confidence interval. This is the probability that the method actually produces an interval that captures the unknown parameter. A 95% confidence interval gives a correct result 95% of the time when we use it repeatedly.

■ A P-value tells us how surprising the observed outcome would be if the null hypothesis were true. That is, P is the probability that the test would produce a result at least as extreme as the observed result if the null hypothesis really were true. Very surprising outcomes (small P-values) are good evidence that the null hypothesis is not true.

Figures 16.3 and 16.4 use the z procedures introduced in Chapter 14 to present in picture form the big ideas of confidence intervals and significance tests. These ideas are the foundation for the rest of this book. We will have much to say about many statistical methods and their use in practice. In every case, the basic reasoning of confidence intervals and significance tests remains the same.

PART II   SUMMARY

Here are the most important skills you should have acquired from reading Chapters 8 to 15.

A. Sampling

1. Identify the population in a sampling situation.

2. Recognize bias due to voluntary response samples and other inferior sampling methods.

3. Use software or Table B of random digits to select a simple random sample (SRS) from a population.

4. Recognize the presence of undercoverage and nonresponse as sources of error in a sample survey. Recognize the effect of the wording of questions on the responses.

5. Use random digits to select a stratified random sample from a population when the strata are identified.

B. Experiments

1. Recognize whether a study is an observational study or an experiment.

2. Recognize bias due to confounding of explanatory variables with lurking variables in either an observational study or an experiment.

3. Identify the factors (explanatory variables), treatments, response variables, and individuals or subjects in an experiment.

4. Outline the design of a completely randomized experiment using a diagram like that in Figure 16.2. The diagram in a specific case should show the sizes of the groups, the specific treatments, and the response variable.

5. Use software or Table B of random digits to carry out the random assignment of subjects to groups in a completely randomized experiment.

6. Recognize the placebo effect. Recognize when the double-blind technique should be used.

7. Explain why randomized comparative experiments can give good evidence for cause-and-effect relationships.

C. Probability

1. Recognize that some phenomena are random. Probability describes the long-run regularity of random phenomena.

2. Understand that the probability of an event is the proportion of times the event occurs in very many repetitions of a random phenomenon. Use the idea of probability as long-run proportion to think about probability.

3. Use basic probability rules to detect illegitimate assignments of probability: any probability must be a number between 0 and 1, and the total probability assigned to all possible outcomes must be 1.

4. Use basic probability rules to find the probabilities of events that are formed from other events. The probability that an event does not occur is 1 minus its probability. If two events are disjoint, the probability that one or the other occurs is the sum of their individual probabilities.

5. Find probabilities in a discrete probability model by adding the probabilities of their outcomes. Find probabilities in a continuous probability model as areas under a density curve.

Icing the kicker

The football team lines up for what they hope will be the winning field goal ... and the other team calls time out. "Make the kicker think about it" is their motto. Does "icing the kicker" really work? That is, does the probability of making a field goal go down when the kicker must wait around during the time out? This isn't a simple question. A detailed statistical study considered the distance, the weather, the kicker's skill, and so on. The conclusion is cheering to coaches: yes, icing the kicker does reduce the probability of success.

6. Use the notation of random variables to make compact statements about random outcomes, such as $P(\bar{x} \leq 4) = 0.3$. Be able to interpret such statements.

D. Sampling Distributions

1. Identify parameters and statistics in a statistical study.

2. Recognize the fact of sampling variability: a statistic will take different values when you repeat a sample or experiment.

3. Interpret a sampling distribution as describing the values taken by a statistic in all possible repetitions of a sample or experiment under the same conditions.

4. Interpret the sampling distribution of a statistic as describing the probabilities of its possible values.

E. The Sampling Distribution of a Sample Mean

1. Recognize when a problem involves the mean $\bar{x}$ of a sample. Understand that $\bar{x}$ estimates the mean μ of the population from which the sample is drawn.

2. Use the law of large numbers to describe the behavior of $\bar{x}$ as the size of the sample increases.

3. Find the mean and standard deviation of a sample mean $\bar{x}$ from an SRS of size n when the mean μ and standard deviation σ of the population are known.

4. Understand that $\bar{x}$ is an unbiased estimator of μ and that the variability of $\bar{x}$ about its mean μ gets smaller as the sample size increases.

5. Understand that $\bar{x}$ has approximately a Normal distribution when the sample is large (central limit theorem). Use this Normal distribution to calculate probabilities that concern $\bar{x}$.

F. General Rules of Probability (Not required for later chapters)

1. Use Venn diagrams to picture relationships among several events.

2. Use the general addition rule to find probabilities that involve overlapping events.

3. Understand the idea of independence. Judge when it is reasonable to assume independence as part of a probability model.

4. Use the multiplication rule for independent events to find the probability that all of several independent events occur.

5. Use the multiplication rule for independent events in combination with other probability rules to find the probabilities of complex events.

6. Understand the idea of conditional probability. Find conditional probabilities for individuals chosen at random from a table of counts of possible outcomes.

7. Use the general multiplication rule to find $P(A \text{ and } B)$ from $P(A)$ and the conditional probability $P(B \mid A)$.

8. Use tree diagrams to organize several-stage probability models.

G. Binomial Distributions (Not required for later chapters)

1. Recognize the binomial setting: a fixed number n of independent success-failure trials with the same probability p of success on each trial.

2. Recognize and use the binomial distribution of the count of successes in a binomial setting.

3. Use the binomial probability formula to find probabilities of events involving the count X of successes in a binomial setting for small values of n.

4. Find the mean and standard deviation of a binomial count X.

5. Recognize when you can use the Normal approximation to a binomial distribution. Use the Normal approximation to calculate probabilities that concern a binomial count X.

H. Confidence Intervals

1. State in nontechnical language what is meant by "95% confidence" or other statements of confidence in statistical reports.

2. Know the four-step process (page 366) for any confidence interval. This process will be used more extensively in later chapters.

3. Calculate a confidence interval for the mean μ of a Normal population with known standard deviation σ, using the formula $\overline{x} \pm z^* \sigma / \sqrt{n}$.

4. Understand how the margin of error of a confidence interval changes with the sample size and the level of confidence C.

5. Find the sample size required to obtain a confidence interval of specified margin of error m when the confidence level and other information are given.

6. Identify sources of error in a study that are *not* included in the margin of error of a confidence interval, such as undercoverage or nonresponse.

I. Significance Tests

1. State the null and alternative hypotheses in a testing situation when the parameter in question is a population mean μ.

2. Explain in nontechnical language the meaning of the P-value when you are given the numerical value of P for a test.

3. Know the four-step process (page 378) for any significance test. This process will be used more extensively in later chapters.

4. Calculate the one-sample z test statistic and the P-value for both one-sided and two-sided tests about the mean μ of a Normal population.

5. Assess statistical significance at standard levels α, either by comparing P with α or by comparing z with standard Normal critical values.

6. Recognize that significance testing does not measure the size or importance of an effect. Explain why a small effect can be significant in a large sample and why a large effect can fail to be significant in a small sample.

7. Recognize that any inference procedure acts as if the data were properly produced. The z confidence interval and test require that the data be an SRS from the population.

R E V I E W   E X E R C I S E S

Review exercises help you solidify the basic ideas and skills in Chapters 8 to 15. In exercises that ask a P-value, you may use either Table A, software, or the P-Value of a Test of Significance applet.

16.1 Spoofing. Online criminals use "spoofing" to direct Internet users to fraudulent Web sites in order to collect information such as passwords. In one study of Internet fraud, students were warned about spoofing and then asked to log in to their university account starting from the university's home page. In some cases, the log-in link led to the genuine dialog box. In others, the box looked genuine but in fact was linked to a different site that recorded the ID and password the student entered. An alert student could detect the fraud by looking at the true Internet address displayed in the browser status bar below the window, but most just entered their ID and password. Is this study an experiment? Why? What are the explanatory and response variables?

16.2 Student counseling. A university offers its students up to 8 sessions per semester of free psychological counseling. To evaluate satisfaction with this service, the counseling office mails questionnaires to 200 of the 1400 students who participated in counseling last semester. Only 93 students return the questionnaire. What is the population in this study? What is the sample?

16.3 How much do students earn? A university's financial aid office wants to know how much it can expect students to earn from summer employment. This information will be used to set the level of financial aid. The population contains 3478 students who have completed at least one year of study but have not yet graduated. The university will send a questionnaire to an SRS of 100 of these students, drawn from an alphabetized list.

(a) Describe how you will label the students in order to select the sample.

(b) Use Table B, beginning at line 105, to select the first 5 students in the sample.

(c) What is the response variable in this study?

16.4 California's endangered animals. The California Department of Fish and Game publishes a list of the state's endangered animals. Here are the reptiles on the list:

Desert tortoise	Green sea turtle	Loggerhead sea turtle
Olive Ridley sea turtle	Leatherback sea turtle	Barefoot banded gecko
Island night lizard	Alameda whipsnake	Coachella Valley fringe-toed lizard
Flat-tailed horned lizard	Southern rubber boa	Blunt-nosed leopard lizard
Giant garter snake	San Francisco garter snake	

© David Paynter/Age fotostock

Your class can't decide which 3 endangered reptiles to choose for special study, so you agree to choose an SRS from the list. Use software, the *Simple Random Sample* applet, or Table B at line 111 to choose an SRS of 3 of these species.

16.5 Elephants and bees. Elephants sometimes damage crops in Africa. It turns out that elephants dislike bees. They recognize beehives in areas where they are common and avoid them. Can this be used to keep elephants away from trees? A group in Kenya placed active beehives in some trees and empty beehives in others. Will elephant damage be less in trees with hives? Will even empty hives keep elephants away?[1]

(a) Outline the design of an experiment to answer these questions using 72 acacia trees (be sure to include a control group).

(b) Use software or the *Simple Random Sample* applet to choose the trees for the active-hive group, or Table B at line 137 to choose the first 4 trees in that group.

(c) What is the response variable in this experiment?

16.6 Does chocolate cause headaches? The evidence linking chocolates to chronic headaches is inconsistent. In one study, 64 women with chronic headaches ate a restricted diet for two weeks. They then ate candy bars containing either chocolate or carob, prepared to taste the same, and reported whether they had a headache in the next 12 hours.[2]

(a) Outline the design of this experiment.

(b) Use software or the *Simple Random Sample* applet to choose the 32 members of the chocolate group (list only the first 10), or use Table B at line 110 to choose the first 5 members of the chocolate group.

16.7 Cash to find work? Will cash bonuses speed the return to work of unemployed people? The Illinois Department of Employment Security designed an experiment to find out. The subjects were 10,065 people aged 20 to 54 who were filing claims for unemployment insurance. Some were offered $500 if they found a job within 11 weeks and held it for at least 4 months. Others could tell potential employers that the state would pay the employer $500 for hiring them. A control group got neither kind of bonus.[3]

(a) Suggest two response variables of interest to the state and outline the design of the experiment.

(b) How will you label the subjects for random assignment? Use Table B at line 127 to choose the first 3 subjects for the first treatment.

16.8 Effects of day care. The Carolina Abecedarian Project investigated the effect of high-quality preschool programs on children from poor families. Children were randomly assigned to two groups. One group participated in a year-round preschool program from the age of three months. The control group received social services but no preschool. At age 21, **35%** of the treatment group and **14%** of the control group were attending a four-year college or had already graduated from college. Is each of the boldface numbers a parameter or a statistic? Why?

16.9 *NOVA* takes a sample. The Web site of the PBS television program *NOVA Science Now* invites viewers to vote on issues such as re-creating the virus responsible for the deadly flu epidemic of 1918. This online poll is unusual in offering detailed arguments for both sides. Of the 790 viewers who read the arguments and voted, 64% said that re-creating the virus was justified.[4] Explain to someone who knows no statistics why these 790 responses probably don't represent the opinions of all American adults.

16.10 Marijuana and driving. Questioning a sample of young people in New Zealand revealed a positive association between use of marijuana (cannabis) and traffic accidents caused by the members of the sample. Both cannabis use and accidents were measured by interviewing the young people themselves. The study report says, "It is unlikely that self reports of cannabis use and accident rates will be perfectly accurate."[5] Is the response bias likely to make the reported association stronger or weaker than the true association? Why?

Emilio Ereza/Alamy

16.11 Fuel economy. The Environmental Protection Agency (EPA) fuel economy ratings say that the Toyota Prius hybrid car gets 48 miles per gallon (mpg) on the highway. Deborah wonders whether the actual long-term average highway mileage of her new Prius is less than 48 mpg. She keeps careful records of gas mileage for 3000 miles of highway driving. Her result is $\overline{x} = 47.2$ mpg. What are her null and alternative hypotheses?

16.12 Did you work hard in high school? The average amount of time that high school students spend on homework is about 5 hours per week. Only 25% of college freshmen say they spent at least 6 hours per week on homework in high school. Your college wonders if the average for its freshmen differs from the national average. A random sample of 500 freshmen claims to have spent an average of $\overline{x} = 6.2$ hours per week on homework in high school. State the null and alternative hypotheses for a comparison of freshmen at your college with national freshmen.

16.13 Estimating blood cholesterol. The distribution of blood cholesterol level in the population of young men aged 20 to 34 years is close to Normal with standard deviation $\sigma = 41$ milligrams per deciliter (mg/dl). You measure the blood cholesterol of 14 cross-country runners. The mean level is $\overline{x} = 172$ mg/dl. Assuming that σ is the same as in the general population, give a 90% confidence interval for the mean level μ among cross-country runners.

16.14 Testing blood cholesterol. The mean blood cholesterol level for all men aged 20 to 34 years is $\mu = 188$ mg/dl. We suspect that the mean for cross-country runners is lower. State hypotheses, use the information in Exercise 16.13 to find the z test statistic, and give the P-value. Is the result significant at the $\alpha = 0.10$ level? At $\alpha = 0.05$? At $\alpha = 0.01$?

16.15 Smaller margin of error. How large a sample is needed to cut the margin of error in Exercise 16.13 in half? How large a sample is needed to cut the margin of error to ± 5 mg/dl?

16.16 More significant results. You increase the sample of cross-country runners in Exercise 16.13 from 14 to 56. Suppose that this larger sample gives the same mean level, $\overline{x} = 172$ mg/dl. Redo the test in Exercise 16.14. What is the P-value now? At which of the levels $\alpha = 0.10$, $\alpha = 0.05$, $\alpha = 0.01$ is the result significant? What general fact about significance tests does comparing your results here and in Exercise 16.14 illustrate?

How many miles per gallon?

As gasoline prices rise, more people pay attention to the government's gas mileage ratings of their vehicles. Until recently these ratings overstated the miles per gallon we can expect in real-world driving. The ratings assumed a top speed of 60 miles per hour, slow acceleration, and no air conditioning. That doesn't resemble what we see around us on the highway. Maybe it doesn't resemble the way we ourselves drive. Starting with 2008 models, the ratings assume higher speeds (80 miles per hour tops), faster acceleration, and air conditioning in warm weather. Mileage ratings of the same vehicle dropped by about 12% in the city and 8% on the highway.

16.17 Pesticides in whale blubber: estimation. The level of pesticides found in the blubber of whales is a measure of pollution of the oceans by runoff from land and can also be used to identify different populations of whales. A sample of 8 male minke whales in the West Greenland area of the North Atlantic found the mean concentration of the insecticide dieldrin to be $\bar{x} = 357$ nanograms per gram of blubber (ng/g).[6] Suppose that the concentration in all such whales varies Normally with standard deviation $\sigma = 50$ ng/g. Use a 95% confidence interval to estimate the mean level. Be sure to state your conclusion in plain language.

16.18 Pesticides in whale blubber: testing. The Food and Drug Administration regulates the amount of dieldrin in raw food. For some foods, no more than 100 ng/g is allowed. Using the information in Exercise 16.17, is there good evidence that the mean concentration in whale blubber is above 100 ng/g? State hypotheses, carry out a test assuming that the "simple conditions" (page 360) hold, and give your conclusion in plain language.

16.19 Other confidence levels. Use the information in Exercise 16.17 to give an 80% confidence interval and a 90% confidence interval for the mean concentration of dieldrin in the whale population. What general fact about confidence intervals do the margins of error of your three intervals illustrate?

16.20 Birth weight and IQ: estimation. Infants weighing less than 1500 grams at birth are classed as "very low birth weight." Low birth weight carries many risks. One study followed 113 male infants with very low birth weight to adulthood. At age 20, the mean IQ score for these men was $\bar{x} = 87.6$.[7] IQ scores vary Normally with standard deviation $\sigma = 15$. Give a 95% confidence interval for the mean IQ score at age 20 for all very-low-birth-weight males.

16.21 Birth weight and IQ: testing. IQ tests are scaled so that the mean score in a large population should be $\mu = 100$. We suspect that the very-low-birth-weight population has mean score less than 100. Does the study described in the previous exercise give good evidence that this is true? State hypotheses, carry out a test assuming that the "simple conditions" (page 360) hold, and give your conclusion in plain language.

16.22 Birth weight and IQ: causation? Very-low-birth-weight babies are more likely to be born to unmarried mothers and to mothers who did not complete high school.

(a) Explain why the study of Exercise 16.20 was not an experiment.

(b) Explain clearly why confounding prevents us from concluding that very low birth weight in itself reduces adult IQ.

16.23 Sample space. A randomly chosen subject arrives for a study of exercise and fitness. Describe a sample space for each of the following. (In some cases, you may have some freedom in your choice of S.)

(a) The subject is either female or male.

(b) After 10 minutes on an exercise bicycle, you ask the subject to rate his or her effort on the Rate of Perceived Exertion (RPE) scale. RPE ranges in whole-number steps from 6 (no exertion at all) to 20 (maximal exertion).

(c) You measure VO2, the maximum volume of oxygen consumed per minute during exercise. VO2 is generally between 2.5 and 6.1 liters per minute.

(d) You measure the maximum heart rate (beats per minute).

16.24 Internet search engines. Internet search sites compete for users because they sell advertising space on their sites and can charge more if they are heavily used. Choose an Internet search attempt at random. Here is the probability distribution for the site the search uses:[8]

Site	Google	Yahoo	MSN	Ask.com	Others
Probability	0.66	0.21	0.07	0.04	?

(a) What is the probability that a search attempt is made at a site other than the leading four?

(b) What is the probability that a search attempt is directed to a site other than Google?

16.25 How many in the house? In government data, a household consists of all occupants of a dwelling unit. Here is the distribution of household size in the United States:

Number of persons	1	2	3	4	5	6	7
Probability	0.26	0.33	0.16	0.15	0.07	0.02	0.01

Choose an American household at random and let the random variable Y be the number of persons living in the household.

(a) Express "more than one person lives in this household" in terms of Y. What is the probability of this event?

(b) What is $P(2 < Y \leq 4)$?

(c) What is $P(Y \neq 2)$?

16.26 How many children? How many children do women give birth to during their childbearing years? Choose at random an American woman who is past childbearing:[9]

Number of children	0	1	2	3	4	5
Probability	0.193	0.174	0.344	0.181	0.074	0.034

(The few women with 6 or more children are included in the "5 children" group.)

(a) Check that this distribution satisfies the two requirements for a legitimate discrete probability model.

(b) Describe in words the event $P(X \leq 2)$. What is the probability of this event?

(c) What is $P(X < 2)$?

(d) Write the event "a woman gives birth to three or more children" in terms of values of X. What is the probability of this event?

16.27 Reaction times. The time that people require to react to a stimulus usually has a right-skewed distribution, as lack of attention or tiredness causes some lengthy reaction times. Reaction times for children with attention deficit hyperactivity disorder (ADHD) are more skewed, as their condition causes more frequent lack of attention. In one study, children with ADHD were asked to press the spacebar on a computer keyboard when any letter other than X appeared on the screen. With 2 seconds between letters, the mean reaction time was 445 milliseconds (ms) and the standard deviation was 82 ms.[10] Take these values to be the population μ and σ for ADHD children.

(a) What are the mean and standard deviation of the mean reaction time $\bar{x}$ for a randomly chosen group of 15 ADHD children? For a group of 150 such children?

(b) The distribution of reaction time is strongly skewed. Explain briefly why we hesitate to regard $\bar{x}$ as Normally distributed for 15 children but are willing to use a Normal distribution for the mean reaction time of 150 children.

(c) What is the approximate probability that the mean reaction time in a group of 150 ADHD children is greater than 450 ms?

16.28 An IQ test. The Wechsler Adult Intelligence Scale (WAIS) is a common "IQ test" for adults. The distribution of WAIS scores for persons over 16 years of age is approximately Normal with mean 100 and standard deviation 15.

(a) What is the probability that a randomly chosen individual has a WAIS score of 105 or higher?

(b) What are the mean and standard deviation of the average WAIS score $\bar{x}$ for an SRS of 60 people?

(c) What is the probability that the average WAIS score of an SRS of 60 people is 105 or higher?

(d) Would your answers to any of (a), (b), or (c) be affected if the distribution of WAIS scores in the adult population were distinctly non-Normal?

16.29 Does chocolate cause headaches, continued. Here are some of the results of the experiment described in Exercise 16.6. There was no significant difference in headaches between the chocolate and carob groups ($P = 0.68$). But subjects who said they had a mild headache before eating the candy bar were more likely to report a headache afterward ($P < 0.001$). Explain carefully why $P = 0.68$ means there is no evidence that chocolate and carob differ in their effects and why $P < 0.001$ is evidence that having a headache before eating the candy bar does increase reports of a headache after eating.

16.30 Brains at work. When our brains store information, complicated chemical changes take place. In trying to understand these changes, researchers blocked some processes in brain cells taken from rats and compared these cells with a control group of normal cells. They say that "no differences were seen" between the two groups in four response variables. They give P-values 0.45, 0.83, 0.26, and 0.84 for these four comparisons.[11]

(a) Say clearly what P-value $P = 0.45$ says about the response that was observed.

(b) It isn't literally true that "no differences were seen." That is, the mean responses were not exactly alike in the two groups. Explain what the researchers mean when they give $P = 0.45$ and say "no differences were seen."

16.31 California brushfires. We often see televised reports of brushfires threatening homes in California. Some people argue that the modern practice of quickly putting out small fires allows fuel to accumulate and so increases the damage done by large fires. A detailed study of historical data suggests that this is wrong—the damage has risen simply because there are more houses in risky areas.[12] As usual, the study report gives statistical information tersely. Here is the summary of a regression of number of fires on decade (9 data points, for the 1910s to the 1990s): "Collectively, since 1910, there has been a highly significant increase ($r^2 = 0.61$, $P < 0.01$) in the number of fires per decade." How would you explain this statement to someone who knows no statistics? Include an explanation of both the description given by r^2 and its statistical significance.

CORBIS

SUPPLEMENTARY EXERCISES

Supplementary exercises apply the skills you have learned in ways that require more thought or more elaborate use of technology.

16.32 Sampling students. You want to investigate the attitudes of students at your school toward the school's policy on sexual harassment. You have a grant that will pay the costs of contacting about 500 students.

(a) Specify the exact population for your study. For example, will you include part-time students?

(b) Describe your sample design. Will you use a stratified sample?

(c) Briefly discuss the practical difficulties that you anticipate. For example, how will you contact the students in your sample?

16.33 The placebo effect. A survey of physicians found that some doctors give a placebo to a patient who complains of pain for which the physician can find no cause. If the patient's pain improves, these doctors conclude that it had no physical basis. The medical school researchers who conducted the survey claimed that these doctors do not understand the placebo effect. Why?

16.34 Informed consent. The requirement that human subjects give their informed consent to participate in an experiment can greatly reduce the number of available subjects. For example, a study of new teaching methods asks the consent of parents for their children to be taught by either a new method or the standard method. Many parents do not return the forms, so their children must continue to follow the standard curriculum. Why is it not correct to consider these children as part of the control group along with children who are randomly assigned to the standard method?

16.35 Fixing health care. The cost of health care and health insurance is the biggest health concern among Americans, even ahead of cancer and other diseases.

Changing to a national government health insurance system is controversial. An opinion poll will give different results depending on the wording of the question asked. For each of the following claims, say whether including it in the question would *increase* or *decrease* the percent of a poll sample who support a government health insurance system.

(a)   A national system would mean that everybody has health insurance.

(b)   A national system would probably require an increase in taxes.

(c)   Eliminating private insurance companies and their profits would reduce insurance costs.

(d)   A national system would limit the medical treatments available in order to contain costs.

16.36 Market research. Stores advertise price reductions to attract customers. What type of price cut is most attractive? Market researchers prepared ads for athletic shoes announcing different levels of discounts (20%, 40%, or 60%). The student subjects who read the ads were also given "inside information" about the fraction of shoes on sale (50% or 100%). Each subject then rated the attractiveness of the sale on a scale of 1 to 7.[13]

(a)   There are two factors. Make a sketch like Figure 9.2 (page 227) that displays the treatments formed by all combinations of levels of the factors.

(b)   Outline a completely randomized design using 60 student subjects. Use software or Table B at line 111 to choose the subjects for the first treatment.

16.37 Making french fries. Few people want to eat discolored french fries. Potatoes are kept refrigerated before being cut for french fries to prevent spoiling and preserve flavor. But immediate processing of cold potatoes causes discoloring due to complex chemical reactions. The potatoes must therefore be brought to room temperature before processing. Design an experiment in which tasters will rate the color and flavor of french fries prepared from several groups of potatoes. The potatoes will be freshly picked or stored for a month at room temperature or stored for a month refrigerated. They will then be sliced and cooked either immediately or after an hour at room temperature.

(a)   What are the factors and their levels, the treatments, and the response variables?

(b)   Describe and outline the design of this experiment.

(c)   It is efficient to have each taster rate fries from all treatments. How will you use randomization in presenting fries to the tasters?

16.38 The addition rule. The addition rule for probabilities, $P(A \text{ or } B) = P(A) + P(B)$, is not always true. Give (in words) an example of real-world events A and B for which this rule is not true.

16.39 Comparing wine tasters. Two wine tasters rate each wine they taste on a scale of 1 to 5. From data on their ratings of a large number of wines, we obtain the following probabilities for both tasters' ratings of a randomly chosen wine:

	TASTER 2				
TASTER 1	**1**	**2**	**3**	**4**	**5**
1	0.03	0.02	0.01	0.00	0.00
2	0.02	0.08	0.05	0.02	0.01
3	0.01	0.05	0.25	0.05	0.01
4	0.00	0.02	0.05	0.20	0.02
5	0.00	0.01	0.01	0.02	0.06

(a) Why is this a legitimate discrete probability model?

(b) What is the probability that the tasters agree when rating a wine?

(c) What is the probability that Taster 1 rates a wine higher than Taster 2? What is the probability that Taster 2 rates a wine higher than Taster 1?

16.40 A 14-sided die. An ancient Korean drinking game involves a 14-sided die. The players roll the die in turn and must submit to whatever humiliation is written on the up-face: something like "Keep still when tickled on face." Six of the 14 faces are squares. Let's call them A, B, C, D, E, and F for short. The other eight faces are triangles, which we will call 1, 2, 3, 4, 5, 6, 7, and 8. Each of the squares is equally likely. Each of the triangles is also equally likely, but the triangle probability differs from the square probability. The probability of getting a square is 0.72. Give the probability model for the 14 possible outcomes.

David Moore

16.41 Distributions: means versus individuals. The z confidence interval and test are based on the sampling distribution of the sample mean $\overline{x}$. Suppose that the distribution of body mass index (BMI) among young women is Normal with mean $\mu = 27$ and standard deviation $\sigma = 7.5$.

(a) You take an SRS of 100 young women. According to the 99.7 part of the 68–95–99.7 rule, about what range of BMI values do you expect to see in your sample?

(b) You look at many SRSs of size 100. About what range of sample mean BMIs $\overline{x}$ do you expect to see?

16.42 Distributions: larger samples. In the setting of the previous exercise, how many women must you sample to cut the range of values of $\overline{x}$ in half? This will also cut the margin of error of a confidence interval for μ in half. Do you expect the range of individual scores in the new sample to also be much less than in a sample of size 100? Why?

16.43 Alcohol and mortality. It appears that people who drink alcohol in moderation have lower death rates than either people who drink heavily or people who do not drink at all. The protection offered by moderate drinking is concentrated among people over 50 and on deaths from heart disease. The Nurses' Health Study played an essential role in establishing these facts for women. This part of the study followed 85,709 female nurses for 12 years, during which time 2658 of the subjects died. The nurses completed a questionnaire that described their diet, including their use of alcohol. They were reexamined every two years. Conclusion: "As compared with nondrinkers and heavy drinkers, light-to-moderate drinkers had a significantly lower risk of death."[14]

(a)   Was this study an experiment? Explain your answer.

(b)   What does "significantly lower risk of death" mean in simple language?

(c)   Suggest some lurking variables that might be confounded with how much a person drinks. The investigators used advanced statistical methods to adjust for many such variables before concluding that the moderate drinkers really have a lower risk of death.

16.44 Time in a restaurant. The owner of a pizza restaurant in France knows that the time customers spend in the restaurant on Saturday evening has mean 90 minutes and standard deviation 15 minutes. He has read that pleasant odors can influence customers, so he spreads a lavender odor throughout the restaurant. Here are the times (minutes) for customers on the next Saturday evening:[15]

92	126	114	106	89	137	93	76	98	108
124	105	129	103	107	109	94	105	102	108
95	121	109	104	116	88	109	97	101	106

(a)   Make a stemplot of the times. The distribution is roughly symmetric and single-peaked, so the distribution of $\bar{x}$ should be close to Normal.

(b)   Suppose that the standard deviation $\sigma = 15$ minutes is not changed by the odor. Is there reason to think that the lavender odor has changed the mean time customers spend in the restaurant? Follow the four-step process for significance tests (page 378).

16.45 Normal body temperature? Here are the daily average body temperatures (degrees Fahrenheit) for 20 healthy adults:[16]

| 98.74 | 98.83 | 96.80 | 98.12 | 97.89 | 98.09 | 97.87 | 97.42 | 97.30 | 97.84 |
| 100.27 | 97.90 | 99.64 | 97.88 | 98.54 | 98.33 | 97.87 | 97.48 | 98.92 | 98.33 |

(a)   Make a stemplot of the data. The distribution is roughly symmetric and single-peaked. There is one mild outlier. We expect the distribution of the sample mean $\bar{x}$ to be close to Normal.

(b)   Do these data give evidence that the mean body temperature for all healthy adults is not equal to the traditional 98.6 degrees? Follow the four-step process for significance tests (page 378). (Suppose that body temperature varies Normally with standard deviation 0.7 degree.)

16.46 Time in a restaurant. Use the data in Exercise 16.44 to estimate the mean time customers spend in this restaurant on Saturday evenings with 95% confidence. Follow the four-step process for confidence intervals (page 366).

16.47 Normal body temperature. Use the data in Exercise 16.45 to estimate mean body temperature with 90% confidence. Follow the four-step process for confidence intervals (page 366).

16.48 Tests from confidence intervals. You read in a Census Bureau report that a 99% confidence interval for the mean income in 2005 of American households headed by a college-educated person at least 25 years old was $100,272 \pm \$1651$. (The median income of these households was lower, $77,179.) Based on this interval, can you reject the null hypothesis that the mean income in this group is $95,000? What is the alternative hypothesis of the test? What is its significance level?

16.49 Low power? (optional) It appears that eating oat bran lowers cholesterol slightly. At a time when oat bran was something of a fad, a paper in the *New England Journal of Medicine* found that it had no significant effect on cholesterol.[17] The paper reported a study with just 20 subjects. Letters to the journal denounced publication of a negative finding from a study with very low power. Explain why lack of significance in a study with low power gives no reason to accept the null hypothesis that oat bran has no effect.

16.50 Type I and Type II errors (optional). Exercise 16.21 asks for a significance test of the null hypothesis that the mean IQ of very-low-birth-weight male babies is 100 against the alternative hypothesis that the mean is less than 100. State in words what it means to make a Type I error and a Type II error in this setting.

O P T I O N A L   E X E R C I S E S

These exercises concern the material in Chapters 12 and 13.

16.51 Is business success just chance? Investors like to think that some companies are consistently successful. Academic researchers looked at data for many companies to determine whether each firm's sales growth was above the median for all firms in each year. They found that a simple "just chance" model fit well: years are independent, and the probability of being above the median in any one year is 1/2.[18] If this model holds, what is the probability that a particular firm is above average for two consecutive years? For all of four years?

16.52 Sharing music online. A sample survey reports that 29% of Internet users download music files online, 21% share music files from their computers, and 12% both download and share music.[19] Make a Venn diagram that displays this information. What percent of Internet users neither download nor share music files?

16.53 Really high incomes. The Internal Revenue Service received 134,372,678 individual income tax returns for 2005. Of these, 303,817 reported an adjusted gross income of $1 million or more and 35,207 reported at least $5 million.[20]

(a) What is the probability that a randomly chosen return shows an income of at least $5 million?

(b) If you know that a randomly chosen return shows an income of $1 million or more, what is the conditional probability that the income is at least $5 million?

16.54 Comparing wine tasters. In the setting of Exercise 16.39, Taster 1's rating for a wine is 3. What is the conditional probability that Taster 2's rating is higher than 3?

16.55 Latinos online. "Latinos comprise 14% of the U.S. adult population and about half of this growing group (56%) goes online."[21] Take *A* to be the event that a randomly chosen adult is a Latino and *B* the event that a randomly chosen adult goes online. Express the two percents given as probabilities. Find the percent of U.S. adults who are Latino and go online.

16.56 A baseball cliché. How often have you heard a baseball radio or TV announcer say something like "Scott has hit safely in 9 of the last 12 games," as if this were an impressive performance? Let's find out how impressive. Major league starting players (leaving out pitchers) hit safely in about 67% of their games.[22]

(a) It's reasonable to take games as independent. What is the distribution of the number of games out of 12 in which a typical player hits safely?

(b) What is the probability that a player hits safely in 9 or more out of 12 games? In 8 or more out of 12?

16.57 Text messaging. Suppose (as is roughly true) that 40% of all cell phone owners have sent or received text messages. A sample survey interviews an SRS of 2000 cell phone owners.

(a) What is the actual distribution of the number X in the sample who have sent or received text messages?

(b) What is the probability that 750 or fewer of the people in the sample have sent or received text messages? (Use software or a suitable approximation.)

16.58 Athletes breaking bones. The intense training needed to reach the Olympics carries risks. About 70% of the members of U.S. Winter Olympics teams have suffered one or more broken bones in the past.

Steve Mason/Age fotostock

(a) The 2006 team had 211 members. What distribution describes the number of team members who had broken bones in the past?

(b) What is the probability that 150 or more of the 211 team members had broken bones in the past? (Use software or an approximation.)

16.59 Airline overbooking. Airlines sell more tickets than a plane has seats because some passengers don't show up and an empty seat is worth nothing once the plane takes off. You are planning ticket sales for a Boeing 737 with 132 seats. You know that about 12% of people who book the flight won't show up. Use binomial distributions (software or the Normal approximation) to answer these questions.

(a) If you sell 145 seats, what is the probability that no more than 132 passengers will show up?

(b) On the average, each passenger without a seat costs the airline $250, either as a reward for voluntarily giving up a seat or as a penalty if a passenger is bumped. If you sell 145 seats, what is the probability that this extra cost is no more than $1000?

16.60 Cystic fibrosis. Cystic fibrosis is a lung disorder that often results in death. It is inherited but can be inherited only if both parents are carriers of an abnormal gene. In 1989, the CF gene that is abnormal in carriers of cystic fibrosis was identified. The probability that a randomly chosen person of European ancestry carries an abnormal CF gene is 1/25. (The probability is less in other ethnic groups.) The CF20m test detects most but not all harmful mutations of the CF gene. The test is positive for 90% of people who are carriers. It is (ignoring human error) never positive for people who are not carriers. What is the probability that a randomly chosen person of European ancestry tests positive?

16.61 Cystic fibrosis, continued. Jason tests positive on the CF20m test. What is the probability that he is a carrier of the abnormal CF gene?

16.62 Smoking and social class. As the dangers of smoking have become more widely known, clear class differences in smoking have emerged. British government statistics classify adult men by occupation as "managerial and professional" (43% of the population), "intermediate" (34%), or "routine and manual" (23%). A survey finds that 20% of men in managerial and professional occupations smoke, 29% of the intermediate group smoke, and 38% in routine and manual occupations smoke.[23] Use a tree diagram to find the percent of all adult British men who smoke.

16.63 Smoking and social class, continued. Use your work from the previous exercise to find the percent of male smokers who have routine and manual occupations. (Start by expressing this as a conditional probability.) This information is used in planning antismoking campaigns.

16.64 Do the rich stay that way? We like to think that anyone can rise to the top. That's possible, but it's easier if you start near the top. Divide families by the income of the parents into the top 20%, the bottom 20%, and the middle 60%. Here are the conditional probabilities that a child of each class of parents ends up in each income class as an adult.[24] For example, a child of parents in the top 20% has probability 0.42 of also being in the top 20%.

	Child's class		
Parents' class	Top 20%	Middle 60%	Bottom 20%
Top 20%	0.42	0.52	0.06
Middle 60%	0.15	0.68	0.17
Bottom 20%	0.07	0.56	0.37

Suppose that these probabilities stay the same for three generations. Draw a tree diagram to show the path of a child and grandchild of parents in the top 20% of incomes. For example, the child might drop to the middle and the grandchild might then rise back to the top. What is the probability that the grandchild of people in the top 20% is also in the top 20%?

Inference about Variables

With the principles in hand, we proceed to practice, that is, to inference in fully realistic settings. In the remaining chapters of this book, you will meet many of the most commonly used statistical procedures. We have grouped these procedures into two classes, corresponding to our division of data analysis into exploring variables and distributions and exploring relationships. The five chapters of Part III concern inference about the distribution of a single variable and inference for comparing the distributions of two variables. Part IV deals with inference for relationships among variables. In Chapters 17 and 18, we analyze data on quantitative variables. We begin with the familiar Normal distribution for a quantitative variable. Chapters 19 and 20 concern categorical variables, so that inference begins with counts and proportions of outcomes. Chapter 21 reviews this part of the text.

The four-step process for approaching a statistical problem can guide much of your work in these chapters. You should review the outlines of the four-step process for a confidence interval (page 366) and for a test of significance (page 378). The statement of an exercise usually does the *State* step for you, leaving the *Plan, Solve,* and *Conclude* steps for you to complete. It is helpful to first summarize the *State* step in your own words to organize your thinking. Many examples and exercises in these chapters involve both carrying out inference and thinking about inference in practice. Remember that any inference method is useful only under certain conditions, and that you must judge these conditions before rushing to inference.

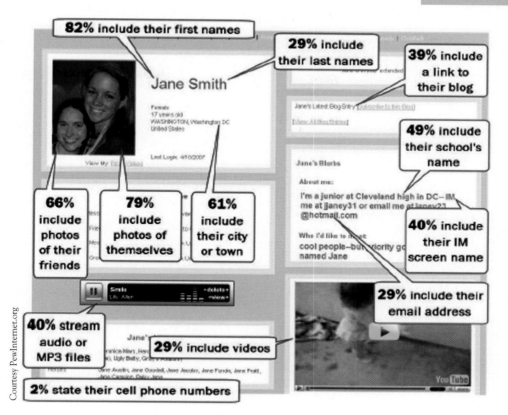

Courtesy PewInternet.org

QUANTITATIVE RESPONSE VARIABLE

CATEGORICAL RESPONSE VARIABLE

Eric Nathan/Alamy

CHAPTER 17

Inference about a Population Mean

This chapter describes confidence intervals and significance tests for the mean μ of a population. We used the z procedures in this same setting to introduce the ideas of confidence intervals and tests. Now we discard the unrealistic condition that we know the population standard deviation σ and present procedures for practical use. We also pay more attention to the real-data setting of our work. The details of confidence intervals and tests change only slightly when you don't know σ. More important, you can interpret your results exactly as before. To emphasize this, Examples 17.2 and 17.3 repeat the most important examples from Chapter 14.

**IN THIS CHAPTER
WE COVER...**

- Conditions for inference about a mean
- The t distributions
- The one-sample t confidence interval
- The one-sample t test
- Using technology
- Matched pairs t procedures
- Robustness of t procedures

Conditions for inference about a mean

Confidence intervals and tests of significance for the mean μ of a Normal population are based on the sample mean $\overline{x}$. Confidence levels and P-values are probabilities calculated from the sampling distribution of $\overline{x}$. Here are the conditions needed for realistic inference about a population mean.

> **CONDITIONS FOR INFERENCE ABOUT A MEAN**
>
> ■ We can regard our data as a **simple random sample** (SRS) from the population. This condition is very important.
>
> ■ Observations from the population have a **Normal distribution** with mean μ and standard deviation σ. In practice, it is enough that the distribution be symmetric and single-peaked unless the sample is very small. Both μ and σ are unknown parameters.

There is another condition that applies to all of the inference methods in this book: *the population must be much larger than the sample, say at least 20 times as large*.[1] All of our examples and exercises satisfy this condition. Practical settings in which the sample is a large part of the population are rather special, and we will not discuss them.

When the conditions for inference are satisfied, the sample mean $\bar{x}$ has the Normal distribution with mean μ and standard deviation $\sigma/\sqrt{n}$. Because we don't know σ, we estimate it by the sample standard deviation s. We then estimate the standard deviation of $\bar{x}$ by $s/\sqrt{n}$. This quantity is called the *standard error* of the sample mean $\bar{x}$.

> **STANDARD ERROR**
>
> When the standard deviation of a statistic is estimated from data, the result is called the **standard error** of the statistic. The standard error of the sample mean $\bar{x}$ is $s/\sqrt{n}$.

APPLY YOUR KNOWLEDGE

17.1   Travel time to work. A study of commuting times reports the travel times to work of a random sample of 20 employed adults in New York State. The mean is $\bar{x} = 31.25$ minutes and the standard deviation is $s = 21.88$ minutes. What is the standard error of the mean?

17.2   Is that light moving? When two lights close together blink alternately, we "see" one light moving back and forth if the time between blinks is short. What is the longest interval of time between blinks that preserves the illusion of motion? Ask subjects to turn a knob that slows the blinking until they "see" two lights rather than one light moving. A report gives the results in the form "mean plus or minus the standard error of the mean."[2] Data for 12 subjects are summarized as 251 ± 45 (in milliseconds). What are $\bar{x}$ and s for these subjects? (This exercise is also a warning to read carefully: that 251 ± 45 is *not* a confidence interval, yet summaries in this form are common in scientific reports.)

© Oote Boe Photography/Alamy

The *t* distributions

If we knew the value of σ, we would base confidence intervals and tests for μ on the one-sample z statistic

$$z = \frac{\overline{x} - \mu}{\sigma/\sqrt{n}}$$

This z statistic has the standard Normal distribution $N(0, 1)$. In practice, we don't know σ, so we substitute the standard error $s/\sqrt{n}$ of $\overline{x}$ for its standard deviation $\sigma/\sqrt{n}$. The statistic that results does not have a Normal distribution. It has a distribution that is new to us, called a *t distribution*.

THE ONE-SAMPLE *t* STATISTIC AND THE *t* DISTRIBUTIONS

Draw an SRS of size n from a large population that has the Normal distribution with mean μ and standard deviation σ. The **one-sample *t* statistic**

$$t = \frac{\overline{x} - \mu}{s/\sqrt{n}}$$

has the **t distribution** with $n - 1$ degrees of freedom.

The t statistic has the same interpretation as any standardized statistic: it says how far $\overline{x}$ is from its mean μ in standard deviation units. There is a different t distribution for each sample size. We specify a particular t distribution by giving its **degrees of freedom.** The degrees of freedom for the one-sample t statistic come from the sample standard deviation s in the denominator of t. We saw in Chapter 2 (page 51) that s has $n - 1$ degrees of freedom. There are other t statistics with different degrees of freedom, some of which we will meet later. We will write the t distribution with $n - 1$ degrees of freedom as $t(n - 1)$ for short.

degrees of freedom

Figure 17.1 compares the density curves of the standard Normal distribution and the t distributions with 2 and 9 degrees of freedom. The figure illustrates these facts about the t distributions:

- The density curves of the t distributions are similar in shape to the standard Normal curve. They are symmetric about 0, single-peaked, and bell-shaped.

- The spread of the t distributions is a bit greater than that of the standard Normal distribution. The t distributions in Figure 17.1 have more probability in the tails and less in the center than does the standard Normal. This is true because substituting the estimate s for the fixed parameter σ introduces more variation into the statistic.

- As the degrees of freedom increase, the t density curve approaches the $N(0, 1)$ curve ever more closely. This happens because s estimates σ more accurately as the sample size increases. So using s in place of σ causes little extra variation when the sample is large.

FIGURE 17.1

Density curves for the t distributions with 2 and 9 degrees of freedom and the standard Normal distribution. All are symmetric with center 0. The t distributions are somewhat more spread out.

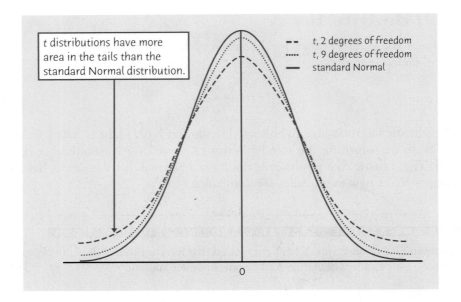

Table C in the back of the book gives critical values for the t distributions. Each row in the table contains critical values for the t distribution whose degrees of freedom appear at the left of the row. For convenience, we label the table entries both by the confidence level C (in percent) required for confidence intervals and by the one-sided and two-sided P-values for each critical value. You have already used the standard Normal critical values in the z^* row at the bottom of Table C. By looking down any column, you can check that the t critical values approach the Normal values as the degrees of freedom increase. If you use statistical software, you don't need Table C.

EXAMPLE 17.1 *t* critical values

Figure 17.1 shows the density curve for the t distribution with 9 degrees of freedom. What point on this distribution has probability 0.05 to its right? In Table C, look in the df = 9 row above one-sided P-value .05 and you will find that this critical value is $t^* = 1.833$. To use software, enter the degrees of freedom and the probability you want to the *left*, 0.95 in this case. Here is Minitab's output:

```
Student's t distribution with 9 DF
P(X<=x)                 x
   0.95           1.83311 ▪
```

APPLY YOUR KNOWLEDGE

17.3 **Critical values.** Use Table C or software to find

(a) the critical value for a one-sided test with level $\alpha = 0.05$ based on the $t(5)$ distribution.

(b)   the critical value for a 98% confidence interval based on the $t(21)$ distribution.

17.4   More critical values. You have an SRS of size 25 and calculate the one-sample *t* statistic. What is the critical value t^* such that

(a)   *t* has probability 0.025 to the right of t^*?

(b)   *t* has probability 0.75 to the left of t^*?

The one-sample *t* confidence interval

To analyze samples from Normal populations with unknown σ, just replace the standard deviation $\sigma/\sqrt{n}$ of $\overline{x}$ by its standard error $s/\sqrt{n}$ in the *z* procedures of Chapters 14 and 15. The confidence interval and test that result are *one-sample t procedures*. Critical values and *P*-values come from the *t* distribution with $n - 1$ degrees of freedom. The one-sample *t* procedures are similar in both reasoning and computational detail to the *z* procedures.

Better statistics, better beer

The *t* distribution and the *t* inference procedures were invented by William S. Gosset (1876–1937). Gosset worked for the Guinness brewery, and his goal in life was to make better beer. He used his new *t* procedures to find the best varieties of barley and hops. Gosset's statistical work helped him become head brewer, a more interesting title than professor of statistics. Because Gosset published under the pen name "Student" you will often see the *t* distribution called "Student's *t*" in his honor.

> ### THE ONE-SAMPLE *t* CONFIDENCE INTERVAL
>
> Draw an SRS of size n from a large population having unknown mean μ. A level C **confidence interval for μ** is
>
> $$\overline{x} \pm t^* \frac{s}{\sqrt{n}}$$
>
> where t^* is the critical value for the $t(n-1)$ density curve with area C between $-t^*$ and t^*. This interval is exact when the population distribution is Normal and is approximately correct for large n in other cases.

EXAMPLE 17.2   Healing of skin wounds

Let's look again at the biological study we met in Example 14.3. We follow the four-step process for a confidence interval, outlined on page 366.

STATE: Biologists studying the healing of skin wounds measured the rate at which new cells closed a razor cut made in the skin of an anesthetized newt. Here are data from 18 newts, measured in micrometers (millionths of a meter) per hour:[3]

$$
\begin{array}{ccccccccc}
29 & 27 & 34 & 40 & 22 & 28 & 14 & 35 & 26 \\
35 & 12 & 30 & 23 & 18 & 11 & 22 & 23 & 33
\end{array}
$$

This is one of several sets of measurements made under different conditions. We want to estimate the mean rate for comparison with rates under other conditions.

PLAN: We will estimate the mean rate μ for all newts of this species by giving a 95% confidence interval.

SOLVE: We must first check the conditions for inference.

David A. Northcott/CORBIS

FIGURE 17.2

Stemplot of the healing rates in Example 17.2.

```
1 | 1 2 4
1 | 8
2 | 2 2 3 3
2 | 6 7 8 9
3 | 0 3 4
3 | 5 5
4 | 0
```

- As in Chapter 14 (page 367), we are willing to regard these newts as an SRS from their species.

- The stemplot in Figure 17.2 does not suggest any strong departures from Normality.

We can proceed to calculation. For these data,

$$\bar{x} = 25.67 \quad \text{and} \quad s = 8.324$$

The degrees of freedom are $n - 1 = 17$. From Table C we find that for 95% confidence $t^* = 2.110$. The confidence interval is

$$\bar{x} \pm t^* \frac{s}{\sqrt{n}} = 25.67 \pm 2.110 \frac{8.324}{\sqrt{18}}$$

$$= 25.67 \pm 4.14$$

$$= 21.53 \text{ to } 29.81 \text{ micrometers per hour}$$

CONCLUDE: We are 95% confident that the mean healing rate for all newts of this species is between 21.53 and 29.81 micrometers per hour. ■

Our work in Example 17.2 is very similar to what we did in Example 14.3 (page 366). To make the inference realistic we replaced the assumed $\sigma = 8$ by $s = 8.324$ calculated from the data and replaced the standard Normal critical value $z^* = 1.960$ by the t critical value $t^* = 2.110$.

The one-sample t confidence interval has the form

$$\text{estimate} \pm t^* \text{SE}_{\text{estimate}}$$

where "SE" stands for "standard error." We will meet a number of confidence intervals that have this common form. In Example 17.2, the estimate is the sample mean $\bar{x}$, and its standard error is

$$\text{SE}_{\bar{x}} = \frac{s}{\sqrt{n}}$$

$$= \frac{8.324}{\sqrt{18}} = 1.962$$

Software will find $\bar{x}$, s, $\text{SE}_{\bar{x}}$, and the confidence interval from the data. Figure 17.5 (page 453) displays typical software output for Example 17.2.

APPLY YOUR KNOWLEDGE

17.5   **Critical values.** What critical value t^* from Table C would you use for a confidence interval for the mean of the population in each of the following situations?

(a)   A 95% confidence interval based on $n = 10$ observations.

(b)   A 99% confidence interval from an SRS of 20 observations.

(c)   A 90% confidence interval from a sample of size 7.

17.6   **To gamble or not to gamble.** Our decisions depend on how the options are presented to us. Here's an experiment that illustrates this phenomenon. Tell 20 subjects that they have been given $50 but can't keep it all. Then present them with a long series of choices between bets they can make with the $50. Scattered among these choices in random order are 64 choices between a fixed amount and an all-or-nothing gamble. The odds for the gamble are always the same, but 32 of the fixed options read "Keep $20" and the other 32 read "Lose $30." These two options are exactly the same except for their wording, but people are more likely to gamble if the fixed option says they lose money. Here are the percent differences ("Lose $30" minus "Keep $20") in the numbers of trials on which the 20 subjects chose to gamble:[4]

37.5	30.8	6.2	17.6	14.3	8.3	16.7	20.0	10.5	21.7
30.8	27.3	22.7	38.5	8.3	10.5	8.3	10.5	25.0	7.7

(a)   Make a stemplot. Is there any sign of a major deviation from Normality?

(b)   All 20 subjects gambled more often when faced with a sure loss than when faced with a sure win. Give a 95% confidence interval for the mean percent increase in gambling when faced with a sure loss.

17.7   **Ancient air.** The composition of the earth's atmosphere may have changed over time. To try to discover the nature of the atmosphere long ago, we can examine the gas in bubbles inside ancient amber. Amber is tree resin that has hardened and been trapped in rocks. The gas in bubbles within amber should be a sample of the atmosphere at the time the amber was formed. Measurements on specimens of amber from the late Cretaceous era (75 to 95 million years ago) give these percents of nitrogen:[5]

63.4	65.0	64.4	63.3	54.8	64.5	60.8	49.1	51.0

Assume (this is not yet agreed on by experts) that these observations are an SRS from the late Cretaceous atmosphere. Use a 90% confidence interval to estimate the mean percent of nitrogen in ancient air. Follow the four-step process as illustrated in Example 17.2.

David Sanger Photography/Alamy

The one-sample *t* test

Like the confidence interval, the *t* test is very similar to the *z* test we met earlier.

Draw an SRS of size n from a large population having unknown mean μ. To **test the hypothesis $H_0: \mu = \mu_0$,** compute the **one-sample t statistic**

$$t = \frac{\bar{x} - \mu_0}{s/\sqrt{n}}$$

In terms of a variable T having the $t(n-1)$ distribution, the P-value for a test of H_0 against

$H_a: \mu > \mu_0$ is $P(T \geq t)$

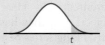

$H_a: \mu < \mu_0$ is $P(T \leq t)$

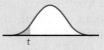

$H_a: \mu \neq \mu_0$ is $2P(T \geq |t|)$

These P-values are exact if the population distribution is Normal and are approximately correct for large n in other cases.

EXAMPLE 17.3 **Sweetening colas**

Here is a more realistic analysis of the cola-sweetening example from Chapter 14. We follow the four-step process for a significance test, outlined on page 378.

STATE: Cola makers test new recipes for loss of sweetness during storage. Trained tasters rate the sweetness before and after storage. Here are the sweetness losses (sweetness before storage minus sweetness after storage) found by 10 tasters for one new cola recipe:

2.0   0.4   0.7   2.0   −0.4   2.2   −1.3   1.2   1.1   2.3

Are these data good evidence that the cola lost sweetness?

PLAN: Tasters vary in their perception of sweetness loss. So we ask the question in terms of the mean loss μ for a large population of tasters. The null hypothesis is "no loss," and the alternative hypothesis says "there is a loss."

$$H_0: \mu = 0$$
$$H_a: \mu > 0$$

SOLVE: First check the conditions for inference. As before, we are willing to regard these 10 carefully trained tasters as an SRS from a large population of all trained tasters.

Figure 17.3 is a stemplot of the data. We can't judge Normality from just 10 observations; there are no outliers but the data are somewhat skewed. *P*-values for the *t* test may be only approximately accurate.

The basic statistics are

$$\bar{x} = 1.02 \quad \text{and} \quad s = 1.196$$

The one-sample *t* statistic is

$$t = \frac{\bar{x} - \mu_0}{s/\sqrt{n}} = \frac{1.02 - 0}{1.196/\sqrt{10}}$$

$$= 2.697$$

The *P*-value for *t* = 2.697 is the area to the right of 2.697 under the *t* distribution curve with degrees of freedom *n* − 1 = 9. Figure 17.4 shows this area. Software (see Figure 17.6, on page 454) tells us that *P* = 0.0123.

−1	3
−0	4
0	4 7
1	1 2
2	0 0 2 3

FIGURE 17.3

Stemplot of the sweetness losses in Example 17.3.

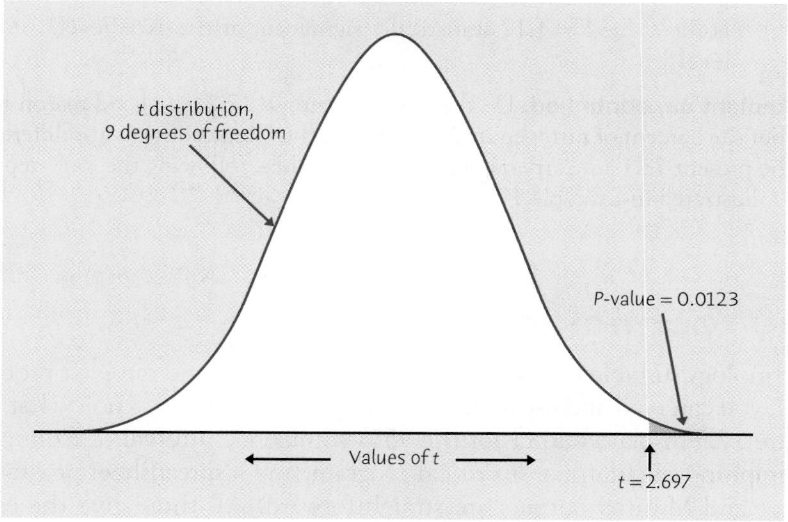

FIGURE 17.4

The *P*-value for the one-sided *t* test in Example 17.3.

Without software, we can pin *P* between two values by using Table C. Search the df = 9 row of Table C for entries that bracket *t* = 2.697. The observed *t* lies between the critical values for one-sided *P*-values 0.02 and 0.01.

CONCLUDE: There is quite strong evidence (*P* < 0.02) for a loss of sweetness. ■

df = 9

*t**	2.398	2.821
One-sided *P*	.02 .	.01

APPLY YOUR KNOWLEDGE

17.8 Is it significant? The one-sample *t* statistic for testing

$$H_0: \mu = 0$$

$$H_a: \mu > 0$$

from a sample of *n* = 15 observations has the value *t* = 1.82.

(a)   What are the degrees of freedom for this statistic?

(b)   Give the two critical values t^* from Table C that bracket t. What are the one-sided P-values for these two entries?

(c)   Is the value $t = 1.82$ significant at the 5% level? Is it significant at the 1% level?

17.9 **Is it significant?** The one-sample t statistic from a sample of $n = 25$ observations for the two-sided test of

$$H_0: \mu = 64$$
$$H_a: \mu \neq 64$$

has the value $t = 1.12$.

(a)   What are the degrees of freedom for t?

(b)   Locate the two critical values t^* from Table C that bracket t. What are the two-sided P-values for these two entries?

(c)   Is the value $t = 1.12$ statistically significant at the 10% level? At the 5% level?

17.10 **Ancient air, continued.** Do the data of Exercise 17.7 give good reason to think that the percent of nitrogen in the air during the Cretaceous era was different from the present 78.1%? Carry out a test of significance, following the four-step process as illustrated in Example 17.3.

Using technology

Any technology suitable for statistics will implement the one-sample t procedures. As usual, you can read and use almost any output now that you know what to look for. Figure 17.5 displays output for the 95% confidence interval of Example 17.2 from a graphing calculator, a statistical program, and a spreadsheet program. The calculator and Minitab outputs are straightforward. All three give the estimate $\overline{x}$ and the confidence interval plus a clearly labeled selection of other information. The confidence interval agrees with our hand calculation in Example 17.2. In general, software results are more accurate because of the rounding in hand calculations. Excel gives several descriptive measures but does not give the confidence interval. The entry labeled "Confidence Level (95.0%)" is the margin of error. You can use this together with $\overline{x}$ to get the interval using either a calculator or the spreadsheet's formula capability.

Figure 17.6 displays output for the t test in Example 17.3. The graphing calculator and Minitab give the sample mean $\overline{x}$, the t statistic, and its P-value. Accurate P-values are the biggest advantage of software for the t procedures. Excel is as usual more awkward than software designed for statistics. It lacks a one-sample t test menu selection but does have a function named TDIST for tail areas under t density curves. The Excel output shows functions for the t statistic and its P-value to the right of the main display, along with their values $t = 2.69669$ and $P = 0.01226$.

Texas Instruments Graphing Calculator

```
TInterval
 (21.527,29.806)
 x̄=25.6667
 Sx=8.3243
 n=18.0000
```

Minitab

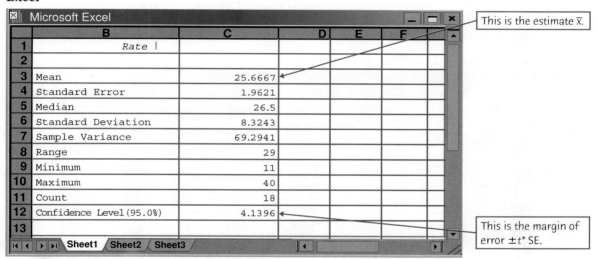

```
Session                                                    _ ◻ ✕

One-Sample T: Rate

Variable      N      Mean    StDev   SE Mean        95.0% CI
Rate         18     25.67     8.32      1.96    (21.53, 29.81)
```

Excel

Microsoft Excel						
	B	**C**	**D**	**E**	**F**	
1	*Rate*					
2						
3	Mean	25.6667				
4	Standard Error	1.9621				
5	Median	26.5				
6	Standard Deviation	8.3243				
7	Sample Variance	69.2941				
8	Range	29				
9	Minimum	11				
10	Maximum	40				
11	Count	18				
12	Confidence Level (95.0%)	4.1396				
13						

This is the estimate x̄.

This is the margin of error ±*t** SE.

FIGURE 17.5

The *t* confidence interval of Example 17.2: output from a graphing calculator, a statistical program, and a spreadsheet program.

Matched pairs *t* procedures

The study of healing in Example 17.2 estimated the mean healing rate for newts under natural conditions, but the researchers then compared results under several conditions. The taste test in Example 17.3 was a matched pairs study in which the same 10 tasters rated before-and-after sweetness. Comparative studies are more convincing than single-sample investigations. For that reason, one-sample inference is less common than comparative inference. One common design to compare two treatments makes use of one-sample procedures. In a **matched pairs design,** subjects are matched in pairs and each treatment is given to one subject in each

matched pairs design

Texas Instruments Graphing Calculator

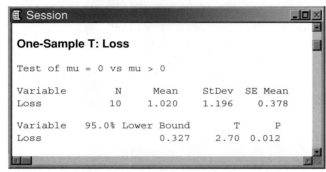

Minitab

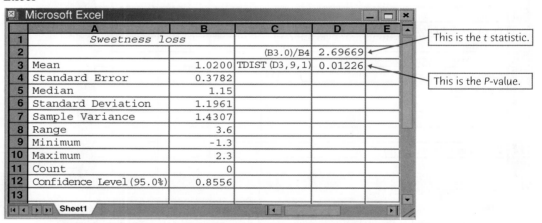

FIGURE 17.6

The *t* test of Example 17.3: output from a graphing calculator, a statistical program, and a spreadsheet program.

pair. Another situation calling for matched pairs is before-and-after observations on the same subjects, as in the taste test of Example 17.3.

> **MATCHED PAIRS *t* PROCEDURES**
>
> To compare the responses to the two treatments in a matched pairs design, find the difference between the responses within each pair. Then apply the one-sample *t* procedures to these differences.

The parameter μ in a matched pairs t procedure is the mean difference in the responses to the two treatments within matched pairs of subjects in the entire population.

EXAMPLE 17.4 Do chimpanzees collaborate?

STATE: Humans often collaborate to solve problems. Will chimpanzees recruit another chimp when solving a problem requires collaboration? Researchers presented chimpanzee subjects with food outside their cage that they could bring within reach by pulling two ropes, one attached to each end of the food tray. If a chimp pulled only one rope, the rope came loose and the food was lost. Another chimp was available as a partner, but only if the subject unlocked a door joining two cages. (Chimpanzees learn these things quickly.) The same 8 chimpanzee subjects faced this problem in two versions: the two ropes were close enough together that one chimp could pull both (no collaboration needed) or the two ropes were too far apart for one chimp to pull both (collaboration needed). Table 17.1 shows how often in 24 trials each subject opened the door to recruit another chimp as partner.[6] Is there evidence that chimpanzees recruit partners more often when a problem requires collaboration?

Manoj Shah/Getty

TABLE 17.1 **Trials (out of 24) on which chimpanzees recruited a partner**

| CHIMPANZEE | COLLABORATION NEEDED | | DIFFERENCE |
	YES	NO	
Namuiska	16	0	16
Kalema	16	1	15
Okech	23	5	18
Baluku	19	3	16
Umugenzi	15	4	11
Indi	20	9	11
Bili	24	16	8
Asega	24	20	4

PLAN: Take μ to be the mean difference (collaboration required minus not) in the number of times a subject recruited a partner. The null hypothesis says that the need for collaboration has no effect, and H_a says that partners are recruited more often when the problem requires collaboration. So we test the hypotheses

$$H_0: \mu = 0$$

$$H_a: \mu > 0$$

SOLVE: The subjects are "semi-free-ranging chimpanzees at Ngamba Island Chimpanzee Sanctuary in Uganda." We are willing to regard them as an SRS from their species. To analyze the data, subtract the "no collaboration needed" count from the "collaboration needed" count for each subject. The 8 differences form a single sample from a population with unknown mean μ. They appear in the "Difference" column in Table 17.1. All of the chimpanzees recruited a partner more often when the ropes were too far apart to be pulled by one chimp.

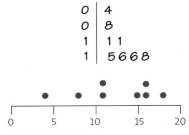

FIGURE 17.7

Stemplot and dotplot of the differences in Example 7.4.

df = 7

t^*	4.785	5.408
One-sided P	.001	.0005

The stemplot in Figure 17.7 creates the impression of a left skew. This is a bit misleading, as the *dotplot* in the bottom part of Figure 17.7 shows. A dotplot simply places the observations on an axis, stacking observations that have the same value. It gives a good picture of distributions with only whole-number values. We can't assess Normality from just 8 observations, but there are no signs of major departures from Normality. The researchers used the matched pairs t test.

The 8 differences have

$$\bar{x} = 12.375 \quad \text{and} \quad s = 4.749$$

The one-sample t statistic is therefore

$$t = \frac{\bar{x} - 0}{s/\sqrt{n}} = \frac{12.375 - 0}{4.749/\sqrt{8}}$$

$$= 7.37$$

Find the P-value from the $t(7)$ distribution. (Remember that the degrees of freedom are 1 less than the sample size.) Table C shows that 7.37 is greater than the critical value for one-sided $P = 0.0005$. The P-value is therefore less than 0.0005. Software says that $P = 0.000077$.

CONCLUDE: The data give very strong evidence ($P < 0.0005$) that chimpanzees recruit a collaborator more often when faced with a problem that requires a collaborator to solve. That is, chimpanzees recognize when collaboration is necessary, a skill that they share with humans. ■

Example 17.4 illustrates how to turn matched pairs data into single-sample data by taking differences within each pair. We are making inferences about a single population, the population of all differences within matched pairs. *It is incorrect to ignore the matching and analyze the data as if we had two samples of chimpanzees, one facing ropes close together and the other facing ropes far apart.* Inference procedures for comparing two samples assume that the samples are selected independently of each other. This condition does not hold when the same subjects are measured twice. The proper analysis depends on the design used to produce the data.

APPLY YOUR KNOWLEDGE

Many exercises from this point on ask you to give the P-value of a t test. If you have suitable technology, give the exact P-value. Otherwise, use Table C to give two values between which P lies.

17.11 The brain responds to sound. The usual way to study the brain's response to sounds is to have subjects listen to "pure tones." The response to recognizable sounds may differ. To compare responses, researchers anesthetized macaque monkeys. They fed pure tones and also monkey calls directly to their brains by inserting electrodes. Response to the stimulus was measured by the firing rate (electrical spikes per second) of neurons in various areas of the brain. Table 17.2 contains the responses for 37 neurons.[7] Researchers suspected that the response to monkey calls would be stronger than the response to a pure tone. Do the data support this idea?

TABLE 17.2 Neuron response to tones and monkey calls

TONE	CALL	TONE	CALL	TONE	CALL	TONE	CALL
474	500	145	42	71	134	35	103
256	138	141	241	68	65	31	70
241	485	129	194	59	182	28	192
226	338	113	123	59	97	26	203
185	194	112	182	57	318	26	135
174	159	102	141	56	201	21	129
176	341	100	118	47	279	20	193
168	85	74	62	46	62	20	54
161	303	72	112	41.	84	19	66
150	208						

Complete the *Plan, Solve,* and *Conclude* steps of the four-step process, following the model of Example 17.4.

17.12 The brain responds, continued. How much more strongly do monkey brains respond to monkey calls than to pure tones? Give a 90% confidence interval to answer this question.

Robustness of *t* procedures

The *t* confidence interval and test are exactly correct when the distribution of the population is exactly Normal. No real data are exactly Normal. The usefulness of the *t* procedures in practice therefore depends on how strongly they are affected by lack of Normality.

> **ROBUST PROCEDURES**
>
> A confidence interval or significance test is called **robust** if the confidence level or *P*-value does not change very much when the conditions for use of the procedure are violated.

The condition that the population is Normal rules out outliers, so the presence of outliers shows that this condition is not fulfilled. The *t* procedures are not robust against outliers unless the sample is large, because $\bar{x}$ and s are not resistant to outliers.

Fortunately, the *t* procedures are quite robust against non-Normality of the population except when outliers or strong skewness are present. (Skewness is more serious than other kinds of non-Normality.) As the size of the sample increases, the central limit theorem ensures that the distribution of the sample mean $\bar{x}$ becomes more nearly Normal and that the *t* distribution becomes more accurate for critical values and *P*-values of the *t* procedures.

Catching cheaters

A certification test for surgeons asks 277 multiple-choice questions. Smith and Jones have 193 common right answers and 53 identical wrong choices. The computer flags their 246 identical answers as evidence of possible cheating. They sue. The court wants to know how unlikely it is that exams this similar would occur just by chance. That is, the court wants a *P*-value. Statisticians offer several *P*-values based on different models for the exam-taking process. They all say that results this similar would almost never happen just by chance. Smith and Jones fail the exam.

Always make a plot to check for skewness and outliers before you use the *t* procedures for small samples. For most purposes, you can safely use the one-sample *t* procedures when $n \geq 15$ unless an outlier or quite strong skewness is present. Here are practical guidelines for inference on a single mean.[8]

USING THE *t* PROCEDURES

- Except in the case of small samples, the condition that the data are an SRS from the population of interest is more important than the condition that the population distribution is Normal.

- *Sample size less than 15:* Use *t* procedures if the data appear close to Normal (roughly symmetric, single peak, no outliers). If the data are clearly skewed or if outliers are present, do not use *t*.

- *Sample size at least 15:* The *t* procedures can be used except in the presence of outliers or strong skewness.

- *Large samples:* The *t* procedures can be used even for clearly skewed distributions when the sample is large, roughly $n \geq 40$.

EXAMPLE 17.5   Can we use *t*?

Figure 17.8 shows plots of several data sets. For which of these can we safely use the *t* procedures?[9]

- Figure 17.8(a) is a histogram of the percent of each state's adult residents who are college graduates. *We have data on the entire population of 50 states, so inference is not needed.* We can calculate the exact mean for the population. There is no uncertainty due to having only a sample from the population, and no need for a confidence interval or test. *If these data were an SRS from a larger population, t inference would be safe despite the mild skewness because n = 50.*

- Figure 17.8(b) is a stemplot of the force required to pull apart 20 pieces of Douglas fir. *The data are strongly skewed to the left with possible low outliers, so we cannot trust the t procedures for n = 20.*

- Figure 17.8(c) is a stemplot of the lengths of 23 specimens of the red variety of the tropical flower *Heliconia*. *The data are mildly skewed to the right and there are no outliers. We can use the t distributions for such data.*

- Figure 17.8(d) is a histogram of the heights of the students in a college class. *This distribution is quite symmetric and appears close to Normal. We can use the t procedures for any sample size.* ■

APPLY YOUR KNOWLEDGE

17.13 Diamonds. A group of earth scientists studied the small diamonds found in a nodule of rock carried up to the earth's surface in surrounding rock. This is an opportunity to examine a sample from a single population of diamonds formed in a single event deep in the earth.[10] Table 17.3 (page 460) presents data on the nitrogen content (parts per million) and the abundance of carbon-13 in these diamonds.

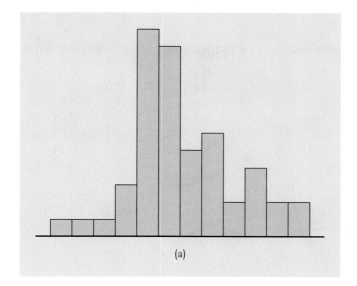

(a)

```
23 | 0
24 | 0
25 |
26 | 5
27 |
28 | 7
29 |
30 | 2 5 9
31 | 3 9 9
32 | 0 3 3 6 7 7
33 | 0 2 3 6
```

(b)

```
37 | 4 8 9
38 | 0 0 1 1 2 2 8 9
39 | 2 6 8
40 | 6 7
41 | 5 7 9 9
42 | 0 2
43 | 1
```

(c)

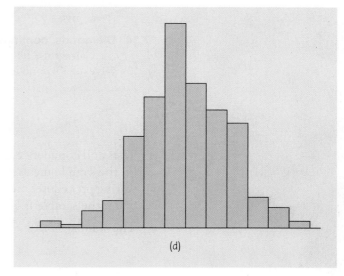

(d)

FIGURE 17.8

Can we use *t* procedures for these data? **(a)** Percent of adult college graduates in the 50 states. *No, this is an entire population, not a sample.* **(b)** Force required to pull apart 20 pieces of Douglas fir. *No, there are just 20 observations and strong skewness.* **(c)** Lengths of 23 tropical flowers of the same variety. *Yes, the sample is large enough to overcome the mild skewness.* **(d)** Heights of college students. *Yes, for any size sample,* because the distribution is close to Normal.

(Carbon has several isotopes, forms with different numbers of neutrons in the nuclei of their atoms. Carbon-12 makes up almost 99% of natural carbon. The abundance of carbon-13 is measured by the ratio of carbon-13 to carbon-12, in parts per thousand more or less than a standard. The minus signs in the data mean that the ratio is smaller in these diamonds than in standard carbon.)

We would like to estimate the mean abundance of both nitrogen and carbon-13 in the population of diamonds represented by this sample. Examine the data for

TABLE 17.3 Nitrogen and carbon-13 in a sample of diamonds

DIAMOND	NITROGEN (ppm)	CARBON-13 RATIO	DIAMOND	NITROGEN (ppm)	CARBON-13 RATIO
1	487	−2.78	13	273	−2.73
2	1430	−1.39	14	94	−2.33
3	60	−4.26	15	69	−3.83
4	244	−1.19	16	262	−2.04
5	196	−2.12	17	120	−2.82
6	274	−2.87	18	302	−0.84
7	41	−3.68	19	75	−3.57
8	54	−3.29	20	242	−2.42
9	473	−3.79	21	115	−3.89
10	30	−4.06	22	65	−3.87
11	98	−1.83	23	311	−1.58
12	41	−4.03	24	61	−3.97

nitrogen. Can we use a t confidence interval for mean nitrogen? Explain your answer. Give a 95% confidence interval if you think the result can be trusted.

17.14 Diamonds, continued. Examine the data in Table 17.3 on abundance of carbon-13. Can we use a t confidence interval for mean carbon-13? Explain your answer. Give a 95% confidence interval if you think the result can be trusted.

CHAPTER 17 SUMMARY

- Tests and confidence intervals for the mean μ of a Normal population are based on the sample mean $\bar{x}$ of an SRS. Because of the central limit theorem, the resulting procedures are approximately correct for other population distributions when the sample is large.

- The standardized sample mean is the **one-sample z statistic**

$$z = \frac{\bar{x} - \mu}{\sigma/\sqrt{n}}$$

If we knew σ, we would use the z statistic and the standard Normal distribution.

- In practice, we do not know σ. Replace the standard deviation $\sigma/\sqrt{n}$ of $\bar{x}$ by the **standard error** $s/\sqrt{n}$ to get the **one-sample t statistic**

$$t = \frac{\bar{x} - \mu}{s/\sqrt{n}}$$

The t statistic has the **t distribution** with $n - 1$ degrees of freedom.

- There is a t distribution for every positive **degrees of freedom.** All are symmetric distributions similar in shape to the standard Normal distribution. The t distribution approaches the $N(0, 1)$ distribution as the degrees of freedom increase.

■ A level C **confidence interval for the mean** μ of a Normal population is

$$\bar{x} \pm t^* \frac{s}{\sqrt{n}}$$

The **critical value** t^* is chosen so that the t curve with $n - 1$ degrees of freedom has area C between $-t^*$ and t^*.

■ **Significance tests** for $H_0: \mu = \mu_0$ are based on the t statistic. Use P-values or fixed significance levels from the $t(n - 1)$ distribution.

■ Use these one-sample procedures to analyze **matched pairs** data by first taking the difference within each matched pair to produce a single sample.

■ The t procedures are quite **robust** when the population is non-Normal, especially for larger sample sizes. The t procedures are useful for non-Normal data when $n \geq 15$ unless the data show outliers or strong skewness.

CHECK YOUR SKILLS

17.15 We prefer the t procedures to the z procedures for inference about a population mean because

(a) z can be used only for large samples.

(b) z requires that you know the population standard deviation σ.

(c) z requires that you can regard your data as an SRS from the population.

17.16 You are testing $H_0: \mu = 10$ against $H_a: \mu < 10$ based on an SRS of 20 observations from a Normal population. The data give $\bar{x} = 8$ and $s = 4$. The value of the t statistic is

(a) -0.5. (b) -10. (c) -2.24.

17.17 You are testing $H_0: \mu = 10$ against $H_a: \mu < 10$ based on an SRS of 20 observations from a Normal population. The t statistic is $t = -2.25$. The degrees of freedom for this statistic are

(a) 19. (b) 20. (c) 21.

17.18 The P-value for the statistic in the previous exercise

(a) falls between 0.01 and 0.02.

(b) falls between 0.02 and 0.04.

(c) is greater than 0.25.

17.19 You have an SRS of 15 observations from a Normally distributed population. What critical value would you use to obtain a 98% confidence interval for the mean μ of the population?

(a) 2.326 (b) 2.602 (c) 2.624

17.20 You are testing $H_0: \mu = 0$ against $H_a: \mu \neq 0$ based on an SRS of 15 observations from a Normal population. What values of the t statistic are statistically significant at the $\alpha = 0.005$ level?

(a) $t < -3.326$ or $t > 3.326$ (b) $t < -3.286$ or $t > 3.286$ (c) $t > 2.977$

17.21 Data on the blood cholesterol levels of 24 rats (milligrams per deciliter of blood) give $\overline{x} = 85$ and $s = 12$. A 95% confidence interval for the mean blood cholesterol of rats under this condition is

(a) 79.9 to 90.1.   (b) 80.2 to 89.8.   (c) 84.0 to 86.0.

17.22 Which of the following would cause the most worry about the validity of the confidence interval you calculated in the previous exercise?

(a) There is a clear outlier in the data.

(b) A stemplot of the data shows a mild right skew.

(c) You do not know the population standard deviation σ.

17.23 Which of these settings does *not* allow use of a matched pairs t procedure?

(a) You interview both the husband and the wife in 64 married couples and ask each about their ideal number of children.

(b) You interview a sample of 64 unmarried male students and another sample of 64 unmarried female students and ask each about their ideal number of children.

(c) You interview 64 female students in their freshman year and again in their senior year and ask each about their ideal number of children.

17.24 Because the t procedures are robust, the most important condition for their safe use is that

(a) the population standard deviation σ is known.

(b) the population distribution is exactly Normal.

(c) the data can be regarded as an SRS from the population.

C H A P T E R 1 7   E X E R C I S E S

17.25 Read carefully. You read in the report of a psychology experiment: "Separate analyses for our two groups of 12 participants revealed no overall placebo effect for our student group (mean = 0.08, SD = 0.37, $t(11) = 0.49$) and a significant effect for our non-student group (mean = 0.35, SD = 0.37, $t(11) = 3.25$, $p < 0.01$)."[11] The null hypothesis is that the mean effect is zero. What are the correct values of the two t statistics based on the means and standard deviations? Compare each correct t-value with the critical values in Table C. What can you say about the two-sided P-value in each case?

17.26 Body mass index of young women. In Example 14.1 (page 360) we developed a 95% z confidence interval for the mean body mass index (BMI) of women aged 20 to 29 years, based on a national random sample of 654 such women. We assumed there that the population standard deviation was known to be $\sigma = 7.5$. In fact, the sample data had mean BMI $\overline{x} = 26.8$ and standard deviation $s = 7.42$. What is the 95% t confidence interval for the mean BMI of all young women?

17.27 Reading scores in Atlanta. The Trial Urban District Assessment (TUDA) is a government-sponsored study of student achievement in large urban school districts. TUDA gives a reading test scored from 0 to 500. A score of 243 is a "basic" reading level and a score of 281 is "proficient." Scores for a random sample of 1470 eighth-graders in Atlanta had $\overline{x} = 240$ with standard error 1.1.[12]

(a) We don't have the 1470 individual scores, but use of the t procedures is surely safe. Why?

(b) Give a 99% confidence interval for the mean score of all Atlanta eighth-graders. (Be careful: the report gives the standard error of $\overline{x}$, not the standard deviation s.)

(c) Urban children often perform below the basic level. Is there good evidence that the mean for all Atlanta eighth-graders is less than the basic level?

David Grossman/The Image Works

17.28 Calcium and blood pressure. In a randomized comparative experiment on the effect of calcium in the diet on blood pressure, researchers divided 54 healthy white males at random into two groups. One group received calcium; the other, a placebo. At the beginning of the study, the researchers measured many variables on the subjects. The paper reporting the study gives $\overline{x} = 114.9$ and $s = 9.3$ for the seated systolic blood pressure of the 27 members of the placebo group.

(a) Give a 95% confidence interval for the mean blood pressure in the population from which the subjects were recruited.

(b) What conditions for the population and the study design are required by the procedure you used in (a)? Which of these conditions are important for the validity of the procedure in this case?

17.29 The placebo effect. The placebo effect is particularly strong in patients with Parkinson's disease. To understand the workings of the placebo effect, scientists measure activity at a key point in the brain when patients receive a placebo that they think is an active drug and also when no treatment is given.[13] The same six patients are measured both with and without the placebo, at different times.

(a) Explain why the proper procedure to compare the mean response to placebo with control (no treatment) is a matched pairs t test.

(b) The six differences (treatment minus control) had $\overline{x} = -0.326$ and $s = 0.181$. Is there significant evidence of a difference between treatment and control?

17.30 The conductivity of glass. How well materials conduct heat matters when designing houses, for example. Conductivity is measured in terms of watts of heat power transmitted per square meter of surface per degree Celsius of temperature difference on the two sides of the material. In these units, glass has conductivity about 1. The National Institute of Standards and Technology (NIST) provides data on properties of materials. Here are 11 NIST measurements of the heat conductivity of a particular type of glass:[14]

1.11	1.07	1.11	1.07	1.12	1.08	1.08	1.18	1.18	1.18	1.12

(a) We can consider this an SRS of all specimens of glass of this type. Make a stemplot. Is there any sign of major deviation from Normality?

(b) Give a 95% confidence interval for the mean conductivity.

(c)   Is there significant evidence at the 5% level that the mean conductivity of this type of glass is not 1?

17.31 Learning Blissymbols. Blissymbols are pictographs (think of Egyptian hieroglyphics) sometimes used to help learning-disabled children. In a study of computer-assisted learning, 12 normal-ability schoolchildren were assigned at random to each of four computer learning programs. After they used the program, they attempted to recognize 24 Blissymbols. Here are the counts correct for one of the programs:[15]

12   22   9   14   20   15   9   10   11   11   15   6

(a)   Make a stemplot (split the stems). Are there outliers or strong skewness that would forbid use of the *t* procedures?

(b)   Give a 90% confidence interval for the mean count correct among all children of this age who use the program.

17.32 A big toe problem. Hallux abducto valgus (call it HAV) is a deformation of the big toe that often requires surgery. Doctors used X-rays to measure the angle (in degrees) of deformity in 38 consecutive patients under the age of 21 who came to a medical center for surgery to correct HAV. The angle is a measure of the seriousness of the deformity. Here are the data:[16]

28   32   25   34   38   26   25   18   30   26   28   13   20
21   17   16   21   23   14   32   25   21   22   20   18   26
16   30   30   20   50   25   26   28   31   38   32   21

It is reasonable to regard these patients as a random sample of young patients who require HAV surgery. Carry out the *Solve* and *Conclude* steps of a 95% confidence interval for the mean HAV angle in the population of all such patients.

17.33 An outlier's effect. Our bodies have a natural electrical field that is known to help wounds heal. Does changing the field strength slow healing? A series of experiments with newts investigated this question. In one experiment, the two hind limbs of 12 newts were assigned at random to either experimental or control groups. This is a matched pairs design. The electrical field in the experimental limbs was reduced to zero by applying a voltage. The control limbs were left alone. Here are the rates at which new cells closed a razor cut in each limb, in micrometers per hour:[17]

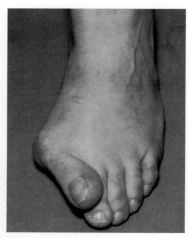

Newt	1	2	3	4	5	6	7	8	9	10	11	12
Control limb	36	41	39	42	44	39	39	56	33	20	49	30
Experimental limb	28	31	27	33	33	38	45	25	28	33	47	23

(a)   Make a stemplot of the differences between limbs of the same newt (control limb minus experimental limb). There is a high outlier.

(b)   A good way to judge the effect of an outlier is to do your analysis twice, once with the outlier and a second time without it. Carry out two *t* tests to see if the mean healing rate is significantly lower in the experimental limbs, one including all 12 newts and another that omits the outlier. What are the test statistics and their *P*-values? Does the outlier have a strong influence on your conclusion?

17.34 An outlier's effect. A good way to judge the effect of an outlier is to do your analysis twice, once with the outlier and a second time without it. The data in Exercise 17.32 follow a Normal distribution quite closely except for one patient with HAV angle 50 degrees, a high outlier.

(a) Find the 95% confidence interval for the population mean based on the 37 patients who remain after you drop the outlier.

(b) Compare your interval in (a) with your interval from Exercise 17.32. What is the most important effect of removing the outlier?

17.35 Genetic engineering for cancer treatment. Here's a new idea for treating advanced melanoma, the most serious kind of skin cancer. Genetically engineer white blood cells to better recognize and destroy cancer cells, then infuse these cells into patients. The subjects in a small initial study were 11 patients whose melanoma had not responded to existing treatments. One question was how rapidly the new cells would multiply after infusion, as measured by the doubling time in days. Here are the doubling times:[18]

> 1.4   1.0   1.3   1.0   1.3   2.0   0.6   0.8   0.7   0.9   1.9

(a) Examine the data. Is it reasonable to use the t procedures?

(b) Give a 90% confidence interval for the mean doubling time. Are you willing to use this interval to make an inference about the mean doubling time in a population of similar patients?

17.36 Genetic engineering for cancer treatment, continued. Another outcome in the cancer experiment described in Exercise 17.35 is measured by a test for the presence of cells that trigger an immune response in the body and so may help fight cancer. Here are data for the 11 subjects: counts of active cells per 100,000 cells before and after infusion of the modified cells. The difference (after minus before) is the response variable.

Before	14	0	1	0	0	0	0	20	1	6	0
After	41	7	1	215	20	700	13	530	35	92	108
Difference	27	7	0	215	20	700	13	510	34	86	108

(a) Examine the data. Is it reasonable to use the t procedures?

(b) If your conclusion in part (a) is Yes, do the data give convincing evidence that the count of active cells is higher after treatment?

17.37 Growing trees faster. The concentration of carbon dioxide (CO_2) in the atmosphere is increasing rapidly due to our use of fossil fuels. Because plants use CO_2 to fuel photosynthesis, more CO_2 may cause trees and other plants to grow faster. An elaborate apparatus allows researchers to pipe extra CO_2 to a 30-meter circle of forest. They selected two nearby circles in each of three parts of a pine forest and randomly chose one of each pair to receive extra CO_2. The response variable is the mean increase in base area for 30 to 40 trees in a circle during a growing season. We measure this in percent increase per year. The following are one year's data.[19]

Pair	Control plot	Treated plot
1	9.752	10.587
2	7.263	9.244
3	5.742	8.675

(a) State the null and alternative hypotheses. Explain clearly why the investigators used a one-sided alternative.

(b) Carry out a test and report your conclusion in simple language.

(c) The investigators used the test you just carried out. Any use of the t procedures with samples this size is risky. Why?

17.38 Fungus in the air. The air in poultry-processing plants often contains fungus spores. Inadequate ventilation can affect the health of the workers. The problem is most serious during the summer. To measure the presence of spores, air samples are pumped to an agar plate and "colony-forming units (CFUs)" are counted after an incubation period. Here are data from two locations in a plant that processes 37,000 turkeys per day, taken on four days in the summer. The units are CFUs per cubic meter of air.[20]

	Day 1	Day 2	Day 3	Day 4
Kill room	3175	2526	1763	1090
Processing	529	141	362	224

(a) Explain carefully why these are matched pairs data.

(b) The spore count is clearly higher in the kill room. Give sample means and a 90% confidence interval to estimate how much higher. Be sure to state your conclusion in plain English.

(c) You will often see the t procedures used for data like these. You should regard the results as only rough approximations. Why?

17.39 Weeds among the corn. Velvetleaf is a particularly annoying weed in cornfields. It produces lots of seeds, and the seeds wait in the soil for years until conditions are right. How many seeds do velvetleaf plants produce? Here are counts from 28 plants that came up in a cornfield when no herbicide was used:[21]

2450	2504	2114	1110	2137	8015	1623	1531	2008	1716
721	863	1136	2819	1911	2101	1051	218	1711	164
2228	363	5973	1050	1961	1809	130	880		

We would like to give a confidence interval for the mean number of seeds produced by velvetleaf plants. Alas, the t interval can't be safely used for these data. Why not?

17.40 How much oil? How much oil wells in a given field will ultimately produce is key information in deciding whether to drill more wells. Here are the estimated total amounts of oil recovered from 64 wells in the Devonian Richmond Dolomite area of the Michigan basin, in thousands of barrels:[22]

21.71	53.2	46.4	42.7	50.4	97.7	103.1	51.9
43.4	69.5	156.5	34.6	37.9	12.9	2.5	31.4
79.5	26.9	18.5	14.7	32.9	196	24.9	118.2
82.2	35.1	47.6	54.2	63.1	69.8	57.4	65.6
56.4	49.4	44.9	34.6	92.2	37.0	58.8	21.3
36.6	64.9	14.8	17.6	29.1	61.4	38.6	32.5
12.0	28.3	204.9	44.5	10.3	37.7	33.7	81.1
12.1	20.1	30.5	7.1	10.1	18.0	3.0	2.0

Take these wells to be an SRS of wells in this area.

(a) Give a 95% t confidence interval for the mean amount of oil recovered from all wells in this area.

(b) Make a graph of the data. The distribution is very skewed, with several high outliers. A computer-intensive method that gives accurate confidence intervals without assuming any specific shape for the distribution gives a 95% confidence interval of 40.28 to 60.32. How does the t interval compare with this? Should the t procedures be used with these data?

The following exercises ask you to answer questions from data without having the details outlined for you. The four-step process is illustrated in Examples 17.2, 17.3, and 17.4. The exercise statements give you the **State** *step. Follow the* **Plan, Solve,** *and* **Conclude** *steps in your work.*

17.41 Natural weed control? Fortunately, we aren't really interested in the number of seeds velvetleaf plants produce (see Exercise 17.39). The velvetleaf seed beetle feeds on the seeds and might be a natural weed control. Here are the total seeds, seeds infected by the beetle, and percent of seeds infected for 28 velvetleaf plants:

Seeds	2450	2504	2114	1110	2137	8015	1623	1531	2008	1716
Infected	135	101	76	24	121	189	31	44	73	12
Percent	5.5	4.0	3.6	2.2	5.7	2.4	1.9	2.9	3.6	0.7
Seeds	721	863	1136	2819	1911	2101	1051	218	1711	164
Infected	27	40	41	79	82	85	42	0	64	7
Percent	3.7	4.6	3.6	2.8	4.3	4.0	4.0	0.0	3.7	4.3
Seeds	2228	363	5973	1050	1961	1809	130	880		
Infected	156	31	240	91	137	92	5	23		
Percent	7.0	8.5	4.0	8.7	7.0	5.1	3.8	2.6		

Do a complete analysis of the percent of seeds infected by the beetle. Include a 90% confidence interval for the mean percent infected in the population of all velvetleaf plants. Do you think that the beetle is very helpful in controlling the weed?

17.42 Does nature heal better? Our bodies have a natural electrical field that is known to help wounds heal. Does changing the field strength slow healing? A series of experiments with newts investigated this question. The data below are the healing rates of cuts (micrometers per hour) in a matched pairs experiment. The pairs are the two hind limbs of the same newt, with the body's natural field in one limb (control) and half the natural value in the other limb (experimental).[23] Is there good evidence that changing the electrical field from its natural level slows healing?

Newt	1	2	3	4	5	6	7	8	9	10	11	12	13	14
Control	25	13	44	45	57	42	50	36	35	38	43	31	26	48
Experimental	24	23	47	42	26	46	38	33	28	28	21	27	25	45

17.43 How much better does nature heal? Give a 90% confidence interval for the difference in healing rates (control minus experimental) in the previous exercise.

 17.44 Mutual-fund performance. Mutual funds often compare their performance with a benchmark provided by an "index" that describes the performance of the class of assets in which the fund invests. For example, the Vanguard International Growth Fund benchmarks its performance against the EAFE (Europe, Australasia, Far East) index. Table 17.4 gives annual returns (percent) for the fund and the index. Does the fund's performance differ significantly from that of its benchmark?

(a) Explain clearly why the matched pairs t test is the proper choice to answer this question.

(b) Do a complete analysis that answers the question posed.

TABLE 17.4 A mutual fund versus its benchmark index

YEAR	FUND RETURN	INDEX RETURN	YEAR	FUND RETURN	INDEX RETURN
1984	−1.02	7.38	1996	14.65	6.05
1985	56.94	56.16	1997	4.12	1.78
1986	56.71	69.44	1998	16.93	20.00
1987	12.48	24.63	1999	26.34	26.96
1988	11.61	28.27	2000	−8.60	−14.17
1989	24.76	10.54	2001	−18.92	−21.44
1990	−12.05	−23.45	2002	−17.79	−15.94
1991	4.74	12.13	2003	34.45	38.59
1992	−5.79	−12.17	2004	18.95	20.25
1993	44.74	32.56	2005	15.00	13.54
1994	0.76	7.78	2006	25.92	26.34
1995	14.89	11.21	2007	15.98	11.17

 17.45 Right versus left. The design of controls and instruments affects how easily people can use them. Timothy Sturm investigated this effect in a course project, asking 25 right-handed students to turn a knob (with their right hands) that moved an indicator by screw action. There were two identical instruments, one with a right-hand thread (the knob turns clockwise) and the other with a left-hand thread (the knob turns counterclockwise). Table 17.5 gives the times in seconds each subject took to move the indicator a fixed distance.[24]

(a) Each of the 25 students used both instruments. Explain briefly how you would use randomization in arranging the experiment.

TABLE 17.5 Performance times (seconds) using right-hand and left-hand threads

SUBJECT	RIGHT THREAD	LEFT THREAD	SUBJECT	RIGHT THREAD	LEFT THREAD
1	113	137	14	107	87
2	105	105	15	118	166
3	130	133	16	103	146
4	101	108	17	111	123
5	138	115	18	104	135
6	118	170	19	111	112
7	87	103	20	89	93
8	116	145	21	78	76
9	75	78	22	100	116
10	96	107	23	89	78
11	122	84	24	85	101
12	103	148	25	88	123
13	116	147			

(b) The project hoped to show that right-handed people find right-hand threads easier to use. Do an analysis that leads to a conclusion about this issue.

17.46 Comparing two drugs. Makers of generic drugs must show that they do not differ significantly from the "reference" drugs that they imitate. One aspect in which drugs

TABLE 17.6 Absorption extent for two versions of a drug

SUBJECT	REFERENCE DRUG	GENERIC DRUG
15	4108	1755
3	2526	1138
9	2779	1613
13	3852	2254
12	1833	1310
8	2463	2120
18	2059	1851
20	1709	1878
17	1829	1682
2	2594	2613
4	2344	2738
16	1864	2302
6	1022	1284
10	2256	3052
5	938	1287
7	1339	1930
14	1262	1964
11	1438	2549
1	1735	3340
19	1020	3050

might differ is their extent of absorption in the blood. Table 17.6 gives data taken from 20 healthy nonsmoking male subjects for one pair of drugs.[25] This is a matched pairs design. Numbers 1 to 20 were assigned at random to the subjects. Subjects 1 to 10 received the generic drug first, and Subjects 11 to 20 received the reference drug first. In all cases, a washout period separated the two drugs so that the first had disappeared from the blood before the subject took the second. Do the drugs differ significantly in absorption?

17.47 Practical significance? Give a 90% confidence interval for the mean time advantage of right-hand over left-hand threads in the setting of Exercise 17.45. Do you think that the time saved would be of practical importance if the task were performed many times—for example, by an assembly-line worker? To help answer this question, find the mean time for right-hand threads as a percent of the mean time for left-hand threads.

CHAPTER 18

Two-Sample Problems

Comparing two populations or two treatments is one of the most common situations encountered in statistical practice. We call such situations *two-sample problems*.

> **TWO-SAMPLE PROBLEMS**
>
> - The goal of inference is to compare the responses to two treatments or to compare the characteristics of two populations.
>
> - We have a separate sample from each treatment or each population.

Two-sample problems

A two-sample problem can arise from a randomized comparative experiment that randomly divides subjects into two groups and exposes each group to a different treatment. Comparing random samples separately selected from two populations is also a two-sample problem. Unlike the matched pairs designs studied earlier, there is no matching of the individuals in the two samples, and the two samples can be of different sizes. Inference procedures for two-sample data differ from those for matched pairs. Here are some typical two-sample problems.

EXAMPLE 18.1 **Two-sample problems**

- Does regular physical therapy help lower back pain? A randomized experiment assigned patients with lower back pain to two groups: 142 received an examination and advice from a physical therapist; another 144 received regular physical therapy for up to five weeks. After a year, the change in their level of disability (0% to 100%) was assessed by a doctor who did not know which treatment the patients had received.

- A psychologist develops a test that measures social insight. He compares the social insight of female college students with that of male college students by giving the test to a sample of female students and a separate sample of male students.

- A bank wants to know which of two incentive plans will most increase the use of its credit cards. It offers each incentive to a random sample of credit card customers and compares the amounts charged during the following six months. ■

APPLY YOUR KNOWLEDGE

Which data design? *Each situation described in Exercises 18.1 to 18.4 requires inference about a mean or means. Identify each as involving (1) a single sample, (2) matched pairs, or (3) two independent samples. The procedures of Chapter 17 apply to designs (1) and (2). We are about to learn procedures for (3).*

Betsie Van Der Meer/Getty

18.1 **Looking back on love.** Choose 40 romantically attached couples in their midtwenties. Interview the man and woman separately about a romantic attachment they had at age 15 or 16. Compare the attitudes of men and women.

18.2 **Whom do you trust?** Companies often place advertisements to improve the image of their brand rather than to promote specific products. In a randomized comparative experiment, business students read ads that cited either the *Wall Street Journal* or the *National Enquirer* for important facts about a fictitious company. The students then rated the trustworthiness of each ad on a 7-point scale. Compare the mean score for the two types of advertisement.

18.3 **Chemical analysis.** To check a new analytical method, a chemist obtains a reference specimen of known concentration from the National Institute of Standards and Technology. She then makes 20 measurements of the concentration of this specimen with the new method and checks for bias by comparing the mean result with the known concentration.

18.4 **Chemical analysis, continued.** Another chemist is checking the same new method. He has no reference specimen, but a familiar analytic method is available. He wants to know if the new and old methods agree. He takes a specimen of unknown concentration and measures the concentration 10 times with the new method and 10 times with the old method.

Comparing two population means

Comparing two populations or the responses to two treatments starts with data analysis: make boxplots, stemplots (for small samples), or histograms (for larger samples) and compare the shapes, centers, and spreads of the two samples. The most common goal of inference is to compare the average or typical responses in the two populations. When data analysis suggests that both population distributions are symmetric, and especially when they are at least approximately Normal, we want to compare the population means. Here are the conditions for inference about means.

CONDITIONS FOR INFERENCE COMPARING TWO MEANS

- We have **two SRSs,** from two distinct populations. The samples are **independent.** That is, one sample has no influence on the other. Matching violates independence, for example. We measure the same response variable for both samples.

- Both populations are **Normally distributed.** The means and standard deviations of the populations are unknown. In practice, it is enough that the distributions have similar shapes and that the data have no strong outliers.

Driving while fasting

Muslims fast from sunrise to sunset during the month of Ramadan. Does this affect the rate of traffic accidents? Fasting can improve alertness, reducing accidents. Or it can cause dehydration, increasing accidents. Data from Turkey show a statistically significant increase, starting two weeks into Ramadan. Ah, but because Ramadan follows a lunar calendar, it cycles through the year. Perhaps accidents go down during a winter Ramadan (alertness) but go up during a summer Ramadan (longer fast and dehydration). Ask the statisticians this question and get their favorite answer: we need more data.

Call the variable we measure x_1 in the first population and x_2 in the second because the variable may have different distributions in the two populations. Here is how we describe the two populations:

Population	Variable	Mean	Standard deviation
1	x_1	μ_1	σ_1
2	x_2	μ_2	σ_2

There are four unknown parameters, the two means and the two standard deviations. The subscripts remind us which population a parameter describes. We want to compare the two population means, either by giving a confidence interval for their difference $\mu_1 - \mu_2$ or by testing the hypothesis of no difference, $H_0: \mu_1 = \mu_2$.

We use the sample means and standard deviations to estimate the unknown parameters. Again, subscripts remind us which sample a statistic comes from. Here is how we describe the samples:

Population	Sample size	Sample mean	Sample standard deviation
1	n_1	$\overline{x}_1$	s_1
2	n_2	$\overline{x}_2$	s_2

To do inference about the difference $\mu_1 - \mu_2$ between the means of the two populations, we start from the difference $\overline{x}_1 - \overline{x}_2$ between the means of the two samples.

EXAMPLE 18.2 Daily activity and obesity

STATE: People gain weight when they take in more energy from food than they expend. James Levine and his collaborators at the Mayo Clinic investigated the link between obesity and energy spent on daily activity.[1]

Choose 20 healthy volunteers who don't exercise. Deliberately choose 10 who are lean and 10 who are mildly obese but still healthy. Attach sensors that monitor the

TABLE 18.1 Time (minutes per day) spent in three different postures by lean and obese subjects

GROUP	SUBJECT	STAND/WALK	SIT	LIE
Lean	1	511.100	370.300	555.500
Lean	2	607.925	374.512	450.650
Lean	3	319.212	582.138	537.362
Lean	4	584.644	357.144	489.269
Lean	5	578.869	348.994	514.081
Lean	6	543.388	385.312	506.500
Lean	7	677.188	268.188	467.700
Lean	8	555.656	322.219	567.006
Lean	9	374.831	537.031	531.431
Lean	10	504.700	528.838	396.962
Obese	11	260.244	646.281	521.044
Obese	12	464.756	456.644	514.931
Obese	13	367.138	578.662	563.300
Obese	14	413.667	463.333	532.208
Obese	15	347.375	567.556	504.931
Obese	16	416.531	567.556	448.856
Obese	17	358.650	621.262	460.550
Obese	18	267.344	646.181	509.981
Obese	19	410.631	572.769	448.706
Obese	20	426.356	591.369	412.919

AP Photo/Toby Talbot

subjects' every move for 10 days. Table 18.1 presents data on the time (in minutes per day) that the subjects spent standing or walking, sitting, and lying down. Do lean and obese people differ in the average time they spend standing and walking?

PLAN: Examine the data and carry out a test of hypotheses. We suspect in advance that lean subjects (Group 1) are more active than obese subjects (Group 2), so we test the hypotheses

$$H_0: \mu_1 = \mu_2$$
$$H_a: \mu_1 > \mu_2$$

SOLVE (first steps): Are the conditions for inference met? The subjects are volunteers, so they are not SRSs from all lean and mildly obese adults. The study tried to recruit comparable groups: all worked in sedentary jobs, none smoked or were taking medication, and so on. Setting clear standards like these helps make up for the fact that we can't reasonably get SRSs for so invasive a study. The subjects were not told that they were chosen from a larger group of volunteers because they did not exercise and were either lean or mildly obese. Because their willingness to volunteer isn't related to the purpose of the experiment, we will treat them as two independent SRSs.

A back-to-back stemplot (Figure 18.1) displays the data in detail. To make the plot, we rounded the data to the nearest 10 minutes and used 100s as stems and 10s as leaves. The distributions are a bit irregular, as we expect with just 10 observations. There are no clear departures from Normality such as extreme outliers or skewness. The lean subjects as a group spend much more time standing and walking than do the obese subjects. Calculating the group means confirms this:

Group	n	Mean $\bar{x}$	Std. dev. s
Group 1 (lean)	10	525.751	107.121
Group 2 (obese)	10	373.269	67.498

The observed difference in mean time per day spent standing or walking is

$$\bar{x}_1 - \bar{x}_2 = 525.751 - 373.269 = 152.482 \text{ minutes}$$

To complete the *Solve* step, we must learn the details of inference comparing two means. ▪

Lean		Obese
	2	6 7
7 2	3	5 6 7
	4	1 1 2 3 6
8 8 6 4 1 0	5	
8 1	6	

FIGURE 18.1

Back-to-back stemplot of the times spent walking or standing, for Example 18.2.

Two-sample *t* procedures

To assess the significance of the observed difference between the means of our two samples, we follow a familiar path. Whether an observed difference is surprising depends on the spread of the observations as well as on the two means. Widely different means can arise just by chance if the individual observations vary a great deal. To take variation into account, we would like to standardize the observed difference $\bar{x}_1 - \bar{x}_2$ by dividing by its standard deviation. This standard deviation is

$$\sqrt{\frac{\sigma_1^2}{n_1} + \frac{\sigma_2^2}{n_2}}$$

This standard deviation gets larger as either population gets more variable, that is, as σ_1 or σ_2 increases. It gets smaller as the sample sizes n_1 and n_2 increase.

Because we don't know the population standard deviations, we estimate them by the sample standard deviations from our two samples. The result is the **standard error,** or estimated standard deviation, of the difference in sample means:

standard error

$$\sqrt{\frac{s_1^2}{n_1} + \frac{s_2^2}{n_2}}$$

When we standardize the estimate by dividing it by its standard error, the result is the **two-sample *t* statistic:**

two-sample t statistic

$$t = \frac{\bar{x}_1 - \bar{x}_2}{\sqrt{\dfrac{s_1^2}{n_1} + \dfrac{s_2^2}{n_2}}}$$

The statistic t has the same interpretation as any z or t statistic: it says how far $\bar{x}_1 - \bar{x}_2$ is from 0 in standard deviation units.

The two-sample t statistic has approximately a t distribution. It does not have exactly a t distribution even if the populations are both exactly Normal. In practice, however, the approximation is very accurate. There are two practical options for using the two-sample t procedures:

Option 1. With software, use the statistic t with accurate critical values from the approximating t distribution. The degrees of freedom are calculated from the data by a somewhat messy formula. Moreover, the degrees of freedom may not be a whole number.

Option 2. Without software, use the statistic t with critical values from the t distribution with *degrees of freedom equal to the smaller of $n_1 - 1$ and $n_2 - 1$*. These procedures are always conservative for any two Normal populations. The confidence interval has a margin of error *as large as or larger than* is needed for the desired confidence level. The significance test gives a P-value *equal to or greater than* the true P-value.

The two options are exactly the same except for the degrees of freedom used for t critical values and P-values. As the sample sizes increase, confidence levels and P-values from Option 2 become more accurate. The gap between what Option 2 reports and the truth is quite small unless the sample sizes are both small and unequal.[2]

THE TWO-SAMPLE t PROCEDURES

Draw an SRS of size n_1 from a large Normal population with unknown mean μ_1, and draw an independent SRS of size n_2 from another large Normal population with unknown mean μ_2. A level C **confidence interval for $\mu_1 - \mu_2$** is given by

$$(\overline{x}_1 - \overline{x}_2) \pm t^* \sqrt{\frac{s_1^2}{n_1} + \frac{s_2^2}{n_2}}$$

Here t^* is the critical value for confidence level C for the t distribution with degrees of freedom from either Option 1 (software) or Option 2 (the smaller of $n_1 - 1$ and $n_2 - 1$).

To **test the hypothesis H_0: $\mu_1 = \mu_2$**, calculate the **two-sample t statistic**

$$t = \frac{\overline{x}_1 - \overline{x}_2}{\sqrt{\dfrac{s_1^2}{n_1} + \dfrac{s_2^2}{n_2}}}$$

Find P-values from the t distribution with degrees of freedom from either Option 1 (software) or Option 2 (the smaller of $n_1 - 1$ and $n_2 - 1$).

EXAMPLE 18.3 **Daily activity and obesity**

We can now complete Example 18.2.

SOLVE (inference): The two-sample *t* statistic comparing the average minutes spent standing and walking in Group 1 (lean) and Group 2 (obese) is

$$t = \frac{\overline{x}_1 - \overline{x}_2}{\sqrt{\dfrac{s_1^2}{n_1} + \dfrac{s_2^2}{n_2}}}$$

$$= \frac{525.751 - 373.269}{\sqrt{\dfrac{107.121^2}{10} + \dfrac{67.498^2}{10}}}$$

$$= \frac{152.482}{40.039} = 3.808$$

Software (Option 1) gives one-sided *P*-value $P = 0.0008$ based on df = 15.174.

Without software, use the conservative Option 2. Because $n_1 - 1 = 9$ and $n_2 - 1 = 9$, there are 9 degrees of freedom. Because H_a is one-sided, the *P*-value is the area to the right of $t = 3.808$ under the $t(9)$ curve. Figure 18.2 illustrates this *P*-value. Table C shows that $t = 3.808$ lies between the critical values t^* for 0.0025 and 0.001. So $0.001 < P < 0.0025$. Option 2 gives a larger (more conservative) *P*-value than Option 1. As usual, the practical conclusion is the same for both versions of the test.

df = 9		
t^*	3.690	4.297
One-sided *P*	.0025	.001

CONCLUDE: There is very strong evidence ($P = 0.0008$) that lean people spend more time walking and standing than do moderately obese people. ■

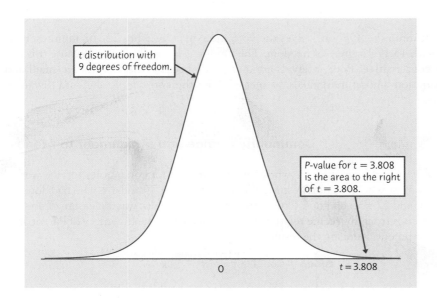

FIGURE 18.2

Using the conservative Option 2, the *P*-value in Example 18.3 comes from the *t* distribution with 9 degrees of freedom.

Does lack of every-day activity *cause* obesity? It may be that some people are naturally more active and are therefore less likely to gain weight. Or it may be that people who gain weight reduce their activity level. The study went on to enroll most of the obese subjects in a weight-reduction program and most of the lean subjects in a supervised program of overeating. After 8 weeks, the obese subjects had lost weight (mean 8 kg) and the lean subjects had gained weight (mean 4 kg). But both groups kept their original allocation of time to the different postures. This suggests that time allocation is biological and influences weight, rather than the other way around. The authors remark: "It should be emphasized that this was a pilot study and that the results need to be confirmed in larger studies."

EXAMPLE 18.4 How much more active are lean people?

PLAN: Give a 90% confidence interval for $\mu_1 - \mu_2$, the difference in average daily minutes spent standing and walking between lean and mildly obese adults.

SOLVE AND CONCLUDE: As in Example 18.3, the conservative Option 2 uses 9 degrees of freedom. Table C shows that the $t(9)$ critical value is $t^* = 1.833$. We are 90% confident that $\mu_1 - \mu_2$ lies in the interval

$$(\bar{x}_1 - \bar{x}_2) \pm t^* \sqrt{\frac{s_1^2}{n_1} + \frac{s_2^2}{n_2}}$$

$$= (525.751 - 373.269) \pm 1.833 \sqrt{\frac{107.121^2}{10} + \frac{67.498^2}{10}}$$

$$= 152.482 \pm 73.390$$

$$= 79.09 \text{ to } 225.87 \text{ minutes}$$

Software using Option 1 gives the 90% interval as 82.35 to 222.62 minutes, based on t with 15.174 degrees of freedom. The Option 2 interval is wider because this method is conservative. Both intervals are quite wide because the samples are small and the variation among individuals, as measured by the two sample standard deviations, is large. ■

EXAMPLE 18.5 Community service and attachment to friends

STATE: Do college students who have volunteered for community service work differ from those who have not? A study obtained data from 57 students who had done service work and 17 who had not. One of the response variables was a measure of attachment to friends (roughly, secure relationships), measured by the Inventory of Parent and Peer Attachment. Here are the results:[3]

Group	Condition	n	$\bar{x}$	s
1	Service	57	105.32	14.68
2	No service	17	96.82	14.26

Meta-analysis

Small samples have large margins of error. Large samples are expensive. Often we can find several studies of the same issue; if we could combine their results, we would have a large sample with a small margin of error. That is the idea of "meta-analysis." Of course, we can't just lump the studies together, because of differences in design and quality. Statisticians have more sophisticated ways of combining the results. Meta-analysis has been applied to issues ranging from the effect of secondhand smoke to whether coaching improves SAT scores.

PLAN: The investigator had no specific direction for the difference in mind before looking at the data, so the alternative is two-sided. We will test the hypotheses

$$H_0: \mu_1 = \mu_2$$
$$H_a: \mu_1 \neq \mu_2$$

SOLVE: The investigator says that the individual scores, examined separately in the two samples, appear roughly Normal. There is a serious problem with the more important condition that the two samples can be regarded as SRSs from two student populations. We will discuss that after we illustrate the calculations.

The two-sample *t* statistic is

$$t = \frac{\overline{x}_1 - \overline{x}_2}{\sqrt{\dfrac{s_1^2}{n_1} + \dfrac{s_2^2}{n_2}}}$$

$$= \frac{105.32 - 96.82}{\sqrt{\dfrac{14.68^2}{57} + \dfrac{14.26^2}{17}}}$$

$$= \frac{8.5}{3.9677} = 2.142$$

Software (Option 1) says that the two-sided *P*-value is $P = 0.0414$.

Without software, use Option 2 to find a conservative *P*-value. There are 16 degrees of freedom, the smaller of

$$n_1 - 1 = 57 - 1 = 56 \quad \text{and} \quad n_2 - 1 = 17 - 1 = 16$$

Figure 18.3 illustrates the *P*-value. Find it by comparing $t = 2.142$ with the two-sided critical values for the $t(16)$ distribution. Table C shows that the *P*-value is between 0.05 and 0.04.

df = 16		
t^*	2.120	2.235
Two-sided *P*	.05	.04

CONCLUDE: The data give moderately strong evidence ($P < 0.05$) that students who have engaged in community service are on the average more attached to their friends. ■

Is the *t* test in Example 18.5 justified? The student subjects were "enrolled in a course on U.S. Diversity at a large mid-western university." Unless this course is required of all students, the subjects cannot be considered a random sample even from this campus. Students were placed in the two groups on the basis of a questionnaire, 39 in the "no service" group and 71 in the "service" group. The data were gathered from a follow-up survey two years later; 17 of the 39 "no service" students responded (44%), compared with 80% response (57 of 71) in the "service" group. Nonresponse is confounded with group: students who had done community service were much more likely to respond. Finally, 75% of the "service" respondents were women, compared with 47% of the "no service" respondents. Gender, which can strongly affect attachment, is badly confounded with the presence or absence of community service. The data are so far from meeting the SRS condition for

FIGURE 18.3

The *P*-value in Example 18.5. Because the alternative is two-sided, the *P*-value is double the area to the left of $t = -2.142$.

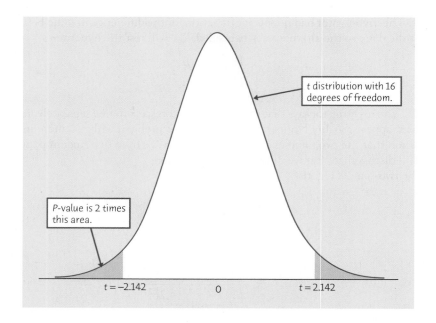

t distribution with 16 degrees of freedom.

P-value is 2 times this area.

$t = -2.142$ 0 $t = 2.142$

inference that the *t* test is meaningless. Difficulties like these are common in social science research, where confounding variables have stronger effects than is usual when biological or physical variables are measured. This researcher honestly disclosed the weaknesses in data production but left it to readers to decide whether to trust her inferences.

APPLY YOUR KNOWLEDGE

In exercises that call for two-sample t procedures, use Option 1 if you have technology that implements that method. Otherwise, use Option 2 (degrees of freedom the smaller of $n_1 - 1$ and $n_2 - 1$).

18.5   **Logging in the rain forest.** "Conservationists have despaired over destruction of tropical rain forest by logging, clearing, and burning." These words begin a report on a statistical study of the effects of logging in Borneo.[4] Here are data on the number of tree species in 12 unlogged forest plots and 9 similar plots logged 8 years earlier:

Unlogged	22	18	22	20	15	21	13	13	19	13	19	15
Logged	17	4	18	14	18	15	15	10	12			

(a)   The study report says, "Loggers were unaware that the effects of logging would be assessed." Why is this important? The study report also explains why the plots can be considered to be randomly assigned.

(b)   Does logging significantly reduce the mean number of species in a plot after 8 years? Follow the four-step process as illustrated in Examples 18.2 and 18.3.

Digital Vision/Getty Images

18.6 **Daily activity and obesity.** We can conclude from Examples 18.2 and 18.3 that mildly obese people spend less time standing and walking (on the average) than lean people. Is there a significant difference between the mean times the two groups spend lying down? Use the four-step process to answer this question from the data in Table 18.1. Follow the model of Examples 18.2 and 18.3.

18.7 **Logging in the rain forest, continued.** Use the data in Exercise 18.5 to give a 90% confidence interval for the difference in mean number of species between unlogged and logged plots.

Using technology

Software should use Option 1 for the degrees of freedom to give accurate confidence intervals and P-values. Unfortunately, there is variation in how well software implements Option 1. Figure 18.4 displays output from a graphing calculator, a statistical program, and a spreadsheet program for the test of Example 18.3. All three claim to use Option 1. The two-sample t statistic is exactly as in Example 18.3, $t = 3.808$. You can find this in all three outputs (Minitab rounds to 3.81; Excel and the graphing calculator give additional decimal places). The different technologies use different methods to find the P-value for $t = 3.808$.

- The calculator gets Option 1 completely right. The accurate approximation uses the t distribution with approximately 15.174 degrees of freedom. The P-value is $P = 0.0008$.

- Minitab uses Option 1, but it *truncates* the exact degrees of freedom to the next smaller whole number to get critical values and P-values. In this example, the exact df $= 15.174$ is truncated to df $= 15$, so that Minitab's results are slightly conservative. That is, Minitab's P-value (rounded to $P = 0.001$ in the output) is slightly larger than the full Option 1 P-value.

- Excel *rounds* the exact degrees of freedom to the nearest whole number, so that df $= 15.174$ becomes df $= 15$. Excel's method agrees with Minitab's in this example. But when rounding moves the degrees of freedom up to the next higher whole number, Excel's P-values are slightly smaller than is correct. This is misleading, another illustration of the fact that Excel is substandard as statistical software.

Excel's label for the test, "Two-Sample Assuming Unequal Variances," is seriously misleading. *The two-sample t procedures we have described work whether or not the two populations have the same variance.* There is an old-fashioned special procedure that works only when the two variances are equal. We discuss this method in an optional section on page 487, but you should never use it.

Although different calculators and software give slightly different P-values, in practice you can just accept what your technology says. The small differences in P don't affect the conclusion. Even "between 0.001 and 0.0025" from Option 2 (Example 18.3) is close enough for practical purposes.

FIGURE 18.4

The two-sample *t* procedures applied to the data on activity and obesity: output from a graphing calculator, a statistical program, and a spreadsheet program.

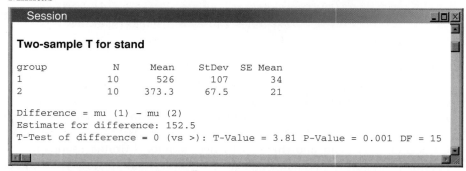

Texas Instruments Graphing Calculator

```
2-SampTTest
 μ1>μ2
 t=3.808375604
 P=8.4101904ε-4
 df=15.17355038
 x̄1=525.7513
↓x̄2=373.2692
```

Minitab

```
Session                                                    _□x

Two-sample T for stand

group       N      Mean    StDev   SE Mean
1          10       526      107        34
2          10     373.3     67.5        21

Difference = mu (1) - mu (2)
Estimate for difference: 152.5
T-Test of difference = 0 (vs >): T-Value = 3.81 P-Value = 0.001 DF = 15
```

Microsoft Excel

	A	B	C
1	t-Test: Two-Sample Assuming Unequal Variances		
2			
3		*Lean*	*Obese*
4	Mean	525.7513	373.2692
5	Variance	11474.8903	4556.019849
6	Observations	10	10
7	Hypothesized Mean	0	
8	df	15	
9	t Stat	3.808375604	
10	P(T<=t) one-tail	0.000856818	
11	t Critical one-tail	1.753051038	
12	P(T<=t) two-tail	0.001713635	
13	P(T<t) two-tail	2.131450856	
14			

Sheet4

Stephen Wilkes/Getty Images

APPLY YOUR KNOWLEDGE

18.8 **Does polyester decay?** How quickly do synthetic fabrics such as polyester decay in landfills? A researcher buried 10 strips of polyester fabric in well-drained soil in the summer. Five of the strips, chosen at random, were dug up after 2 weeks; the other 5 were dug up after 16 weeks. Breaking strength is easy to measure and is a good indicator of decay. Lower strength means the fabric has decayed. Here are the breaking strengths in pounds:[5]

Sample 1 (2 weeks)	118	126	126	120	129
Sample 2 (16 weeks)	124	98	110	140	110

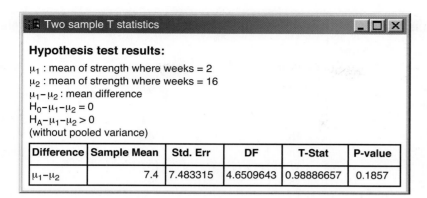

FIGURE 18.5
Two-sample t output from CrunchIt! for Exercise 18.8.

Figure 18.5 shows output for the two-sample t test using Option 1. (This output is from CrunchIt!, software that does Option 1 without rounding or truncating the degrees of freedom.) Do polyester strips buried for 16 weeks have significantly lower mean breaking strength than those buried for 2 weeks? Using the output in Figure 18.5, write a summary in a sentence or two, including t, df, P, and a conclusion.

Robustness again

The two-sample t procedures are more robust than the one-sample t methods, particularly when the distributions are not symmetric. When the sizes of the two samples are equal and the two populations being compared have distributions with similar shapes, probability values from the t table are quite accurate for a broad range of distributions when the sample sizes are as small as $n_1 = n_2 = 5$.[6] When the two population distributions have different shapes, larger samples are needed.

As a guide to practice, adapt the guidelines given on page 458 for the use of one-sample t procedures to two-sample procedures by replacing "sample size" with the "sum of the sample sizes," $n_1 + n_2$. These guidelines err on the side of safety, especially when the two samples are of equal size. *In planning a two-sample study, choose equal sample sizes whenever possible. The two-sample t procedures are most robust against non-Normality in this case, and the conservative Option 2 probability values are most accurate.*

APPLY YOUR KNOWLEDGE

18.9 **Do good smells bring good business?** Businesses know that customers often respond to background music. Do they also respond to odors? One study of this question took place in a small pizza restaurant in France on Saturday evenings in May. On one of these evenings, a relaxing lavender odor was spread through the restaurant. Table 18.2 gives the time (minutes) that two samples of 30 customers spent in the restaurant and the amount they spent (in euros).[7] The two evenings were comparable in many ways (weather, customer count, and so on), so we are

---

TABLE 18.2 Time (minutes) and spending (euros) by restaurant customers

NO ODOR		LAVENDER	
MINUTES	EUROS SPENT	MINUTES	EUROS SPENT
103	15.9	92	21.9
68	18.5	126	18.5
79	15.9	114	22.3
106	18.5	106	21.9
72	18.5	89	18.5
121	21.9	137	24.9
92	15.9	93	18.5
84	15.9	76	22.5
72	15.9	98	21.5
92	15.9	108	21.9
85	15.9	124	21.5
69	18.5	105	18.5
73	18.5	129	25.5
87	18.5	103	18.5
109	20.5	107	18.5
115	18.5	109	21.9
91	18.5	94	18.5
84	15.9	105	18.5
76	15.9	102	24.9
96	15.9	108	21.9
107	18.5	95	25.9
98	18.5	121	21.9
92	15.9	109	18.5
107	18.5	104	18.5
93	15.9	116	22.8
118	18.5	88	18.5
87	15.9	109	21.9
101	25.5	97	20.7
75	12.9	101	21.9
86	15.9	106	22.5

willing to regard the data as independent SRSs from spring Saturday evenings at this restaurant. The authors say, "Therefore at this stage it would be impossible to generalize the results to other restaurants."

(a) Does a lavender odor encourage customers to stay longer in the restaurant? Examine the time data and explain why they are suitable for two-sample t procedures. Use the two-sample t test to answer the question posed.

(b) Does a lavender odor encourage customers to spend more while in the restaurant? Examine the spending data. In what ways do these data deviate from Normality? With 30 observations, the t procedures are nonetheless reasonably accurate. Use the two-sample t test to answer the question posed.

18.10 Compressing soil. Farmers know that driving heavy equipment on wet soil compresses the soil and injures future crops. Here are data on the "penetrability" of the same type of soil at two levels of compression.[8] Penetrability is a measure of how much resistance plant roots will meet when they try to grow through the soil.

Compressed Soil									
2.86	2.68	2.92	2.82	2.76	2.81	2.78	3.08	2.94	2.86
3.08	2.82	2.78	2.98	3.00	2.78	2.96	2.90	3.18	3.16

Intermediate Soil									
3.14	3.38	3.10	3.40	3.38	3.14	3.18	3.26	2.96	3.02
3.54	3.36	3.18	3.12	3.86	2.92	3.46	3.44	3.62	4.26

(a) Make stemplots to investigate the shape of the distributions. The penetrabilities for intermediate soil are skewed to the right and have a high outlier. Returning to the source of the data shows that the outlying sample had unusually low soil density, so that it belongs in the "loose soil" class. We are justified in removing the outlier.

(b) We suspect that the penetrability of compressed soil is less than that of intermediate soil. Do the data (with the outlier removed) support this suspicion?

18.11 Weeds among the corn. Exercise 7.32 (page 189) gives these corn yields (bushels per acre) for experimental plots controlled to have 1 weed per meter of row and 3 weeds per meter of row:

1 weed/meter	166.2	157.3	166.7	161.1
3 weeds/meter	158.6	176.4	153.1	156.0

Explain carefully why a two-sample *t* confidence interval for the difference in mean yields may not be accurate.

18.12 Compressing soil, continued. Use the data in Exercise 18.10, omitting the outlier, to give a 95% confidence interval for the decrease in penetrability of compressed soil relative to intermediate soil.

Details of the *t* approximation*

The exact distribution of the two-sample *t* statistic is not a *t* distribution. Moreover, the distribution changes as the unknown population standard deviations σ_1 and σ_2 change. However, an excellent approximation is available. We call this Option 1 for *t* procedures.

* This section can be omitted unless you are using software and wish to understand what the software does.

APPROXIMATE DISTRIBUTION OF THE TWO-SAMPLE *t* STATISTIC

The distribution of the two-sample *t* statistic is very close to the *t* distribution with degrees of freedom df given by

$$df = \frac{\left(\dfrac{s_1^2}{n_1} + \dfrac{s_2^2}{n_2}\right)^2}{\dfrac{1}{n_1 - 1}\left(\dfrac{s_1^2}{n_1}\right)^2 + \dfrac{1}{n_2 - 1}\left(\dfrac{s_2^2}{n_2}\right)^2}$$

This approximation is accurate when both sample sizes n_1 and n_2 are 5 or larger.

EXAMPLE 18.6 **Daily activity and obesity**

In the experiment of Examples 18.2 and 18.3, the data on minutes per day spent standing and walking give

Group	n	$\bar{x}$	s
Group 1 (lean)	10	525.751	107.121
Group 2 (obese)	10	373.269	67.498

The two-sample *t* test statistic calculated from these values is $t = 3.808$.

   The one-sided *P*-value is the area to the right of 3.808 under a *t* density curve, as in Figure 18.2. The conservative Option 2 uses the *t* distribution with 9 degrees of freedom. Option 1 finds a very accurate *P*-value by using the *t* distribution with degrees of freedom df given by

$$df = \frac{\left(\dfrac{107.121^2}{10} + \dfrac{67.498^2}{10}\right)^2}{\dfrac{1}{9}\left(\dfrac{107.121^2}{10}\right)^2 + \dfrac{1}{9}\left(\dfrac{67.498^2}{10}\right)^2}$$

$$= \frac{2{,}569{,}894}{169{,}367.2} = 15.1735$$

These degrees of freedom appear in the graphing calculator output in Figure 18.4. Because the formula is messy and roundoff errors are likely, we don't recommend calculating df by hand. ■

   The degrees of freedom df is generally not a whole number. It is always at least as large as the smaller of $n_1 - 1$ and $n_2 - 1$. The larger degrees of freedom that result from Option 1 give slightly shorter confidence intervals and slightly smaller *P*-values than the conservative Option 2 produces. There is a *t* distribution for any positive degrees of freedom, even though Table C contains entries only for whole-number degrees of freedom.

The difference between the _t_ procedures using Options 1 and 2 is rarely of practical importance. That is why we recommend the simpler, conservative Option 2 for inference without software. With software, the more accurate Option 1 procedures are painless.

APPLY YOUR KNOWLEDGE

18.13 Students' self-concept. A study of the self-concept of seventh-grade students asked if male and female students differ in mean score on the Piers-Harris Children's Self-Concept Scale.[9] Software that uses Option 1 gives these summary results:

```
Gender  n     Mean    Std dev  Std err       t     df       P
F       31   55.5161  12.6961  2.2803   -0.8276  62.8  0.4110
M       47   57.9149  12.2649  1.7890
```

Starting from the sample means and standard deviations, verify each of these entries: the standard errors of the means; the degrees of freedom for two-sample _t_; the value of _t_.

18.14 Does polyester decay? Figure 18.5 gives output for the breaking strength data in Exercise 18.8 from software that does Option 1 with the correct degrees of freedom. What are $\bar{x}_i$ and s_i for the two samples? Starting from these values, find the _t_ test statistic and its degrees of freedom. Your work should agree with Figure 18.5.

18.15 Students' self-concept, continued. Write a sentence or two summarizing the comparison of female and male students in Exercise 18.13, as if you were preparing a report for publication. Use the output in Exercise 18.13.

Avoid the pooled two-sample _t_ procedures*

Most software, including all three illustrated in Figure 18.4, offers a choice of two-sample _t_ statistics. One is often labeled for "unequal" variances, the other for "equal" variances. The "unequal" variance procedure is our two-sample _t_. _This test is valid whether or not the population variances are equal._ The other choice is a special version of the two-sample _t_ statistic that assumes that the two populations have the same variance. This procedure averages (the statistical term is "pools") the two sample variances to estimate the common population variance. The resulting statistic is called the _pooled two-sample t statistic_. It is equal to our _t_ statistic if the two sample sizes are the same, but not otherwise. We could choose to use the pooled _t_ for tests and confidence intervals.

The pooled _t_ statistic has exactly the _t_ distribution with $n_1 + n_2 - 2$ degrees of freedom _if_ the two population variances really are equal and the population distributions are exactly Normal. The pooled _t_ was in common use before software

* The two short sections that follow offer advice on what not to do. They are not needed to read the rest of the book.

made it easy to use Option 1 for our two-sample *t* statistic. Of course, in the real world distributions are not exactly Normal and population variances are not exactly equal. In practice, the Option 1 two-sample *t* procedures are almost always more accurate than the pooled procedures. Our advice: *Never use the pooled t procedures if you have software that will implement Option 1*.

Avoid inference about standard deviations*

Two basic features of a distribution are its center and spread. In a Normal population, we measure center by the mean and spread by the standard deviation. We use the *t* procedures for inference about population means for Normal populations, and we know that *t* procedures are widely useful for non-Normal populations as well. It is natural to turn next to inference about the standard deviations of Normal populations. Our advice here is short and clear: Don't do it without expert advice.

There are methods for inference about the standard deviations of Normal populations. The most common such method is the "*F* test" for comparing the standard deviations of two Normal populations. You will find this test in the menus of most statistical software. *Unlike the t procedures for means, the F test for standard deviations is extremely sensitive to non-Normal distributions.* This lack of robustness does not improve in large samples. It is difficult in practice to tell whether a significant test result is evidence of unequal population spreads or simply a sign that the populations are not Normal. Because this test is of little use in practice, we don't give its details.

The deeper difficulty underlying the very poor robustness of Normal population procedures for inference about spread already appeared in our work on describing data. The standard deviation is a natural measure of spread for Normal distributions but not for distributions in general. In fact, because skewed distributions have unequally spread tails, no single numerical measure does a good job of describing the spread of a skewed distribution. In summary, the standard deviation is not always a useful parameter, and even when it is (for symmetric distributions), the results of inference are not trustworthy. Consequently, *we do not recommend trying to do inference about population standard deviations in basic statistical practice*.[10]

C H A P T E R 1 8   S U M M A R Y

■ The data in a **two-sample problem** are two independent SRSs, each drawn from a separate population.

■ Tests and confidence intervals for the difference between the means μ_1 and μ_2 of two Normal populations start from the difference $\overline{x}_1 - \overline{x}_2$ between the two sample means. Because of the central limit theorem, the resulting procedures are approximately correct for other population distributions when the sample sizes are large.

- Draw independent SRSs of sizes n_1 and n_2 from two Normal populations with parameters μ_1, σ_1 and μ_2, σ_2. The **two-sample t statistic** is

$$t = \frac{(\bar{x}_1 - \bar{x}_2) - (\mu_1 - \mu_2)}{\sqrt{\dfrac{s_1^2}{n_1} + \dfrac{s_2^2}{n_2}}}$$

The statistic t has approximately a t distribution.

- There are two choices for the **degrees of freedom** of the two-sample t statistic. Option 1: software produces accurate probability values using degrees of freedom calculated from the data. Option 2: for conservative inference procedures, use degrees of freedom equal to the smaller of $n_1 - 1$ and $n_2 - 1$.

- The **confidence interval for $\mu_1 - \mu_2$** is

$$(\bar{x}_1 - \bar{x}_2) \pm t^* \sqrt{\dfrac{s_1^2}{n_1} + \dfrac{s_2^2}{n_2}}$$

The critical value t^* from Option 1 gives a confidence level very close to the desired level C. Option 2 produces a margin of error at least as wide as is needed for the desired level C.

- **Significance tests for H_0: $\mu_1 = \mu_2$** are based on

$$t = \frac{\bar{x}_1 - \bar{x}_2}{\sqrt{\dfrac{s_1^2}{n_1} + \dfrac{s_2^2}{n_2}}}$$

P-values calculated from Option 1 are very accurate. Option 2 P-values are always at least as large as the true P.

- The two-sample t procedures are quite **robust** against departures from Normality. Guidelines for practical use are similar to those for one-sample t procedures. Equal sample sizes are recommended.

- Procedures for inference about the standard deviations of Normal populations are very sensitive to departures from Normality. Avoid inference about standard deviations unless you have expert advice.

CHECK YOUR SKILLS

18.16 The 2005 National Assessment of Educational Progress (NAEP) gave a mathematics test to a random sample of eighth-graders in Texas. The mean score was 281 out of 500. To give a confidence interval for the mean score of all Texas eighth-graders, you would use

(a) the one-sample t interval.

(b) the matched pairs t interval.

(c) the two-sample t interval.

18.17 In the 2005 NAEP sample of Texas eighth-graders, the mean mathematics scores were 279 for female students and 283 for male students. To see if there is a significant difference between the mean scores of all female and male students in Texas, you would use

(a) the one-sample t test.

(b) the matched pairs t test.

(c) the two-sample t test.

18.18 There are two common methods for measuring the concentration of a pollutant in fish tissue. Do the two methods differ on the average? You apply both methods to a sample of 18 carp and use

(a) the one-sample t test.

(b) the matched pairs t test.

(c) the two-sample t test.

18.19 One major reason that the two-sample t procedures are widely used is that they are quite *robust*. This means that

(a) t procedures do not require that we know the standard deviations of the populations.

(b) confidence levels and P-values from the t procedures are quite accurate even if the population distribution is not exactly Normal.

(c) t procedures compare population means, a comparison that answers many practical questions.

18.20 A study of the effects of exercise used rats bred to have high or low capacity for exercise. There were 8 high-capacity and 8 low-capacity rats. To compare the mean blood pressure of the two types of rats using the two-sample t procedures with the conservative Option 2, the correct degrees of freedom is

(a) 7.   (b) 14.   (c) 15.

18.21 The 8 high-capacity rats had mean blood pressure 89 with standard deviation 9; the 8 low-capacity rats had mean blood pressure 105 with standard deviation 13. (Blood pressure is measured in millimeters of mercury.) The two-sample t statistic for comparing the population means has value

(a) 0.5.   (b) 2.86.   (c) 9.65.

18.22 A study of road rage asked samples of 596 men and 523 women about their behavior while driving. Based on their answers, each subject was assigned a road rage score on a scale of 0 to 20. The subjects were chosen by random digit dialing of telephone numbers. Are the conditions for two-sample t inference satisfied?

(a) Maybe: the SRS condition is OK but we need to look at the data to check Normality.

(b) No: scores in a range between 0 and 20 can't be Normal.

(c) Yes: the SRS condition is OK and large sample sizes make the Normality condition unnecessary.

18.23 We suspect that men are more prone to road rage than women. To see if this is true, test these hypotheses for the mean road rage scores of all male and female drivers:

(a) $H_0: \mu_M = \mu_F$ versus $H_a: \mu_M > \mu_F$.

(b) $H_0: \mu_M = \mu_F$ versus $H_a: \mu_M \neq \mu_F$.

(c) $H_0: \mu_M = \mu_F$ versus $H_a: \mu_M < \mu_F$.

18.24 The two-sample t statistic for the road rage study (male mean minus female mean) is $t = 3.18$. The P-value for testing the hypotheses from the previous exercise satisfies

(a) $0.001 < P < 0.005$.   (b) $0.0005 < P < 0.001$.

(c) $0.001 < P < 0.002$.

C H A P T E R 1 8   E X E R C I S E S

Exercises 18.25 to 18.34 are based on summary statistics rather than raw data. This information is typically all that is presented in published reports. You can calculate inference procedures by hand from the summaries. Use the conservative Option 2 (degrees of freedom the smaller of $n_1 - 1$ and $n_2 - 1$) for two-sample t confidence intervals and P-values. You must trust that the authors understood the conditions for inference and verified that they apply. This isn't always true.

18.25 Do women talk more than men? Equip male and female students with a small device that secretly records sound for a random 30 seconds during each 12.5-minute period over two days. Count the words each subject speaks during each recording period, and from this, estimate how many words per day each subject speaks. The published report includes a table summarizing six such studies.[11] Here are two of the six:

Study	Sample Size		Estimated Average Number (SD) of Words Spoken per Day	
	Women	Men	Women	Men
1	56	56	16,177 (7520)	16,569 (9108)
2	27	20	16,496 (7914)	12,867 (8343)

Readers are supposed to understand that, for example, the 56 women in the first study had $\bar{x} = 16,177$ and $s = 7520$. It is commonly thought that women talk more than men. Does either of the two samples support this idea? For each study:

(a) State hypotheses in terms of the population means for men (μ_M) and women (μ_F).

(b) Find the two-sample t statistic.

(c) What degrees of freedom does Option 2 use to get a conservative P-value?

(d) Compare your value of t with the critical values in Table C. What can you say about the P-value of the test?

(e) What do you conclude from the results of these two studies?

18.26 Eating potato chips. Give healthy women aged 18 to 40 bags of potato chips and bottled water and invite them to snack freely. Some women were trying to restrain their diet out of concern about their weight. How much effect did these good intentions have on their eating habits? The table below summarizes data on grams of potato chips consumed.[12] (The study report gave the standard error of the mean $s/\sqrt{n}$, abbreviated as SEM, rather than the standard deviation s.)

Group	n	$\bar{x}$	SEM
Unrestrained	9	59	7
Restrained	11	32	10

(a) What are the two sample standard deviations?

(b) What degrees of freedom does the conservative Option 2 use for two-sample t procedures for these samples?

(c) Using Option 2, give a 90% confidence interval for the mean difference between the two populations of women.

18.27 Whelks on the Pacific coast. In a study of the presence of whelks along the Pacific coast, investigators put down a frame that covers 0.25 square meter and counted the whelks on the sea bottom inside the frame. They did this at 7 locations in California and 6 locations in Oregon. The report says that whelk densities "were twice as high in Oregon as in California (mean ± SEM, 26.9 ± 1.56 versus 11.9 ± 2.68 whelks per 0.25 m², Oregon versus California, respectively; Student's t test, $P < 0.001$)."[13]

Peter Egerton

(a) SEM stands for the standard error of the mean, $s/\sqrt{n}$. Fill in the values in this summary table:

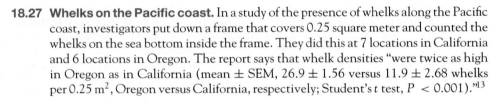

Group	Location	n	$\bar{x}$	s
1	Oregon	?	?	?
2	California	?	?	?

(b) What degrees of freedom would you use in the conservative two-sample t procedures to compare Oregon and California?

(c) What is the two-sample t test statistic for comparing the mean densities of whelks in Oregon and California?

(d) Test the null hypothesis of no difference between the two population means against the two-sided alternative. Use your statistic from part (c) with degrees of freedom from part (b). Does your conclusion agree with the published report?

18.28 Is Montessori preschool beneficial? Do education programs for preschool children that follow the Montessori method perform better than other programs? A study compared 5-year-old children in Milwaukee, Wisconsin, who had been enrolled in preschool programs from the age of 3.[14]

(a) Explain why comparing children whose parents chose a Montessori school with children of other parents would not show whether Montessori schools perform better than other programs. (In fact, all of the children in the study applied to the Montessori school. The school district assigned students to Montessori or other preschools by a random lottery.)

(b) In all, 54 children were assigned to the Montessori school and 112 to other schools at age 3. When the children were 5, parents of 39 of the Montessori children and 31 of the others could be located and agreed to participate in testing. This information reveals a possible source of bias in the comparison of outcomes. Explain why.

(c) One of the many response variables was score on a test of ability to apply basic mathematics to solve problems. Here are summaries for the children who took this test:

Group	n	$\bar{x}$	s
Montessori	30	19	3.11
Control	25	17	4.19

Is there evidence of a difference in the population mean scores? (The researchers used two-sided alternative hypotheses.)

18.29 Ginkgo extract and the post-lunch dip. The post-lunch dip is the drop in mental alertness after a midday meal. Does an extract of the leaves of the ginkgo tree reduce the post-lunch dip? Assign healthy people aged 18 to 40 to take either ginkgo extract or a placebo pill. After lunch, ask them to read seven pages of random letters and place an X over every e. Count the number of misses per line read.[15]

(a) What is a placebo and why was one group given a placebo?

(b) What is the double-blind method and why should it be used in this experiment?

(c) Here are summaries of performance after 13 weeks of either ginkgo extract or placebo:

Group	Group size	Mean	Std. dev.
Ginkgo	21	0.06383	0.01462
Placebo	18	0.05342	0.01549

Emilio Ereza/Alamy

Is there a significant difference between the two groups? What do these data show about the effect of ginkgo extract?

18.30 Hispanic customers and Anglo customers. As the presence of Hispanics in the United States has grown, businesses try to understand what Hispanics like. One study sampled customers leaving a bank. Customers were classified as Hispanic if they preferred to be interviewed in Spanish or as Anglo if they preferred English. Each customer rated the importance of several aspects of bank service on a 10-point scale.[16] Here are summary results for the importance of "reliability" (the accuracy of account records and so on):

Group	n	$\bar{x}$	s
Anglo	92	6.37	0.60
Hispanic	86	5.91	0.93

Another aspect of service quality is "empathy," the relationship that bank employees have with customers. The summary data are

Group	n	$\bar{x}$	s
Anglo	92	6.00	0.89
Hispanic	86	6.43	0.70

Do Hispanic and Anglo bank customers differ in the importance they place on either or both of these qualities? Write a one-sentence description of the differences in what the two groups find important.

18.31 Immigrant mothers at a university. The SAFE (Social, Attitudinal, Familial, and Environmental Stress) scale measures the stress level of adults adjusting to a different culture. Scores range from 1 (not stressful) to 5 (extremely stressful). In a study of stress among immigrant mothers in a university community, mothers of children between 2 and 10 years of age whose families had come to the United States for professional or academic reasons took the SAFE questionnaire. Here are summaries for mothers from Asia and Europe:[17]

Origin	Sample size	Mean	Std. dev.
Asian	12	1.92	0.60
European	9	1.74	0.57

Is there evidence of a difference in mean stress levels between mothers from Asia and Europe?

18.32 Coaching and SAT scores. Coaching companies claim that their courses can raise the SAT scores of high school students. Of course, students who retake the SAT without paying for coaching generally raise their scores. A random sample of students who took the SAT twice found 427 who were coached and 2733 who were uncoached.[18] Starting with their verbal scores on the first and second tries, we have these summary statistics:

		Try 1		Try 2		Gain	
	n	$\bar{x}$	s	$\bar{x}$	s	$\bar{x}$	s
Coached	427	500	92	529	97	29	59
Uncoached	2733	506	101	527	101	21	52

Let's first ask if students who are coached increased their scores significantly.

(a) You could use the information on the Coached line to carry out either a two-sample t test comparing Try 1 with Try 2 for coached students or a matched pairs t test using Gain. Which is the correct test? Why?

(b) Carry out the proper test. What do you conclude?

(c) Give a 99% confidence interval for the mean gain of all students who are coached.

18.33 Coaching and SAT scores, continued. What we really want to know is whether coached students improve more than uncoached students, and whether any advantage is large enough to be worth paying for. Use the information in the previous exercise to answer these questions:

(a)   Is there good evidence that coached students gained more on the average than uncoached students?

(b)   How much more do coached students gain on the average? Give a 99% confidence interval.

(c)   Based on your work, what is your opinion: do you think coaching courses are worth paying for?

18.34 Coaching and SAT scores: critique. The data you used in the previous two problems came from a random sample of students who took the SAT twice. The response rate was 63%, which is pretty good for nongovernment surveys, so let's accept that the respondents do represent all students who took the exam twice. Nonetheless, we can't be sure that coaching actually *caused* the coached students to gain more than the uncoached students. Explain briefly but clearly why this is so.

Exercises 18.35 to 18.41 include the actual data. To apply the two-sample t procedures, use Option 1 if you have technology that implements that method. Otherwise, use Option 2.

18.35 IQ scores for boys and girls. Here are the IQ test scores of 31 seventh-grade girls in a Midwest school district:[19]

114	100	104	89	102	91	114	114	103	105
108	130	120	132	111	128	118	119	86	72
111	103	74	112	107	103	98	96	112	112	93

The IQ test scores of 47 seventh-grade boys in the same district are

111	107	100	107	115	111	97	112	104	106	113
109	113	128	128	118	113	124	127	136	106	123
124	126	116	127	119	97	102	110	120	103	115
93	123	79	119	110	110	107	105	105	110	77
90	114	106								

(a)   Make stemplots or histograms of both sets of data. Because the distributions are reasonably symmetric with no extreme outliers, the *t* procedures will work well.

(b)   Treat these data as SRSs from all seventh-grade students in the district. Is there good evidence that girls and boys differ in their mean IQ scores? State hypotheses, carry out a two-sample *t* test, and report your conclusions.

18.36 Do good smells bring good business, continued. In Exercise 18.9 (page 483) you examined the effects of a lavender odor on customer behavior in a small restaurant. Lavender is a relaxing odor. The researchers also looked at the effects of lemon, a stimulating odor. The design of the study is described in Exercise 18.9. Here are the times in minutes that customers spent in the restaurant when no odor was present:

103	68	79	106	72	121	92	84	72	92
85	·69	73	87	109	115	91	84	76	96
107	98	92	107	93	118	87	101	75	86

When a lemon odor was present, customers lingered for these times:

78	104	74	75	112	88	105	97	101	89
88	73	94	63	83	108	91	88	83	106
108	60	96	94	56	90	113	97		

(a) Examine both samples. Does it appear that use of two-sample t procedures is justified? Do the sample means suggest that a lemon odor changes the average length of stay?

(b) Does a lemon odor influence the length of time customers stay in the restaurant? State hypotheses, carry out a t test, and report your conclusions.

18.37 IQ scores for boys and girls, continued. Use the data in Exercise 18.35 to give a 95% confidence interval for the difference between the mean IQ scores of all boys and girls in the district.

18.38 How strong are durable press fabrics? "Durable press" cotton fabrics are treated to improve their recovery from wrinkles after washing. Unfortunately, the treatment also reduces the strength of the fabric. A study compared the breaking strength of fabrics treated by two commercial durable press processes. Five swatches of the same fabric were assigned at random to each process. Here are the data, in pounds of pull needed to tear the fabric:[20]

Permafresh	29.9	30.7	30.0	29.5	27.6
Hylite	28.8	23.9	27.0	22.1	24.2

Is there good evidence that the two processes result in different mean breaking strengths?

(a) Do the sample means suggest that one of the processes is superior in breaking strength?

(b) Make stemplots for both samples. The Permafresh sample contains a mild outlier. With just 5 observations per group, we worry that this outlier will affect our conclusions.

(c) Test the hypothesis H_0: $\mu_1 = \mu_2$ against the two-sided alternative twice: once using all the data and again without the outlier in the Permafresh sample. Do the two tests lead to similar conclusions? Can we safely conclude that one treatment has significantly higher mean breaking strength than the other?

18.39 Reducing wrinkles. Of course, the reason for durable press treatment is to reduce wrinkling. "Wrinkle recovery angle" measures how well a fabric recovers from wrinkles. Higher is better. Here are data on the wrinkle recovery angle (in degrees) for the same fabric swatches discussed in the previous exercise:

Permafresh	136	135	132	137	134
Hylite	143	141	146	141	145

Is there a significant difference in wrinkle resistance?

(a) Do the sample means suggest that one process has better wrinkle resistance?

(b) Make stemplots for both samples. There are no obvious deviations from Normality.

(c) Test the hypothesis $H_0: \mu_1 = \mu_2$ against the two-sided alternative. What do you conclude from part (a) and from the result of your test?

18.40 How much stronger? Continue your work from Exercise 18.38. A fabric manufacturer wants to know how large an advantage in strength fabrics treated by the Permafresh method have over fabrics treated by the Hylite process. Give a 90% confidence interval for the difference in mean breaking strengths. (Use all 5 fabric swatches.)

18.41 How much less wrinkling? In Exercise 18.39, you found that the Hylite process results in significantly greater wrinkle resistance than the Permafresh process. How large is the difference in mean wrinkle recovery angle? Give a 90% confidence interval.

Do birds learn to time their breeding? *Blue titmice eat caterpillars. The birds would like lots of caterpillars around when they have young to feed, but they breed earlier than peak caterpillar season. Do the birds learn from one year's experience when they time breeding the next year? Researchers randomly assigned 7 pairs of birds to have the natural caterpillar supply supplemented while feeding their young and another 6 pairs to serve as a control group relying on natural food supply. The next year, they measured how many days after the caterpillar peak the birds produced their nestlings.[21] Exercises 18.42 to 18.44 are based on this experiment.*

Hugh Clark/Frank Lane Picture Agency/CORBIS

18.42 Did the randomization produce similar groups? First, compare the two groups in the first year. The only difference should be the chance effect of the random assignment. The study report says: "In the experimental year, the degree of synchronization did not differ between food-supplemented and control females." For this comparison, the report gives $t = -1.05$. What type of t statistic (paired or two-sample) is this? What are the degrees of freedom for this statistic? Show that this t leads to the quoted conclusion.

18.43 Did the treatment have an effect? The investigators expected the control group to adjust their breeding date the next year, whereas the well-fed supplemented group had no reason to change. The report continues: "but in the following year food-supplemented females were more out of synchrony with the caterpillar peak than the controls." Here are the data (days behind the caterpillar peak):

Control	4.6	2.3	7.7	6.0	4.6	−1.2	
Supplemented	15.5	11.3	5.4	16.5	11.3	11.4	7.7

Carry out a t test and show that it leads to the quoted conclusion.

18.44 Year-to-year comparison. Rather than comparing the two groups in each year, we could compare the behavior of each group in the first and second years. The study report says: "Our main prediction was that females receiving additional food in the nestling period should not change laying date the next year, whereas controls, which (in our area) breed too late in their first year, were expected to advance their laying date in the second year."

Comparing days behind the caterpillar peak in Years 1 and 2 gave $t = 0.63$ for the control group and $t = -2.63$ for the supplemented group. Are these paired or two-sample t statistics? What are the degrees of freedom for each t? Show that these t-values do *not* agree with the prediction.

The remaining exercises ask you to answer questions from data without having the details outlined for you. The exercise statements give you the **State** *step of the four-step process. Follow the* **Plan, Solve,** *and* **Conclude** *steps as illustrated in Examples 18.2 and 18.3 for tests and Example 18.4 for confidence intervals. Remember that examining the data and discussing the conditions for inference are part of the* **Solve** *step.*

18.45 Thinking about money changes behavior. Kathleen Vohs of the University of Minnesota and her coworkers carried out several randomized comparative experiments on the effects of thinking about money. Here's part of one such experiment.[22] Ask student subjects to unscramble 30 sets of five words to make a meaningful phrase from four of the five words. The control group unscrambled phrases like "cold it desk outside is" into "it is cold outside." The treatment group unscrambled phrases that lead to thinking about money, turning "high a salary desk paying" into "a high-paying salary." Then each subject worked a hard puzzle, knowing that he or she could ask for help. Here are the times in seconds until subjects asked for help, for the treatment group,

609	444	242	199	174	55	251	466	443
531	135	241	476	482	362	69	160	

and for the control group,

118	272	413	291	140	104	55	189	126
400	92	64	88	142	141	373	156	

The researchers suspected that money is connected with self-sufficiency, so that the treatment group will ask for help less quickly on the average. Do the data support this idea?

18.46 Active versus passive learning. A study of computer-assisted learning examined the learning of "Blissymbols" by children. Blissymbols are pictographs (think of Egyptian hieroglyphs) that are sometimes used to help learning-impaired children communicate. The researcher designed two computer lessons that taught the same content using the same examples. One lesson required the children to interact with the material, while in the other the children controlled only the pace of the lesson. Call these two styles "Active" and "Passive." Children were assigned at random to Active and Passive groups. After the lesson, the computer presented a quiz that asked the children to identify 56 Blissymbols. Here are the numbers of correct identifications by the 24 children in the Active group:[23]

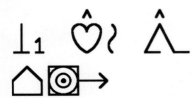

29	28	24	31	15	24	27	23	20	22	23	21
24	35	21	24	44	28	17	21	21	20	28	16

The 24 children in the Passive group had these counts of correct identifications:

16	14	17	15	26	17	12	25	21	20	18	21
20	16	18	15	26	15	13	17	21	19	15	12

Is there good evidence that active learning is superior to passive learning?

18.47 Each day I am getting better in math. A "subliminal" message is below our threshold of awareness but may nonetheless influence us. Can subliminal messages help students learn math? A group of students who had failed the mathematics part of the City University of New York Skills Assessment Test agreed to participate in a study to find out.

All received a daily subliminal message, flashed on a screen too rapidly to be consciously read. The treatment group of 10 students (chosen at random) was exposed to "Each day I am getting better in math." The control group of 8 students was exposed to a neutral message, "People are walking on the street." All students participated in a summer program designed to raise their math skills, and all took the assessment test again at the end of the program. Table 18.3 gives data on the subjects' scores before and after the program.[24] Is there good evidence that the treatment brought about a greater improvement in math scores than the neutral message? How large is the mean difference in gains between treatment and control? (Use 90% confidence.)

TABLE 18.3 Mathematics skills scores before and after a subliminal message

TREATMENT GROUP		CONTROL GROUP	
BEFORE	AFTER	BEFORE	AFTER
18	24	18	29
18	25	24	29
21	33	20	24
18	29	18	26
18	33	24	38
20	36	22	27
23	34	15	22
23	36	19	31
21	34		
17	27		

18.48 Active versus passive learning, continued.

(a) Use the data in Exercise 18.46 to give a 90% confidence interval for the difference in mean number of Blissymbols identified correctly by children after active and passive lessons.

(b) Give a 90% confidence interval for the mean number of Blissymbols identified correctly by children after the active lesson.

18.49 Tropical flowers. Different varieties of the tropical flower *Heliconia* are fertilized by different species of hummingbirds. Over time, the lengths of the flowers and the forms of the hummingbirds' beaks have evolved to match each other. Here are data

on the lengths in millimeters of two color varieties of the same species of flower on the island of Dominica:[25]

			H. caribaea red				
41.90	42.01	41.93	43.09	41.47	41.69	39.78	40.57
39.63	42.18	40.66	37.87	39.16	37.40	38.20	38.07
38.10	37.97	38.79	38.23	38.87	37.78	38.01	

			H. caribaea yellow				
36.78	37.02	36.52	36.11	36.03	35.45	38.13	37.1
35.17	36.82	36.66	35.68	36.03	34.57	34.63	

Is there good evidence that the mean lengths of the two varieties differ? Estimate the difference between the population means. (Use 95% confidence.)

18.50 Student drinking. A professor asked her sophomore students, "How many drinks do you typically have per session? (A drink is defined as one 12 oz beer, one 4 oz glass of wine, or one 1 oz shot of liquor.)" Some of the students didn't drink. Table 18.4 gives the responses of the female and male students who did drink.[26] It is likely that some of the students exaggerated a bit. The sample is all students in one large sophomore-level class. The class is popular, so we are tentatively willing to regard its members as an SRS of sophomore students at this college. Do a complete analysis that reports on

(a) the drinking behavior claimed by sophomore women.

(b) the drinking behavior claimed by sophomore men.

(c) a comparison of the behavior of women and men.

TABLE 18.4 Drinks per session claimed by female and male students

					FEMALE	STUDENTS						
2.5	9	1	3.5	2.5	3	1	3	3	3	3	2.5	2.5
5	3.5	5	1	2	1	7	3	7	4	4	6.5	4
3	6	5	3	8	6	6	3	6	8	3	4	7
4	5	3.5	4	2	1	5	5	3	3	6	4	2
7	7	7	3.5	3	2.5	10	5	4	9	8	1	6
2	5	2.5	3	4.5	9	5	4	4	3	4	6	7
4	5	1	5	3	4	10	7	3	4	4	4	4
2	1	2.5	2.5									

					MALE	STUDENTS						
7	7.5	8	15	3	4	1	5	11	4.5	6	4	10
16	4	8	5	9	7	7	3	5	6.5	1	12	4
6	8	8	4.5	10.5	8	6	10	1	9	8	7	8
15	3	10	7	4	6	5	2	10	7	9	5	8
7	3	7	6	4	5	2	5	5.5	9	10	10	4
8	4	2	4	12.5	3	15	2	6	3	4	3	10
6	4.5	5										

Kolvenbach/Alamy

CHAPTER 19

Inference about a Population Proportion

Our discussion of statistical inference to this point has concerned making inferences about population *means*. Now we turn to questions about the *proportion* of some outcome in a population. Here are some examples that call for inference about population proportions.

EXAMPLE 19.1 Risky behavior in the age of AIDS

How common is behavior that puts people at risk of AIDS? The National AIDS Behavioral Surveys interviewed a random sample of 2673 adult heterosexuals. Of these, 170 had more than one sexual partner in the past year. That's 6.36% of the sample.[1] Based on these data, what can we say about the percent of all adult heterosexuals who have multiple partners? We want to *estimate a single population proportion*. This chapter concerns inference about one proportion. ■

EXAMPLE 19.2 Young adults living at home

A surprising number of young adults (ages 19 to 25) still live at home with their parents. A random sample of 2253 men and 2629 women in this age group found that 44% of the men but only 35% of the women lived at home. Is this significant evidence that the proportions living at home differ in the populations of all young men and all young women? We want to *compare two population proportions*. This is the topic of Chapter 20. ■

To do inference about a population mean μ, we use the mean $\bar{x}$ of a random sample from the population. The reasoning of inference starts with the sampling distribution of $\bar{x}$. Now we follow the same pattern, replacing means by proportions.

The sample proportion $\hat{p}$

We are interested in the unknown proportion p of a population that has some outcome. For convenience, call the outcome we are looking for a "success." In Example 19.1, the population is adult heterosexuals, and the parameter p is the proportion who have had more than one sexual partner in the past year. To esti-

sample proportion

mate p, the National AIDS Behavioral Surveys used random dialing of telephone numbers to contact a sample of 2673 people. Of these, 170 said they had multiple sexual partners. The statistic that estimates the parameter p is the **sample proportion**

$$\hat{p} = \frac{\text{number of successes in the sample}}{\text{total number of individuals in the sample}}$$

$$= \frac{170}{2673} = 0.0636$$

Read the sample proportion $\hat{p}$ as "p-hat."

How good is the statistic $\hat{p}$ as an estimate of the parameter p? To find out, we ask, "What would happen if we took many samples?" The sampling distribution of $\hat{p}$ answers this question. Here are the facts.[2]

SAMPLING DISTRIBUTION OF A SAMPLE PROPORTION

Draw an SRS of size n from a large population that contains proportion p of successes. Let $\hat{p}$ be the **sample proportion** of successes,

$$\hat{p} = \frac{\text{number of successes in the sample}}{n}$$

Then:

- The **mean** of the sampling distribution is p.
- The **standard deviation** of the sampling distribution is

$$\sqrt{\frac{p(1-p)}{n}}$$

- As the sample size increases, the sampling distribution of $\hat{p}$ becomes **approximately Normal.** That is, for large n, $\hat{p}$ has approximately the $N(p, \sqrt{p(1-p)/n})$ distribution.

Figure 19.1 summarizes these facts in a form that helps you recall the big idea of a sampling distribution. The behavior of sample proportions $\hat{p}$ is similar to the behavior of sample means $\bar{x}$, except that the distribution of $\hat{p}$ is only approximately Normal. The mean of the sampling distribution of $\hat{p}$ is the true value of

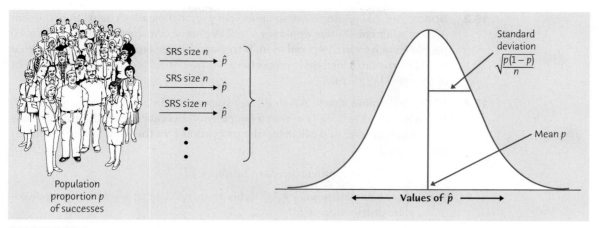

FIGURE 19.1

Select a large SRS from a population in which the proportion p are successes. The sampling distribution of the proportion $\hat{p}$ of successes in the sample is approximately Normal. The mean is p and the standard deviation is $\sqrt{p(1-p)/n}$.

the population proportion p. That is, $\hat{p}$ is an unbiased estimator of p. The standard deviation of $\hat{p}$ gets smaller as the sample size n gets larger, so that estimation is likely to be more accurate when the sample is larger. As is the case for $\bar{x}$, the standard deviation gets smaller only at the rate $\sqrt{n}$. We need four times as many observations to cut the standard deviation in half.

EXAMPLE 19.3 Asking about risky behavior

Suppose that in fact 6% of all adult heterosexuals had more than one sexual partner in the past year (and would admit it when asked). The National AIDS Behavioral Surveys interviewed a random sample of 2673 people from this population. In many such samples, the proportion $\hat{p}$ of the 2673 people in the sample who had more than one partner would vary according to (approximately) the Normal distribution with mean 0.06 and standard deviation

$$\sqrt{\frac{p(1-p)}{n}} = \sqrt{\frac{(0.06)(0.94)}{2673}}$$

$$= \sqrt{0.0000211} = 0.00459 \ \blacksquare$$

APPLY YOUR KNOWLEDGE

19.1   **Do college students pray?** A study of religious practices among college students interviewed a sample of 127 students; 107 of the students said that they prayed at least once in a while.

(a)   Describe the population and explain in words what the parameter p is.

(b)   Give the numerical value of the statistic $\hat{p}$ that estimates p.

19.2   **Spam.** Are you getting more spam in your personal email account? Suppose that 40% of adult email users would say "Yes." A polling firm contacts an SRS of 1500 people chosen from this population. If the sample were repeated many times, what would be the range of sample proportions who say "Yes," according to the 99.7 part of the 68–95–99.7 rule?

19.3   **Watching online video.** About 75% of young adult Internet users (ages 18 to 29) watch online video. Suppose that a sample survey contacts an SRS of 1000 young adult Internet users and calculates the proportion $\hat{p}$ in this sample who watch online video.

(a)   What is the approximate distribution of $\hat{p}$?

(b)   If the sample size were 4000 rather than 1000, what would be the approximate distribution of $\hat{p}$?

Large-sample confidence intervals for a proportion

We can follow the same path from sampling distribution to confidence interval as we did for $\bar{x}$ in Chapter 14. To obtain a level C confidence interval for p, we start by capturing the central probability C in the distribution of $\hat{p}$. To do this, go out z^* standard deviations from the mean p, where z^* is the critical value that captures the central area C under the standard Normal curve. Figure 19.2 shows the result. The confidence interval is

$$\hat{p} \pm z^* \sqrt{\frac{p(1-p)}{n}}$$

FIGURE 19.2

With probability C, $\hat{p}$ lies within $\pm z^* \sqrt{p(1-p)/n}$ of the unknown population proportion p. That is to say that in these samples p lies within $\pm z^* \sqrt{p(1-p)/n}$ of $\hat{p}$.

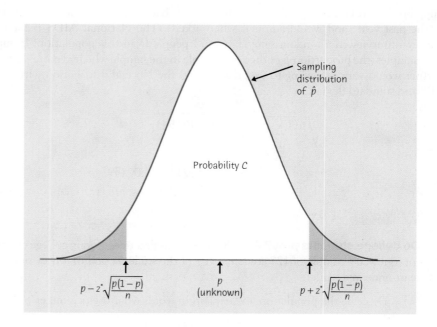

Sampling distribution of $\hat{p}$

Probability C

$p - z^* \sqrt{\dfrac{p(1-p)}{n}}$ p (unknown) $p + z^* \sqrt{\dfrac{p(1-p)}{n}}$

This won't do, because we don't know the value of p. So we replace the standard deviation by the **standard error of $\hat{p}$**

standard error of $\hat{p}$

$$SE = \sqrt{\frac{\hat{p}(1-\hat{p})}{n}}$$

to get the confidence interval

$$\hat{p} \pm z^*\sqrt{\frac{\hat{p}(1-\hat{p})}{n}}$$

This interval has the form

$$\text{estimate} \pm z^*SE_{\text{estimate}}$$

We can trust this confidence interval only for large samples. Because the number of successes must be a whole number, using a continuous Normal distribution to describe the behavior of $\hat{p}$ may not be accurate unless n is large. Because the approximation is least accurate for populations that are almost all successes or almost all failures, we require that the sample have both enough successes and enough failures rather than that the overall sample size be large. *Pay attention to both conditions for inference in the box below that summarizes the confidence interval: we must as usual be willing to regard the sample as an SRS from the population, and the sample must have both enough successes and enough failures.*

LARGE-SAMPLE CONFIDENCE INTERVAL FOR A POPULATION PROPORTION

Draw an SRS of size n from a large population that contains an unknown proportion p of successes. An approximate level C **confidence interval for p** is

$$\hat{p} \pm z^*\sqrt{\frac{\hat{p}(1-\hat{p})}{n}}$$

where z^* is the critical value for the standard Normal density curve with area C between $-z^*$ and z^*.

Use this interval only when the numbers of successes and failures in the sample are both at least 15.[3]

Why not t? Notice that we *don't* change z^* to t^* when we replace the standard deviation by the standard error. When the sample mean $\overline{x}$ estimates the population mean μ, a separate parameter σ describes the spread of the distribution of $\overline{x}$. We separately estimate σ, and this leads to a t distribution. When the sample proportion $\hat{p}$ estimates the population proportion p, the spread depends on p, not on a separate parameter. There is no t distribution—we just make the Normal approximation a bit less accurate when we replace p in the standard deviation by $\hat{p}$.

EXAMPLE 19.4 **Estimating risky behavior**

The four-step process for any confidence interval is outlined on page 366.

STATE: The National AIDS Behavioral Surveys found that 170 of a sample of 2673 adult heterosexuals had multiple partners. That is,

$$\hat{p} = \frac{170}{2673} = 0.0636$$

What can we say about the population of all adult heterosexuals?

PLAN: We will give a 99% confidence interval to estimate the proportion p of all adult heterosexuals who have multiple partners.

SOLVE: First verify the conditions for inference:

- The sampling design was a complex stratified sample, and the survey used inference procedures for that design. The overall effect is close to an SRS, however.

- The sample is large enough: the numbers of successes (170) and failures (2503) in the sample are both much larger than 15.

The sample size condition is easily satisfied. The condition that the sample be an SRS is only approximately met.

A 99% confidence interval for the proportion p of all adult heterosexuals with multiple partners uses the standard Normal critical value $z^* = 2.576$. The confidence interval is

$$\hat{p} \pm z^* \sqrt{\frac{\hat{p}(1-\hat{p})}{n}} = 0.0636 \pm 2.576 \sqrt{\frac{(0.0636)(0.9364)}{2673}}$$

$$= 0.0636 \pm 0.0122$$

$$= 0.0514 \text{ to } 0.0758$$

CONCLUDE: We are 99% confident that the percent of adult heterosexuals who have had more than one sexual partner in the past year lies between about 5.1% and 7.6%. ■

As usual, the practical problems of a large sample survey weaken our confidence in the AIDS survey's conclusions. Only people in households with landline telephones could be reached. This is acceptable for surveys of the general population, because about 89% of American households have landline telephones. However, some groups at high risk for AIDS, such as people who inject illegal drugs, often don't live in settled households and therefore are underrepresented in the sample. About 30% of the people reached refused to cooperate. A nonresponse rate of 30% is not unusual in large sample surveys, but it may cause some bias if those who refuse differ systematically from those who cooperate. The survey used statistical methods that adjust for unequal response rates in different groups. Finally, some respondents may not have told the truth when asked about their sexual behavior. The survey team tried to make respondents feel comfortable. For example, Hispanic women were interviewed only by Hispanic women, and Spanish speakers were interviewed by Spanish speakers with the same regional accent (Cuban, Mexican, or Puerto Rican). Nonetheless, the survey report says that some bias is probably present:

> *It is more likely that the present figures are underestimates; some respondents may underreport their numbers of sexual partners and intravenous drug use because of embarrassment and fear of reprisal, or they may forget or not know details of their own or of their partner's HIV risk and their antibody testing history.*[4]

Reading the report of a large study like the National AIDS Behavioral Surveys reminds us that statistics in practice involves much more than formulas for inference.

Who is a smoker?

When estimating a proportion p, be sure you know what counts as a "success." The news says that 20% of adolescents smoke. Shocking. It turns out that this is the percent who smoked at least once in the past month. If we say that a smoker is someone who smoked on at least 20 of the past 30 days and smoked at least half a pack on those days, fewer than 4% of adolescents qualify.

APPLY YOUR KNOWLEDGE

19.4 No confidence interval. In the National AIDS Behavioral Surveys sample of 2673 adult heterosexuals, 0.2% (that's 0.002 as a decimal fraction) had both received a blood transfusion and had a sexual partner from a group at high risk of AIDS. Explain why we can't use the large-sample confidence interval to estimate the proportion p in the population who share these two risk factors.

19.5 Canadian attitudes toward guns. Canada has much stronger gun control laws than the United States, and Canadians support gun control more strongly than do Americans. A sample survey asked a random sample of 1505 adult Canadians, "Do you agree or disagree that all firearms should be registered?" Of the 1505 people in the sample, 1288 answered either "Agree strongly" or "Agree somewhat."[5]

    (a)   The survey dialed residential telephone numbers at random in all ten Canadian provinces (omitting the sparsely populated northern territories). Based on what you know about sample surveys, what is likely to be the biggest weakness in this survey?

    (b)   Nonetheless, act as if we have an SRS from adults in the Canadian provinces. Give a 95% confidence interval for the proportion who support registration of all firearms.

19.6 How common is SAT coaching? A random sample of students who took the SAT college entrance examination twice found that 427 of the respondents had paid for coaching courses and that the remaining 2733 had not.[6] Give a 99% confidence interval for the proportion of coaching among students who retake the SAT. Follow the four-step process as illustrated in Example 19.4.

Accurate confidence intervals for a proportion

The confidence interval $\hat{p} \pm z^*\sqrt{\hat{p}(1 - \hat{p})/n}$ for a sample proportion p is easy to calculate. It is also easy to understand because it rests directly on the approximately Normal distribution of $\hat{p}$. Unfortunately, confidence levels from this interval are often quite inaccurate unless the sample is very large. The actual confidence level is usually *less* than the confidence level you asked for in choosing the critical value z^*. That's bad. What is worse, accuracy does not consistently get better as the sample size n increases. There are "lucky" and "unlucky" combinations of the sample size n and the true population proportion p.

Fortunately, there is a simple modification that is almost magically effective in improving the accuracy of the confidence interval. We call it the "plus four" method because all you need to do is *add four imaginary observations, two successes and two failures*. With the added observations, the **plus four estimate** of p is

plus four estimate

$$\tilde{p} = \frac{\text{number of successes in the sample} + 2}{n + 4}$$

The formula for the confidence interval is exactly as before, with the new sample size and number of successes.[7] You do not need software that offers the plus four interval—just enter the new sample size (actual size $+ 4$) and number of successes (actual number $+ 2$) into the large-sample procedure.

PLUS FOUR CONFIDENCE INTERVAL FOR A PROPORTION

Draw an SRS of size n from a large population that contains an unknown proportion p of successes. To get the **plus four confidence interval for p,** add four imaginary observations, two successes and two failures. Then use the large-sample confidence interval with the new sample size ($n + 4$) and number of successes (actual number $+ 2$).

Use this interval when the confidence level is at least 90% and the sample size n is at least 10, with any counts of successes and failures.

EXAMPLE 19.5 Cocaine traces in Spanish currency

STATE: Cocaine users commonly snort the powder up the nose through a rolled-up paper currency bill. Spain has a high rate of cocaine use, so it's not surprising that euro paper currency in Spain often contains traces of cocaine. Researchers collected 20 euro bills in each of several Spanish cities. In Madrid, 17 out of 20 contained traces of cocaine.[8] The researchers note that we can't tell whether the bills had been used to snort cocaine or had been contaminated in currency-sorting machines. Estimate the proportion of all euro bills in Madrid that have traces of cocaine.

PLAN: Take p to be the proportion of bills that contain cocaine traces. Give a 95% confidence interval for p.

SOLVE: The conditions for use of the large-sample interval are not met because there are only 3 failures. To apply the plus four method, add two successes and two failures to the original data. The plus four estimate of p is

$$\tilde{p} = \frac{17 + 2}{20 + 4} = \frac{19}{24} = 0.7917$$

Kolvenbach/Alamy

The plus four confidence interval is the same as the large-sample interval based on 19 successes in 24 observations. Here it is:

$$\tilde{p} \pm z^* \sqrt{\frac{\tilde{p}(1-\tilde{p})}{n+4}} = 0.7917 \pm 1.960 \sqrt{\frac{(0.7917)(0.2083)}{24}}$$

$$= 0.7917 \pm 0.1625$$

$$= 0.6292 \text{ to } 0.9542$$

CONCLUDE: We estimate with 95% confidence that between about 63% and 95% of all euro bills in Madrid contain traces of cocaine. ∎

For comparison, the ordinary sample proportion is

$$\hat{p} = \frac{17}{20} = 0.85$$

The plus four estimate $\tilde{p} = 0.7917$ in Example 19.5 is farther away from 1 than $\hat{p} = 0.85$. The plus four estimate gains its added accuracy by always moving toward 0.5 and away from 1 or 0, whichever is closer. This is particularly helpful when the sample contains only a few successes or a few failures. The numerical difference between a large-sample interval and the corresponding plus four interval is often small. Remember that the confidence level is the probability that the interval will catch the true population proportion *in very many uses*. Small differences every time add up to accurate confidence levels from plus four versus inaccurate levels from the large-sample interval.

How much more accurate is the plus four interval? Computer studies have asked how large n must be to guarantee that the actual probability that a 95% confidence interval covers the true parameter value is at least 0.94 for all samples of size n or larger. If $p = 0.1$, for example, the answer is $n = 646$ for the large-sample interval and $n = 11$ for the plus four interval.[9] The consensus of computational and theoretical studies is that plus four is very much better than the large-sample interval for many combinations of n and p. (If you use software such as Minitab, you may find an "exact method" on the menu. Despite the appealing name, this method is often less accurate than plus four.) **We recommend that you always use the plus four interval for estimating a proportion.**

APPLY YOUR KNOWLEDGE

19.7   Whelks and mussels. Sample surveys usually contact large samples, so we can use the large-sample confidence interval if the sample design is close to an SRS. Scientific studies often use smaller samples that require the plus four method. For example, the small round holes you often see in sea shells were drilled by other sea creatures, who ate the former owners of the shells. Whelks often drill into mussels, but this behavior appears to be more or less common in different locations. Investigators collected whelk eggs from the coast of Oregon, raised the whelks in the laboratory, then put each whelk in a container with some delicious mussels. Only 9 of 98 whelks drilled into a mussel.[10]

(a)   Why can't we use the large-sample confidence interval for the proportion p of Oregon whelks that will spontaneously drill mussels?

(b)   The plus four method adds four observations, two successes and two failures. What are the sample size and the number of successes after you do this? What is the plus four estimate $\tilde{p}$ of p?

(c)   Give the plus four 90% confidence interval for the proportion of Oregon whelks that will spontaneously drill mussels.

19.8   Teens' MySpace profiles. Over half of all American teens (ages 12 to 17 years) have an online profile, mainly on MySpace. A random sample of 487 teens with profiles found that 385 included photos of themselves.[11]

(a)   Give the 95% large-sample confidence interval for the proportion p of all teens with profiles who include photos of themselves.

(b)   Give the plus four 95% confidence interval for p. If you express the two intervals in percents, rounded to the nearest tenth of a percent, how do they differ? (The plus four interval always pulls the results away from 0% or 100%, whichever is closer. Even though the condition for using the large-sample interval is met, the plus four interval is more trustworthy.)

19.9   Cocaine traces in Spanish currency, continued. The plus four method is particularly useful when there are *no* successes or *no* failures in the data. The study of Spanish currency described in Example 19.5 found that in Seville, all 20 of a sample of 20 euro bills had cocaine traces.

(a)   What is the sample proportion $\hat{p}$ of contaminated bills? What is the large-sample 95% confidence interval for p? It's not plausible that *every* bill in Seville has cocaine traces, as this interval says.

(b)   Find the plus four estimate $\tilde{p}$ and the plus four 95% confidence interval for p. These results are more reasonable.

Choosing the sample size

In planning a study, we may want to choose a sample size that will allow us to estimate the parameter within a given margin of error. We saw earlier (page 405) how to do this for a population mean. The method is similar for estimating a population proportion.

The margin of error in the large-sample confidence interval for p is

$$m = z^* \sqrt{\frac{\hat{p}(1 - \hat{p})}{n}}$$

Here z^* is the standard Normal critical value for the level of confidence we want. Because the margin of error involves the sample proportion of successes $\hat{p}$, we need to guess this value when choosing n. Call our guess p^*. Here are two ways to get p^*:

Courtesy Pewinternet.org

New York, New York

New York City, they say, is bigger, richer, faster, ruder. Maybe there's something to that. The sample survey firm Zogby International says that as a national average it takes 5 telephone calls to reach a live person. When calling to New York, it takes 12 calls. Survey firms assign their best interviewers to make calls to New York and often pay them bonuses to cope with the stress.

1. Use a guess p^* based on a pilot study or on past experience with similar studies. You can do several calculations to cover the range of values of $\hat{p}$ you might get.

2. Use $p^* = 0.5$ as the guess. The margin of error m is largest when $\hat{p} = 0.5$, so this guess is conservative in the sense that if we get any other $\hat{p}$ when we do our study, we will get a margin of error smaller than planned.

Once you have a guess p^*, the recipe for the margin of error can be solved to give the sample size n needed. Here is the result for the large-sample confidence interval. For simplicity, use this result even if you plan to use the plus four interval.

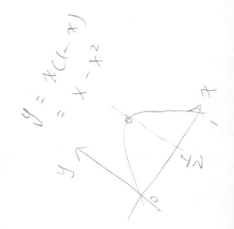

SAMPLE SIZE FOR DESIRED MARGIN OF ERROR

The level C confidence interval for a population proportion p will have margin of error approximately equal to a specified value m when the sample size is

$$n = \left(\frac{z^*}{m}\right)^2 p^*(1 - p^*)$$

where p^* is a guessed value for the sample proportion. The margin of error will always be less than or equal to m if you take the guess p^* to be 0.5.

Which method for finding the guess p^* should you use? The n you get doesn't change much when you change p^* as long as p^* is not too far from 0.5. You can use the conservative guess $p^* = 0.5$ if you expect the true $\hat{p}$ to be roughly between 0.3 and 0.7. If the true $\hat{p}$ is close to 0 or 1, using $p^* = 0.5$ as your guess will give a sample much larger than you need. Try to use a better guess from a pilot study when you suspect that $\hat{p}$ will be less than 0.3 or greater than 0.7.

EXAMPLE 19.6 Planning a poll

STATE: Gloria Chavez and Ronald Flynn are the candidates for mayor in a large city. You are planning a sample survey to determine what percent of the voters intend to vote for Chavez. You will contact an SRS of registered voters in the city. You want to estimate the proportion p of Chavez voters with 95% confidence and a margin of error no greater than 3%, or 0.03. How large a sample do you need?

PLAN: Find the sample size n needed for margin of error $m = 0.03$ and 95% confidence. The winner's share in all but the most lopsided elections is between 30% and 70% of the vote. You can use the guess $p^* = 0.5$.

SOLVE: The sample size you need is

$$n = \left(\frac{1.96}{0.03}\right)^2 (0.5)(1 - 0.5) = 1067.1$$

Round the result up to $n = 1068$. (Rounding down would give a margin of error slightly greater than 0.03.)

CONCLUDE: An SRS of 1068 registered voters is adequate for margin of error ±3%. ■

VOTE CHAVEZ

Colin Anderson/BrandX/Age fotostock

If you want a 2.5% margin of error rather than 3%, then (after rounding up)

$$n = \left(\frac{1.96}{0.025}\right)^2 (0.5)(1 - 0.5) = 1537$$

For a 2% margin of error the sample size you need is

$$n = \left(\frac{1.96}{0.02}\right)^2 (0.5)(1 - 0.5) = 2401$$

As usual, smaller margins of error call for larger samples.

APPLY YOUR KNOWLEDGE

19.10 Canadians and doctor-assisted suicide. A Gallup Poll asked a sample of Canadian adults if they thought the law should allow doctors to end the life of a patient who is in great pain and near death if the patient makes a request in writing. The poll included 270 people in Québec, 221 of whom agreed that doctor-assisted suicide should be allowed.[12]

(a)   What is the margin of error of the large-sample 95% confidence interval for the proportion of all Québec adults who would allow doctor-assisted suicide?

(b)   How large a sample is needed to get the common ±3 percentage point margin of error? Use the previous sample as a pilot study to get p^*.

19.11 Can you taste PTC? PTC is a substance that has a strong bitter taste for some people and is tasteless for others. The ability to taste PTC is inherited. About 75% of Italians can taste PTC, for example. You want to estimate the proportion of Americans with at least one Italian grandparent who can taste PTC. Starting with the 75% estimate for Italians, how large a sample must you collect in order to estimate the proportion of PTC tasters within ±0.04 with 90% confidence?

Significance tests for a proportion

The test statistic for the null hypothesis H_0: $p = p_0$ is the sample proportion $\hat{p}$ standardized using the value p_0 specified by H_0,

$$z = \frac{\hat{p} - p_0}{\sqrt{\dfrac{p_0(1 - p_0)}{n}}}$$

This z statistic has approximately the standard Normal distribution when H_0 is true. P-values therefore come from the standard Normal distribution. Because H_0 fixes a value of p, the inaccuracy that plagues the large-sample confidence interval does not affect tests. Here is the procedure for tests.

SIGNIFICANCE TESTS FOR A PROPORTION

Draw an SRS of size n from a large population that contains an unknown proportion p of successes. To **test the hypothesis** $H_0: p = p_0$, compute the z statistic

$$z = \frac{\hat{p} - p_0}{\sqrt{\dfrac{p_0(1 - p_0)}{n}}}$$

In terms of a variable Z having the standard Normal distribution, the approximate P-value for a test of H_0 against

$H_a: p > p_0$ is $P(Z \geq z)$

$H_a: p < p_0$ is $P(Z \leq z)$

$H_a: p \neq p_0$ is $2P(Z \geq |z|)$

Use this test when the sample size n is so large that both np_0 and $n(1 - p_0)$ are 10 or more.[13]

EXAMPLE 19.7   Are boys more likely?

The four-step process for any significance test is outlined on page 378.

STATE: We hear that newborn babies are more likely to be boys than girls, presumably to compensate for higher mortality among boys in early life. Is this true? A random sample found 13,173 boys among 25,468 firstborn children.[14] The sample proportion of boys was

$$\hat{p} = \frac{13{,}173}{25{,}468} = 0.5172$$

Boys do make up more than half of the sample, but of course we don't expect a perfect 50-50 split in a random sample. Is this sample evidence that boys are more common than girls in the entire population?

PLAN: Take p to be the proportion of boys among all firstborn children of American mothers. (Biology says that this should be the same as the proportion among all children, but the survey data concern first births.) We want to test the hypotheses

$$H_0: p = 0.5$$
$$H_a: p > 0.5$$

Blaine Harrington III/CORBIS

SOLVE: The conditions for inference are met, so we can go on to the z test statistic:

$$z = \frac{\hat{p} - p_0}{\sqrt{\dfrac{p_0(1 - p_0)}{n}}}$$

$$= \frac{0.5172 - 0.5}{\sqrt{\dfrac{(0.5)(0.5)}{25,468}}} = 5.49$$

The P-value is the area under the standard Normal curve to the right of $z = 5.49$. We know that this is very small; Table C shows that $P < 0.0005$. Minitab (Figure 19.3) says that P is 0 to three decimal places.

CONCLUDE: There is very strong evidence that more than half of newborns are boys ($P < 0.001$). ∎

FIGURE 19.3

Minitab output for the significance test of Example 19.7. Roundoff error in Example 19.7 explains the small difference (5.49 versus 5.50) in the values of the z statistic.

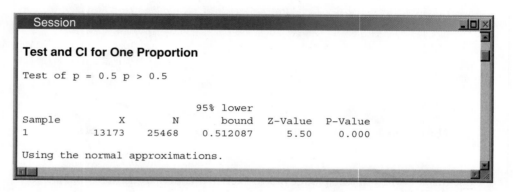

EXAMPLE 19.8 **Estimating the chance of a boy**

With 13,173 successes in 25,468 trials, the large-sample and plus four estimates of p are almost identical. Both are 0.5172 to four decimal places. So both methods give the 99% confidence interval

$$\hat{p} \pm z^* \sqrt{\frac{\hat{p}(1 - \hat{p})}{n}} = 0.5172 \pm 2.576 \sqrt{\frac{(0.5172)(0.4828)}{25,468}}$$

$$= 0.5172 \pm 0.0081$$

$$= 0.5091 \text{ to } 0.5253$$

We are 99% confident that between about 51% and 52.5% of first children are boys.

The confidence interval is more informative than the test in Example 19.7, which tells us only that more than half are boys. ∎

APPLY YOUR KNOWLEDGE

19.12 Spinning pennies. Spinning a coin, unlike tossing it, may not give heads and tails equal probabilities. I spun a penny 200 times and got 83 heads. How significant is this evidence against equal probabilities? Follow the four-step process as illustrated in Example 19.7.

19.13 Vote for the best face? We often judge other people by their faces. It appears that some people judge candidates for elected office by their faces. Psychologists showed head-and-shoulders photos of the two main candidates in 32 races for the U.S. Senate to many subjects (dropping subjects who recognized one of the candidates) to see which candidate was rated "more competent" based on nothing but the photos. On election day, the candidates whose faces looked more competent won 22 of the 32 contests.[15] If faces don't influence voting, half of all races in the long run should be won by the candidate with the better face. Is there evidence that the candidate with the better face wins more than half the time? Follow the four-step process as illustrated in Example 19.7.

19.14 No test. Explain why we can't use the z test for a proportion in these situations:

(a) You toss a coin 10 times in order to test the hypothesis H_0: $p = 0.5$ that the coin is balanced.

(b) A college president says, "99% of the alumni support my firing of Coach Boggs." You contact an SRS of 200 of the college's 15,000 living alumni to test the hypothesis H_0: $p = 0.99$.

CHAPTER 19 SUMMARY

- Tests and confidence intervals for a population proportion p when the data are an SRS of size n are based on the **sample proportion $\hat{p}$.**

- When n is large, $\hat{p}$ has approximately the Normal distribution with mean p and standard deviation $\sqrt{p(1-p)/n}$.

- The level C **large-sample confidence interval for p** is

$$\hat{p} \pm z^* \sqrt{\frac{\hat{p}(1-\hat{p})}{n}}$$

where z^* is the critical value for the standard Normal curve with area C between $-z^*$ and z^*.

- The true confidence level of the large-sample interval can be substantially less than the planned level C unless the sample is very large. We recommend using the plus four interval instead.

- To get a more accurate confidence interval, add four imaginary observations, two successes and two failures, to your sample. Then use the same formula for the confidence interval. This is the **plus four confidence interval.** Use this interval in practice for confidence level 90% or higher and sample size n at least 10.

- The **sample size** needed to obtain a confidence interval with approximate margin of error m for a population proportion is

$$n = \left(\frac{z^*}{m}\right)^2 p^*(1-p^*)$$

where p^* is a guessed value for the sample proportion $\hat{p}$, and z^* is the standard Normal critical point for the level of confidence you want. If you use $p^* = 0.5$ in this formula, the margin of error of the interval will be less than or equal to m no matter what the value of $\hat{p}$ is.

■ **Significance tests for H_0: $p = p_0$** are based on the z statistic

$$z = \frac{\hat{p} - p_0}{\sqrt{\dfrac{p_0(1 - p_0)}{n}}}$$

with P-values calculated from the standard Normal distribution. Use this test in practice when $np_0 \geq 10$ and $n(1 - p_0) \geq 10$.

Kids on bikes

In the most recent year for which data are available, 77% of children killed in bicycle accidents were boys. You might take these data as a sample and start from $\hat{p} = 0.77$ to do inference about bicycle deaths in the near future. What you should not do is conclude that boys on bikes are in greater danger than girls. We don't know how many boys and girls ride bikes—it may be that most fatalities are boys because most riders are boys.

CHECK YOUR SKILLS

19.15 *Sports Illustrated* asked a random sample of 757 Division I college athletes, "Do you believe performance-enhancing drugs are a problem in college sports?" Suppose that in fact 30% of all Division I athletes think that drugs are a problem. In repeated samples, the sample proportion $\hat{p}$ would follow a Normal distribution with mean

(a) 227.   (b) 0.3.   (c) 0.017.

19.16 The standard deviation of the distribution of $\hat{p}$ in the previous exercise is about

(a) 0.00028.   (b) 0.033.   (c) 0.017.

19.17 In fact, 273 of the 757 athletes in the *Sports Illustrated* sample said "Yes." The sample proportion $\hat{p}$ who said "Yes" is

(a) 36.   (b) 2.77.   (c) 0.36.

19.18 Based on the *Sports Illustrated* sample, the 95% large-sample confidence interval for the proportion of all Division I athletes who think performance-enhancing drugs are a problem is

(a) 0.36 ± 0.017.   (b) 0.36 ± 0.034.   (c) 0.36 ± 0.00030.

19.19 How many athletes must be interviewed to estimate the proportion concerned about use of drugs within ± 0.02 with 95% confidence? Use 0.5 as the conservative guess for p.

(a) $n = 25$ (b) $n = 1225$ (c) $n = 2401$

19.20 An opinion poll asks an SRS of 100 college seniors how they view their job prospects. In all, 53 say "Good." The plus four 95% confidence interval for estimating the proportion of all college seniors who think their job prospects are good is

(a) 0.529 ± 0.096.   (b) 0.529 ± 0.098.   (c) 0.529 ± 0.049.

19.21 The sample survey in Exercise 19.20 actually called 130 seniors, but 30 of the seniors refused to answer. This nonresponse could cause the survey result to be in error. The error due to nonresponse

(a) is in addition to the margin of error found in Exercise 19.20.

(b) is included in the margin of error found in Exercise 19.20.

(c) can be ignored because it isn't random.

19.22 Does the poll in Exercise 19.20 give reason to conclude that more than half of all seniors think their job prospects are good? The hypotheses for a test to answer this question are

(a) H_0: $p = 0.5$, H_a: $p > 0.5$.

(b) H_0: $p > 0.5$, H_a: $p = 0.5$.

(c) H_0: $p = 0.5$, H_a: $p \neq 0.5$.

19.23 The value of the z statistic for the test of the previous exercise is about

(a) $z = 12$.   (b) $z = 6$.   (c) $z = 0.6$.

19.24 A Gallup Poll found that only 28% of American adults expect to inherit money or valuable possessions from a relative. The poll's margin of error was 3%. This means that

(a) the poll used a method that gets an answer within 3% of the truth about the population 95% of the time.

(b) we can be sure that the percent of all adults who expect an inheritance is between 25% and 31%.

(c) if Gallup takes another poll using the same method, the results of the second poll will lie between 25% and 31%.

CHAPTER 19   EXERCISES

We recommend using the plus four method for all confidence intervals for a proportion. However, the large-sample method is acceptable when the guidelines for its use are met.

19.25 Do smokers know that smoking is bad for them? The Harris Poll asked a sample of smokers, "Do you believe that smoking will probably shorten your life, or not?" Of the 1010 people in the sample, 848 said "Yes."

(a) Harris called residential telephone numbers at random in an attempt to contact an SRS of smokers. Based on what you know about national sample surveys, what is likely to be the biggest weakness in the survey?

(b) We will nonetheless act as if the people interviewed are an SRS of smokers. Give a 95% confidence interval for the percent of smokers who agree that smoking will probably shorten their lives.

19.26 Reporting cheating. Students are reluctant to report cheating by other students. A student project put this question to an SRS of 172 undergraduates at a large university: "You witness two students cheating on a quiz. Do you go to the professor?" Only 19 answered "Yes."[16] Give a 95% confidence interval for the proportion of all undergraduates at this university who would report cheating.

19.27 Harris announces a margin of error. Exercise 19.25 describes a Harris Poll survey of smokers in which 848 of a sample of 1010 smokers agreed that smoking would probably shorten their lives. Harris announces a margin of error of ±3 percentage points for all samples of about this size. Opinion polls announce the margin of error for 95% confidence.

(a)   What is the actual margin of error (in percent) for the large-sample confidence interval from this sample?

(b)   The margin of error is largest when $\hat{p} = 0.5$. What would the margin of error (in percent) be if the sample had resulted in $\hat{p} = 0.5$?

(c)   Why do you think that Harris announces a ±3% margin of error for all samples of about this size?

19.28 Do college students pray? Social scientists asked 127 undergraduate students "from courses in psychology and communications" about prayer and found that 107 prayed at least a few times a year.[17]

(a)   To use any inference procedure, we must be willing to regard these 127 students, as far as their religious behavior goes, as an SRS from the population of all undergraduate students. Do you think it is reasonable to do this? Why or why not?

(b)   If we act as if the sample is an SRS, what is the large-sample 99% confidence interval for the proportion p of all students who pray?

(c)   Give the plus four 99% confidence interval for p. If you express the two intervals in percents and round to the nearest tenth of a percent, how do they differ? (As always, the plus four method pulls results away from 0% or 100%, whichever is closer. Although the condition for the large-sample interval is met, the plus four interval is more trustworthy.)

19.29 I don't like my life. The Pew Research Center asked a random sample of 1128 adult women, "How satisfied are you with your life overall?" Of these women, 56 said either "Mostly dissatisfied" or "Very dissatisfied."[18]

(a)   Pew dialed residential telephone numbers at random in the continental United States in an attempt to contact a random sample of adults. Based on what you know about national sample surveys, what is likely to be the biggest weakness in the survey?

(b)   Act as if the sample is an SRS. Give a large-sample 90% confidence interval for the proportion p of all adult women who are mostly or very dissatisfied with their lives.

(c)   Give the plus four confidence interval for p. If you express the two confidence intervals in percents and round to the nearest tenth of a percent, how do they differ? (As always, the plus four method pulls results away from 0% or 100%, whichever is closer. Although the condition for the large-sample interval is met, we can place more trust in the plus four interval.)

19.30 Which font? Plain type fonts such as Times New Roman are easier to read than fancy fonts such as Gigi. A group of 25 volunteer subjects read the same text in both fonts. (This is a matched pairs design. One-sample procedures for proportions, like those for means, are used to analyze data from matched pairs designs.) Of the 25 subjects, 17 said that they preferred Times New Roman for Web use. But 20 said that Gigi was more attractive.[19]

(a) Because the subjects were volunteers, conclusions from this sample can be challenged. Show that the sample size condition for the large-sample confidence interval is not met, but that the condition for the plus four interval is met.

(b) Give a 95% confidence interval for the proportion of all adults who prefer Times New Roman for Web use. Give a 90% confidence interval for the proportion of all adults who think Gigi is more attractive.

19.31 Detecting genetically modified soybeans. Most soybeans grown in the United States are genetically modified to, for example, resist pests and so reduce use of pesticides. Because some nations do not accept genetically modified (GM) foods, grain-handling facilities routinely test soybean shipments for the presence of GM beans. In a study of the accuracy of these tests, researchers submitted shipments of soybeans containing 1% of GM beans to 23 randomly selected facilities. Eighteen detected the GM beans.[20]

Wesley Hitt/Alamy

(a) Show that the conditions for the large-sample confidence interval are not met. Show that the conditions for the plus four interval are met.

(b) Use the plus four method to give a 90% confidence interval for the percent of all grain-handling facilities that will correctly detect 1% of GM beans in a shipment.

19.32 Running red lights. A random digit dialing telephone survey of 880 drivers asked, "Recalling the last ten traffic lights you drove through, how many of them were red when you entered the intersections?" Of the 880 respondents, 171 admitted that at least one light had been red.[21]

(a) Give a 95% confidence interval for the proportion of all drivers who ran one or more of the last ten red lights they met.

(b) Nonresponse is a practical problem for this survey—only 21.6% of calls that reached a live person were completed. Another practical problem is that people may not give truthful answers. What is the likely direction of the bias: do you think more or fewer than 171 of the 880 respondents really ran a red light? Why?

19.33 The IRS plans an SRS. The Internal Revenue Service plans to examine an SRS of individual federal income tax returns from each state. One variable of interest is the proportion of returns claiming itemized deductions. The total number of tax returns in a state varies from more than 15 million in California to fewer than 250,000 in Wyoming.

(a) Will the margin of error for estimating the population proportion change from state to state if an SRS of 2000 tax returns is selected in each state? Explain your answer.

(b) Will the margin of error change from state to state if an SRS of 1% of all tax returns is selected in each state? Explain your answer.

19.34 Customer satisfaction. An automobile manufacturer would like to know what proportion of its customers are not satisfied with the service provided by the local dealer. The customer relations department will survey a random sample of customers and compute a 99% confidence interval for the proportion who are not satisfied.

(a) Past studies suggest that this proportion will be about 0.2. Find the sample size needed if the margin of error of the confidence interval is to be about 0.015.

(b) When the sample is actually contacted, 10% of the sample say they are not satisfied. What is the margin of error of the 99% confidence interval?

19.35 Surveying students. You are planning a survey of students at a large university to determine what proportion favor an increase in student fees to support an expansion of the student newspaper. Using records provided by the registrar, you can select a random sample of students. You will ask each student in the sample whether he or she is in favor of the proposed increase. Your budget will allow a sample of 100 students.

(a) For a sample of size 100, construct a table of the margins of error for 95% confidence intervals when $\hat{p}$ takes the values 0.1, 0.2, 0.3, 0.4, 0.5, 0.6, 0.7, 0.8, and 0.9.

(b) A former editor of the student newspaper offers to provide funds for a sample of size 500. Repeat the margin of error calculations in (a) for the larger sample size. Then write a short thank-you note to the former editor describing how the larger sample size will improve the results of the survey.

In responding to Exercises 19.36 to 19.44, follow the **Plan, Solve,** *and* **Conclude** *steps of the four-step process.*

19.36 Student drinking. The College Alcohol Study interviewed a sample of 14,941 college students about their drinking habits. The sample was stratified using 140 colleges as strata, but the overall effect is close to an SRS of students. The response rate was between 60% and 70% at most colleges. This is quite good for a national sample, though nonresponse is as usual the biggest weakness of this survey. Of the students in the sample, 10,010 supported cracking down on underage drinking.[22] Estimate with 99% confidence the proportion of all college students who feel this way.

19.37 Shrubs that survive fires. Some shrubs have the useful ability to resprout from their roots after their tops are destroyed. Fire is a particular threat to shrubs in dry climates, as it can injure the roots as well as destroy the aboveground material. One study of resprouting took place in a dry area of Mexico.[23] The investigators clipped the tops of samples of several species of shrubs. In some cases, they also applied a propane torch to the stumps to simulate a fire. Of 12 specimens of the shrub *Krameria cytisoides*, 5 resprouted after fire. Estimate with 90% confidence the proportion of all shrubs of this species that will resprout after fire.

19.38 Condom usage. The National AIDS Behavioral Surveys (Example 19.1) also interviewed a sample of adults in the cities where AIDS is most common. This sample included 803 heterosexuals who reported having more than one sexual partner in the past year. We can consider this an SRS of size 803 from the population of all heterosexuals in high-risk cities who have multiple partners. These people risk infection with the AIDS virus. Yet 304 of the respondents said they never use condoms. Is this strong evidence that more than one-third of this population never use condoms?

19.39 Opinions about evolution. A sample survey funded by the National Science Foundation asked a random sample of American adults about biological evolution.[24] One question asked subjects to answer "True," "False," or "Not sure" to the statement "Human beings, as we know them today, developed from earlier species of animals." Of the 1484 respondents, 594 said "True." What can you say with 95% confidence about the percent of all American adults who think that humans developed from earlier species of animals?

19.40 Seat belt use. The proportion of drivers who use seat belts depends on things like age, gender, ethnicity, and local law. As part of a broader study, investigators observed a random sample of 117 female Hispanic drivers in Boston; 68 of these drivers were wearing seat belts.[25] Give a 95% confidence interval for the proportion of all female Hispanic drivers in Boston who wear seat belts.

19.41 Opinions about evolution, continued. Does the sample in Exercise 19.39 give good evidence to support the claim "Fewer than half of American adults think that humans developed from earlier species of animals"?

19.42 Seat belt use, continued. Do the data in Exercise 19.40 give good reason to conclude that more than half of Hispanic female drivers in Boston wear seat belts?

19.43 Does the double-blind method work? Many medical trials randomly assign patients to either an active treatment or a placebo. These trials are always double-blind. Sometimes the patients can tell whether or not they are getting the active treatment. This defeats the purpose of blinding. Reports of medical research usually ignore this problem. Investigators looked at a random sample of 97 articles reporting on placebo-controlled randomized trials in the top five general medical journals. Only 7 of the 97 discussed the success of blinding—and in 5 of these the blinding was imperfect.[26] What proportion of all such studies discuss the success of blinding? (Use 95% confidence.)

19.44 Seat belt use: planning a study. How large a sample would be needed to obtain margin of error ±0.05 in the study of seat belt use among Hispanic females? Use the $\hat{p}$ from Exercise 19.40 as your guess for the unknown p.

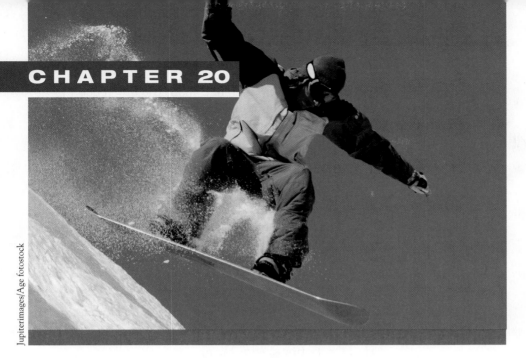

Jupiterimages/Age fotostock

Comparing Two Proportions

In a **two-sample problem,** we want to compare two populations or the responses to two treatments based on two independent samples. When the comparison involves the *means* of two populations, we use the two-sample t methods of Chapter 18. Now we turn to methods to compare the *proportions* of successes in two populations.

Two-sample problems: proportions

We will use notation similar to that used in our study of two-sample t statistics. The groups we want to compare are Population 1 and Population 2. We have a separate SRS from each population or responses from two treatments in a randomized comparative experiment. A subscript shows which group a parameter or statistic describes. Here is our notation:

Population	Population proportion	Sample size	Sample proportion
1	p_1	n_1	$\hat{p}_1$
2	p_2	n_2	$\hat{p}_2$

523

We compare the populations by doing inference about the difference $p_1 - p_2$ between the population proportions. The statistic that estimates this difference is the difference between the two sample proportions, $\hat{p}_1 - \hat{p}_2$.

EXAMPLE 20.1 Young adults living with their parents

STATE: A surprising number of young adults (ages 19 to 25) still live in their parents' home. A random sample by the National Institutes of Health included 2253 men and 2629 women in this age group.[1] The survey found that 986 of the men and 923 of the women lived with their parents. Is this good evidence that different proportions of young men and young women live with their parents? How large is the difference between the proportions of young men and young women who live with their parents?

PLAN: Take young men to be Population 1 and young women to be Population 2. The population proportions who live in their parents' home are p_1 for men and p_2 for women. We want to test the hypotheses

$$H_0: p_1 = p_2 \text{ (the same as } H_0: p_1 - p_2 = 0)$$

$$H_a: p_1 \neq p_2 \text{ (the same as } H_a: p_1 - p_2 \neq 0)$$

We also want to give a confidence interval for the difference $p_1 - p_2$.

SOLVE: Inference about population proportions is based on the sample proportions

$$\hat{p}_1 = \frac{986}{2253} = 0.4376 \quad \text{(men)}$$

$$\hat{p}_2 = \frac{923}{2629} = 0.3511 \quad \text{(women)}$$

We see that about 44% of the men but only about 35% of the women lived with their parents. Because the samples are large and the sample proportions are quite different, we expect that a test will be highly significant (in fact, $P < 0.0001$). So we concentrate on the confidence interval. To estimate $p_1 - p_2$, start from the difference of sample proportions

$$\hat{p}_1 - \hat{p}_2 = 0.4376 - 0.3511 = 0.0865$$

To complete the *Solve* step, we must know how this difference behaves. ■

The sampling distribution of a difference between proportions

To use $\hat{p}_1 - \hat{p}_2$ for inference, we must know its sampling distribution. Here are the facts we need:

■ When the samples are large, the distribution of $\hat{p}_1 - \hat{p}_2$ is **approximately Normal.**

■ The **mean** of the sampling distribution is $p_1 - p_2$. That is, the difference between sample proportions is an unbiased estimator of the difference between population proportions.

■ The **standard deviation** of the distribution is

$$\sqrt{\frac{p_1(1 - p_1)}{n_1} + \frac{p_2(1 - p_2)}{n_2}}$$

Figure 20.1 displays the distribution of $\hat{p}_1 - \hat{p}_2$. The standard deviation of $\hat{p}_1 - \hat{p}_2$ involves the unknown parameters p_1 and p_2. Just as in the previous chapter, we must replace these by estimates in order to do inference. And just as in the previous chapter, we do this a bit differently for confidence intervals and for tests.

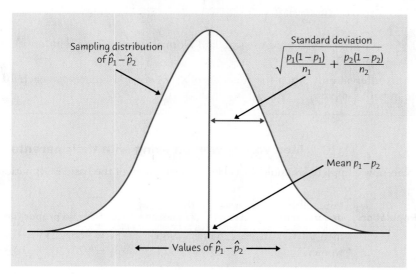

FIGURE 20.1

Select independent SRSs from two populations having proportions of successes p_1 and p_2. The proportions of successes in the two samples are $\hat{p}_1$ and $\hat{p}_2$. When the samples are large, the sampling distribution of the difference $\hat{p}_1 - \hat{p}_2$ is approximately Normal.

Large-sample confidence intervals for comparing proportions

To obtain a confidence interval, replace the population proportions p_1 and p_2 in the standard deviation by the sample proportions. The result is the **standard error** of the statistic $\hat{p}_1 - \hat{p}_2$:

standard error

$$\text{SE} = \sqrt{\frac{\hat{p}_1(1 - \hat{p}_1)}{n_1} + \frac{\hat{p}_2(1 - \hat{p}_2)}{n_2}}$$

The confidence interval has the same form we met in the previous chapter

$$\text{estimate} \pm z^* \text{SE}_{\text{estimate}}$$

LARGE-SAMPLE CONFIDENCE INTERVAL FOR COMPARING TWO PROPORTIONS

Draw an SRS of size n_1 from a large population having proportion p_1 of successes and draw an independent SRS of size n_2 from another large population having proportion p_2 of successes. When n_1 and n_2 are large, an approximate level C **confidence interval for $p_1 - p_2$** is

$$(\hat{p}_1 - \hat{p}_2) \pm z^*\text{SE}$$

In this formula the standard error SE of $\hat{p}_1 - \hat{p}_2$ is

$$\text{SE} = \sqrt{\frac{\hat{p}_1(1 - \hat{p}_1)}{n_1} + \frac{\hat{p}_2(1 - \hat{p}_2)}{n_2}}$$

and z^* is the critical value for the standard Normal density curve with area C between $-z^*$ and z^*.

Use this interval only when the numbers of successes and failures are each 10 or more in both samples.

EXAMPLE 20.2 Men versus women living with their parents

We can now complete Example 20.1. Here is a summary of the basic information:

Population	Population description	Sample size	Number of successes	Sample proportion
1	men	$n_1 = 2253$	986	$\hat{p}_1 = 986/2253 = 0.4376$
2	women	$n_2 = 2629$	923	$\hat{p}_2 = 923/2629 = 0.3511$

SOLVE: We will give a 95% confidence interval for $p_1 - p_2$, the difference between the proportions of young men and young women who live with their parents. To check that the large-sample confidence interval is safe, look at the counts of successes and failures in the two samples. All of these four counts are much larger than 10, so the large-sample method will be accurate. The standard error is

$$\text{SE} = \sqrt{\frac{\hat{p}_1(1 - \hat{p}_1)}{n_1} + \frac{\hat{p}_2(1 - \hat{p}_2)}{n_2}}$$

$$= \sqrt{\frac{(0.4376)(0.5624)}{2253} + \frac{(0.3511)(0.6489)}{2629}}$$

$$= \sqrt{0.0001959} = 0.01400$$

The 95% confidence interval is

$$(\hat{p}_1 - \hat{p}_2) \pm z^*\text{SE} = (0.4376 - 0.3511) \pm (1.960)(0.01400)$$

$$= 0.0865 \pm 0.0274$$

$$= 0.059 \text{ to } 0.114$$

> CONCLUDE: We are 95% confident that the percent of young men living with their parents is between 5.9 and 11.4 percentage points higher than the percent of young women who live with their parents. ∎

The sample survey in this example selected a single random sample of young adults, not two separate random samples of young men and young women. To get two samples, we divided the single sample by sex. This means that we did not know the two sample sizes n_1 and n_2 until after the data were in hand. The two-sample z procedures for comparing proportions are valid in such situations. This is an important fact about these methods.

Using technology

Figure 20.2 displays software output for Example 20.2 from a graphing calculator and a statistical software program. As usual, you can understand the output even without knowledge of the program that produced it. Minitab gives the test as well as the confidence interval, confirming that the difference between men and women is highly significant. Excel spreadsheet output is not shown because Excel lacks menu items for inference about proportions. You must use the spreadsheet's formula capability to program the confidence interval or test statistic and then to find the P-value of a test.

Texas Instruments Graphing Calculator

```
2-PropZInt
 (.05912,.11399)
 p̂1=.437638704
 p̂2=.3510840624
 n1=2253
 n2=2629
```

Minitab

```
Session                                                    _ □ ×

Test and CI for Two Proportions

Sample   X    N    Sample p
1       986  2253  0.437639
2       923  2629  0.351084

Difference = p (1) - p (2)
Estimate for difference: 0.0865546
95% CI for difference: (0.0591225, 0.113987)
Test for difference = 0 (vs not = 0): Z = 6.18   P-Value = 0.000
```

FIGURE 20.2

Output from a graphing calculator and Minitab for the 95% confidence interval of Example 20.2.

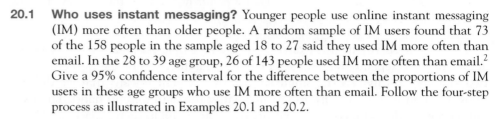

20.1   **Who uses instant messaging?** Younger people use online instant messaging (IM) more often than older people. A random sample of IM users found that 73 of the 158 people in the sample aged 18 to 27 said they used IM more often than email. In the 28 to 39 age group, 26 of 143 people used IM more often than email.[2] Give a 95% confidence interval for the difference between the proportions of IM users in these age groups who use IM more often than email. Follow the four-step process as illustrated in Examples 20.1 and 20.2.

20.2   **Listening to rap.** Rap music is more popular among young blacks than among young whites. A sample survey compared 634 randomly chosen blacks aged 15 to 25 with 567 randomly selected whites in the same age group. It found that 368 of the blacks and 130 of the whites listened to rap music every day.[3] Give a 90% confidence interval for the difference between the proportions of black and white young people who listen to rap every day. Follow the four-step process as illustrated in Examples 20.1 and 20.2.

20.3   **High school students in action.** A government survey randomly selected 6889 female high school students and 7028 male high school students.[4] Of these students, 1915 females and 3078 males met recommended levels of physical activity. (These levels are quite high: at least 60 minutes of activity that makes you breathe hard on at least 5 of the past 7 days.) Give a 99% confidence interval for the difference between the proportions of all female and male high school students who meet the recommended levels of activity.

Dennis MacDonald/Age fotostock

Accurate confidence intervals for comparing proportions

Like the large-sample confidence interval for a single proportion p, the large-sample interval for $p_1 - p_2$ generally has true confidence level less than the level you asked for. The inaccuracy is not as serious as in the one-sample case, at least if our guidelines for use are followed. Once again, adding imaginary observations greatly improves the accuracy.[5]

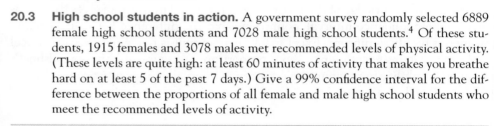

> **PLUS FOUR CONFIDENCE INTERVAL FOR COMPARING TWO PROPORTIONS**
>
> Draw independent SRSs from two large populations with population proportions of successes p_1 and p_2. To get the **plus four confidence interval for the difference $p_1 - p_2$,** add four imaginary observations, one success and one failure in each of the two samples. Then use the large-sample confidence interval with the new sample sizes (actual sample sizes + 2) and counts of successes (actual counts + 1).
>
> Use this interval when the sample size is at least 5 in each group, with any counts of successes and failures.

If your software does not offer the plus four method, just enter the new plus four sample sizes and success counts into the large-sample procedure.

EXAMPLE 20.3 **Shrubs that withstand fire**

STATE: Fire is a serious threat to shrubs in dry climates. Some shrubs can resprout from their roots after their tops are destroyed. One study of resprouting took place in a dry area of Mexico.[6] The investigators randomly assigned shrubs to treatment and control groups. They clipped the tops of all the shrubs. They then applied a propane torch to the stumps of the treatment group to simulate a fire. A shrub is a success if it resprouts. Here are the data for the shrub *Xerospirea hartwegiana*:

Population	Population description	Sample size	Number of successes	Sample proportion
1	control	$n_1 = 12$	12	$\hat{p}_1 = 12/12 = 1.000$
2	treatment	$n_2 = 12$	8	$\hat{p}_2 = 8/12 = 0.667$

How much does burning reduce the proportion of shrubs of this species that resprout?

PLAN: Give a 90% confidence interval for the difference of population proportions, $p_1 - p_2$.

SOLVE: The conditions for the large-sample interval are not met. In fact, there are *no* failures in the control group. We will use the plus four method. Add four imaginary observations. The new data summary is

Population	Population description	Sample size	Number of successes	Plus four sample proportion
1	control	$n_1 + 2 = 14$	$12 + 1 = 13$	$\tilde{p}_1 = 13/14 = 0.9286$
2	treatment	$n_2 + 2 = 14$	$8 + 1 = 9$	$\tilde{p}_2 = 9/14 = 0.6429$

The standard error based on the new facts is

$$\text{SE} = \sqrt{\frac{\tilde{p}_1(1 - \tilde{p}_1)}{n_1 + 2} + \frac{\tilde{p}_2(1 - \tilde{p}_2)}{n_2 + 2}}$$

$$= \sqrt{\frac{(0.9286)(0.0714)}{14} + \frac{(0.6429)(0.3571)}{14}}$$

$$= \sqrt{0.02113} = 0.1454$$

The plus four 90% confidence interval is

$$(\tilde{p}_1 - \tilde{p}_2) \pm z^*\text{SE} = (0.9286 - 0.6429) \pm (1.645)(0.1454)$$

$$= 0.2857 \pm 0.2392$$

$$= 0.047 \text{ to } 0.525$$

CONCLUDE: We are 90% confident that burning reduces the percent of these shrubs that resprout by between 4.7% and 52.5%. ■

Computer-assisted interviewing

The days of the interviewer with a clipboard are past. Interviewers now read questions from a computer screen and use the keyboard to enter responses. The computer skips irrelevant items—once a woman says that she has no children, further questions about her children never appear. The computer can even present questions in random order to avoid bias due to always following the same order. Software keeps records of who has responded and prepares a file of data from the responses. The tedious process of transferring responses from paper to computer, once a source of errors, has disappeared.

The plus four interval may be conservative (that is, the true confidence level may be *higher* than you asked for) for very small samples and population p's close to 0 or 1, as in this example. It is generally much more accurate than the large-sample interval when the samples are small. Nevertheless, the plus four interval in Example 20.3 cannot save us from the fact that small samples produce wide confidence intervals.

Jupiterimages/Comstock Images/Alamy

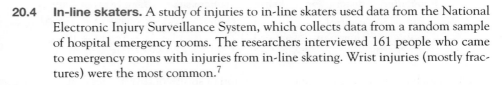

APPLY YOUR KNOWLEDGE

20.4 **In-line skaters.** A study of injuries to in-line skaters used data from the National Electronic Injury Surveillance System, which collects data from a random sample of hospital emergency rooms. The researchers interviewed 161 people who came to emergency rooms with injuries from in-line skating. Wrist injuries (mostly fractures) were the most common.[7]

   (a) The interviews found that 53 people were wearing wrist guards and 6 of these had wrist injuries. Of the 108 who did not wear wrist guards, 45 had wrist injuries. Why should we not use the large-sample confidence interval for these data?

   (b) The plus four method adds one success and one failure in each sample. What are the sample sizes and counts of successes after you do this?

   (c) Give the plus four 95% confidence interval for the difference between the two population proportions of wrist injuries. State carefully what populations your inference compares.

20.5 **Broken crackers.** We don't like to find broken crackers when we open the package. How can makers reduce breaking? One idea is to microwave the crackers for 30 seconds right after baking them. Breaks start as hairline cracks called "checking." Assign 65 newly baked crackers to the microwave and another 65 to a control group that is not microwaved. After one day, none of the microwave group and 16 of the control group show checking.[8] Give the 95% plus four confidence interval for the amount by which microwaving reduces the proportion of checking. The plus four method is particularly helpful when, as here, a count of successes is zero. Follow the four-step process as illustrated in Example 20.3.

Significance tests for comparing proportions

An observed difference between two sample proportions can reflect an actual difference between the populations, or it may just be due to chance variation in random sampling. Significance tests help us decide if the effect we see in the samples is really there in the populations. The null hypothesis says that there is no difference between the two populations:

$$H_0: p_1 = p_2$$

The alternative hypothesis says what kind of difference we expect.

EXAMPLE 20.4 Choosing a mate

STATE: "Would you marry a person from a lower social class than your own?" Researchers asked this question of a sample of 385 black, never-married students at two historically black colleges in the South. We will consider this to be an SRS of black students at historically black colleges. Of the 149 men in the sample, 91 said "Yes." Among the 236 women, 117 said "Yes."[9] Is there reason to think that different proportions of men and women in this student population would be willing to marry beneath their class?

PLAN: Call the population proportions p_1 for men and p_2 for women. We had no direction for the difference in mind before looking at the data, so we have a two-sided alternative:

$$H_0: p_1 = p_2$$
$$H_a: p_1 \neq p_2$$

SOLVE: The men and women in a single SRS can be treated as if they were separate SRSs of men and women students. The sample proportions who would marry someone from a lower social class are

$$\hat{p}_1 = \frac{91}{149} = 0.611 \quad \text{(men)}$$

$$\hat{p}_2 = \frac{117}{236} = 0.496 \quad \text{(women)}$$

That is, about 61% of the men but only about 50% of the women would marry beneath their class. Is this apparent difference statistically significant? To continue the solution, we must learn the proper test. ■

To do a test, standardize the difference between the sample proportions $\hat{p}_1 - \hat{p}_2$ to get a z statistic. If H_0 is true, both samples come from populations in which the same unknown proportion p would marry someone from a lower social class. We take advantage of this by combining the two samples to estimate this single p instead estimating p_1 and p_2 separately. Call this the **pooled sample proportion.** It is

$$\hat{p} = \frac{\text{number of successes in both samples combined}}{\text{number of individuals in both samples combined}}$$

Use $\hat{p}$ in place of both $\hat{p}_1$ and $\hat{p}_2$ in the expression for the standard error SE of $\hat{p}_1 - \hat{p}_2$ to get a z statistic that has the standard Normal distribution when H_0 is true. Here is the test.

The cookie strikes

How many different people clicked on your business Web site last month? Technology tries to help: when someone visits your site, a little piece of code called a cookie is left on their computer. When the same person clicks again, the cookie says not to count them as a "unique visitor" because this isn't their first visit. But lots of Web users delete cookies, either by hand or automatically with software. These people get counted again when they visit your site again. That's bias: your counts of unique visitors are systematically too high. One study found that unique visitor counts were as much as 50% too high.

pooled sample proportion

SIGNIFICANCE TEST FOR COMPARING TWO PROPORTIONS

Draw an SRS of size n_1 from a large population having proportion p_1 of successes and draw an independent SRS of size n_2 from another large population having proportion p_2 of successes. To **test the hypothesis** $H_0: p_1 = p_2$, first find the pooled proportion $\hat{p}$ of successes in both samples combined. Then compute the z statistic

$$z = \frac{\hat{p}_1 - \hat{p}_2}{\sqrt{\hat{p}(1 - \hat{p})\left(\dfrac{1}{n_1} + \dfrac{1}{n_2}\right)}}$$

In terms of a variable Z having the standard Normal distribution, the P-value for a test of H_0 against

$H_a: p_1 > p_2$ is $P(Z \geq z)$

$H_a: p_1 < p_2$ is $P(Z \leq z)$

$H_a: p_1 \neq p_2$ is $2P(Z \geq |z|)$

Use this test when the counts of successes and failures are each 5 or more in both samples.[10]

EXAMPLE 20.5 **Choosing a mate, continued**

SOLVE: The data come from an SRS and the counts of successes and failures are all much larger than 5. The pooled proportion of students who would marry beneath their own social class is

$$\hat{p} = \frac{\text{number of "Yes" responses among men and women combined}}{\text{number of men and women combined}}$$

$$= \frac{91 + 117}{149 + 236}$$

$$= \frac{208}{385} = 0.5403$$

The z test statistic is

$$z = \frac{\hat{p}_1 - \hat{p}_2}{\sqrt{\hat{p}(1-\hat{p})\left(\dfrac{1}{n_1} + \dfrac{1}{n_2}\right)}}$$

$$= \frac{0.611 - 0.496}{\sqrt{(0.5403)(0.4597)\left(\dfrac{1}{149} + \dfrac{1}{236}\right)}}$$

$$= \frac{0.115}{0.05215} = 2.205$$

The two-sided P-value is the area under the standard Normal curve more than 2.205 distant from 0. Figure 20.3 shows this area. Software tells us that $P = 0.0275$.

Without software, you can compare $z = 2.205$ with the bottom row of Table C (standard Normal critical values) to approximate P. It lies between the critical values 2.054 and 2.326 for two-sided P-values 0.04 and 0.02.

z^*	2.054	2.326
Two-sided P	.04	.02

CONCLUDE: There is good evidence ($P < 0.04$) that men are more likely than women to say they will marry someone from a lower social class. ■

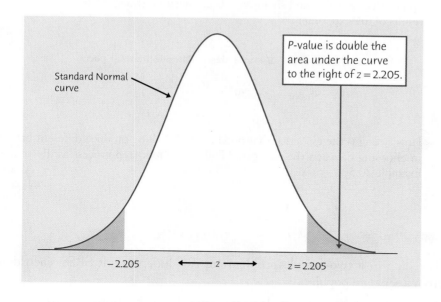

FIGURE 20.3

The P-value for the two-sided test of Example 20.5.

Standard Normal curve

P-value is double the area under the curve to the right of $z = 2.205$.

-2.205 $\longleftarrow z \longrightarrow$ $z = 2.205$

APPLY YOUR KNOWLEDGE

20.6 **Did the random assignment work?** A large clinical trial of the effect of diet on breast cancer assigned women at random to either a normal diet or a low-fat diet. To check that the random assignment did produce comparable groups, we can compare the two groups at the start of the study. Ask if there is a family history of breast cancer: 3396 of the 19,541 women in the low-fat group and 4929 of the 29,294 women in the control group said "Yes."[11] If the random assignment worked well, there should *not* be a significant difference in the proportions with a family history of breast cancer. How significant is the observed difference? Follow the four-step process as illustrated in Example 20.5.

Jupiterimages/Age fotostock

20.7   Protecting skiers and snowboarders. Most alpine skiers and snowboarders do not use helmets. Do helmets reduce the risk of head injuries? A study in Norway compared skiers and snowboarders who suffered head injuries with a control group who were not injured. Of 578 injured subjects, 96 had worn a helmet. Of the 2992 in the control group, 656 wore helmets.[12] Is helmet use less common among skiers and snowboarders who have head injuries? Follow the four-step process as illustrated in Example 20.5. (Note that this is an observational study that compares injured and uninjured subjects. An experiment that assigned subjects to helmet and no-helmet groups would be more convincing.)

20.8   Shake your head. Our bodies affect our behavior in strange ways. Psychologists asked student subjects to either nod their heads up and down or shake their heads side to side while listening to music through headphones. The subjects thought that they were evaluating the headphones. An experimenter placed a pen on the table before the music played. At the end of the session, each subject was offered a choice of two pens as a gift: one identical to the pen on the table and another in a different color. The "nodding" group more often chose the familiar pen and the "shaking" group more often chose the new pen. It seems that nodding calls up positive thoughts and shaking calls up negative thoughts. Here are the actual results of the experiment:[13]

Group	Sample size	Choose familiar pen
Nod	61	45
Shake	59	15

How strong is the evidence that nodding and shaking produce different behavior in choosing between the two pens? Follow the four-step process as illustrated in Example 20.5.

CHAPTER 20   SUMMARY

- The data in a **two-sample problem** are two independent SRSs, each drawn from a separate population.

- Tests and confidence intervals to compare the proportions p_1 and p_2 of successes in the two populations are based on the difference $\hat{p}_1 - \hat{p}_2$ between the sample proportions of successes in the two SRSs.

- When the sample sizes n_1 and n_2 are large, the sampling distribution of $\hat{p}_1 - \hat{p}_2$ is close to Normal with mean $p_1 - p_2$.

- The level C **large-sample confidence interval for $p_1 - p_2$** is

$$(\hat{p}_1 - \hat{p}_2) \pm z^* \text{SE}$$

where the **standard error** of $\hat{p}_1 - \hat{p}_2$ is

$$SE = \sqrt{\frac{\hat{p}_1(1 - \hat{p}_1)}{n_1} + \frac{\hat{p}_2(1 - \hat{p}_2)}{n_2}}$$

and z^* is a standard Normal critical value.

- The true confidence level of the large-sample interval can be substantially less than the planned level C. Use this interval only if the counts of successes and failures in both samples are 10 or greater.

- To get a more accurate confidence interval, add four imaginary observations, one success and one failure in each sample. Then use the same formula for the confidence interval. This is the **plus four confidence interval.** You can use it whenever both samples have 5 or more observations.

- **Significance tests for H_0: $p_1 = p_2$** use the **pooled sample proportion**

$$\hat{p} = \frac{\text{number of successes in both samples combined}}{\text{number of individuals in both samples combined}}$$

and the z statistic

$$z = \frac{\hat{p}_1 - \hat{p}_2}{\sqrt{\hat{p}(1 - \hat{p})\left(\dfrac{1}{n_1} + \dfrac{1}{n_2}\right)}}$$

P-values come from the standard Normal distribution. Use this test when there are 5 or more successes and 5 or more failures in both samples.

C H E C K   Y O U R   S K I L L S

A sample survey interviews SRSs of 500 female college students and 550 male college students. Each student is asked if he or she worked for pay last summer. In all, 410 of the women and 484 of the men say "Yes." Exercises 20.9 to 20.13 are based on this survey.

20.9 Take p_M and p_F to be the proportions of all college males and females who worked last summer. We conjectured before seeing the data that men are more likely to work. The hypotheses to be tested are

(a) H_0: $p_M = p_F$ versus H_a: $p_M \neq p_F$.

(b) H_0: $p_M = p_F$ versus H_a: $p_M > p_F$.

(c) H_0: $p_M = p_F$ versus H_a: $p_M < p_F$.

20.10 The sample proportions of college males and females who worked last summer are about

(a) $\hat{p}_M = 0.88$ and $\hat{p}_F = 0.82$.

(b) $\hat{p}_M = 0.82$ and $\hat{p}_F = 0.88$.

(c) $\hat{p}_M = 0.75$ and $\hat{p}_F = 0.97$.

20.11 The pooled sample proportion who worked last summer is about

(a) $\hat{p} = 1.70$.   (b) $\hat{p} = 0.89$.   (c) $\hat{p} = 0.85$.

20.12 The z statistic for a test comparing the proportions of college men and women who worked last summer is about

(a) $z = 2.66$.   (b) $z = 2.72$.   (c) $z = 3.10$.

20.13 The 95% large-sample confidence interval for the difference $p_M - p_F$ in the proportions of college men and women who worked last summer is about

(a) 0.06 ± 0.00095.   (b) 0.06 ± 0.043.   (c) 0.06 ± 0.036.

20.14 In an experiment to learn if substance M can help restore memory, the brains of 20 rats were treated to damage their memories. The rats were trained to run a maze. After a day, 10 rats were given M and 7 of them succeeded in the maze; only 2 of the 10 control rats were successful. The z test for "no difference" against "a higher proportion of the M group succeeds" has

(a) $z = 2.25$, $P < 0.02$.

(b) $z = 2.60$, $P < 0.005$.

(c) $z = 2.25$, $P < 0.04$ but not < 0.02.

20.15 The z test in the previous exercise

(a) may be inaccurate because the populations are too small.

(b) may be inaccurate because some counts of successes and failures are too small.

(c) is reasonably accurate because the conditions for inference are met.

20.16 The plus four 90% confidence interval for the difference between the proportion of rats that succeed when given M and the proportion that succeed without it is

(a) 0.455 ± 0.312.   (b) 0.417 ± 0.304.   (c) 0.417 ± 0.185.

CHAPTER 20   EXERCISES

We recommend using the plus four method for all confidence intervals for proportions. However, the large-sample method is acceptable when the guidelines for its use are met.

20.17 Truthfulness in online profiles. Many teens have posted profiles on sites such as MySpace. A sample survey asked random samples of teens with online profiles if they included false information in their profiles. Of 170 younger teens (ages 12 to 14), 117 said "Yes." Of 317 older teens (ages 15 to 17), 152 said "Yes."[14]

(a) Do these samples satisfy the guidelines for the large-sample confidence interval?

(b) Give a 95% confidence interval for the difference between the proportions of younger and older teens who include false information in their online profiles.

20.18 Drug testing in schools. In 2002 the Supreme Court ruled that schools could require random drug tests of students participating in competitive after-school activities such as athletics. Does drug testing reduce use of illegal drugs? A study compared two similar high schools in Oregon. Wahtonka High School tested athletes at random and Warrenton High School did not. In a confidential survey, 7 of 135 athletes at Wahtonka and 27 of 141 athletes at Warrenton said they were using drugs.[15] Regard these athletes as SRSs from the populations of athletes at similar schools with and without drug testing.

(a) You should not use the large-sample confidence interval. Why not?

(b) The plus four method adds two observations, a success and a failure, to each sample. What are the sample sizes and the numbers of drug users after you do this?

(c) Give the plus four 95% confidence interval for the difference between the proportion of athletes using drugs at schools with and without testing.

20.19 Genetically altered mice. Genetic influences on cancer can be studied by manipulating the genetic makeup of mice. One of the processes that turn genes on or off (so to speak) in particular locations is called "DNA methylation." Do low levels of this process help cause tumors? Compare mice altered to have low levels with normal mice. Of 33 mice with lowered levels of DNA methylation, 23 developed tumors. None of the control group of 18 normal mice developed tumors in the same time period.[16]

(a) Explain why we cannot safely use either the large-sample confidence interval or the test for comparing the proportions of normal and altered mice that develop tumors.

(b) The plus four method adds two observations, a success and a failure, to each sample. What are the sample sizes and the numbers of mice with tumors after you do this?

(c) Give a 99% confidence interval for the difference in the proportions of the two populations that develop tumors.

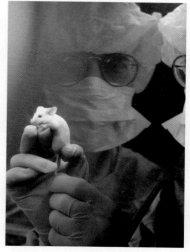

Mark Harmel/Getty Images

20.20 Drug testing in schools, continued. Exercise 20.18 describes a study that compared the proportions of athletes who use illegal drugs in two similar high schools, one that tests for drugs and one that does not. Drug testing is intended to reduce use of drugs. Do the data give good reason to think that drug use among athletes is lower in schools that test for drugs? State hypotheses, find the test statistic, and use either software or the bottom row of Table C for the *P*-value. Be sure to state your conclusion. (Because the study is not an experiment, the conclusion depends on the condition that athletes in these two schools can be considered SRSs from all similar schools.)

Call a statistician. *Does involving a statistician to help with statistical methods improve the chance that a medical research paper will be published? A study of papers submitted to two medical journals found that 135 of 190 papers that lacked statistical assistance were rejected without even being reviewed in detail. In contrast, 293 of the 514 papers with statistical help were sent back without review.*[17] *Exercises 20.21 to 20.23 are based on this study.*

20.21 Does statistical help make a difference? Is there a significant difference in the proportions of papers with and without statistical help that are rejected without review? State hypotheses, find the test statistic, use software or the bottom row of Table C to get a *P*-value, and give your conclusion. (This observational study does not establish causation, because studies that include statistical help may also be better in other ways than those that do not.)

20.22 **How often are statisticians involved?** Give a 95% confidence interval for the proportion of papers submitted to these journals that include help from a statistician.

20.23 **How big a difference?** Give a 95% confidence interval for the difference between the proportions of papers rejected without review when a statistician is and is not involved in the research.

20.24 **The design of the study matters.** How accurate are the tests that grain-handling facilities make to detect the presence of genetically modified (GM) soybeans in shipments to countries that do not allow GM beans? Batches of soybeans containing some genetically modified (GM) beans were submitted to 23 grain-handling facilities. When batches contained 1% of GM beans, 18 of the facilities detected the presence of GM beans. Only 7 of the facilities detected GM beans when they made up one-tenth of 1% of the beans in the batches.[18] Explain why we *cannot* use the methods of this chapter to compare the proportions of facilities that will detect the two levels of GM soybeans.

20.25 **Significant does not mean important.** Never forget that even small effects can be statistically significant if the samples are large. To illustrate this fact, consider a sample of 148 small businesses. During a three-year period, 15 of the 106 headed by men and 7 of the 42 headed by women failed.[19]

(a) Find the proportions of failures for businesses headed by women and businesses headed by men. These sample proportions are quite close to each other. Give the P-value for the z test of the hypothesis that the same proportion of women's and men's businesses fail. (Use the two-sided alternative.) The test is very far from being significant.

(b) Now suppose that the same sample proportions came from a sample 30 times as large. That is, 210 out of 1260 businesses headed by women and 450 out of 3180 businesses headed by men fail. Verify that the proportions of failures are exactly the same as in (a). Repeat the z test for the new data, and show that it is now significant at the $\alpha = 0.05$ level.

(c) It is wise to use a confidence interval to estimate the size of an effect rather than just giving a P-value. Give 95% confidence intervals for the difference between the proportions of women's and men's businesses that fail for the settings of both (a) and (b). What is the effect of larger samples on the confidence interval?

In responding to Exercises 20.26 to 20.35, follow the **Plan, Solve,** *and* **Conclude** *steps of the four-step process.*

20.26 **Are urban students more successful?** North Carolina State University looked at the factors that affect the success of students in a required chemical engineering course. Students must get a C or better in the course in order to continue as chemical engineering majors, so a "success" is a grade of C or better. There were 65 students from urban or suburban backgrounds, and 52 of these students succeeded. Another 55 students were from rural or small-town backgrounds; 30 of these students succeeded in the course.[20] Is there good evidence that the proportion of students who succeed is different for urban/suburban versus rural/small-town backgrounds?

20.27 Female and male students. The North Carolina State University study in the previous exercise also looked at possible differences in the proportions of female and male students who succeeded in the course. They found that 23 of the 34 women and 60 of the 89 men succeeded. Is there evidence of a difference between the proportions of women and men who succeed?

20.28 Are urban students more successful, continued. Continue your work from Exercise 20.26. Estimate the difference between the success rates for all urban/ suburban and rural/small-town students who plan to study chemical engineering at North Carolina State. (Use 90% confidence.)

20.29 How to quit smoking. Nicotine patches are often used to help smokers quit. Does giving medicine to fight depression help? A randomized double-blind experiment assigned 244 smokers who wanted to stop to receive nicotine patches and another 245 to receive both a patch and the antidepression drug bupropion. After a year, 40 subjects in the nicotine patch group and 87 in the patch-plus-drug group had abstained from smoking.[21] Give a 99% confidence interval for the difference (treatment minus control) in the proportion of smokers who quit.

20.30 The Gold Coast. A historian examining British colonial records for the Gold Coast in Africa suspects that the death rate was higher among African miners than among European miners. In the year 1936, there were 223 deaths among 33,809 African miners and 7 deaths among 1541 European miners on the Gold Coast.[22] (The Gold Coast became the independent nation of Ghana in 1957.)

Consider this year as a random sample from the colonial era in West Africa. Is there good evidence that the proportion of African miners who died was higher than the proportion of European miners who died?

Michael S. Lewis/CORBIS

20.31 I refuse! Do our emotions influence economic decisions? One way to examine the issue is to have subjects play an "ultimatum game" against other people and against a computer. Your partner (person or computer) gets $10, on the condition that it be shared with you. The partner makes you an offer. If you refuse, neither of you gets anything. So it's to your advantage to accept even the unfair offer of $2 out of the $10. Some people get mad and refuse unfair offers. Here are data on the responses of 76 subjects randomly assigned to receive an offer of $2 from either a person they were introduced to or a computer:[23]

	Accept	Reject
Human offers	20	18
Computer offers	32	6

We suspect that emotion will lead to offers from another person being rejected more often than offers from an impersonal computer. Do a test to assess the evidence for this conjecture.

20.32 Seat belt use. The proportion of drivers who use seat belts depends on things like age (young people are more likely to go unbelted) and gender (women are more likely to use belts). It also depends on local law. In New York City, police can stop a driver who is not belted. In Boston at the time of the survey, police could cite a driver for not wearing a seat belt only if the driver had been stopped for some other violation. Here are data from observing random samples of female Hispanic drivers in these two cities:[24]

City	Drivers	Belted
New York	220	183
Boston	117	68

(a) Is this an experiment or an observational study? Why?

(b) Comparing local laws suggests the hypothesis that a smaller proportion of drivers wear seat belts in Boston than in New York. Do the data give good evidence that this is true for female Hispanic drivers?

20.33 Lyme disease. Lyme disease is spread in the northeastern United States by infected ticks. The ticks are infected mainly by feeding on mice, so more mice result in more infected ticks. The mouse population in turn rises and falls with the abundance of acorns, their favored food. Experimenters studied two similar forest areas in a year when the acorn crop failed. They added hundreds of thousands of acorns to one area to imitate an abundant acorn crop, while leaving the other area untouched. The next spring, 54 of the 72 mice trapped in the first area were in breeding condition, versus 10 of the 17 mice trapped in the second area.[25] Estimate the difference between the proportions of mice ready to breed in good acorn years and bad acorn years. (Use 90% confidence. Be sure to justify your choice of confidence interval.)

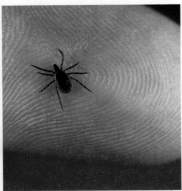

Scott Camazine/Photo Researchers

20.34 Does preschool help? To study the long-term effects of preschool programs for poor children, the High/Scope Educational Research Foundation has followed two groups of Michigan children since early childhood.[26] One group of 62 attended preschool as 3- and 4-year-olds. A control group of 61 children from the same area and similar backgrounds did not attend preschool. Over a ten-year period as adults, 38 of the preschool sample and 49 of the control sample needed social services (mainly welfare). Does the study provide significant evidence that children who attend preschool have less need for social services as adults? How large is the difference between the proportions of the preschool and no-preschool populations that require social services? Do inference to answer both questions. Be sure to explain exactly what inference you choose to do.

20.35 Using credit cards. Are shoppers more or less likely to use credit cards for "impulse purchases" that they decide to make on the spot, as opposed to purchases that they had in mind when they went to the store? Stop every third person leaving a department store with a purchase. (This is in effect a random sample of people who buy at that store.) A few questions allow us to classify the purchase as impulse or not. Here are the data on how the customer paid:[27]

	Credit Card?	
	Yes	No
Impulse purchases	13	18
Planned purchases	35	31

Estimate with 95% confidence the percent of all customers at this store who use a credit card. Give numerical summaries to describe the difference in credit card use between impulse and planned purchases. Is this difference statistically significant?

Kelly Kalhoefer/Jupiter Images

CHAPTER 21

Inference about Variables: Part III Review

The procedures of Chapters 17 to 20 are among the most common of all statistical inference methods. Now that you have mastered important ideas and practical methods for inference, it's time to review the big ideas of statistics in outline form. Here is a summary of Parts I and II of this book, leading up to Part III. The outline contains some important warnings: look for the Caution icon.

1. **Data Production**
    - Data basics:
        Individuals (subjects).
        Variables: categorical versus quantitative, units of measurement, explanatory versus response.
        Purpose of study.
    - Data production basics:
        Observation versus experiment.
        Simple random samples.
        Completely randomized experiments.
    - Beware: really bad data production (voluntary response, confounding) can make interpretation impossible.
    - Beware: weaknesses in data production (for example, sampling students at only one campus) can make generalizing conclusions difficult.

IN THIS CHAPTER
WE COVER...

- Part III Summary
- Review Exercises
- Supplementary Exercises

541

2. **Data Analysis**
   - Plot your data. Look for overall pattern and striking deviations.
   - Add numerical descriptions based on what you see.
   - Beware: averages and other simple descriptions can miss the real story.
   - One quantitative variable:
     Graphs: stemplot, histogram, boxplot.
     Pattern: distribution shape, center, spread. Outliers?
     Density curves (such as Normal curves) to describe overall pattern.
     Numerical descriptions: five-number summary or $\overline{x}$ and s.
   - Relationships between two quantitative variables:
     Graph: scatterplot.
     Pattern: relationship form, direction, strength. Outliers? Influential
       observations?
     Numerical description for linear relationships: correlation, regression
       line.
     Beware the lurking variable: correlation does not imply causation.
   - Beware the effects of outliers and influential observations.

3. **The Reasoning of Inference**
   - Inference uses data to infer conclusions about a wider population.
   - When you do inference, you are acting as if your data come from random
     samples or randomized comparative experiments. Beware: if they don't,
     you may have "garbage in, garbage out."
   - Always examine your data before doing inference. Inference often requires
     a regular pattern, such as roughly Normal with no strong outliers.
   - Key idea: "What would happen if we did this many times?"
   - Confidence intervals: estimate a population parameter.
     95% confidence: I used a method that captures the true parameter 95%
       of the time in repeated use.
     Beware: the margin of error of a confidence interval does not include
       the effects of practical errors such as undercoverage and nonresponse.
   - Significance tests: assess evidence against H_0 in favor of H_a.
     P-value: If H_0 were true, how often would I get an outcome favoring
       the alternative this strongly? Smaller P = stronger evidence
       against H_0.
     Statistical significance at the 5% level, $P < 0.05$, means that an
       outcome this extreme would occur less than 5% of the time if H_0
       were true.
     Beware: $P < 0.05$ is not sacred.
     Beware: statistical significance is not the same as practical
       significance.
     Large samples can make small effects significant. Small samples can fail
       to declare large effects significant.
     Always try to estimate the size of an effect (for example, with a
       confidence interval), not just its significance.

4.   **Methods of Inference**
   - Choose the right inference procedure.
   - Carry out the details.
   - State your conclusion.

   Part III of this book introduces the fourth and last part of this outline. To actually do inference, you must choose the right procedure and carry out the details.
The Statistics in Summary flowchart given below offers a brief guide. It is important

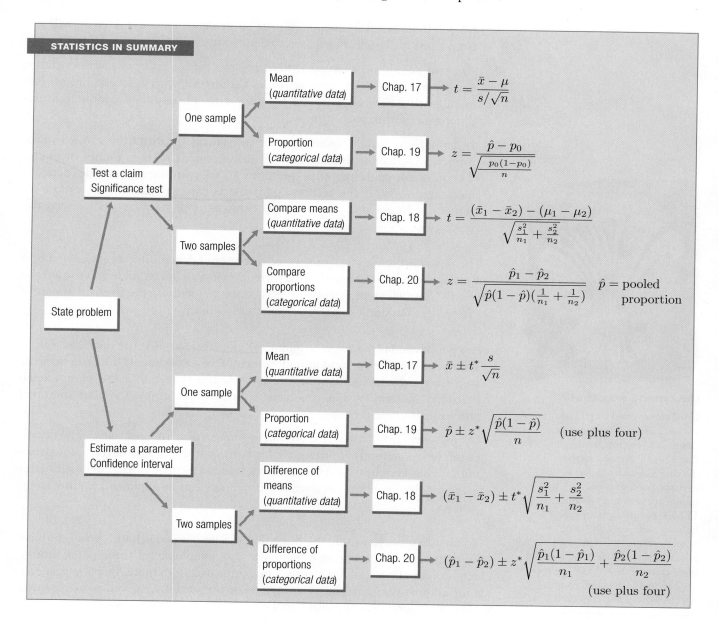

to do some of the review exercises because now, for the first time, you must decide which of several inference procedures to use. Learning to recognize problem settings in order to choose the right type of inference is a key step in advancing your mastery of statistics. This is the *Plan* step in the four step process, in which you translate the real-world problem from the *State* step into a specific inference procedure.

The flowchart organizes one way of planning inference problems. Let's go through it from left to right.

1. *Do you want to test a claim or estimate an unknown quantity?* That is, will you need a test of significance or a confidence interval?

2. *Are your data a single sample representing one population or two samples chosen to compare two populations or responses to two treatments in an experiment?* Remember that to work with *matched pairs* data you form one sample from the differences within pairs.

3. *Is the response variable quantitative or categorical?* Quantitative variables take numerical values with some unit of measurement such as inches or grams. The most common inference questions about quantitative variables concern *mean* responses. If the response variable is categorical, inference most often concerns the *proportion* of some category (call it a "success") among the responses.

The flowchart leads you to a specific test or confidence interval, indicated by a formula at the end of each path. The formula is just an aid to guide you toward the *Solve* and *Conclude* steps. You (or your technology) will use the formula as part of the *Solve* step, but don't forget that you must do more.

■ *Are the conditions for this procedure met?* Can you act as if the data come from a random sample or randomized comparative experiment? Does data analysis show extreme outliers or strong skewness that forbid use of inference based on Normality? Do you have enough observations for your intended procedure?

■ *Do your data come from an experiment or from an observational study?* The details of inference methods are the same for both. But the design of the study determines what conclusions you can reach, because experiments give much better evidence that an effect uncovered by inference can be explained by direct causation.

You may ask, as you study the Statistics in Summary flowchart, "What if I have an experiment comparing four treatments, or samples from three populations?" The flowchart allows only one or two, not three or four or more. Be patient: methods for comparing more than two means or proportions, as well as some other settings for inference, appear in Part IV.

How many was that?

Good causes often breed bad statistics. An advocacy group claims, without much evidence, that 150,000 Americans suffer from the eating disorder anorexia nervosa. Soon someone misunderstands and says that 150,000 people *die* from anorexia nervosa each year. This wild number gets repeated in countless books and articles. It really is a wild number: only about 55,000 women aged 15 to 44 (the main group affected) die of *all causes* each year.

PART III   SUMMARY

Here are the most important skills you should have acquired from reading Chapters 17 to 20.

A. Recognition

1. Recognize when a problem requires inference about population means (quantitative response variable) or population proportions (usually categorical response variable).

2. Recognize from the design of a study whether one-sample, matched pairs, or two-sample procedures are needed.

3. Based on recognizing the problem setting, choose among the one- and two-sample t procedures for means and the one- and two-sample z procedures for proportions.

B. Inference about One Mean

1. Verify that the t procedures are appropriate in a particular setting. Check the study design and the distribution of the data and take advantage of robustness against lack of Normality.

2. Recognize when poor study design, outliers, or a small sample from a skewed distribution make the t procedures risky.

3. Use the one-sample t procedure to obtain a confidence interval at a stated level of confidence for the mean μ of a population.

4. Carry out a one-sample t test for the hypothesis that a population mean μ has a specified value against either a one-sided or a two-sided alternative. Use software to find the P-value or Table C to get an approximate value.

5. Recognize matched pairs data and use the t procedures to obtain confidence intervals and to perform tests of significance for such data.

C. Comparing Two Means

1. Verify that the two-sample t procedures are appropriate in a particular setting. Check the study design and the distribution of the data and take advantage of robustness against lack of Normality.

2. Give a confidence interval for the difference between two means. Use software if you have it. Use the two-sample t statistic with conservative degrees of freedom and Table C if you do not have statistical software.

3. Test the hypothesis that two populations have equal means against either a one-sided or a two-sided alternative. Use software if you have it. Use the two-sample t test with conservative degrees of freedom and Table C if you do not have statistical software.

4. Know that procedures for comparing the standard deviations of two Normal populations are available, but that these procedures are risky because they are not at all robust against non-Normal distributions.

D. Inference about One Proportion

1. Verify that you can safely use either the large-sample or the plus four z procedures in a particular setting. Check the study design and the guidelines for sample size.

2. Use the large-sample z procedure to give a confidence interval for a population proportion p. Understand that the true confidence level may be substantially less than you ask for unless the sample is very large and the true p is not close to 0 or 1.

3. Use the plus four modification of the z procedure to give a confidence interval for p that is accurate even for small samples and for any value of p.

4. Use the z statistic to carry out a test of significance for the hypothesis H_0: $p = p_0$ about a population proportion p against either a one-sided or a two-sided alternative. Use software or Table A to find the P-value, or Table C to get an approximate value.

E. Comparing Two Proportions

1. Verify that you can safely use either the large-sample or the plus four z procedures in a particular setting. Check the study design and the guidelines for sample sizes.

2. Use the large-sample z procedure to give a confidence interval for the difference $p_1 - p_2$ between proportions in two populations based on independent samples from the populations. Understand that the true confidence level may be less than you ask for unless the samples are quite large.

3. Use the plus four modification of the z procedure to give a confidence interval for $p_1 - p_2$ that is accurate even for very small samples and for any values of p_1 and p_2.

4. Use a z statistic to test the hypothesis H_0: $p_1 = p_2$ that proportions in two distinct populations are equal. Use software or Table A to find the P-value, or Table C to get an approximate value.

R E V I E W　　 E X E R C I S E S

Review exercises are short and straightforward exercises that help you solidify the basic ideas and skills from Chapters 17 to 20.

In review exercises that call for inference, you can assume that the conditions for inference are met. For tests of significance, use Table C to approximate P-values unless you use software that reports the P-value. For confidence intervals for proportions, we recommend the plus four procedures unless the sample sizes are very large. Be sure to state your conclusions in plain language that identifies the populations to which they apply.

21.1 **Wikipedia.** A sample survey of 1497 adult Internet users found that 36% consult the online collaborative encyclopedia Wikipedia.[1] Give a 95% confidence interval for the proportion of all adult Internet users who refer to Wikipedia.

21.2 Game players. A government survey randomly selected 6889 female high school students and 7028 male high school students.[2] Of these students, 1020 females and 1926 males played video or computer games for 3 or more hours a day. Use a 99% confidence interval to estimate the proportion of all male high school students who play games at least 3 hours a day.

21.3 Spinning euros. When the new euro coins were introduced throughout Europe in 2002, curious people tried all sorts of things. Two Polish mathematicians spun a Belgian euro (one side of the coin has a different design for each country) 250 times. They got 140 heads. Newspapers reported this result widely. Is it significant evidence that the coin is not balanced when spun?

21.4 Game players, female and male. Use the information in Exercise 21.2 to answer these questions.

(a) What are the sample proportions of females and males who played games at least 3 hours a day?

(b) The samples are large and the sample proportions are quite different, so the difference is sure to be highly significant. Give a 99% confidence interval for the difference between the proportions of all male and female high school students who play games at least 3 hours a day.

Matthias Kulka/CORBIS

21.5 Men and muscle. Ask young men to estimate their own degree of body muscle by choosing from a set of 100 photos. Then ask them to choose what they think women prefer. The researchers know the actual degree of muscle, measured as kilograms per square meter of fat-free mass, for each of the photos. They can therefore measure the difference between what a subject thinks women prefer and the subject's own self-image. Call this difference the "muscle gap." Here are summary statistics for the muscle gap from two samples, one of American and European young men and the other of Chinese young men from Taiwan:[3]

Group	n	$\bar{x}$	s
American/European	200	2.35	2.5
Chinese	55	1.20	3.2

Give a 95% confidence interval for the mean size of the muscle gap for all American and European young men. On the average, men think they need this much more muscle to match what women prefer.

21.6 Are Chinese men different? Continue to use the data in Exercise 21.5. Is there a significant difference between the mean size of the muscle gap for American/European men and Chinese men?

21.7 Butterflies mating. Here's how butterflies mate: a male passes to a female a packet of sperm called a spermatophore. Females may mate several times. Will they remate sooner if the first spermatophore they receive is small? Among 20 females who received a large spermatophore (greater than 25 milligrams), the mean time to the next mating was 5.15 days, with standard deviation 0.18 day. For 21 females who received a small spermatophore (about 7 milligrams), the mean was 4.33 days and the standard deviation was 0.31 day.[4] Is the observed difference in means statistically significant?

Listening to rap. The Black Youth Project of the University of Chicago interviewed random samples of black, Hispanic, and white young people aged 15 to 25. We can consider this a stratified sample or three separate random samples of 634 blacks, 314 Hispanics, and 567 whites. The survey found that 58% of black youth listen to rap music every day, compared with 45% of Hispanics and 23% of whites. But attitudes were quite similar in the three groups. For example, 72% of blacks, 72% of Hispanics, and 68% of whites agreed that "rap music videos contain too many references to sex."[5] Exercises 21.8 to 21.10 are based on this study.

21.8 Black young people listening to rap. Give a 90% confidence interval for the proportion of all black young people who listen to rap every day.

21.9 Listening to rap: Hispanics and whites. Give a 90% confidence interval for the difference between the proportions of all Hispanic and all white young people who listen to rap every day.

21.10 Too much sex in rap music? Is there a significant difference between the proportions of black and white young people who think that rap videos contain too much sex?

21.11 Mouse endurance. A study of the inheritance of speed and endurance in mice found a trade-off between these two characteristics, both of which help mice survive. To test endurance, mice were made to swim in a bucket with a weight attached to their tails. (The mice were rescued when exhausted.) Here are data on endurance in minutes for female and male mice:[6]

Group	n	Mean	Standard Deviation
Female	162	11.4	26.09
Male	135	6.7	6.69

(a) Both sets of endurance data are skewed to the right. Why are t procedures nonetheless reasonably accurate for these data?

(b) Do the data show that female mice have significantly higher endurance on the average than male mice?

21.12 Female mouse endurance. Use the information in the previous exercise to give a 95% confidence interval for the mean endurance of female mice swimming.

21.13 More on mouse endurance. Use the information in Exercise 21.11 to give a 95% confidence interval for the mean difference (female minus male) in endurance times.

The National Assessment of Educational Progress (NAEP) includes a "long-term trend" study that tracks reading and mathematics skills over time in a way that allows comparisons between results from different years. Exercises 21.14 to 22.18 are based on information on 17-year-old students from the report on the latest long-term trend study, carried out in 2004.[7] The NAEP sample used a multistage design, but the overall effect is quite similar to an SRS of 17-year-olds who are still in school.

21.14 Better-educated parents. In the 1978 sample of 17,554 students, 5617 had at least one parent who was a college graduate. In the 2004 sample of 2158 students, 1014 had at least one college graduate parent. Give a 99% confidence interval for the increase in the proportion of students with a college graduate parent between 1978 and 2004.

21.15 College-educated parents. Use the information in the previous exercise to give a 99% confidence interval for the proportion of all students in 2004 who had at least one parent who graduated from college. (The sample excludes 17-year-olds who had dropped out of school, so your estimate is valid for students but is probably too high for all 17-year-olds.)

21.16 The effect of parents' education. The mean NAEP mathematics score (on a scale of 0 to 500) for the 1014 students in the 2004 sample with at least one parent who graduated from college was 317, with standard deviation 28.6. The 2004 sample contained 410 students whose parents' highest level of education was high school graduate. The mean score for these students was 295, with standard deviation 22.3. Is there a significant difference in mean score between these two groups of students? Estimate the size of the difference in the entire student population (use 95% confidence).

21.17 Students with college-educated parents. Use the information in the previous exercise to give a 95% confidence interval for the mean NAEP mathematics score of all 17-year-old students with at least one parent who graduated from college.

21.18 Men versus women. The 2004 NAEP sample contained 1122 female students and 1036 male students. The women had a mean mathematics score (on a scale of 0 to 500) of 305, with standard error 0.9. The male mean was 308, with standard error 1.0. Is there evidence that the mean mathematics scores of men and women differ in the population of all 17-year-old students?

21.19 Very-low-birth-weight babies. Starting in the 1970s, medical technology allowed babies with very low birth weight (VLBW, less than 1500 grams, about 3.3 pounds) to survive without major handicaps. It was noticed that these children nonetheless had difficulties in school and as adults. A long-term study has followed 242 VLBW babies to age 20 years, along with a control group of 233 babies from the same population who had normal birth weight.[8]

(a) Is this an experiment or an observational study? Why?

(b) At age 20, 179 of the VLBW group and 193 of the control group had graduated from high school. Is the graduation rate among the VLBW group significantly lower than for the normal-birth-weight controls?

Knut Mueller/Peter Arnold

21.20 Very low birth weight and IQ. IQ scores were available for 113 men in the VLBW group. The mean IQ was 87.6, and the standard deviation was 15.1. The 106 men in the control group had mean IQ 94.7, with standard deviation 14.9. Is there good evidence that mean IQ is lower among VLBW men than among controls from similar backgrounds?

21.21 Very low birth weight, drug use, and IQ. Of the 126 women in the VLBW group, 37 said they had used illegal drugs; 52 of the 124 control group women had done so. The IQ scores for the VLBW women had mean 86.2 (standard deviation 13.4), and the normal-birth-weight controls had mean IQ 89.8 (standard deviation 14.0). Is there a statistically significant difference between the two groups in either proportion using drugs or mean IQ?

21.22 Favoritism for college athletes? *Sports Illustrated* surveyed a random sample of 757 Division I college athletes in 36 sports. One question asked was "Have you ever received preferential treatment from a professor because of your status as an athlete?" Of the athletes polled, 225 said "Yes." Give a 99% confidence interval for the proportion of all Division I college athletes who believe they have received preferential treatment.

21.23 Breast cancer. The Women's Health Initiative is a randomized, controlled clinical trial designed to see if a low-fat diet reduces the incidence of breast cancer. In all, 19,541 women were assigned at random to a low-fat diet and a control group of 29,294 women were assigned to a normal diet. All the subjects were between ages 50 and 79 and had no prior breast cancer. After 8 years, 655 of the women in the low-fat group and 1072 of the women in the control group had developed breast cancer.[9] Does this clinical trial give evidence that a low-fat diet reduces breast cancer?

21.24 Do fruit flies sleep? Mammals and birds sleep. Fruit flies show a daily cycle of rest and activity, but does the rest qualify as sleep? Researchers looking at brain activity and behavior finally concluded that fruit flies do sleep. A small part of the study used an infrared motion sensor to see if flies moved in response to vibrations. Here are results for low levels of vibration:[10]

	Response to Vibration?	
	No	**Yes**
Fly was walking	10	54
Fly was resting	28	4

Analyze these results. Is there good reason to think that resting flies respond differently than flies that are walking? (That's a sign that the resting flies may actually be sleeping.)

21.25 Cholesterol in dogs. High levels of cholesterol in the blood are not healthy in either humans or dogs. Because a diet rich in saturated fats raises the cholesterol level, it is plausible that dogs owned as pets have higher cholesterol levels than dogs owned by a veterinary research clinic. "Normal" levels of cholesterol based on the clinic's dogs would then be misleading. A clinic compared healthy dogs it owned with healthy pets brought to the clinic to be neutered. The summary statistics for blood cholesterol levels (milligrams per deciliter of blood) appear below.[11]

Group	n	$\bar{x}$	s
Pets	26	193	68
Clinic	23	174	44

Is there strong evidence that pets have a higher mean cholesterol level than clinic dogs?

21.26 Pets versus clinic dogs. Using the information in the previous exercise, give a 95% confidence interval for the difference in mean cholesterol levels between pets and clinic dogs.

21.27 Cholesterol in pets. Continue your work with the information in Exercise 21.25. Give a 95% confidence interval for the mean cholesterol level in pets.

21.28 Conditions for inference. What conditions must be satisfied to justify the procedures you used in Exercise 21.25? In Exercise 21.26? In Exercise 21.27? Assuming that the cholesterol measurements have no outliers and are not strongly skewed, what is the chief threat to the validity of the results of this study?

Choosing an inference procedure. *In each of Exercises 21.29 to 21.35, say which type of inference procedure from the Statistics in Summary flowchart (page 543) you would use, or explain why none of these procedures fits the problem. You do not need to carry out any procedures.*

21.29 Driving too fast. How seriously do people view speeding in comparison with other annoying behaviors? A large random sample of adults was asked to rate a number of behaviors on a scale of 1 (no problem at all) to 5 (very severe problem). Do speeding drivers get a higher average rating than noisy neighbors?

21.30 Preventing drowning. Drowning in bathtubs is a major cause of death in children less than 5 years old. A random sample of parents was asked many questions related to bathtub safety. Overall, 85% of the sample said they used baby bathtubs for infants. Estimate the percent of all parents of young children who use baby bathtubs.

21.31 Acid rain? You have data on rainwater collected at 16 locations in the Adirondack Mountains of New York State. One measurement is the acidity of the water, measured by pH on a scale of 0 to 14 (the pH of distilled water is 7.0). Estimate the average acidity of rainwater in the Adirondacks.

21.32 Athletes' salaries. Looking online, you find the salaries of all 27 players for the Chicago Cubs as of opening day of the 2008 baseball season. The team total was $118.6 million, seventh highest in the major leagues. Estimate the average salary of the Cubs players.

21.33 Looking back on love. How do young adults look back on adolescent romance? Investigators interviewed 40 couples in their midtwenties. The female and male partners were interviewed separately. Each was asked about his or her current relationship and also about a romantic relationship that lasted at least two months when they were aged 15 or 16. One response variable was a measure on a numerical scale of how much the attractiveness of the adolescent partner mattered. You want to compare the men and women on this measure.

21.34 Dropping out. You have data from interviews with a random sample of students who failed to graduate from your college in 7 years and also from a random sample of students who entered at the same time and did graduate. You will use these data to

Kyodo via AP Images

(a) estimate the percent of dropouts who transferred to another school.

(b) compare the first-year grade point averages of dropouts and graduates.

(c) compare the percents of students from rural backgrounds among dropouts and graduates.

21.35 Preventing AIDS through education. The Multisite HIV Prevention Trial was a randomized comparative experiment to compare the effects of twice-weekly small-group AIDS discussion sessions (the treatment) with a single one-hour session (the control). Compare the effects of treatment and control on each of the following response variables:

(a) A subject does or does not use condoms 6 months after the education sessions.

(b) The number of unprotected intercourse acts by a subject between 4 and 8 months after the sessions.

(c) A subject is or is not infected with a sexually transmitted disease 6 months after the sessions.

S U P P L E M E N T A R Y   E X E R C I S E S

Supplementary exercises apply the skills you have learned in ways that require more thought or more use of technology. Some of these exercises start from actual data rather than from data summaries. Many of these exercises ask you to follow the **Plan, Solve,** *and* **Conclude** *steps of the four step process. Remember that the* **Solve** *step includes checking the conditions for the inference you plan.*

21.36 Do you have confidence? A report of a survey distributed to randomly selected email addresses at a large university says: "We have collected 427 responses from our sample of 2,100 as of April 30, 2004. This number of responses is large enough to achieve a 95% confidence interval with ±5% margin of sampling error in generalizing the results to our study population."[12] Why would you be reluctant to trust a confidence interval based on these data?

21.37 Pain from a rubber hand. People who have had limbs amputated sometimes feel sensations from the limb that is no longer there. To study this effect, psychologists asked subjects to place their right arm on a table. They then put a rubber arm and hand next to the real arm, with a high partition arranged so that the subject could see only the rubber arm. After a few minutes during which the real and fake hand were both tapped by an experimenter, the subjects felt the taps coming from the location of the rubber hand they could see, not from the real hand they couldn't see. Now the experiment begins: bend back a finger of the fake hand in a way that would cause pain, while merely lifting a real finger. Do electrical measurements show a response to pain? Because there would be some response from the surprise of being touched, a control treatment delayed the touch to the real hand to separate surprise from "pain." Here are summary data for 16 undergraduate students who were subject to both stimuli:[13]

Stimulus	$\bar{x}$	s
Treatment	0.39	0.28
Control	0.18	0.20

(a) Which t procedures are correct for comparing the mean response to treatment and control: one-sample, matched pairs, or two-sample?

(b) The data summary given is not enough information to carry out the correct t procedures. Explain why not.

21.38 Monkeys and music. Humans generally prefer music to silence. What about monkeys? Allow a tamarin monkey to enter a V-shaped cage with food in both arms of the V. After the monkey eats the food, which arm will it prefer? The monkey's location determines what it hears, a lullaby played by a flute in one arm and silence in the other. Each of 4 monkeys was tested 6 times, on different days and with the music arm alternating between left and right (in case a monkey prefers one direction). The monkeys chose silence for about 65% of their time in the cage. The researchers reported a one-sample t test for the mean percent of time spent in the music arm, $H_0: \mu = 50\%$ against the two-sided alternative, $t = -5.26$, df $= 23$, $P < 0.0001$.[14]

Although the result is interesting, the statistical analysis is not correct. The degrees of freedom df $= 23$ show that the researchers assumed that they had 24 independent observations. Explain why the results of the 24 trials are not independent.

21.39 Drug-detecting rats? Dogs are big and expensive. Rats are small and cheap. Might rats be trained to replace dogs in sniffing out illegal drugs? A first study of this idea trained rats to rear up on their hind legs when they smelled simulated cocaine. To see how well rats performed after training, they were let loose on a surface with many cups sunk in it, one of which contained simulated cocaine. Four out of six trained rats succeeded in 80 out of 80 trials.[15] How should we estimate the long-term success rate p of a rat that succeeds in every one of 80 trials?

Holt Studios International/Alamy

(a) What is the rat's sample proportion $\hat{p}$? What is the large-sample 95% confidence interval for p? It's not plausible that the rat will *always* be successful, as this interval says.

(b) Find the plus four estimate $\tilde{p}$ and the plus four 95% confidence interval for p. These results are more reasonable.

21.40 A new vaccine. In 2006, the pharmaceutical company Merck released a vaccine named Gardasil for human papilloma virus, the most common cause of cervical cancer in young women. The Merck Web site gives results from "four placebo-controlled, double-blind, randomized clinical studies" with women 16 to 26 years of age, as follows:[16]

	n	Cervical Cancer	n	Genital Warts
Gardasil	8487	0	7897	1
Placebo	8460	32	7899	91

(a) Give a 99% confidence interval for the difference in the proportions of young women who develop cervical cancer with and without the vaccine.

(b) Do the same for the proportions who develop genital warts.

(c) What do you conclude about the overall effectiveness of the vaccine?

21.41 Starting to talk. At what age do infants speak their first word of English? Here are data on 20 children (ages in months):[17]

$$15 \quad 26 \quad 10 \quad 9 \quad 15 \quad 20 \quad 18 \quad 11 \quad 8 \quad 20$$
$$7 \quad 9 \quad 10 \quad 11 \quad 11 \quad 10 \quad 12 \quad 17 \quad 11 \quad 10$$

(In fact, the sample contained one more child, who began to speak at 42 months. Child development experts consider this abnormally late, so we dropped the outlier to get a sample of "normal" children. The investigators are willing to treat these data as an SRS.) Is there good evidence that the mean age at first word among all normal children is greater than one year?

Kelly Kalhoefer/Jupiter Images

21.42 Fertilizing a tropical plant. Bromeliads are tropical flowering plants. Many are epiphytes that attach to trees and obtain moisture and nutrients from air and rain. Their leaf bases form cups that collect water and are home to the larvae of many insects. In an experiment in Costa Rica, Jacqueline Ngai and Diane Srivastava studied whether added nitrogen increases the productivity of bromeliad plants. Bromeliads were randomly assigned to nitrogen or control groups. Here are data on the number of new leaves produced over a 7-month period:[18]

Control	11	13	16	15	15	11	12	
Nitrogen	15	14	15	16	17	18	17	13

Is there evidence that adding nitrogen increases the mean number of new leaves formed?

21.43 Starting to talk, continued. Use the data in Exercise 21.41 to give a 90% confidence interval for the mean age at which children speak their first word.

21.44 Dyeing fabrics. Different fabrics respond differently when dyed. This matters to clothing manufacturers, who want the color of the fabric to be just right. A researcher dyed fabrics made of cotton and of ramie with the same "procion blue" dye applied in the same way. Then she used a colorimeter to measure the lightness of the color on a scale in which black is 0 and white is 100. Here are the data for 8 pieces of each fabric:[19]

Cotton	48.82	48.88	48.98	49.04	48.68	49.34	48.75	49.12
Ramie	41.72	41.83	42.05	41.44	41.27	42.27	41.12	41.49

Is there a significant difference between the fabrics? Which fabric is darker when dyed in this way?

21.45 More on dyeing fabrics. The color of a fabric depends on the dye used and also on how the dye is applied. This matters to clothing manufacturers, who want the color of the fabric to be just right. The study discussed in the previous exercise went on to dye fabric made of ramie with the same "procion blue" dye applied in two different ways. Here are the lightness scores for 8 pieces of identical fabric dyed in each way:

Method B	40.98	40.88	41.30	41.28	41.66	41.50	41.39	41.27
Method C	42.30	42.20	42.65	42.43	42.50	42.28	43.13	42.45

(a) This is a randomized comparative experiment. Outline the design.

(b) A clothing manufacturer wants to know which method gives the darker color (lower lightness score). Use sample means to answer this question. Is the difference between the two sample means statistically significant? Can you tell from just the P-value whether the difference is large enough to be important in practice?

21.46 Do parents matter? A professor asked her sophomore students, "Does either of your parents allow you to drink alcohol around him or her?" and "How many drinks do you typically have per session? (A drink is defined as one 12 oz beer, one 4 oz glass of wine, or one 1 oz shot of liquor)." Table 21.1 contains the responses of the

TABLE 21.1 Drinks per session by female students

				Parent Allows Student to Drink								
2.5	1	2.5	3	1	3	3	3	2.5	2.5	3.5	5	2
7	7	6.5	4	8	6	6	3	6	3	4	7	5
3.5	2	1	5	3	3	6	4	2	7	5	8	1
6	5	2.5	3	4.5	9	5	4	4	3	4	6	4
5	1	5	3	10	7	4	4	4	4	2	2.5	2.5

				Parent Does Not Allow Student to Drink								
9	3.5	3	5	1	1	3	4	4	3	6	5	3
8	4	4	5	7	7	3.5	3	10	4	9	2	7
4	3	1										

female students who are not abstainers.[20] The sample is all students in one large sophomore-level class. The class is popular, so we are tentatively willing to regard its members as an SRS of sophomore students at this college. Does the behavior of parents make a significant difference in how many drinks students have on the average?

21.47 Parents' behavior. We wonder what proportion of female students have at least one parent who allows them to drink around him or her. Table 21.1 contains information about a sample of 94 students. Use this sample to give a 95% confidence interval for this proportion.

21.48 Diabetic mice. The body's natural electrical field helps wounds heal. If diabetes changes this field, that might explain why people with diabetes heal slowly. A study of this idea compared normal mice and mice bred to spontaneously develop diabetes. The investigators attached sensors to the right hip and front feet of the mice and measured the difference in electrical potential (millivolts) between these locations. Here are the data:[21]

	Diabetic Mice						**Normal Mice**			
14.70	13.60	7.40	1.05	10.55	16.40	13.80	9.10	4.95	7.70	9.40
10.00	22.60	15.20	19.60	17.25	18.40	7.20	10.00	14.55	13.30	6.65
9.80	11.70	14.85	14.45	18.25	10.15	9.50	10.40	7.75	8.70	8.85
10.85	10.30	10.45	8.55	8.85	19.20	8.40	8.55	12.60		

(a) Make a stemplot of each sample of potentials. There is a low outlier in the diabetic group. Does it appear that potentials in the two groups differ in a systematic way?

(b) Is there significant evidence of a difference in mean potentials between the two groups?

(c) Repeat your inference without the outlier. Does the outlier affect your conclusion?

21.49 Keeping crackers from breaking. We don't like to find broken crackers when we open the package. How can makers reduce breaking? One idea is to microwave the crackers for 30 seconds right after baking them. Analyze the following results from two experiments intended to examine this idea.[22] Does microwaving significantly improve indicators of future breaking? How large is the improvement? What do you conclude about the idea of microwaving crackers?

(a) The experimenter randomly assigned 65 newly baked crackers to be microwaved and another 65 to a control group that is not microwaved. Fourteen days after baking, 3 of the 65 microwaved crackers and 57 of the 65 crackers in the control group showed visible checking, which is the starting point for breaks.

(b) The experimenter randomly assigned 20 crackers to be microwaved and another 20 to a control group. After 14 days, he broke the crackers. Here are summaries of the pressure needed to break them, in pounds per square inch:

	Microwave	Control
Mean	139.6	77.0
Standard deviation	33.6	22.6

2006 Bill Watkins/AlaskaStock.com

21.50 Falling through the ice. Table 7.3 (page 192) gives the dates on which a wooden tripod fell through the ice of the Tanana River in Alaska, thus deciding the winner of the Nenana Ice Classic contest, for the years 1917 to 2007. Give a 95% confidence interval for the mean date on which the tripod falls through the ice. After calculating the interval in the scale used in the table (days from April 20, which is Day 1), translate your result into calendar dates and hours within the dates. (Each hour is 1/24, or 0.042, of a day.)

21.51 A case for the Supreme Court. In 1986, a Texas jury found a black man guilty of murder. The prosecutors had used "peremptory challenges" to remove 10 of the 11 blacks and 4 of the 31 whites in the pool from which the jury was chosen.[23] The law says that there must be a plausible reason (that is, a reason other than race) for different treatment of blacks and whites in the jury pool. When the case reached the Supreme Court 17 years later, the Court said that "happenstance is unlikely to produce this disparity." Explain why the methods we know can't be safely used to do the inference that lies behind the Court's finding that chance is unlikely to produce so large a black-white difference.

21.52 Mouse genes. A study of genetic influences on diabetes compared normal mice with similar mice genetically altered to remove a gene called $aP2$. Mice of both types were allowed to become obese by eating a high-fat diet. The researchers then measured the levels of insulin and glucose in their blood plasma. Here are some excerpts from their findings.[24] The normal mice are called "wild-type" and the altered mice are called "$aP2^{-/-}$."

Each value is the mean ± SEM of measurements on at least 10 mice. Mean values of each plasma component are compared between $aP2^{-/-}$ mice and wild-type controls by Student's t test (*$P < 0.05$ and **$P < 0.005$).

Parameter	Wild Type	$aP2^{-/-}$
Insulin (ng/ml)	5.9 ± 0.9	$0.75 \pm 0.2^{**}$
Glucose (mg/dl)	230 ± 25	$150 \pm 17^{*}$

Despite much greater circulating amounts of insulin, the wild-type mice had higher blood glucose than the aP2$^{-/-}$ animals. These results indicate that the absence of aP2 interferes with the development of dietary obesity-induced insulin resistance.

Other biologists are supposed to understand the statistics reported so tersely.

(a)   What does "SEM" mean? What is the expression for SEM based on n, $\bar{x}$, and s from a sample?

(b)   Which of the tests we have studied did the researchers apply?

(c)   Explain to a biologist who knows no statistics what $P < 0.05$ and $P < 0.005$ mean. Which is stronger evidence of a difference between the two types of mice?

21.53 Mouse genes, continued. The report quoted in the previous exercise says only that the sample sizes were "at least 10." Suppose that the results are based on exactly 10 mice of each type. Use the values in the table to find $\bar{x}$ and s for the insulin concentrations in the two types of mice. Carry out a test to assess the significance of the difference in mean insulin concentration. Does your P-value confirm the claim in the report that $P < 0.005$?

Inference about Relationships

Statistical inference offers more methods than anyone can know well, as a glance at the offerings of any large statistical software package demonstrates. In an introductory text, we must be selective. Parts I to III have laid a foundation for understanding statistics:

- The nature and purpose of data analysis.
- The central ideas of designs for data production.
- The reasoning behind confidence intervals and significance tests.
- Experience applying these ideas in practice.

Each of the three chapters of Part IV offers an introduction to a more advanced topic in statistical inference. You may choose to read any or all of them, in any order.

What makes a statistical method "more advanced"? More complex data, for one thing. In Part III, we looked only at methods for inference about a single population parameter and for comparing two parameters. All of the chapters in Part IV present methods for studying relationships between two variables. In Chapter 22, both variables are categorical, with data given as a two-way table of counts of outcomes. Chapter 23 considers inference in the setting of regressing a response variable on an explanatory variable. This is an important type of relationship between two quantitative variables. In Chapter 24 we meet methods for comparing the mean response in more than two groups. Here, the explanatory variable (group) is categorical and the response variable is quantitative. These chapters together bring our knowledge of inference to the same point that our study of data analysis reached in Chapters 1 to 7.

With greater complexity comes greater reliance on technology. In these final three chapters you will more often be interpreting the output of statistical software or using software yourself. With effort, you can do the calculations needed in Chapter 22 with a basic calculator. In Chapters 23 and 24, the pain is too great and the contribution to learning too small. Fortunately, you can grasp the ideas without step-by-step arithmetic.

Another aspect of "more advanced" methods is new concepts and ideas. This is where we draw the line in deciding what statistical topics we can master in a first course. Part IV builds elaborate methods on the foundation we have laid without introducing fundamentally new concepts. You can see that statistical practice does need additional big ideas by reading the sections on "the problem of multiple comparisons" in Chapters 22 and 24. But the ideas you already know place you among the world's statistical sophisticates.

INFERENCE ABOUT RELATIONSHIPS

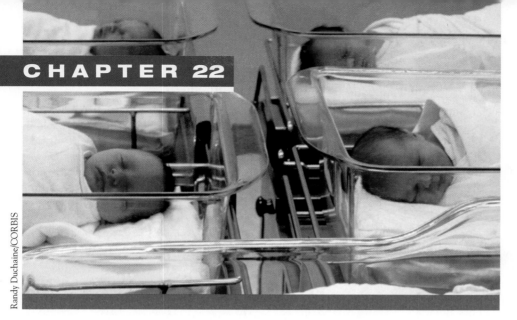

Randy Duchaine/CORBIS

Two Categorical Variables: The Chi-Square Test

The two-sample z procedures of Chapter 20 allow us to compare the proportions of successes in two groups, either two populations or two treatment groups in an experiment. In the first example in Chapter 20 (page 524), we compared young men and young women by looking at whether or not they lived with their parents. That is, we looked at a relationship between two categorical variables, gender (female or male) and "Where do you live?" (with parents or not). In fact, the data include three more outcomes for "Where do you live?": in another person's home, in your own place, and in group quarters such as a dormitory. When there are more than two outcomes, or when we want to compare more than two groups, we need a new statistical test. The new test addresses a general question: *is there a relationship between two categorical variables?*

Two-way tables

We saw in Chapter 6 that we can present data on two categorical variables in a **two-way table** of counts. That's our starting point. Let's continue our exploration of where college-age young people live.

two-way table

EXAMPLE 22.1 **Where do young people live?**

A sample survey asked a random sample of young adults, "Where do you live now? That is, where do you stay most often?" Table 22.1 is a two-way table of all 2984 people in the sample (both men and women) classified by their age and by where they lived.[1] Living arrangement is a categorical variable. Even though age is quantitative, the two-way table treats age as dividing young adults into four categories. Table 22.1 gives the counts for all 20 combinations of age and living arrangement. Each of the 20 counts occupies a **cell** of the table. ■

cell

TABLE 22.1 **Young adults by age and living arrangement**

| | AGE | | | | |
	19	20	21	22	TOTAL
Parents' home	324	378	337	318	1357
Another person's home	37	47	40	38	162
Your own place	116	279	372	487	1254
Group quarters	58	60	49	25	192
Other	5	2	3	9	19
Total	540	766	801	877	2984

As usual, we prepare for inference by first doing data analysis. Because we think that age helps explain where young people live, find the percents of people in each age group who have each living arrangement. The percents appear in Table 22.2. Each column adds to 100% (up to roundoff error) because we are looking at each age group separately. In the language of Chapter 6 (page 165), Table 22.2 shows the four *conditional distributions* of living arrangements given a specific age.

TABLE 22.2 **Percents of each age group who have each living arrangement (read down columns)**

| | AGE | | | |
	19	20	21	22
Parents' home	60.0	49.3	42.1	36.3
Another person's home	6.9	6.1	5.0	4.3
Your own place	21.5	36.4	46.4	55.5
Group quarters	10.7	7.8	6.1	2.9
Other	0.9	0.3	0.4	1.0
Total	100.0	99.9	100.0	100.0

Figure 22.1 is Minitab's bar graph comparing the four conditional distributions. The graph shows a strong relationship between age and living arrangement. As young adults age from 19 to 22, the percent living with their parents drops and the percent living in their own place rises. The percent living in group quarters also declines with age as college students move out of dormitories. Are these differences among the four age groups large enough to be statistically significant?

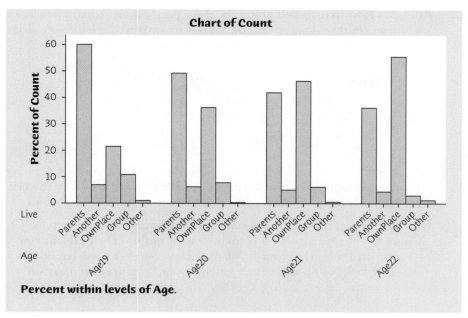

Chart of Count

Percent within levels of Age.

FIGURE 22.1

Minitab bar graph comparing the four conditional distributions of living arrangements given age, for Example 22.1.

APPLY YOUR KNOWLEDGE

22.1   Facebook at Penn State. The Pennsylvania State University has its main campus in University Park and more than 20 smaller "commonwealth campuses" around the state. The Penn State Division of Student Affairs polled a random sample of undergraduates about their use of online social networking. (The response rate was only about 20%, which casts some doubt on the usefulness of the data.) Facebook was the most popular site, with more than 80% of students having an account. Here is a comparison of Facebook use by undergraduates at the University Park and commonwealth campuses:[2]

Courtesy Pennsylvania State University

	University Park	Commonwealth
Do not use Facebook	68	248
Several times a month or less	55	76
At least once a week	215	157
At least once a day	640	394

(a)   What percent of University Park students fall in each Facebook category? What percent of commonwealth campus students fall in each category? Each column should add to 100% (up to roundoff error). These are the conditional distributions of Facebook use given campus setting.

(b)   Make a bar graph that compares the two conditional distributions. What are the most important differences in Facebook use between the two campus settings?

22.2 Attitudes toward recycled products. Some people think recycled products are lower in quality than other products, a fact that makes recycling less practical. Here are data on attitudes toward coffee filters made of recycled paper.[3]

	Think the Quality of the Recycled Product Is		
	Higher	Same	Lower
Buyers	20	7	9
Nonbuyers	29	25	43

(a) It appears that people who have bought the recycled filters have more positive opinions than those who have not. Give percents to back up this claim. Make a bar graph that compares your percents for buyers and nonbuyers.

(b) Association does not prove causation. Explain how buying recycled filters might improve a person's opinion of their quality. Then explain how the opinion a person holds might influence his or her decision to buy or not. You see that the cause-and-effect relationship might go in either direction.

The problem of multiple comparisons

The null hypothesis in Example 22.1 is that in the population of all American young adults there is *no difference* among the four distributions of living arrangements for people aged 19, 20, 21, and 22. If the null hypothesis is true, the differences in the sample are just accidents due to random selection of the sample. Put more generally, the null hypothesis is that there is *no relationship* between two categorical variables,

H_0: there is no relationship between age and where young people live

The alternative hypothesis says that there *is* a relationship but does not specify any particular kind of relationship,

H_a: there is some relationship between age and living arrangement

Any difference among the four distributions of living arrangements in the population of all young adults means that the null hypothesis is false and the alternative hypothesis is true. The alternative hypothesis is not one-sided or two-sided. We might call it "many-sided" because it allows any kind of difference.

With only the methods we already know, we might start by comparing the proportions of people aged 19 and 22 who live with their parents. We could similarly compare other pairs of proportions, ending up with many tests and many P-values. This is a bad idea. The P-values belong to each test separately, not to the collection of all the tests together. Think of the distinction between the probability that a basketball player makes a free throw and the probability that she makes all of her free throws in a game. *When we do many individual tests or confidence intervals, the individual P-values and confidence levels don't tell us how confident we can be in all of the inferences taken together.*

He started it!

A study of deaths in bar fights showed that in 90% of the cases, the person who died started the fight. You shouldn't believe this. If you killed someone in a fight, what would you say when the police ask you who started the fight? After all, dead men tell no tales.

Because of this, it's cheating to pick out one large difference from Table 22.2 and then test its significance as if it were the only comparison we had in mind. For example, the percents of people aged 19 and 22 who live with their parents are significantly different ($z = 8.92$, $P < 0.001$) if we make just this one comparison. But we could also pick a comparison that is not significant; for example, the proportions of people aged 21 and 22 who live in another person's home do not differ significantly ($z = 0.64$, $P = 0.522$). Individual comparisons can't tell us whether the four distributions, each with five outcomes, are significantly different.

The problem of how to do many comparisons at once with an overall measure of confidence in all our conclusions is common in statistics. This is the problem of **multiple comparisons.** Statistical methods for dealing with multiple comparisons usually have two steps:

multiple comparisons

1. An *overall test* to see if there is good evidence of *any* differences among the parameters that we want to compare.

2. A detailed *follow-up analysis* to decide which of the parameters differ and to estimate how large the differences are.

The overall test, though more complex than the tests we met earlier, is reasonably straightforward. The follow-up analysis can be quite elaborate. We will concentrate on the overall test and use data analysis to describe in detail the nature of the differences.

APPLY YOUR KNOWLEDGE

22.3 **Facebook at Penn State.** In the setting of Exercise 22.1, we might do several significance tests to compare University Park with the commonwealth campuses.

(a) Is there a significant difference between the proportions of students in the two locations who do not use Facebook? Give the P-value.

(b) Is there a significant difference between the proportions of students in the two locations who are in the "At least once a week" category? Give the P-value.

(c) Explain clearly why P-values for individual outcomes like these can't tell us whether the two distributions for all four outcomes in the two locations differ significantly.

22.4 **Is astrology scientific?** The University of Chicago's General Social Survey (GSS) is the nation's most important social science sample survey. The GSS asked a random sample of adults their opinion about whether astrology is very scientific, sort of scientific, or not at all scientific. Here is a two-way table of counts for people in the sample who had three levels of higher education:[4]

	Degree Held		
	Associate's	Bachelor's	Master's
Not at all scientific	169	256	114
Very or sort of scientific	65	65	18

Courtesy University of Chicago General Social Survey

(a) Give three 95% confidence intervals, for the percents of people with each degree who think that astrology is not at all scientific.

(b) Explain clearly why we are *not* 95% confident that *all three* of these intervals capture their respective population proportions.

Expected counts in two-way tables

Our general null hypothesis H_0 is that there is *no relationship* between the two categorical variables that label the rows and columns of a two-way table. To test H_0, we compare the observed counts in the table with the *expected counts*, the counts we would expect—except for random variation—if H_0 were true. If the observed counts are far from the expected counts, that is evidence against H_0. It is easy to find the expected counts.

EXPECTED COUNTS

The **expected count** in any cell of a two-way table when H_0 is true is

$$\text{expected count} = \frac{\text{row total} \times \text{column total}}{\text{table total}}$$

EXAMPLE 22.2 Where young people live: expected counts

Let's find the expected counts for the study of where young people live. Look back at the two-way table of counts, Table 22.1. That table includes the row and column totals. The expected count of 19-year-olds who live in their parents' home is

$$\frac{\text{row 1 total} \times \text{column 1 total}}{\text{table total}} = \frac{(1357)(540)}{2984} = 245.57$$

The expected count of 22-year-olds who live with their parents is

$$\frac{\text{row 1 total} \times \text{column 4 total}}{\text{table total}} = \frac{(1357)(877)}{2984} = 398.82$$

The actual counts are 324 and 318. More younger people and fewer older people live with their parents than we would expect if there were no relationship between age and living arrangement. Table 22.3 shows all 20 expected counts.

TABLE 22.3 Young adults by age and living arrangement: expected cell counts

| | AGE | | | | |
	19	20	21	22	TOTAL
Parents' home	245.57	348.35	364.26	398.82	1357
Another person's home	29.32	41.59	43.49	47.61	162
Your own place	226.93	321.90	336.61	368.55	1254
Group quarters	34.75	49.29	51.54	56.43	192
Other	3.44	4.88	5.10	5.58	19
Total	540	766	801	877	2984

As this table shows, *the expected counts have exactly the same row and column totals (up to roundoff error) as the observed counts.* That's a good way to check your work. Comparing the actual counts (Table 22.1) and the expected counts (Table 22.3) shows in what ways the data diverge from the null hypothesis. ■

Why the formula works Where does the formula for an expected count come from? Think of a basketball player who makes 70% of her free throws in the long run. If she shoots 10 free throws in a game, we expect her to make 70% of them, or 7 of the 10. Of course, she won't make exactly 7 every time she shoots 10 free throws in a game. There is chance variation from game to game. But in the long run, 7 of 10 is what we expect. In more formal language, if we have n independent tries and the probability of a success on each try is p, we expect np successes.

Now go back to the count of 19-year-olds living in their parents' home. The proportion of all 2984 subjects who live with their parents is

$$\frac{\text{count of successes}}{\text{table total}} = \frac{\text{row 1 total}}{\text{table total}} = \frac{1357}{2984}$$

Think of this as p, the overall proportion of successes. If H_0 is true, we expect (except for random variation) this same proportion of successes in all four age groups. So the expected count of successes among the 540 19-year-olds is

$$np = (540)\left(\frac{1357}{2984}\right) = 245.57$$

That's the formula in the Expected Counts box.

APPLY YOUR KNOWLEDGE

22.5 Facebook at Penn State. The two-way table in Exercise 22.1 displays data on use of Facebook by two groups of Penn State students. It's clear that nonusers are much more frequent at the commonwealth campuses. Let's look just at students who have Facebook accounts:

Use Facebook	University Park	Commonwealth
Several times a month or less	55	76
At least once a week	215	157
At least once a day	640	394
Total Facebook users	910	627

The null hypothesis is that there is no relationship between campus and Facebook use.

(a) If this hypothesis is true, what are the expected counts for Facebook use among commonwealth campus students? This is one column of the two-way table of expected counts. Find the column total and verify that it agrees with the column total for the observed counts.

(b) Commonwealth campus students as a group are older and more likely to be married and employed than University Park students. What does comparing the observed and expected counts in this column show about Facebook use by these students?

22.6 **Attitudes toward recycled products.** Exercise 22.2 describes a comparison of the attitudes of people who do and don't buy coffee filters made of recycled paper. The null hypothesis "no relationship" says that in the population of all consumers, the proportions who hold each attitude are the same for buyers and nonbuyers.

(a) Find the expected cell counts if this hypothesis is true and display them in a two-way table. Add the row and column totals to your table and check that they agree with the totals for the observed counts.

(b) Are there any large deviations between the observed counts and the expected counts? What kind of relationship between the two variables do these deviations point to?

The chi-square test statistic

To test whether the observed differences among the four distributions of living arrangements given age are statistically significant, we compare the observed and expected counts. The test statistic that makes the comparison is the *chi-square statistic*.

> **CHI-SQUARE STATISTIC**
>
> The **chi-square statistic** is a measure of how far the observed counts in a two-way table are from the expected counts. The formula for the statistic is
>
> $$\chi^2 = \sum \frac{(\text{observed count} - \text{expected count})^2}{\text{expected count}}$$
>
> The sum is over all cells in the table.

As you might guess, the symbol χ in the box is the Greek letter chi. The chi-square statistic is a sum of terms, one for each cell in the table.

EXAMPLE 22.3 Where young people live: the test statistic

In the study of where young people live, 324 19-year-olds lived with their parents. The expected count for this cell is 245.57. So the term of the chi-square statistic from this cell is

$$\frac{(\text{observed count} - \text{expected count})^2}{\text{expected count}} = \frac{(324 - 245.57)^2}{245.57}$$

$$= \frac{6151.26}{245.57} = 25.05$$

The chi-square statistic χ^2 is the sum of 20 terms like this one. Here they are, arranged to match the layout of the two-way table:

$$\begin{aligned} \chi^2 &= 25.05 + 2.53 + 2.04 + 16.38 \\ &+ \ 2.01 + 0.71 + 0.28 + \ 1.94 \\ &+ 54.23 + 5.72 + 3.72 + 38.07 \\ &+ 15.56 + 2.33 + 0.13 + 17.51 \\ &+ \ 0.71 + 1.70 + 0.87 + \ 2.09 \\ &= 193.58 \end{aligned}$$

To find the value $\chi^2 = 193.58$, we had to calculate the 20 expected counts in Table 22.3 and then the 20 terms of the sum. Moreover, rounding each term to two decimal places creates roundoff error in the sum. Software is very handy in finding χ^2. ∎

Think of χ^2 as a measure of the distance of the observed counts from the expected counts. Like any distance, it is always zero or positive, and it is zero only when the observed counts are exactly equal to the expected counts. Large values of χ^2 are evidence against H_0 because they say that the observed counts are far from what we would expect if H_0 were true. *Although the alternative hypothesis H_a is many-sided, the chi-square test is one-sided* because any violation of H_0 tends to produce a large value of χ^2. Small values of χ^2 are not evidence against H_0.

Cell counts required for the chi-square test

The chi-square test, like the z procedures for comparing two proportions, is an approximate method that becomes more accurate as the counts in the cells of the table get larger. We must therefore check that the counts are large enough to allow us to trust the P-value. Fortunately, the chi-square approximation is accurate for quite modest counts. Here is a practical guideline.[5]

> ### CELL COUNTS REQUIRED FOR THE CHI-SQUARE TEST
>
> You can safely use the chi-square test with critical values from the chi-square distribution when no more than 20% of the expected counts are less than 5 and all individual expected counts are 1 or greater. In particular, all four expected counts in a 2 × 2 table should be 5 or greater.

Note that the guideline uses *expected* cell counts. The expected counts for the living arrangements study of Example 22.1 appear in Table 22.3. Only 2 of the 20 expected counts (that's 10%) are less than 5 and all are greater than 1, so the data meet the guideline for safe use of chi-square.

Using technology

Calculating the expected counts and then the chi-square statistic by hand is time-consuming. As usual, software saves time and always gets the arithmetic right. Figure 22.2 shows output for the chi-square test for the living arrangements data from a graphing calculator and a statistical program.

> **EXAMPLE 22.4** **Where young people live: chi-square output**
>
> Both outputs tell us that the chi-square statistic is $\chi^2 = 193.55$, with very small P-value. Minitab reports $P = 0.000$, rounded to three decimal places. The graphing calculator says that $P = 6.98 \times 10^{-35}$, a very small number indeed. This P-value comes from an approximation to the sampling distribution of χ^2. The approximation is less accurate far out in the tail than near the center of the distribution, so you should not take 6.98×10^{-35} literally. Just read it as "P is very small." The sample gives very strong evidence that living arrangements differ among the four age groups.
>
> Statistical software generally offers additional information on request. We asked Minitab to show the observed counts and expected counts and also the term in the chi-square statistic for each cell, called the "chi-square contribution." The top left cell has expected count 245.57 and contributes 25.049 to the chi-square statistic, as we calculated earlier. (Roundoff errors are smaller with software than in hand calculation.) The graphing calculator also displays the observed and expected cell counts on request. We told the calculator to display the expected counts rounded to two decimal places. To see the remaining columns of expected counts on the calculator's small screen, you must scroll to the right.
>
> What about the Excel spreadsheet program? Excel is in general a poor choice for statistics. It is particularly awkward for chi-square because its Data Analysis tool pack omits this common test. You will need add-in modules to use Excel effectively for chi-square. ■

The chi-square test is an overall test for detecting relationships between two categorical variables. If the test is significant, it is important to look at the data to learn the nature of the relationship. We have three ways to look at the living arrangements data:

■ **Compare selected percents:** which living arrangements occur in quite different percents of the four age groups? This is the method we learned in Chapter 6.

■ **Compare observed and expected cell counts:** which cells have more or fewer observations than we would expect if H_0 were true?

■ **Look at the terms of the chi-square statistic:** which cells contribute the most to the value of χ^2?

Texas Instruments Graphing Calculator

Minitab

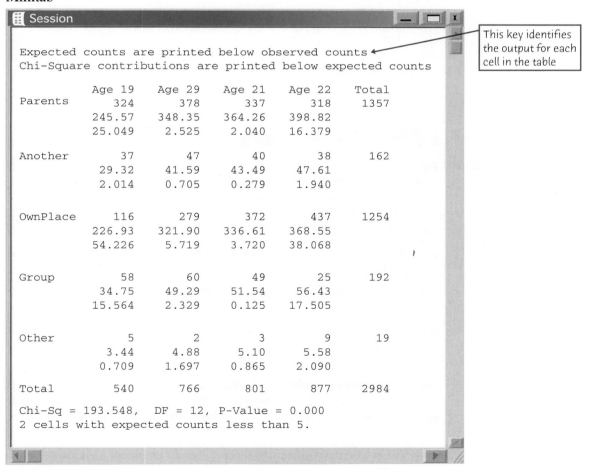

FIGURE 22.2

Output from a graphing calculator and Minitab for the two-way table in the study of where young people live, for Example 22.4.

EXAMPLE 22.5   **Where young people live: conclusion**

There is very strong evidence ($\chi^2 = 193.55$, $P < 0.001$) that living arrangements of young people are not the same for ages 19, 20, 21, and 22. Comparing selected percents—specifically, the four conditional distributions of living arrangements for each age in Table 22.2 and Figure 22.1—shows how young people become more independent as they grow older.

The additional information provided by programs like Minitab shows what differences among the age groups explain the large value of the chi-square statistic. Look at the 20 terms in the chi-square statistic in the Minitab output and compare the observed and expected counts in the cells that contribute most to chi-square. Just 6 of the 20 cells contribute 166.79 of the total chi-square $\chi^2 = 193.55$. These 6 cells occur in pairs:

- 54.226 and 38.068: fewer 19-year-olds than expected and more 22-year-olds than expected live in their own place.

- 25.049 and 16.379: more 19-year-olds than expected and fewer 22-year-olds than expected live in their parents' home.

- 15.564 and 17.505: more 19-year-olds than expected and fewer 22-year-olds than expected live in group quarters.

These three trends display the increase in independent living between age 19 and age 22. ■

APPLY YOUR KNOWLEDGE

22.7   **Facebook at Penn State.** Figure 22.3 displays Minitab output for how frequently students at the University Park and commonwealth campuses of Penn State University who have Facebook accounts make use of their accounts. The output includes the two-way table of observed counts, the expected counts, and each cell's contribution to the chi-square statistic.

(a)   Verify from the output that the data meet the cell count requirement for use of chi-square.

(b)   What hypotheses does chi-square test? What are the test statistic and its P-value?

(c)   Which cells contribute the most to χ^2? Compare the observed and expected counts in these cells and comment on the most important differences in Facebook use between students at the two locations.

22.8   **Attitudes toward recycled products.** Your data analysis in Exercise 22.2 found that people who have bought recycled coffee filters tend to have more positive opinions about the filters than those who have not. Figure 22.4 gives Minitab output for the two-way table in Exercise 22.2.

(a)   Verify from the output that the data meet the cell count requirement for use of chi-square.

(b)   What are the chi-square statistic and its P-value? Explain in simple language what it means to reject H_0 in this setting.

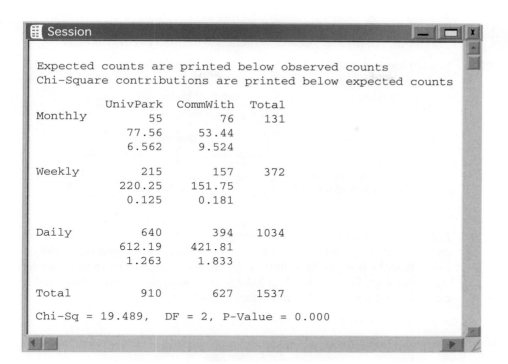

FIGURE 22.3

Minitab output for the two-way table of Facebook use by Penn State campus, for Exercise 22.7.

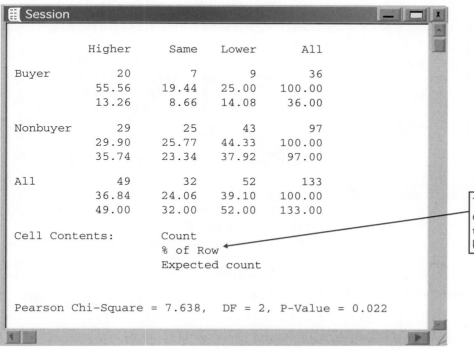

FIGURE 22.4

Minitab output for the study of consumer attitudes toward recycled products, for Exercise 22.8.

(c) Give an overall conclusion that refers to row percents to describe the nature of the relationship between attitude and decision to buy or not.

22.9 Is astrology scientific? The General Social Survey asked a random sample of adults about their education and about their view of astrology as scientific or not. Here are the data for people with three levels of higher education:

	Degree Held		
	Associate's	Bachelor's	Master's
Not at all scientific	169	256	114
Very or sort of scientific	65	65	18

Figure 22.5 gives Minitab chi-square output for these data. Follow the *Plan*, *Solve*, and *Conclude* steps of the four-step process in using the information in the output to describe how people with these levels of education differ in their opinions about astrology. Be sure that your *Solve* step includes data analysis and checking conditions for inference as well as a formal test.

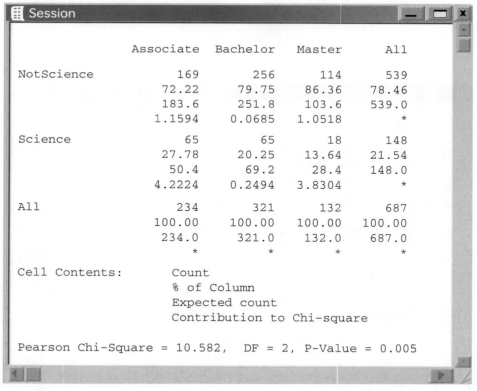

FIGURE 22.5

Minitab output for the two-way table of opinion about astrology by degree held, for Exercise 22.9.

Uses of the chi-square test

Two-way tables can arise in several ways. Most commonly, the subjects in a single sample are classified by two categorical variables. For example, we classified young adults by their age group and where they lived. The next example illustrates a different setting, in which we compare two separate samples. "Which sample" is now one of the variables for a two-way table.

EXAMPLE 22.6 Are cell-only telephone users different?

STATE: Random digit dialing telephone surveys do not call cell phone numbers. If the opinions of people who have only cell phones differ from those of people who still have landline service, the poll results may not represent the entire adult population. The Pew Research Center interviewed separate random samples of cell-only and landline telephone users. We will compare the 96 cell-only users and the 104 landline users who were less than 30 years old. Here's what the Pew survey found about how these people describe their political party affiliation:[6]

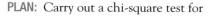

	Cell-only sample	Landline sample
Democrat or lean Democratic	49	47
Refuse to lean either way	15	27
Republican or lean Republican	32	30
Total	96	104

AB/Getty Images

PLAN: Carry out a chi-square test for

H_0: no relationship; that is, the distribution of party affiliation is the same in both populations

H_a: there is some relationship; that is, the party distribution in the cell-only population differs from that of landline users

Compare column percents or observed versus expected cell counts or terms of chi-square to see the nature of the relationship.

SOLVE: The Minitab output in Figure 22.6 includes the column percents. These give the conditional distributions of party given telephone use. Cell-only users are less likely to have no party affiliation (15.63% versus 25.96% of landline users). The party affiliations among the people who prefer one party are nearly the same in both groups, 60% Democrat for cell-only, 61% Democrat for landline.

To see if the differences are significant, first check the guidelines for use of chi-square. The samples are reasonably close to SRSs, though nonresponse was higher for the cell phone calls. The Minitab output shows that all expected cell counts are greater than 5. The chi-square test shows that there is no significant difference between the party affiliations of the two groups of young adults ($\chi^2 = 3.22$, $P = 0.200$). Comparing observed and expected cell counts again shows that cell-only young adults are less likely to have no party preference than would be expected if there were no relationship, and that landline users are more likely to have no preference. The two "refuse to lean" cells contribute 2.54 of the total chi-square $\chi^2 = 3.22$. But the overall comparison is not significant.

FIGURE 22.6

Minitab output for the two-way table of political party affiliation and telephone use, for Example 22.6.

```
┌─ Session ──────────────────────────────────────── _ □ X ─┐
│                                                          │
│                  Cell-only   Landline      All           │
│                                                          │
│   Democrat            49         47          96          │
│                    51.04      45.19       48.00          │
│                    46.08      49.92       96.00          │
│                   0.1850     0.1708           *          │
│                                                          │
│   RefuseToLean        15         27          42          │
│                    15.63      25.96       21.00          │
│                    20.16      21.84       42.00          │
│                   1.3207     1.2191           *          │
│                                                          │
│   Republican          32         30          62          │
│                    33.33      28.85       31.00          │
│                    29.76      32.24       62.00          │
│                   0.1686     0.1556           *          │
│                                                          │
│   All                 96        104         200          │
│                   100.00     100.00      100.00          │
│                    96.00     104.00      200.00          │
│                        *          *           *          │
│                                                          │
│   Cell Contents:      Count                              │
│                       % of Column                        │
│                       Expected count                     │
│                       Contribution to Chi-square         │
│                                                          │
│   Pearson Chi-Square = 3.220,  DF = 2, P-Value = 0.200   │
│                                                          │
└──────────────────────────────────────────────────────────┘
```

More chi-square tests

There are other chi-square tests for hypotheses more specific than "no relationship." A sociologist places people in classes by social status, waits ten years, then classifies the same people again. The row and column variables are the classes at the two times. She might test the hypothesis that there has been no change in the overall distribution of social status in the group. Or she might ask if moves up in status are balanced by matching moves down. These and other null hypotheses can be tested by variations of the chi-square test.

CONCLUDE: There is no significant difference between the political party affiliations of young people who have a landline telephone and those who rely entirely on cell phones. The data do suggest that cell-only users are more likely to have some affiliation, so that a larger sample might find a significant difference. The Pew study found "little difference" to be true for all adults and for a variety of political questions. Traditional telephone sample surveys will live on, at least for a while. ■

One of the most useful properties of chi-square is that it tests the null hypothesis "the row and column variables are not related to each other" whenever this hypothesis makes sense for a two-way table. It makes sense when we are comparing a categorical response in two or more samples, as when we compared people who have only a cell phone with people who have a landline phone. The hypothesis also makes sense when we have data on two categorical variables for the individuals in a single sample, as when we examined age group and living arrangement for a sample of young adults. Statistical significance has the same meaning in both settings: "A relationship this strong is not likely to happen just by chance."

USES OF THE CHI-SQUARE TEST

Use the chi-square test to test the null hypothesis

H_0: there is no relationship between two categorical variables

when you have a two-way table from one of these situations:

- Independent SRSs from two or more populations, with each individual classified according to one categorical variable. (The other variable says which sample the individual comes from.)

- A single SRS, with each individual classified according to both of two categorical variables.

APPLY YOUR KNOWLEDGE

22.10 Cell-only versus landline users. We suspect that people who rely entirely on cell phones will as a group be younger than those who have a landline telephone. Do data confirm this guess? Here is a two-way table that breaks down both of Pew's samples (see Example 22.6) by age group:

	Landline sample	Cell-only sample
Age 18–29	108	96
Age 30–49	264	70
Age 50–64	202	26
Age 65 or older	178	8
Total	752	200

Do a complete analysis of these data, following the four-step process as illustrated in Example 22.6.

The chi-square distributions

Software usually finds P-values for us. The P-value for a chi-square test comes from comparing the value of the chi-square statistic with critical values for a *chi-square distribution*.

THE CHI-SQUARE DISTRIBUTIONS

The **chi-square distributions** are a family of distributions that take only positive values and are skewed to the right. A specific chi-square distribution is specified by giving its **degrees of freedom.**

The chi-square test for a two-way table with r rows and c columns uses critical values from the chi-square distribution with $(r-1)(c-1)$ degrees of freedom. The P-value is the area under the density curve of this chi-square distribution to the right of the value of test statistic.

Density curves for the chi-square distributions with 1, 4, and 8 degrees of freedom. Chi-square distributions take only positive values and are right-skewed.

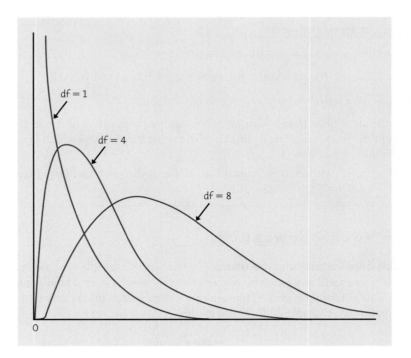

Figure 22.7 shows the density curves for three members of the chi-square family of distributions. As the degrees of freedom increase, the density curves become less skewed and larger values become more probable. Table D in the back of the book gives critical values for chi-square distributions. You can use Table D if you do not have software that gives you P-values for a chi-square test.

EXAMPLE 22.7 **Using the chi-square table**

The two-way table of 5 outcomes by 4 age groups for the living arrangements study (Table 22.1) study has 5 rows and 4 columns. That is, $r = 5$ and $c = 4$. The chi-square statistic therefore has degrees of freedom

$$(r - 1)(c - 1) = (5 - 1)(4 - 1) = (4)(3) = 12$$

Both outputs in Figure 22.2 give 12 as the degrees of freedom.

The observed value of the chi-square statistic is $\chi^2 = 193.55$. Look in the df = 12 row of Table D. The value $\chi^2 = 193.55$ falls above the largest critical value in the table, for $P = 0.0005$. Remember that the chi-square test is always one-sided. So the P-value of $\chi^2 = 193.55$ is less than 0.0005. ■

df = 12

p	.001	.0005
x^*	32.91	34.82

We know that all z and t statistics measure the size of an effect in the standard scale centered at zero. We can roughly assess the size of any z or t statistic by the 68–95–99.7 rule, though this is exact only for z. The chi-square statistic does not have any such natural interpretation. But here is a helpful fact: *the mean of any chi-square distribution is equal to its degrees of freedom.* In Example 22.7, χ^2 would have mean 12 if the null hypothesis were true. The observed value $\chi^2 = 193.55$ is so much larger than 12 that we suspect it is significant even before we look at Table D.

APPLY YOUR KNOWLEDGE

22.11 Facebook at Penn State. The Minitab output in Figure 22.3 gives the degrees of freedom for a table of Facebook use by students at two campus locations as DF = 2.

(a) Show that this is correct for a table with 3 rows and 2 columns.

(b) Minitab gives the chi-square statistic as Chi-Sq = 19.489. Where does this value fall when compared with critical values of the chi-square distribution with 2 degrees of freedom in Table D? How does Minitab's result P-Value = 0.000 compare with the P-value from the table?

(c) The table included only students who have Facebook accounts. The original table in Exercise 22.1 had 4 rows and 2 columns. What is the proper degrees of freedom for that table?

22.12 Attitudes toward recycled products. The Minitab output in Figure 22.4 gives 2 degrees of freedom for the table in Exercise 22.2.

(a) Verify that this is correct.

(b) The computer gives the value of the chi-square statistic as $\chi^2 = 7.638$. Between what two entries in Table D does this value lie? Verify that Minitab's P-value does fall between the tail probabilities p for these two entries.

(c) What is the mean value of the statistic χ^2 if the null hypothesis is true? How does the observed value of χ^2 compare with this mean?

The chi-square test for goodness of fit*

The most common and most important use of the chi-square statistic is to test the hypothesis that there is *no relationship between two categorical variables.* A variation of the statistic can be used to test a different kind of null hypothesis: that *a categorical variable has a specified distribution.* Here is an example that illustrates this use of chi-square.

EXAMPLE 22.8 Never on Sunday?

Births are not evenly distributed across the days of the week. Fewer babies are born on Saturday and Sunday than on other days, probably because doctors find weekend births inconvenient.

---

* This special topic is optional.

A random sample of 140 births from local records shows this distribution across the days of the week:

Day	Sun.	Mon.	Tue.	Wed.	Thu.	Fri.	Sat.
Births	13	23	24	20	27	18	15

Sure enough, the two smallest counts of births are on Saturday and Sunday. Do these data give significant evidence that local births are not equally likely on all days of the week? ■

The chi-square test answers the question of Example 22.8 by comparing observed counts with expected counts under the null hypothesis. The null hypothesis for births says that they *are* evenly distributed. To state the hypotheses carefully, write the discrete probability distribution for days of birth:

Day	Sun.	Mon.	Tue.	Wed.	Thu.	Fri.	Sat.
Probability	p_1	p_2	p_3	p_4	p_5	p_6	p_7

The null hypothesis says that the probabilities are the same on all days. In that case, all 7 probabilities must be 1/7. So the null hypothesis is

$$H_0: p_1 = p_2 = p_3 = p_4 = p_5 = p_6 = p_7 = \frac{1}{7}$$

The alternative hypothesis says that days are *not* all equally probable:

$$H_a: \text{ not all } p_i = \frac{1}{7}$$

As usual in chi-square tests, H_a is a "many-sided" hypothesis that simply says that H_0 is not true. The chi-square statistic is also as usual:

$$\chi^2 = \sum \frac{(\text{observed count} - \text{expected count})^2}{\text{expected count}}$$

The expected count for an outcome with probability p is np, as we saw in the discussion following Example 22.2. Under the null hypothesis, all the probabilities p_i are the same, so all 7 expected counts are equal to

$$np_i = 140 \times \frac{1}{7} = 20$$

These expected counts easily satisfy our guidelines for using chi-square. The chi-square statistic is

$$\chi^2 = \sum \frac{(\text{observed count} - 20)^2}{20}$$

$$= \frac{(13 - 20)^2}{20} + \frac{(23 - 20)^2}{20} + \cdots + \frac{(15 - 20)^2}{20}$$

$$= 7.6$$

This new use of χ^2 requires a different degrees of freedom. To find the P-value, compare χ^2 with critical values from the chi-square distribution with degrees of freedom one less than the number of values the birth day can take. That's $7 - 1 = 6$ degrees of freedom. From Table D, we see that $\chi^2 = 7.6$ is smaller than the smallest entry in the df $= 6$ row, which is the critical value for tail area 0.25. The P-value is therefore greater than 0.25 (software gives the more exact value $P = 0.269$). These 140 births don't give convincing evidence that births are not equally likely on all days of the week.

The chi-square test applied to the hypothesis that a categorical variable has a specified distribution is called the test for *goodness of fit*. The idea is that the test assesses whether the observed counts "fit" the distribution. The chi-square statistic is the same as for the two-way table test, but the expected counts and degrees of freedom are different. Here are the details.

df $= 6$		
p	.25	.20
x^*	7.84	8.56

THE CHI-SQUARE TEST FOR GOODNESS OF FIT

A categorical variable has k possible outcomes, with probabilities $p_1, p_2, p_3, \ldots, p_k$. That is, p_i is the probability of the ith outcome. We have n independent observations from this categorical variable.

To test the null hypothesis that the probabilities have specified values

$$H_0: p_1 = p_{10}, \quad p_2 = p_{20}, \quad \ldots, \quad p_k = p_{k0}$$

find the **expected count** for the ith possible outcome as np_{i0} and use the **chi-square statistic**

$$\chi^2 = \sum \frac{(\text{observed count} - \text{expected count})^2}{\text{expected count}}$$

The sum is over all the possible outcomes.

The P-value is the area to the right of χ^2 under the density curve of the chi-square distribution with $k - 1$ degrees of freedom.

In Example 22.8, the outcomes are days of the week, with $k = 7$. The null hypothesis says that the probability of a birth on the ith day is $p_{i0} = 1/7$ for all days. We observe $n = 140$ births and count how many fall on each day. These are the counts used in the chi-square statistic.

Chi-square in the casino

Gambling devices such as slot machines and roulette wheels are supposed to have a fixed and known distribution of outcomes. Here's a job for the chi-square test of goodness of fit: state gambling regulators use it to verify that casino devices are honest. How much deviation a casino can get away with depends on the state. Nevada cracks down if chi-square is significant at the 5% level. Mississippi gives more leeway, acting only when the 1% level is reached.

APPLY YOUR KNOWLEDGE

22.13 Saving birds from windows. Many birds are injured or killed by flying into windows. It appears that birds don't see windows. Can tilting windows down so that they reflect earth rather than sky reduce bird strikes? Place six windows at the edge of a woods: two vertical, two tilted 20 degrees, and two tilted 40 degrees. During the next four months, there were 53 bird strikes, 31 on the vertical windows, 14 on the 20-degree windows, and 8 on the 40-degree windows.[7] If the tilt has no effect, we expect strikes on windows with all three tilts to have equal probability. Test this null hypothesis. What do you conclude?

Randy Duchaine/CORBIS

22.14 More on birth days. Births really are not evenly distributed across the days of the week. The data in Example 22.8 failed to reject this null hypothesis because of random variation in a quite small number of births. Here are data on 700 births in the same locale:

Day	Sun.	Mon.	Tue.	Wed.	Thu.	Fri.	Sat.
Births	84	110	124	104	94	112	72

(a) The null hypothesis is that all days are equally probable. What are the probabilities specified by this null hypothesis? What are the expected counts for each day in 700 births?

(b) Calculate the chi-square statistic for goodness of fit.

(c) What are the degrees of freedom for this statistic? Do these 700 births give significant evidence that births are not equally probable on all days of the week?

22.15 Police harassment? Police may use minor violations such as not wearing a seat belt to stop motorists for other reasons. A large study in Michigan first studied the population of drivers not wearing seat belts during daylight hours by observation at more than 400 locations around the state. Here is the population distribution of seat belt violators by age group:[8]

Age group	16 to 29	30 to 59	60 or older
Proportion	0.328	0.594	0.078

The researchers then looked at court records and called a random sample of 803 drivers who had actually been cited by police for not wearing a seat belt. Here are the counts:

Age group	16 to 29	30 to 59	60 or older
Count	401	382	20

Does the age distribution of people cited differ significantly from the distribution of ages of all seat belt violators? Which age groups have the largest contributions to chi-square? Are these age groups cited more or less frequently than is justified? (The study found that males, blacks, and younger drivers were all over-cited.)

22.16 Course grades. Most students in a large statistics course are taught by teaching assistants (TAs). One section is taught by the course supervisor, a senior professor. The distribution of grades for the hundreds of students taught by TAs this semester was

Grade	A	B	C	D/F
Probability	0.32	0.41	0.20	0.07

The grades assigned by the professor to students in his section were

Grade	A	B	C	D/F
Count	22	38	20	11

(These data are real. We won't say when and where, but the professor was not the author of this book.)

(a)   What percents of each grade did students in the professor's section earn? In what ways does this distribution of grades differ from the TA distribution?

(b)   Because the TA distribution is based on hundreds of students, we are willing to regard it as a fixed probability distribution. If the professor's grading follows this distribution, what are the expected counts of each grade in his section?

(c)   Does the chi-square test for goodness of fit give good evidence that the professor's grades follow a different distribution? (State hypotheses, check the guidelines for using chi-square, give the test statistic and its P-value, and state your conclusion.)

22.17 What's your sign? For reasons known only to social scientists, the General Social Survey (GSS) regularly asks its subjects their astrological sign. Here are the counts of responses for the most recent GSS:

Sign	Aries	Taurus	Gemini	Cancer	Leo	Virgo
Count	321	360	367	374	383	402

Sign	Libra	Scorpio	Sagittarius	Capricorn	Aquarius	Pisces
Count	392	329	331	354	376	355

If births are spread uniformly across the year, we expect all 12 signs to be equally likely. Are they? Follow the four-step process in your answer.

CHAPTER 22   SUMMARY

■ The **chi-square test** for a two-way table tests the null hypothesis H_0 that there is no relationship between the row variable and the column variable. The alternative hypothesis H_a says that there is some relationship but does not say what kind.

■ The test compares the observed counts of observations in the cells of the table with the counts that would be expected if H_0 were true. The **expected count** in any cell is

$$\text{expected count} = \frac{\text{row total} \times \text{column total}}{\text{table total}}$$

■ The **chi-square statistic** is

$$\chi^2 = \sum \frac{(\text{observed count} - \text{expected count})^2}{\text{expected count}}$$

- The chi-square test compares the value of the statistic χ^2 with critical values from the **chi-square distribution** with $(r-1)(c-1)$ **degrees of freedom.** Large values of χ^2 are evidence against H_0, so the P-value is the area under the chi-square density curve to the right of χ^2.

- The chi-square distribution is an approximation to the distribution of the statistic χ^2. You can safely use this approximation when all expected cell counts are at least 1 and no more than 20% are less than 5.

- If the chi-square test finds a statistically significant relationship between the row and column variables in a two-way table, do data analysis to describe the nature of the relationship. You can do this by comparing well-chosen percents, comparing the observed counts with the expected counts, and looking for the largest **terms of the chi-square statistic.**

STATISTICS IN SUMMARY

Here are the most important skills you should have acquired from reading this chapter.

A. Two-Way Tables

1. Understand that the data for a chi-square test must be presented as a two-way table of counts of outcomes.

2. Use percents to describe the relationship between any two categorical variables, starting from the counts in a two-way table.

B. Interpreting Chi-Square Tests

1. Locate the chi-square statistic, its P-value, and other useful facts (row or column percents, expected counts, terms of chi-square) in output from your software or calculator.

2. Use the expected counts to check whether you can safely use the chi-square test.

3. Explain what null hypothesis the chi-square statistic tests in a specific two-way table.

4. If the test is significant, compare percents, compare observed with expected cell counts, or look for the largest terms of the chi-square statistic to see what deviations from the null hypothesis are most important.

C. Doing Chi-Square Tests by Hand

1. Calculate the expected count for any cell from the observed counts in a two-way table. Check whether you can safely use the chi-square test.

2. Calculate the term of the chi-square statistic for any cell, as well as the overall statistic.

3. Give the degrees of freedom of a chi-square statistic. Make a quick assessment of the significance of the statistic by comparing the observed value with the degrees of freedom.

4. Use the chi-square critical values in Table D to approximate the P-value of a chi-square test.

The National Longitudinal Study of Adolescent Health interviewed several thousand teens (grades 7 to 12). One question asked was "What do you think are the chances you will be married in the next ten years?" Here is a two-way table of the responses by gender:[9]

	Female	Male
Almost no chance	119	103
Some chance, but probably not	150	171
A 50-50 chance	447	512
A good chance	735	710
Almost certain	1174	756

22.18 The number of female teenagers in the sample is

(a) 4877.   (b) 2625.   (c) 2252.

22.19 The percent of the females in the sample who responded "almost certain" is about

(a) 44.7%.   (b) 39.6%.   (c) 33.6%.

22.20 The percent of the females in the sample who responded "almost certain" is

(a) higher than the percent of males who felt this way.

(b) about the same as the percent of males who felt this way.

(c) lower than the percent of males who felt this way.

22.21 The expected count of females who respond "almost certain" is about

(a) 464.6.   (b) 891.2.   (c) 1038.8.

22.22 The term in the chi-square statistic for the cell of females who respond "almost certain" is about

(a) 17.6.   (b) 15.6.   (c) 0.1.

22.23 The degrees of freedom for the chi-square test for this two-way table are

(a) 4.   (b) 8.   (c) 20.

22.24 The null hypothesis for the chi-square test for this two-way table is

(a) Equal proportions of female and male teenagers are almost certain they will be married in ten years.

(b) There is no difference between female and male teenagers in their distributions of opinions about marriage.

(c) There are equal numbers of female and male teenagers.

22.25 The alternative hypothesis for the chi-square test for this two-way table is

(a) Female and male teenagers do not have the same distribution of opinions about marriage.

(b) Female teenagers are more likely than male teenagers to think it is almost certain they will be married in ten years.

(c) Female teenagers are less likely than male teenagers to think it is almost certain they will be married in ten years.

22.26 Software gives chi-square statistic $\chi^2 = 69.8$ for this table. From the table of critical values, we can say that the P-value is

(a) between 0.0025 and 0.001.

(b) between 0.001 and 0.0005.

(c) less than 0.0005.

22.27 The most important fact that allows us to trust the results of the chi-square test is that

(a) the sample is large, 4877 teenagers in all.

(b) the sample is close to an SRS of all teenagers.

(c) all of the cell counts are greater than 100.

C H A P T E R 2 2   E X E R C I S E S

If you have access to software or a graphing calculator, use it to speed your analysis of the data in these exercises. Exercises 22.28 to 23.33 are suitable for hand calculation if necessary.

22.28 Got broadband? A sample survey by the Pew Internet and American Life Project asked a random sample of adults about use of the Internet. One question was whether the subject had a broadband Internet connection at home. Here is a two-way table of home broadband use by type of community:[10]

	Community Type		
	Urban	**Suburban**	**Rural**
Have broadband	300	521	174
No broadband	276	542	387

(a) Give a 95% confidence interval for the difference between the proportions of all rural and urban adults who have a home broadband connection.

(b) What proportion of each of the three groups in the sample have home broadband? Are there statistically significant differences among these proportions? State hypotheses and give a test statistic and its P-value.

22.29 Free speech for racists? The General Social Survey (GSS) asked this question: "Consider a person who believes that Blacks are genetically inferior. If such a person wanted to make a speech in your community claiming that Blacks are inferior, should he be allowed to speak, or not?" Here are the responses, broken down by the race of the respondent:

	Black	**White**	**Other**
Allowed	140	976	121
Not allowed	129	480	131

(a) Because the GSS is essentially an SRS of all adults, we can combine the races in these data and give a 99% confidence interval for the proportion of all adults who would allow a racist to speak. Do this.

(b) Find the column percents and use them to compare the attitudes of the three racial groups. How significant are the differences found in the sample?

22.30 Do you use cocaine? Sample surveys on sensitive issues can give different results depending on how the question is asked. A University of Wisconsin study divided 2400 respondents into 3 groups at random. All were asked if they had ever used cocaine. One group of 800 was interviewed by phone; 21% said they had used cocaine. Another 800 people were asked the question in a one-on-one personal interview; 25% said "Yes." The remaining 800 were allowed to make an anonymous written response; 28% said "Yes."[11] Are there statistically significant differences among these proportions? State the hypotheses, convert the information given into a two-way table of counts, give the test statistic and its P-value, and state your conclusions.

22.31 Did the randomization work? After randomly assigning subjects to treatments in a randomized comparative experiment, we can compare the treatment groups to see how well the randomization worked. We hope to find no significant differences among the groups. A study of how to provide premature infants with a substance essential to their development assigned infants at random to receive one of four types of supplement, called PBM, NLCP, PL-LCP, and TG-LCP.[12]

(a) The subjects were 77 premature infants. Outline the design of the experiment if 20 are assigned to the PBM group and 19 to each of the other treatments.

(b) The random assignment resulted in 9 females in the TG-LCP group and 11 females in each of the other groups. Make a two-way table of group by gender and do a chi-square test to see if there are significant differences among the groups. What do you find?

22.32 Opinions about the death penalty. The data for comparing two sample proportions can be presented in a two-way table containing the counts of successes and failures in both samples, with two rows and two columns. Here is an example from the General Social Survey. The question was "Do you favor or oppose the death penalty for persons convicted of murder?" The table gives the responses of people whose highest education was a high school degree and of people with a bachelor's degree.

	Favor	Oppose
High school	1010	369
Bachelor's	319	185

(a) Is there evidence that the proportions of all people at these levels of education who favor the death penalty differ? Find the two sample proportions, the z statistic, and its P-value.

(b) Is there evidence that the opinions of all people at these levels of education differ? Find the chi-square statistic χ^2 and its P-value.

(c) Show that (up to roundoff error) your χ^2 is the same as z^2. The two P-values are also the same. These facts are always true, so you will often see chi-square for 2×2 tables used to compare two proportions.

22.33 Unhappy rats and tumors. Some people think that the attitude of cancer patients can influence the progress of their disease. We can't experiment with humans, but here is a rat experiment on this theme. Inject 60 rats with tumor cells and then divide them at random into two groups of 30. All the rats receive electric shocks, but rats in Group 1 can end the shock by pressing a lever. (Rats learn this sort of thing quickly.) The rats in Group 2 cannot control the shocks, which presumably makes them feel helpless and unhappy. We suspect that the rats in Group 1 will develop fewer tumors. The results: 11 of the Group 1 rats and 22 of the Group 2 rats developed tumors.[13]

(a) Make a two-way table of tumors by group. State the null and alternative hypotheses for this investigation.

(b) Although we have a two-way table, the chi-square test can't test a one-sided alternative. Carry out the z test and report your conclusion.

22.34 I think I'll be rich by age 30. A sample survey asked young adults (aged 19 to 25), "What do you think are the chances you will have much more than a middle-class income at age 30?" The Minitab output in Figure 22.8 shows the two-way table and related information, omitting a few subjects who refused to respond or who said they were already rich.[14] Use the output as the basis for a discussion of the differences between young men and young women in assessing their chances of being rich by age 30.

22.35 Sexy magazine ads? Look at full-page ads in magazines with a young adult readership. Classify ads that show a model as "not sexual" or "sexual" depending on how the model is dressed (or not dressed). Here are data on 1509 ads in magazines aimed at young men, at young women, or at young adults in general:[15]

	Readers		
	Men	Women	General
Sexual	105	225	66
Not sexual	514	351	248

Figure 22.9 displays Minitab chi-square output. Use the information in the output to describe the relationship between the target audience and the sexual content of ads in magazines for young adults.

Mistakes in using the chi-square test are unusually common. Exercises 22.36 to 22.39 illustrate several kinds of mistake.

22.36 Sorry, no chi-square. We would prefer to learn from teachers who know their subject. Perhaps even preschool children are affected by how knowledgeable they think teachers are. Assign 48 3- and 4-year-olds at random to be taught the name of a new toy by either an adult who claims to know about the toy or an adult who claims not to know about it. Then ask the children to pick out a picture of the new

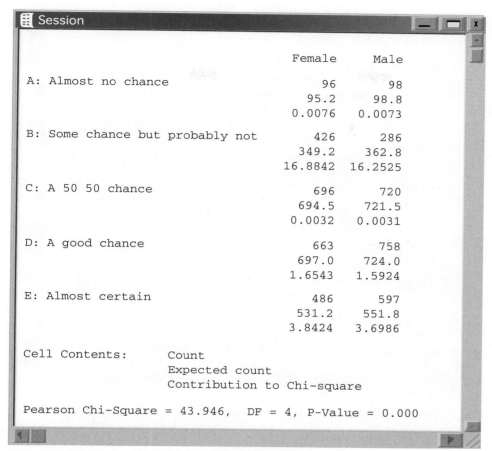

FIGURE 22.8

Minitab output for the sample survey responses of Exercise 22.34.

toy in a set of pictures of other toys and say its name. The response variable is the count of right answers in four tries. Here are the data:[16]

	Correct Answers				
	0	**1**	**2**	**3**	**4**
Knowledgeable teacher	5	1	6	3	9
Ignorant teacher	20	0	3	0	1

The researchers report that children taught by the teacher who claimed to be knowledgeable did significantly better ($\chi^2 = 20.04$, $P < 0.05$).

(a) The data do show that children taught by the knowledgeable teacher did better. Show this by comparing suitable percents.

(b) Why can't the chi-square test be trusted to assess significance for these data?

(c) If you use software, does the chi-square output for these data warn you against using the test?

FIGURE 22.9

Minitab output for a study of ads in magazines, for Exercise 22.35.

```
┌─────────────────────────────────────────────────────────────────────┐
│ ▦ Session                                            ─  □  x          │
│                                                                       │
│                 Men     Women    General     All                      │
│                                                                       │
│  Sex           105       225        66        396                     │
│                16.96     39.06     21.02      26.24                    │
│                162.4     151.2     82.4       396.0                    │
│                20.312    36.074    3.265        *                      │
│                                                                       │
│  notsexy       514       351       248        1113                    │
│                83.04     60.94     78.98      73.76                    │
│                456.6     424.8     231.6      1113.0                   │
│                7.227     12.835    1.162        *                      │
│                                                                       │
│  All           619       576       314        1509                    │
│                100.00    100.00    100.00     100.00                   │
│                619.0     576.0     314.0      1509.0                   │
│                  *         *         *          *                      │
│                                                                       │
│  Cell Contents:         Count                                         │
│                         % of Column                                   │
│                         Expected count                                │
│                         Contribution to Chi-square                    │
│                                                                       │
│  Pearson Chi-Square = 80.874,  DF = 2, P-Value = 0.00                  │
│                                                                       │
└─────────────────────────────────────────────────────────────────────┘
```

22.37 Sorry, no chi-square. How do U.S. residents who travel overseas for leisure differ from those who travel for business? The following is the breakdown by occupation.[17]

	Leisure travelers	Business travelers
Professional/technical	36%	39%
Manager/executive	23%	48%
Retired	14%	3%
Student	7%	3%
Other	20%	7%
Total	100%	100%

Explain why we don't have enough information to use the chi-square test to learn whether these two distributions differ significantly.

22.38 Sorry, no chi-square. Here is more information about Internet use by students at Penn State, based on a random sample of 1852 undergraduates. Explain why it is not correct to use a chi-square test on this table to compare the University Park and commonwealth campuses.

	University Park	Commonwealth
Viewed a video on YouTube or similar site	875	700
Legally purchased music or videos online	514	348
Downloaded a podcast	235	145
Participated in Internet gambling	114	93

22.39 Sorry, no chi-square. Does eating chocolate trigger headaches? To find out, women with chronic headaches followed the same diet except for eating chocolate bars and carob bars that looked and tasted the same. Each subject ate both chocolate and carob bars in random order with at least three days between. Each woman then reported whether or not she had a headache within 12 hours of eating the bar. Here is a two-way table of the results for the 64 subjects:[18]

	No headache	Headache
Chocolate	53	11
Carob (placebo)	38	26

The researchers carried out a chi-square test on this table to see if the two types of bar differ in triggering headaches. Explain why this test is incorrect. (*Hint:* There are 64 subjects. How many observations appear in the two-way table?)

*The remaining exercises concern larger tables that require software for easy analysis. In many cases, you should follow the **Plan, Solve,** and **Conclude** steps of the four-step process in your answers.*

22.40 Animal testing. "It is right to use animals for medical testing if it might save human lives." The General Social Survey asked 1152 adults to react to this statement. Here is the two-way table of their responses:

	Male	Female
Strongly agree	76	59
Agree	270	247
Neither agree nor disagree	87	139
Disagree	61	123
Strongly disagree	22	68

(a) Compare the conditional distributions of opinion among men and among women using both a table and a graph. What are the most important differences?

(b) Carry out the chi-square test for the hypothesis of no difference between the opinions of men and women. What would be the mean of the test statistic if the null hypothesis were true? The value of the statistic is so far above this mean that you can see at once that it must be highly significant. What is the approximate P-value?

(c) Look at the terms of the chi-square statistic and compare observed and expected counts in the cells that contribute the most to chi-square. Based on this and your findings in part (a), write a short comparison of the opinions of men and women about animal testing.

22.41 Smoking among French men. In the United States, there is a strong relationship between education and smoking: well-educated people are less likely to smoke. Does a similar relationship hold in France? Here is a two-way table of the level of education and smoking status (nonsmoker, former smoker, moderate smoker, heavy smoker) of a sample of 459 French men aged 20 to 60 years.[19] The subjects are a

Lisl Dennis/Getty Images

random sample of men who visited a health center for a routine checkup. We are willing to consider them an SRS of men from their region of France.

Education	Smoking Status			
	Nonsmoker	**Former**	**Moderate**	**Heavy**
Primary school	56	54	41	36
Secondary school	37	43	27	32
University	53	28	36	16

(a) Find the conditional distributions of smoking status for each education level. Make a graph that compares the three conditional distributions. Use your work to describe the overall relationship between education and smoking in this sample.

(b) Do French men with different levels of education differ significantly in their smoking status? State hypotheses, give the chi-square statistic and its P-value, and state your conclusion.

22.42 Students and catalog shopping. What is the most important reason that students buy from catalogs? The answer may differ for different groups of students. Here are results for samples of American and East Asian students at a large midwestern university:[20]

	American	**Asian**
Save time	29	10
Easy	28	11
Low price	17	34
Live far from stores	11	4
No pressure to buy	10	3
Other reason	20	7
Total	115	69

Describe the most important differences between American and Asian students. Is there a significant overall difference between the two distributions of responses?

22.43 How are schools doing? The nonprofit group Public Agenda conducted telephone interviews with a stratified sample of parents of high school children. There were 202 black parents, 202 Hispanic parents, and 201 white parents. One question asked was "Are the high schools in your state doing an excellent, good, fair or poor job, or don't you know enough to say?" Here are the survey results:[21]

	Black parents	**Hispanic parents**	**White parents**
Excellent	12	34	22
Good	69	55	81
Fair	75	61	60
Poor	24	24	24
Don't know	22	28	14
Total	202	202	201

Are the differences in the distributions of responses for the three groups of parents statistically significant? What departures from the null hypothesis "no relationship between group and response" contribute most to the value of the chi-square statistic? Write a brief conclusion based on your analysis.

22.44 The Mediterranean diet. Cancer of the colon and rectum is less common in the Mediterranean region than in other Western countries. The Mediterranean diet contains little animal fat and lots of olive oil. Italian researchers compared 1953 patients with colon or rectal cancer with a control group of 4154 patients admitted to the same hospitals for unrelated reasons. They estimated consumption of various foods from a detailed interview, then divided the patients into three groups according to their consumption of olive oil. Here are some of the data:[22]

| | Olive Oil | | | |
	Low	Medium	High	Total
Colon cancer	398	397	430	1225
Rectal cancer	250	241	237	728
Controls	1368	1377	1409	4154

Hugh Burden/SuperStock

(a) Is this study an experiment? Explain your answer.

(b) The investigators report that "less than 4% of cases or controls refused to participate." Why does this fact strengthen our confidence in the results?

(c) The researchers conjectured that high olive oil consumption would be more common among patients without cancer than among patients with colon cancer or rectal cancer. What do the data say?

22.45 Market research. Before bringing a new product to market, firms carry out extensive studies to learn how consumers react to the product and how best to advertise its advantages. Here are data from a study of a new laundry detergent.[23] The subjects are people who don't currently use the established brand that the new product will compete with. Give subjects free samples of both detergents. After they have tried both for a while, ask which they prefer. The answers may depend on other facts about how people do laundry.

| | Laundry Practices | | | |
	Soft water, warm wash	Soft water, hot wash	Hard water, warm wash	Hard water, hot wash
Prefer standard product	53	27	42	30
Prefer new product	63	29	68	42

How do laundry practices (water hardness and wash temperature) influence the choice of detergent? In which settings does the new detergent do best? Are the differences between the detergents statistically significant?

Support for political parties. *Political parties want to know what groups of people support them. The General Social Survey (GSS) asked its 2006 sample, "Generally speaking, do you usually think of yourself as a Republican, Democrat, Independent, or what?" The GSS is essentially an SRS of American adults. Here is a large two-way table breaking down the responses by the highest degree the subject held:*

	None	High school	Jr. college	Bachelor	Graduate
Strong Democrat	97	347	54	110	91
Not strong Democrat	115	384	52	116	69
Independent, near Democrat	67	265	50	87	58
Independent	263	503	86	92	53
Independent, near Republican	39	168	28	60	32
Not strong Republican	56	307	64	158	52
Strong Republican	40	256	37	118	44
Other party	9	32	3	18	3

Exercises 22.46 to 22.48 are based on this table.

22.46 Other parties. Give a 95% confidence interval for the proportion of adults who support "other parties."

22.47 Party support in brief. Make a 2 × 5 table by combining the counts in the three rows that mention Democrat and in the three rows that mention Republican and ignoring strict independents and supporters of other parties. We might think of this table as comparing all adults who lean Democrat and all adults who lean Republican. How does support for the two major parties differ among adults with different levels of education?

22.48 Party support in full. Use the full table to analyze the differences in political party support among levels of education. The sample is so large that the differences are bound to be highly significant, but give the chi-square statistic and its P-value nonetheless. The main challenge is in seeing what the data say. Does the full table yield any insights not found in the compressed table you analyzed in the previous exercise?

CHAPTER 23

Millard H. Sharp/Photo Researchers

Inference for Regression

When a scatterplot shows a linear relationship between a quantitative explanatory variable x and a quantitative response variable y, we can use the least-squares line fitted to the data to predict y for a given value of x. When the data are a sample from a larger population, we need statistical inference to answer questions like these about the population:

- Is there really a linear relationship between x and y in the population, or might the pattern we see in the scatterplot plausibly arise just by chance?

- What is the slope (rate of change) that relates y to x in the population, including a margin of error for our estimate of the slope?

- If we use the least-squares line to predict y for a given value of x, how accurate is our prediction (again, with a margin of error)?

This chapter shows you how to answer these questions. Here is an example we will explore.

EXAMPLE 23.1 Crying and IQ

STATE: Infants who cry easily may be more easily stimulated than others. This may be a sign of higher IQ. Child development researchers explored the relationship between the crying of infants four to ten days old and their later IQ test scores. A snap of a rubber band on the sole of the foot caused the infants to cry. The researchers recorded the crying and measured its intensity by the number of peaks in the most active 20 seconds. They later measured the children's IQ at age three years using the Stanford-Binet IQ test. Table 23.1 contains data on 38 infants.[1] Do children with higher crying counts tend to have higher IQ?

Benelux Press/Index Stock
Imagery/PictureQuest

TABLE 23.1 Infants' crying and IQ scores

CRYING	IQ	CRYING	IQ	CRYING	IQ	CRYING	IQ
10	87	20	90	17	94	12	94
12	97	16	100	19	103	12	103
9	103	23	103	13	104	14	106
16	106	27	108	18	109	10	109
18	109	15	112	18	112	23	113
15	114	21	114	16	118	9	119
12	119	12	120	19	120	16	124
20	132	15	133	22	135	31	135
16	136	17	141	30	155	22	157
33	159	13	162				

PLAN: Make a scatterplot. If the relationship appears linear, use correlation and regression to describe it. Finally, ask whether there is a *statistically significant* linear relationship between crying and IQ.

SOLVE (first steps): Chapters 4 and 5 introduced the data analysis that must come before inference. The first steps we take are a review of this data analysis. Figure 23.1 is a *scatterplot* of the crying data. Plot the explanatory variable (count of crying peaks) horizontally and the response variable (IQ) vertically. Look for the form, direction, and strength of the relationship as well as for outliers or other deviations. There is a moderate positive linear relationship, with no extreme outliers or potentially influential observations.

scatterplot

Because the scatterplot shows a roughly linear (straight-line) pattern, the **correlation** describes the direction and strength of the relationship. The correlation between crying and IQ is $r = 0.455$. We are interested in predicting the response from information about the explanatory variable. So we find the **least-squares regression line** for predicting IQ from crying. The equation of the regression line is

correlation

least-squares line

$$\hat{y} = a + bx$$
$$= 91.27 + 1.493x$$

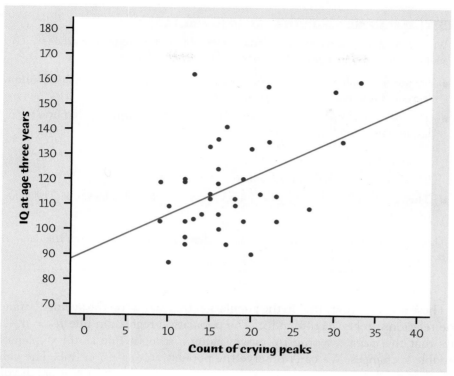

FIGURE 23.1
Scatterplot of the IQ score of infants at age three years against the intensity of their crying soon after birth, with the least-squares regression line, for Example 23.1.

CONCLUDE (first steps): Children who cry more vigorously do tend to have higher IQs. Because $r^2 = 0.207$, only about 21% of the variation in IQ scores is explained by crying intensity. Prediction of IQ will not be very accurate. It is nonetheless impressive that behavior soon after birth can even partly predict IQ three years later. Is this observed relationship statistically significant? We must now develop tools for inference in the regression setting.

Conditions for regression inference

We can fit a regression line to *any* data relating two quantitative variables, though the results are useful only if the scatterplot shows a linear pattern. Statistical inference requires more detailed conditions. Because the conclusions of inference always concern some *population*, the conditions describe the population and how the data are produced from it. The slope *b* and intercept *a* of the least-squares line are *statistics*. That is, we calculated them from the sample data. These statistics would take somewhat different values if we repeated the study with different infants. To do inference, think of *a* and *b* as estimates of unknown *parameters* that describe the population of all infants.

CONDITIONS FOR REGRESSION INFERENCE

We have n observations on an explanatory variable x and a response variable y. Our goal is to study or predict the behavior of y for given values of x.

- For any fixed value of x, the response y varies according to a **Normal distribution.** Repeated responses y are **independent** of each other.

- The mean response μ_y has a **straight-line relationship** with x given by a **population regression line**

$$\mu_y = \alpha + \beta x$$

The slope β and intercept α are unknown parameters.

- The **standard deviation** of y (call it σ) is the same for all values of x. The value of σ is unknown.

There are thus three population parameters that we must estimate from the data: α, β, and σ.

These conditions say that in the population there is an "on the average" straight-line relationship between y and x. The population regression line $\mu_y = \alpha + \beta x$ says that the *mean* response μ_y moves along a straight line as the explanatory variable x changes. We can't observe the population regression line. The values of y that we do observe vary about their means according to a Normal distribution. If we hold x fixed and take many observations on y, the Normal pattern will eventually appear in a stemplot or histogram. In practice, we observe y for many different values of x, so that we see an overall linear pattern formed by points scattered about the population line. The standard deviation σ determines whether the points fall close to the population regression line (small σ) or are widely scattered (large σ).

Figure 23.2 shows the conditions for regression inference in picture form. The line in the figure is the population regression line. The mean of the response y

FIGURE 23.2

The nature of regression data when the conditions for inference are met. The line is the population regression line, which shows how the mean response μ_y changes as the explanatory variable x changes. For any fixed value of x, the observed response y varies according to a Normal distribution having mean μ_y and standard deviation σ.

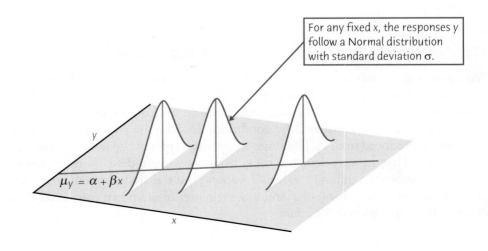

For any fixed x, the responses y follow a Normal distribution with standard deviation σ.

moves along this line as the explanatory variable x takes different values. The Normal curves show how y will vary when x is held fixed at different values. All of the curves have the same σ, so the variability of y is the same for all values of x. You should check the conditions for inference when you do inference about regression. We will see later how to do that.

Estimating the parameters

The first step in inference is to estimate the unknown parameters α, β, and σ.

ESTIMATING THE POPULATION REGRESSION LINE

When the conditions for regression are met and we calculate the least-squares line $\hat{y} = a + bx$, the slope b of the least-squares line is an unbiased estimator of the population slope β, and the intercept a of the least-squares line is an unbiased estimator of the population intercept α.

EXAMPLE 23.2 Crying and IQ: slope and intercept

The data in Figure 23.1 satisfy the condition of scatter about an invisible population regression line reasonably well. The least-squares line is $\hat{y} = 91.27 + 1.493x$. The slope is particularly important. *A slope is a rate of change.* The population slope β says how much higher average IQ is for children with one more peak in their crying measurement. Because $b = 1.493$ estimates the unknown β, we estimate that, on the average, IQ is about 1.5 points higher for each added crying peak.

We need the intercept $a = 91.27$ to draw the line, but it has no statistical meaning in this example. No child had fewer than 9 crying peaks, so we have no data near $x = 0$. We suspect that all normal children would cry when snapped with a rubber band, so that we will never observe $x = 0$. ■

The remaining parameter is the standard deviation σ, which describes the variability of the response y about the population regression line. The least-squares line estimates the population regression line. So the **residuals** estimate how much y varies about the population line. Recall that the residuals are the vertical deviations of the data points from the least-squares line:

residuals

$$\text{residual} = \text{observed } y - \text{predicted } y$$
$$= y - \hat{y}$$

There are n residuals, one for each data point. Because σ is the standard deviation of responses about the population regression line, we estimate it by a sample standard deviation of the residuals. We call this sample standard deviation the *regression standard error* to emphasize that it is estimated from data. The residuals from a least-squares line always have mean zero. That simplifies their standard error.

REGRESSION STANDARD ERROR

The **regression standard error** is

$$s = \sqrt{\frac{1}{n-2}\sum \text{residual}^2}$$

$$= \sqrt{\frac{1}{n-2}\sum (y - \hat{y})^2}$$

Use s to estimate the standard deviation σ of responses about the mean given by the population regression line.

---

degrees of freedom

Because we use the regression standard error so often, we just call it s. The quantity $\sum(y - \hat{y})^2$ is the sum of the squared deviations of the data points from the line. We average the squared deviations by dividing by $n - 2$, the number of data points less 2. It turns out that if we know $n - 2$ of the n residuals, the other two are determined. That is, $n - 2$ are the **degrees of freedom** of s. We first met the idea of degrees of freedom in the case of the ordinary sample standard deviation of n observations, which has $n - 1$ degrees of freedom. Now we observe two variables rather than one, and the proper degrees of freedom are $n - 2$ rather than $n - 1$.

Calculating s is unpleasant. You must find the predicted response for each x in your data set, then the residuals, and then s. In practice you will use software that does this arithmetic instantly. Nonetheless, here is an example to help you understand the standard error s.

The jinx!

Athletes are often jinxed. We read of "the rookie of the year jinx," the "cover of *Sports Illustrated* jinx," and many others. That is, athletes who are recognized for an outstanding performance often fail to do as well in the future. No, nature isn't retaliating against them. It's just random variation about their long-term mean performance. They were recognized because they randomly varied above their typical performance, and in the future they return to the mean or randomly vary down from it. If they randomly vary down, they can hope for a "comeback" award the next year.

EXAMPLE 23.3 **Crying and IQ: residuals and standard error**

Table 23.1 shows that the first infant studied had 10 crying peaks and a later IQ of 87. The predicted IQ for $x = 10$ is

$$\hat{y} = 91.27 + 1.493x$$

$$= 91.27 + 1.493(10) = 106.2$$

The residual for this observation is

$$\text{residual} = y - \hat{y}$$

$$= 87 - 106.2 = -19.2$$

That is, the observed IQ for this infant lies 19.2 points below the least-squares line on the scatterplot.

Repeat this calculation 37 more times, once for each subject. The 38 residuals are

−19.20	−31.13	−22.65	−15.18	−12.18	−15.15	−16.63	−6.18
−1.70	−22.60	−6.68	−6.17	−9.15	−23.58	−9.14	2.80
−9.14	−1.66	−6.14	−12.60	0.34	−8.62	2.85	14.30
9.82	10.82	0.37	8.85	10.87	19.34	10.89	−2.55
20.85	24.35	18.94	32.89	18.47	51.32		

Check the calculations by verifying that the sum of the residuals is zero. It is 0.04, not quite zero, because of roundoff error. Another reason to use software in regression is that roundoff errors in hand calculation can accumulate to make the results inaccurate.

The variance about the line is

$$s^2 = \frac{1}{n-2} \sum \text{residual}^2$$

$$= \frac{1}{38-2}[(-19.20)^2 + (-31.13)^2 + \cdots + (51.32)^2]$$

$$= \frac{1}{36}(11023.3) = 306.20$$

Finally, the regression standard error is

$$s = \sqrt{306.20} = 17.50$$ ■

We will study several kinds of inference in the regression setting. The regression standard error s is the key measure of the variability of the responses in regression. It is part of the standard error of all the statistics we will use for inference.

APPLY YOUR KNOWLEDGE

Millard H. Sharp/Photo Researchers

23.1 Ebola and gorillas. An outbreak of the deadly Ebola virus in 2002 and 2003 killed 91 of the 95 gorillas in 7 home ranges in the Congo. To study the spread of the virus, measure "distance" by the number of home ranges separating a group of gorillas from the first group infected. Here are data on distance and number of days until deaths began in each later group:[2]

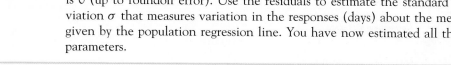

Distance x	1	3	4	4	4	5
Days y	4	21	33	41	43	46

(a) Examine the data. Make a scatterplot with distance as the explanatory variable and find the correlation. There is a strong linear relationship.

(b) Explain in words what the slope β of the population regression line would tell us if we knew it. Based on the data, what are the estimates of β and the intercept α of the population regression line?

(c) Calculate by hand the residuals for the six data points. Check that their sum is 0 (up to roundoff error). Use the residuals to estimate the standard deviation σ that measures variation in the responses (days) about the means given by the population regression line. You have now estimated all three parameters.

Using technology

Basic "two-variable statistics" calculators will find the slope b and intercept a of the least-squares line from keyed-in data. Inference about regression requires in addition the regression standard error s. At this point, software or a graphing calculator that includes procedures for regression inference becomes almost essential for practical work.

Figure 23.3 shows regression output for the data of Table 23.1 from a graphing calculator, a statistical program, and a spreadsheet program. When we entered the data into the programs, we called the explanatory variable "Crycount." The software outputs use that label. The graphing calculator just uses "x" and "y" to label

FIGURE 23.3

Regression of IQ on crying peaks: output from a graphing calculator, a statistical program, and a spreadsheet program.

Texas Instruments Graphing Calculator

Minitab

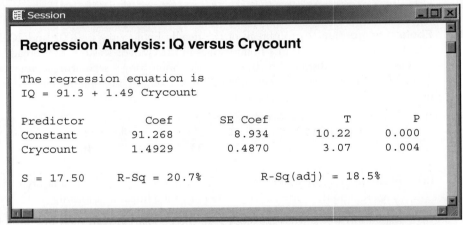

the explanatory and response variables. You can locate the basic information in all of the outputs. The regression slope is $b = 1.4929$ and the regression intercept is $a = 91.268$. The equation of the least-squares line is therefore (after rounding) just as given in Example 23.1. The regression standard error is $s = 17.4987$ and the squared correlation is $r^2 = 0.207$. Both of these results reflect the rather wide scatter of the points in Figure 23.1 about the least-squares line.

Each output contains other information, some of which we will need shortly and some of which we don't need. In fact, we left out some output to save space. Once you know what to look for, you can find what you want in almost any output and ignore what doesn't interest you.

APPLY YOUR KNOWLEDGE

23.2 **How fast do icicles grow?** The rate at which an icicle grows depends on temperature, water flow, and wind. The data below are for an icicle grown in a cold chamber at $-11°C$ with no wind and a water flow of 11.9 milligrams per second.[3]

Kristjan Fridriksson/Getty Images

Time (min)	10	20	30	40	50	60	70	80	90
Length (cm)	0.6	1.8	2.9	4.0	5.0	6.1	7.9	10.1	10.9
Time (min)	100	110	120	130	140	150	160	170	180
Length (cm)	12.7	14.4	16.6	18.1	19.9	21.0	23.4	24.7	27.8

We want to predict length from time. Figure 23.4 shows Minitab regression output for these data.

(a) Make a scatterplot suitable for predicting length from time. The pattern is very linear. What is the squared correlation r^2? Time explains almost all of the change in length.

(b) For regression inference, we must estimate the three parameters α, β, and σ. From the output, what are the estimates of these parameters?

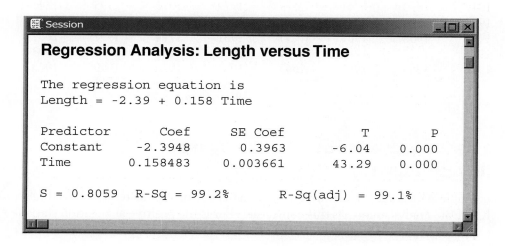

```
Session                                                    _ |□| X|

Regression Analysis: Length versus Time


The regression equation is
Length = -2.39 + 0.158 Time

Predictor        Coef      SE Coef           T          P
Constant       -2.3948      0.3963       -6.04      0.000
Time          0.158483     0.003661       43.29      0.000

S = 0.8059   R-Sq = 99.2%        R-Sq(adj) = 99.1%
```

FIGURE 23.4

Minitab output for the icicle growth data, for Exercise 23.2.

(c)   What is the equation of the least-squares regression line of length on time? Add this line to your plot. We will continue the analysis of these data in later exercises.

23.3   **Great Arctic rivers.** One effect of global warming is to increase the flow of water into the Arctic Ocean from rivers. Such an increase may have major effects on the world's climate. Six rivers (Yenisey, Lena, Ob, Pechora, Kolyma, and Severnaya Dvina) drain two-thirds of the Arctic in Europe and Asia. Several of these are among the largest rivers on earth. Table 23.2 presents the total discharge from these rivers each year from 1936 to 1999.[4] Discharge is measured in cubic kilometers of water. Use software to analyze these data.

(a)   Make a scatterplot of river discharge against time. Is there a clear increasing trend? Calculate r^2 and briefly interpret its value. There is considerable year-to-year variation, so we wonder if the trend is statistically significant.

(b)   As a first step, find the least-squares line and draw it on your plot. Then find the regression standard error s, which measures scatter about this line. We will continue the analysis in later exercises.

TABLE 23.2 Arctic river discharge (cubic kilometers), 1936 to 1999

YEAR	DISCHARGE	YEAR	DISCHARGE	YEAR	DISCHARGE	YEAR	DISCHARGE
1936	1721	1952	1829	1968	1713	1984	1823
1937	1713	1953	1652	1969	1742	1985	1822
1938	1860	1954	1589	1970	1751	1986	1860
1939	1739	1955	1656	1971	1879	1987	1732
1940	1615	1956	1721	1972	1736	1988	1906
1941	1838	1957	1762	1973	1861	1989	1932
1942	1762	1958	1936	1974	2000	1990	1861
1943	1709	1959	1906	1975	1928	1991	1801
1944	1921	1960	1736	1976	1653	1992	1793
1945	1581	1961	1970	1977	1698	1993	1845
1946	1834	1962	1849	1978	2008	1994	1902
1947	1890	1963	1774	1979	1970	1995	1842
1948	1898	1964	1606	1980	1758	1996	1849
1949	1958	1965	1735	1981	1774	1997	2007
1950	1830	1966	1883	1982	1728	1998	1903
1951	1864	1967	1642	1983	1920	1999	1970

Testing the hypothesis of no linear relationship

Example 23.1 asked, "Do children with higher crying counts tend to have higher IQ?" Data analysis supports this conjecture. But is the positive association statistically significant? That is, is it too strong to often occur just by chance? To answer this question, test hypotheses about the slope β of the population regression line:

$$H_0: \beta = 0$$

$$H_a: \beta > 0$$

A regression line with slope 0 is horizontal. That is, the mean of y does not change at all when x changes. So H_0 says that there is *no linear relationship* between x and y in the population. Put another way, H_0 says that *linear regression of y on x is of no value for predicting y*.

The test statistic is just the standardized version of the least-squares slope b, using the hypothesized value $\beta = 0$ for the mean of b. It is another t statistic. Here are the details.

SIGNIFICANCE TEST FOR REGRESSION SLOPE

To **test the hypothesis $H_0: \beta = 0$,** compute the t statistic

$$t = \frac{b}{\mathrm{SE}_b}$$

In this formula, the standard error of the least-squares slope b is

$$\mathrm{SE}_b = \frac{s}{\sqrt{\sum(x - \overline{x})^2}}$$

The sum runs over all observations on the explanatory variable x. In terms of a random variable T having the $t(n-2)$ distribution, the P-value for a test of H_0 against

$H_a: \beta > 0$ is $P(T \geq t)$

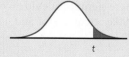

$H_a: \beta < 0$ is $P(T \leq t)$

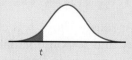

$H_a: \beta \neq 0$ is $2P(T \geq |t|)$

As advertised, the standard error of b is a multiple of the regression standard error s. The degrees of freedom $n-2$ are the degrees of freedom of s. Although we give the formula for this standard error, you should not try to calculate it by hand. Regression software gives the standard error SE_b along with b itself.

Is regression garbage?

No—but garbage can be the setting for regression. The Census Bureau once asked if weighing a neighborhood's garbage would help count its people. So 63 households had their garbage sorted and weighed. It turned out that pounds of plastic in the trash gave the best garbage prediction of the number of people in a neighborhood. The margin of error for a 95% prediction interval in a neighborhood of about 100 households, based on five weeks' worth of garbage, was about ±2.5 people. Alas, that is not accurate enough to help the Census Bureau.

EXAMPLE 23.4 **Crying and IQ: is the relationship significant?**

The hypothesis $H_0: \beta = 0$ says that crying has no straight-line relationship with IQ. We conjecture that there is a positive relationship, so we use the one-sided alternative $H_a: \beta > 0$.

Figure 23.1 shows that there is a positive relationship, so it is not surprising that all of the outputs in Figure 23.3 give $t = 3.07$ with two-sided P-value 0.004. The P-value for the one-sided test is half of this, $P = 0.002$. There is very strong evidence that IQ increases as the intensity of crying increases. ■

APPLY YOUR KNOWLEDGE

23.4 **Ebola and gorillas: testing.** Exercise 23.1 gives data on the distance of gorilla groups from the origin of an Ebola outbreak and the number of days until deaths from Ebola began. Software tells us that the least-squares slope is $b = 11.263$ with standard error $\text{SE}_b = 1.591$.

(a) What is the t statistic for testing $H_0: \beta = 0$?

(b) How many degrees of freedom does t have? Use Table C to approximate the P-value of t against the one-sided alternative $H_a: \beta > 0$. What do you conclude?

23.5 **Great Arctic rivers: testing.** The most important question we ask of the data in Table 23.2 is this: is the increasing trend visible in your plot (Exercise 23.3) statistically significant? If so, changes in the Arctic may already be affecting the earth's climate. Use software to answer this question. Give a test statistic, its P-value, and the conclusion you draw from the test.

23.6 **Does fast driving waste fuel?** Exercise 4.8 (page 102) gives data on the fuel consumption of a small car at various speeds from 10 to 150 kilometers per hour. Is there significant evidence of straight-line dependence between speed and fuel use? Make a scatterplot and use it to explain the result of your test.

Testing lack of correlation

The least-squares slope b is closely related to the correlation r between the explanatory and response variables x and y. In the same way, the slope β of the population regression line is closely related to the correlation between x and y in the population. In particular, the slope is 0 exactly when the correlation is 0.

Testing the null hypothesis $H_0: \beta = 0$ is therefore exactly the same as testing that there is *no correlation* between x and y in the population from which we drew our data. You can use the test for zero slope to test the hypothesis of zero correlation between any two quantitative variables. That's a useful trick.

Because correlation also makes sense when there is no explanatory-response distinction, it is handy to be able to test correlation without doing regression. Table E in the back of the book gives critical values of the sample correlation r under the null hypothesis that the correlation is 0 in the population. Use this table when both variables have at least approximately Normal distributions or when the sample size is large.

EXAMPLE 23.5 Testing lack of correlation

Figure 23.5 displays two scatterplots that we will use to illustrate testing lack of correlation and also to illustrate once again the need for formal statistical tests. On the left are data from an experiment on the healing of cuts in the limbs of newts. The data are the healing rates (micrometers per hour) for the two front limbs of 18 newts. The right-hand scatterplot shows the first- and second-round scores for the 96 golfers in the 2007 Masters Tournament. (There are fewer than 96 points because of duplicate scores.)

AP Photo/Elise Amendola

We will test the hypotheses

$$H_0: \text{population correlation} = 0$$

$$H_a: \text{population correlation} \neq 0$$

for both sets of data. (The Masters scores are all whole numbers, but with $n = 96$ the robustness of t procedures allows their use.) Software gives

```
newts     r=0.3581    t=1.5342    P=0.1445
masters   r=0.1924    t=1.9010    P=0.0604
```

The two-sided P-values for the t statistic for testing slope 0 are also the two-sided P-values for testing correlation 0.

Without software, compare the correlation $r = 0.3581$ for newts with the critical values in the $n = 18$ row of Table E. It falls between the table entries for one-tail

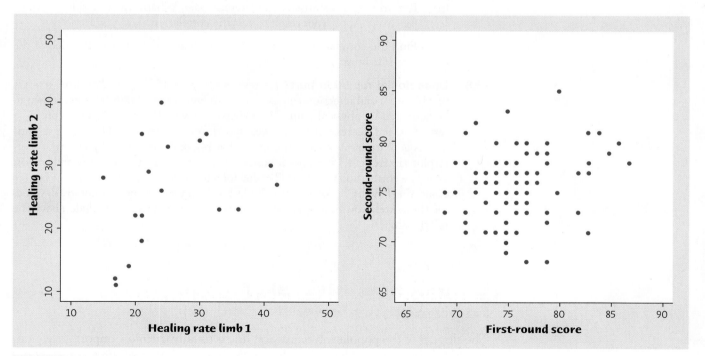

FIGURE 23.5

Two scatterplots for inference about the population correlation, for Example 23.5. (a) Healing rates for the two front limbs of 18 newts. (b) Scores on the first two rounds of the 2007 Masters Tournament.

probabilities 0.05 and 0.10, so the two-sided P-value lies between 0.10 and 0.20. For the Masters data, use the $n = 80$ row of Table E. (There is no table entry for sample size $n = 96$, so we use the next smaller sample size.) The two-sided P-value lies between 0.05 and 0.10. ■

Although the evidence for nonzero correlation is not strong for either set of data, it is stronger for Masters scores ($t = 1.9$, $P = 0.06$) than for newts ($t = 1.5$, $P = 0.14$). Yet the correlation for the newts is almost double that for the Masters, and the scatterplots suggest a stronger linear relationship for the newts. What happened? The larger sample size for the Masters data is largely responsible. The same r will have a smaller P-value for $n = 96$ than for $n = 18$. Our eyeball impression, even aided by calculating r, can't assess significance. We need the P-value from a formal test to guide us.

APPLY YOUR KNOWLEDGE

23.7   **Ebola and gorillas: testing correlation.** Exercise 23.1 gives data showing that the delay in deaths from an Ebola outbreak in groups of gorillas increases linearly with distance from the origin of the outbreak. There are only 6 observations, so we worry that the apparent relationship may be just chance. Is the correlation significantly greater than 0? Answer this question in two ways.

  (a)   Return to your t statistic from Exercise 23.4. What is the one-sided P-value for this t? Apply your result to test the correlation.

  (b)   Find the correlation r and use Table E to approximate the P-value of the one-sided test.

23.8   **Does social rejection hurt?** Exercise 4.45 (page 122) gives data from a study of whether social rejection causes activity in areas of the brain that are known to be activated by physical pain. The explanatory variable is a subject's score on a test of "social distress" after being excluded from an activity. The response variable is activity in an area of the brain that responds to physical pain. Your scatterplot (Exercise 4.45) shows a positive linear relationship. The research report gives the correlation r and the P-value for a test that r is greater than 0. What are r and the P-value? (You can use Table E or you can get more accurate P-values for the correlation from regression software.) What do you conclude about the relationship?

Confidence intervals for the regression slope

The slope β of the population regression line is usually the most important parameter in a regression problem. The slope is the rate of change of the mean response as the explanatory variable increases. We often want to estimate β. The slope b of the least-squares line is an unbiased estimator of β. A confidence interval is more useful because it shows how accurate the estimate b is likely to be. The confidence interval for β has the familiar form

$$\text{estimate} \pm t^*\text{SE}_{\text{estimate}}$$

Because b is our estimate, the confidence interval is $b \pm t^*\text{SE}_b$. Here are the details.

CONFIDENCE INTERVAL FOR REGRESSION SLOPE

A level C **confidence interval for the slope** β of the population regression line is

$$b \pm t^*\text{SE}_b$$

Here t^* is the critical value for the $t(n-2)$ density curve with area C between $-t^*$ and t^*. The formula for SE_b appears in the box on page 605.

EXAMPLE 23.6 **Crying and IQ: estimating the slope**

The two software outputs in Figure 23.3 give the slope $b = 1.4929$ and also the standard error $\text{SE}_b = 0.4870$. The outputs use a similar arrangement, a table in which each regression coefficient is followed by its standard error. Excel also gives the lower and upper endpoints of the 95% confidence interval for the population slope β, 0.505 and 2.481.

Once we know b and SE_b, it is easy to find the confidence interval. There are 38 data points, so the degrees of freedom are $n - 2 = 36$. Because Table C does not have a row for df $= 36$, we must use either software or the next smaller degrees of freedom in the table, df $= 30$. To use software, enter 36 degrees of freedom. For 95% confidence, enter the cumulative proportion 0.975 that corresponds to upper tail area 0.025. Minitab gives

```
Student's t distribution with 36 DF
P(X<=x)        x
 0.975    2.02809
```

The 95% confidence interval for the population slope β is

$$b \pm t^*\text{SE}_b = 1.4929 \pm (2.02809)(0.4870)$$

$$= 1.4929 \pm 0.9877$$

$$= 0.505 \text{ to } 2.481$$

This agrees with Excel's result. We are 95% confident that mean IQ increases by between about 0.5 and 2.5 points for each additional peak in crying. ■

You can find a confidence interval for the intercept α of the population regression line in the same way, using a and SE_a from the "Constant" line of the Minitab output or the "Intercept" line in Excel. We rarely need to estimate α.

23.9 Ebola and gorillas: estimating slope. Exercise 23.1 presents data on distance and days until an Ebola outbreak reached six groups of gorillas. Software tells us that the least-squares slope is $b = 11.263$ with standard error $SE_b = 1.591$. Because there are only 6 observations, the observed slope b may not be an accurate estimate of the population slope β. Give a 90% confidence interval for β.

23.10 Growth of icicles: estimating slope. Exercise 23.2 gives data on the growth of an icicle. We want a 95% confidence interval for the slope of the population regression line. Starting from the information in the Minitab output in Figure 23.4, find this interval. Say in words what the slope of the population regression line tells us about the growth of icicles under the conditions of this experiment.

23.11 Great Arctic rivers: estimating slope. Use the data in Table 23.2 to give a 90% confidence interval for the slope of the population regression of Arctic river discharge on year. Does this interval convince you that discharge is actually increasing over time? Explain your answer.

Inference about prediction

One of the most common reasons to fit a line to data is to predict the response to a particular value of the explanatory variable. This is another setting for regression inference: we want, not simply a prediction, but a prediction with a margin of error that describes how accurate the prediction is likely to be.

EXAMPLE 23.7 **Beer and blood alcohol**

STATE: The EESEE story "Blood Alcohol Content" describes a study in which 16 student volunteers at the Ohio State University drank a randomly assigned number of cans of beer. Thirty minutes later, a police officer measured their blood alcohol content (BAC) in grams of alcohol per deciliter of blood. Here are the data:[5]

Student	1	2	3	4	5	6	7	8
Beers	5	2	9	8	3	7	3	5
BAC	0.10	0.03	0.19	0.12	0.04	0.095	0.07	0.06
Student	9	10	11	12	13	14	15	16
Beers	3	5	4	6	5	7	1	4
BAC	0.02	0.05	0.07	0.10	0.085	0.09	0.01	0.05

Jame Shaffer/The Image Works

The students were equally divided between men and women and differed in weight and usual drinking habits. Because of this variation, many students don't believe that number of drinks predicts blood alcohol well. Steve thinks he can drive legally 30 minutes after he finishes drinking 5 beers. The legal limit for driving is BAC 0.08 in all states. We want to predict Steve's blood alcohol content, using no information except that he drinks 5 beers.

PLAN: Regress BAC on number of beers. Use the regression line to predict Steve's BAC. Give a margin of error that allows us to have 95% confidence in our prediction.

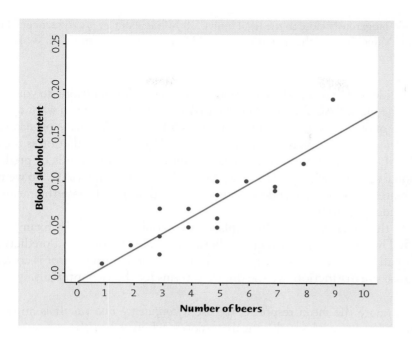

FIGURE 23.6
Scatterplot of students' blood alcohol content against the number of cans of beer consumed, with the least-squares regression line, for Example 23.7.

SOLVE: The scatterplot in Figure 23.6 and the regression output in Figure 23.7 show that student opinion is wrong: number of beers predicts blood alcohol content quite well. In fact, $r^2 = 0.80$, so that number of beers explains 80% of the observed variation in BAC. To predict Steve's BAC after 5 beers, use the equation of the regression line:

$$\hat{y} = -0.0127 + 0.0180x$$
$$= -0.0127 + 0.0180(5) = 0.077$$

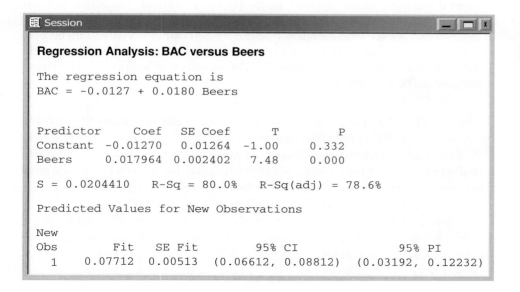

FIGURE 23.7
Minitab regression output for the blood alcohol content data, for Example 23.7.

That's dangerously close to the legal limit 0.08. What about 95% confidence? The "Predicted Values" part of the output in Figure 23.7 shows *two* 95% intervals. Which should we use? ■

To decide which interval to use, you must answer this question: do you want to predict the *mean* BAC for *all students* who drink 5 beers, or do you want to predict the BAC of *one individual student* who drinks 5 beers? *Both of these predictions may be interesting, but they are two different problems.* The actual prediction is the same, $\hat{y} = 0.077$. But the margin of error is different for the two kinds of prediction. Individual students who drink 5 beers don't all have the same BAC. So we need a larger margin of error to pin down Steve's result than to estimate the mean BAC for all students who have 5 beers.

Write the given value of the explanatory variable x as x^*. In Example 23.7, $x^* = 5$. The distinction between predicting a single outcome and predicting the mean of all outcomes when $x = x^*$ determines what margin of error is correct. To emphasize the distinction, we use different terms for the two intervals.

- To estimate the *mean* response, we use a *confidence interval*. It is an ordinary confidence interval for the mean response when x has the value x^*, which is $\mu_y = \alpha + \beta x^*$. This is a parameter, a fixed number whose value we don't know.

prediction interval

- To estimate an *individual* response y, we use a **prediction interval.** A prediction interval estimates a single random response y rather than a parameter like μ_y. The response y is not a fixed number. If we took more observations with $x = x^*$, we would get different responses.

EXAMPLE 23.8 **Beer and blood alcohol: conclusion**

Steve is one individual, so we must use the prediction interval. The output in Figure 23.7 helpfully labels the confidence interval as "95% CI" and the prediction interval as "95% PI." We are 95% confident that Steve's BAC after 5 beers will lie between 0.032 and 0.122. The upper part of that range will get him arrested if he drives. The 95% confidence interval for the mean BAC of all students who drink 5 beers is much narrower, 0.066 to 0.088. ■

The meaning of a prediction interval is very much like the meaning of a confidence interval. A 95% prediction interval, like a 95% confidence interval, is right 95% of the time in repeated use. "Repeated use" now means that we take an observation on y for each of the n values of x in the original data and then take one more observation y with $x = x^*$. Form the prediction interval from the n observations, then see if it covers the one more y. It will in 95% of all repetitions.

The interpretation of prediction intervals is a minor point. The main point is that it is harder to predict one response than to predict a mean response. Both intervals have the usual form

$$\hat{y} \pm t^* SE$$

but the prediction interval is wider than the confidence interval because individuals are more variable than averages. You will rarely need to know the details, because software automates the calculation, but here they are.

> **CONFIDENCE AND PREDICTION INTERVALS FOR REGRESSION RESPONSE**
>
> A level C **confidence interval for the mean response** μ_y when x takes the value x^* is
>
> $$\hat{y} \pm t^* SE_{\hat{\mu}}$$
>
> The standard error $SE_{\hat{\mu}}$ is
>
> $$SE_{\hat{\mu}} = s\sqrt{\frac{1}{n} + \frac{(x^* - \overline{x})^2}{\sum(x - \overline{x})^2}}$$
>
> A level C **prediction interval for a single observation y** when x takes the value x^* is
>
> $$\hat{y} \pm t^* SE_{\hat{y}}$$
>
> The standard error for prediction $SE_{\hat{y}}$ is
>
> $$SE_{\hat{y}} = s\sqrt{1 + \frac{1}{n} + \frac{(x^* - \overline{x})^2}{\sum(x - \overline{x})^2}}$$
>
> In both intervals, t^* is the critical value for the $t(n - 2)$ density curve with area C between $-t^*$ and t^*.

May the longer name win!

Regression is far from perfect, but it beats most other ways of predicting. A writer in the early 1960s noted a simple method for predicting presidential elections: just choose the candidate with the longer name. In the 22 elections from 1876 to 1960, this method failed only once. Let's hope that the writer didn't bet the family silver on this idea. The 11 elections from 1964 to 2004 presented 9 tests of the "long name wins" method (the 1980 candidates and 2000 candidates had names of the same length). The longer name lost 6 of the 9.

There are two standard errors: $SE_{\hat{\mu}}$ for estimating the mean response μ_y and $SE_{\hat{y}}$ for predicting an individual response y. The only difference between the two standard errors is the extra 1 under the square root sign in the standard error for prediction. The extra 1 makes the prediction interval wider. Both standard errors are multiples of the regression standard error s. The degrees of freedom are again $n - 2$, the degrees of freedom of s.

APPLY YOUR KNOWLEDGE

23.12 Ebola and gorillas: prediction. Exercise 23.1 presents data on distance and days until an Ebola outbreak reached six groups of gorillas. If another group were located 2 ranges from the original outbreak, predict the number of days until deaths begin in this group.

(a) Figure 23.8 is part of the output from Minitab for prediction when $x^* = 2$. Which interval in the output is the proper 95% interval for predicting the number of days until deaths begin in this group?

(b) Minitab gives only one of the two standard errors used in prediction. It is $SE_{\hat{\mu}}$, the standard error for estimating the mean response. Use this fact along

FIGURE 23.8

Partial Minitab output for regressing days until Ebola outbreak on distance (number of ranges) from initial outbreak, for Exercise 23.12.

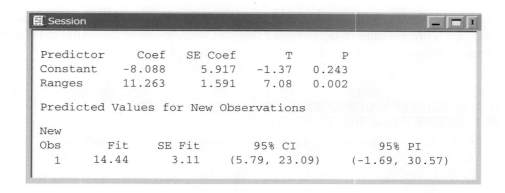

with the output to give a 90% confidence interval for the mean number of days the Ebola outbreak takes to reach gorilla groups 2 ranges from the origin.

23.13 Growth of icicles: prediction. Analysis of the data in Exercise 23.2 shows that growth of icicles is very linear. We might want to predict the mean length of icicles after 200 minutes under the same conditions of temperature, wind, and water flow. Here is the Minitab output for prediction when $x^* = 200$ minutes:

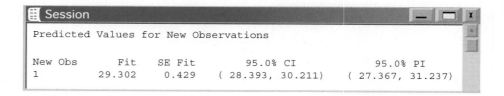

(a)   Use the regression line from Figure 23.4 to verify that "Fit" is the predicted value for $x^* = 200$. (Start with the results in the "Coef" column of Figure 23.4 to reduce roundoff error.)

(b)   What is the 95% interval we want?

Checking the conditions for inference

You can fit a least-squares line to any set of explanatory-response data when both variables are quantitative. If the scatterplot doesn't show a roughly linear pattern, the fitted line may be almost useless. But it is still the line that fits the data best in the least-squares sense. To use regression inference, however, the data must satisfy additional conditions. *Before you can trust the results of inference, you must check the conditions for inference one by one.* There are ways to deal with violations of any of the conditions. If you see a clear violation, get expert advice.

Although the conditions for regression inference are a bit elaborate, it is not hard to check for major violations. The conditions involve the population regression line and the deviations of responses from this line. We can't observe the population line, but the least-squares line estimates it and the residuals estimate the deviations from the population line. *You can check all the conditions for regression inference by looking at graphs of the residuals.* Most regression software will calculate and save the residuals for you. Start by making a stemplot or histogram of the residuals and also a **residual plot,** a plot of the residuals against the explanatory variable x, with a horizontal line at the "residual $= 0$" position. The "residual $= 0$" line represents the position of the least-squares line in the scatterplot of y against x. Let's look at each condition in turn.

residual plot

- **The relationship is linear in the population.** Look for curved patterns or other departures from a straight-line overall pattern in the residual plot. You can also use the original scatterplot, but the residual plot magnifies any effects.

- **The response varies Normally about the population regression line.** Because different y-values usually come from different x-values, the responses themselves need not be Normal. It is the deviations from the population line—estimated by the residuals—that must be Normal. Check for clear skewness or other major departures from Normality in your stemplot or histogram of the residuals.

- **Observations are independent.** In particular, repeated observations on the same individual are not allowed. You should not use ordinary regression to make inferences about the growth of a single child over time, for example. Signs of dependence in the residual plot are a bit subtle, so we usually rely on common sense.

- **The standard deviation of the responses is the same for all values of x.** Look at the scatter of the residuals above and below the "residual $= 0$" line in the residual plot. The scatter should be roughly the same from one end to the other. You will sometimes find that, as the response y gets larger, so does the scatter of the residuals. Rather than remaining fixed, the standard deviation σ about the line changes with x as the mean response changes with x. There is no fixed σ for s to estimate. You cannot trust the results of inference when this happens.

You will always see some irregularity when you look for Normality and fixed standard deviation in the residuals, especially when you have few observations. Don't overreact to minor violations of the conditions. Like other t procedures, inference for regression is (with one exception) not very sensitive to lack of Normality, especially when we have many observations. Do beware of influential observations, which can greatly affect the results of inference.

The exception is the prediction interval for a single response y. This interval relies on Normality of individual observations, not just on the approximate Normality of statistics like the slope a and intercept b of the least-squares line. The statistics a and b become more Normal as we take more observations. This contributes to the robustness of regression inference, but it isn't enough for the prediction interval. We will not study methods that carefully check Normality of the residuals, so *you should regard prediction intervals as rough approximations.*

Dave Harasti

TABLE 23.3 Winter temperature and anglerfish latitude, 1977 to 2001

YEAR	TEMP.	LATITUDE	YEAR	TEMP.	LATITUDE	YEAR	TEMP.	LATITUDE
1977	6.26	57.20	1986	6.52	57.72	1994	7.02	58.71
1978	6.26	57.96	1987	6.68	57.83	1995	7.09	58.07
1979	6.27	57.65	1988	6.76	57.87	1996	7.13	58.49
1980	6.31	57.59	1989	6.78	57.48	1997	7.15	58.28
1981	6.34	58.01	1990	6.89	58.13	1998	7.29	58.49
1982	6.32	59.06	1991	6.90	58.52	1999	7.34	58.01
1983	6.37	56.85	1992	6.93	58.48	2000	7.57	58.57
1984	6.39	56.87	1993	6.98	57.89	2001	7.65	58.90
1985	6.42	57.43						

EXAMPLE 23.9 **Climate change chases fish north**

STATE: As the climate grows warmer, we expect many animal species to move toward the poles in an attempt to maintain their preferred temperature range. Do data on fish in the North Sea confirm this expectation? Table 23.3 gives data for 25 years on mean winter temperatures at the bottom of the North Sea (degrees Celsius) and the center of the distribution of anglerfish in degrees of north latitude.[6]

PLAN: Regress latitude on temperature. Look for a positive linear relationship and assess its significance. Be sure to check the conditions for regression inference.

SOLVE: The scatterplot in Figure 23.9 shows a clear positive linear relationship. The solid line in the plot is the least-squares regression line of the center of the fish distribution (north latitude) on winter ocean temperature. Software shows that the slope

FIGURE 23.9

Plot of the latitude of the center of the distribution of anglerfish in the North Sea against mean winter temperature at the bottom of the sea, for Example 23.9. The two regression lines are for the data with (solid) and without (dashed) Observation 6.

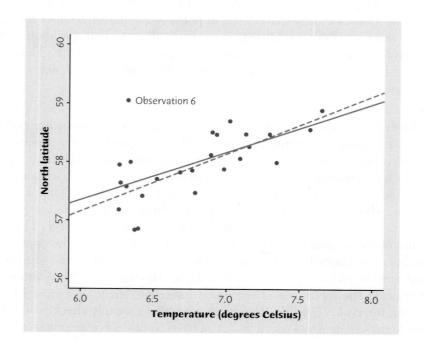

is $b = 0.818$. That is, each degree of ocean warming moves the fish about 0.8 degree of latitude farther north. The t statistic for testing H_0: $\beta = 0$ is $t = 3.6287$ with one-sided P-value $P = 0.0007$. There is very strong evidence that the population slope is positive, $\beta > 0$.

CONCLUDE: The data give highly significant evidence that anglerfish have moved north as the ocean has grown warmer. Before relying on this conclusion, we must check the conditions for inference. ▪

The software that did the regression calculations also finds the 25 residuals. In the same order as the observations in Example 23.9, they are

```
-0.3731   0.3869   0.0687  -0.0240   0.3714   1.4378  -0.8131
-0.8095  -0.2740  -0.0658  -0.0867  -0.1121  -0.5185   0.0415
 0.4234   0.3588  -0.2721   0.5152  -0.1821   0.2052  -0.0211
 0.0743  -0.4466  -0.0747   0.1899
```

Begin by making two graphs of the residuals. Figure 23.10 is a histogram of the residuals. Figure 23.11 is the residual plot, a plot of the residuals against the explanatory variable, sea-bottom temperature. The "residual = 0" line marks the position of the regression line. Notice that the vertical scale in Figure 23.11 is wider than is necessary to simply show the points. Patterns in residual plots are often easier to see if you use a wider vertical scale than your software's default plot.

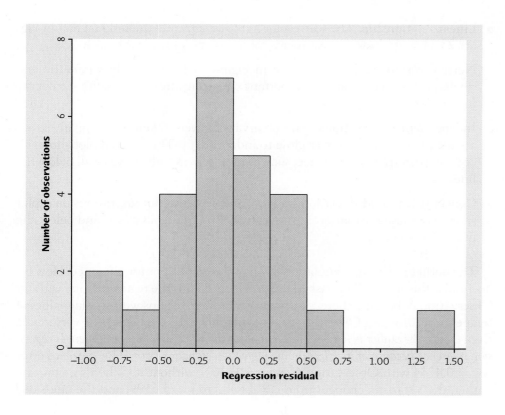

FIGURE 23.10

Histogram of the residuals from the regression of latitude on temperature in Example 23.9.

FIGURE 23.11

Residual plot for the regression of latitude on temperature in Example 23.9.

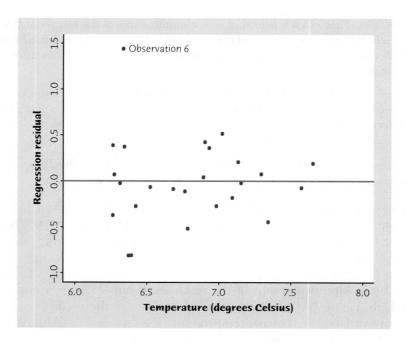

Both graphs show that Observation 6 is a high outlier. Let's check the conditions for regression inference.

- **Linear relationship.** The scatterplot in Figure 23.9 and the residual plot in Figure 23.11 both show a linear relationship except for the outlier.

- **Normal residuals.** The histogram in Figure 23.10 is roughly symmetric and single-peaked. There are no important departures from Normality except for the outlier.

- **Independent observations.** The observations were taken a year apart, so we are willing to regard them as close to independent. The residual plot shows no obvious pattern of dependence, such as runs of points all above or all below the line.

- **Constant standard deviation.** Again excepting the outlier, the residual plot shows no unusual variation in the scatter of the residuals above and below the line as x varies.

The outlier is the only serious violation of the conditions for inference. How influential is the outlier? The dashed line in Figure 23.9 is the regression line without Observation 6. Because there are several other observations with similar values of temperature, dropping Observation 6 does not move the regression line very much. *Even though the outlier is not very influential for the regression line, it influences regression inference because of its effect on the regression standard error.* The standard error is $s = 0.4734$ with Observation 6 and $s = 0.3622$ without it. When we omit the outlier, the t statistic changes from $t = 3.6287$ to $t = 5.5599$, and the one-sided

P-value changes from $P = 0.0007$ to $P < 0.00001$. Fortunately, the outlier does not affect the conclusion we drew from the data. Dropping Observation 6 makes the test for the population slope *more* significant and *increases* the percent of variation in fish location explained by ocean temperature.

One more caution about inference in this example: as usual in an observational study, the possibility of lurking variables makes us hesitant to conclude that rising temperature is *causing* anglerfish to move north. Ocean temperature was steadily rising during these years. The effect on fish latitude of any lurking variable that increased over time—perhaps increased commercial fishing—is confounded with the effect of temperature.

APPLY YOUR KNOWLEDGE

23.14 Crying and IQ: residuals. The residuals for the study of crying and IQ appear in Example 23.3.

    (a)   Make a stemplot to display the distribution of the residuals (round to the nearest whole number). Are there strong outliers or other signs of departures from Normality?

    (b)   Make a residual plot, residuals against crying peaks. Try a vertical scale of -60 to 60 to show patterns more clearly. Draw the "residual $= 0$" line. Does the residual plot show clear deviations from a linear pattern or clearly unequal spread about the line?

    (c)   Using the information given in Example 23.1, explain why the 38 observations are independent.

23.15 Growth of icicles: residuals. Figure 23.4 gives part of the Minitab output for the data on growth of icicles in Exercise 23.2. Figure 23.12 comes from another part of the output. It gives x, y, the predicted response $\hat{y}$, the residual $y - \hat{y}$, and related quantities for each of the 18 observations. This table is stored in the data file *ex23-15.dat* on the text CD and Web site. Most statistical software provides similar output. Examine the conditions for regression inference one by one. This example illustrates mild violations of the conditions that did not prevent the researchers from doing inference.

    (a)   **Linear relationship.** Your scatterplot and r^2 from Exercise 23.2 show that the relationship is very linear. Residual plots magnify effects. Plot the residuals against time. What kind of deviation from a straight line is now visible? (The deviation is clear in the residual plot, but it is very small in the original scale.)

    (b)   **Normal variation about the line.** Make a stemplot of the residuals (round to the nearest tenth, use split stems, and don't forget that -0 and 0 are separate stems). With only 18 observations, the small right skew is not disturbing. Minitab suggests that Observation 18 may be an outlier. Does your plot confirm or refute this suggestion?

    (c)   **Independent observations.** The data come from the growth of a single icicle, not from a different icicle at each time. Explain why this would violate the independence condition if we had data on the growth of a child rather than

```
┌─────────────────────────────────────────────────────────────────────────────┐
│ ▦ Session                                                    ─ □ ▪           │
├─────────────────────────────────────────────────────────────────────────────┤
│                                                                              │
│  Obs    Time    Length     Fit   SE Fit    Residual    St Resid              │
│   1      10     0.600    -0.810   0.365      1.410       1.96                 │
│   2      20     1.800     0.775   0.334      1.025       1.40                 │
│   3      30     2.900     2.360   0.304      0.540       0.72                 │
│   4      40     4.000     3.945   0.277      0.055       0.07                 │
│   5      50     5.000     5.529   0.251     -0.529      -0.69                 │
│   6      60     6.100     7.114   0.229     -1.014      -1.31                 │
│   7      70     7.900     8.699   0.211     -0.799      -1.03                 │
│   8      80    10.100    10.284   0.198     -0.184      -0.24                 │
│   9      90    10.900    11.869   0.191     -0.969      -1.24                 │
│  10     100    12.700    13.454   0.191     -0.754      -0.96                 │
│  11     110    14.400    15.038   0.198     -0.638      -0.82                 │
│  12     120    16.600    16.623   0.211     -0.023      -0.03                 │
│  13     130    18.100    18.208   0.229     -0.108      -0.14                 │
│  14     140    19.900    19.793   0.251      0.107       0.14                 │
│  15     150    21.000    21.378   0.277     -0.378      -0.50                 │
│  16     160    23.400    22.963   0.304      0.437       0.59                 │
│  17     170    24.700    24.547   0.334      0.153       0.21                 │
│  18     180    27.800    26.132   0.365      1.668       2.32R                │
│                                                                              │
│  R denotes an observation with a large standardized residual.                │
│                                                                              │
└─────────────────────────────────────────────────────────────────────────────┘
```

FIGURE 23.12

Residuals from Minitab for Exercise 23.15. The table gives the predicted value ("Fit") and the residual for each observation.

of an icicle. (The researchers decided that all icicles, unlike all children, grow at the same rate if the conditions are held fixed. So one icicle can stand in for a separate icicle at each time.)

(d) **Spread about the line stays the same.** Your plot in (a) suggests that spread may be higher at both ends. (Once again, the plot greatly magnifies small deviations.)

C H A P T E R 2 3   S U M M A R Y

- **Least-squares regression** fits a straight line to data in order to predict a response variable y from an explanatory variable x. Inference about regression requires more conditions.

- The **conditions for regression inference** say that there is a **population regression line** $\mu_y = \alpha + \beta x$ that describes how the mean response varies as x changes. The observed response y for any x has a Normal distribution with mean given by the population regression line and with the same standard deviation σ for any value of x. Observations on y are independent.

- The **parameters to be estimated** are the intercept α and the slope β of the population regression line, and also the standard deviation σ. The slope a and intercept b of the least-squares line estimate α and β. Use the **regression standard error** s to estimate σ.

- The regression standard error s has $n - 2$ **degrees of freedom.** All t procedures in regression inference have $n - 2$ degrees of freedom.

- To test **the hypothesis that the slope is zero in the population,** use the t statistic $t = b/\text{SE}_b$. This null hypothesis says that straight-line dependence on x has no value for predicting y. In practice, use software to find the slope b of the least-squares line, its standard error SE_b, and the t statistic.

- The t test for regression slope is also a test for **the hypothesis that the population correlation between x and y is zero.** To do this test without software, use the sample correlation r and Table E.

- **Confidence intervals for the slope** of the population regression line have the form $b \pm t^*\text{SE}_b$.

- **Confidence intervals for the mean response** when x has value x^* have the form $\hat{y} \pm t^*\text{SE}_{\hat{\mu}}$. **Prediction intervals** for an individual future response y have a similar form with a larger standard error, $\hat{y} \pm t^*\text{SE}_{\hat{y}}$. Software often gives these intervals.

S T A T I S T I C S   I N   S U M M A R Y

Here are the most important skills you should have acquired from reading this chapter.

A. Preliminaries

1. Make a scatterplot to show the relationship between an explanatory and a response variable.

2. Use a calculator or software to find the correlation and the equation of the least-squares regression line.

3. Recognize which type of inference you need in a particular regression setting.

B. Inference Using Software Output

1. Explain in any specific regression setting the meaning of the slope β of the population regression line.

2. Understand software output for regression. Find in the output the slope and intercept of the least-squares line, their standard errors, and the regression standard error.

3. Use that information to carry out tests of H_0: $\beta = 0$ and calculate confidence intervals for β.

4. Explain the distinction between a confidence interval for the mean response and a prediction interval for an individual response.

5. If software gives output for prediction, use that output to give either confidence or prediction intervals.

C. Checking the Conditions for Regression Inference

1. Make a stemplot or histogram of the residuals and look for strong departures from Normality.
2. Make a residual plot and look for departures from a linear pattern or unequal spread about the "residual = 0" line.
3. Ask whether the study design suggests that observations are independent.

Franz Marc Frei/CORBIS

CHECK YOUR SKILLS

Florida reappraises real estate every year, so the county appraiser's Web site lists the current "fair market value" of each piece of property. Property usually sells for somewhat more than the appraised market value. Here are the appraised market values and actual selling prices (in thousands of dollars) of condominium units sold in a beachfront building over a 19-month period:[7]

Selling Price	Appraised Value	Month	Selling Price	Appraised Value	Month
850	758.0	0	790	605.9	13
900	812.7	1	700	483.8	14
625	504.0	2	715	585.8	14
1075	956.7	2	825	707.6	14
890	747.9	8	675	493.9	17
810	717.7	8	1050	802.6	17
650	576.6	9	1325	1031.8	18
845	648.3	12	845	586.7	19

Here is part of the Minitab output for regressing selling price on appraised value, along with prediction for a unit with appraised value $802,600:

```
Predictor   Coef    SE Coef     T      P
Constant  127.27      79.49   1.60  0.132
appraisal 1.0466     0.1126   9.29  0.000

S = 69.7299   R-Sq = 86.1%   R-Sq(adj) = 85.1%

Predicted Values for New Observations

New
Obs    Fit   SE Fit      95% CI              95% PI
1    967.3     21.6  (920.9, 1013.7)  (810.7, 1123.9)
```

Exercises 23.16 to 23.24 are based on this information.

23.16 The equation of the least-squares regression line for predicting selling price from appraised value is

 (a) price $= 79.49 + 0.1126 \times$ appraised value.

 (b) price $= 127.27 + 1.0466 \times$ appraised value.

 (c) price $= 1.0466 + 127.27 \times$ appraised value.

23.17 What is the correlation between selling price and appraised value?

 (a) 0.1126   (b) 0.861   (c) 0.928

23.18 The slope β of the population regression line describes

 (a) the exact increase in the selling price of an individual unit when its appraised value increases by $1000.

 (b) the average increase in selling price in a population of units when appraised value increases by $1000.

 (c) the average selling price in a population of units when a unit's appraised value is 0.

23.19 Is there significant evidence that selling price increases as appraised value increases? To answer this question, test the hypotheses

 (a) $H_0: \beta = 0$ versus $H_a: \beta > 0$.

 (b) $H_0: \beta = 0$ versus $H_a: \beta \neq 0$.

 (c) $H_0: \alpha = 0$ versus $H_a: \alpha > 0$.

23.20 Minitab shows that the P-value for this test is

 (a) 0.132.   (b) less than 0.001.   (c) 0.861.

23.21 The regression standard error for these data is

 (a) 0.1126.   (b) 69.7299.   (c) 79.49.

23.22 Confidence intervals and tests for these data use the t distribution with degrees of freedom

 (a) 14.   (b) 15.   (c) 16.

23.23 A 95% confidence interval for the population slope β is

 (a) 1.0466 ± 0.2415.   (b) 1.0466 ± 149.5706.   (c) 1.0466 ± 0.2387.

23.24 Hamada owns a unit in this building appraised at $802,600. The Minitab output includes prediction for this appraised value. She can be 95% confident that her unit would sell for between

 (a) $920,900 and $1,013,700.

 (b) $810,700 and $1,123,900.

 (c) $945,700 and $988,900.

23.25 Too much nitrogen? Burning fossil fuels deposits extra nitrogen on the land. Too much nitrogen can reduce the variety of plants by favoring rapid growth of some species—think of putting fertilizer on your lawn to help grass choke out weeds. A study of 68 grassland sites in Britain measured nitrogen deposited (kilograms of nitrogen per hectare of land area per year) and also the "richness" of plant species (based on number of species and how abundant each species is). The authors reported a regression analysis as follows:[8]

$$\text{plant species richness} = 23.3 - 0.408 \times \text{nitrogen deposited}$$

$$r^2 = 0.55 \quad P < 0.0001$$

(a) What does the slope $b = -0.408$ say about the effect of increased nitrogen deposits on species richness?

(b) What does $r^2 = 0.55$ add to the information given by the equation of the least-squares line?

(c) What null and alternative hypotheses do you think the P-value refers to? What does this P-value tell you?

Exercise 7.46 (page 195) gives data from a study of the "gate velocity" of molten metal that experienced foundry workers choose based on the thickness of the aluminum piston being cast. Gate velocity is measured in feet per second, and the piston wall thickness is in inches. A scatterplot (you need not make one) shows a moderate positive linear relationship. Figure 23.13 displays part of the Minitab regression output. Exercises 23.26 to 23.28 analyze these data.

23.26 Casting aluminum: is there a relationship? Figure 23.13 leaves out the t statistics and their P-values. Based on the information in the output, test the hypothesis that there is no straight-line relationship between thickness and gate velocity. State hypotheses, give a test statistic and its approximate P-value, and state your conclusion.

23.27 Casting aluminum: intervals. The output in Figure 23.13 includes prediction for piston wall thickness $x^* = 0.5$ inch. Use the output to give 90% intervals for

(a) the slope of the population regression line of gate velocity on piston thickness.

(b) the average gate velocity for a type of piston with thickness 0.5 inch.

23.28 Casting aluminum: residuals. The output in Figure 23.13 includes a table of the the x and y variables, the fitted values $\hat{y}$ for each x, the residuals, and some related quantities. (This table is stored as *ex23-28.dat* on the text CD and Web site.)

(a) Plot the residuals against thickness (the explanatory variable). Use vertical scale −200 to 200 so that the pattern is clearer. Add the "residual = 0" line. Does your plot show a systematically nonlinear relationship? Does it show systematic change in the spread about the regression line?

(b) Make a histogram of the residuals. Minitab identifies the residual for Observation 9 as a suspected outlier. Does your histogram agree?

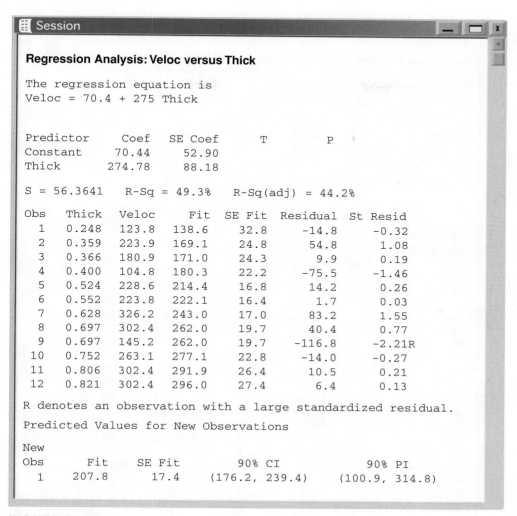

FIGURE 23.13
Minitab output for the regression of gate velocity on piston thickness in casting aluminum parts, for Exercises 23.26 to 23.28.

(c)   Redoing the regression without Observation 9 gives regression standard error $s = 42.4725$ and predicted mean velocity 216 feet per second (90% confidence interval 191.4 to 240.6) for piston walls 0.5 inch thick. Compare these values with those in Figure 23.13. Is Observation 9 influential for inference?

Table 4.1 gives 30 years' data on boats registered in Florida and manatees killed by boats. Figure 4.2 (page 102) shows a strong linear relationship. The correlation is $r = 0.953$. Figure 23.14 shows part of the Minitab regression output. Exercises 23.29 to 23.31 analyze the manatee data.

23.29 Manatees: conditions for inference. We know that there is a strong linear relationship. Let's check the other conditions for inference. Figure 23.14 includes a table of the two variables, the predicted values $\hat{y}$ for each x in the data, the residuals,

FIGURE 23.14

Minitab output for the regression of number of manatees killed by boats on the number of boats (in thousands) registered in Florida, for Exercises 23.29 to 23.31.

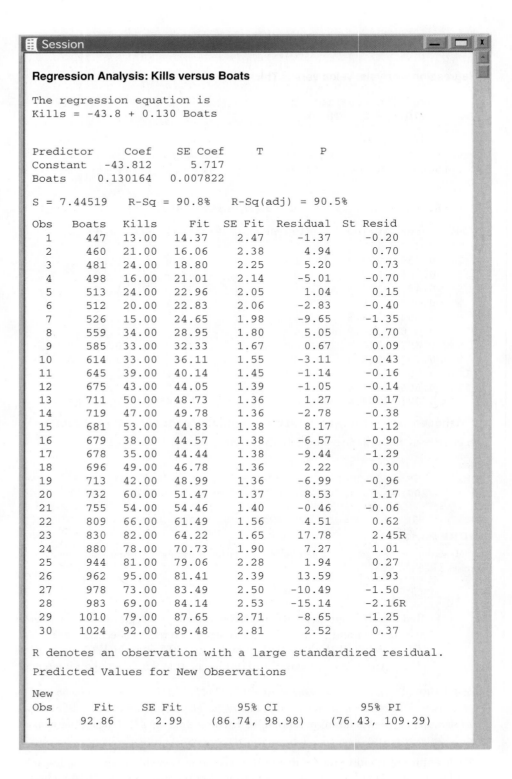

```
Session                                              _  □  X

Regression Analysis: Kills versus Boats

The regression equation is
Kills = -43.8 + 0.130 Boats


Predictor      Coef     SE Coef      T         P
Constant    -43.812       5.717
Boats      0.130164    0.007822

S = 7.44519   R-Sq = 90.8%   R-Sq(adj) = 90.5%

Obs    Boats    Kills      Fit   SE Fit   Residual  St Resid
  1      447    13.00    14.37     2.47      -1.37     -0.20
  2      460    21.00    16.06     2.38       4.94      0.70
  3      481    24.00    18.80     2.25       5.20      0.73
  4      498    16.00    21.01     2.14      -5.01     -0.70
  5      513    24.00    22.96     2.05       1.04      0.15
  6      512    20.00    22.83     2.06      -2.83     -0.40
  7      526    15.00    24.65     1.98      -9.65     -1.35
  8      559    34.00    28.95     1.80       5.05      0.70
  9      585    33.00    32.33     1.67       0.67      0.09
 10      614    33.00    36.11     1.55      -3.11     -0.43
 11      645    39.00    40.14     1.45      -1.14     -0.16
 12      675    43.00    44.05     1.39      -1.05     -0.14
 13      711    50.00    48.73     1.36       1.27      0.17
 14      719    47.00    49.78     1.36      -2.78     -0.38
 15      681    53.00    44.83     1.38       8.17      1.12
 16      679    38.00    44.57     1.38      -6.57     -0.90
 17      678    35.00    44.44     1.38      -9.44     -1.29
 18      696    49.00    46.78     1.36       2.22      0.30
 19      713    42.00    48.99     1.36      -6.99     -0.96
 20      732    60.00    51.47     1.37       8.53      1.17
 21      755    54.00    54.46     1.40      -0.46     -0.06
 22      809    66.00    61.49     1.56       4.51      0.62
 23      830    82.00    64.22     1.65      17.78      2.45R
 24      880    78.00    70.73     1.90       7.27      1.01
 25      944    81.00    79.06     2.28       1.94      0.27
 26      962    95.00    81.41     2.39      13.59      1.93
 27      978    73.00    83.49     2.50     -10.49     -1.50
 28      983    69.00    84.14     2.53     -15.14     -2.16R
 29     1010    79.00    87.65     2.71      -8.65     -1.25
 30     1024    92.00    89.48     2.81       2.52      0.37

R denotes an observation with a large standardized residual.

Predicted Values for New Observations

New
Obs      Fit    SE Fit      95% CI              95% PI
  1    92.86      2.99   (86.74, 98.98)   (76.43, 109.29)
```

and related quantities. (This table is stored as *ex23-29.dat* on the text CD and Web site.)

(a) Round the residuals to the nearest whole number and make a stemplot. The distribution is single-peaked and symmetric and appears close to Normal.

(b) Make a residual plot, residuals against boats registered. Use a vertical scale from −25 to 25 to show the pattern more clearly. Add the "residual = 0" line. There is no clearly nonlinear pattern. The spread about the line may be a bit greater for larger values of the explanatory variable, but the effect is not large.

(c) It is reasonable to regard the number of manatees killed by boats in successive years as independent. The number of boats grew over time. Someone says that pollution also grew over time and may explain more manatee deaths. How would you respond to this idea?

23.30 Manatees: do more boats bring more kills? The output in Figure 23.14 omits the t statistics and their P-values. Based on the information in the output, is there good evidence that the number of manatees killed increases as the number of boats registered increases? State hypotheses and give a test statistic and its approximate P-value. What do you conclude?

23.31 Manatees: estimation. The output in Figure 23.14 includes prediction of manatees killed when there are 1,050,000 boats registered in Florida. Give 95% intervals for

(a) the increase in manatees killed for each additional 1000 boats registered.

(b) the number of manatees that will be killed next year if there are 1,050,000 boats registered next year.

23.32 Fidgeting keeps you slim: inference. Our first example of regression (Example 5.1, page 126) presented data showing that people who increased their nonexercise activity (NEA) when they were deliberately overfed gained less fat than other people. Use software to add formal inference to the data analysis for these data.

(a) Based on 16 subjects, the correlation between NEA increase and fat gain was $r = -0.7786$. Is this significant evidence that people with higher NEA increase gain less fat? (Report a t statistic from regression output and give the one-sided P-value.)

(b) The slope of the least-squares regression line was $b = -0.00344$, so that fat gain decreased by 0.00344 kilogram for each added calorie of NEA. Give a 90% confidence interval for the slope of the population regression line. This rate of change is the most important parameter to be estimated.

(c) Sam's NEA increases by 400 calories. His predicted fat gain is 2.13 kilograms. Give a 95% interval for predicting Sam's fat gain.

23.33 Predicting tropical storms. Exercise 5.53 (page 158) gives data on William Gray's predictions of the number of named tropical storms in Atlantic hurricane seasons from 1984 to 2007. Use these data for regression inference as follows.

(a) Does Professor Gray do better than random guessing? That is, is there a significantly positive correlation between his forecasts and the actual number of storms? (Report a t statistic from regression output and give the one-sided P-value.)

(b) Give a 95% confidence interval for the mean number of storms in years when Professor Gray forecasts 16 storms.

23.34 Sparrowhawk colonies. One of nature's patterns connects the percent of adult birds in a colony that return from the previous year and the number of new adults that join the colony. Exercise 4.28 (page 116) gives data for 13 colonies of sparrowhawks. In Exercises 4.28 and 5.36 (page 153), you showed that there is a moderately strong linear relationship.

(a) Biologists conjecture that the number of new adults joining colonies goes down as higher percents of adults return from the previous year. Do the data show this effect? Is it statistically significant?

(b) Use the data to predict with 95% confidence the average number of new birds in colonies to which 60% of the past year's adults return.

23.35 Predicting tropical storms: residuals. Make a stemplot of the residuals (round to the nearest tenth) from your regression in Exercise 23.33. Explain why your plot suggests that we should not use these data to get a prediction interval for the number of storms in a single year.

23.36 Sparrowhawk colonies: residuals. The regression of number of new birds that join a sparrowhawk colony on the percent of adult birds in the colony that return from the previous year is an example of data that satisfy the conditions for regression inference well. Here are the residuals for the 13 colonies in Exercise 23.34:

Percent return	74	66	81	52	73	62	52
Residual	−4.44	−5.87	0.69	−5.13	2.26	1.92	−0.13
Percent return	45	62	46	60	46	38	
Residual	−1.25	4.92	0.05	5.31	2.05	−0.38	

(a) **Linear relationship.** A plot of the residuals against the explanatory variable x magnifies the deviations from the least-squares line. Does the plot show any systematic deviation from a roughly linear pattern?

(b) **Normal variation about the line.** Make a histogram of the residuals. With only 13 observations, no clear shape emerges. Do strong skewness or outliers suggest lack of Normality?

(c) **Independent observations.** Why are the 13 observations independent?

(d) **Spread about the line stays the same.** Does your plot in (a) show any systematic change in spread as x changes?

23.37 Our brains don't like losses. Exercise 4.29 (page 116) describes an experiment that showed a linear relationship between how sensitive people are to monetary losses ("behavioral loss aversion") and activity in one part of their brains ("neural loss aversion").

(a) Make a scatterplot with neural loss aversion as x and behavioral loss aversion as y. One point is a high outlier in both the x and y directions. In Exercise 5.37

(page 153) you found that this outlier is not influential for the least-squares line.

(b) The research report says that $r = 0.85$ and that the test for regression slope has $P < 0.001$. Verify these results, using all of the observations.

(c) The report recognizes the outlier and says, "However, this regression also remained highly significant ($P = 0.004$) when the extreme data point (top right corner) was removed from the analysis." Repeat your analysis omitting the outlier. Show that the outlier influences regression inference by comparing the t statistic for testing slope with and without the outlier. Then verify the report's claim about the P-value of this test.

23.38 Time at the table. Does how long young children remain at the lunch table help predict how much they eat? Here are data on 20 toddlers observed over several months at a nursery school.[9] "Time" is the average number of minutes a child spent at the table when lunch was served. "Calories" is the average number of calories the child consumed during lunch, calculated from careful observation of what the child ate each day.

Time	21.4	30.8	37.7	33.5	32.8	39.5	22.8	34.1	33.9	43.8
Calories	472	498	465	456	423	437	508	431	479	454
Time	42.4	43.1	29.2	31.3	28.6	32.9	30.6	35.1	33.0	43.7
Calories	450	410	504	437	489	436	480	439	444	408

(a) Make a scatterplot. Find the correlation and the least-squares regression line. (Be sure to save the regression residuals.) Based on your work, describe the direction, form, and strength of the relationship.

(b) Check the conditions for regression inference. Parts (a) to (d) of Exercise 23.36 provide a handy outline. Use vertical limits -100 to 100 in your plot of the residuals against time to help you see the pattern. What do you conclude?

(c) Is there significant evidence that more time at the table is associated with more calories consumed? Give a 95% confidence interval to estimate how rapidly calories consumed changes as time at the table increases.

23.39 DNA on the ocean floor. We think of DNA as the stuff that stores the genetic code. It turns out that DNA occurs, mainly outside living cells, on the ocean floor. It is important in nourishing seafloor life. Scientists think that this DNA comes from organic matter that settles to the bottom from the top layers of the ocean. "Phytopigments," which come mainly from algae, are a measure of the amount of organic matter that has settled to the bottom. The data file *ex23-39.dat* on the text CD and Web site contains data on concentrations of DNA and phytopigments (both in grams per square meter) in 116 ocean locations around the world.[10] Look first at DNA alone. Describe the distribution of DNA concentration and give a confidence interval for the mean concentration. Be sure to explain why your confidence interval is trustworthy in the light of the shape of the distribution. The data show surprisingly high DNA concentration, and this by itself was an important finding.

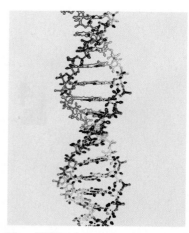

Minoru Toi/Getty Images

23.40 Time at the table: prediction. Rachel is another child at the nursery school of Exercise 23.38. Over several months, Rachel averages 40 minutes at the lunch table. Give a 95% interval to predict Rachel's average calorie consumption at lunch.

Exercises 23.41 to 23.45 ask practical questions of regression inference without step-by-step instructions. Do complete regression analyses, using the **Plan, Solve,** *and* **Conclude** *steps of the four-step process to organize your answers. Follow the model of Example 23.9 and the following discussion, with checking the conditions as part of the* **Solve** *step.*

23.41 Squirrels and their food supply. Exercise 7.25 (page 188) gives data on the abundance of the pine cones that red squirrels feed on and the mean number of offspring per female squirrel over 16 years. The strength of the relationship is remarkable because females produce young before the food is available. How significant is the evidence that more cones leads to more offspring? (Use a vertical scale from −2 to 2 in your residual plot to show the pattern more clearly.)

23.42 A big toe problem. Table 7.4 (page 194) and Exercises 7.42 and 7.44 describe the relationship between two deformities of the feet in young patients. Metatarsus abductus (MA) may help predict the severity of hallux abducto valgus (HAV). The paper that reports this study says, "Linear regression analysis, using the hallux abductus angle as the response variable, demonstrated a significant correlation between the metatarsus abductus and hallux abductus angles."[11] Do a suitable analysis to verify this finding. The study authors note that the scatterplot suggests that the variation in y may change as x changes, so they offer a more elaborate analysis as well.

23.43 Beavers and beetles. Exercise 5.51 (page 157) describes a study that found that the number of stumps from trees felled by beavers predicts the abundance of beetle larvae. Is there good evidence that more beetle larvae clusters are present when beavers have left more tree stumps? Estimate how many more clusters accompany each additional stump, with 95% confidence.

23.44 Sulfur, the ocean, and the sun. Sulfur in the atmosphere affects climate by influencing formation of clouds. The main natural source of sulfur is dimethylsulfide (DMS) produced by small organisms in the upper layers of the oceans. DMS production is in turn influenced by the amount of energy the upper ocean receives from sunlight. Exercise 4.30 (page 117) gives monthly data on solar radiation dose (SRD, in watts per square meter) and surface DMS concentration (in nanomolars) for a region in the Mediterranean. Do the data provide convincing evidence that DMS increases as SRD increases? We also want to estimate the rate of increase, with 90% confidence.

23.45 DNA on the ocean floor. Another conclusion of the study introduced in Exercise 23.39 was that organic matter settling down from the top layers of the ocean is the main source of DNA on the seafloor. An important piece of evidence is the relationship between DNA and phytopigments. Do the data in the file *ex23-39.dat* give good reason to think that phytopigment concentration helps explain DNA concentration? (Try vertical limits −1 to 1 to make the pattern of your residual plot clearer.)

23.46 A lurking variable (optional). Return to the data on selling price versus appraised value for beachfront condominiums that are the basis for the Check Your Skills Exercises 23.16 to 23.24. Prices for beachfront property were rising rapidly during this period. Because property is reassessed just once a year, selling prices might pull away from appraised values over time. The data are in order by date of the sale, and the data table includes the number of months from the start of the data period. Here are the residuals from the regression of selling price on appraised value (rounded):

−70.60	−77.85	−29.76	−53.56	−20.03	−68.42	−80.75	39.21
28.59	66.38	−25.38	−42.85	30.81	82.72	117.83	103.68

(a) Plot the residuals against the explanatory variable (appraised value). To make the pattern clearer, use vertical limits −200 to 200. Explain why the pattern you see agrees with the conditions of linear relationship and constant standard deviation needed for regression inference.

(b) Make a stemplot of the residuals. The distribution has a bit of a cluster at the left, but there are no outliers or other strong deviations from Normality that would prevent regression inference.

(c) Next, plot the residuals against month. Explain why the pattern fits the fact that selling prices were rising rapidly. Is your prediction in Exercise 23.24 likely to be too low or too high? (*Comment:* As this example illustrates, it is often wise to plot residuals against important lurking variables as well as against the explanatory variable. The margin of error for prediction includes the effect of rising prices. Your plot shows that the prediction could be improved by using month as a second explanatory variable. This is *multiple regression,* using more than one explanatory variable to predict a response.)

23.47 Standardized residuals (optional). Software often calculates **standardized residuals** as well as the actual residuals from regression. Because the standardized residuals have the standard z-score scale, it is easier to judge whether any are extreme. Figure 23.13 and the data file *ex23-28.dat* include the standardized residuals for the regression of gate velocity on piston wall thickness.

standardized residuals

(a) Find the mean and standard deviation of the standardized residuals. Why do you expect values close to those you obtain?

(b) Make a stemplot of the standardized residuals. Are there any striking deviations from Normality?

(c) The most extreme standardized residual is $z = −2.21$. Minitab flags this as "large." What is the probability that a standard Normal variable takes a value this extreme (that is, less than −2.21 or greater than 2.21)? Your result suggests that a residual this extreme would be a bit unusual when there are only 12 observations. That's why we examined Observation 9 in Exercise 23.28.

23.48 Tests for the intercept (optional). Figure 23.7 (page 611) gives Minitab output for the regression of blood alcohol content (BAC) on number of beers consumed. The t test for the hypothesis that the population regression line has *slope* $\beta = 0$ has $P < 0.001$. The data show a positive linear relationship between BAC and beers. We might expect the *intercept* α of the population regression line to be 0, because no beers ($x = 0$) should produce no alcohol in the blood ($y = 0$). To test

$$H_0: \alpha = 0$$
$$H_a: \alpha \neq 0$$

we use a t statistic formed by dividing the least-squares intercept a by its standard error SE_a. Locate this statistic in the output of Figure 23.7 and verify that it is in fact a divided by its standard error. What is the P-value? Do the data suggest that the intercept is not 0?

23.49 Confidence intervals for the intercept (optional). The output in Figure 23.7 allows you to calculate confidence intervals for both the slope β and the intercept α of the population regression line of BAC on beers in the population of all students. Confidence intervals for the intercept α have the familiar form $a \pm t^*SE_a$ with degrees of freedom $n - 2$. What is the 95% confidence interval for the intercept? Does it contain 0, the value we might guess for α?

Getty Images/AsiaPix

One-Way Analysis of Variance: Comparing Several Means

The two-sample *t* procedures of Chapter 18 compare the means of two populations or the mean responses to two treatments in an experiment. Of course, studies don't always compare just two groups. We need a method for comparing any number of means.

EXAMPLE 24.1 Comparing tropical flowers

STATE: Ethan Temeles and W. John Kress of Amherst College studied the relationship between varieties of the tropical flower *Heliconia* on the island of Dominica and the different species of hummingbirds that fertilize the flowers.[1] Over time, the researchers believe, the lengths of the flowers and the forms of the hummingbirds' beaks have evolved to match each other. If that is true, flower varieties fertilized by different hummingbird species should have distinct distributions of length.

Table 24.1 gives length measurements (in millimeters) for samples of three varieties of *Heliconia*, each fertilized by a different species of hummingbird. Do the three varieties display distinct distributions of length? In particular, are the mean lengths of their flowers different?

PLAN: Use graphs and numerical descriptions to describe and compare the three distributions of flower length. Finally, ask whether the differences among the mean lengths of the three varieties are *statistically significant*.

TABLE 24.1 Flower lengths (millimeters) for three *Heliconia* varieties

H. bihai

47.12	46.75	46.81	47.12	46.67	47.43	46.44	46.64
48.07	48.34	48.15	50.26	50.12	46.34	46.94	48.36

H. caribaea red

41.90	42.01	41.93	43.09	41.47	41.69	39.78	40.57
39.63	42.18	40.66	37.87	39.16	37.40	38.20	38.07
38.10	37.97	38.79	38.23	38.87	37.78	38.01	

H. caribaea yellow

36.78	37.02	36.52	36.11	36.03	35.45	38.13	37.10
35.17	36.82	36.66	35.68	36.03	34.57	34.63	

SOLVE (first steps): We first met these data in Chapter 2 (page 56), where we compared the distributions. Figure 24.1 repeats a side-by-side stemplot from Chapter 2. The lengths have been rounded to the nearest tenth of a millimeter. Here are the summary measures we will use in further analysis:

Sample	Variety	Sample size	Mean length	Standard deviation
1	*bihai*	16	47.60	1.213
2	red	23	39.71	1.799
3	yellow	15	36.18	0.975

CONCLUDE (first steps): The three varieties differ so much in flower length that there is little overlap among them. In particular, the flowers of *bihai* are longer than either red or yellow. The mean lengths are 47.6 mm for H. *bihai*, 39.7 mm for H. *caribaea* red, and 36.2 mm for H. *caribaea* yellow. Are these observed differences in sample means statistically significant? We must develop a test for comparing more than two population means. ■

Kevin Schafer/Alamy

FIGURE 24.1

Side-by-side stemplots comparing the lengths in millimeters of samples of flowers from three varieties of *Heliconia*, from Table 24.1.

```
     bihai            red               yellow
 34 |            34 |              34 | 6 6
 35 |            35 |              35 | 2 5 7
 36 |            36 |              36 | 0 0 1 5 7 8 8
 37 |            37 | 4 8 9        37 | 0 1
 38 |            38 | 0 0 1 1 2 2 8 9   38 | 1
 39 |            39 | 2 6 8        39 |
 40 |            40 | 6 7          40 |
 41 |            41 | 5 7 9 9      41 |
 42 |            42 | 0 2          42 |
 43 |            43 | 1            43 |
 44 |            44 |              44 |
 45 |            45 |              45 |
 46 | 3 4 6 7 8 8 9  46 |          46 |
 47 | 1 1 4       47 |            47 |
 48 | 1 2 3 4     48 |            48 |
 49 |            49 |              49 |
 50 | 1 3         50 |            50 |
```

Comparing several means

Call the mean lengths for the three populations of flowers μ_1 for *bihai*, μ_2 for red, and μ_3 for yellow. The subscript reminds us which group a parameter or statistic describes. To compare these three population means, we might use the two-sample *t* test several times:

- Test H_0: $\mu_1 = \mu_2$ to see if the mean length for *bihai* differs from the mean for red.

- Test H_0: $\mu_1 = \mu_3$ to see if *bihai* differs from yellow.

- Test H_0: $\mu_2 = \mu_3$ to see if red differs from yellow.

The weakness of doing three tests is that we get three *P*-values, one for each test alone. That doesn't tell us how likely it is that *three* sample means are spread apart as far as these are. It may be that $\bar{x}_1 = 47.60$ and $\bar{x}_3 = 36.18$ are significantly different if we look at just two groups but not significantly different if we know that they are the largest and the smallest means in three groups. As we look at more groups, we expect the gap between the largest and smallest sample mean to get larger. (Think of comparing the tallest and shortest person in larger and larger groups of people.) *We can't safely compare many parameters by doing tests or confidence intervals for two parameters at a time.*

The problem of how to do many comparisons at once with an overall measure of confidence in all our conclusions is common in statistics. This is the problem of **multiple comparisons.** Statistical methods for dealing with multiple compar-   *multiple comparisons*
isons usually have two steps:

1. An *overall test* to see if there is good evidence of *any* differences among the parameters that we want to compare.

2. A detailed *follow-up* analysis to decide which of the parameters differ and to estimate how large the differences are.

The overall test, though more complex than the tests we met in Chapters 17 to 20, is reasonably straightforward. Formal follow-up analysis can be quite elaborate. We will concentrate on the overall test and use data analysis to describe in detail the nature of the differences. Companion Chapter 28, on the text CD and Web site, presents some details of follow-up inference.

The analysis of variance *F* test

We want to test the null hypothesis that there are *no differences* among the mean lengths for the three populations of flowers:

$$H_0: \mu_1 = \mu_2 = \mu_3$$

The basic *conditions for inference* (more detail later) are that we have random samples from the three populations and that flower lengths are Normally distributed in each population.

The alternative hypothesis is that there is *some difference*. That is, not all three population means are equal:

$$H_a: \text{not all of } \mu_1, \ \mu_2, \text{ and } \mu_3 \text{ are equal}$$

analysis of variance F test

The alternative hypothesis is no longer one-sided or two-sided. It is "many-sided," because it allows any relationship other than "all three equal." For example, H_a includes the case in which $\mu_2 = \mu_3$ but μ_1 has a different value. The test of H_0 against H_a is called the **analysis of variance F test.** Analysis of variance is usually abbreviated as ANOVA. The ANOVA F test is almost always carried out by software that reports the test statistic and its P-value.

EXAMPLE 24.2 Comparing tropical flowers: ANOVA

SOLVE (inference): Software tells us that for the flower length data in Table 24.1, the test statistic is $F = 259.12$ with P-value $P < 0.0001$. There is very strong evidence that the three varieties of flowers do not all have the same mean length.

The F test does not say *which* of the three means are significantly different. It appears from our preliminary data analysis that *bihai* flowers are distinctly longer than either red or yellow. Red and yellow are closer together, but the red flowers tend to be longer.

CONCLUDE: There is strong evidence ($P < 0.0001$) that the population means are not all equal. The most important difference among the means is that the *bihai* variety has longer flowers than the red and yellow varieties. ■

Example 24.2 illustrates our approach to comparing means. The ANOVA F test (done by software) assesses the evidence for *some* difference among the population means. Formal follow-up analysis would allow us to say which means differ and by how much, with (say) 95% confidence that *all* our conclusions are correct. We rely instead on examination of the data to show what differences are present and whether they are large enough to be interesting.

APPLY YOUR KNOWLEDGE

24.1   Choose your parents wisely. To live long, it helps to have long-lived parents. One study of this "inheritance of longevity" looked at records of families in a small mountain valley in France where the style of life changed little over time. Married couples in which both husband and wife were born between 1745 and 1849 were divided into four groups as follows:

	Short-lived wife	Long-lived wife
Short-lived husband	Group A	Group C
Long-lived husband	Group B	Group D

"Long-lived" parents were in the top 70% of the distribution of age at death, "short-lived" parents were in the bottom 30%. To examine extended life, the investigators looked at the ages at death of the children of these parents who lived to at least age 55. The report includes a graph like Figure 24.2 and concludes: "The mean

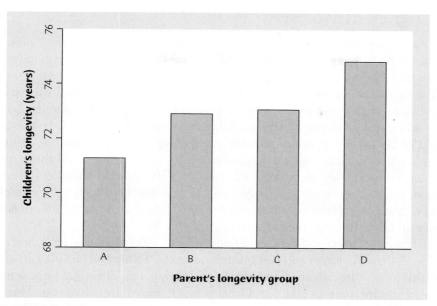

FIGURE 24.2

Bar graph comparing the longevity of children from four groups of parents, for Exercise 24.1.

life spans varied significantly among the four groups, indicating an influence of parental phenotype on offspring longevity (ANOVA, $F = 9.2$, d.f. $= 3, 1085$, $P = 0.0001$)."[2]

(a)   What are the null and alternative hypotheses for the ANOVA F test? Be sure to explain what means the test compares.

(b)   Based on the graph and the F test, what do you conclude?

24.2   **Road rage.** "The phenomenon of road rage has been frequently discussed but infrequently examined." So begins a report based on interviews with 1382 randomly selected drivers.[3] The respondents' answers to interview questions produced scores on an "angry/threatening driving scale" with values between 0 and 19. What driver characteristics go with road rage? There were no significant differences among races or levels of education. What about the effect of the driver's age? Here are the mean responses for three age groups:

<30 yr	30–55 yr	>55 yr
2.22	1.33	0.66

The report says that $F = 34.96$, with $P < 0.01$.

(a)   What are the null and alternative hypotheses for the ANOVA F test? Be sure to explain what means the test compares.

(b)   Based on the sample means and the F test, what do you conclude?

Using technology

Any technology used for statistics should perform analysis of variance. Figure 24.3 displays ANOVA output for the data of Table 24.1 from a graphing calculator, a statistical program, and a spreadsheet program.

The two software outputs give the sizes of the three samples and their means. These agree with those in Example 24.1. Minitab also gives the standard deviations and Excel gives the variances. All three outputs report the F test statistic, $F = 259.12$, and its P-value. Minitab sensibly reports the P-value as 0 to three decimal places, which is all we need to know in practice. Excel and the graphing calculator offer the specific value 1.92×10^{-27}. (This would be correct if the population distributions were exactly Normal. In practice, read such values simply as "P is very small.") There is very strong evidence that the three varieties of flowers do not all have the same mean length.

All three outputs report degrees of freedom (df), sums of squares (SS), and mean squares (MS). We don't need this information now. Minitab also gives confidence intervals for all three means that help us see which means differ and by how much. None of the intervals overlap, and *bihai* is much above the other two. These are 95% confidence intervals for each mean separately. We are *not* 95% confident that *all three* intervals cover the three means. This is another example of the peril of multiple comparisons.

APPLY YOUR KNOWLEDGE

24.3 **Logging in the rain forest.** How does logging in a tropical rain forest affect the forest in later years? Researchers compared forest plots in Borneo that had never been logged (Group 1) with similar plots nearby that had been logged 1 year earlier (Group 2) and 8 years earlier (Group 3). Although the study was not an experiment, the authors explain why we can consider the plots to be randomly selected. The data appear in Table 24.2 (page 640). The variable Trees is the count of trees in a plot; Species is the count of tree species in a plot. The variable Richness is Species/Trees, the number of species divided by the number of individual trees.[4]

    (a)   Make side-by-side stemplots of Trees for the three groups. Use stems 0, 1, 2, and 3 and split the stems (see page 21). What effects of logging are visible?

    (b)   Figure 24.4 (page 640) shows Excel ANOVA output for Trees. What do the group means show about the effects of logging?

    (c)   What are the values of the ANOVA F statistic and its P-value? What hypotheses does F test? What conclusions about the effects of logging on number of trees do the data lead to?

24.4 **Do good smells bring good business?** Businesses know that customers often respond to background music. Do they also respond to odors? Nicolas Guéguen and his colleagues studied this question in a small pizza restaurant in France on Saturday evenings in May. On one evening, a relaxing lavender odor was spread through the restaurant; on another evening, a stimulating lemon odor; a third evening served as a control, with no odor. The three evenings were comparable in many ways (weather, customer count, and so on), so we are willing to regard the data as

Texas Instruments Graphing Calculator

```
One-way ANOVA
 F=259.1192995
 P=1.918818E-27
 Factor
  df=2
  SS=1082.87237
↓ MS=541.436183
```

```
One-way ANOVA
↑ MS=541.436183
 Error
  df=51
  SS=106.565761
  MS=2.08952472
  SxP=1.44551884
```

FIGURE 24.3
ANOVA for the flower length data: output from a graphing calculator, a statistical program, and a spreadsheet program.

Minitab

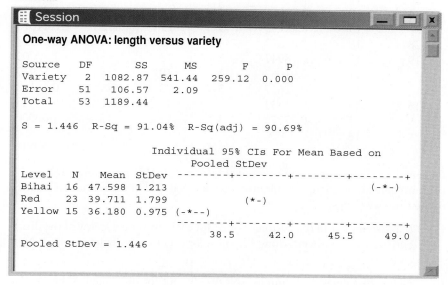

```
Session                                               _ □ x

One-way ANOVA: length versus variety

Source    DF       SS      MS       F       P
Variety    2  1082.87  541.44  259.12   0.000
Error     51   106.57    2.09
Total     53  1189.44


S = 1.446   R-Sq = 91.04%   R-Sq(adj) = 90.69%

                      Individual 95% CIs For Mean Based on
                                  Pooled StDev
Level      N    Mean  StDev  --------+--------+--------+--------+
Bihai     16  47.598  1.213                               (-*-)
Red       23  39.711  1.799                  (*-)
Yellow    15  36.180  0.975  (-*--)
                             --------+--------+--------+--------+
                                 38.5     42.0     45.5     49.0

Pooled StDev = 1.446
```

Excel

Microsoft Excel - ta25-01.dat						
	A	B	C	D	E	F	G
1	Anova: Single Factor						
2							
3	SUMMARY						
4	*Groups*	*Count*	*Sum*	*Average*	*Variance*		
5	bihai	16	761.56	47.5975	1.471073		
6	red	23	913.36	39.7113	3.235548		
7	yellow	15	542.7	36.18	0.951257		
8							
9							
10	ANOVA						
11	*Source of variation*	*SS*	*df*	*MS*	*F*	*P-value*	*F crit*
12	Between Groups	1082.872	2	541.4362	259.1193	1.92E-27	3.178799
13	Within Groups	106.5658	51	2.089525			
14							
15	Total	1189.438	53				

TABLE 24.2 Data from a study of logging in Borneo

GROUP	TREES	SPECIES	RICHNESS	GROUP	TREES	SPECIES	RICHNESS
1	27	22	0.81481	2	18	15	0.83333
1	22	18	0.81818	2	17	15	0.88235
1	29	22	0.75862	2	14	12	0.85714
1	21	20	0.95238	2	14	13	0.92857
1	19	15	0.78947	2	2	2	1.00000
1	33	21	0.63636	2	17	15	0.88235
1	16	13	0.81250	2	19	8	0.42105
1	20	13	0.65000	3	18	17	0.94444
1	24	19	0.79167	3	4	4	1.00000
1	27	13	0.48148	3	22	18	0.81818
1	28	19	0.67857	3	15	14	0.93333
1	19	15	0.78947	3	18	18	1.00000
2	12	11	0.91667	3	19	15	0.78947
2	12	11	0.91667	3	22	15	0.68182
2	15	14	0.93333	3	12	10	0.83333
2	9	7	0.77778	3	12	12	1.00000
2	20	18	0.90000				

independent SRSs from spring Saturday evenings at this restaurant. Table 24.3 contains data on how long (in minutes) customers stayed in the restaurant on each of the three evenings.[5]

(a)   Make stemplots of the customer times for each evening. Do any of the distributions show outliers, strong skewness, or other clear deviations from Normality?

(b)   Figure 24.5 gives the Minitab ANOVA output for these data. What do the mean times in the restaurant say about the effects of the two odors?

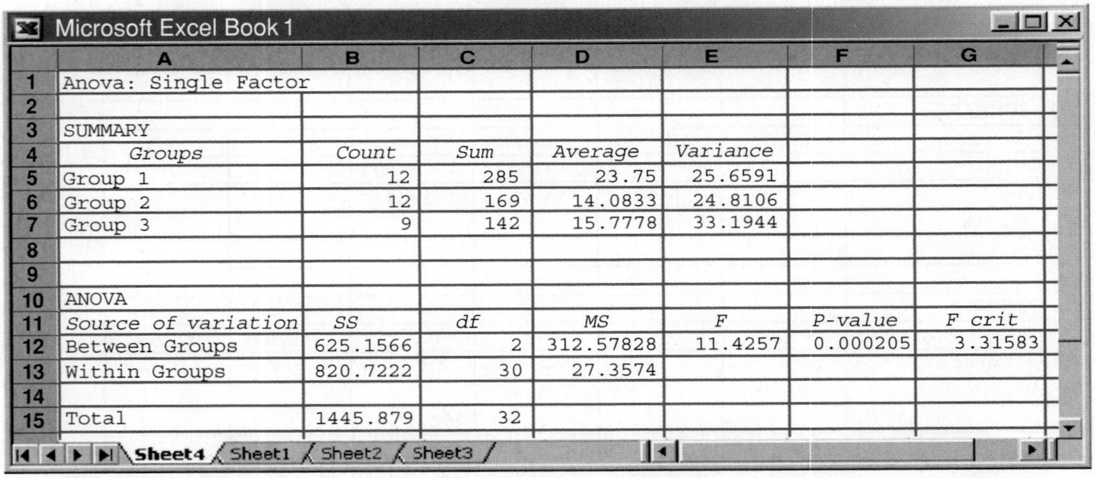

	A	B	C	D	E	F	G
1	Anova: Single Factor						
2							
3	SUMMARY						
4	Groups	Count	Sum	Average	Variance		
5	Group 1	12	285	23.75	25.6591		
6	Group 2	12	169	14.0833	24.8106		
7	Group 3	9	142	15.7778	33.1944		
8							
9							
10	ANOVA						
11	Source of variation	SS	df	MS	F	P-value	F crit
12	Between Groups	625.1566	2	312.57828	11.4257	0.000205	3.31583
13	Within Groups	820.7222	30	27.3574			
14							
15	Total	1445.879	32				

Microsoft Excel Book 1

Sheet4 / Sheet1 / Sheet2 / Sheet3 /

FIGURE 24.4

Excel output for analysis of variance on the number of trees in forest plots, for Exercise 24.3.

TABLE 24.3 Time (minutes) that customers remain in a restaurant when exposed to odors

LAVENDER ODOR

92	126	114	106	89	137	93	76	98	108
124	105	129	103	107	109	94	105	102	108
95	121	109	104	116	88	109	97	101	106

LEMON ODOR

78	104	74	75	112	88	105	97	101	89
88	73	94	63	83	108	91	88	83	106
108	60	96	94	56	90	113	97		

NO ODOR

103	68	79	106	72	121	92	84	72	92
85	69	73	87	109	115	91	84	76	96
107	98	92	107	93	118	87	101	75	86

(c)   What are the values of the ANOVA F statistic and its P-value? What hypotheses does F test? Briefly describe the conclusions you draw from these data. Did you find anything surprising?

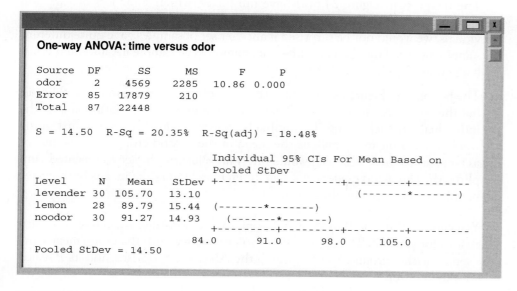

FIGURE 24.5
Minitab output for the data in Table 24.3 on time in minutes that customers spend in a restaurant, for Exercise 24.4.

The idea of analysis of variance

The details of ANOVA are a bit daunting (they appear in an optional section at the end of this chapter). The main idea of ANOVA is more accessible and much more important. Here it is: when we ask if a set of sample means gives evidence for differences among the population means, what matters is not how far apart the sample means are but how far apart they are *relative to the variability of individual observations*.

Look at the two sets of boxplots in Figure 24.6. For simplicity, these distributions are all symmetric, so that the mean and median are the same. The center line in each boxplot is therefore the sample mean. Both sets of boxplots compare three samples with the same three means. Could differences this large easily arise just due to chance, or are they statistically significant?

FIGURE 24.6

Boxplots for two sets of three samples each. The sample means are the same in (a) and (b). Analysis of variance will find a more significant difference among the means in (b) because there is less variation among the individuals within those samples.

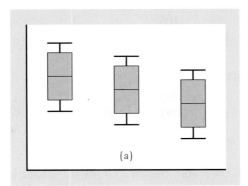
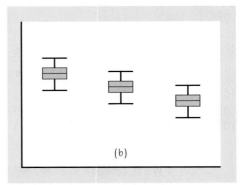

(a)   (b)

- The boxplots in Figure 24.6(a) have tall boxes, which show lots of variation among the individuals in each group. With this much variation among individuals, we would not be surprised if another set of samples gave quite different sample means. The observed differences among the sample means could easily happen just by chance.

- The boxplots in Figure 24.6(b) have the same centers as those in Figure 24.6(a), but the boxes are much shorter. That is, there is much less variation among the individuals in each group. It is unlikely that any sample from the first group would have a mean as small as the mean of the second group. Because means as far apart as those observed would rarely arise just by chance in repeated sampling, they are good evidence of real differences among the means of the three populations we are sampling from.

You can use the *One-Way ANOVA* applet to demonstrate the analysis of variance idea for yourself. The applet allows you to change both the group means and the spread within groups. You can watch the ANOVA F statistic and its P-value change as you work.

This comparison of the two parts of Figure 24.6 is too simple in one way. It ignores the effect of the sample sizes, an effect that boxplots do not show. *Small differences among sample means can be significant if the samples are large. Large differences among sample means can fail to be significant if the samples are small.* All we can be sure of is that for the same sample size, Figure 24.6(b) will give a much smaller P-value than Figure 24.6(a). Despite this qualification, the big idea remains: if sample means are far apart relative to the variation among individuals in the same groups, that's evidence that something other than chance is at work.

> ### THE ANALYSIS OF VARIANCE IDEA
>
> **Analysis of variance** compares the variation due to specific sources with the variation among individuals who should be similar. In particular, ANOVA tests whether several populations have the same mean by comparing how far apart the sample means are with how much variation there is within the samples.

It is one of the oddities of statistical language that methods for comparing means are named after the variance. The reason is that the test works by comparing two kinds of variation. Analysis of variance is a general method for studying sources of variation in responses. Comparing several means is the simplest form of ANOVA, called **one-way ANOVA.**

one-way ANOVA

> ### THE ANOVA *F* STATISTIC
>
> The **analysis of variance *F* statistic** for testing the equality of several means has this form:
>
> $$F = \frac{\text{variation among the sample means}}{\text{variation among individuals in the same sample}}$$

If you want more detail, read the optional section at the end of this chapter. The *F* statistic can take only values that are zero or positive. It is zero only when all the sample means are identical and gets larger as they move farther apart. Large values of *F* are evidence against the null hypothesis H_0 that all population means are the same. Although the alternative hypothesis H_a is many-sided, the ANOVA *F* test is one-sided because any violation of H_0 tends to produce a large value of *F*.

APPLY YOUR KNOWLEDGE

24.5 ANOVA compares several means. The *One-Way* ANOVA applet displays the observations in three groups, with the group means highlighted by black dots. When you open or reset the applet, the scale at the bottom of the display shows that for these groups the ANOVA *F* statistic is $F = 31.74$, with $P < 0.001$. (The *P*-value is marked by a red dot that moves along the scale.)

(a) The middle group has larger mean than the other two. Grab its mean point with the mouse. How small can you make *F*? What did you do to the mean to make *F* small? Roughly how significant is your small *F*?

(b) Starting with the three means aligned from your configuration at the end of (a), drag any one of the group means either up or down. What happens to *F*? What happens to the *P*-value? Convince yourself that the same thing happens if you move any one of the means, or if you move one slightly and then another slightly in the opposite direction.

24.6   **ANOVA uses within-group variation.** Reset the *One-Way ANOVA* applet to its original state. As in Figure 24.6(b), the differences among the three means are highly significant (large F, small P-value) because the observations in each group cluster tightly about the group mean.

(a)   Use the mouse to slide the Pooled Standard Error at the top of the display to the right. You see that the group means do not change, but the spread of the observations in each group increases. What happens to F and P as the spread among the observations in each group increases? What are the values of F and P when the slider is all the way to the right? This is similar to Figure 24.6(a): variation within groups hides the differences among the group means.

(b)   Leave the Pooled Standard Error slider at the extreme right of its scale, so that spread within groups stays fixed. Use the mouse to move the group means apart. What happens to F and P as you do this?

We weren't working anyway

A "consultant" estimated that the annual NCAA men's basketball tournament costs employers \$3.8 billion in time wasted by workers participating in office pools, checking game scores, and so on. That's unlikely. Most of the games are played outside work hours, at night and on weekends. More to the point, economists note that workers waste lots of time every workday, by talking to other employees, chatting on the phone, shopping online, and so on. The dollar value of time spent on the basketball tournament probably comes in large part from time we were wasting anyway.

robustness

Conditions for ANOVA

Like all inference procedures, ANOVA is valid only in some circumstances. Here are the conditions under which we can use ANOVA to compare population means.

CONDITIONS FOR ANOVA INFERENCE

■ We have *I* **independent SRSs,** one from each of *I* populations. We measure the same response variable for each sample.

■ The *i*th population has a **Normal distribution** with unknown mean μ_i. One-way ANOVA tests the null hypothesis that all the population means are the same.

■ All of the populations have the **same standard deviation** σ, whose value is unknown.

The first two conditions are familiar from our study of the two-sample t procedures for comparing two means. As usual, the design of the data production is the most important condition for inference. Biased sampling or confounding can make any inference meaningless. *If we do not actually draw separate SRSs from each population or carry out a randomized comparative experiment, it may be unclear to what population the conclusions of inference apply.* ANOVA, like other inference procedures, is often used when random samples are not available. You must judge each use on its merits, a judgment that usually requires some knowledge of the subject of the study in addition to some knowledge of statistics.

Because no real population has exactly a Normal distribution, the usefulness of inference procedures that assume Normality depends on how sensitive they are to departures from Normality. Fortunately, procedures for comparing means are not very sensitive to lack of Normality. The ANOVA F test, like the t procedures, is **robust.** What matters is Normality of the sample means, so ANOVA becomes safer as the sample sizes get larger, because of the central limit theorem effect. Remember to check for outliers that change the value of sample means and for extreme skewness. When there are no outliers and the distributions are roughly symmetric, you can safely use ANOVA for sample sizes as small as 4 or 5.

The third condition is annoying: ANOVA assumes that the variability of observations, measured by the standard deviation, is the same in all populations. The *t* test for comparing two means (Chapter 18) does not require equal standard deviations. Unfortunately, the ANOVA *F* for comparing more than two means is less broadly valid. It is not easy to check the condition that the populations have equal standard deviations. Statistical tests for equality of standard deviations are very sensitive to lack of Normality, so much so that they are of little practical value. You must either seek expert advice or rely on the robustness of ANOVA.

How serious are unequal standard deviations? ANOVA is not too sensitive to violations of the condition, especially when all samples have the same or similar sizes and no sample is very small. When designing a study, try to take samples of about the same size from all the groups you want to compare. The sample standard deviations estimate the population standard deviations, so check before doing ANOVA that the sample standard deviations are similar to each other. We expect some variation among them due to chance. Here is a rule of thumb that is safe in almost all situations.

CHECKING STANDARD DEVIATIONS IN ANOVA

The results of the ANOVA *F* test are approximately correct when the largest sample standard deviation is no more than twice as large as the smallest sample standard deviation.

EXAMPLE 24.3 **Comparing tropical flowers: conditions for ANOVA**

The study of *Heliconia* blossoms is based on three independent samples that the researchers consider to be random samples from all flowers of these varieties in Dominica. The stemplots in Figure 24.1 show that the *bihai* and red varieties have slightly skewed distributions, but the sample means of samples of sizes 16 and 23 will have distributions that are close to Normal. The sample standard deviations for the three varieties are

$$s_1 = 1.213 \quad s_2 = 1.799 \quad s_3 = 0.975$$

These standard deviations satisfy our rule of thumb:

$$\frac{\text{largest } s}{\text{smallest } s} = \frac{1.799}{0.975} = 1.85 \quad \text{(less than 2)}$$

We can safely use ANOVA to compare the mean lengths for the three populations. ■

EXAMPLE 24.4 **Thinking about money changes behavior**

STATE: Kathleen Vohs of the University of Minnesota and her coworkers carried out several randomized comparative experiments on the effects of thinking about money. Here's an outline of one of the experiments. Ask student subjects to unscramble

TABLE 24.4 Time (seconds) until subjects ask for help with a puzzle

GROUP	TIME	GROUP	TIME	GROUP	TIME
Prime	609	Play	455	Control	118
Prime	444	Play	100	Control	272
Prime	242	Play	238	Control	413
Prime	199	Play	243	Control	291
Prime	174	Play	500	Control	140
Prime	55	Play	570	Control	104
Prime	251	Play	231	Control	55
Prime	466	Play	380	Control	189
Prime	443	Play	222	Control	126
Prime	531	Play	71	Control	400
Prime	135	Play	232	Control	92
Prime	241	Play	219	Control	64
Prime	476	Play	320	Control	88
Prime	482	Play	261	Control	142
Prime	362	Play	290	Control	141
Prime	69	Play	495	Control	373
Prime	160	Play	600	Control	156
		Play	67		

30 sets of five words to make a meaningful phrase from four of the five. The control group unscrambled phrases like "cold it desk outside is" into "it is cold outside." The "play money" group unscrambled similar sets of words, but a stack of Monopoly money was placed nearby. The "money prime" group unscrambled phrases that lead to thinking about money, turning "high a salary desk paying" into "a high-paying salary." Then each subject worked a hard puzzle, knowing that they could ask for help. Table 24.4 shows the time in seconds that each subject worked on the puzzle before asking for help.[6] Psychologists think that money tends to make people self-sufficient. If so, the two groups that were encouraged in different ways to think about money should take longer on the average to ask for help. Do the data support this idea?

PLAN: Examine the data to compare the effect of the treatments and check that we can safely use ANOVA. If the data allow ANOVA, assess the significance of observed differences in mean times to ask for help.

SOLVE: Figure 24.7 shows side-by-side stemplots of the data in the three groups. We expect some irregularity in small samples, but there are no outliers or strong skewness that would hinder use of ANOVA. The Minitab ANOVA output in Figure 24.8 shows that the group standard deviations easily satisfy our rule of thumb. The control group subjects asked for help much sooner (mean 186.1 seconds) than did subjects in the two money groups (means 305.2 seconds and 314.1 seconds). The three means are significantly different ($F = 3.73$, $P = 0.031$).

CONCLUDE: The experiment gives good evidence that reminding people of money in either of two ways does make them less willing to ask others for help. This is consistent with the idea that money makes people feel more self-sufficient. ■

```
     Prime           Play            Control
 0 | 6 7         0 | 7 7         0 | 6 6 9 9
 1 | 4 6 7       1 | 0           1 | 0 2 3 4 4 4 6 9
 2 | 0 4 4 5     2 | 2 2 3 3 4 4 6 9   2 | 7 9
 3 | 6           3 | 2 8         3 | 7
 4 | 4 4 7 8 8   4 | 6           4 | 0 1
 5 | 3           5 | 0 0 7       5 |
 6 | 1           6 | 0           6 |
```

FIGURE 24.7

Side-by-side stemplots comparing the time until subjects asked for help with a puzzle, for Example 24.4.

```
 _ □ X

One-way ANOVA: time versus group

Source    DF       SS      MS      F      P
Group      2   174912   87456   3.73  0.031
Error     49  1149568   23461
Total     51  1234479

S = 153.2   R-Sq = 13.21%   R-Sq(adj) = 9.66%

                        Individual 95% CIs For Mean Based on
                        Pooled StDev
Level      N    Mean   StDev  ----+---------+---------+---------+----
control   17   186.1   118.1  (----------*----------)
play      18   305.2   162.5                (----------*----------)
prime     17   314.1   172.8                   (----------*----------)
                              ----+---------+---------+---------+----
                              140       210       280       350

Pooled StDev = 153.2
```

FIGURE 24.8

Minitab ANOVA output for comparing the three treatments in Example 24.4.

24.7 **Checking standard deviations.** Verify that the sample standard deviations for these sets of data do allow use of ANOVA to compare the population means.

(a) The counts of trees in Exercise 24.3 and Figure 24.4.

(b) The restaurant times of Exercise 24.4 and Figure 24.5.

24.8 **Species richness after logging.** Table 24.2 gives data on the species richness in rain forest plots, defined as the number of tree species in a plot divided by the number of trees in the plot. ANOVA may not be trustworthy for the richness data. Do data analysis: make side-by-side stemplots to examine the distributions of the response variable in the three groups, and also compare the standard deviations. What characteristic of the data makes ANOVA risky?

24.9 **Fertilizing bromeliads.** Bromeliads are tropical flowering plants. Many are epiphytes that attach to trees and obtain moisture and nutrients from air and rain. Their leaf bases form cups that collect water and are home to the larvae of many insects. As a preliminary to a study of changes in the nutrient cycle, Jacqueline Ngai and Diane Srivastava examined the effects of adding nitrogen, phosphorus, or both to the cups. They randomly assigned 8 bromeliads growing in Costa Rica to each of four treatment groups, including an unfertilized control group. A monkey destroyed one of the plants in the control group, leaving 7 bromeliads in that group. Here are the numbers of new leaves on each plant over the 7 months following fertilization:[7]

Nitrogen	Phosphorus	Both	Neither
15	14	14	11
14	14	16	13
15	14	15	16
16	11	14	15
17	13	14	15
18	12	13	11
17	15	17	12
13	15	14	

Analyze these data and discuss the results. Does nitrogen or phosphorus have a greater effect on the growth of bromeliads? Follow the four-step process as illustrated in Example 24.4.

F distributions and degrees of freedom

The ANOVA F statistic is

$$F = \frac{\text{variation among the sample means}}{\text{variation among individuals in the same sample}}$$

F distribution

To find the P-value for this statistic, we must know the sampling distribution of F when the null hypothesis (all population means equal) is true. This sampling distribution is an **F distribution.**

The F distributions are a family of right-skewed distributions that take only values greater than 0. The density curves in Figure 24.9 illustrate their shapes. A specific F distribution is determined by the *degrees of freedom* of the numerator

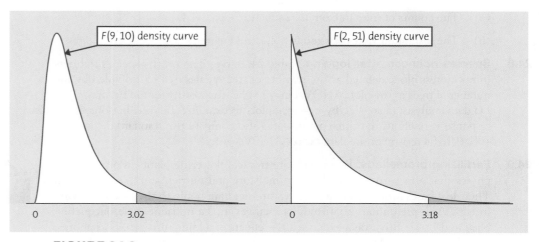

FIGURE 24.9

Density curves for two F distributions. Both are right-skewed and take only positive values. The upper 5% critical values are marked under the curves.

and denominator of the *F* statistic. You may have noticed that all of our software outputs include degrees of freedom, labeled either "df" or "DF." The optional section "Some Details of ANOVA" shows where the degrees of freedom come from. When describing an *F* distribution, always give the numerator degrees of freedom first. Our brief notation will be $F(\text{df1}, \text{df2})$ for the *F* distribution with df1 degrees of freedom in the numerator and df2 in the denominator. *Interchanging the degrees of freedom changes the distribution, so the order is important.*

Tables of *F* critical points are awkward, because we need a separate table for every pair of degrees of freedom df1 and df2. Fortunately, software gives you *P*-values for the ANOVA *F* test without the need for a table.

EXAMPLE 24.5 **Comparing flowers: the *F* distribution**

Look again at the software output in Figure 24.3 for the flower length data. All three outputs give the degrees of freedom for the *F* test, labeled "df" or "DF." There are 2 degrees of freedom in the numerator and 51 in the denominator. *P*-values for the *F* test therefore come from the *F* distribution $F(2, 51)$ with 2 and 51 degrees of freedom. The right-hand curve in Figure 24.9 is the density curve of this distribution. The 5% critical value marked on that curve is 3.18, and the 1% critical value is 5.05. The observed value $F = 259.12$ of the ANOVA *F* statistic lies far to the right of these values, so the *P*-value is extremely small. ■

The degrees of freedom of the ANOVA *F* statistic depend on the number of means we are comparing and the number of observations in each sample. That is, the *F* test takes into account the number of observations. Here are the details.

DEGREES OF FREEDOM FOR THE *F* TEST

We want to compare the means of *I* populations. We have an SRS of size n_i from the *i*th population, so that the total number of observations in all samples combined is

$$N = n_1 + n_2 + \cdots + n_I$$

If the null hypothesis that all population means are equal is true, the ANOVA *F* statistic has the *F* distribution with $I - 1$ degrees of freedom in the numerator and $N - I$ degrees of freedom in the denominator.

EXAMPLE 24.6 **Degrees of freedom for *F***

In Examples 24.1 and 24.2, we compared the mean lengths for three varieties of flowers, so $I = 3$. The three sample sizes are

$$n_1 = 16 \quad n_2 = 23 \quad n_3 = 15$$

The total number of observations is therefore

$$N = 16 + 23 + 15 = 54$$

The ANOVA *F* test has numerator degrees of freedom

$$I - 1 = 3 - 1 = 2$$

and denominator degrees of freedom

$$N - I = 54 - 3 = 51$$

These are the degrees of freedom given in the outputs in Figure 24.3. ■

APPLY YOUR KNOWLEDGE

24.10 Logging in the rain forest, continued. Exercise 24.3 compares the number of tree species in rain forest plots that had never been logged (Group 1) with similar plots nearby that had been logged 1 year earlier (Group 2) and 8 years earlier (Group 3).

    (a)   What are I, the n_i, and N for these data? Identify these quantities in words and give their numerical values.

    (b)   Find the degrees of freedom for the ANOVA F statistic. Check your work against the Excel output in Figure 24.4.

24.11 What music will you play? People often match their behavior to their social environment. One study of this idea first established that the type of music most preferred by black college students is R&B and that whites' most preferred music is rock. Will students hosting a small group of other students choose music that matches the makeup of the people attending? Assign 90 black business students at random to three equal-sized groups. Do the same for 96 white students. Each student sees a picture of the people he or she will host. Group 1 sees 6 blacks, Group 2 sees 3 whites and 3 blacks, and Group 3 sees 6 whites. Ask how likely the host is to play the type of music preferred by the other race. Use ANOVA to compare the three groups to see whether the racial mix of the gathering affects the choice of music.[8]

    (a)   For the white subjects, $F = 16.48$. What are the degrees of freedom?

    (b)   For the black subjects, $F = 2.47$. What are the degrees of freedom?

Robert Glenn/Getty

Some details of ANOVA*

Now we will give the actual formula for the ANOVA F statistic. We have SRSs from each of I populations. Subscripts from 1 to I tell us which sample a statistic refers to:

Population	Sample size	Sample mean	Sample std. dev.
1	n_1	$\overline{x}_1$	s_1
2	n_2	$\overline{x}_2$	s_2
.	.	.	.
.	.	.	.
.	.	.	.
I	n_I	$\overline{x}_I$	s_I

---

* This more advanced section is optional if you are using software to find the F statistic.

You can find the F statistic from just the sample sizes n_i, the sample means $\overline{x}_i$, and the sample standard deviations s_i. You don't need to go back to the individual observations.

The ANOVA F statistic has the form

$$F = \frac{\text{variation among the sample means}}{\text{variation among individuals in the same sample}}$$

The measures of variation in the numerator and denominator of F are called **mean squares.** A mean square is a more general form of a sample variance. An ordinary sample variance s^2 is an average (or mean) of the squared deviations of observations from their mean, so it qualifies as a "mean square."

mean squares

Call the overall mean response $\overline{x}$. That is, $\overline{x}$ is the mean of all N observations together. You can find $\overline{x}$ from the I sample means by

$$\overline{x} = \frac{\text{sum of all observations}}{N} = \frac{n_1\overline{x}_1 + n_2\overline{x}_2 + \cdots + n_I\overline{x}_I}{N}$$

(This expression works because multiplying a group mean $\overline{x}_i$ by the number of observations n_i it represents gives the sum of the observations in that group.)

The numerator of F is a mean square that measures variation among the I sample means $\overline{x}_1, \overline{x}_2, \ldots, \overline{x}_I$. To measure this variation, look at the I deviations of the means of the samples from $\overline{x}$,

$$\overline{x}_1 - \overline{x}, \overline{x}_2 - \overline{x}, \ldots, \overline{x}_I - \overline{x}$$

The mean square in the numerator of F is an average of the squares of these deviations. We call it the **mean square for groups,** abbreviated as MSG:

MSG

$$\text{MSG} = \frac{n_1(\overline{x}_1 - \overline{x})^2 + n_2(\overline{x}_2 - \overline{x})^2 + \cdots + n_I(\overline{x}_I - \overline{x})^2}{I - 1}$$

Each squared deviation is weighted by n_i, the number of observations it represents.

The mean square in the denominator of F measures variation among individual observations in the same sample. For any one sample, the sample variance s_i^2 does this job. For all I samples together, we use an average of the individual sample variances. It is another weighted average, in which each s_i^2 is weighted by its degrees of freedom $n_i - 1$. The resulting mean square is called the **mean square for error,** MSE:

MSE

$$\text{MSE} = \frac{(n_1 - 1)s_1^2 + (n_2 - 1)s_2^2 + \cdots + (n_I - 1)s_I^2}{N - I}$$

"Error" doesn't mean a mistake has been made. It's a traditional term for chance variation. Here is a summary of the ANOVA test.

> **THE ANOVA _F_ TEST**
>
> Draw an independent SRS from each of I Normal populations that have a common standard deviation but may have different means. The sample from the ith population has size n_i, sample mean $\bar{x}_i$, and sample standard deviation s_i.
>
> To test the null hypothesis that all I populations have the same mean against the alternative hypothesis that not all the means are equal, calculate the **ANOVA _F_ statistic**
>
> $$F = \frac{\text{MSG}}{\text{MSE}}$$
>
> The numerator of F is the **mean square for groups**
>
> $$\text{MSG} = \frac{n_1(\bar{x}_1 - \bar{x})^2 + n_2(\bar{x}_2 - \bar{x})^2 + \cdots + n_I(\bar{x}_I - \bar{x})^2}{I - 1}$$
>
> The denominator of F is the **mean square for error**
>
> $$\text{MSE} = \frac{(n_1 - 1)s_1^2 + (n_2 - 1)s_2^2 + \cdots + (n_I - 1)s_I^2}{N - I}$$
>
> When H_0 is true, F has the **_F_ distribution** with $I - 1$ and $N - I$ degrees of freedom.

sums of squares

ANOVA table

The denominators in the formulas for MSG and MSE are the two degrees of freedom $I - 1$ and $N - I$ of the F test. The numerators are called **sums of squares,** from their algebraic form. It is usual to present the results of ANOVA in an **ANOVA table.** Output from software usually includes an ANOVA table.

EXAMPLE 24.7 ANOVA calculations: software

Look again at the three outputs in Figure 24.3. The two software outputs give the ANOVA table. The calculator, with its small screen, gives the degrees of freedom, sums of squares, and mean squares separately. Each output uses slightly different language to identify the two sources of variation. The basic ANOVA table is

Source of variation	df	SS	MS	F statistic
Variation among samples	2	1082.87	MSG = 541.44	259.12
Variation within samples	51	106.57	MSE = 2.09	

You can check that each mean square MS is the corresponding sum of squares SS divided by its degrees of freedom df. The F statistic is MSG divided by MSE. ■

pooled standard deviation

Because MSE is an average of the individual sample variances, it is also called the *pooled sample variance*, written as s_p^2. When all I populations have the same population variance σ^2, as ANOVA assumes that they do, s_p^2 estimates the common variance σ^2. The square root of MSE is the **pooled standard deviation** s_p. It estimates the common standard deviation σ of observations in each group. The Minitab and calculator outputs in Figure 24.3 give the value $s_p = 1.446$.

The pooled standard deviation s_p is a better estimator of the common σ than any individual sample standard deviation s_i because it combines (pools) the information in all I samples. We can get a confidence interval for any one of the means μ_i from the usual form

$$\text{estimate} \pm t^* \text{SE}_{\text{estimate}}$$

using s_p to estimate σ. The confidence interval for μ_i is

$$\bar{x}_i \pm t^* \frac{s_p}{\sqrt{n_i}}$$

Use the critical value t^* from the t distribution with $N - I$ degrees of freedom because s_p has $N - I$ degrees of freedom. These are the confidence intervals that appear in Minitab ANOVA output.

EXAMPLE 24.8 ANOVA calculations: without software

We can do the ANOVA test comparing the mean lengths of *bihai*, red, and yellow flower varieties using only the sample sizes, sample means, and sample standard deviations. These appear in Example 24.1, but it is easy to find them with a calculator. There are $I = 3$ groups with a total of $N = 54$ flowers.

The overall mean of the 54 lengths in Table 24.1 is

$$\bar{x} = \frac{n_1 \bar{x}_1 + n_2 \bar{x}_2 + n_3 \bar{x}_3}{N}$$

$$= \frac{(16)(47.598) + (23)(39.711) + (15)(36.180)}{54}$$

$$= \frac{2217.621}{54} = 41.067$$

The mean square for groups is

$$\text{MSG} = \frac{n_1(\bar{x}_1 - \bar{x})^2 + n_2(\bar{x}_2 - \bar{x})^2 + n_3(\bar{x}_3 - \bar{x})^2}{I - 1}$$

$$= \frac{1}{3 - 1}[(16)(47.598 - 41.067)^2 + (23)(39.711 - 41.067)^2$$

$$+ (15)(36.180 - 41.067)^2]$$

$$= \frac{1082.996}{2} = 541.50$$

The mean square for error is

$$\text{MSE} = \frac{(n_1 - 1)s_1^2 + (n_2 - 1)s_2^2 + (n_3 - 1)s_3^2}{N - I}$$

$$= \frac{(15)(1.213^2) + (22)(1.799^2) + (14)(0.975^2)}{51}$$

$$= \frac{106.580}{51} = 2.09$$

Finally, the ANOVA test statistic is

$$F = \frac{MSG}{MSE} = \frac{541.50}{2.09} = 259.09$$

Our work differs slightly from the output in Figure 24.3 because of roundoff error. We don't recommend doing these calculations, because tedium and roundoff errors cause frequent mistakes. ∎

APPLY YOUR KNOWLEDGE

The calculations of ANOVA use only the sample sizes n_i, the sample means $\bar{x}_i$, and the sample standard deviations s_i. You can therefore re-create the ANOVA calculations when a report gives these summaries but does not give the actual data. These optional exercises ask you to do the ANOVA calculations starting with the summary statistics. P-values require either a table or software for the F distributions.

24.12 Road rage. Exercise 24.2 describes a study of road rage. Here are the means and standard deviations for a measure of "angry/threatening driving" for random samples of drivers in three age groups:

Age group	n	$\bar{x}$	s
Less than 30 years	244	2.22	3.11
30 to 55 years	734	1.33	2.21
Over 55 years	364	0.66	1.60

(a)   The distributions of responses are somewhat right-skewed. ANOVA is nonetheless safe for these data. Why?

(b)   Check that the standard deviations satisfy the guideline for ANOVA inference.

(c)   Calculate the overall mean response $\bar{x}$, the mean squares MSG and MSE, and the ANOVA F statistic.

(d)   Which F distribution would you use to find the P-value of the ANOVA F test? Software gives $P < 0.001$. Write a brief conclusion based on the sample means and the ANOVA F test.

24.13 Exercise and weight loss. What conditions help overweight people exercise regularly? Subjects were randomly assigned to three treatments: a single long exercise period 5 days per week; several 10-minute exercise periods 5 days per week; and several 10-minute periods 5 days per week on a home treadmill that was provided to the subjects. The study report contains the following information about weight loss (in kilograms) after six months of treatment:[9]

Treatment	n	$\bar{x}$	s
Long exercise periods	37	10.2	4.2
Short exercise periods	36	9.3	4.5
Short periods with equipment	42	10.2	5.2

(a)   Do the standard deviations satisfy the rule of thumb for safe use of ANOVA?

(b)   Calculate the overall mean response $\bar{x}$, the mean squares MSG and MSE, and the F statistic.

(c)   Which F distribution would you use to find the P-value of the ANOVA F test? Software says that $P = 0.634$. What do you conclude from this study?

24.14 Attitudes toward math. Do high school students from different racial/ethnic groups have different attitudes toward mathematics? Measure the level of interest in mathematics on a 5-point scale for a national random sample of students. Here are summaries for students who were taking math at the time of the survey:[10]

Racial/ethnic group	n	$\bar{x}$	s
African American	809	2.57	1.40
White	1860	2.32	1.36
Asian/Pacific Islander	654	2.63	1.32
Hispanic	883	2.51	1.31
Native American	207	2.51	1.28

(a)   The conditions for ANOVA are clearly satisfied. Explain why.

(b)   Calculate the ANOVA table and the F statistic.

(c)   Software gives $P < 0.001$. What explains the small P-value? Do you think the differences are large enough to be important?

CHAPTER 24   SUMMARY

- **One-way analysis of variance (ANOVA)** compares the means of several populations. The **ANOVA F test** tests the null hypothesis that all the populations have the same mean. If the F test shows significant differences, examine the data to see where the differences lie and whether they are large enough to be important.

- The **conditions for ANOVA** state that we have an **independent SRS** from each population; that each population has a **Normal distribution;** and that all populations have the **same standard deviation.**

- In practice, ANOVA inference is relatively **robust** when the populations are non-Normal, especially when the samples are large. Before doing the F test, check the observations in each sample for outliers or strong skewness. Also verify that the largest sample standard deviation is no more than twice as large as the smallest standard deviation.

- When the null hypothesis is true, the **ANOVA F statistic** for comparing I means from a total of N observations in all samples combined has the **F distribution** with $I - 1$ and $N - I$ degrees of freedom.

- ANOVA calculations are reported in an **ANOVA table** that gives sums of squares, mean squares, and degrees of freedom for variation among groups and for variation within groups. In practice, we use software to do the calculations.

Here are the most important skills you should have acquired from reading this chapter.

A. Recognition

1. Recognize when testing the equality of several means is helpful in understanding data.

2. Recognize that the statistical significance of differences among sample means depends on the sizes of the samples and on how much variation there is within the samples.

3. Recognize when you can safely use ANOVA to compare means. Check the data production, the presence of outliers, and the sample standard deviations for the groups you want to compare.

B. Interpreting ANOVA

1. Explain what null hypothesis F tests in a specific setting.

2. Locate the F statistic and its P-value on the output of analysis of variance software.

3. Find the degrees of freedom for the F statistic from the number and sizes of the samples.

4. If the test is significant, use graphs and descriptive statistics to see what differences among the means are most important.

C H E C K   Y O U R   S K I L L S

24.15 The purpose of analysis of variance is to compare

(a) the variances of several populations.

(b) the proportions of successes in several populations.

(c) the means of several populations.

24.16 A study of the effects of smoking classifies subjects as nonsmokers, moderate smokers, or heavy smokers. The investigators interview a sample of 200 people in each group. Among the questions is "How many hours do you sleep on a typical night?" The degrees of freedom for the ANOVA F statistic comparing mean hours of sleep are

(a) 2 and 197.   (b) 2 and 597.   (c) 3 and 597.

24.17 The alternative hypothesis for the ANOVA F test in the previous exercise is

(a) the mean hours of sleep in the groups are all the same.

(b) the mean hours of sleep in the groups are all different.

(c) the mean hours of sleep in the groups are not all the same.

24.18 The F distributions are

(a)   a family of distributions with bell-shaped density curves centered at 0.

(b)   a family of distributions that are right-skewed and take only values greater than 0.

(c)   a family of distributions that are left-skewed and take values between 0 and 1.

The air in poultry processing plants often contains fungus spores. Large spore concentrations can affect the health of the workers. To measure the presence of spores, air samples are pumped to an agar plate and "colony forming units (CFUs)" are counted after an incubation period. Here are data from the "kill room" of a plant that slaughters 37,000 turkeys per day, taken at four seasons of the year. Each observation was made on a different day. The units are CFUs per cubic meter of air.[11]

Fall	Winter	Spring	Summer
1231	384	2105	3175
1254	104	701	2526
752	251	2947	1763
1088	97	842	1090

Here is Minitab output for ANOVA to compare mean CFUs in the four seasons:

```
Source   DF       SS        MS    F      P
Season    3   8236359   2745453  5.38  0.014
Error    12   6124211    510351
Total    15  14360570

           Individual 95% CIs For Mean Based on Pooled StDev
Level   N   Mean    StDev ------+---------+---------+---------+---
Fall    4  1081.3   231.5              (-------*-------)
Spring  4  1648.8  1071.2                  (------*-------)
Summer  4  2138.5   906.4                     (------*-------)
Winter  4   209.0   136.6  (-------*-------)
                      ------+---------+---------+---------+---
                          0      1000      2000      3000
```

Exercises 24.19 to 24.23 are based on this study.

24.19 The most striking conclusion from the numerical summaries for the turkey processing plant is that

(a)   there appears to be little difference among the seasons.

(b)   on the average, CFUs are much lower in winter than in other seasons.

(c)   the air in the plant is clearly unhealthy.

24.20 We might use the two-sample t procedures to compare summer and winter. The conservative 90% confidence interval for the difference in the two population means is

(a) 1929.5 ± 458.3.   (b) 1929.5 ± 1078.4.   (c) 1929.5 ± 1458.4.

24.21 In all, we would have to give 6 two-sample confidence intervals to compare all pairs of seasons. The weakness of doing this is that

    (a) we don't know how confident we can be that all 6 intervals cover the true differences in means.

    (b) 90% confidence is OK for one comparison, but it isn't high enough for 6 comparisons done at once.

    (c) the conditions for two-sample t inference are not met for all 6 pairs of seasons.

24.22 The conclusion of the ANOVA test is that

    (a) there is quite strong evidence ($P = 0.014$) that the mean CFUs are not the same in all four seasons.

    (b) there is quite strong evidence ($P = 0.014$) that the mean CFUs are much lower in winter than in any other season.

    (c) the data give no evidence ($P = 0.014$) to suggest that mean CFUs differ from season to season.

24.23 The P-value 0.014 in the output may not be accurate because the conditions for ANOVA are not satisfied. The most serious violation of the conditions is that

    (a) the sample standard deviations are too different.

    (b) there is an extreme outlier in the data.

    (c) the data can't be regarded as random samples.

C H A P T E R 2 4   E X E R C I S E S

Exercises 24.24 to 24.27 describe situations in which we want to compare the mean responses in several populations. For each setting, identify the populations and the response variable. Then give I, the n_i, and N. Finally, give the degrees of freedom of the ANOVA F statistic.

24.24 Morning or evening? Are you a morning person, an evening person, or neither? Does this personality trait affect how well you perform? A sample of 100 students took a psychological test that found 16 morning people, 30 evening people, and 54 who were neither. All the students then took a test of their ability to memorize at 8 A.M. and again at 9 P.M. The response variable is the score at 8 A.M. minus the score at 9 P.M.

24.25 Writing essays. Do strategies such as preparing a written outline help students write better essays? College students were divided at random into four groups of 20 students each, then asked to write an essay on an assigned topic. Group A (the control group) received no additional instruction. Group B was required to prepare a written outline. Group C was given 15 ideas that might be relevant to the essay topic. Group D was given the ideas and also required to prepare an outline. An expert scored the quality of the essays on a scale of 1 to 7.

24.26 Test accommodations. Many states require schoolchildren to take regular statewide tests to assess their progress. Children with learning disabilities who read poorly may not do well on mathematics tests because they can't read the problems. Most states allow "accommodations" for learning-disabled children. Randomly assign 100 learning-disabled children in equal numbers to three types of accommodation and a control group: math problems are read by a teacher; by a computer; by a computer that also shows a video; and standard test conditions. Compare the mean scores on the state mathematics assessment.

James Woodson/Age fotostock

24.27 A medical study. The Québec (Canada) Cardiovascular Study recruited men aged 34 to 64 at random from towns in the Québec City metropolitan area. Of these, 1824 met the criteria (no diabetes, free of heart disease, and so on) for a study of the relationship between being overweight and medical risks. The 719 normal-weight men had mean triglyceride level 1.5 millimoles per liter (mmol/l); the 885 over-weight men had mean 1.7 mmol/l; and the 220 obese men had mean 1.9 mmol/l.[12]

24.28 Plants defend themselves. When some plants are attacked by leaf-eating insects, they release chemical compounds that attract other insects that prey on the leaf-eaters. A study carried out on plants growing naturally in the Utah desert demon-strated both the release of the compounds and that they not only repel the leaf-eaters but attract predators that act as the plants' bodyguards.[13] The investigators chose 8 plants attacked by each of three leaf-eaters and 8 more that were undam-aged, 32 plants of the same species in all. They then measured emissions of several compounds during seven hours. Here are data (mean ± standard error of the mean for eight plants) for one compound. The emission rate is measured in nanograms (ng) per hour.

Group	Emission rate (ng/hr)
Control	9.22 ± 5.93
Hornworm	31.03 ± 8.75
Leaf bug	18.97 ± 6.64
Flea beetle	27.12 ± 8.62

(a)   Make a graph that compares the mean emission rates for the four groups. Does it appear that emissions increase when the plant is attacked?

(b)   What hypotheses does ANOVA test in this setting?

(c)   We do not have the full data. What would you look for in deciding whether you can safely use ANOVA?

(d)   What is the relationship between the standard error of the mean (SEM) and the standard deviation for a sample? What are the four sample standard devi-ations? Do they satisfy our rule of thumb for safe use of ANOVA?

24.29 Can you hear these words? To test whether a hearing aid is right for a patient, audiologists play a tape on which words are pronounced at low volume. The patient tries to repeat the words. There are several different lists of words that are supposed to be equally difficult. Are the lists equally difficult when there is background noise? To find out, an experimenter had subjects with normal hearing listen to four lists with a noisy background. The response variable was the percent of the 50 words in a list that the subject repeated correctly. The data set contains 96 responses.[14] Here are two study designs that could produce these data:

Phanie/Photo Researchers

Design A. The experimenter assigns 96 subjects to 4 groups at random. Each group of 24 subjects listens to one of the lists. All individuals listen and respond separately.

Design B. The experimenter has 24 subjects. Each subject listens to all four lists in random order. All individuals listen and respond separately.

Does Design A allow use of one-way ANOVA to compare the lists? Does Design B allow use of one-way ANOVA to compare the lists? Briefly explain your answers.

24.30 More rain for California? The changing climate will probably bring more rain to California, but we don't know whether the additional rain will come during the winter wet season or extend into the long dry season in spring and summer. Kenwyn Suttle of the University of California at Berkeley and his coworkers randomly assigned plots of open grassland to three treatments: added water equal to 20% of annual rainfall either during January to March (winter) or during April to June (spring), and no added water (control). Here are some of the data, for plant biomass (in grams per square meter) produced by each plot in a single year.[15]

Winter	Spring	Control
264.1514	318.4182	129.0538
187.7312	281.6830	144.6578
291.1431	288.8433	172.7772
176.2879	382.6673	113.2813
141.7525	326.8877	142.1562
169.9737	293.8502	117.9808

Figure 24.10 shows Minitab ANOVA output for these data.

(a) Make side-by-side stemplots of plant biomass for the three treatments, as well as a table of the sample means and standard deviations. What do the data appear to show about the effect of extra water in winter and in spring on biomass? Do these data satisfy the conditions for ANOVA?

(b) State H_0 and H_a for the ANOVA F test, and explain in words what ANOVA tests in this setting.

(c) Report your overall conclusions about the effect of added water on plant growth in California.

FIGURE 24.10

Minitab ANOVA output for comparing the total plant biomass of grassland plots under different water conditions, for Exercise 24.30.

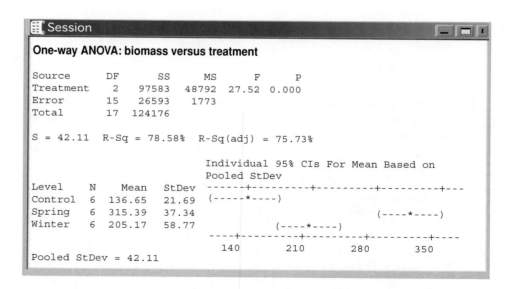

```
Session                                                          _ □ X

One-way ANOVA: biomass versus treatment

Source       DF      SS      MS       F      P
Treatment     2    97583   48792   27.52  0.000
Error        15    26593    1773
Total        17   124176


S = 42.11   R-Sq = 78.58%   R-Sq(adj) = 75.73%


                              Individual 95% CIs For Mean Based on
                              Pooled StDev
Level     N    Mean   StDev  ------+---------+---------+---------+---
Control   6   136.65  21.69  (-----*----)
Spring    6   315.39  37.34                              (----*----)
Winter    6   205.17  58.77            (----*----)
                              ----+---------+---------+---------+----
                              140       210       280       350
Pooled StDev = 42.11
```

24.31 Can you hear these words? Figure 24.11 displays the Minitab output for one-way ANOVA applied to the hearing data described in Exercise 24.29. The response

```
┌─────────────────────────────────────────────────────────────────────┐
│  ▣ Session                                              ─ ■ x         │
├─────────────────────────────────────────────────────────────────────┤
│  Analysis of Variance for Percent                               ▲    │
│  Source    DF       SS       MS       F       P                      │
│  List       3     920.5    306.8    4.92   0.003                     │
│  Error     92    5738.2     62.4                                     │
│  Total     95    6658.6                                              │
│                                                                      │
│                           Individual 95% CIs For Mean                │
│                           Based on Pooled StDev                      │
│  Level      N     Mean     StDev ----+---------+---------+---------+----│
│  1          24   32.750    7.409                    (------*------)   │
│  2          24   29.667    8.058               (------*------)        │
│  3          24   25.250    8.316        (------*------)               │
│  4          24   25.583    7.779         (------*------)              │
│                                                                      │
│                                  ----+---------+---------+---------+----│
│  Pooled StDev =    7.898         24.0      28.0     32.0     36.0   ▼ │
└─────────────────────────────────────────────────────────────────────┘
```

FIGURE 24.11

Minitab ANOVA output for comparing the percents heard correctly in four lists of words, for Exercise 25.31.

variable is "Percent," and "List" identifies the four lists of words. Based on this analysis, is there good reason to think that the four lists are not all equally difficult? Write a brief summary of the study findings.

24.32 Which blue is most blue? The color of a fabric depends on the dye used and also on how the dye is applied. This matters to clothing manufacturers, who want the color of the fabric to be just right. Dye fabric made of ramie with the same "procion blue" die applied in four different ways. Then use a colorimeter to measure the lightness of the color on a scale in which black is 0 and white is 100. Here are the data for 8 pieces of fabric dyed in each way:[16]

Method A	41.72	41.83	42.05	41.44	41.27	42.27	41.12	41.49
Method B	40.98	40.88	41.30	41.28	41.66	41.50	41.39	41.27
Method C	42.30	42.20	42.65	42.43	42.50	42.28	43.13	42.45
Method D	41.68	41.65	42.30	42.04	42.25	41.99	41.72	41.97

(a) This is a randomized comparative experiment. Outline the design.

(b) A clothing manufacturer wants to know which method gives the darkest color. Follow the four-step process in answering this question.

24.33 Does nature heal best? Our bodies have a natural electrical field that helps wounds heal. Might higher or lower levels speed healing? An experiment with newts investigated this question. Newts were randomly assigned to five groups. In four of the groups, an electrode applied to one hind limb (chosen at random) changed the natural field, while the other hind limb was not manipulated. Both limbs in the fifth (control) group remained in their natural state.[17]

Table 24.5 gives data from this experiment. The "Group" variable shows the field applied as a multiple of the natural field for each newt. For example, "0.5" is half the natural field, "1" is the natural level (the control group), and "1.5" indicates a field 1.5 times natural. "Diff" is the response variable, the difference in the healing rate (in micrometers per hour) of cuts made in the experimental and control limbs of that newt. Negative values mean that the experimental limb healed more slowly.

TABLE 24.5 Effect of electrical field on healing rate in newts

GROUP	DIFF	GROUP	DIFF	GROUP	DIFF	GROUP	DIFF	GROUP	DIFF
0	−10	0.5	−1	1	−7	1.25	1	1.5	−13
0	−12	0.5	10	1	15	1.25	8	1.5	−49
0	−9	0.5	3	1	−4	1.25	−15	1.5	−16
0	−11	0.5	−3	1	−16	1.25	14	1.5	−8
0	−1	0.5	−31	1	−2	1.25	−7	1.5	−2
0	6	0.5	4	1	−13	1.25	−1	1.5	−35
0	−31	0.5	−12	1	5	1.25	11	1.5	−11
0	−5	0.5	−3	1	−4	1.25	8	1.5	−46
0	13	0.5	−7	1	−2	1.25	11	1.5	−22
0	−2	0.5	−10	1	−14	1.25	−4	1.5	2
0	−7	0.5	−22	1	5	1.25	7	1.5	10
0	−8	0.5	−4	1	11	1.25	−14	1.5	−4
		0.5	−1	1	10	1.25	0	1.5	−10
		0.5	−3	1	3	1.25	5	1.5	2
				1	6	1.25	−2	1.5	−5
				1	−1				
				1	13				
				1	−8				

The investigators conjectured that nature heals best, so that changing the field from the natural state (the "1" group) will slow healing.

Do a complete analysis to see whether the groups differ in the effect of the electrical field level on healing. Follow the four-step process in your work.

24.34 Does polyester decay? How quickly do synthetic fabrics such as polyester decay in landfills? A researcher buried polyester strips in the soil for different lengths of time, then dug up the strips and measured the force required to break them. Breaking strength is easy to measure and is a good indicator of decay; lower strength means the fabric has decayed.

Part of the study buried 20 polyester strips in well-drained soil in the summer. Five of the strips, chosen at random, were dug up after each of 2 weeks, 4 weeks, and 8 weeks. Here are the breaking strengths in pounds:[18]

2 weeks	118	126	126	120	129
4 weeks	130	120	114	126	128
8 weeks	122	136	128	146	140

The investigator conjectured that buried polyester loses strength over time. Do the data support this conjecture? Follow the four-step process in data analysis and ANOVA. Be sure to check the conditions for ANOVA.

24.35 Durable press fabrics are weaker. "Durable press" cotton fabrics are treated to improve their recovery from wrinkles after washing. Unfortunately, the treatment also reduces the strength of the fabric. A study compared the breaking strength of untreated fabric with that of fabrics treated by three commercial durable press processes. Five specimens of the same fabric were assigned at random to each group. Here are the data, in pounds of pull needed to tear the fabric:[19]

Untreated	60.1	56.7	61.5	55.1	59.4
Permafresh 55	29.9	30.7	30.0	29.5	27.6
Permafresh 48	24.8	24.6	27.3	28.1	30.3
Hylite LF	28.8	23.9	27.0	22.1	24.2

The untreated fabric is clearly much stronger than any of the treated fabrics. We want to know if there is a significant difference in breaking strength among the three durable press treatments. Analyze the data for the three processes and write a clear summary of your findings. Which process do you recommend if breaking strength is a main concern? Use the four-step process to guide your discussion. (Although the standard deviations do not quite satisfy our rule of thumb, that rule is conservative and many statisticians would use ANOVA for these data.)

24.36 Durable press fabrics wrinkle less. The data in Exercise 24.35 show that durable press treatment greatly reduces the breaking strength of cotton fabric. Of course, durable press treatment also reduces wrinkling. How much? "Wrinkle recovery angle" measures how well a fabric recovers from wrinkles. Higher is better. Here are data on the wrinkle recovery angle (in degrees) for the same fabric specimens discussed in the previous exercise:

Untreated	79	80	78	80	78
Permafresh 55	136	135	132	137	134
Permafresh 48	125	131	125	145	145
Hylite LF	143	141	146	141	145

The untreated fabric once again stands out, this time as inferior to the treated fabrics in wrinkle resistance. Examine the data for the three durable press processes and summarize your findings. How does the ranking of the three processes by wrinkle resistance compare with their ranking by breaking strength in Exercise 24.35? Explain why we can't trust the ANOVA F test.

24.37 Logging in the rain forest: species counts. Table 24.2 gives data on the number of trees per forest plot, the number of species per plot, and species richness. Exercise 24.3 analyzed the effect of logging on number of trees. Exercise 24.8 concludes that it would be risky to use ANOVA to analyze richness. Use software to analyze the effect of logging on the number of species.

(a) Make a table of the group means and standard deviations. Do the standard deviations satisfy our rule of thumb for safe use of ANOVA? What do the means suggest about the effect of logging on the number of species?

(b) Carry out the ANOVA. Report the F statistic and its P-value and state your conclusion.

More rain for California? Exercise 24.30 describes a randomized experiment carried out by Kenwyn Suttle and his coworkers to examine the effects of additional water on California grassland. The experimental units are 18 plots of grassland, assigned at random among three treatments: added water in the winter wet season, added water in the spring dry season, and no added water (control group). Field experiments, unlike laboratory experiments, are exposed to variations in the natural environment. The experiment therefore continued over 5 years, 2001 to 2005.

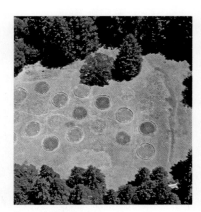

Courtesy Blake Suttle

TABLE 24.6 **Plant biomass for three water conditions over five years**

				YEAR		
TREATMENT	PLOT	2001	2002	2003	2004	2005
Winter	3	136.8358	228.0717	264.1514	254.6453	344.3933
Winter	8	151.4154	189.9505	187.7312	233.8155	203.3908
Winter	14	136.1536	209.0485	291.1431	253.4506	331.9724
Winter	20	121.6323	189.6755	176.2879	228.5882	388.1056
Winter	27	124.1459	188.0090	141.7525	158.6675	382.8617
Winter	32	125.2986	215.2174	169.9737	212.3232	346.3042
Spring	4	338.1301	422.7411	318.4182	517.6650	344.0489
Spring	11	291.8597	339.8243	281.6830	342.2825	261.8016
Spring	18	244.8727	398.7296	288.8433	270.5785	262.7238
Spring	22	234.6599	400.6878	382.6673	212.5324	316.9683
Spring	25	197.5830	326.9497	326.8877	213.9879	224.1109
Spring	35	239.0122	444.1556	293.8502	240.1927	328.2783
Control	6	73.4288	148.8907	129.0538	178.9988	237.6596
Control	7	110.6306	182.6762	144.6578	205.5165	281.1442
Control	17	95.3405	196.8303	172.7772	242.6795	313.7242
Control	24	83.0584	186.1953	113.2813	231.7639	258.3631
Control	28	30.5886	154.0401	142.1562	134.9847	235.8320
Control	33	96.9709	213.2537	117.9808	212.4862	217.5060

Table 24.6 gives data on the total plant biomass (grams per square meter) that grew on each plot during each year.[20] The "Plot" column shows how the random assignment of 18 of the 36 available plots worked. Exercises 24.38 to 24.40 are based on this information.

24.38 Plot the means. Starting from the data in Table 24.6, you can calculate the mean plant biomass for each treatment in each year as follows:

	Year				
	2001	2002	2003	2004	2005
Winter	132.58	203.33	205.17	223.58	332.84
Spring	257.69	388.85	315.39	299.54	289.66
Control	81.67	180.31	136.65	201.07	257.37

Plot the means for each of the three treatments against year, connecting the yearly means for each treatment by lines to show the pattern over time. Use the same plot for all three treatments, with a different color for each treatment. From this plot, you can get an overall picture of the experiment's results.

(a) Across all 5 years, does more water in the wet season increase plant growth? What about more water in the dry season? Which seasonal addition of water has the larger effect?

(b) One-way ANOVAs comparing the mean plant biomass separately in each year find significant differences in 3 years and no significant difference in 2 years. Based on your plot, in which three years do you think the treatment means differ significantly?

(c)   In 2005, there were unusual late rains during the spring. How does the effect of this natural rainfall show up in your plot? (You see that it would not be wise to do an experiment like this in just one year.)

24.39 The results for 2001. Your work in Exercise 24.30 shows that there were significant differences in mean plant biomass among the three treatments in 2003. Do a complete analysis of the data for 2001 and report your conclusions.

24.40 Conditions for ANOVA. Examine the data for the year 2004. The conditions for ANOVA inference are not met. In what way do these data fail to meet the conditions? (It is not very surprising that in 5 ANOVAs one will fail to satisfy our quite conservative conditions.)

24.41 Which test? Example 24.4 describes one of the experiments done by Kathleen Vohs and her coworkers to demonstrate that even being reminded of money makes people more self-sufficient and less involved with other people. Here are three more of these experiments. For each experiment, which statistical test from Chapters 17 to 24 would you use, and why?

(a)   Randomly assign student subjects to money and control groups. The control group unscrambles neutral phrases, and the money group unscrambles money-oriented phrases, as described in Example 24.4. Then ask the subjects to volunteer to help the experimenter by coding data sheets, about 5 minutes per sheet. Subjects said how many sheets they would volunteer to code. "Participants in the money condition volunteered to help code fewer data sheets than did participants in the control condition."

(b)   Randomly assign student subjects to high-money, low-money, and control groups. After playing Monopoly for a short time, the high-money group is left with $4000 in Monopoly money, the low-money group with $200, and the control group with no money. Each subject is asked to imagine a future with lots of money (high-money group), a little money (low-money group), or just their future plans (control group). Another student walks in and spills a box of 27 pencils. How many pencils does the subject pick up? "Participants in the high-money condition gathered fewer pencils" than subjects in the other two groups.

(c)   Randomly assign student subjects to three groups. All do paperwork while a computer on the desk shows a screensaver of currency floating underwater (Group 1), a screensaver of fish swimming underwater (Group 2), or a blank screen (Group 3). Each subject must now develop an advertisement and can choose whether to work alone or with a partner. Count how many in each group make each choice. "Choosing to perform the task with a coworker was reduced among money condition participants."

"About This Book" Notes

1. D. S. Moore and discussants, "New pedagogy and new content: the case of statistics," *International Statistical Review*, 65 (1997), pp. 123–165. Richard Scheaffer's comment appears on p. 156.

2. This summary of the committee's report was unanimously endorsed by the Board of Directors of the American Statistical Association. The full report is George Cobb, "Teaching statistics," in L. A. Steen (ed.), *Heeding the Call for Change: Suggestions for Curricular Action*, Mathematical Association of America, 1990, pp. 3–43.

3. A summary of the GAISE "Introductory College Course Guidelines," which have also been endorsed by the ASA Board of Directors, appears in *Amstat News*, June 2006, p. 31. See www.amstat.org/education/gaise for details.

4. Lawrence D. Brown, Tony Cai, and Anirban DasGupta, "Interval estimation for a binomial proportion," *Statistical Science*, 16 (2001), pp. 101–133.

"Statistical Thinking" Notes

1. Parts of this essay are shared with David S. Moore, "Introduction: learning from data," in Roxy Peck, et al. (eds.), *Statistics: A Guide to the Unknown*, 4th ed., Thomson, 2006.

2. See, for example, Martin Enserink, "The vanishing promises of hormone replacement," *Science*, 297 (2002), pp. 325–326; and Brian Vastag, "Hormone replacement therapy falls out of favor with expert committee," *Journal of the American Medical Association*, 287 (2002), pp. 1923–1924. A National Institutes of Health panel's comprehensive report is *International Position Paper on Women's Health and Menopause*, NIH Publication 02-3284, 2002.

3. A. C. Nielsen, Jr., "Statistics in marketing," in *Making Statistics More Effective in Schools of Business*, Graduate School of Business, University of Chicago, 1986.

4. The data in Figure 2 are based on a component of the Consumer Price Index, from the Bureau of Labor Statistics Web site: www.bls.gov. I converted the index number into cents per gallon using retail price information from the Automobile Association of America, www.fuelgaugereport.com.

5. FUTURE II Study Group, "Quadrivalent vaccine against human papillomavirus to prevent high-grade cervical lesions," *New England Journal of Medicine*, 356 (2007), pp. 1915–1927. I have simplified the conclusions so that students with as yet no statistics background can better follow the essay.

Chapter 1 Notes

1. Data for 2005 from the 2007 *Statistical Abstract of the United States* at the Census Bureau Web site, www.census.gov.

2. *Arbitron Internet and Media 2006*, at www.arbitron.com.

3. *Arbitron Radio Today 2007 Edition*, at www.arbitron.com.

4. Penn State Division of Student Affairs, "Student Drinking March 2006," at www.sa.psu.edu/sara/pulse/140-StudentDrinking.pdf.

5. Centers for Disease Control and Prevention, *Births: Final Data for 2005*, National Vital Statistics Reports, 56, No. 6, 2007, at www.cdc.gov. These are the most recent data available as of early 2008, but the numbers change only slightly from year to year.

6. From the 2006 American Community Survey, at factfinder.census.gov.

7. Our eyes do respond to area, but not quite linearly. It appears that we perceive the ratio of two bars to be about the 0.7 power of the ratio of their actual areas. See W. S. Cleveland, *The Elements of Graphing Data*, Wadsworth, 1985, pp. 278–284.

8. See Note 6.

9. From the Gary Community School Corporation, courtesy of Celeste Foster, Purdue University.

10. From the College Board Web site, `www.collegeboard.com`.

11. United Nations data, at `hdr.undp.org/en/statistics`. All amounts are in U.S. dollars at purchasing power parity. That is, the exchange rate between each currency and the dollar is set not at the fluctuating market rate but at the rate that gives a dollar the same buying power in each country.

12. Raymond W. Schaffranek and Ami L. Riscassi, *Flow Velocity, Water Temperature, and Conductivity at Selected Locations in Shark River Slough, Everglades National Park, Florida; July 1999–July 2003*, Data Series 110, U.S. Geological Survey, 2004, `water.usgs.gov`.

13. College Entrance Examination Board, *Trends in College Pricing, 2007*, at `www.collegeboard.com`. The averages are "enrollment weighted," so that they give average tuition over *students* rather than over *colleges*. Data for years before 1997 are from past editions, updated to June 2007 price levels by the Consumer Price Index.

14. *DuPont 2007 Color Popularity Report*, at `www2.dupont.com`.

15. See Note 2.

16. National Center for Health Statistics, *Deaths: Preliminary Data for 2005*, at `www.cdc.gov/nchs`.

17. *U.S. Hispanic Population 2006*, at `www.census.gov`, based on the March 2006 Current Population Survey, Annual Social and Economic Supplement.

18. Found at `spam-filter-review.toptenreviews.com`, which claims to have compiled data "from a number of different reputable sources."

19. Tom Lloyd et al., "Fruit consumption, fitness, and cardiovascular health in female adolescents: the Penn State Young Women's Health Study," *American Journal of Clinical Nutrition*, 67 (1998), pp. 624–630.

20. Data provided by Darlene Gordon from her PhD thesis, "Relationships among academic self-concept, academic achievement, and persistence with self-attribution, study habits, and perceived school environment," Purdue University, 1997.

21. Monthly stock returns from the Web site of Professor Kenneth French of Dartmouth, `mba.tuck.dartmouth.edu/pages/faculty/ken.french`. A fine point: the data are actually the "excess returns" on stocks, the actual returns less the small monthly returns on Treasury bills.

22. National Institutes of Health, Essential Fatty Acids Education site, `efaeducation.nih.gov`.

23. *2007 Statistical Abstract of the United States*, Table 154, at `www.census.gov`.

24. International Energy Agency, *Key World Energy Statistics 2007*, at `www.iea.org`.

25. National Oceanic and Atmospheric Administration, at `www.noaa.gov`.

26. David M. Fergusson and L. John Horwood, "Cannabis use and traffic accidents in a birth cohort of young adults," *Accident Analysis and Prevention*, 33 (2001), pp. 703–711.

27. From a plot in K. Krishna Kumar et al., "Unraveling the mystery of Indian monsoon failure during El Niño," *Science*, 314 (2006), pp. 115–119.

28. Census Bureau, New Residential Construction page, at `www.census.gov`. These are monthly data not seasonally adjusted.

29. Florida Fish and Wildlife Conservation Commission, at `myfwc.com/gators`.

Chapter 2 Notes

1. From the 2003 American Community Survey, at the Census Bureau Web site, `www.census.gov`. The data are a subsample of the 13,194 individuals in the ACS North Carolina sample who had travel times greater than zero.

2. This isn't a mathematical theorem. The mean can be less than the median in right-skewed distributions that take only a few values, many of which lie exactly at the median. The rule almost never fails for distributions taking many values, and most counterexamples don't appear clearly skewed in graphs even though they may be slightly skewed according to technical measures of skewness. See Paul T. von Hippel, "Mean, median, and skew: correcting a textbook rule," *Journal of Statistics Education*, 13, No. 2 (2005), online journal,

3. National Association of College and University Business Officers, 2007 Endowment Study, at `www.nacubo.org`.

4. From the National Association of Realtors, at `www.realtor.org`.

5. U.S. Census Bureau, *Income, Poverty, and Health Insurance Coverage in the United States: 2006*, August 2007, at `www.census.org`.

6. Figure 2.2 displays the daily returns for the year ending November 8, 2007, for the CREF Equity Index Fund and the TIAA Real Estate Fund. Daily price data for these funds are at `www.tiaa-cref.org`. Returns can be easily calculated because dividends are incorporated in the daily prices rather than given separately.

7. Michael W. Peugh, "Field investigation of ventilation and air quality in duck and turkey slaughter plants," MS thesis, Purdue University, 1996.

8. Ethan J. Temeles and W. John Kress, "Adaptation in a plant-hummingbird association," *Science*, 300 (2003), pp. 630–633. I thank Ethan J. Temeles for providing the data.

9. C. H. Cannon, D. R. Peart, and M. Leighton, "Tree species diversity in commercially logged Bornean rainforest," *Science*, 281 (1998), pp. 1366–1367. I thank Charles Cannon for providing the data.

10. Raymond Fisman and Edward Miguel, "Cultures of corruption: evidence from diplomatic parking tickets," National Bureau of Economic Research working paper 12312, June 2006, at `www.nber.org`.

11. Patrick J. Purcell, *Retirement Savings and Household Wealth: A Survey of Recent Data*, Congressional Research Service, 2004, available at several online repositories. The data are for 2001. To make the amounts more realistic, I have updated them to 2007 dollars using the Consumer Price Index.

12. See Note 5 for Chapter 1.

13. T. Bjerkedal, "Acquisition of resistance in guinea pigs infected with different doses of virulent tubercle bacilli," *American Journal of Hygiene*, 72 (1960), pp. 130–148.

14. Data for 1986 from David Brillinger, University of California, Berkeley. See David R. Brillinger, "Mapping aggregate birth data," in A. C. Singh and P. Whitridge (eds.), *Analysis of Data in Time*, Statistics Canada, 1990, pp. 77–83. A boxplot similar to Figure 2.6 appears in David R. Brillinger, "Some examples of random process environmental data analysis," in P. K. Sen and C. R. Rao (eds.), *Handbook of Statistics*, Vol. 18, *Bioenvironmental and Public Health Statistics*, North Holland, 2000.

15. Nicolas Guéguen and Christine Petr, "Odors and consumer behavior in a restaurant," *Journal of Hospitality Management*, 25 (2006), pp. 335–339. I thank Nicolas Guéguen for providing the data.

16. James A. Levine et al., "Inter-individual variation in posture allocation: possible role in human obesity," *Science*, 307 (2005), pp. 584–586. I thank James Levine for providing the data.

17. Parmeshwar S. Gupta, "Reaction of plants to the density of soil," *Journal of Ecology*, 21 (1933), pp. 452–474.

18. M. Ann Laskey et al., "Bone changes after 3 mo of lactation: influence of calcium intake, breast-milk output, and vitamin D–receptor genotype," *American Journal of Clinical Nutrition*, 67 (1998), pp. 685–692.

Chapter 3 Notes

1. See Note 9 for Chapter 1.

2. Thomas K. Cureton et al., *Endurance of Young Men*, Monographs of the Society for Research in Child Development, Vol. 10, No. 1, 1945.

3. Monsoon rainfall from B. Parthasarathy, Indian Institute of Tropical Meterology, at `www.iges.org`. The data cover the years 1871 to 2000.

4. Based on the National Health and Nutrition Examination Surveys, 1988–1994. From the Web site of the National Center for Health Statistics, `www.cdc.gov/nchs`.

5. Detailed data appear in P. S. Levy et al., *Total Serum Cholesterol Values for Youths 12–17 Years*, Vital and Health Statistics, Series 11, No. 155, National Center for Health Statistics, 1976.

6. See Note 16 for Chapter 2.

7. All ACT facts are from the ACT Web site, `www.act.org`.

8. All SAT facts are from the College Board Web site, `www.collegeboard.com`.

9. See Note 20 for Chapter 1.

10. The data were provided by Nicolas Fisher.

11. From a graph in L. Partridge and M. Farquhar, "Sexual activity reduces lifespan of male fruitflies," *Nature*, 294 (1981), pp. 580–582. Provided by Brigitte Baldi.

12. See Note 3.

Chapter 4 Notes

1. Data for 2007 graduates from the College Board Web site, www.collegeboard.com.

2. Government data for 2005, presented in "An accident waiting to happen?" *Consumer Reports*, March 2007, pp. 16–19.

3. The Florida Boating Accident Statistical Report for each year (at myfwc.com/law/boating) gives the number of registered vessels. The Florida Wildlife Commission maintains a manatee death data base at research.myfwc.com/manatees.

4. Based on T. N. Lam, "Estimating fuel consumption from engine size," *Journal of Transportation Engineering*, 111 (1985), pp. 339–357. The data for 10 to 50 km/h are measured; those for 60 and higher are calculated from a model given in the paper and are therefore smoothed.

5. A careful study of this phenomenon is W. S. Cleveland, P. Diaconis, and R. McGill, "Variables on scatterplots look more highly correlated when the scales are increased," *Science*, 216 (1982), pp. 1138–1141.

6. From a graph in Magdalena Bermejo et al., "Ebola outbreak killed 5000 gorillas," *Science*, 314 (2006), p. 1564.

7. Data for Fig. 4.6(b) come from William Gray's Web site, at hurricane.atmos.colostate.edu. Data for Fig. 4.6(c) were provided by Drina Iglesia, Purdue University, from a study reported in D. D. S. Iglesia, E. J. Cragoe, Jr., and J. W. Vanable, "Electric field strength and epithelization in the newt (*Notophthalmus viridescens*)," *Journal of Experimental Zoology*, 274 (1996), pp. 56–62. Data for Fig. 4.6(d) are for the Wilshire 5000 stock index. As a fine point, plots (b), (c), and (d) are square with the same scales on both axes because both variables measure similar quantities in the same units.

8. This exercise is motivated by Scott Berry, "Statistical fallacies in sports," *Chance*, 19, No. 4 (2006), pp. 50–56, where scores from the 2006 Masters are analyzed.

9. James T. Fleming, "The measurement of children's perception of difficulty in reading materials," *Research in the Teaching of English*, 1 (1967), pp. 136–156.

10. From a graph in Timothy G. O'Brien and Margaret F. Kinnaird, "Caffeine and conservation," *Science*, 300 (2003), p. 587.

11. From a graph in Bernt-Erik Saether, Steiner Engen, and Erik Mattysen, "Demographic characteristics and population dynamical patterns of solitary birds," *Science*, 295 (2002), pp. 2070–2073.

12. From a graph in Sabrina M. Tom et al., "The neural basis of loss aversion in decision-making under risk," *Science*, 315 (2007), pp. 515–518.

13. From a graph in Sergio M. Vallina and Rafel Simó, "Strong relationship between DMS and the solar radiation dose over the global surface ocean," *Science*, 315 (2007), pp. 506–508.

14. N. Maeno et al., "Growth rates of icicles," *Journal of Glaciology*, 40 (1994), pp. 319–326.

15. Simplified from W. L. Colville and D. P. McGill, "Effect of rate and method of planting on several plant characters and yield of irrigated corn," *Agronomy Journal*, 54 (1962), pp. 235–238.

16. Modified from M. C. Wilson and R. E. Shade, "Relative attractiveness of various luminescent colors to the cereal leaf beetle and the meadow spittlebug," *Journal of Economic Entomology*, 60 (1967), pp. 578–580.

17. From a graph in Martin Wild et al., "From dimming to brightening: decadal changes in solar radiation at Earth's surface," *Science*, 308 (2005), pp. 847–850.

18. Data provided by Robert Dale, Purdue University.

19. From a graph in Christer G. Wiklund, "Food as a mechanism of density-dependent regulation of breeding numbers in the merlin *Falco columbarius*," *Ecology*, 82 (2001), pp. 860–867.

20. From a graph in Naomi I. Eisenberger, Matthew D. Lieberman, and Kipling D. Williams, "Does rejection hurt? An fMRI study of social exclusion," *Science*, 302 (2003), pp. 290–292.

21. Justin S. Brashares et al., "Bushmeat hunting, wildlife declines, and fish supply in West Africa," *Science*, 306 (2004), pp. 1180–1183. The data used here are found in the online supplementary material. The published

analysis omits data for 1999, an extreme low outlier, without explanation.

Chapter 5 Notes

1. From a graph in James A. Levine, Norman L. Eberhardt, and Michael D. Jensen, "Role of nonexercise activity thermogenesis in resistance to fat gain in humans," *Science*, 283 (1999), pp. 212–214.

2. See Note 6 for Chapter 4.

3. From a graph in Tania Singer et al., "Empathy for pain involves the affective but not sensory components of pain," *Science*, 303 (2004), pp. 1157–1162. Data for other brain regions showed a stronger correlation and no outliers.

4. Contributed by Marigene Arnold, Kalamazoo College.

5. Gannett News Service article appearing in the *Lafayette (Ind.) Journal and Courier*, April 23, 1994.

6. P. Goldblatt (ed.), *Longitudinal Study: Mortality and Social Organization*, Her Majesty's Stationery Office, 1990. At least, so claims Richard Conniff, *The Natural History of the Rich*, Norton, 2002, p. 45. The Goldblatt report is not available to me.

7. Laura L. Calderon et al., "Risk factors for obesity in Mexican-American girls: dietary factors, anthropometric factors, physical activity, and hours of television viewing," *Journal of the American Dietetic Association*, 96 (1996), pp. 1177–1179.

8. *The Health Consequences of Smoking: 1983*, Public Health Service, Washington, D.C., 1983.

9. A. K. Yousafzai et al., "Comparison of armspan, arm length and tibia length as predictors of actual height of disabled and nondisabled children in Dharavi, Mumbai, India," *European Journal of Clinical Nutrition*, 57 (2003), pp. 1230–1234. In fact, $r^2 = 0.93$. I changed the value in Exercise 5.25 to avoid confusion with $b = 0.93$.

10. G. L. Kooyman et al., "Diving behavior and energetics during foraging cycles in king penguins," *Ecological Monographs*, 62 (1992), pp. 143–163.

11. T. Constable and E. McBean, "BOD/TOC correlations and their application to water quality evaluation," *Water, Air, and Soil Pollution*, 11 (1979), pp. 363–375.

12. Karl Pearson and A. Lee, "On the laws of inheritance in man," *Biometrika*, 2 (1902), p. 357. These data also appear in D. J. Hand et al., *A Handbook of Small Data Sets*, Chapman & Hall, 1994. This book offers more than 500 data sets that can be used in statistical exercises.

13. From a presentation by Charles Knauf, Monroe County (N.Y.) Environmental Health Laboratory.

14. See Note 11 for Chapter 4.

15. Frank J. Anscombe, "Graphs in statistical analysis," *The American Statistician*, 27 (1973), pp. 17–21.

16. Debora L. Arsenau, "Comparison of diet management instruction for patients with non-insulin dependent diabetes mellitus: learning activity package vs. group instruction," MS thesis, Purdue University, 1993.

17. Gary Smith, "Do statistics test scores regress toward the mean?" *Chance*, 10, No. 4 (1997), pp. 42–45.

18. From a graph in G. D. Martinsen, E. M. Driebe, and T. G. Whitham, "Indirect interactions mediated by changing plant chemistry: beaver browsing benefits beetles," *Ecology*, 79 (1998), pp. 192–200.

19. P. Velleman, *ActivStats 2.0*, Addison Wesley Interactive, 1997.

20. From William Gray's Web site, `hurricane.atmos.colostate.edu`.

21. From a graph in Joaquim I. Goes et al., "Warming of the Eurasian landmass is making the Arabian Sea more productive," *Science*, 308 (2005), pp. 545–547.

22. See Note 11 for Chapter 1.

Chapter 6 Notes

1. The National Longitudinal Study of Adolescent Health interviewed a stratified random sample of 27,000 adolescents, then reinterviewed many of the subjects six years later, when most were aged 19 to 25. These data are from the Wave III reinterviews in 2000 and 2001, found at the Web site of the Carolina Population Center, `www.cpc.unc.edu`.

2. Lien-Ti Bei, "Consumers' purchase behavior toward recycled products: an acquisition-transaction utility theory perspective," MS thesis, Purdue University, 1993.

3. From the October 2005 Current Population Survey, at www.census.gov.

4. Siem Oppe and Frank De Charro, "The effect of medical care by a helicopter trauma team on the probability of survival and the quality of life of hospitalized victims," *Accident Analysis and Prevention*, 33 (2001), pp. 129–138. The authors give the data in Example 6.4 as a "theoretical example" to illustrate the need for their more elaborate analysis of actual data using severity scores for each victim.

5. These data, from reports submitted by airlines to the Department of Transportation, appear in A. Barnett, "How numbers can trick you," *Technology Review*, October 1994, pp. 38–45.

6. See Note 1.

7. Data for Purdue University, fall 2007, omitting small programs and social and behavioral science departments.

8. D. M. Barnes, "Breaking the cycle of addiction," *Science*, 241 (1988), pp. 1029–1030.

9. Sanders Korenman and David Neumark, "Does marriage really make men more productive?" *Journal of Human Resources*, 26 (1991), pp. 282–307.

10. See P. J. Bickel and J. W. O'Connell, "Is there a sex bias in graduate admissions?" *Science*, 187 (1975), pp. 398–404.

11. The General Social Survey exercises in this chapter present tables constructed using the search function at the GSS archive, sda.berkeley.edu/archive.htm.

12. Michael Gurian, "Where have the men gone? No place good," *Washington Post*, December 4, 2005, at www.washingtonpost.com. The data are from the 2006 *Digest of Education Statistics* at the Web site of the National Center for Education Statistics, nces.ed.gov.

13. Claudia Braga et al., "Olive oil, other seasoning fats, and the risk of colorectal carcinoma," *Cancer*, 82 (1998), pp. 448–453.

14. Janice E. Williams et al., "Anger proneness predicts coronary heart disease risk," *Circulation*, 101 (2000), pp. 2034–2039.

15. R. Shine, T. R. L. Madsen, M. J. Elphick, and P. S. Harlow, "The influence of nest temperatures and maternal brooding on hatchling phenotypes in water pythons," *Ecology*, 78 (1997), pp. 1713–1721.

Chapter 7 Notes

1. Pew Research Center for the People and the Press, *Trends in Political Values and Core Attitudes: 1987–2007*, March 2007, at people-press.org.

2. See Note 1.

3. Table 1 of E. Thomassot et al., "Methane-related diamond crystallization in the earth's mantle: stable isotopes evidence from a single diamond-bearing xenolith," *Earth and Planetary Science Letters*, 257 (2007), pp. 362–371.

4. Environmental Protection Agency, *Municipal Solid Waste Generation, Recycling, and Disposal in the United States: Facts and Figures for 2006*, November 2007, at www.epa.gov.

5. Data provided by Brigitte Baldi, University of California at Irvine.

6. Data for Cohort 2 in Richard A. Morgan et al., "Cancer regression in patients after transfer of genetically engineered lymphocytes," *Science*, 314 (2006), pp. 126–129. The doubling time data are given in the paper and the immune response data appear in the supplementary online material.

7. Margaret A. McDowell et al., "Anthropometric reference data for children and adults: U.S. population, 1999–2002," National Center for Health Statistics, Advance Data from Vital and Health Statistics, No. 361, 2005, at www.cdc.gov/nchs.

8. From the MCAT Web site, www.aamc.org/students/mcat.

9. From a plot in Jon J. Ramsey et al., "Energy expenditure, body composition, and glucose metabolism in lean and obese rhesus monkeys treated with ephedrine and caffeine," *American Journal of Clinical Nutrition*, 68 (1998), pp. 42–51.

10. Centers for Disease Control and Prevention, *Cigarette Smoking among Adults—United States, 2006*, and related publications at www.cdc.gov.

11. From a graph in Stan Boutin et al., "Anticipatory reproduction and population growth in seed predators," *Science*, 314 (2006), pp. 1928–1930.

12. Data from a plot in Josef P. Rauschecker, Biao Tian, and Marc Hauser, "Processing of complex sounds in the macaque nonprimary auditory cortex," *Science*, 268 (1995), pp. 111–114. The paper states that there are $n = 41$ observations, but only $n = 37$ can be read accurately from the plot.

13. J. T. Dwyer et al., "Memory of food intake in the distant past," *American Journal of Epidemiology*, 130 (1989), pp. 1033–1046.

14. "Dancing in step," *Economist*, March 22, 2001.

15. Data provided by Samuel Phillips, Purdue University.

16. Mei-Hui Chen, "An exploratory comparison of American and Asian consumers' catalog patronage behavior," MS thesis, Purdue University, 1994.

17. From a graph in Peter A. Raymond and Jonathan J. Cole, "Increase in the export of alkalinity from North America's largest river," *Science*, 301 (2003), pp. 88–91.

18. From the Nenana Ice Classic Web site, `www.nenanaakiceclassic.com`. See Raphael Sagarin and Fiorenza Micheli, "Climate change in nontraditional data sets," *Science*, 294 (2001), p. 811, for a careful discussion.

19. Data for 2004 from Alan Heston, Robert Summers, and Bettina Aten, *Penn World Table Version 6.2*, Center for International Comparisons of Production, Income, and Prices at the University of Pennsylvania, September 2006, at `pwt.econ.upenn.edu`.

20. Louie H. Yang, "Periodical cicadas as resource pulses in North American forests," *Science*, 306 (2004), pp. 1565–1567. The data are simulated Normal values that match the means and standard deviations reported in this article.

21. Alan S. Banks et al., "Juvenile hallux abducto valgus association with metatarsus adductus," *Journal of the American Podiatric Medical Association*, 84 (1994), pp. 219–224.

22. Todd W. Anderson, "Predator responses, prey refuges, and density-dependent mortality of a marine fish," *Ecology*, 81 (2001), pp. 245–257.

23. From a graph in Craig Packer et al., "Ecological change, group territoriality, and population dynamics in Serengeti lions," *Science*, 307 (2005), pp. 390–393.

24. Peter H. Chen, Neftali Herrera, and Darren Christiansen, "Relationships between gate velocity and casting features among aluminum round castings," no date. Provided by Darren Christiansen.

25. Data compiled from a table of percents in "Americans view higher education as key to the American dream," press release by the National Center for Public Policy and Higher Education, at `www.highereducation.org`, May 3, 2000.

Chapter 8 Notes

1. From the index of recent *New York Times* surveys at `nytimes.com/polls`. The methodological statement is similar for most polls listed.

2. Gary S. Foster and Craig M. Eckert, "Up from the grave: a sociohistorical reconstruction of an African American community from cemetary data in the rural Midwest," *Journal of Black Studies*, 33 (2003), pp. 468–489.

3. Pew Forum on Religion and Public Life, *Spirit and Power: A 10-Country Survey of Pentecostals*, October 2006, at `www.pewforum.org`.

4. The regulations that govern seat belt survey design can be found at `www-nrd.nhtsa.dot.gov`. Details on the Hawaii survey are in Karl Kim et al., *Results of the 2002 Highway Seat Belt Use Survey*, at `www.state.hi.us/dot`.

5. Donald L. McCabe, Linda Klebe Trevino, and Kenneth D. Butterfield, "Dishonesty in academic environments," *Journal of Higher Education*, 72 (2001), pp. 29–45.

6. For information on the 2006 American Community Survey of households (there is a separate sample of group quarters), go to `www.census.gov/acs`.

7. The Pew press release and the full report "Polls face growing resistance, but still representative" (dated April 20, 2004) are at `people-press.org/reports`.

8. For more detail on the limits of memory in surveys, see N. M. Bradburn, L. J. Rips, and S. K. Shevell, "Answering autobiographical questions: the impact of memory and inference on surveys," *Science*, 236 (1987), pp. 157–161.

9. The immigration questions are from the *New York Times*/CBS News Poll taken May 18 to 23, 2007, found

at www.pollingreport.com. The responses on welfare are from a *New York Times*/CBS News Poll reported in the *New York Times*, July 5, 1992. Many other examples appear in T. W. Smith, "That which we call welfare by any other name would smell sweeter," *Public Opinion Quarterly*, 51 (1987), pp. 75–83. The example on the effect of question order is cited in Daniel Kahnemann et al., "Would you be happier if you were richer? A focusing illusion," *Science*, 312 (2006), pp. 1908–1910.

10. Giuliana Coccia, "An overview of non-response in Italian telephone surveys," *Proceedings of the 99th Session of the International Statistical Institute*, 1993, Book 3, pp. 271–272.

11. Information from various articles in the special issue on cell phone surveys, *Public Opinion Quarterly*, 71, No. 5 (2007). See also the Pew study cited in Note 7 for a comparison of a standard RDD survey with a rigorous survey that reduced nonresponse from 73% to 49%.

12. See Mick P. Couper, "Web surveys: A review of issues and approaches," *Public Opinion Quarterly*, 64 (2000), pp. 464–494.

13. Rachel Sherman and John Hickner, "Academic physicians use placebos in clinical practice and believe in the mind-body connection," *Journal of General Internal Medicine*, 23 (2008), pp. 7–10.

14. From the Web site of the Gallup Organization, www.gallup.com. Individual poll reports remain on this site for only a limited time.

15. Robert C. Parker and Patrick A. Glass, "Preliminary results of double-sample forest inventory of pine and mixed stands with high- and low-density LiDAR," in Kristina F. Connoe (ed.), *Proceedings of the 12th Biennial Southern Silvicultural Research Conference*, U.S. Department of Agriculture, Forest Service, Southern Research Station, 2004. The researchers actually sampled every 10th plot. This is a systematic sample; see Exercise 8.41.

16. Bryan E. Porter and Thomas D. Berry, "A nationwide survey of self-reported red light running: measuring prevalence, predictors, and perceived consequences," *Accident Analysis and Prevention*, 33 (2001), pp. 735–741.

17. Mario A. Parada et al., "The validity of self-reported seatbelt use: Hispanic and non-Hispanic drivers in El Paso," *Accident Analysis and Prevention*, 33 (2001), pp. 139–143.

18. Clyde O. McDaniel, Jr., "Dating roles and reasons for dating," *Journal of Marriage and the Family*, 31 (1969), pp. 97–107.

19. Susan B. Sorenson, "Regulating firearms as a consumer product," *Science*, 286 (1999), pp. 1481–1482.

20. Information from Warren McIsaac and Vivek Goel, "Is access to physician services in Ontario equitable?" Institute for Clinical Evaluative Sciences in Ontario, October 18, 1993.

21. Adam Nagourney and Janet Elder, "New York Times CBS Poll: What Hispanics Believe," at www.Hispanic.cc.

Chapter 9 Notes

1. I. J. Goldberg et al., "Wine and your heart: a science advisory for healthcare professionals from the Nutrition Committee, Council on Epidemiology and Prevention, and Council on Cardiovascular Nursing of the American Heart Association," *Circulation*, 103 (2001), pp. 472–475.

2. J. E. Muscat et al., "Handheld cellular telephone use and risk of brain cancer," *Journal of the American Medical Association*, 284 (2000), pp. 3001–3007.

3. Charles A. Nelson III et al., "Cognitive recovery in socially deprived young children: the Bucharest Early Intervention Project," *Science*, 318 (2007), pp. 1937–1940.

4. Martin J. Bergee and Lecia Cecconi-Roberts, "Effects of small-group peer interaction on self-evaluation of music performance," *Journal of Research in Music Education*, 50 (2002), pp. 256–268.

5. K. B. Suttle, Meredith A. Thomsen, and Mary E. Power, "Species interactions reverse grassland responses to changing climate," *Science*, 315 (2007), pp. 640–642. See Chapter 24 for an analysis of some data from this experiment.

6. How-Ran Guo, "Arsenic level in drinking water and mortality of lung cancer (Taiwan)," *Cancer Causes and Control*, 15 (2004), pp. 171–177.

7. Marielle H. Emmelot-Vonk et al., "Effect of testosterone supplementation on functional mobility, cognition, and other parameters in older men," *Journal of the American Medical Association*, 299 (2008), pp. 39–52.

8. David L. Strayer, Frank A. Drews, and William A. Johnston, "Cell phone–induced failures of visual attention during simulated driving," *Journal of Experimental Psychology: Applied*, 9 (2003), pp. 23–32.

9. Sterling C. Hilton et al., "A randomized controlled experiment to assess technological innovations in the classroom on student outcomes: an overview of a clinical trial in education," manuscript, no date. A brief report is Sterling C. Hilton and Howard B. Christensen, "Evaluating the impact of multimedia lectures on student learning and attitudes," *Proceedings of the 6th International Conference on the Teaching of Statistics*, at www.stat.aukland.ac.nz.

10. K. J. Mukamal et al., "Prior alcohol consumption and mortality following acute myocardial infarction," *Journal of the American Medical Association*, 285 (2001), pp. 1965–1970.

11. Rita F. Redburg, "Vitamin E and cardiovascular health," *Journal of the American Medical Association*, 294 (2005), pp. 107–109.

12. Jo Phelan et al., "The stigma of homelessness: the impact of the label 'homeless' on attitudes towards poor persons," *Social Psychology Quarterly*, 60 (1997), pp. 323–337.

13. Esther Duflo, Rema Hanna, and Stephan Ryan, "Monitoring works: getting teachers to come to school," report dated November 21, 2007, at econ-mit.edu/files/2066.

14. John H. Kagel, Raymond C. Battalio, and C. G. Miles, "Marijuana and work performance: results from an experiment," *Journal of Human Resources*, 15 (1980), pp. 373–395.

15. Shailja V. Nigdikar et al., "Consumption of red wine polyphenols reduces the susceptibility of low-density lipoproteins to oxidation in vivo," *American Journal of Clinical Nutrition*, 68 (1998), pp. 258–265. (There were in fact only 30 subjects, some of whom received more than one treatment with a four-week period intervening.)

16. Ian G. Williamson et al., "Antibiotics and topical nasal steroid for treatment of acute maxillary sinusitis," *Journal of the American Medical Association*, 298 (2007), pp. 2487–2496.

17. Based on Evan H. DeLucia et al., "Net primary production of a forest ecosystem with experimental CO_2 enhancement," *Science*, 284 (1999), pp. 1177–1179. The investigators used the block design.

18. E. M. Peters et al., "Vitamin C supplementation reduces the incidence of postrace symptoms of upper-respiratory tract infection in ultramarathon runners," *American Journal of Clinical Nutrition*, 57 (1993), pp. 170–174.

19. The study is described in Gina Kolata, "New study finds vitamins are not cancer preventers," *New York Times*, July 21, 1994. Look in the *Journal of the American Medical Association* of the same date for the details.

20. R. C. Shelton et al., "Effectiveness of St. John's wort in major depression," *Journal of the American Medical Association*, 285 (2001), pp. 1978–1986.

Data Ethics Notes

1. John C. Bailar III, "The real threats to the integrity of science," *Chronicle of Higher Education*, April 21, 1995, pp. B1–B2.

2. See the details on the Web site of the Office for Human Research Protections of the Department of Health and Human Services, www.hhs.gov/ohrp.

3. The difficulties of interpreting guidelines for informed consent and for the work of institutional review boards in medical research are a main theme of Beverly Woodward, "Challenges to human subject protections in U.S. medical research," *Journal of the American Medical Association*, 282 (1999), pp. 1947–1952. The references in this paper point to other discussions. Updated regulations and guidelines appear on the OHRP Web site (see Note 2).

4. Quotation from the *Report of the Tuskegee Syphilis Study Legacy Committee*, May 20, 1996. A detailed history is James H. Jones, *Bad Blood: The Tuskegee Syphilis Experiment*, Free Press, 1993.

5. Dr. Hennekens's words are from an interview in the Annenberg/Corporation for Public Broadcasting video series *Against All Odds: Inside Statistics*. The lack of certainty that Dr. Hennekens refers to is now called "clinical equipoise" in discussions of ethics.

6. R. D. Middlemist, E. S. Knowles, and C. F. Matter, "Personal space invasions in the lavatory: suggestive evidence for arousal," *Journal of Personality and Social Psychology*, 33 (1976), pp. 541–546.

7. For a review of domestic violence experiments, see C. D. Maxwell et al., *The Effects of Arrest on Intimate Partner Violence: New Evidence from the Spouse Assault Replication Program*, U.S. Department of Justice, NCH188199, 2001. Available online at www.ojp.usdoj.gov/nij/pubs-sum/188199.htm.

8. Joseph Millum and Ezekial J. Emanuel, "The ethics of international research with abandoned children," *Science*, 318 (2007), pp. 1874–1875. This paper has some useful comments on international research in general.

Chapter 10 Notes

1. Note that pennies have rims that make spinning more stable. The probability of a head in spinning a coin depends on the type of coin and also on the surface. See Exercise 21.3 for an account of 56% of heads in spinning a Belgian 1-euro coin. *Chance News 11.02* at www.dartmouth.edu/~chance reports about 45% heads in more than 20,000 spins of American pennies by Robin Lock's students at Saint Lawrence University.

2. Data for 2006 from the Web site of Statistics Canada, www.statcan.ca.

3. You can find a mathematical explanation of Benford's law in Ted Hill, "The first-digit phenomenon," *American Scientist*, 86 (1996), pp. 358–363; and Ted Hill, "The difficulty of faking data," *Chance*, 12, No. 3 (1999), pp. 27–31. Applications in fraud detection are discussed in the second paper by Hill and in Mark A. Nigrini, "I've got your number," *Journal of Accountancy*, May 1999, available online at www.aicpa.org/pubs/jofa/joaiss.htm.

4. See Note 1 for Chapter 6.

5. Information from www-records.ncsu.edu/cgi-bin/grddist.pl.

6. See Note 2 for Chapter 3.

7. See Note 14 for Chapter 1.

8. National population estimates for July 1, 2006, at the Census Bureau Web site www.census.gov. The table omits people who consider themselves to belong to more than one race.

9. See Note 1 for Chapter 6.

Chapter 11 Notes

1. U.S. Census Bureau, *Income, Poverty, and Health Insurance in the United States: 2006*, Current Population Reports P60-233, August 2007.

2. Strictly speaking, the formula $\sigma/\sqrt{n}$ for the standard deviation of $\bar{x}$ assumes that we draw an SRS of size n from an *infinite* population. If the population has finite size N, this standard deviation is multiplied by $\sqrt{1 - (n-1)/(N-1)}$. This "finite population correction" approaches 1 as N increases. When the population is at least 20 times as large as the sample, the correction factor is between about 0.97 and 1. It is reasonable to use the simpler form $\sigma/\sqrt{n}$ in these settings.

3. Earnings for all 98,105 households were downloaded using the Census Bureau's Data Ferret software. The histograms in Figure 11.4 were produced from the downloaded data.

4. Sherri A. Buzinski, "The effect of position of methylation on the performance properties of durable press treated fabrics," CSR490 honors paper, Purdue University, 1985.

5. Elroy Dimson, Paul Marsh, and Mike Staunton, *Triumph of the Optimists: 101 Years of Global Investment Returns*, Princeton University Press, 2002. Sophisticates will note that for compounding over several years we want the geometric mean return, which was 6.7%.

Chapter 12 Notes

1. This is one of several tests discussed in Bernard M. Branson, "Rapid HIV testing: 2005 update," a presentation by the Centers for Disease Control and Prevention, at www.cdc.gov. The Malawi clinic result is reported by Bernard M. Branson, "Point-of-care rapid tests for HIV antibody," *Journal of Laboratory Medicine*, 27 (2003), pp. 288–295.

2. Robert P. Dellavalle et al., "Going, going, gone: lost Internet references," *Science*, 302 (2003), pp. 787–788.

3. Ward's Automotive, *U.S. Light Vehicle Sales Summary*, December 2007, at wardsauto.com.

4. Sales in 2006 from the Web site of the Entertainment Software Association, at www.theesa.com.

5. Information about Internet users comes from sample surveys carried out by the Pew Internet and American Life Project, at `www.pewinternet.org`.

6. S. H. Sicherer, "Prevalence of peanut and tree nut allergy in the US determined by random digit dial telephone survey," *Journal of Allergy and Clinical Immunolgy*, 103 (1999), pp. 559–562.

7. Probabilities from trials with 2897 people known to be free of HIV antibodies and 673 people known to be infected, reported in J. Richard George, "Alternative specimen sources: methods for confirming positives," 1998 Conference on the Laboratory Science of HIV, found online at the Centers for Disease Control and Prevention, `www.cdc.gov`.

8. Projections from U.S. Department of Education, *Projections of Education Statistics to 2016*, December 2007, at `nces.ed.gov`.

9. Data provided by Patricia Heithaus and the Department of Biology at Kenyon College.

10. F. J. G. M. Klaassen and J. R. Magnus, "How to reduce the service dominance in tennis? Empirical results from four years at Wimbledon," in S. J. Haake and A. O. Coe (eds.), *Tennis Science and Technology*, Blackwell, 2000, pp. 277–284.

11. From the National Institutes of Health's National Digestive Diseases Information Clearinghouse, found at `wrongdiagnosis.com`.

12. The probabilities given are realistic, according to the fundraising firm SCM Associates, at `scmassoc.com`.

Chapter 13 Notes

1. Matthew A. Carlton and William D. Stansfield, "Making babies by the flip of a coin?" *American Statistician*, 59 (2005), pp. 180–182.

2. From a Gallup Poll taken in 2003, at `www.gallup.com`.

3. The survey question is reported in Trish Hall, "Shop? Many say 'Only if I must,' " *New York Times*, November 28, 1990. In fact, 66% (1650 of 2500) in the sample said "Agree."

4. John Schwartz, "Leisure pursuits of today's young men," *New York Times*, March 29, 2004. The source cited is comScore Media Matrix.

5. N. Ranjit et al., "Contraceptive failure in the first two years of use: differences across socioeconomic subgroups," *Family Planning Perspectives*, 33 (2001), pp. 19–27.

6. Results for the full 2007 season, at `www.pgatour.com`.

7. Associated Press news item dated December 9, 2007, found at `www.msnbc.msn.com`.

8. See `demonstrations.wolfram.com/MonteCarloEstimateForPi/` for an online demonstration of this idea.

Chapter 14 Notes

1. See Note 7 for Chapter 7.

2. Francisco L. Rivera-Batiz, "Quantitative literacy and the likelihood of employment among young adults," *Journal of Human Resources*, 27 (1992), pp. 313–328.

3. Data provided by Drina Iglesia, Purdue University. The data are part of a larger study reported in D. D. S. Iglesia, E. J. Cragoe, Jr., and J. W. Vanable, "Electric field strength and epithelization in the newt (*Notophthalmus viridescens*)," *Journal of Experimental Zoology*, 274 (1996), pp. 56–62.

4. See Note 20 for Chapter 1.

5. E. M. Peters et al., "Vitamin C supplementation reduces the incidence of postrace symptoms of upper-respiratory tract infection in ultramarathon runners," *American Journal of Clinical Nutrition*, 57 (1993), pp. 170–174.

6. Data simulated from a Normal distribution with the mean and standard deviation reported by Sarah Morrison and Jan Noyes, "A comparison of two computer fonts: serif versus ornate sans serif," *Usability News*, 5.2 (2003), at `psychology.wichita.edu/surl/usability_news.html`.

7. Chi-Fu Jeffrey Yang, Peter Gray, Harrison G. Pope, Jr., "Male body image in Taiwan versus the West," *American Journal of Psychiatry*, 162 (2005), pp. 263–269.

8. Ajay Ghei, "An empirical analysis of psychological androgeny in the personality profile of the successful hotel manager," MS thesis, Purdue University, 1992.

9. Seung-Ok Kim, "Burials, pigs, and political prestige in Neolithic China," *Current Anthropology*, 35 (1994), pp. 119–141.

10. Louie H. Yang, "Periodical cicadas as resource pulses in North American forests," *Science*, 306 (2004), pp. 1565–1567.

11. Sara L. Webb and Sara E. Scanga, "Windstorm disturbance without patch dynamics: twelve years of change in a Minnesota forest," *Ecology*, 82 (2001), pp. 893–897.

12. Mario A. Parada et al., "The validity of self-reported seatbelt use: Hispanic and non-Hispanic drivers in El Paso," *Accident Analysis and Prevention*, 33 (2001), pp. 139–143.

13. M. Ann Laskey et al., "Bone changes after 3 mo of lactation: influence of calcium intake, breast-milk output, and vitamin D–receptor genotype," *American Journal of Clinical Nutrition*, 67 (1998), pp. 685–692.

14. Data simulated from a Normal distribution based on information in Brian M. DeBroff and Patricia J. Pahk, "The ability of periorbitally applied antiglare products to improve contrast sensitivity in conditions of sunlight exposure," *Archives of Ophthamology*, 121 (2003), pp. 997–1001.

Chapter 15 Notes

1. See www.cdc.gov/nchs/tutorials/currentnhanes/SurveyDesign/SampleDesign/sample_intro.htm.

2. Bryan E. Porter and Thomas D. Berry, "A nationwide survey of self-reported red light running: measuring prevalence, predictors, and perceived consequences," *Accident Analysis and Prevention*, 33 (2001), pp. 735–741.

3. From the Gallup Web site, www.gallup.com. The poll was taken in December 2006.

4. For a discussion of statistical significance in the legal setting, see D. H. Kaye, "Is proof of statistical significance relevant?" *Washington Law Review*, 61 (1986), pp. 1333–1365. Kaye argues: "Presenting the P-value without characterizing the evidence by a significance test is a step in the right direction. Interval estimation, in turn, is an improvement over P-values."

5. From a press release from the Harvard School of Public Health College Alcohol Study, April 12, 2001, at www.hsph.harvard.edu/cas/.

6. Warren E. Leary, "Cell phones: questions but no answers," *New York Times*, October 26, 1999.

7. Poll for June 7, 2007, at retention.harrisblackintl.com/harris_poll. A note at the bottom of the page says: "Because the sample is based on those who agreed to participate in the Harris Interactive panel, no estimates of theoretical sampling error can be calculated."

8. Justin S. Brashares et al., "Bushmeat hunting, wildlife declines, and fish supply in West Africa," *Science*, 306 (2004), pp. 1180–1183. The data used here (and in Figure 1B of the article) are found in the online supplementary material.

9. Gabriel Gregoratos et al., "ACC/AHA guidelines for implantation of cardiac pacemakers and antiarrhythmia devices: executive summary," *Circulation*, 97 (1998), pp. 1325–1335.

10. From the commentary by Frank J. Sulloway, "Birth order and intelligence," *Science*, 316 (2007), pp. 1711–1712. The study report appears in the same issue: Petter Kristensen and Tor Bjerkedal, "Explaining the relation between birth order and intelligence," *Science*, 316 (2007), p. 1717.

11. C. Kopp et al., "Modulation of rhythmic brain activity by diazepam: GABAA receptor subtype and state specificity," *Proceedings of the National Academy of Sciences*, 101 (2004), pp. 3674–3679.

12. J. Bruce Moseley et al., "A controlled trial of arthroscopic surgery for osteoarthritis of the knee," *New England Journal of Medicine*, 347, No. 2 (2002), pp. 81–88.

Chapter 16 Notes

1. Based on a news item "Bee off with you," *Economist*, November 2, 2002, p. 78.

2. Simplified from D. A. Marcus et al., "A double-blind provocative study of chocolate as a trigger of headache," *Cephalalgia*, 17 (1997), pp. 855–862.

3. Based on Stephen A. Woodbury and Robert G. Spiegelman, "Bonuses to workers and employers to reduce

unemployment: randomized trials in Illinois," *American Economic Review*, 77 (1987), pp. 513–530.

4. Votes as of June 27, 2007, at `www.pbs.org/wgbh/nova/sciencenow`.

5. David M. Fergusson and L. John Horwood, "Cannabis use and traffic accidents in a birth cohort of young adults," *Accident Analysis and Prevention*, 33 (2001), pp. 703–711.

6. K. E. Hobbs et al., "Levels and patterns of persistent organochlorines in minke whale (*Balaenoptera acutorostrata*) stocks from the North Atlantic and European Arctic," *Environmental Pollution*, 121 (2003), pp. 239–252.

7. Maureen Hack et al., "Outcomes in young adulthood for very-low-birth-weight infants," *New England Journal of Medicine*, 346 (2002), pp. 149–157.

8. Data for U.S. searches in February 2008 from Hitwise, at `www.hitwise.com`.

9. U.S. Census Bureau, *Fertility of American Women: June 2004*, at `www.census.gov`.

10. Aaron S. Hervey et al., "Reaction time distribution analysis of neuropsychological performance in an ADHD sample," *Child Neuropsychology*, 12 (2006), pp. 125–140.

11. Mikyoung Park et al., "Recycling endosomes supply AMPA receptors for LTP," *Science*, 305 (2004), pp. 1972–1975.

12. Jon E. Keeley, C. J. Fotheringham, and Marco Morais, "Reexamining fire suppression impacts on brushland fire regimes," *Science*, 284 (1999), pp. 1829–1831.

13. Simplified from Sanjay K. Dhar, Claudia González-Vallejo, and Dilip Soman, "Modeling the effects of advertised price claims: tensile versus precise pricing," *Marketing Science*, 18 (1999), pp. 154–177.

14. Charles S. Fuchs et al., "Alcohol consumption and mortality among women," *New England Journal of Medicine*, 332 (1995), pp. 1245–1250.

15. See Note 15 for Chapter 2. Later chapters will suggest more suitable analyses.

16. Data simulated from a Normal distribution with $\mu = 98.2$ and $\sigma = 0.7$. These values are based on P. A. Mackowiak, S. S. Wasserman, and M. M. Levine, "A critical appraisal of 98.6 degrees F, the upper limit of the normal body temperature, and other legacies of Carl Reinhold August Wunderlich," *Journal of the American Medical Association*, 268 (1992), pp. 1578–1580.

17. J. F. Swain et al., "Comparison of the effects of oat bran and low-fiber wheat on serum lipoprotein levels and blood pressure," *New England Journal of Medicine*, 322 (1990), pp. 147–152.

18. Research by Louis Chan et al., reported by Robert Schiller, *Irrational Exuberance*, Broadway Books, 2001, p. 253.

19. Amanda Lenhart and Mary Madden, "Music downloading, file-sharing and copyright," Pew Internet and American Life Project, 2003, at `www.pewinternet.org`.

20. From the Internal Revenue Service Web site, `www.irs.gov/taxstats`.

21. Susannah Fox and Gretchen Livingston, "Latinos online," Pew Hispanic Center and Pew Internet Project, 2007, at `www.pewinternet.org`.

22. Alan Schwarz, "In a game of statistics, some numbers have little meaning," *New York Times*, April 3, 2005.

23. Population base from the 2006 Labor Force Survey, at `www.statistics.gov.uk`. Smoking data from Action on Smoking and Health, *Smoking and Health Inequality*, at `www.ash.org.uk`.

24. Based on a study reported by Alan B. Krueger, "Economic scene" column, *New York Times*, November 14, 2002.

Chapter 17 Notes

1. Note 2 for Chapter 11 explains the reason for this condition in the case of inference about a population mean.

2. Helen E. Staal and D. C. Donderi, "The effect of sound on visual apparent movement," *American Journal of Psychology*, 96 (1983), pp. 95–105.

3. See Note 3 for Chapter 14.

4. From a graph in Benedetto De Martino et al., "Frames, biases, and rational decision-making in the human brain," *Science*, 313 (2006), pp. 684–687. I simplified the design a bit for easier comprehension: the starting amounts and gambles offered differed from trial to trial,

though still matched in pairs; 32 very unbalanced "catch trials" were mixed with the 64 experimental trials to be sure subjects were paying attention; and all money amounts were in British pounds, not dollars.

5. R. A. Berner and G. P. Landis, "Gas bubbles in fossil amber as possible indicators of the major gas composition of ancient air," *Science*, 239 (1988), pp. 1406–1409. The 95% t confidence interval is 54.78 to 64.40. A bootstrap BCa interval is 55.03 to 62.63. So t is reasonably accurate despite the skew and the small sample.

6. Alice P. Melis, Brian Hare, and Michael Tomasello, "Chimpanzees recruit the best collaborators," *Science*, 311 (2006), pp. 1297–1300. A Normal quantile plot does not show major lack of Normality and a saddle-point approximation that allows for skew gives $P = 0.0039$. So the t test is reasonably accurate despite the skew and small sample size.

7. Josef P. Rauschecker, Biao Tian, and Marc Hauser, "Processing of complex sounds in the macaque nonprimary auditory cortex," *Science*, 268 (1995), pp. 111–114.

8. For a qualitative discussion explaining why skewness is the most serious violation of the Normal shape condition, see Dennis D. Boos and Jacqueline M. Hughes-Oliver, "How large does n have to be for the Z and t intervals?" *American Statistician*, 54 (2000), pp. 121–128. Our recommendations are based on extensive computer work. See, for example, Harry O. Posten, "The robustness of the one-sample t-test over the Pearson system," *Journal of Statistical Computation and Simulation*, 9 (1979), pp. 133–149; and E. S. Pearson and N. W. Please, "Relation between the shape of population distribution and the robustness of four simple test statistics," *Biometrika*, 62 (1975), pp. 223–241.

9. For more advanced users, a good way to ascertain if the t procedures are safe is to compare the 95% confidence interval produced by t with the BCa interval from a bootstrap with at least 1000 resamples. For (b) the t interval is 29,428 to 32,254 and a BCa interval is 29,106 to 31,894. For (c), on the other hand, t gives 38.93 to 40.49 and BCa gives 38.97 to 40.44. These results confirm the judgment that t is safe for (c) but not for (b).

10. See Note 3 for Chapter 7.

11. From the online supplement to Tor D. Wager et al., "Placebo-induced changes in fMRI in the anticipation

and experience of pain," *Science*, 303 (2004), pp. 1162–1167.

12. TUDA results for 2003 from the National Center for Education Statistics, at nces.ed.gov/ nationsreportcard.

13. Raul de la Fuente-Fernandez et al., "Expectation and dopamine release: mechanism of the placebo effect in Parkinson's disease," *Science*, 293 (2001), pp. 1164–1166.

14. From the National Institute of Standards and Technology Web site, www.nist.gov/srd/online.htm.

15. Orit E. Hetzroni, "The effects of active versus passive computer-assisted instruction on the acquisition, retention, and generalization of Blissymbols while using elements for teaching compounds," PhD thesis, Purdue University, 1995.

16. See Note 21 for Chapter 7.

17. See Note 3 for Chapter 14.

18. See Note 6 for Chapter 7.

19. I thank Jason Hamilton, University of Illinois, for providing the data. The study is reported in Evan H. DeLucia et al., "Net primary production of a forest ecosystem with experimental CO_2 enhancement," *Science*, 284 (1999), pp. 1177–1179. No method for inference can be trusted with $n = 3$. In this study, each observation is very costly, so the small n is inevitable.

20. See Note 7 for Chapter 2.

21. Harry B. Meyers, "Investigations of the life history of the velvetleaf seed beetle, *Althaeus folkertsi* Kingsolver," MS thesis, Purdue University, 1996. The 95% t interval is 1227.9 to 2507.6. A 95% bootstrap BCa interval is 1444 to 2718, confirming that t inference is inaccurate for these data.

22. J. Marcus Jobe and Hutch Jobe, "A statistical approach for additional infill development," *Energy Exploration and Exploitation*, 18 (2000), pp. 89–103. The comparison interval is the BCa interval based on 1000 bootstrap resamples.

23. See Note 3 for Chapter 14.

24. Data provided by Timothy Sturm.

25. Lianng Yuh, "A biopharmaceutical example for undergraduate students," manuscript, no date.

Chapter 18 Notes

1. See Note 16 for Chapter 2.

2. Detailed information about the conservative t procedures can be found in Paul Leaverton and John J. Birch, "Small sample power curves for the two sample location problem," *Technometrics*, 11 (1969), pp. 299–307; Henry Scheffé, "Practical solutions of the Behrens-Fisher problem," *Journal of the American Statistical Association*, 65 (1970), pp. 1501–1508; and D. J. Best and J. C. W. Rayner, "Welch's approximate solution for the Behrens-Fisher problem," *Technometrics*, 29 (1987), pp. 205–210.

3. Kathleen G. McKinney, "Engagement in community service among college students: is it affected by significant attachment relationships?" *Journal of Adolescence*, 25 (2002), pp. 139–154.

4. See Note 9 for Chapter 2.

5. Sapna Aneja, "Biodeterioration of textile fibers in soil," MS thesis, Purdue University, 1994.

6. See the extensive simulation studies in Harry O. Posten, "The robustness of the two-sample t-test over the Pearson system," *Journal of Statistical Computation and Simulation*, 6 (1978), pp. 295–311; and Harry O. Posten, H. Yeh, and Donald B. Owen, "Robustness of the two-sample t-test under violations of the homogeneity assumption," *Communications in Statistics*, 11 (1982), pp. 109–126.

7. See Note 15 for Chapter 2. Although the spending data are quite discrete, a bootstrap BCa 95% confidence interval for the difference in means based on 1000 resamples is 2.394 to 4.826, close to the Option 1 95% interval 2.209 to 4.736. So the sample means are sufficiently Normal to allow use of t procedures.

8. See Note 17 for Chapter 2.

9. See Note 20 for Chapter 1.

10. The problem of comparing spreads is difficult even with advanced methods. Common distribution-free procedures do not offer a satisfactory alternative to the F test, because they are sensitive to unequal shapes when comparing two distributions. A recent survey of possible approaches is Dennis D. Boos and Cavell Brownie, "Comparing variances and other measures of dispersion," *Statistical Science*, 19 (2005), pp. 571–578.

11. Matthias R. Mehl et al., "Are women really more talkative than men?" *Science*, 317 (2007), p. 82.

12. Debra L. Miller et al., "Effect of fat-free potato chips with and without nutrition labels on fat and energy intakes," *American Journal of Clinical Nutrition*, 68 (1998), pp. 282–290.

13. Eric Sanford et al., "Local selection and latitudinal variation in a marine predator-prey interaction," *Science*, 300 (2003), pp. 1135–1137.

14. Angeline Lillard and Nicole Else-Quest, "Evaluating Montessori education," *Science*, 313 (2006), pp. 1893–1894. Many of the details appear in the supporting online material.

15. Mary K. Pawlik, "The effect of ginkgo biloba on the post-lunch dip and chemosensory function," MS thesis, Purdue University, 2002.

16. Gabriela S. Castellani, "The effect of cultural values on Hispanics' expectations about service quality," MS thesis, Purdue University, 2000.

17. Lynette C. Magna, "Immigrant parents and acculturation stress effects on maternal perceptions and child behaviors," MS thesis, Purdue University, 2001.

18. Wayne J. Camera and Donald Powers, "Coaching and the SAT I," *TIP* (online journal at www.siop.org/tip), July 1999.

19. See Note 20 for Chapter 1.

20. See Note 4 for Chapter 11.

21. Fabrizio Grieco, Arie J. van Noordwijk, and Marcel E. Visser, "Evidence for the effect of learning on timing of reproduction in blue tits," *Science*, 296 (2002), pp. 136–138. The data in Exercise 18.48 are from a graph in this paper.

22. Kathleen D. Vohs, Nicole L. Mead, and Miranda R. Goode, "The psychological consequences of money," *Science*, 314 (2006), pp. 1154–1156. I thank Kathleen Vohs for supplying the data.

23. See Note 15 for Chapter 17.

24. Data provided by Warren Page, New York City Technical College, from a study done by John Hudesman.

25. See Note 8 for Chapter 2.

26. Data provided by Marigene Arnold, Kalamazoo College.

Chapter 19 Notes

1. Joseph H. Catania et al., "Prevalence of AIDS-related risk factors and condom use in the United States," *Science*, 258 (1992), pp. 1101–1106.

2. Strictly speaking, the formula $\sqrt{p(1-p)/n}$ for the standard deviation of $\hat{p}$ assumes that we draw an SRS of size n from an *infinite* population. If the population has finite size N, this standard deviation is multiplied by $\sqrt{1-(n-1)/(N-1)}$. This "finite population correction" approaches 1 as N increases. When the population is at least 20 times as large as the sample, the correction factor is between about 0.97 and 1. It is reasonable to use the simpler form $\sqrt{p(1-p)/n}$ in these settings. See also Note 2 for Chapter 11.

3. This rule of thumb is based on study of computational results in the papers cited in Note 7 and discussion with Alan Agresti. I recommend using the plus four interval.

4. The quotation is from page 1104 of the article cited in Note 1.

5. G. A. Mauser and H. Taylor Buckner, "Canadian attitudes toward gun control: the real story," The Mackenzie Institute, 1997, at `teapot.usask.ca/cdn-firearms/Mauser/gunstory.html`.

6. See Note 18 for Chapter 18.

7. This interval is proposed by Alan Agresti and Brent A. Coull, "Approximate is better than 'exact' for interval estimation of binomial proportions," *The American Statistician*, 52 (1998), pp. 119–126. Note in particular that the plus four interval is often more accurate than the Clopper-Pearson "exact interval" based on the binomial distribution of the sample count and implemented by, for example, Minitab.

    There are several even more accurate but considerably more complex intervals for p that might be used in professional practice. See Lawrence D. Brown, Tony Cai, and Anirban DasGupta, "Interval estimation for a binomial proportion," *Statistical Science*, 16 (2001), pp. 101–133. A detailed theoretical study that uncovers the reason the large-sample interval is inaccurate is Lawrence D. Brown, Tony Cai, and Anirban DasGupta, "Confidence intervals for a binomial proportion and asymptotic expansions," *Annals of Statistics*, 30 (2002), pp. 160–201.

8. BBC News, December 25, 2006, at `news.bbc.co.uk`.

9. From Alan Agresti and Brian Caffo, "Simple and effective confidence intervals for proportions and differences of proportions result from adding two successes and two failures," *The American Statistician*, 45 (2000), pp. 280–288. When can the plus four interval be safely used? The answer depends on just how much accuracy you insist on. Brown and coauthors (see Note 7) recommend $n \geq 40$. Agresti and Coull (see Note 7) demonstrate that performance is almost always satisfactory in their eyes when $n \geq 5$. My rule of thumb $n \geq 10$ allows for confidence levels C other than 95% and fits our philosophy of not insisting on more exact results than practice requires. The big point is that plus four is very much more accurate than the standard interval for most values of p and all but very large n.

10. See Note 13 for Chapter 18.

11. Amanda Lenhart and Mary Madden, "Teens, Privacy and Online Social Networks," Pew Internet and American Life Project, 2007, at `www.pewinternet.org`.

12. Gary Edwards and Josephine Mazzuca, "Three quarters of Canadians support doctor-assisted suicide," Gallup Poll press release, March 24, 1999, at `www.gallup.com`.

13. In fact, P-values for two-sided tests are more accurate than those for one-sided tests. Our rule of thumb is a compromise to avoid the confusion of too many rules.

14. Matthew A. Carlton and William D. Stansfield, "Making babies by the flip of a coin?" *The American Statistician*, 59 (2005), pp. 180–182.

15. Alexander Todorov et al., "Inferences of competence from faces predict election outcomes," *Science*, 308 (2005), pp. 1623–1626.

16. Michele L. Head, "Examining college students' ethical values," Consumer Science and Retailing honors project, Purdue University, 2003.

17. John Paul McKinney and Kathleen G. McKinney, "Prayer in the lives of late adolescents," *Journal of Adolescence*, 22 (1999), pp. 279–290.

18. Paul Taylor, Cary Funk, and April Clark, "As marriage and parenthood drift apart, public is concerned about social impact," Pew Research Center, 2007, at `pewresearch.org`.

19. See Note 6 for Chapter 14.

20. John Fagan et al., "Performance assessment under field conditions of a rapid immunological test for transgenic soybeans," *International Journal of Food Science and Technology*, 36 (2001), pp. 357–367.

21. See Note 2 for Chapter 15.

22. Henry Wechsler et al., *Binge Drinking on America's College Campuses*, Harvard School of Public Health, 2001.

23. Francisco Lloret et al., "Fire and resprouting in Mediterranean ecosystems: insights from an external biogeographical region, the Mexican shrubland," *American Journal of Botany*, 88 (1999), pp. 1655–1661.

24. Jon D. Miller, Eugenie C. Scott, and Shinji Okamoto, "Public acceptance of evolution," *Science*, 313 (2006), pp. 765–766. The information in the exercise appears in the supplementary online material.

25. JoAnn K. Wells, Allan F. Williams, and Charles M. Farmer, "Seat belt use among African Americans, Hispanics, and whites," *Accident Analysis and Prevention*, 34 (2002), pp. 523–529.

26. Dean Fergusson et al., "Turning a blind eye: the success of blinding reported in a random sample of randomised, placebo-controlled trials," *British Medical Journal*, 328 (2004), pp. 432–436.

Chapter 20 Notes

1. See Note 1 for Chapter 6.

2. Eulynn Shiu and Amanda Lenhart, "How Americans use instant messaging," Pew Internet and American Life Project, 2004, at www.pewinternet.org.

3. From the Web site of the Black Youth Project, blackyouthproject.uchicago.edu.

4. *Youth Risk Behavior Surveillance—United States 2005*, Centers for Disease Control and Prevention, Morbidity and Mortality Weekly Report, Vol. 55, No. 66-5, 2007, at www.cdc.gov/mmwr. The data are from a complex multistage sample, so that acting as if we have SRSs is oversimplified.

5. The plus four method is due to Alan Agresti and Brian Caffo. See Note 9 for Chapter 19.

6. See Note 23 for Chapter 19.

7. Modified from Richard A. Schieber et al., "Risk factors for injuries from in-line skating and the effectiveness of safety gear," *New England Journal of Medicine*, 335 (1996), Internet summary at content.nejm.org.

8. Saiyad S. Ahmed, "Effects of microwave drying on checking and mechanical strength of low-moisture baked products," MS thesis, Purdue University, 1994.

9. Louie E. Ross, "Mate selection preferences among African American college students," *Journal of Black Studies*, 27 (1997), pp. 554–569.

10. This rule of thumb is quite conservative. It is in fact safe to arrange the data as a 2×2 table and apply the rule of thumb from Chapter 22 that all four *expected* counts must be 5 or greater. I give the conservative rule here because expected counts are messy to explain in the present contxt.

11. Ross L. Prentice et al., "Low-fat dietary pattern and risk of invasive breast cancer," *Journal of the American Medical Association*, 295 (2006), pp. 629–642.

12. Steiner Sulheim et al., "Helmet use and risk of head injuries in alpine skiers and snowboarders," *Journal of the American Medical Association*, 295 (2006), pp. 919–924.

13. Gail Tom et al., "The role of overt head movement in the formation of affect," *Basic and Applied Social Psychology*, 12 (1991), pp. 281–289.

14. See Note 11 for Chapter 19.

15. From an Associated Press dispatch appearing on December 30, 2002. The study report appeared in the *Journal of Adolescent Health*.

16. François Gaudet et al., "Induction of tumors in mice by genomic hypomethylation," *Science*, 300 (2003), pp. 489–492.

17. Douglas G. Altman, Steven N. Goodman, and Sara Schroter, "How statistical expertise is used in medical research," *Journal of the American Medical Association*, 287 (2002), pp. 2817–2820.

18. See Note 20 for Chapter 19.

19. Arne L. Kalleberg and Kevin T. Leicht, "Gender and organizational performance: determinants of small business survival and success," *The Academy of Management Journal*, 34 (1991), pp. 136–161.

20. Richard M. Felder et al., "Who gets it and who doesn't: a study of student performance in an introductory chemical engineering course," *1992 ASEE Annual Conference Proceedings*, American Society for Engineering Education, Washington, D.C., 1992, pp. 1516–1519.

21. Douglas E. Jorenby et al., "A controlled trial of sustained-release bupropion, a nicotine patch, or both for smoking cessation," *New England Journal of Medicine*, 340 (1999), pp. 685–691.

22. Data courtesy of Raymond Dumett, Purdue University.

23. Based on Alan G. Sanfey et al., "The neural basis of economic decision-making in the ultimatum game," *Science*, 300 (2003), pp. 1755–1758. The paper reports a chi-square test (equivalent to a two-sided z test). This analysis is incorrect for the paper's data, as there were in fact only 19 participants, each appearing twice in each row of the table given in the exercise. Exercise 20.32 therefore amends the data, assuming 76 participants, so that the elementary analysis is correct.

24. See Note 25 for Chapter 19.

25. Clive G. Jones et al., "Chain reactions linking acorns to gypsy moth outbreaks and Lyme disease risk," *Science*, 279 (1998), pp. 1023–1026.

26. The study is reported in William Celis III, "Study suggests Head Start helps beyond school," *New York Times*, April 20, 1993. See www.highscope.org.

27. Karen M. Herbert, "Does impulse buying vary by mode of payment?" MS thesis, Purdue University, 1994.

Chapter 21 Notes

1. Lee Rainie and Bill Tancer, "36% of online American adults consult Wikipedia," Pew Internet and American Life Project, 2007, at www.pewinternet.org.

2. See Note 4 for Chapter 20.

3. See Note 7 for Chapter 14.

4. K. S. Oberhauser, "Fecundity, lifespan and egg mass in butterflies: effects of male-derived nutrients and female size," *Functional Ecology*, 11 (1997), pp. 166–175.

5. See Note 3 for Chapter 20.

6. Michael R. Dohm, Jack P. Hayes, and Theodore Garland, Jr., "Quantitative genetics of sprint running speed and swimming endurance in laboratory house mice (*Mus domesticus*)," *Evolution*, 50 (1996), pp. 1688–1701.

7. Marianne Perle, Rebecca Moran, and Anthony D. Lutkus, NAEP 2004 *Trends in Academic Progress: Three Decades of Student Performance in Reading and Mathematics*, National Center for Education Statistics, 2005, at nces.ed.gov. The data given are approximate due to rounding in the study report.

8. See Note 7 for Chapter 16. The exercises are simplified, in that the measures reported in this paper have been statistically adjusted for "sociodemographic status."

9. See Note 11 for Chapter 20.

10. Based on the online supplement to Paul J. Shaw et al., "Correlates of sleep and waking in *Drosophila melanogaster*," *Science*, 287 (2000), pp. 1834–1837.

11. V. D. Bass, W. E. Hoffmann, and J. L. Dorner, "Normal canine lipid profiles and effects of experimentally induced pancreatitis and hepatic necrosis on lipids," *American Journal of Veterinary Research*, 37 (1976), pp. 1355–1357.

12. Jin Ha Lee and J. Stephen Downie, "Survey of music information needs, uses, and seeking behaviors: preliminary findings," online Proceedings of the 5th International Conference on Music Information Retrieval, 2004, at ismir2004.ismir.net.

13. K. Carrie Armel and V. S. Ramachandran, "Projecting sensations to external objects: evidence from skin conductance response," *Proceedings of the Royal Society of London, Series B*, 270 (2003), pp. 1499–1506.

14. Josh McDermott and Marc D. Hauser, "Nonhuman primates prefer slow tempos but dislike music overall," *Cognition*, 104 (2007), pp. 654–668. Failure to take account of repeated measures on the same subjects is one of the most common errors observed in statistical analysis.

15. James Otto, Michael F. Brown, and William Long III, "Training rats to search and alert on contraband odors," *Applied Animal Behaviour Science*, 77 (2002), pp. 217–232.

16. From the Merck Web site, www.merckvaccines.com/gardasilProductPage_frmst.html.

17. These data were originally collected by L. M. Linde of UCLA but were first published by M. R. Mickey, O. J. Dunn, and V. Clark, "Note on the use of stepwise

regression in detecting outliers,"*Computers and Biomedical Research,* 1 (1967), pp. 105–111. The data have been used by several authors. I found them in N. R. Draper and J. A. John, "Influential observations and outliers in regression," *Technometrics,* 23 (1981), pp. 21–26.

18. Jacqueline T. Ngai and Diane S. Srivastava, "Predators accelerate nutrient cycling in a bromeliad ecosystem," *Science,* 314 (2006), p. 963. I thank Jacqueline Ngai for providing the data.

19. Yvan R. Germain, "The dyeing of ramie with fiber reactive dyes using the cold pad-batch method," MS thesis, Purdue University, 1988.

20. Data provided by Marigene Arnold, Kalamazoo College.

21. Data provided by Corinne Lim, Purdue University, from a student project supervised by Professor Joseph Vanable.

22. See Note 8 for Chapter 20.

23. Michael O. Finkelstein and Bruce Levin, "Statistical proof of discrimination in peremptory challenges," *Chance,* 17, No. 1 (2004), pp. 35–38.

24. G. S. Hotamisligil et al., "Uncoupling of obesity from insulin resistance through a targeted mutation in *aP*2, the adipocyte fatty acid binding protein," *Science,* 274 (1996), pp. 1377–1379.

Chapter 22 Notes

1. See Note 1 for Chapter 6.

2. Pennsylvania State University Division of Student Affairs, "Net behaviors November 2006," *Penn State Pulse,* at www.sa.psu.edu.

3. See Note 2 for Chapter 6.

4. All General Social Survey exercises in this chapter present tables constructed using the search function at the GSS archive, sda.berkeley.edu/archive.htm. Most concern data from the 2006 GSS.

5. There are many computer studies of the accuracy of chi-square critical values for X^2. Our guideline goes back to W. G. Cochran (1954). Later work has shown that it is often conservative in the sense that, if the expected cell counts are all similar and the degrees of freedom exceed 1, the chi-square approximation works well for an average expected count as small as 1 or 2. Our guideline protects against dissimilar expected counts. It has the added advantage that it is safe in the 2×2 case, where the chi-square approximation is least good. So our guideline is helpful for beginners—there is no single condition that is not conservative and applies to 2×2 and larger tables with similar and dissimilar expected cell counts. There are exact procedures that (with software) should be used for tables that do not satisfy our guideline. For a survey, see Alan Agresti, "A survey of exact inference for contingency tables," *Statistical Science,* 7 (1992), pp. 131–177.

6. Pew Research Center for the People and the Press, "The cell phone challenge to survey research," news release for May 15, 2006, at www.people-press.org.

7. Based on a news item in *Science,* 305 (2004), p. 1560. The study, by Daniel Klem, appeared in the *Wilson Journal.*

8. David W. Eby et al., "The effect of changing from secondary to primary safety belt enforcement on police harassment," *Accident Analysis and Prevention,* 36 (2000), pp. 819–828.

9. Public use data available on the Web site of the Carolina Population Center, www.cpc.unc.edu.

10. Based on John B. Horrigan and Aaron Smith, "Home broadband adoption 2007," Pew Internet and American Life Project, 2007, at www.pewinternet.org. The exercise is slightly modified because the data in the paper are not consistent: the authors claim both that 47% have home broadband and that 966 in their sample of 2200 have home broadband.

11. Modified from Felicity Barringer, "Measuring sexuality through polls can be shaky," *New York Times,* April 25, 1993.

12. Virgilio P. Carnielli et al., "Intestinal absorption of long-chain polyunsaturated fatty acids in preterm infants fed breast milk or formula," *American Journal of Clinical Nutrition,* 67 (1998), pp. 97–103.

13. Adapted from M. A. Visintainer, J. R. Volpicelli, and M. E. P. Seligman, "Tumor rejection in rats after inescapable or escapable shock," *Science,* 216 (1982), pp. 437–439.

14. See Note 1 for Chapter 6.

15. Tom Reichert, "The prevalence of sexual imagery in ads targeted to young adults," *Journal of Consumer Affairs*, 37 (2003), pp. 403–412.

16. Mark A. Sabbagh and Dare A. Baldwin, "Learning words from knowledgeable versus ignorant speakers: links between preschoolers' theory of mind and semantic development," *Child Development*, 72 (2001), pp. 1054–1070. Many statistical software packages offer "exact tests" that are valid even when there are small expected counts.

17. U.S. Department of Commerce, Office of Travel and Tourism Industries, in-flight survey, 2007, at `tinet.ita.doc.gov`.

18. See Note 2 for Chapter 16. I have simplified slightly: the table in the paper is exactly as in the exercise but contains data for 63 subjects plus data from one type of bar for 3 subjects who dropped out. Although the authors say that their chi-square refers to this table, they give a nonsignificant value that contradicts what the table shows.

19. Karine Marangon et al., "Diet, antioxidant status, and smoking habits in French men," *American Journal of Clinical Nutrition*, 67 (1998), pp. 231–239.

20. Mei-Hui Chen, "An exploratory comparison of American and Asian consumers' catalog patronage behavior," MS thesis, Purdue University, 1994.

21. Data compiled from a table of percents in "Americans view higher education as key to the American dream," press release by the National Center for Public Policy and Higher Education, May 3, 2000, at `www.highereducation.org`.

22. See Note 13 for Chapter 6.

23. Data produced by Ries and Smith, found in William D. Johnson and Gary G. Koch, "A note on the weighted least squares analysis of the Ries-Smith contingency table data," *Technometrics*, 13 (1971), pp. 438–447.

Chapter 23 Notes

1. Samuel Karelitz et al., "Relation of crying activity in early infancy to speech and intellectual development at age three years," *Child Development*, 35 (1964), pp. 769–777.

2. See Note 6 for Chapter 4.

3. See Note 14 for Chapter 4.

4. From a graph in Bruce J. Peterson et al., "Increasing river discharge to the Arctic Ocean," *Science*, 298 (2002), pp. 2171–2173.

5. Electronic Encyclopedia of Statistical Examples and Exercises (EESEE) at the text Web site, `www.whfreeman.com/bps`.

6. From a graph in Allison L. Perry et al., "Climate change and distribution shifts in marine fishes," *Science*, 308 (2005), pp. 1912–1915. The explanatory variable is the five-year running mean of winter (December to March) sea-bottom temperature.

7. Data for the building at 1800 Ben Franklin Drive, Sarasota, Florida, starting in March 2003. From the Web site of the Sarasota County Property Appraiser, `www.sarasotaproperty.net`.

8. Carly J. Stevens et al., "Impact of nitrogen deposition on the species richness of grasslands," *Science*, 303 (2004), pp. 1876–1879.

9. Based on Marion E. Dunshee, "A study of factors affecting the amount and kind of food eaten by nursery school children," *Child Development*, 2 (1931), pp. 163–183. This article gives the means, standard deviations, and correlation for 37 children, from which the data in the exercise are simulated.

10. From Table S2 in the online supplement to Antonio Dell'Anno and Roberto Danovaro, "Extracellular DNA plays a key role in deep-sea ecosystem functioning," *Science*, 309 (2005), p. 2179.

11. See Note 21 for Chapter 7.

Chapter 24 Notes

1. See Note 8 for Chapter 2.

2. Amandine Cournil, Jean-Marie Legay, and François Schächter, "Evidence of sex-linked effects on the inheritance of human longevity: a population-based study in the Valserine valley (French Jura), 18–20th centuries," *Proceedings of the Royal Society of London: Biological Sciences*, 267 (2000), pp. 1021–1025.

3. Elisabeth Wells-Parker et al., "An exploratory study of the relationship between road rage and crash experience in a representative sample of US drivers," *Accident Analysis and Prevention*, 34 (2002), pp. 271–278.

4. See Note 9 for Chapter 2.

5. See Note 15 for Chapter 2.

6. See Note 22 for Chapter 18.

7. See Note 18 for Chapter 21.

8. David B. Wooten, "One-of-a-kind in a full house: some consequences of ethnic and gender distinctiveness," *Journal of Consumer Psychology*, 4 (1995), 205–224.

9. John M. Jakicic et al., "Effects of intermittent exercise and use of home exercise equipment on adherence, weight loss, and fitness in overweight women," *Journal of the American Medical Association*, 282 (1999), pp. 1554–1560.

10. John P. Thomas, "Influences on mathematics learning and attitudes among African American high school students," *Journal of Negro Education*, 69 (2000), pp. 165–183.

11. See Note 7 for Chapter 2.

12. Annie C. St.-Pierre et al., "Insulin resistance syndrome, body mass index and the risk of ischemic heart disease," *Canadian Medical Association Journal*, 172 (2005), pp. 1301–1305.

13. Data from the online supplement to André Kessler and Ian T. Baldwin, "Defensive function of herbivore-induced plant volatile emissions in nature," *Science*, 291 (2001), pp. 2141–2144.

14. The data and the full story can be found in the Data and Story Library at lib.stat.cmu.edu. The original study is by Faith Loven, "A study of interlist equivalency of the CID W-22 word list presented in quiet and in noise," MS thesis, University of Iowa, 1981.

15. See Note 5 for Chapter 9. I thank Kenwyn Suttle for providing these data, for the year 2003.

16. See Note 19 for Chapter 21.

17. See Note 3 for Chapter 14.

18. See Note 5 for Chapter 18.

19. See Note 4 for Chapter 11.

20. See Note 5 for Chapter 9.

Table entry for C is the critical value t^* required for confidence level C. To approximate one- and two-sided P-values, compare the value of the t statistic with the critical values of t^* that match the P-values given at the bottom of the table.

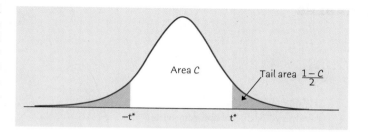

Area C

Tail area $\frac{1-C}{2}$

$-t^*$ t^*

TABLE C *t* distribution critical values

DEGREES OF FREEDOM	CONFIDENCE LEVEL C											
	50%	60%	70%	80%	90%	95%	96%	98%	99%	99.5%	99.8%	99.9%
1	1.000	1.376	1.963	3.078	6.314	12.71	15.89	31.82	63.66	127.3	318.3	636.6
2	0.816	1.061	1.386	1.886	2.920	4.303	4.849	6.965	9.925	14.09	22.33	31.60
3	0.765	0.978	1.250	1.638	2.353	3.182	3.482	4.541	5.841	7.453	10.21	12.92
4	0.741	0.941	1.190	1.533	2.132	2.776	2.999	3.747	4.604	5.598	7.173	8.610
5	0.727	0.920	1.156	1.476	2.015	2.571	2.757	3.365	4.032	4.773	5.893	6.869
6	0.718	0.906	1.134	1.440	1.943	2.447	2.612	3.143	3.707	4.317	5.208	5.959
7	0.711	0.896	1.119	1.415	1.895	2.365	2.517	2.998	3.499	4.029	4.785	5.408
8	0.706	0.889	1.108	1.397	1.860	2.306	2.449	2.896	3.355	3.833	4.501	5.041
9	0.703	0.883	1.100	1.383	1.833	2.262	2.398	2.821	3.250	3.690	4.297	4.781
10	0.700	0.879	1.093	1.372	1.812	2.228	2.359	2.764	3.169	3.581	4.144	4.587
11	0.697	0.876	1.088	1.363	1.796	2.201	2.328	2.718	3.106	3.497	4.025	4.437
12	0.695	0.873	1.083	1.356	1.782	2.179	2.303	2.681	3.055	3.428	3.930	4.318
13	0.694	0.870	1.079	1.350	1.771	2.160	2.282	2.650	3.012	3.372	3.852	4.221
14	0.692	0.868	1.076	1.345	1.761	2.145	2.264	2.624	2.977	3.326	3.787	4.140
15	0.691	0.866	1.074	1.341	1.753	2.131	2.249	2.602	2.947	3.286	3.733	4.073
16	0.690	0.865	1.071	1.337	1.746	2.120	2.235	2.583	2.921	3.252	3.686	4.015
17	0.689	0.863	1.069	1.333	1.740	2.110	2.224	2.567	2.898	3.222	3.646	3.965
18	0.688	0.862	1.067	1.330	1.734	2.101	2.214	2.552	2.878	3.197	3.611	3.922
19	0.688	0.861	1.066	1.328	1.729	2.093	2.205	2.539	2.861	3.174	3.579	3.883
20	0.687	0.860	1.064	1.325	1.725	2.086	2.197	2.528	2.845	3.153	3.552	3.850
21	0.686	0.859	1.063	1.323	1.721	2.080	2.189	2.518	2.831	3.135	3.527	3.819
22	0.686	0.858	1.061	1.321	1.717	2.074	2.183	2.508	2.819	3.119	3.505	3.792
23	0.685	0.858	1.060	1.319	1.714	2.069	2.177	2.500	2.807	3.104	3.485	3.768
24	0.685	0.857	1.059	1.318	1.711	2.064	2.172	2.492	2.797	3.091	3.467	3.745
25	0.684	0.856	1.058	1.316	1.708	2.060	2.167	2.485	2.787	3.078	3.450	3.725
26	0.684	0.856	1.058	1.315	1.706	2.056	2.162	2.479	2.779	3.067	3.435	3.707
27	0.684	0.855	1.057	1.314	1.703	2.052	2.158	2.473	2.771	3.057	3.421	3.690
28	0.683	0.855	1.056	1.313	1.701	2.048	2.154	2.467	2.763	3.047	3.408	3.674
29	0.683	0.854	1.055	1.311	1.699	2.045	2.150	2.462	2.756	3.038	3.396	3.659
30	0.683	0.854	1.055	1.310	1.697	2.042	2.147	2.457	2.750	3.030	3.385	3.646
40	0.681	0.851	1.050	1.303	1.684	2.021	2.123	2.423	2.704	2.971	3.307	3.551
50	0.679	0.849	1.047	1.299	1.676	2.009	2.109	2.403	2.678	2.937	3.261	3.496
60	0.679	0.848	1.045	1.296	1.671	2.000	2.099	2.390	2.660	2.915	3.232	3.460
80	0.678	0.846	1.043	1.292	1.664	1.990	2.088	2.374	2.639	2.887	3.195	3.416
100	0.677	0.845	1.042	1.290	1.660	1.984	2.081	2.364	2.626	2.871	3.174	3.390
1000	0.675	0.842	1.037	1.282	1.646	1.962	2.056	2.330	2.581	2.813	3.098	3.300
z^*	0.674	0.841	1.036	1.282	1.645	1.960	2.054	2.326	2.576	2.807	3.091	3.291
One-sided P	.25	.20	.15	.10	.05	.025	.02	.01	.005	.0025	.001	.0005
Two-sided P	.50	.40	.30	.20	.10	.05	.04	.02	.01	.005	.002	.001

Table entry for p is the critical value χ^* with probability p lying to its right.

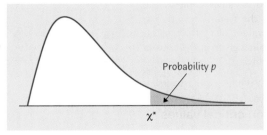

Probability p

χ^*

TABLE D **Chi-square distribution critical values**

df	.25	.20	.15	.10	.05	.025	.02	.01	.005	.0025	.001	.0005
1	1.32	1.64	2.07	2.71	3.84	5.02	5.41	6.63	7.88	9.14	10.83	12.12
2	2.77	3.22	3.79	4.61	5.99	7.38	7.82	9.21	10.60	11.98	13.82	15.20
3	4.11	4.64	5.32	6.25	7.81	9.35	9.84	11.34	12.84	14.32	16.27	17.73
4	5.39	5.99	6.74	7.78	9.49	11.14	11.67	13.28	14.86	16.42	18.47	20.00
5	6.63	7.29	8.12	9.24	11.07	12.83	13.39	15.09	16.75	18.39	20.51	22.11
6	7.84	8.56	9.45	10.64	12.59	14.45	15.03	16.81	18.55	20.25	22.46	24.10
7	9.04	9.80	10.75	12.02	14.07	16.01	16.62	18.48	20.28	22.04	24.32	26.02
8	10.22	11.03	12.03	13.36	15.51	17.53	18.17	20.09	21.95	23.77	26.12	27.87
9	11.39	12.24	13.29	14.68	16.92	19.02	19.68	21.67	23.59	25.46	27.88	29.67
10	12.55	13.44	14.53	15.99	18.31	20.48	21.16	23.21	25.19	27.11	29.59	31.42
11	13.70	14.63	15.77	17.28	19.68	21.92	22.62	24.72	26.76	28.73	31.26	33.14
12	14.85	15.81	16.99	18.55	21.03	23.34	24.05	26.22	28.30	30.32	32.91	34.82
13	15.98	16.98	18.20	19.81	22.36	24.74	25.47	27.69	29.82	31.88	34.53	36.48
14	17.12	18.15	19.41	21.06	23.68	26.12	26.87	29.14	31.32	33.43	36.12	38.11
15	18.25	19.31	20.60	22.31	25.00	27.49	28.26	30.58	32.80	34.95	37.70	39.72
16	19.37	20.47	21.79	23.54	26.30	28.85	29.63	32.00	34.27	36.46	39.25	41.31
17	20.49	21.61	22.98	24.77	27.59	30.19	31.00	33.41	35.72	37.95	40.79	42.88
18	21.60	22.76	24.16	25.99	28.87	31.53	32.35	34.81	37.16	39.42	42.31	44.43
19	22.72	23.90	25.33	27.20	30.14	32.85	33.69	36.19	38.58	40.88	43.82	45.97
20	23.83	25.04	26.50	28.41	31.41	34.17	35.02	37.57	40.00	42.34	45.31	47.50
21	24.93	26.17	27.66	29.62	32.67	35.48	36.34	38.93	41.40	43.78	46.80	49.01
22	26.04	27.30	28.82	30.81	33.92	36.78	37.66	40.29	42.80	45.20	48.27	50.51
23	27.14	28.43	29.98	32.01	35.17	38.08	38.97	41.64	44.18	46.62	49.73	52.00
24	28.24	29.55	31.13	33.20	36.42	39.36	40.27	42.98	45.56	48.03	51.18	53.48
25	29.34	30.68	32.28	34.38	37.65	40.65	41.57	44.31	46.93	49.44	52.62	54.95
26	30.43	31.79	33.43	35.56	38.89	41.92	42.86	45.64	48.29	50.83	54.05	56.41
27	31.53	32.91	34.57	36.74	40.11	43.19	44.14	46.96	49.64	52.22	55.48	57.86
28	32.62	34.03	35.71	37.92	41.34	44.46	45.42	48.28	50.99	53.59	56.89	59.30
29	33.71	35.14	36.85	39.09	42.56	45.72	46.69	49.59	52.34	54.97	58.30	60.73
30	34.80	36.25	37.99	40.26	43.77	46.98	47.96	50.89	53.67	56.33	59.70	62.16
40	45.62	47.27	49.24	51.81	55.76	59.34	60.44	63.69	66.77	69.70	73.40	76.09
50	56.33	58.16	60.35	63.17	67.50	71.42	72.61	76.15	79.49	82.66	86.66	89.56
60	66.98	68.97	71.34	74.40	79.08	83.30	84.58	88.38	91.95	95.34	99.61	102.7
80	88.13	90.41	93.11	96.58	101.9	106.6	108.1	112.3	116.3	120.1	124.8	128.3
100	109.1	111.7	114.7	118.5	124.3	129.6	131.1	135.8	140.2	144.3	149.4	153.2

Table entry for p is the critical value r^* of the correlation coefficient r with probability p lying to its right.

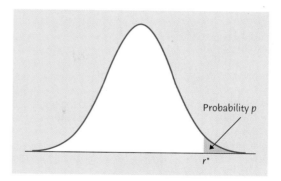

Probability p

r^*

TABLE E Critical values of the correlation r

	UPPER TAIL PROBABILITY p									
n	.20	.10	.05	.025	.02	.01	.005	.0025	.001	.0005
3	0.8090	0.9511	0.9877	0.9969	0.9980	0.9995	0.9999	1.0000	1.0000	1.0000
4	0.6000	0.8000	0.9000	0.9500	0.9600	0.9800	0.9900	0.9950	0.9980	0.9990
5	0.4919	0.6870	0.8054	0.8783	0.8953	0.9343	0.9587	0.9740	0.9859	0.9911
6	0.4257	0.6084	0.7293	0.8114	0.8319	0.8822	0.9172	0.9417	0.9633	0.9741
7	0.3803	0.5509	0.6694	0.7545	0.7766	0.8329	0.8745	0.9056	0.9350	0.9509
8	0.3468	0.5067	0.6215	0.7067	0.7295	0.7887	0.8343	0.8697	0.9049	0.9249
9	0.3208	0.4716	0.5822	0.6664	0.6892	0.7498	0.7977	0.8359	0.8751	0.8983
10	0.2998	0.4428	0.5494	0.6319	0.6546	0.7155	0.7646	0.8046	0.8467	0.8721
11	0.2825	0.4187	0.5214	0.6021	0.6244	0.6851	0.7348	0.7759	0.8199	0.8470
12	0.2678	0.3981	0.4973	0.5760	0.5980	0.6581	0.7079	0.7496	0.7950	0.8233
13	0.2552	0.3802	0.4762	0.5529	0.5745	0.6339	0.6835	0.7255	0.7717	0.8010
14	0.2443	0.3646	0.4575	0.5324	0.5536	0.6120	0.6614	0.7034	0.7501	0.7800
15	0.2346	0.3507	0.4409	0.5140	0.5347	0.5923	0.6411	0.6831	0.7301	0.7604
16	0.2260	0.3383	0.4259	0.4973	0.5177	0.5742	0.6226	0.6643	0.7114	0.7419
17	0.2183	0.3271	0.4124	0.4821	0.5021	0.5577	0.6055	0.6470	0.6940	0.7247
18	0.2113	0.3170	0.4000	0.4683	0.4878	0.5425	0.5897	0.6308	0.6777	0.7084
19	0.2049	0.3077	0.3887	0.4555	0.4747	0.5285	0.5751	0.6158	0.6624	0.6932
20	0.1991	0.2992	0.3783	0.4438	0.4626	0.5155	0.5614	0.6018	0.6481	0.6788
21	0.1938	0.2914	0.3687	0.4329	0.4513	0.5034	0.5487	0.5886	0.6346	0.6652
22	0.1888	0.2841	0.3598	0.4227	0.4409	0.4921	0.5368	0.5763	0.6219	0.6524
23	0.1843	0.2774	0.3515	0.4132	0.4311	0.4815	0.5256	0.5647	0.6099	0.6402
24	0.1800	0.2711	0.3438	0.4044	0.4219	0.4716	0.5151	0.5537	0.5986	0.6287
25	0.1760	0.2653	0.3365	0.3961	0.4133	0.4622	0.5052	0.5434	0.5879	0.6178
26	0.1723	0.2598	0.3297	0.3882	0.4052	0.4534	0.4958	0.5336	0.5776	0.6074
27	0.1688	0.2546	0.3233	0.3809	0.3976	0.4451	0.4869	0.5243	0.5679	0.5974
28	0.1655	0.2497	0.3172	0.3739	0.3904	0.4372	0.4785	0.5154	0.5587	0.5880
29	0.1624	0.2451	0.3115	0.3673	0.3835	0.4297	0.4705	0.5070	0.5499	0.5790
30	0.1594	0.2407	0.3061	0.3610	0.3770	0.4226	0.4629	0.4990	0.5415	0.5703
40	0.1368	0.2070	0.2638	0.3120	0.3261	0.3665	0.4026	0.4353	0.4741	0.5007
50	0.1217	0.1843	0.2353	0.2787	0.2915	0.3281	0.3610	0.3909	0.4267	0.4514
60	0.1106	0.1678	0.2144	0.2542	0.2659	0.2997	0.3301	0.3578	0.3912	0.4143
80	0.0954	0.1448	0.1852	0.2199	0.2301	0.2597	0.2864	0.3109	0.3405	0.3611
100	0.0851	0.1292	0.1654	0.1966	0.2058	0.2324	0.2565	0.2786	0.3054	0.3242
1000	0.0266	0.0406	0.0520	0.0620	0.0650	0.0736	0.0814	0.0887	0.0976	0.1039

Chapter 1

1.1 (a) Vehicles. (b) Make/model, vehicle type, transmission type (all categorical); number of cylinders, city MPG, highway MPG (all quantitative).

1.3 (a) 61.6%; 38.4% listen to other formats. (c) A pie chart could be used.

1.5 A pie graph could be used. Perhaps induced or C-section births are often scheduled for weekdays.

1.9 (a) Women in DC are more likely to be career-oriented. (b) The midpoint is between 24% and 26%; the spread is from 20% to 34% (ignoring the outlier).

1.11 The distribution is somewhat right-skewed, although with the United States (a high outlier) removed, the remaining data are relatively symmetric. The center (median) is $1954 per person; aside from the U.S., health care spending ranges from $419 to $3809 per person.

1.13 (a) **1.14** (c) **1.15** (c) **1.16** (b) **1.17** (b)

1.18 (a) **1.19** (a) **1.20** (b) **1.21** (a) **1.22** (c)

1.23 (a) Medical students. (b) Five variables; medical school, sex, and specialty are categorical, while age and USMLE are quantitative.

1.25 5% are other colors. Either a bar or pie chart could be used.

1.27 (b) To make a pie chart, we need to know the total number of deaths in this age group.

1.31 (a) Fairly symmetric (perhaps slightly left-skewed); center near 110, spread from the mid 80s to the high 130s. (b) 79.49%.

1.33 (a) 4, (b) 2, (c) 1, (d) 3.

1.35 (a) States with large populations need more doctors to serve that population. (b) Right-skewed, with D.C. a high outlier.

1.37 The distribution is right skewed with a high outlier. The center is around 700 million, and the spread is from 173 million to 4700 million (2809 million, if we omit the outlier).

1.39 The number of recruits peaked in the mid-1980s, and in recent years has fallen back to levels similar to those in the 1970s.

1.41 Newer coins are more common than older coins.

1.43 With most software that can create such plots, the mouse can be used to resize the plot.

1.45 (a) The midpoint is 13 people. (b) The number of bites has exceeded the midpoint 15 times since 1986.

Chapter 2

2.1 30,841 pounds; only 6 pieces of wood had breaking strengths below the mean. The mean is made smaller by the left-skew of the distribution.

2.3 The mean (31.25 minutes) is larger than the median (22.5 minutes), as we expect for a right-skewed distribution.

2.5 The mean (0.7607) is larger than the median (0.075) because the sharp right skew.

2.7 (a) About −3.5% and 3.0%. (b) About 0% or 0.1%. (c) The stock fund is much more variable.

2.9 Yes; the IQR is 8.8, and anything above 25.8% is considered an outlier.

2.11 Both sets have $\overline{x} \doteq 7.50$ and $s \doteq 2.03$; set A is left-skewed, while set B has a high outlier.

2.13 The means and the stemplots appear to suggest that logging reduces the number of trees per plot and that recovery is slow (the 1-year-after and 8-years-after means and stemplots are similar).

2.15 (c)   **2.16** (b)   **2.17** (a)   **2.18** (c)   **2.19** (b)

2.20 (c)   **2.21** (b)   **2.22** (a)   **2.23** (b)   **2.24** (a)

2.25 Median $46,453, mean $58,886.

2.27 Median is at position 393, Q_1 is at position 196.5, and Q_3 is at position 589.5.

2.29 The five-number summaries (in mm) are *bihai*: 46.34, 46.71, 47.12, 48.245, 50.26; red: 37.40, 38.07, 39.16, 41.69, 43.09; yellow: 34.57, 35.45, 36.11, 36.82, 38.13.

2.31 (a) The total number of births in a year will vary greatly from one country to another; it would be difficult to compare counts for a small country with those of a large country. (b) Divide each count by 4,134,370 to compute the percents. (c) Q_1 is in position 1,033,593, in the 2,500 to 2,999 grams weight class. The median (position 2,067,185.5) is between 3,000 and 3,499 g, and Q_3 (3,100,778) is between 3,500 and 3,999 g.

2.33 (a) The mean moves in the same direction as the moving point, while the median points to the rightmost nonmoving point. (b) The mean follows the moving point. As the moving point crosses the other two, the median slides along with it, then stays at the leftmost fixed point.

2.35 (a) The main peak occurs from 50 to 150 days; the distribution is right-skewed. (b) The five-number summary is 43, 82.5, 102.5, 151.5, and 598 days.

2.37 The mean is 8.4%; the smaller states are given too much weight in this average.

2.39 (a) Any set of four identical numbers works. (b) 0, 0, 10, 10 is the only possible answer.

2.41 To begin, note the five numbers must add to 35.

2.43 The distribution is right-skewed; the five-number summary is $380,000, $424,500, $2,800,000, $8,625,000, $17,016,381.

2.45 Lemon had little or no effect, but spending was noticeably higher with the lavender odor.

2.47 The means are 2.9075 (compressed), 3.3360 (intermediate), and 4.2315 (loose). Five number summaries would also be a good choice. Soil penetrability is greatest for loose soil and least for compressed soil.

2.49 **(a)** The five number summary is 6.8%, 12.1%, 12.8%, 13.4%, 16.8%. **(b)** The highest and five lowest observations are flagged as outliers.

2.51 Outliers are those salaries above $20,925,750; there are none.

Chapter 3

3.3 Both are 1.5.

3.7 **(a)** 688 to 1016 mm. **(b)** Less than 688 mm.

3.9 Women: $z = 2.96$; men: $z = 0.96$.

3.11 **(a)** About 2.94%. **(b)** About 96.11%.

3.13 **(a)** $z \doteq -0.84$. **(b)** $z \doteq 0.25$.

3.15 (a)   **3.16** (a)   **3.17** (b)   **3.18** (b)   **3.19** (b)

3.20 (c)   **3.21** (a)   **3.22** (c)   **3.23** (b)   **3.24** (b)

3.25 The peak at 0 should be "tall and skinny." Near 1, the curve should be "short and fat."

3.27 About 2.5%.

3.29 **(a)** $z \doteq 0.84$. **(b)** $z \doteq 0.39$.

3.31 About 0.2296.

3.33 About 0.9876.

3.35 About 92.92%.

3.37 About 15.8 and 21.6 mpg.

3.39 15th percentile.

3.41 About 2.5%.

3.43 **(a)** About 3.07%. **(b)** About 1.13%.

3.45 **(a)** About 0.6%. **(b)** About 31%.

3.47 **(a)** 11.47 **(b)** 15.34 **(c)** About 12.3%.

3.49 **(b)** $\bar{x} \doteq 0.8004$, $M = 0.8$, $s \doteq 0.0782$, $Q_1 = 0.76$, and $Q_3 = 0.86$. $\bar{x}$ and M are close, and the distances from the median to each quartile are similar. **(c)** For a Normal distribution, the proportion is 0.4748. For the 49 observations, the proportions is 0.5306.

3.51 The distributions appear to be slightly skewed to the right.

3.53 $Q_1 \doteq -0.67$ and $Q_3 \doteq 0.67$.

Chapter 4

4.1 (a) Explanatory: time spent studying; response: grade. (b) Explore the relationship. (c) Explanatory: time spent on extracurricular activities; response: GPA. (d) Explore the relationship.

4.3 For example, weight and gender.

4.5 Delays are generally higher for heavily-outsourced airlines.

4.7 The association is positive. Aside from Hawaiian Airlines (the outlier), the relationship is roughly linear, but only moderately strong.

4.9 (b) The association is linear and positive, stronger for women. Males typically have larger values for both variables.

4.11 No; units do not affect correlation.

4.13 These variables do not have a straight-line relationship; the association is neither positive nor negative.

4.14 (a)   **4.15** (a)   **4.16** (c)   **4.17** (a)   **4.18** (c)

4.19 (c)   **4.20** (b)   **4.21** (a)   **4.22** (b)   **4.23** (a)

4.25 (a) The association is (weakly) positive. (b) The estimate is 4, which is an overestimate; that child had the lowest score on the test.

4.27 (a) Price is explanatory; the plot shows a positive linear association. (b) $r \doteq 0.9552$. (c) No; units do not affect correlation.

4.29 (b) The association is moderately strong, positive, and linear. (c) For all points, $r \doteq 0.8486$. Without the outlier, $r \doteq 0.7015$, because the scatter (relative to the length of the line) is greater.

4.31 (b) Icicles seem to grow faster when the water runs more slowly.

4.33 (a) Yellow appears to be best. (b) Neither "positive or negative association" nor "correlation" make sense, because color is not quantitative.

4.35 (b) $r \doteq 0.8770$ for both sets; units of measurement have no effect on the correlation.

4.37 (a) Small-cap stocks. (b) A negative correlation.

4.39 (a) Gender is not quantitative. (b) r cannot exceed 1. (c) r has no units.

4.41 (a) $r = 1$ for a line. (c) Leave some space above your vertical stack. (d) The curve must be higher at the right than at the left.

4.43 The new points are generally slightly lower than the old points, suggesting that gas usage has dropped slightly.

4.45 A fairly strong, positive, linear association ($r \doteq 0.8782$); social exclusion does appear to trigger a pain response.

Chapter 5

5.1   (a) The slope is 1.109. On average, highway mileage increases by 1.109 mpg for each 1 mpg change in city mileage. (b) The intercept is 4.62 mpg; this is the highway mileage for a nonexistent car that gets 0 mpg in the city. (c) 22.36 and 35.67 mpg.

5.3   (a) $\bar{x} = 3.5$, $s_x \doteq 1.3784$, $\bar{y} = 31.3333$, $s_y \doteq 16.1328$, and $r \doteq 0.9623$. $\hat{y} \doteq -8.088 + 11.26x$.

5.5   A correlation close to 1 (or -1) means a strong linear relationship.

5.7   (a) The residuals are 0.8246, -4.7018, -3.9649, 4.0351, 6.0351, and -2.2281.

5.9   (a) Any point that falls exactly on the regression line will not increase the sum of squared vertical distances. (b) Influential points are those whose x coordinates are outliers.

5.11   (b) $r \doteq 0.4765$ with all points, and 0.4838 with the outlier removed. (c) With all points, $\hat{y} \doteq 4.73 + 0.3868x$, and the prediction is 34.13%. With Hawaiian Airlines removed, $\hat{y} \doteq 10.88 + 0.2495x$, and the prediction is 29.84%.

5.13   E.g., a student's general intelligence, or problem-solving ability, or self-confidence.

5.15   For example, social status and ethnic background.

5.17   (b)   **5.18**   (c)   **5.19**   (c)   **5.20**   (a)   **5.21**   (b)

5.22   (a)   **5.23**   (b)   **5.24**   (a)   **5.25**   (c)   **5.26**   (b)

5.27   (a) 0.0138 minutes per meter. On the average, if the depth of the dive is increased by 1 meter, it adds 0.0138 minutes (about 0.83 seconds) to the time spent underwater. (b) 5.45 minutes. (c) To plot the line, compute DD (y) when $x = 40$ and when $x = 300$, then connect those points.

5.29   (a) $\hat{y} = -0.126 + 0.0608x$; $\hat{y} = -0.0044$ when $x = 2.0$. (b) 77.1%. (c) $r \doteq 0.88$; it is positive because the slope is positive.

5.31   (a) Slope 0.5185, intercept 36.115. (b) The prediction is about 70.85 inches. (c) This regression only explains 25% of the variation in the husband's height.

5.33 $r = 0.40$.

5.35   (a) $r \doteq 0.9999$, so recalibration is not necessary. (b) $\hat{y} = 8.825x - 14.52$; when $x = 40$, $\hat{y} \doteq 338.5$ mg/l. (c) The relationship is strong, so the prediction should be very accurate.

5.37   (b) $\hat{y} = 0.586 + 0.00891x$. (c) The line will not change much because the outlier fits the pattern of the other points; r changes because the scatter (relative to the length of the line) is greater with the outlier removed. (d) The correlation changes from 0.8486 to 0.7015. With all points, $\hat{y} = 0.585 + 0.00879x$.

5.39   (b) For all points, $r \doteq 0.4819$. Removing Subject 15 increases r to 0.5684, because its presence makes the scatterplot less linear. Removing Subject 18 decreases r to 0.3837, because its presence decreases the relative scatter about the linear pattern.

5.41 $\hat{y} \doteq 66.4 + 10.4x$ (all observations), $\hat{y} \doteq 69.5 + 8.92x$ (without #15), $\hat{y} \doteq 52.3 + 12.1x$ (without #18). Neither point is particularly influential. Outliers in the y direction (like Subject 15) are typically not very influential; Subject 18 is consistent with the linear pattern suggested by the other points.

5.43 For example, students might differ in self-motivation or computer skills.

5.45 For example, suppose each worker's salary is his/her age (in thousands of dollars per year), and most workers are currently 30 to 50 years old. Over the next 10 years, all workers age, and their salaries increase. If each salary increases by between $4000 and $8000, then every worker will be making *more* money than he/she did 10 years before, but *less* money than a worker of that same age 10 years before.

5.47 The predictions are 76.33 and 74.53.

5.51 A scatterplot shows a positive linear association. The regression line is $\hat{y} = -1.286 + 11.89x$; $r^2 \doteq 83.9\%$.

5.53 There is a positive association, strengthened by the 2005 season, but weakened by the last two years of data. The regression line $\hat{y} = 1.803 + 0.9031x$ explains 28% of the variation in tropical storms. If Dr. Gray forecasts 16 tropical storms, we expect 16.25.

5.55 The regression lines are $\hat{y} = 1.09 + 0.189x$ (before) and $\hat{y} = 0.853 + 0.157x$ (after). The predictions are 9.59 and 7.91 hundred cubic feet, for an estimated savings of about 168 cubic feet.

Chapter 6

6.1 (a) We have the opinions of 133 people, of which 36 buy the recycled product. (b) 36.8% (higher), 24.1% (the same), 39.1% (lower). 60.9% think that the recycled product is no worse than the other filters.

6.3 Those who use the product have higher opinions of it: Among buyers, opinions were 55.6% "higher," 19.4% "same," and 25.0% "lower." Among nonbuyers, the corresponding numbers were 29.9%, 25.8%, and 44.3%.

6.5 Start by setting a equal to any number from 10 to 50.

6.7 (a) At Hospital A, 99% of good-condition patients and 96.2% of poor-condition patients survive, compared to 98.67% and 96% at Hospital B. (b) At Hospital A, 97% (2037/2100) survived, compared to 98% (784/800) at Hospital B. (c) More than 70% of Hospital A's patients arrive in poor condition, compared to 25% at Hospital B.

6.8 (b)   **6.9** (a)   **6.10** (b)   **6.11** (a)   **6.12** (b)

6.13 (c)   **6.14** (c)   **6.15** (b)   **6.16** (b)   **6.17** (b)

6.19 (a) For example, the first (desipramine) column should have counts 10 and 14. (b) The percents who stayed cocaine-free are 58.3% (desipramine), 25.0% (lithium), and 16.7% (placebo). Because random assignment was used, we have evidence that desipramine is more effective.

6.21 17.2% of single men hold Grade 1 jobs. 6.07% of Grade 1 jobs are held by single men.

6.23 (a) We need to account for the fact that the study included many more married men than single men, so that we would expect their numbers to be higher in every job grade even if marital status and job level were unrelated. (b) For example, single and widowed men had higher percents of Grade 1 jobs (17.2% and 19.0%). Single men were least likely (2.1%), and widowed men most likely (9.5%) to hold Grade 4 jobs.

6.25 (a) Males: 490 admitted, 210 not. Females: 280 admitted, 220 not. (b) Males: 70% admitted. Females: 56% admitted. (c) Business school: 80% of males, 90% of females. Law school: 10% of males, 33.3% of females. (d) Most male applicants apply to the business school, where admission is easier. A majority of women apply to the law school, which is more selective.

6.27 Generally, more education means more freedom. For example, 45% of those with bachelor's degrees felt free to organize their work, compared to 29.7% of those who finished high school, and 24.4% of those who did not finish high school.

6.29 Women constitute a substantial majority of associate's, bachelor's, and master's degrees, and a scant majority of professional and doctor's degrees. Those percents are 62.5%, 59.2%, 61.3%, 52.7%, and 51.0%.

6.31 The risk of CHD is 1.7% for the low anger group, 2.3% for the moderate anger group, and 4.3% for the high anger group.

Chapter 7

7.1 (a) Quantitative. (b) Categorical. (c) Categorical. (d) Quantitative.

7.3 A bar graph is the best choice.

7.5 Use either a histogram or a stemplot. The distribution is sharply right-skewed, with several possible high outliers. The five number summary is 0.1, 3.5, 5.4, 9, 33.8.

7.7 (b) A pie chart cannot be used because each percent represents a different whole.

7.9 The five number summaries are 7.2, 8.5, 9.3, 10.9, 12.8 for Aleppo pines, and 21.2, 23.7, 26.7, 29.7, 33.7 for Torrey pines. It should be easy to tell which species a needle comes from.

7.11 Neither distribution has an outlier.

7.13 (a) The Aleppo distribution looks more Normal. (b) For Aleppo pines, $\bar{x} = 9.593$ and $M = 9.3$ cm, and for Torrey pines, $\bar{x} = 27$ and $M = 26.7$ cm. These are similar because both distributions are relatively symmetric.

7.15 95% are between 6.4 and 12.8 cm; 2.5% are less than 6.4 cm.

7.17 (a) About 6%. (b) About 73%.

7.19 The plot shows a strong negative association; $r \doteq -0.998$.

7.21 -56.1 grams; prediction outside the range of the available data is risky.

7.23 (a) Lean monkeys: $\bar{x} = 8.6833$ kg. Obese monkeys: $\bar{x} = 10.5167$ kg. (b) Lean monkeys: $\hat{y} = 0.541 + 0.0826x$. Obese monkeys: $\hat{y} = 0.371 + 0.0852x$. Energy increases at about the same rate for both types of monkeys, but obese monkeys have a lower intercept, meaning that they expend less energy overall.

7.25 A scatterplot shows a moderately strong positive association (correlation 0.7565).

7.27 (a) 31 neurons. (b) The correlation is 0.6386; the relationship is moderately strong.

7.29 (a) Fidelity Technology Fund. (b) No.

7.31 Incorrect; it should be $r^2 = 64\%$.

7.33 (a) $\hat{y} = 166 - 1.10x$. Each additional lamb's-quarter per meter decreases yield by about 1.1 bushels/acre. (b) We predict 159.4 bushels/acre.

7.35 (a) Slightly right-skewed; one observation is somewhat low, but not really an outlier. (b) $\bar{x} = 563.1$ and $M = 560$ km^3 of water. (c) $s = 136.5$ km^3 of water; the five-number summary is Min $= 290$, $Q_1 = 445$, $M = 560$, $Q_3 = 670$, Max $= 900$ km^3 of water.

7.37 The distribution is roughly Normal. $\bar{x} \doteq 15.374$; $s \doteq 5.979$; Min $= 1$, $Q_1 = 11$, $M = 16$, $Q_3 = 20$, Max $= 31$ days. The median date is May 5.

7.39 The five-number summaries are
Min $= 7$, $Q_1 = 12$, $M = 19$, $Q_3 = 22$, Max $= 26$;
Min $= 1$, $Q_1 = 11$, $M = 16$, $Q_3 = 20$, Max $= 27$;
Min $= 9$, $Q_1 = 11$, $M = 17$, $Q_3 = 20$, Max $= 31$;
Min $= 1$, $Q_1 = 8$, $M = 10.5$, $Q_3 = 16$, Max $= 25$.
There is no clearly discernible pattern in the boxes, but the minimum values suggest a cyclic pattern.

7.41 Stemplots or boxplots show little difference; the means (0.2426 mg for cicada, 0.2221 mg for control) and medians (0.2380 mg for cicada, 0.2410 mg for control) are also similar.

7.43 A scatterplot shows a moderate positive association ($r \doteq 0.6821$), which supports the idea that the proportion of perch killed rises with the number of perch present.

7.45 The claim is supported: The scatterplot suggests a negative linear association, $r \doteq -0.8035$, and $\hat{y} = 92.29 - 0.05762x$. Each additional 1000 wildebeest decreases burned area by about 0.058% on the average.

7.47 Compute and compare the percents of each group giving each rating; for example, 5.9% of black parents rated schools as excellent.

7.49 (a) $\hat{y} = 93.9 + 0.778x$. The third point is A; the first point is B. (b) The correlation drops only slightly (from 0.6386 to 0.6101) when A is removed; it drops more drastically (to 0.4793) without B. (c) Without A: $\hat{y} = 98.4 + 0.679x$. Without B: $\hat{y} = 101 + 0.693x$.

Chapter 8

8.1 (a) All college students. (b) 104 students.

8.3 Population: all 45,000 people who made credit card purchases. Sample: the 137 people who returned the survey form.

8.5 Possible answers: (a) A call-in poll. (b) Interviewing students as they enter the student center.

8.7 Number from 01 to 28 and choose 04, 10, 17, 19, 12, 13.

8.9 Larger samples provide better information about the population.

8.11 Number from 01 to 30 and choose 19, 22, 05, 13, 25, 28. Then number from 1 to 8 and choose 3, 8, 4, 7.

8.13 The higher no-answer was probably the second period—more families are likely to be gone for vacations, etc. High nonresponse rates are problematic because we do not know what information we are missing.

8.15 Online polls rely on voluntary response.

8.16 (b)   **8.17** (a)   **8.18** (b)   **8.19** (a)   **8.20** (c)

8.21 (a)   **8.22** (b)   **8.23** (b)   **8.24** (a)

8.25 Population: Adult U.S. residents. Sample: The 1027 adults who were interviewed.

8.27 From line 117 we choose 38, 16, 32, 18, 37, 06, 23, 19, 03, 25.

8.29 (a) 0001 through 1410. (b) Table B gives 0769, 1315, 0094, 0720, and 0906.

8.31 The results would not be random.

8.33 (a) Adult residents of the U.S. (b) 58.45%. (c) Response bias (giving inaccurate numbers).

8.35 89.1%.

8.37 Some people who don't wear their seat belts may be embarrassed or ashamed to admit it.

8.39 Assign labels 001 through 290 to the men, and 001 through 110 to the women. From Table B, the first three men are 174, 095, and 178, and the first three women are 019, 007, and 041.

8.41 (a) Select 35, 75, 115, 155, 195. (Only the first number is from Table B; the others are 40, 60, 120, and 160 places down the list.) (b) Each of the first 40 addresses has chance 1/40 of being selected; selections from the other four groups of 40 addresses are tied to the first choice. However, the only possible samples have exactly one address from the first 40, one address from the second 40, and so on; while an SRS could contain any five of the 200 addresses in the population.

8.43 (a) Dialing a randomly chosen phone number can conceivably contact any person who has a telephone. (b) "Phone-answerers" and "non-phone-answerers" might have different characteristics; a good survey should fairly represent both groups.

8.47 **(a)** Smaller sample sizes give less information about the population. **(b)** The margin of error was so large that the results could not be viewed as an accurate reflection of the population of Cubans.

Chapter 9

9.1 This is an observational study. Explanatory variable: cell phone usage. Response variable: presence/absence of brain cancer.

9.3 For ideas, ask: What else might binge-drinking students do that might lead to lower grades?

9.5 Individuals: pine seedlings. Factor: amount of light. Treatments: full light, 25% light, or 5% light. Response variable: dry weight at the end of the study.

9.7 Over a year, many things can change: the state of the economy, hiring costs (due to an increasing minimum wage or the cost of employee benefits), etc.

9.9 **(a)** Assign 6 plots to each treatment. **(b)** If using Table B, label 01 to 36 and take two digits at a time.

9.11 The effects of factors other than the nonphysical treatment have been eliminated or accounted for, so that the differences in improvement observed between the subjects can be attributed to the differences in treatments.

9.13 **(a)** The researchers simply observed the existing arsenic levels; they did not alter them. **(b)** For example: The increased lung cancer mortality rate is greater than would be likely to occur by chance if arsenic had no effect.

9.15 The experimenter knows which subjects were taught to meditate; if he or she has some expectations about the effect of meditation, this could influence the diagnosis.

9.17 **(a)** Randomly assign 15 students to Group 1 (easy mazes) and the other 15 to Group 2 (hard mazes). Compare each group's time estimates. **(b)** Each student does the activity twice, once with the easy mazes, and once with the hard mazes, in a randomly determined order. Compare each student's "easy" and "hard" time estimate.

9.19 (a)   **9.20** (b)   **9.21** (b)   **9.22** (a)   **9.23** (c)

9.24 (a)   **9.25** (b)   **9.26** (a)   **9.27** (b)

9.29 This is an experiment, because the interviewer chooses a treatment. The explanatory variable is the level of identification, and the response variable is whether or not the interview is completed.

9.31 **(a)** In an observational study, we observe subjects who have chosen to take supplements and compare them with others who do not take supplements. In an experiment, we assign some subjects to take supplements and others to take a placebo. **(b)** Treatments are assigned at random, and a control group is used as a basis for comparison to observe the effects of the treatment. **(c)** Subjects who choose to take supplements are more likely to make healthy lifestyle choices. When random assignment is used, some of those subjects will take the supplement, and some will take the placebo.

9.33 (a) Randomly assign 60 schools to each group, then compare teacher attendance. (b) From Table B, the first 10 schools are 090, 009, 067, 092, 041, 059, 040, 080, 029, 091.

9.35 Use a completely randomized design. From Table B, we assign 20, 11, 38, 31, 07, 24, 17, 09, 06; 36, 15, 23, 34, 16, 19, 18, 33, 39; 08, 30, 27, 12, 04, 35; 02, 32, 25, 14, 29, 03, 22, 26, 10; the rest drink vodka and lemonade.

9.37 (a) Assign 10 subjects to each treatment. (b) From Table B, the first group is 05, 16, 17, 40, 20, 19, 32, 04, 25, 29.

9.39 The factors are pill type and spray type. See the definitions in this chapter.

9.41 (a) The subjects are randomly chosen Starbucks customers. Each subject tastes both drinks, from identical unlabeled cups, in random order, and is asked which he or she prefers. (b) Assign 10 customers to get regular coffee first; using Table B, we choose 12, 16, 02, 08, 17, 10, 05, 09, 19, 06.

9.43 Each player will be put through the running sequence twice—once with oxygen and once without. For each player, randomly determine whether to use oxygen on the first or second trial. Allow ample time between trials for full recovery.

9.45 Randomly assign 100 subjects to each group.

9.47 All the women are assigned to one treatment and all the men to the other. If women and men respond differently to the treatment, the experiment will be strongly biased.

9.49 (a) See the definitions in this chapter.

Commentary: Data Ethics

1. Opinions may vary. (a) seems to qualify as minimal risk, while (c) seems to be excessive.

3. It is good to state the *purpose* of the research plainly, but refrain from stating the research *thesis*.

7. This offers anonymity, because names are never revealed.

11. For example, informed consent is lacking.

Chapter 10

10.1 Considering many hands in which you are dealt a pair, the proportion of times that you get four of a kind is about 0.088. This does not mean that, if you've been dealt a pair 999 times and gotten four of a kind 87 times, you are guaranteed to get four of a kind from your next pair.

10.3 (a) There are 4 zeros among the first 50, for a proportion of 0.08.

10.5 (a) $S = \{$male, female$\}$. (b) Upper and lower limits will vary. (c) $S = \{0, 0.01, 0.02, 0.03, \ldots\}$. (d) $S = \{$A, B, C, D, F$\}$.

10.7   Count how many ways each intelligence level can occur; the probabilities are counts/16.

10.9   **(a)** Event B specifically rules out obese subjects. **(b)** "The person chosen is overweight or obese." $P(A \text{ or } B) = 0.66$. **(c)** 0.34.

10.11   Only Model 2 is legitimate.

10.13   **(a)** The eight probabilities have sum 1. **(b)** "The subject worked out fewer than 7 days in the past week." $P(X < 7) = 0.98$. **(c)** $P(X > 0) = 0.32$.

10.15   **(a)** The density curve is a triangle. **(b)** 0.5. **(c)** 0.125.

10.17   **(a)** The student's grade is either an A or a B; $P(X \geq 3) = 0.68$.
(b) $P(X < 2) = P(X \leq 1) = 0.12$.

10.19   **(b)** A personal probability might take into account specific information about your driving habits. **(c)** Most people believe that they are better-than-average drivers.

10.21   (a)   **10.22**   (b)   **10.23**   (b)   **10.24**   (b)   **10.25**   (c)

10.26   (a)   **10.27**   (b)   **10.28**   (c)   **10.29**   (a)   **10.30**   (c)

10.31   **(a)** 16 possible outcomes (all possible sequences of hits and misses).
(b) {0,1,2,3,4}.

10.33   **(a)** 0.28. **(b)** 0.87.

10.35   **(a)** All probabilities are between 0 and 1, and add to 1. **(b)** 0.41. **(c)** 0.38.

10.37   **(a)** 3/7. **(b)** 2/7. **(c)** 5/7.

10.39   The probability is 42/90.

10.41   **(a)** Every person must fall into exactly one category, and the probabilities add to 1. **(b)** 0.04. **(c)** 0.18. **(d)** 0.42.

10.43   **(a)** 0.56. **(b)** 0.54.

10.45   **(a)** Each digit 1–9 has probability 1/9. **(b)** $P(W \geq 6) = \frac{4}{9}$—twice as big as the Benford's law probability.

10.47   **(a)** Each arrangement has probability 1/8. **(b)** 0.375. **(c)** X can be 0,1,2,3 with probabilities 0.125, 0.375, 0.375, 0.125.

10.49   **(a)** Continuous; the set of possible values is an interval. **(b)** The height is $\frac{1}{2}$ so that the area is 1. **(c)** $P(Y \leq 1) = \frac{1}{2}$.

10.51   **(a)** 0.9652. **(b)** $P(Z \geq 8.42)$, which is basically 0.

10.53   **(a)** 1/10,000. **(b)** 24/10,000.

10.55   Results will vary.

Chapter 11

11.1   Both are statistics.

11.3   Both are statistics.

11.5 If 1 of the 12 homes were lost, it would cost more than the collected premiums. For many policies, the average claim should be close to $250.

11.7 (a) 69.4. (b) Answers will vary. (c) Answers will vary.

11.9 (a) $N(188, 4.1)$. $P(185 < \bar{x} < 191) \doteq 0.5346$. (b) $P(185 < \bar{x} < 191) \doteq 0.9792$.

11.11 No. The histogram of the sample values will look like the population distribution. The histogram of *sample means* (from many large samples) will look more and more Normal.

11.13 We can be about 99.4% certain that average losses will not exceed $275 per policy.

11.14 (c)   **11.15** (b)   **11.16** (b)   **11.17** (a)   **11.18** (c)

11.19 (a)   **11.20** (c)   **11.21** (b)

11.23 Both are statistics.

11.25 Mean 6 strikes/km^2, standard deviation 0.7589 strikes/km^2.

11.27 (a) 0.0668. (b) 0.0013.

11.29 About 133.2 mg/dl.

11.31 (a) Approximately $N(2.2, 0.1941)$. (b) 0.1515. (c) 0.0764.

11.33 $P(\bar{x} > 10\%) = 0.3409$ and $P(\bar{x} < 5\%) = 0.1230$.

11.35 $n = 32$.

11.37 On the average, Joe loses 40 cents each time he plays.

11.39 (a) Mean $0.40, standard deviation $0.049. (b) About 0.9586.

11.41 The mean is 10.5.

Chapter 12

12.1 Independence is not a reasonable assumption.

12.3 About 0.38 (0.3773).

12.5 (b) For example, "A and not B" is "student is at least 25 and not local." (c) The first is given; the other three are 0.20, 0.65, and 0.10.

12.7 2/3.

12.9 4% of adults go to health clubs at least twice per week.

12.11 (a) 0.25; 0.2353. (b) 0.22; 0.2041; 0.1875. (c) The general multiplication rule applies; the probability is about 0.0004952. (d) About 0.001981.

12.13 $P(X = 0) \doteq 0.9510$, $P(X = 1) \doteq 0.0480$, and $P(X = 2) \doteq 0.0010$; the other three probabilities are less than 0.00001.

12.15 About 0.9800.

12.17 (b) **12.18** (b) **12.19** (c) **12.20** (a) **12.21** (b)

12.22 (c) **12.23** (c) **12.24** (b) **12.25** (c) **12.26** (b)

12.27 0.1001.

12.29 (a) 0.001125. (b) The other symbol can show up on the middle wheel (probability 0.001375), or on either of the outside wheels (probability 0.021375 each). (c) 0.044125.

12.31 (a) 0.3. (b) 0.3. (c) Yes.

12.33 $P(D) = 0.4$.

12.35 (a) 0.321. (b) 0.3302.

12.37 1/4.

12.39 (a) 0.5910. (b) 0.5424. (c) No.

12.41 (a) 0.2400. (b) 0.2844, 0.3248, 0.1991, and 0.1415. (c) The probability of damage decreases noticeably when thorny cover is 1/3 or more.

12.43 26.59%.

12.45 0.1.

12.47 (a) 1/6. (b) 5/36. (c) 25/216. (d) $(5/6)^3(1/6)$; $(5/6)^4(1/6)$; $(5/6)^k(1/6)$.

12.49 The black candidate expects to get 58% of the vote.

12.51 $P(B \mid F) \doteq 62.07\%$.

12.53 (a) 0.224. (b) 0.7143.

12.55 (a) Either B or O. (b) $P(B) = 0.75$, $P(O) = 0.25$.

12.57 (a) 0.25. (b) 0.015625. (c) 0.140625.

Chapter 13

13.1 Yes.

13.3 No; p does not remain fixed.

13.5 (a) Caught: binomial, $n = 10$, $p = 0.7$. Missed: binomial, $n = 10$, $p = 0.3$. (b) 0.2668; 0.6172.

13.7 (a) 10. (b) "500 choose 2" returns 124,750. "500 choose 100" returns $2.04169424 \times 10^{107}$ (c) 0.387420489.

13.9 (a) X is binomial with $n = 10$ and $p = 0.3$; Y is binomial with $n = 10$ and $p = 0.7$ (b) 7 errors caught, 3 missed. (c) 1.4491 errors.

13.11 (a) Mean 414.45, standard deviation 17.3939 students. (b) 0.4641. (c) 0.4742.

13.13 (b) **13.14** (b) **13.15** (c) **13.16** (b) **13.17** (c)

13.18 (b) **13.19** (c) **13.20** (a) **13.21** (b)

13.23 (a) p varies from one attempt to another. (b) The binomial distribution is appropriate.

13.25 (a) $n = 5$, $p = 0.65$. (b) 0, 1, ..., 5. (c) 0.00525, 0.04877, 0.18115, 0.33642, 0.31239, 0.11603. (d) $\mu = 3.25$ and $\sigma \doteq 1.0665$ years.

13.27 (a) All woman are independent, and each has the same probability of getting pregnant. (b) Ideal: 0.1821. Typical: 0.6415.

13.29 (a) Normal: 0.5, exact: 0.5286. (b) $np = 5$ is too small.

13.31 (a) 0.3115. (b) $\mu = 60$ red-blossomed plants. (c) Normal: 0.5000; exact: 0.5597.

13.33 (a) 5250 dropouts. (b) 0.9999.

13.35 (a) 0.7498. (b) 0.9328.

13.39 The probability is nearly 1.

13.41 (a) $P(1$ of the vaccinated children$) \doteq 0.3741$. $P(1$ of the unvaccinated children$) \doteq 0.0960$. $P(1$ of each$) \doteq 0.0359$. (b) Total probability: 0.1977.

13.43 (a) $n(1 - p) = 10$. (b) 0.5. (c) 0.5793.

Chapter 14

14.1 (a) 2.0702. (b) 4.1404. (c) About 268 to 276.

14.3 $z^* = 2.24$.

14.5 (a) Aside from two low outliers, the distribution is close to Normal. (b) 98.90 to 112.78 IQ points.

14.7 (a) $N(5, 0.0816)$. (b) 4.98 is less than 0.25 standard deviations below the presumed mean, while 4.7 is about 3.67 standard deviations below.

14.9 H_0: $\mu = 5$; H_a: $\mu \neq 5$.

14.11 H_0: $\mu = 64.2$ in; H_a: $\mu \neq 64.2$ in.

14.13 (a) $N(0, 0.3162)$. (b) $P = 0.1711$.

14.15 (a) Such results would rarely occur if vitamin C were ineffective. (b) Such results would occur less than 1% of the time if vitamin C were ineffective.

14.17 (a) $P = 0.8104$; not significant. (b) $P = 0.0002$; significant. (c) The second outcome would almost never happen if H_0 were true.

14.19 $\bar{x} \doteq 4.9883$, $P = 0.8886$; not significant.

14.21 Significant at 5% but not at 1%.

14.23 (a) $z \doteq -2.20$. (b) Significant at the 5% level. (c) Not significant at the 1% level. (d) z is between 2.054 and 2.326, so P is between 0.02 and 0.04.

14.24 (b)   **14.25** (b)   **14.26** (c)   **14.27** (a)   **14.28** (b)

14.29 (c)   **14.30** (c)   **14.31** (a)   **14.32** (b)   **14.33** (c)

14.35 2.0592 to 2.6408 kg/m^2.

14.37   No: The interval refers to the mean BMI, not to individual BMIs.

14.39   Other surveys should be close to the truth—not necessarily close to the results of this survey.

14.41   **(a)** $H_0: \mu = 0$; $H_a: \mu > 0$. **(b)** $z \doteq 13.29$. **(c)** This is far outside the range we would expect from a $N(0, 1)$ distribution.

14.43 $P = 0.03$ means that *if* H_0 is true, we have observed outcomes that occur about 3% of the time.

14.45   Such an increase only has a 3.1% chance of happening by accident.

14.47   Something that occurs "less than once in 100 repetitions" also occurs "less than 5 times in 100 repetitions," but not vice versa.

14.49 $P \leq 1$ always; it should be $P = 0.0918$.

14.51   **(b)** -4.526% to -2.648%.

14.53 $H_0: \mu = 0\%$; $H_a: \mu < 0\%$; $z = -9.84$, so $P \doteq 0$.

14.55   **(a)** $H_0: \mu = 0$; $H_a: \mu > 0$. μ is the mean difference in the population. **(b)** $z \doteq 1.84$, $P = 0.0329$.

14.57   **(a)** 123.16 to 128.98. **(b)** $z \doteq -1.09$, $P = 0.2757$. **(c)** $z \doteq -1.66$, $P = 0.0969$.

Chapter 15

15.1   The most important reason is (c).

15.3   Depending on the time of day or the day of the week, certain types of shoppers would or would not be present.

15.5   **(a)** 1.47. **(b)** $n = 400$: 0.735. $n = 1600$: 0.3675. **(c)** Margin of error decreases as n increases. (Specifically, it is halved each time n quadruples.)

15.7   **(a)** $z = 1.61$ is not significant. **(b)** $z = 1.67$ is significant.

15.9   The confidence intervals are 4.362 to 5.238, 4.547 to 5.053, and 4.645 to 4.955.

15.11   Use $n = 217$.

15.13   **(a)** If $\mu = 5.1$, the probability of (correctly) rejecting H_0 is 0.23. **(b)** The risk of mistakenly concluding that $\mu = 5$ is too high.

15.15   **(a)** The powers are 0.227, 0.405, and 0.684. **(b)** The powers are 0.227, 0.684, and 0.955. **(c)** The powers are 0.227, 0.336, and 0.528.

15.17   **(a)** H_0: Patient is healthy; H_a: Patient is ill. Type I error: sending a healthy patient to the doctor. Type II error: clearing a patient who is ill. (These hypotheses, and error types, could be switched.)

15.18 (a)   **15.19** (c)   **15.20** (b)   **15.21** (c)   **15.22** (a)

15.23 (a)   **15.24** (c)   **15.25** (a)   **15.26** (b)   **15.27** (c)

15.29   Were these random samples? How big were the samples?

15.31 (a) This is not an SRS from the population of all women. (b) It may be reasonable to view this sample as an SRS of women who shop at large suburban malls.

15.33 Many people might be reluctant to discuss such intimate details of their personal lives. The margin of error allows only for random sampling error, not bias.

15.35 The effect is greater if the sample is small.

15.37 B, A, C

15.39 (a) Margin of error decreases. (b) The P-value decreases. (c) The power increases.

15.41 (a) $z \doteq 1.70$, $P = 0.0891$. (b) The sample size is small, so the test has low power.

15.43 (a) The differences observed might occur by chance even if SES had no effect. (b) This tells us that the test was not insignificant merely because of a small sample size.

15.45 (a) This test has a 20% chance of rejecting H_0 when the alternative is true. (b) Such a test will fail to reject H_0 80% of the time. (c) The sample sizes are very small.

15.47 The power is 0.606.

15.49 (a) $z = (\overline{x} - 5)/(0.2/\sqrt{6})$. Reject H_0 if $z > 1.96$ or $z < -1.96$. (b) Reject H_0 if $\overline{x} > 5.16$ or $\overline{x} < 4.84$. (c) The power is about 0.23.

15.51 0.86.

15.53 (a) In the long run, this probability should be 0.05. (b) In the long run, this probability should be 0.192.

Chapter 16

16.1 This is an experiment; students were randomly (at the time they visited the site) assigned to a treatment. The explanatory variable is the login box (genuine or not), and the response is the student's action (logging in or not).

16.3 (a) Label from 0001 to 3478. (b) 2940, 0769, 1481, 2975, 1315. (c) The response variable is how much money the students earn in the summer.

16.5 (a) The control group should have 24 trees with no beehives. (b) Table B gives 53, 64, 56, and 68. (c) The response variable is elephant damage.

16.7 (a) For example: Did subject get a job? Did subject keep the job? For the design, randomly assign about one-third of the group to each treatment, and observe the chosen response variables after a suitable amount of time. (b) Label from 00001 through 10065 and choose 06565, 00795, and 08727.

16.9 This is a convenience sample.

16.11 $H_0: \mu = 48$; $H_a: \mu < 48$.

16.13 154 to 190 mg/dl.

16.15 Cut in half: $n = 56$. ± 5 mg/dl: $n = 182$.

16.17 322.35 to 391.65 ng/g.

16.19 80%: 334.35 to 379.65 ng/g; 90%: 327.92 to 386.08 ng/g; margin of error grows with increasing confidence.

16.21 $H_0: \mu = 100$; $H_a: \mu < 100$. $z \doteq -8.79$, $P \doteq 0$.

16.23 (a) {F,M}. (b) {6,7,...,20}. (c) All numbers between 2.5 and 6 liters/min. (d) Upper limits will vary.

16.25 (a) $P(Y > 1) = 0.74$. (b) 0.31. (c) 0.67.

16.27 (a) Mean 445 ms, standard deviation 21.1723 ($n = 15$) or 6.6953 ($n = 150$) ms. (b) Large sample sizes overcome skewness. (c) 0.2266.

16.29 The first result would happen frequently when chocolate and carob have identical effects; the second would rarely happen in that event.

16.31 The increase in fires over time would occur less than 1% of the time by chance. This straight-line relationship explains 61% of the variation in the number of fires.

16.33 Placebos can provide genuine pain relief. (Believing that one will experience relief can lead to actual relief.)

16.35 (a) Increase. (b) Decrease. (c) Increase. (d) Decrease.

16.37 (a) Factors: storage method (fresh, room temperature for one month, refrigerated for one month) and preparation method (cooked immediately or after one hour). This makes six treatments (storage-preparation combinations). Response variables: tasters' color and flavor ratings. (b) Randomly allocate n potatoes to each of the 6 groups, then compare ratings. (c) For each taster, randomly choose the order in which the fries are tasted.

16.39 (a) All probabilities are greater than or equal to 0, and their sum is 1. (b) 0.62. (c) Both probabilities are 0.19.

16.41 (a) Nearly all should be in the range 4.5 to 49.5. (b) $\bar{x}$ will nearly always fall in the range 24.75 to 29.25.

16.43 (a) An observational study; no treatment is imposed. (b) This means unlikely to happen by chance, if alcohol consumption were unrelated to death rate. (c) For example, some nondrinkers might avoid drinking because of other health concerns.

16.45 (b) $z \doteq -2.54$, $P = 0.0110$.

16.47 97.95° to 98.46°.

16.49 When we use a low-power test, we run a fairly high risk of making a Type II error for the low-power alternatives.

16.51 Two consecutive years: 0.25. Four consecutive years: 0.0625.

16.53 (a) 0.000262. (b) 0.1159.

16.55 $P(A) = 0.14$ and $P(B|A) = 0.56$. $P(A \text{ and } B) = 0.0784 = 7.84\%$.

16.57 (a) Binomial with $n = 2000$ and $p = 0.4$. (b) $P(X \leq 750) \doteq 0.0113$.

16.59   (a) 0.8989. (b) 0.9930.

16.61   1.

16.63   32.13%.

Chapter 17

17.1   4.8925 minutes.

17.3   (a) 2.015. (b) 2.518.

17.5   (a) 2.262. (b) 2.861. (c) 1.943.

17.7   The distribution is slightly, but not unreasonably, left-skewed, with no outliers. The interval is 55.71% to 63.47%.

17.9   (a) 24. (b) t is between 1.059 and 1.318, so $0.20 < P < 0.30$. (c) $t = 1.12$ is not significant at either level.

17.11   This is a matched pairs setting; compute the call minus pure tone differences. $\bar{x} \doteq 70.38$, $s \doteq 88.47$, $t \doteq 4.84$, df $= 36$, $P < 0.0001$.

17.13   The distribution of nitrogen concentration is too skewed—and the sample size too small—to use t procedures.

17.15   (b)   **17.16**   (c)   **17.17**   (a)   **17.18**   (a)   **17.19**   (c)

17.20   (a)   **17.21**   (a)   **17.22**   (a)   **17.23**   (b)   **17.24**   (c)

17.25   Student group: $t = 0.749$, for which Table C gives $0.4 < P < 0.5$. Non-student group: $t = 3.277$, for which Table C gives $0.005 < P < 0.01$.

17.27   (a) Because scores range from 0 to 500, there is a limit to how skewed the distribution could be. (b) 237.2 to 242.8. (c) We can reject H_0.

17.29   (a) A subject's responses to the two treatments would not be independent. (b) Yes ($t = -4.41$, $P = 0.0069$).

17.31   (a) The use of t procedures should be fairly safe. (b) 10.423 to 15.243 correct answers.

17.33   (b) All points: $t = 2.08$, $P = 0.031$. Without the outlier: $t = 1.79$, $P = 0.052$. The evidence is weaker with the outlier removed.

17.35   (a) A stemplot shows no major causes for concern. (b) 0.9211 to 1.4244 days.

17.37   (a) $H_0: \mu = 0$; $H_a: \mu > 0$. Researchers have reason to believe that CO_2 will increase growth rate. (b) $t \doteq 3.16$, df $= 2$, $P = 0.044$. (c) For small samples, the t procedures should be used only for samples from a Normal population.

17.39   The data contain two extreme high outliers, 5973 and 8015. These may distort the t statistic.

17.41   A stemplot shows a single-peaked and roughly symmetric distribution. The 90% confidence interval is 3.43% to 4.73%.

17.43   0.71 to 10.71 μm/hr.

17.45 (a) For each subject, randomly select which knob should be used first. (b) μ is the mean of (right-hand-thread time minus left-hand-thread time); H_0: $\mu = 0$ sec; H_a: $\mu < 0$ sec. $t = -2.90$, df $= 24$, $P = 0.0039$; right-hand-thread times are less.

17.47 -21.2 to -5.5 sec. $\bar{x}_{RH}/\bar{x}_{LH} = 88.7\%$.

Chapter 18

18.1 Matched pairs.

18.3 Single sample.

18.5 (a) If the loggers had known that a study would be done, they might have (consciously or subconsciously) altered their behavior. (b) With all points: $t \doteq 2.11$, for which $0.25 < P < 0.05$ (df $= 7$) or $P = 0.026$ (df $= 14.8$). With the low outlier removed from the logged data: $t \doteq 1.832$, for which $0.05 < P < 0.1$ (df $= 7$) or $P = 0.042$ (df $= 17.2$).

18.7 0.46 to 7.20 species (df $= 8$) or 0.65 to 7.02 species (df $= 14.8$).

18.9 (a) The time data are reasonably Normal. $t = -3.98$, for which $P < 0.0005$ (df $= 29$) or $P = 0.0001$ (df $= 57.041$). (b) The spending data are skewed, and have many gaps. $t = -5.945$, for which $P < 0.0005$ (df $= 29$) or P is very small (df $= 57.998$).

18.11 With such small samples, t procedures are not reliable unless both distributions are Normal.

18.15 There was no significant difference in mean Self-Concept Scale scores for men and women ($t = -0.8276$, df $= 62.8$, and $P = 0.4110$).

18.16 (a)   **18.17** (c)   **18.18** (b)   **18.19** (b)   **18.20** (a)

18.21 (b)   **18.22** (c)   **18.23** (a)   **18.24** (b)

18.25 (a) H_0: $\mu_M = \mu_F$; H_a: $\mu_M < \mu_F$. (b) $t = -0.248$ (Study 1); $t = 1.507$ (Study 2). (c) df $= 55$ (Study 1); df $= 19$ (Study 2). (d) $P > 0.5$ (Study 1); $0.05 < P < 0.10$ (Study 2). (e) The first study gives no support to the belief that women talk more than men; the second study gives only moderate (not significant) support.

18.27 (a) Oregon: $n = 6$, $\bar{x} = 26.9$, $s = 3.82$. California: $n = 7$, $\bar{x} = 11.9$, $s = 7.09$. (b) Use conservative df $= 5$. (c) $t \doteq 4.837$. (d) $0.002 < P < 0.005$.

18.29 (a) A placebo allows researchers to account for any psychological benefit (or detriment) subject might get from taking a pill. (b) Neither the subjects nor the researchers who worked with them knew who was getting ginkgo extract; this prevents expectations or prejudices from affecting the evaluation of the effectiveness of the treatment. (c) $t \doteq 2.147$, for which $0.04 < P < 0.05$ (df $= 17$) or $P = 0.0387$ (df $\doteq 35.35$). Those who took gingko extract had significantly more misses per line.

18.31 No evidence of a difference: $t = 0.7$, for which $P > 0.5$ (df $= 8$) or $P = 0.4930$ (df $= 17.85$).

18.33 **(a)** H_0: $\mu_1 = \mu_2$; H_a: $\mu_1 > \mu_2$; $t \doteq 2.646$; either df $= 426$ or 534.45 gives $P < 0.005$. **(b)** 0.06 to 15.94 points (df $= 100$) or 0.184 to 15.816 points (df $= 534.45$). **(c)** A 16-point gain on the SAT is unlikely to make a practical difference.

18.35 **(b)** H_0: $\mu_G = \mu_B$; H_a: $\mu_G \neq \mu_B$; $t \doteq 1.64$, for which $0.1 < P < 0.2$ (df $= 30$) or $P = 0.1057$ (df $= 56.93$).

18.37 -1.24 to 11.48 (df $= 30$) or -1.12 to 11.35 (df $= 56.9$).

18.39 **(a)** The Hylite mean (143.2) is greater than the Permafresh mean (134.8). **(c)** $t \doteq -6.296$, for which $0.002 < P < 0.005$ (df $= 4$) or $P = 0.0003$ (df $= 7.779$); Hylite is significantly better.

18.41 $-11.244°$ to $-5.556°$ (df $= 4$) or $-10.891°$ to $-5.909°$ (df $= 7.779$).

18.43 $t = -3.74$; $0.01 < P < 0.02$ (df $= 5$) or 0.0033 (df $= 10.9$).

18.45 $t = 2.521$, for which $0.01 < P < 0.02$ (df $= 16$) or $P = 0.0088$ (df $= 28.27$).

18.47 H_0: $\mu_1 = \mu_2$; H_a: $\mu_1 > \mu_2$; $t \doteq 1.914$, $0.025 < P < 0.05$ (df $= 7$) or 0.0382 (df $= 13.92$). The confidence interval is 0.03 to 6.27 points (df $= 7$) or 0.25 to 6.05 points (df $= 13.92$).

18.49 $t \doteq 7.817$; with either df $= 14$ or df $= 35.1$, $P < 0.001$. The confidence interval is either 2.562 to 4.500 mm, or 2.614 to 4.448 mm.

Chapter 19

19.1 **(a)** p is the proportion of the population (all college students) who say they pray at least once in a while. **(b)** $\hat{p} = 0.8425$.

19.3 **(a)** $N(0.75, 0.01369)$. **(b)** $N(0.75, 0.00685)$.

19.5 **(a)** The survey excludes residents of the northern territories, as well as those who have no phones or have only cell phone service. **(b)** 0.8381 to 0.8736.

19.7 **(a)** The number of "successes" is only 9. **(b)** The sample size is 102, with 11 successes; $\tilde{p} \doteq 0.1078$. **(c)** 0.0573 to 0.1584.

19.9 **(a)** $\hat{p} = 1$; the confidence interval is 1 to 1. **(b)** $\tilde{p} = 0.9167$; 0.8061 to 1.0272 (use 1 for the high end).

19.11 $n = 318$.

19.13 $z \doteq 2.12$ and $P = 0.0169$.

19.15 (b)   **19.16** (c)   **19.17** (c)   **19.18** (b)   **19.19** (c)

19.20 (a)   **19.21** (a)   **19.22** (a)   **19.23** (c)   **19.24** (a)

19.25 **(a)** The survey excludes those who have no phones or have only cell phone service. **(b)** 0.8170 to 0.8622.

19.27 **(a)** 2.26%. **(b)** 3.08%. **(c)** For samples of about this size, the margin of error is no more than about $\pm 3\%$ no matter what $\hat{p}$ is.

19.29 (a) The survey excludes residents of Alaska and Hawaii, and those who have no phones or have only cell phone service. (b) 0.0390 to 0.0603. (c) 0.0405 to 0.0620; both the lower and upper limits of this interval are slightly higher than those in (b).

19.31 (a) The failure count (5) is too small. (b) 0.6020 to 0.8795.

19.33 (a) The margin of error will be the same for all states. (b) There would be less margin of error for states with larger samples.

19.35 (a) 0.0588, 0.0784, 0.0898, 0.0960, 0.0980, 0.0960, 0.0898, 0.0784, 0.0588. (b) 0.0263, 0.0351, 0.0402, 0.0429, 0.0438, 0.0429, 0.0402, 0.0351, 0.0263. The new margins of error are less than half their former size.

19.37 The plus four interval is 0.2335 to 0.6415.

19.39 The large sample interval is 0.3753 to 0.4252.

19.41 Yes: $z = -7.68$, $P < 0.0001$.

19.43 The plus four interval is 0.0335 to 0.1447.

Chapter 20

20.1 0.1800 to 0.3804.

20.3 The male proportion is higher by 0.1394 to 0.1806.

20.5 0.1306 to 0.3470.

20.7 Yes: $z = -2.87$, $P = 0.0021$.

20.9 (b)   **20.10** (a)   **20.11** (c)   **20.12** (b)   **20.13** (b)

20.14 (a)   **20.15** (b)   **20.16** (b)

20.17 (a) Yes, all counts are large enough. (b) 0.1200 to 0.2975.

20.19 (a) One count is 0. (b) Altered mice: sample size 35, 24 with tumors. Control: sample size 20, 1 with tumors. (c) The plus four interval for $p_1 - p_2$ is 0.3978 to 0.8736.

20.21 $H_0: p_1 = p_2$; $H_a: p_1 \neq p_2$; $z \doteq 3.39$, $P = 0.0006$.

20.23 0.0631 to 0.2179 (plus four: 0.0614 to 0.2158).

20.25 (a) $\hat{p}_1 \doteq 0.1415$, $\hat{p}_2 \doteq 0.1667$; $z \doteq -0.39$, $P = 0.6966$. (b) $z \doteq -2.12$ and $P = 0.0340$. (c) For (a): -0.1559 to 0.1056 (plus four: -0.1659 to 0.0985). For (b): -0.04904 to -0.001278 (plus four: -0.0493 to -0.0016). Larger samples make the margin of error smaller.

20.27 $H_0: p_1 = p_2$; $H_a: p_1 \neq p_2$; $z \doteq 0.02$, $P = 0.9840$—no evidence of a difference.

20.29 0.0915 to 0.2908.

20.31 $H_0: p_1 = p_2$; $H_a: p_1 > p_2$; $z \doteq 2.96$, $P = 0.0015$.

20.33 Use the plus four interval, because one count is only 7: -0.0399 to 0.3685.

20.35 The confidence interval for the proportion of all customers who pay by credit card is 0.3953 to 0.5943. The difference is not significant: $z \doteq -1.02$, $P = 0.3078$.

Chapter 21

21.1 0.3357 to 0.3843.

21.3 $H_0: p = 0.5$; $H_a: p \neq 0.5$; $z \doteq 1.90$, $P = 0.0578$.

21.5 2.00 to 2.70 kg/m^2.

21.7 $H_0: \mu_1 = \mu_2$; $H_a: \mu_1 > \mu_2$; $t \doteq 10.4$, $P < 0.0001$ (df = 19 or 32.39).

21.9 0.1654 to 0.2746.

21.11 (a) With large samples, the t procedures are fairly safe even with non-Normal data. (b) Yes: $t \doteq 2.207$, for which $0.01 < P < 0.02$ (df = 134) or $P = 0.0143$ (df = 186.02).

21.13 Either 0.48 to 8.92 minutes (df = 100) or 0.50 to 8.90 minutes (df = 186.02).

21.15 0.4422 to 0.4976.

21.17 315.24 to 318.76 points (using either df = 100 or 1013).

21.19 (a) This is an observational study (one cannot "assign" a baby's birthweight). (b) Yes: $z \doteq -2.34$, $P = 0.0096$.

21.21 Both differences are significant. For drug use, $z \doteq -2.08$, $P = 0.0376$. For IQ scores, $t \doteq -2.08$, $0.04 < P < 0.05$ (df = 100) or $P = 0.0389$ (df = 247.1).

21.23 $z = -1.80$, $P = 0.0357$.

21.25 $t \doteq 1.17$, so $0.10 < P < 0.15$ (df = 22) or $P = 0.1234$ (df = 43.3).

21.27 165.5 to 220.5 mg/dl.

21.29 Compare two means (line 3).

21.31 Estimate one mean (line 5).

21.33 Use a matched pairs test for a single mean (line 1).

21.35 (a) Two-sample z for proportions (line 4). (b) Two-sample t for means (line 3) (c) Two-sample z for proportions (line 4).

21.37 (a) Matched pairs. (b) We need to know the standard deviation of the differences, not the two individual standard deviations.

21.39 (a) $\hat{p} = 1$; the interval is 1 to 1. (b) $\tilde{p} \doteq 0.9762$; the interval is 0.9436 to 1.0088.

21.41 $t \doteq 0.907$, df = 19, $P = 0.1879$.

21.43 10.74 to 13.90 months.

21.45 (a) Randomly assign 8 pieces of fabric to each method, then compare color. (b) $t \doteq -8.79$, for which P is very small (using either df = 7 or df = 13.73). However, we cannot tell if the difference is important in practice.

21.47 0.5981 to 0.7849.

21.49 (a) The large-sample significance test is not reliable, but with such a large difference between the sample proportions, the difference is clearly significant. A plus four 95% confidence interval for the difference in checking is 0.7065 to 0.9054. (b) $t \doteq 6.914$, P is very small (using either df $= 19$ or df $= 33.27$). The 95% confidence interval is either 43.65 to 81.55 psi (df $= 19$) or 44.18 to 81.02 psi (df $= 33.27$).

21.51 Two of the counts are too small to perform a significance test safely.

21.53 $t \doteq 5.59$; with either df $= 9$ or df $= 9.89$, $P < 0.001$.

Chapter 22

22.1 (a) University Park: 7.0%, 5.6%, 22.0%, and 65.4%. Commonwealth: 28.3%, 8.7%, 17.9%, and 45.0%. (b) Students on the main campus are much more likely to use Facebook at least daily, while commonwealth campus students are more likely not to use it at all.

22.3 (a) $z \doteq -12.22$, P is very small. (b) $z \doteq 2.17$, $P = 0.0300$. (c) If we did four individual tests, we would not know how confident we could be in all four results taken together.

22.5 (a) The expected counts are 53.44, 151.75, and 421.81. (b) Commonwealth use Facebook less than weekly more often than we would expect, and use it daily less often than we expect.

22.7 (a) All expected counts are well above 5. (b) H_0: there is no relationship between setting and Facebook use; H_a: there is some relationship. $\chi^2 = 19.489$, $P < 0.0005$. (c) The largest contributions come from the first row, reflecting the fact that monthly use is lower among University Park students, and higher among commonwealth students.

22.9 There is a significant relationship between education level and attitudes about astrology ($\chi^2 = 10.582$, $P = 0.005$), with the greatest contributions coming from first and last entries on the second row.

22.11 (a) df $= (r - 1)(c - 1) = (3 - 1)(2 - 1) = 2$. (b) $\chi^2 > 15.20$, $P < 0.0005$. (c) df $= 3$.

22.13 H_0: $p_1 = p_2 = p_3 = \frac{1}{3}$; H_a: not all three are equally likely; $\chi^2 = 16.11$, df $= 2$, $P < 0.0005$.

22.15 The difference is significant: $\chi^2 \doteq 119.84$, df $= 2$, P is very small. The largest contribution comes from the youngest age group (which is cited more frequently than we would expect); the other two age groups (which are cited less frequently than expected) also have large contributions.

22.17 The observed data differs significantly from the expected uniform distribution: $\chi^2 = 19.76$, df $= 11$, $P = 0.0487$.

22.18 (b)   **22.19** (a)   **22.20** (a)   **22.21** (c)   **22.22** (a)

22.23 (a)   **22.24** (b)   **22.25** (a)   **22.26** (c)   **22.27** (b)

22.29   **(a)** 0.5977 to 0.6537 (plus four: 0.5974 to 0.6535). **(b)** $\hat{p}_b \doteq 52.0\%$, $\hat{p}_w \doteq 67.0\%$, and $\hat{p}_o \doteq 48.0\%$. To test H_0: $p_b = p_w = p_o$; H_a: some proportion is different, we find $\chi^2 \doteq 47.899$, df $= 2$, $P < 0.0005$.

22.31   **(a)** Label 01 to 77, and choose pairs of random digits. **(b)** $\chi^2 = 0.568$, df $= 3$, $P = 0.904$.

22.33   **(a)** The four numbers in the two-way table are: Group 1: 11 with tumors, 19 without; Group 2: 22 with tumors, 8 without. H_0: $p_1 = p_2$; H_a: $p_1 < p_2$. **(b)** $z \doteq -2.85$, $P = 0.0022$.

22.35   Magazines aimed at women are much more likely to have sexual depictions of models than the other two types of magazines.

22.37   We need cell counts, not just percents.

22.39   In order to do a chi-square test, each subject can only be counted once.

22.41   **(a)** 29.9%, 28.9%, 21.9%, 19.3%; 26.6%, 30.9%, 19.4%, 23.0%; 39.8%, 21.1%, 27.1%, 12.0%. **(b)** There is a significant relationship ($\chi^2 \doteq 13.305$, df $= 6$, $P = 0.039$). University-educated men are more likely to be nonsmokers or moderate smokers.

22.43 $\chi^2 = 22.426$, df $= 8$, $P = 0.004$. Blacks are less likely, and Hispanics more likely, to consider schools excellent, while Hispanics and whites differ in percent considering schools good (whites are higher) and percent who "don't know" (Hispanics are higher).

22.45   The percents in each group who prefer the new detergent are 54%, 52%, 62%, and 58%. The differences are small; the "hard water, warm wash" group is most likely to prefer the new product. For testing H_0: no relationship between laundry habits and preference; H_a: at least one proportion is different, we have $\chi^2 = 2.058$, df $= 3$, $P = 0.560$.

22.47   The counts of Democrats for each education level are 279, 996, 156, 313, and 218. The counts for Republicans are 135, 731, 129, 336, and 128. The percents leaning Democrat are 67.4%, 57.7%, 54.7%, 48.2%, and 63.0%. Support for the Democrats is highest among the least and most educated.

Chapter 23

23.1   **(a)** $r \doteq 0.9623$. **(b)** β is the average number of additional days before the first death for each unit increase in distance. We estimate $b \doteq 11.263$ and $a \doteq -8.088$. **(c)** The residuals are 0.825, -4.702, -3.965, 4.035, 6.035, and -2.228. The estimated standard deviation is $s \doteq 4.903$.

23.3   **(a)** $r^2 = 11.2\%$. **(b)** $\hat{y} = -2057 + 1.97x$, $s = 104.0$ km^3 of water.

23.5 $t \doteq 2.79$, df $= 62$, $P = 0.007$.

23.7   **(a)** $t \doteq 7.08$, df $= 4$, $0.001 < P < 0.0025$. **(b)** $r \doteq 0.9623$, $0.001 < P < 0.0025$.

23.9   With df $= 4$, the 90% confidence interval is 7.87 to 14.66.

23.11   0.79 to 3.14 km^3/year; this interval does not contain 0.

23.13   (a) $\hat{y} = -2.3948 + 0.158483(200) = 29.3018$ cm. (b) We want the 95% confidence interval (CI): 28.393 to 30.211 cm.

23.15   (a) and (d) The residuals are first positive, then negative, then positive again. (b) There is slight right-skewness, but no clear deviations from Normality. Observation 18 does not appear to be an outlier. (c) A child who is, say, above average in height at one age is likely to be tall at other ages, so that repeated measurements on the same child cannot be considered independent.

23.16   (b)   **23.17**   (c)   **23.18**   (b)   **23.19**   (a)   **23.20**   (b)

23.21   (b)   **23.22**   (a)   **23.23**   (a)   **23.24**   (b)

23.25   (a) For every additional kilogram of nitrogen per hectare, the species richness measure decreases by 0.408 on the average. (b) The straight-line relationship with nitrogen accounts for 55% of the observed variation in species richness. (c) H_0: $\beta = 0$; H_a: $\beta \neq 0$ (or $\beta < 0$). There is very strong evidence that β is negative, so species richness does decrease as nitrogen deposited increases.

23.27   (a) 115.0 to 434.6 fps/inch. (b) 176.2 to 239.4 fps.

23.29   (a) For best results, split stems. Don't forget that -0 and 0 are separate stems. (c) The table shows counts of manatees killed by boats. Perhaps some deaths due to pollution that were mistakenly attributed to boats, but we assume that number is small.

23.31   (a) 0.1141 to 0.1462 additional manatee deaths. (b) 76.43 to 109.29 manatee deaths.

23.33   (a) $r \doteq 0.5289$ is significantly greater than 0 ($t \doteq 2.92$, $P = 0.008$). (b) 12.77 to 19.73 storms.

23.35   The stemplot suggests that the residuals do not follow a Normal distribution.

23.37   (b) $r \doteq 0.8486$. For testing the slope, $t \doteq 6.00$, for which P is *much* smaller than 0.001. (c) Without the outlier, $t \doteq 3.55$, $P = 0.004$.

23.39   The distribution is skewed to the right, but the sample is large, so t procedures should be safe. $\bar{x} = 0.2781$ and $s = 0.1803$, and the 95% confidence interval for μ is 0.2449 to 0.3113 g/m^2.

23.41   The evidence for a relationship is very strong: $t = 4.33$, $P = 0.001$.

23.43   The evidence of a relationship is strong ($t \doteq 10.47$, $P < 0.0005$). The 95% confidence interval for the slope is 9.53 to 14.26 clusters per stump.

23.45   The scatterplot shows a positive association; the equation is $\hat{y} = 0.1523 + 8.1676x$, with $r^2 = 0.606$. The slope is significantly different from 0 ($t = 13.25$, $P < 0.001$).

23.47   (a) $\bar{x} \doteq -0.00333$, $s \doteq 1.0233$. "Standardized" typically means $\bar{x} \doteq 0$ and $s \doteq 1$. (b) A stemplot is not strikingly non-Normal (for such a small sample). (c) The probability that a standard Normal variable is as extreme as this is about 0.0272.

23.49 -0.0398 to 0.0144; it does contain 0.

Chapter 24

24.1   **(a)** H_0: $\mu_A = \mu_B = \mu_C = \mu_D$; H_a: at least one mean is different. **(b)** A child's life expectancy increases by about 2 years for each long-lived parent.

24.3   **(a)** The stemplots show no extreme outliers or strong skewness (given the small sample sizes). **(b)** The means suggest that logging reduces the number of trees per plot, and that recovery is slow. **(c)** $F = 11.43$, $P = 0.000205$; H_0: $\mu_1 = \mu_2 = \mu_3$; H_a: not all means are the same.

24.5   **(a)** When the three means are similar, F is very small, and P is near 1. **(b)** By moving any mean, F increases and P decreases.

24.7   **(a)** The ratio of standard deviations is about 1.16. **(b)** The ratio of standard deviations is about 1.18.

24.9   The differences among the groups were significant ($F = 3.44$, df $= 3$ and 27, $P = 0.031$). Nitrogen had a positive effect, the phosphorus and control groups were similar, and the plants that got both nutrients fell between the others.

24.11   **(a)** df $= 2$ and 93. **(b)** df $= 2$ and 87.

24.13   **(a)** Yes: the ratio of standard deviations is 1.24. **(b)** $\bar{x} \doteq 9.92$, MSG $\doteq 10.0$, MSE $\doteq 21.9$, and $F \doteq 0.46$. **(c)** df $= 2$ and 112 (use 2 and 100 in the table). The difference in means is not significant.

24.15   (c)   **24.16**   (b)   **24.17**   (c)   **24.18**   (b)   **24.19**   (b)

24.20   (b)   **24.21**   (a)   **24.22**   (a)   **24.23**   (a)

24.25   Populations: students with no additional instruction, those required to prepare a written outline, those given 15 relevant ideas, and those given the ideas and required to prepare an outline. Response variable: essay quality scores. $I = 4$, $n_1 = n_2 = n_3 = n_4 = 20$, $N = 80$; df $= 3$ and 76.

24.27   Populations: normal-weight men, overweight men, obese men. Response variable: triglyceride level. $I = 3$, $n_1 = 719$, $n_2 = 885$, $n_3 = 220$, $N = 1824$; df $= 2$ and 1821.

24.29   Only Design A would allow use of one-way ANOVA.

24.31   There is strong evidence ($F = 4.92$, df $= 3$ and 92, $P = 0.003$) that the means are not all the same. List 1 seems to be the easiest, and lists 3 and 4 are the most difficult.

24.33   The differences in mean healing rate are significant ($F = 4.04$, df $= 4$ and 69, $P = 0.005$). The natural group and the 1.25-times groups showed the fastest healing, the 0- and 0.5-times groups were slower, and the 1.5-times group had the slowest healing.

24.35   There is a highly significant ($F = 5.02$, df $= 2$ and 12, $P = 0.026$) difference among the mean breaking strengths. Fabrics treated with Permafresh 55 have considerably higher strength than fabrics treated with Permafresh 48 or Hylite.

24.37   **(a)** The means are 17.5, 11.75, and 13.67; the standard deviations are 3.529, 4.372, and 4.500. The standard deviations satisfy the rule of thumb (ratio 1.28). The means suggest that logging reduces the number of species per plot and that recovery takes more than 8 years. **(b)** $F = 6.02$, df $= 2$ and 30, $P = 0.006$. The differences are significant; the number of species per plot really is lower in logged areas.

24.39 ANOVA for 2001 shows significant differences ($F = 43.79$, df $= 2$ and 15), $P < 0.0005$). The standard deviation ratio (4.42) does not meet our guidelines.

24.41 **(a)** t test (comparing two means). **(b)** ANOVA (comparing three means). **(c)** Chi-square (comparing three proportions).

Table entry for C is the critical value t^* required for confidence level C. To approximate one- and two-sided P-values, compare the value of the t statistic with the critical values of t^* that match the P-values given at the bottom of the table.

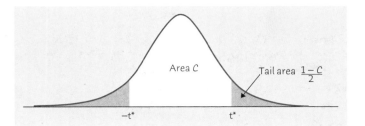

Area C   Tail area $\frac{1-C}{2}$

$-t^*$ t^*

TABLE C *t* distribution critical values

DEGREES OF FREEDOM	CONFIDENCE LEVEL C											
	50%	60%	70%	80%	90%	95%	96%	98%	99%	99.5%	99.8%	99.9%
1	1.000	1.376	1.963	3.078	6.314	12.71	15.89	31.82	63.66	127.3	318.3	636.6
2	0.816	1.061	1.386	1.886	2.920	4.303	4.849	6.965	9.925	14.09	22.33	31.60
3	0.765	0.978	1.250	1.638	2.353	3.182	3.482	4.541	5.841	7.453	10.21	12.92
4	0.741	0.941	1.190	1.533	2.132	2.776	2.999	3.747	4.604	5.598	7.173	8.610
5	0.727	0.920	1.156	1.476	2.015	2.571	2.757	3.365	4.032	4.773	5.893	6.869
6	0.718	0.906	1.134	1.440	1.943	2.447	2.612	3.143	3.707	4.317	5.208	5.959
7	0.711	0.896	1.119	1.415	1.895	2.365	2.517	2.998	3.499	4.029	4.785	5.408
8	0.706	0.889	1.108	1.397	1.860	2.306	2.449	2.896	3.355	3.833	4.501	5.041
9	0.703	0.883	1.100	1.383	1.833	2.262	2.398	2.821	3.250	3.690	4.297	4.781
10	0.700	0.879	1.093	1.372	1.812	2.228	2.359	2.764	3.169	3.581	4.144	4.587
11	0.697	0.876	1.088	1.363	1.796	2.201	2.328	2.718	3.106	3.497	4.025	4.437
12	0.695	0.873	1.083	1.356	1.782	2.179	2.303	2.681	3.055	3.428	3.930	4.318
13	0.694	0.870	1.079	1.350	1.771	2.160	2.282	2.650	3.012	3.372	3.852	4.221
14	0.692	0.868	1.076	1.345	1.761	2.145	2.264	2.624	2.977	3.326	3.787	4.140
15	0.691	0.866	1.074	1.341	1.753	2.131	2.249	2.602	2.947	3.286	3.733	4.073
16	0.690	0.865	1.071	1.337	1.746	2.120	2.235	2.583	2.921	3.252	3.686	4.015
17	0.689	0.863	1.069	1.333	1.740	2.110	2.224	2.567	2.898	3.222	3.646	3.965
18	0.688	0.862	1.067	1.330	1.734	2.101	2.214	2.552	2.878	3.197	3.611	3.922
19	0.688	0.861	1.066	1.328	1.729	2.093	2.205	2.539	2.861	3.174	3.579	3.883
20	0.687	0.860	1.064	1.325	1.725	2.086	2.197	2.528	2.845	3.153	3.552	3.850
21	0.686	0.859	1.063	1.323	1.721	2.080	2.189	2.518	2.831	3.135	3.527	3.819
22	0.686	0.858	1.061	1.321	1.717	2.074	2.183	2.508	2.819	3.119	3.505	3.792
23	0.685	0.858	1.060	1.319	1.714	2.069	2.177	2.500	2.807	3.104	3.485	3.768
24	0.685	0.857	1.059	1.318	1.711	2.064	2.172	2.492	2.797	3.091	3.467	3.745
25	0.684	0.856	1.058	1.316	1.708	2.060	2.167	2.485	2.787	3.078	3.450	3.725
26	0.684	0.856	1.058	1.315	1.706	2.056	2.162	2.479	2.779	3.067	3.435	3.707
27	0.684	0.855	1.057	1.314	1.703	2.052	2.158	2.473	2.771	3.057	3.421	3.690
28	0.683	0.855	1.056	1.313	1.701	2.048	2.154	2.467	2.763	3.047	3.408	3.674
29	0.683	0.854	1.055	1.311	1.699	2.045	2.150	2.462	2.756	3.038	3.396	3.659
30	0.683	0.854	1.055	1.310	1.697	2.042	2.147	2.457	2.750	3.030	3.385	3.646
40	0.681	0.851	1.050	1.303	1.684	2.021	2.123	2.423	2.704	2.971	3.307	3.551
50	0.679	0.849	1.047	1.299	1.676	2.009	2.109	2.403	2.678	2.937	3.261	3.496
60	0.679	0.848	1.045	1.296	1.671	2.000	2.099	2.390	2.660	2.915	3.232	3.460
80	0.678	0.846	1.043	1.292	1.664	1.990	2.088	2.374	2.639	2.887	3.195	3.416
100	0.677	0.845	1.042	1.290	1.660	1.984	2.081	2.364	2.626	2.871	3.174	3.390
1000	0.675	0.842	1.037	1.282	1.646	1.962	2.056	2.330	2.581	2.813	3.098	3.300
z^*	0.674	0.841	1.036	1.282	1.645	1.960	2.054	2.326	2.576	2.807	3.091	3.291
One-sided P	.25	.20	.15	.10	.05	.025	.02	.01	.005	.0025	.001	.0005
Two-sided P	.50	.40	.30	.20	.10	.05	.04	.02	.01	.005	.002	.001